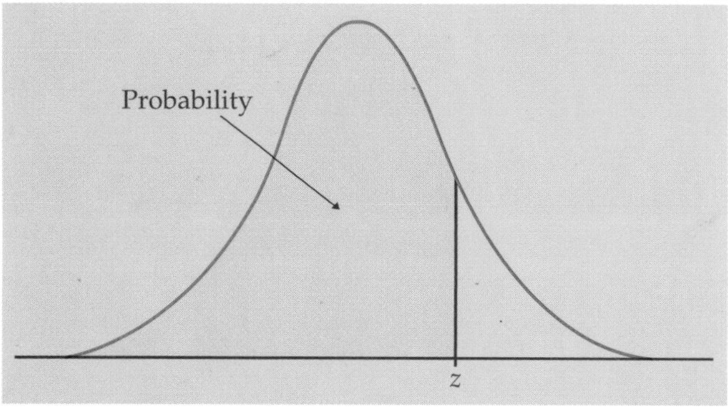

Table entry for z is the area under the standard normal curve to the left of z.

TABLE A Standard normal probabilities (*continued*)

z	.00	.01	.02	.03	.04	.05	.06	.07	.08	.09
0.0	.5000	.5040	.5080	.5120	.5160	.5199	.5239	.5279	.5319	.5359
0.1	.5398	.5438	.5478	.5517	.5557	.5596	.5636	.5675	.5714	.5753
0.2	.5793	.5832	.5871	.5910	.5948	.5987	.6026	.6064	.6103	.6141
0.3	.6179	.6217	.6255	.6293	.6331	.6368	.6406	.6443	.6480	.6517
0.4	.6554	.6591	.6628	.6664	.6700	.6736	.6772	.6808	.6844	.6879
0.5	.6915	.6950	.6985	.7019	.7054	.7088	.7123	.7157	.7190	.7224
0.6	.7257	.7291	.7324	.7357	.7389	.7422	.7454	.7486	.7517	.7549
0.7	.7580	.7611	.7642	.7673	.7704	.7734	.7764	.7794	.7823	.7852
0.8	.7881	.7910	.7939	.7967	.7995	.8023	.8051	.8078	.8106	.8133
0.9	.8159	.8186	.8212	.8238	.8264	.8289	.8315	.8340	.8365	.8389
1.0	.8413	.8438	.8461	.8485	.8508	.8531	.8554	.8577	.8599	.8621
1.1	.8643	.8665	.8686	.8708	.8729	.8749	.8770	.8790	.8810	.8830
1.2	.8849	.8869	.8888	.8907	.8925	.8944	.8962	.8980	.8997	.9015
1.3	.9032	.9049	.9066	.9082	.9099	.9115	.9131	.9147	.9162	.9177
1.4	.9192	.9207	.9222	.9236	.9251	.9265	.9279	.9292	.9306	.9319
1.5	.9332	.9345	.9357	.9370	.9382	.9394	.9406	.9418	.9429	.9441
1.6	.9452	.9463	.9474	.9484	.9495	.9505	.9515	.9525	.9535	.9545
1.7	.9554	.9564	.9573	.9582	.9591	.9599	.9608	.9616	.9625	.9633
1.8	.9641	.9649	.9656	.9664	.9671	.9678	.9686	.9693	.9699	.9706
1.9	.9713	.9719	.9726	.9732	.9738	.9744	.9750	.9756	.9761	.9767
2.0	.9772	.9778	.9783	.9788	.9793	.9798	.9803	.9808	.9812	.9817
2.1	.9821	.9826	.9830	.9834	.9838	.9842	.9846	.9850	.9854	.9857
2.2	.9861	.9864	.9868	.9871	.9875	.9878	.9881	.9884	.9887	.9890
2.3	.9893	.9896	.9898	.9901	.9904	.9906	.9909	.9911	.9913	.9916
2.4	.9918	.9920	.9922	.9925	.9927	.9929	.9931	.9932	.9934	.9936
2.5	.9938	.9940	.9941	.9943	.9945	.9946	.9948	.9949	.9951	.9952
2.6	.9953	.9955	.9956	.9957	.9959	.9960	.9961	.9962	.9963	.9964
2.7	.9965	.9966	.9967	.9968	.9969	.9970	.9971	.9972	.9973	.9974
2.8	.9974	.9975	.9976	.9977	.9977	.9978	.9979	.9979	.9980	.9981
2.9	.9981	.9982	.9982	.9983	.9984	.9984	.9985	.9985	.9986	.9986
3.0	.9987	.9987	.9987	.9988	.9988	.9989	.9989	.9989	.9990	.9990
3.1	.9990	.9991	.9991	.9991	.9992	.9992	.9992	.9992	.9993	.9993
3.2	.9993	.9993	.9994	.9994	.9994	.9994	.9994	.9995	.9995	.9995
3.3	.9995	.9995	.9995	.9996	.9996	.9996	.9996	.9996	.9996	.9997
3.4	.9997	.9997	.9997	.9997	.9997	.9997	.9997	.9997	.9997	.9998

INTRODUCTION TO THE PRACTICE OF STATISTICS

FOURTH EDITION

David S. Moore
George P. McCabe
Purdue University

W. H. Freeman and Company
New York

Senior Acquisitions Editor: Patrick Farace
Associate Editor: Danielle Swearengin
Publisher: Michelle Julet
Marketing Manager: Jeffrey Rucker
Media/Supplements Editor: Brian Donnellan
Project Editor: Mary Louise Byrd
Cover and Text Designer: Vicki Tomaselli
Illustrations: Publication Services
Illustration Coordinator: Bill Page
Photo Editor: Patricia Marx
Production Coordinator: Paul W. Rohloff
Composition: Publication Services
Manufacturing: RR Donnelly & Sons Company

Cataloging-in-Publication Data available from the Library of Congress

Printed in the United States of America

Second printing, 2003

CONTENTS

*Sections marked with asterisks are optional.

Additional chapters are on CD-ROM and available in a separate print supplement.

TO TEACHERS: ABOUT THIS BOOK

Statistics is the science of data. ***Introduction to the Practice of Statistics (IPS)*** is an introductory text based on this principle. We present the most-used methods of basic statistics in a way that emphasizes working with data and mastering statistical reasoning. ***IPS*** is elementary in mathematical level, but conceptually rich in statistical ideas and serious in its aim to help students think about data and use statistical methods with understanding.

Some schematic history will help place ***IPS*** in the universe of texts for a first course in statistics for students from a variety of disciplines. Traditional texts were almost entirely devoted to methods of inference, with quick coverage of means, medians, and histograms as a preliminary. No doubt this reflected the fact that inference is the only part of statistics that has a mathematical theory behind it. Several innovative books aimed at nontraditional audiences pioneered a broader approach that paid more attention to design of samples and experiments, the messiness of real data, and discussion of real-world statistical studies and controversies. All were written by widely known statisticians whose main business was not writing textbooks. *The Nature of Statistics* (Wallis and Roberts) has passed away, but *Statistics* (Freedman and collaborators) and *Statistics Concepts and Controversies* (Moore) remain alive and well. None of these books tried to meet the needs of a typical first course because their audiences did not need full coverage of standard statistical methods.

IPS was the first book to successfully combine attention to broader content and reasoning with comprehensive presentation of the most-used statistical methods. It reflects the consensus among statisticians—even stronger now than when the first edition appeared—concerning the content of an introduction to our discipline. This consensus is expressed in a report from the joint curriculum committee of the American Statistical Association and the Mathematical Association of America[1] and in discussions in leading journals.[2] (Notes and data sources are collected at the end of the book.) ***IPS*** has been successful for several reasons:

1. ***IPS*** reflects at a level suitable for beginners the nature of modern statistical practice. Attention to data analysis and data production as well as to probability and inference is "new" only in the world of textbooks. Users of statistical methods have always paid attention to all of these. Contemporary research in statistics, driven by advances in computing, puts more emphasis on sophisticated "looking at data" and

on data-analytic ways of thinking. Formal inference remains important, and receives careful treatment, but it appears as part of a larger picture.

2. ***IPS*** has a logical overall progression, so that data analysis and data production strengthen the presentation of inference rather than stand apart from it. We stress that data analysis is an essential preliminary to inference because inference requires clean data. The most useful "goodness of fit" procedure, for example, is the normal quantile plot presented in Chapter 1 and used frequently in the inference chapters. We emphasize that when you do formal statistical inference, you are acting as if your data come from properly randomized data production. We use random samples and experimental randomization to motivate the need for probability as a language for inference.
3. ***IPS*** presents data analysis as more than a collection of techniques for exploring data. We integrate techniques with discussion of systematic ways of thinking about data. We also work hard to make data-analytic thinking accessible to beginners by presenting a series of simple principles: always plot your data; look for overall patterns and deviations from them; the overall pattern of a distribution for one variable should consider shape, center, and spread; for relations between two variables, consider form, direction, and strength; always ask whether a relationship between variables is influenced by other variables lurking in the background. Inference is similarly treated as more than a collection of methods. We warn students about pitfalls in clear cautionary discussions—about regression and correlation (Section 2.4), experiments (page 237), sample surveys (page 252), confidence intervals (page 426), and significance tests (Section 6.3). Our goal throughout ***IPS*** is to present principles and techniques together in a way that is accessible to beginners and lays a foundation for those who will go on to more advanced study.
4. ***IPS*** integrates discussion of techniques, reasoning, and practice using real examples to drive the exposition. Students learn the technique of least-squares regression and how to interpret regression slope. But they also learn the conceptual ties between regression and correlation, the importance of looking for influential observations (always plot your data), and to beware of averaged data and the restricted-range effect.
5. ***IPS*** is aware of current developments both in statistical science and in teaching statistics. For example, the first edition already favored the version of the two-sample t procedures that does not assume equal population variances and discussed the great difference in robustness between standard tests for means and for variances. In this fourth edition, we present the modified ("add two successes and two failures") confidence intervals for proportions that are now supported both by extensive simulation and by theory. Brief optional "Beyond the Basics" sections give overviews of contemporary topics such as density estimation, scatterplot smoothers, the bootstrap, and (new in this edition) data mining.

The title of the book expresses our intent to introduce readers to statistics as it is used in practice. Statistics in practice is concerned with gaining understanding from data; it focuses on problem solving rather than on methods that may be useful in specific settings. A text cannot fully imitate practice, because it must teach specific methods in a logical order and must use data that are not the reader's own. Nonetheless, our interest and experience in applying statistics have influenced the nature of ***IPS*** in several ways.

Statistical Thinking Statistics is interesting and useful because it provides strategies and tools for using data to gain insight into real problems. As the continuing revolution in computing automates most tiresome details, an emphasis on statistical concepts and on insight from data becomes both more practical for students and teachers and more important for users who must supply what is not automated. No student should complete a first statistics course, for example, without a firm grasp of the distinction between observational studies and experiments and of why randomized comparative experiments are the gold standard for evidence of causation. We have seen many statistical mistakes, but few that involved simply getting a calculation wrong. We therefore ask students to learn to explore data, always starting with plots, to think about the context of the data and the design of the study that produced the data, the possible influence of wild observations on conclusions, and the reasoning that lies behind standard methods of inference. Users of statistics who form these habits from the beginning are well prepared to learn and use more advanced methods.

Data Data are numbers with a context, as we say in our Introduction for students. A newborn that weighs 10.3 pounds is a big baby, and the birth weight could not plausibly be 10.3 ounces or 10.3 kilograms. Because context makes numbers meaningful, our examples and exercises use real data with real contexts that we briefly describe. Calculating the mean of five numbers is arithmetic, not statistics. We hope that the presence of background information, even in exercises intended for routine drill, will encourage students to always consider the meaning of their calculations as well as the calculations themselves. Note in this connection that a calculation or a graph or "reject H_0" is rarely a full answer to a statistical problem. We strongly encourage requiring students to always state a brief conclusion in the context of the problem. This helps build data sense as well as the communications skills that employers value.

Mathematics Although statistics is a mathematical science, it is not a field of mathematics and should not be taught as if it were. A fruitful mathematical theory (based on probability, which *is* a field of mathematics) underlies some parts of basic statistics, but by no means all. The distinction between observation and experiment, for example, is a core statistical idea that is ignored by the theory.[3] Mathematically trained teachers, rightly resisting a formula-based approach, sometimes identify conceptual understanding with mathematical understanding. When teaching statistics, we must emphasize statistical ideas and recognize that mathematics is not the only vehicle for conceptual understanding. ***IPS*** requires only the ability to read and use equations without having each step parsed. We require no algebraic derivations, let alone calculus. Because this is a *statistics* text, it is richer in ideas and requires more thought than the low mathematical level suggests.

Calculators and Computers Statistical calculations and graphics are, in practice, automated by software. We encourage instructors to use software of their choice or a graphing calculator that includes functions for both data analysis and basic inference. ***IPS*** includes some topics that reflect the dominance of software in practice, such as normal quantile plots and the version of the two-sample t procedures that does not require equal variances. We several times display the output of multiple software systems for the same problem—the regression output on page 142 is an example. The point is that a student who knows the basics of regression can interpret almost any regression output. Students like this reassurance, and it helps focus their attention on understanding rather than reading output.

Nonetheless, much statistical work can be done with just a "two-variable statistics" calculator with functions for correlation and the least-squares regression line as well as for the mean and standard deviation. Although not all exercises are feasible for students with such a calculator, we have taken care to make the book usable by students without access to computing.

Judgment Statistics in practice requires judgment. It is easy to list the mathematical assumptions that justify use of a particular procedure, but not so easy to decide when the procedure can be safely used in practice. Because judgment develops through experience, an introductory course should present clear guidelines and not make unreasonable demands on the judgment of students. We have given guidelines—for example, on using the t procedures for comparing two means but avoiding the F procedures for comparing two variances—that we follow ourselves. Similarly, many exercises require students to use some judgment and (equally important) to explain their choices in words. Many students would prefer to stick to calculating, and many statistics texts allow them to. Requiring more will do them much good in the long run.

Teaching Experiences We have successfully used ***IPS*** in courses taught to quite diverse student audiences. For general undergraduates from mixed disciplines, we cover Chapters 1 to 8 and one of Chapters 9, 10, and 12, omitting all optional material. For a quantitatively strong audience—sophomores planning to major in actuarial science or statistics—we move more quickly. We add Chapters 10 and 11 to the core material in Chapters 1 to 8 and include most optional content. We deemphasize Chapter 4 (probability) because these students will take a probability course later in their program, though we make intensive use of software for simulating probabilities as well as for statistical analysis. The third group we teach contains beginning graduate students in such fields as education, family studies, and retailing. These mature but sometimes quantitatively unprepared students read the entire text (Chapters 11 and 13 lightly), again with reduced emphasis on Chapter 4 and some parts of Chapter 5. In all cases, beginning with data analysis and data production (Part I) helps students overcome their fear of statistics and builds a sound base for studying inference. We find that ***IPS*** can be flexibly adapted to quite varied audiences by paying attention to our clear designation of some material as optional and by varying the chapters assigned.

The Fourth Edition

In revising ***Introduction to the Practice of Statistics*** after three successful editions, we have concentrated on new exercises and on new material (some of it optional) that matches changes in statistical practice and the increasing power of software typically available to students.

Examples and exercises

- Over one-fourth of all examples are new or revised.
- The fourth edition has over 150 more exercises than the third edition, and more than half of all exercises are new or revised. Even some old favorites, such as the Florida manatee, have been retired to make room for new material. Long-time users of ***IPS*** (we are among them) will appreciate the need to present students with a different set of challenges. For those who like the old favorites, several hundred exercises from the third edition are available as "Supplementary Exercises" online and on the CD that accompanies each book. We have, in addition, carefully updated all the exercises that depend on current data.
- Each chapter now begins with a Prelude, a short, informal (and sometimes amusing) case study that illustrates the content of the chapter.
- In general, the Section Exercises follow the order of material in each section. Straightforward problems precede those that require more thought or several steps, and more elaborate exercises come at the end. The Chapter Exercises allow review without the clues that come from placement in a specific section. What is more, we have expanded the set of exercises available by placing online and on the CD many of the exercises from the previous edition that we replaced in preparing this new edition.

Integrated technology

applet

- A set of online applets built to our specifications is freely available on the ***IPS*** Web site `www.whfreeman.com/ips`. These interactive graphical programs are by far the most effective way to help students grasp the sensitivity of correlation and regression to outliers, the idea of a confidence interval, the way ANOVA responds to both within-group and among-group variation, and many other statistical fundamentals. Exercises using these applets appear throughout the text, marked by a distinctive icon. We urge you to assign some of these, and we suggest that if your classroom is suitably equipped the applets are very helpful tools for classroom presentation as well.

- We include more exercises based on moderate-to-large sets of data that are best done using software. Full data sets are available in several formats (ASCII, Excel, JMP, Minitab, SPSS, and TI calculator) both online (`www.whfreeman.com/ips`) and on the CD that accompanies each copy of the text. We have added SPSS to a number of displays of computer output to better match software popular for teaching.
- "Exploring the Web" boxes appear regularly. These point students to Web sites relevant to material in ***IPS.*** Some sites can directly aid instruction, while others are interesting byways.

Additional coverage

- Chapter 2 now includes a thorough optional presentation of transformations of data, with special attention to exponential growth and to the power models that are so common in biology.
- The discussion of two-way data tables and relations among categorical variables has been moved from Chapter 2 to Chapter 9, where it accompanies the discussion of inference about two-way tables. This allows Chapter 2 to concentrate on quantitative variables and makes room for the new material on transformations.
- Chapter 4 now discusses variances for sums of correlated random variables. One motivation for this is that "mean-variance analysis" of investments now appears even in mutual fund reports and other material for a broad public. One of our applets automates the calculation for two variables, in the guise of mean-variance analysis of portfolios that combine two investments.
- The standard z procedures for confidence intervals for one and two population proportions have been replaced in Chapter 8 by new methods that replace the usual sample proportion by the proportion that results from adding one success and one failure to each sample. The resulting procedures are as easy to use as the traditional intervals and are very much more accurate. This has been demonstrated both by extensive computational studies[4] and by mathematical analysis.[5]

Acknowledgments

We are pleased that the first three editions of ***Introduction to the Practice of Statistics*** have helped move the teaching of introductory statistics in a direction supported by most statisticians. We are grateful to the many colleagues and students who have provided helpful comments, and we hope that they will find this new edition another step forward. In particular, we would like to thank the following colleagues who offered specific comments on the new edition:

Robert A. Berman, USDA Graduate School

Patricia M. Buchanan, Pennsylvania State University

Joseph Chang, Yale University

Smiley Cheng, University of Manitoba

DeWayne Derrberry, University of Puget Sound

Michael Evans, University of Toronto

K. Renee Fister, Murray State University

Amaris Franz, Oregon State University

Betsy S. Greenberg, University of Texas–Austin

Lance E. Hemlow, Raritan Valley Community College

Patricia B. Humphrey, Georgia Southern University

John Z. Imbrie, University of Virginia

Ron Loser, Adams State College

Marilou Mendel, University of Arizona

Ron Millard, Shawnee Mission South High School

David A. Palkovich, Oklahoma City Community College

Charles M. Parks, Ohio University

Matthew G. Pickard, University of Puget Sound

David M. Reineke, University of Wisconsin–La Crosse

Lawrence D. Ries, University of Missouri–Columbia

Larry J. Ringer, Texas A&M University

Nancy Roper, Portland Community College

Kenneth A. Ross, University of Oregon

Nathan Smith, University of Texas–Tyler

Patty Solomon, University of Adelaide

Nathan Taback, University of Toronto

Augustin M. Vukov, University of Toronto

Nathan Wetzel, University of Wisconsin–Stevens Point

Dan Williams, Baruch College

The professionals at W. H. Freeman, in particular, Mary Louise Byrd, Patrick Farace, and Danielle Swearengin, have contributed greatly to the success of ***IPS***. Most of all, we are grateful to the many people in varied disciplines and occupations with whom we have worked to gain understanding from data. They have provided both material for this book and the experience that enabled us to write it. What the eminent statistician John Tukey called "the real problems experience and the real data experience" has shaped our view of statistics. It has convinced us of the need for beginning instruction to focus on data and concepts, building intellectual skills that transfer to more elaborate settings and remain essential when all details are automated. We hope that users and potential users of statistical techniques will find this emphasis helpful.

David S. Moore
George P. McCabe

SUPPLEMENTS AND MEDIA FOR STUDENTS

A full range of supplements and media is available to help students get the most out of ***IPS:***

SUPPLEMENTS

- **Study Guide** (0-7167-4912-2) prepared by Michael A. Fligner and William I. Notz of The Ohio State University offers students explanations of crucial concepts in each section of ***IPS***, plus detailed solutions to key text problems and step-through models of important statistical techniques.
- **Statistical Software Manuals** will guide students in the use of particular statistical software with ***IPS.*** The chapters of each manual correlate to those of ***IPS*** and include a set of exercises specific to that chapter's concepts. These manuals are:
  - **Excel Manual** (0-7167-4915-7) prepared by Fred Hoppe of McMaster University. Excel Macros can be accessed on our Web site, `www.whfreeman.com/ips`.
  - **JMP Manual** (0-7167-9631-7) prepared by Thomas Devlin of Montclair State University.
  - **Minitab Manual** (0-7167-4913-0) prepared by Michael Evans of University of Toronto.
  - **S-PLUS Manual** (0-7167-9619-8) prepared by Greg Snow of Brigham Young University and Laura Chihara of Carleton College.
  - **SPSS Manual** (0-7167-4914-9) prepared by Paul Stephenson, Neal Rogness, and Patricia Stephenson of Grand Valley State University.
  - **TI-83 Graphing Calculator Manual** (0-7167-4911-4) prepared by David K. Neal of Western Kentucky University.
- **Chapters 14 and 15** (0-7167-9810-7) written by David S. Moore and George P. McCabe contain optional instruction on Nonparametric Tests and Logistic Regression. These chapters are available both as a printed supplement and on the CD. Refer to the table of contents for a complete listing of ***IPS*** chapters.
- **Telecourse Study Guide** (0-7167-4916-5) for "Against All Odds" guides students' learning of basic statistics by linking the television programs in *Against All Odds: Inside Statistics* with a telecourse text. *Against All Odds* is a series of 26 half-hour television shows that introduces modern statistical methods in the context of a wide variety of real-world applications. The series is available through Annenberg/CPB at `www.learner.org` in both VHS and DVD formats.
- **Projects Book** (0-7167-9809-3) prepared by Ron Millard of Shawnee Mission South High School and John C. Turner of the U.S. Naval Academy is a new supplement that offers students, as well as instructors, ideas for hands-on explorations that prompt students to think critically about statistics.

MEDIA

- ***IPS* Web Site** www.whfreeman.com/ips seamlessly integrates topics from the text. On the Web site students can find:
  - **Interactive Statistical Applets** that allow students to manipulate data and see results graphically. These applets automate calculations and graphics in a way that nicely complements classroom work.
  - **Data Sets** in ASCII, Excel, JMP, Minitab, SPSS, and TI formats.
  - **Self Quizzes** to help students prepare for tests. Each quiz contains 20 multiple-choice questions that include automatic feedback for incorrect answers, reinforcement for correct answers, and text section references to aid in additional study for incorrect answers.
  - **Student version of the Electronic Encyclopedia of Statistical Examples and Exercises (EESEE)** is a rich repository of case studies that apply the concepts of the text to various real-world venues such as the mass media, sports, natural sciences, social sciences, and medicine. Each case study is accompanied by practice problems, and most include full data sets that are exportable to various statistical software packages. EESEE was developed by a consortium at The Ohio State University dedicated to statistical education.
  - **Excel Macros** prepared by Fred Hoppe of McMaster University are new to the fourth edition of ***IPS*** and facilitate use of spreadsheet operations by students and instructors, allowing for easier and more accurate data and statistical analyses. Adopters of the **Excel Manual** will be able to download updates to these macros.
  - **The Practice of Business Statistics (PBS) Chapter 12** entitled "Statistics for Quality Control and Capability" by David S. Moore is available as an online supplement.
- **Interactive Student CD-ROM** is included with every new copy of ***IPS.*** The CD contains:
  - **EESEE** (Electronic Encyclopedia of Statistical Examples and Exercises) case studies.
  - **Chapters 14 and 15** written by David S. Moore and George McCabe, additional chapters that cover Nonparametric Tests and Logistic Regression.
  - **Self Quizzes** in multiple-choice format for each chapter constructed by experienced instructors to anticipate typical errors.
  - **Data Sets** in ASCII, Excel, JMP, Minitab, SPSS and TI formats.
  - **Interactive Statistical Applets** that allow students to manipulate data and see results graphically. These applets automate calculations and graphics in a way that nicely complements classroom work.
- **Student versions of JMP, Minitab, SPLUS, and SPSS** can be packaged with ***IPS.***

SUPPLEMENTS AND MEDIA SUPPORT FOR INSTRUCTORS

A full range of supplements and media support is available to help instructors better teach from ***IPS:***

INSTRUCTOR'S CD

Instructor's Resource CD-ROM contains all the student CD material plus the following:

- **Instructor's Guide with Solutions** in Adobe .pdf electronic format.
- **Instructor's version of the Electronic Encyclopedia of Statistical Examples and Exercises (EESEE),** with solutions to the exercises in the student version.
- **All *IPS* figures** in an exportable presentation format, JPEG for Windows and PICT for Macintosh users.
- **Presentation Manager Pro,** which creates presentations for all the figures and selected tables from the text. No extra software is necessary to save, print, and present electronic presentations using the materials from the text. Extra material can be imported to the presentation list from locally saved files and the Web.
- **PowerPoint Slides** that can be used directly or customized for your particular course. Every image and table from the textbook is formatted for your lecture needs.

ASSESSMENT TOOLS

- **Instructor's Guide with Solutions** prepared by John C. Turner of the U.S. Naval Academy includes worked out solutions to all exercises, teaching suggestions, and chapter comments.
- **Test Bank** (0-7167-4881-9) prepared by Michael Fligner and William Notz of The Ohio State University is an easy-to-use CD that includes Windows and MAC versions on a single disc. The format lets you add, edit, and resequence questions to suit your needs.
- **Online Testing** powered by Diploma from the Brownstone Research Group offers instructors the ability to easily create and administer secure exams over a network and over the Internet, with questions that incorporate multimedia and interactive exercises. The program lets you restrict tests to specific computers or blocks of time, and includes an impressive suite of gradebook and result-analysis features.
- **Online Quizzing** powered by Questionmark and accessed via the ***IPS*** Web site uses Question Mark's Perception to enable instructors to easily and securely quiz students online using prewritten, multiple-choice questions for each text chapter, separate from those appearing in the Test Bank. Students receive instant feedback and can take the quizzes multiple times. Instructors can view the results by quiz, student, or question or can get weekly results via e-mail.

IPS WEB SITE

The following ***IPS*** **Web Site** for Instructors, `www.whfreeman.com/ips`, contains all features available to students along with:

- **Instructor's version of the Electronic Encyclopedia of Statistical Examples and Exercises (EESEE),** with solutions to the exercises in the student version.
- **Instructor's Guide with Solutions** available in Adobe .pdf electronic format.

COURSE MANAGEMENT

- **Online Course Materials (WebCT, Blackboard)** can be provided as a service for adopters. We offer electronic content of ***IPS,*** including the complete Test Bank in either the WebCT or Blackboard format.

STATISTICAL SOFTWARE PACKAGES

- **Student versions of JMP, Minitab, SPLUS, and SPSS** can be bundled with ***IPS*** for those instructors who wish to use a statistical software package in the course.

APPLICATIONS

Introduction to the Practice of Statistics presents a wide variety of applications from diverse disciplines. The list below indicates the number of exercises and examples that relate to different fields.

Exercises by application

Biology and environmental sciences: 129
Business and management: 82
Demographics and population characteristics: 102
Economics and investment: 79
Education and child development: 171
Manufacturing and product development: 79
Marketing and consumer behavior: 86
Medicine and clinical studies: 77
Paleontology: 3
Physical sciences: 35
Psychology and human behavior: 68
Public health and nutrition: 101
Public policy and political science: 34
Sports and leisure: 51

Examples by application

Biology and environmental sciences: 31
Business and management: 9
Demographics and population characteristics: 22
Economics and investment: 9
Education and child development: 46
Manufacturing and product development: 17
Marketing and consumer behavior: 22
Medicine and clinical studies: 27
Paleontology: 1
Physical sciences: 12
Psychology and human behavior: 2
Public health and nutrition: 40
Public policy and political science: 6
Sports and leisure: 22

For a complete list of examples and exercises by chapter and number, please see the *Instructor's Guide* or the ***IPS*** Web site: `www.whfreeman.com/ips`.

TO STUDENTS: WHAT IS STATISTICS?

Statistics is the science of collecting, organizing, and interpreting numerical facts, which we call *data*.

We are bombarded by data in our everyday life. The news mentions imported car sales, the latest poll of the president's popularity, and the average high temperature for today's date. Advertisements claim that data show the superiority of the advertiser's product. All sides in public debates about economics, education, and social policy argue from data. A knowledge of statistics helps separate sense from nonsense in the flood of data.

The study and collection of data are also important in the work of many professions, so that training in the science of statistics is valuable preparation for a variety of careers. Each month, for example, government statistical offices release the latest numerical information on unemployment and inflation. Economists and financial advisors as well as policy makers in government and business study these data in order to make informed decisions. Doctors must understand the origin and trustworthiness of the data that appear in medical journals. Politicians rely on data from polls of public opinion. Business decisions are based on market research data that reveal consumer tastes. Engineers gather data on the quality and reliability of manufactured products. Most areas of academic study make use of numbers, and therefore also make use of the methods of statistics.

Understanding from data

The goal of statistics is to gain understanding from data. To gain understanding, we often operate on a set of numbers—we average or graph them, for example. But we must do more, because data are not just numbers; they are numbers that have some context that helps us understand them.

You read that low birth weight is a major reason why infant mortality in the United States is higher than in most other advanced nations. The report goes on to say that 7.6% of children born in the United States have low birth weight and that 13.1% of black infants have low birth weight.[1] To make sense of these numbers you must know what counts as low birth weight (less than 2500 grams, or 5.5 pounds) and have some feeling for the weights of babies. You probably recognize that 5.5 pounds is small, that 7.5 pounds (3400 grams) is about average, and that 10 pounds (4500 grams) is a big baby.

Another part of the context is the source of the data. How do we know that 7.6% of American babies have low birth weight, or that the average weight of newborns is about 3400 grams? The data come from the National Center for Health Statistics, a government office to which the states report information from all birth certificates issued each month. These are the most complete data available about births in the United States.

When you do statistical problems—even simple textbook problems—don't just graph or calculate. Think about the context and state your conclusions in the specific setting of the problem. As you are learning how to do statistical calculations and graphs, remember that the goal of statistics is not calculation for its own sake but gaining understanding from numbers. The calculations and graphs can be automated by a calculator or software, but you must supply the understanding. This book presents only the most common specific procedures for statistical analysis. A thorough grasp of the principles of statistics will enable you to quickly learn more advanced methods as needed. On the other hand, a fancy computer analysis carried out without attention to basic principles will often produce elaborate nonsense. As you read, seek to understand the principles as well as the necessary details of methods and recipes.

The rise of statistics

Historically, the ideas and methods of statistics developed gradually as society grew interested in collecting and using data for a variety of applications. The earliest origins of statistics lie in the desire of rulers to count the number of inhabitants or measure the value of taxable land in their domains. As the physical sciences developed in the seventeenth and eighteenth centuries, the importance of careful measurements of weights, distances, and other physical quantities grew. Astronomers and surveyors striving for exactness had to deal with variation in their measurements. Many measurements should be better than a single measurement, even though they vary among themselves. How can we best combine many varying observations? Statistical methods that are still important were invented in order to analyze scientific measurements.

By the nineteenth century, the agricultural, life, and behavioral sciences also began to rely on data to answer fundamental questions. How are the heights of parents and children related? Does a new variety of wheat produce higher yields than the old, and under what conditions of rainfall and fertilizer? Can a person's mental ability and behavior be measured just as we measure height and reaction time? Effective methods for dealing with such questions developed slowly and with much debate.[2]

As methods for producing and understanding data grew in number and sophistication, the new discipline of statistics took shape in the twentieth century. Ideas and techniques that originated in the collection of government data, in the study of astronomical or biological measurements, and in the attempt to understand heredity or intelligence came together to form a unified "science of data." That science of data—statistics—is the topic of this text.

The organization of this book

Part I of this book, called simply "Data," concerns data analysis and data production. The first two chapters deal with statistical methods for organizing and describing data. These chapters progress from simpler to more complex data. Chapter 1 examines data on a single variable; Chapter 2 is devoted to relationships among two or more variables. You will learn both how to examine data produced by others and how to organize and summarize your own data. These summaries will be first graphical, then numerical, then when appropriate in the form of a mathematical model that gives a compact description of the overall pattern of the data. Chapter 3 outlines arrangements (called designs) for producing data that answer specific questions. The principles presented in this chapter will help you to design proper samples and experiments and to evaluate such investigations in your field of study.

Part II, consisting of Chapters 4 to 8, introduces statistical inference—formal methods for drawing conclusions from properly produced data. Statistical inference uses the language of probability to describe how reliable its conclusions are, so some basic facts about probability are needed to understand inference. Probability is the subject of Chapters 4 and 5. Chapter 6, perhaps the most important chapter in the text, introduces the reasoning of statistical inference. Effective inference is based on good procedures for producing data (Chapter 3), careful examination of the data (Chapters 1 and 2), and an understanding of the nature of statistical inference as discussed in Chapter 6. Chapters 7 and 8 describe some of the most common specific methods of inference, for drawing conclusions about means and proportions from one and two samples.

The five shorter chapters in Part III and two short chapters on CD-ROM introduce somewhat more advanced methods of inference, dealing with relations in categorical data, regression and correlation, and analysis of variance.

What lies ahead

Introduction to the Practice of Statistics is full of data from many different areas of life and study. Many exercises ask you to express briefly some understanding gained from the data. In practice, you would know much more about the background of the data you work with and about the questions you hope the data will answer. No textbook can be fully realistic. But it is important to form the habit of asking, "What do the data tell me?" rather than just concentrating on making graphs and doing calculations.

You should have some help in automating many of the graphs and calculations. You should certainly have a calculator with basic statistical functions. Look for keywords such as "two-variable statistics" or "regression" when you shop for a calculator. More advanced (and much more expensive) calculators will do much more, including some statistical graphs. You may be asked to use software as well. There are many different kinds of statistical software, from spreadsheets to large programs for advanced users of statistics. The kind of computing available to learners varies a great deal from place to place—but the big ideas of statistics don't depend on any particular level of access to computing.

Because graphing and calculating are automated in statistical practice, the most important assets you can gain from the study of statistics are an understanding of the big ideas and the beginnings of good judgment in working with data. Ideas and judgment can't (at least yet) be automated. They guide you in telling the computer what to do and in interpreting its output. This book tries to explain the most important ideas of statistics, not just teach methods. Some examples of big ideas that you will meet are "always plot your data," "randomized comparative experiments," and "statistical significance."

You learn statistics by doing statistical problems. "Practice, practice, practice." Be prepared to work problems. The basic principle of learning is persistence. Being organized and persistent is more helpful in reading this book than knowing lots of math. The main ideas of statistics, like the main ideas of any important subject, took a long time to discover and take some time to master. The gain will be worth the pain.

Notes

1. S. J. Ventura et al., *Births: Final Data for 1999,* National Vital Statistics Reports, Volume 49, No. 1, National Center for Health Statistics, 2001, table 43.

2. The rise of statistics from the physical, life, and behavioral sciences is described in detail by S. M. Stigler, *The History of Statistics: The Measurement of Uncertainty before 1900,* Harvard-Belknap, Cambridge, Mass., 1986. Much of the information in the brief historical notes appearing throughout the text is drawn from this book.

ABOUT THE AUTHORS

David S. Moore is Shanti S. Gupta Distinguished Professor of Statistics at Purdue University and was 1998 president of the American Statistical Association. He received his A.B. from Princeton and the Ph.D. from Cornell, both in mathematics. He has written many research papers in statistical theory and served on the editorial boards of several major journals. Professor Moore is an elected fellow of the American Statistical Association and of the Institute of Mathematical Statistics and an elected member of the International Statistical Institute. He has served as program director for statistics and probability at the National Science Foundation.

In recent years, Professor Moore has devoted his attention to the teaching of statistics. He was the content developer for the Annenberg/Corporation for Public Broadcasting college-level telecourse *Against All Odds: Inside Statistics* and for the series of video modules *Statistics: Decisions Through Data,* intended to aid the teaching of statistics in schools. He is the author of influential articles on statistics education and of several leading texts. Professor Moore has served as president of the International Association for Statistical Education and has received the Mathematical Association of America's national award for distinguished college or university teaching of mathematics.

George P. McCabe is Professor of Statistics and Head of the Statistical Consulting Service at Purdue University. In 1966 he received a B.S. degree in mathematics from Providence College and in 1970 a Ph.D. in mathematical statistics from Columbia University. His entire professional career has been spent at Purdue with sabbaticals at Princeton, the Commonwealth Scientific and Industrial Research Organization (CSIRO) in Melbourne, Australia, the University of Berne (Switzerland), and the National Institute of Standards and Technology (NIST) in Boulder, Colorado. Professor McCabe is an elected fellow of the American Statistical Association and was 1998 chair of its section on Statistical Consulting. He has served on the editorial boards of several statistics journals. He has consulted with many major corporations and has testified as an expert witness on the use of statistics in several cases.

Professor McCabe's research interests have focused on applications of statistics. He is author or co-author of over 125 publications in many different journals. Since working on a U.S. Agency for International Development project on functional implications of malnutrition in Egypt, his collaborations on problems in human nutrition have increased. Other recent interests include problems associated with metal fatigue in aging aircraft structures, epidemiology, and industrial applications of statistics.

PART 1

Data

Prelude: the case of the missing vans

Auto manufacturers lend their dealers money to help them keep vehicles on their lots. The loans are repaid when the vehicles are sold. A Long Island auto dealer named John McNamara borrowed over $6 billion from General Motors between 1985 and 1991. In December 1990 alone, Mr. McNamara borrowed $425 million to buy 17,000 GM vans customized by an Indiana company for sale overseas. GM happily lent McNamara the money because he always repaid the loans.

Let's pause to consider the numbers, as GM should have done, but didn't. The entire van-customizing industry produces only about 17,000 customized vans a month. So McNamara was claiming to buy an entire month's production. These large, luxurious, and gas-guzzling vehicles are designed for U.S. interstate highways. The recreational vehicle trade association says that only 1.35% were exported in 1990. It's not plausible to claim that 17,000 vans in a single month are being bought for export.

Having looked at the numbers, you can guess the rest. McNamara admitted in federal court in 1992 that he was defrauding GM on a massive scale. The Indiana company was a shell set up by McNamara, its invoices were phony, and the vans didn't exist. McNamara borrowed vast sums from GM, used most of each loan to pay off the previous loan, and skimmed off a bit for himself. The bit he skimmed amounted to over $400 million. GM set aside $275 million to cover its losses. Two executives, who should have looked at the numbers relevant to their business, were fired.[1]

CHAPTER 1

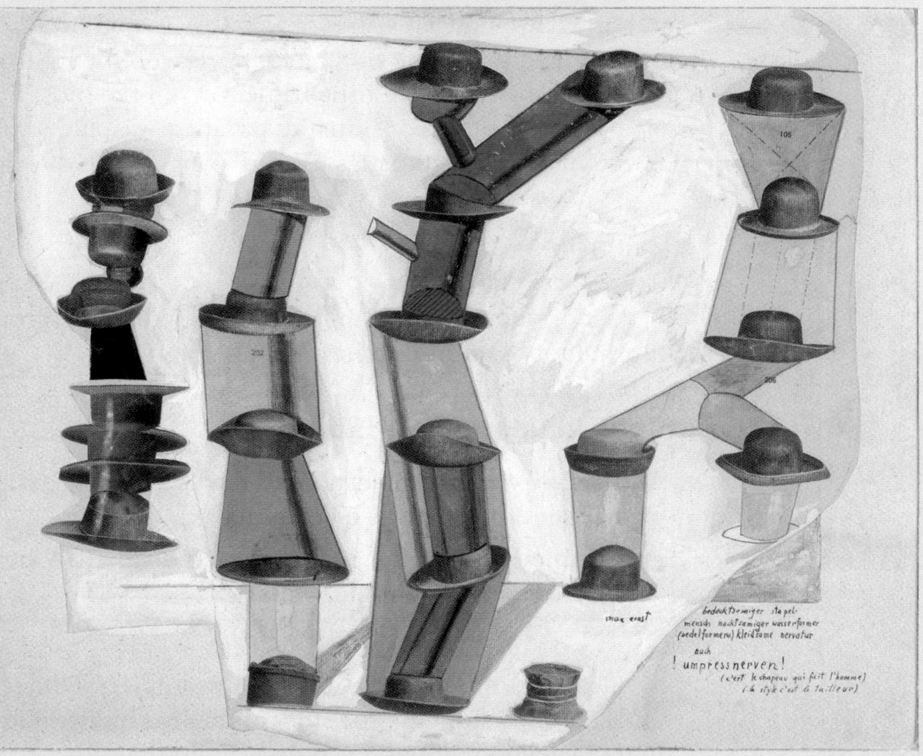

Max Ernst, *The Hat Makes the Man*, 1920.

Looking at Data—Distributions

1.1 Displaying Distributions with Graphs
1.2 Describing Distributions with Numbers
1.3 The Normal Distributions

Introduction

Statistics uses data to gain understanding. Understanding comes from combining knowledge of the background of the data with the ability to use graphs and calculations to see what the data tell us. The numbers in a medical study, for example, mean little without some knowledge of the goals of the study and of what blood pressure, heart rate, and other measurements contribute to those goals. On the other hand, measurements from the study's several hundred subjects are of little value to even the most knowledgeable medical expert until the tools of statistics organize, display, and summarize them. We begin our study of statistics by mastering the art of examining data.

Variables

Any set of data contains information about some group of *individuals*. The information is organized in *variables*.

> **Individuals and Variables**
>
> **Individuals** are the objects described by a set of data. Individuals may be people, but they may also be animals or things.
>
> A **variable** is any characteristic of an individual. A variable can take different values for different individuals.

A college's student data base, for example, includes data about every currently enrolled student. The students are the individuals described by the data set. For each individual, the data contain the values of variables such as date of birth, gender (female or male), choice of major, and grade point average. In practice, any set of data is accompanied by background information that helps us understand the data. When you plan a statistical study or explore data from someone else's work, ask yourself the following questions:

1. **Why? What purpose** do the data have? Do we hope to answer some specific questions? Do we want to draw conclusions about individuals other than those for whom we actually have data?
2. **Who?** What **individuals** do the data describe? **How many** individuals appear in the data?
3. **What?** How many **variables** do the data contain? What are the **exact definitions** of these variables? In what **units of measurement** is each variable recorded? Weights, for example, might be recorded in pounds, in thousands of pounds, or in kilograms.

Some variables, like gender and college major, simply place individuals into categories. Others, like height and grade point average, take numerical values for which we can do arithmetic. It makes sense to give an average income for a company's employees, but it does not make sense to give an "average" gender. We can, however, count the numbers of female and male employees and do arithmetic with these counts.

Categorical and Quantitative Variables

A **categorical variable** places an individual into one of several groups or categories.

A **quantitative variable** takes numerical values for which arithmetic operations such as adding and averaging make sense.

The **distribution** of a variable tells us what values it takes and how often it takes these values.

EXAMPLE 1.1

Here is a small part of the data set in which CyberStat Corporation records information about its employees:

	A	B	C	D	E	F
1	Name	Job Type	Age	Gender	Race	Salary
2	Cedillo, Jose	Technical	27	Male	White	52,300
3	Chambers, Tonia	Management	42	Female	Black	112,800
4	Childers, Amanda	Clerical	39	Female	White	27,500
5	Chen, Huabang	Technical	51	Male	Asian	83,600
6						
Ready				NUM		

case

spreadsheet

Each row records data on one individual. You will often see each row of data called a **case.** Each column contains the values of one variable for all the individuals. In addition to the person's name, there are 5 variables. Gender, race, and job type are categorical variables; age in years and salary in dollars are quantitative variables.

Most statistical software uses this format to enter data—each row is an individual, and each column is a variable. This data set appears in a **spreadsheet** program that has rows and columns ready for your use. Spreadsheets are commonly used to enter and transmit data. Most statistical software can read data from the major spreadsheet programs.

1.1 Displaying Distributions with Graphs

exploratory data analysis

Statistical tools and ideas help us examine data in order to describe their main features. This examination is called **exploratory data analysis.** Like an explorer crossing unknown lands, we want first to simply describe what we see. Here are two basic strategies that help us organize our exploration of a set of data:

- Begin by examining each variable by itself. Then move on to study the relationships among the variables.
- Begin with a graph or graphs. Then add numerical summaries of specific aspects of the data.

We will follow these principles in organizing our learning. This chapter presents methods for describing a single variable. We study relationships among several variables in Chapter 2. Within each chapter, we begin with graphical displays, then add numerical summaries for more complete description.

Graphs for categorical variables

The values of a categorical variable are labels for the categories, such as "male" and "female." The distribution of a categorical variable lists the categories and gives either the **count** or the **percent** of individuals who fall in each category. For example, how well educated are 30-something young adults? Here is the distribution of the highest level of education for people aged 25 to 34 years:[2]

Education	Count (millions)	Percent
Less than high school	4.7	12.3
High school graduate	11.8	30.7
Some college	10.9	28.3
Bachelor's degree	8.5	22.1
Advanced degree	2.5	6.6

bar graph

The graphs in Figure 1.1 display these data. The **bar graph** in Figure 1.1(a) quickly compares the sizes of the five education groups. The heights

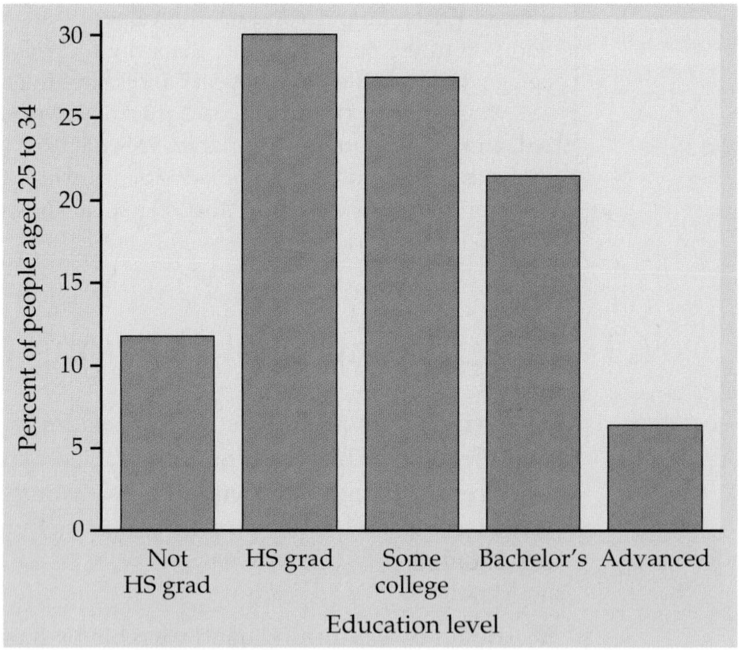

FIGURE 1.1(a) Bar graph of the educational attainment of people aged 25 to 34 years.

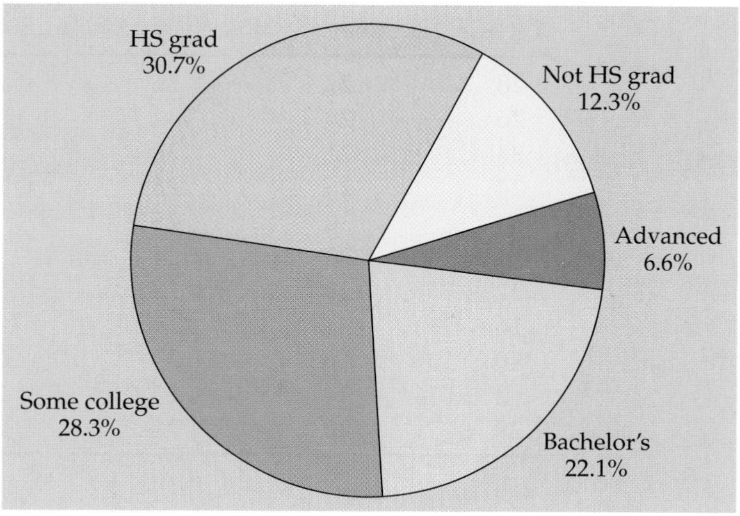

FIGURE 1.1(b) Pie chart of the education data.

pie chart

of the bars show the counts in the five categories. The **pie chart** in Figure 1.1(b) helps us see what part of the whole each group forms. For example, the "HS grad" slice makes up 30.7% of the pie because 30.7% of young adults have only a high school education. Because pie charts lack a scale, we have added the percents to the labels for the wedges. Pie charts require that you include all the categories that make up a whole. Bar graphs are more flexible. For example, you can use a bar graph to compare the numbers of students at your college majoring in biology, business, and political science. A pie chart cannot make this comparison because not all students fall into one of these three majors.

Bar graphs and pie charts help an audience grasp the distribution quickly. They are, however, of limited use for data analysis because it is easy to understand categorical data on a single variable such as educational attainment without a graph. We can move on to quantitative variables, where graphs are essential tools.

Measuring the speed of light

How do we begin to examine the distribution of a single quantitative variable? Let's look at an important scientific study.

EXAMPLE 1.2

Light travels fast, but it is not transmitted instantaneously. Light takes over a second to reach us from the moon and over 10 billion years to reach us from the most distant objects yet observed in the expanding universe. Because radio and radar also travel at the speed of light, an accurate value for that speed is important in communicating with astronauts and orbiting satellites. An accurate value for the speed of light is also important to computer designers because electrical signals travel at light speed. The first reasonably accurate measurements of the speed of light were made over 100 years ago by A. A. Michelson and Simon Newcomb. Table 1.1 contains 66 measurements made by Newcomb between July and September 1882.[3]

TABLE 1.1 Newcomb's measurements of the passage time of light

28	22	36	26	28	28
26	24	32	30	27	24
33	21	36	32	31	25
24	25	28	36	27	32
34	30	25	26	26	25
−44	23	21	30	33	29
27	29	28	22	26	27
16	31	29	36	32	28
40	19	37	23	32	29
−2	24	25	27	24	16
29	20	28	27	39	23

A set of numbers such as those in Table 1.1 is meaningless without some background information. The *individuals* here are Newcomb's 66 repetitions of his experiment. We need to know exactly *what variable* he measured, and in *what units*. Newcomb measured the time in seconds that a light signal took to pass from his laboratory on the Potomac River to a mirror at the base of the Washington Monument and back, a total distance of about 7400 meters. Just as you can compute the speed of a car from the time required to drive a mile, Newcomb could compute the speed of light from the passage time. Newcomb's first measurement of the passage time of light was 0.000024828 second, or 24,828 nanoseconds. (There are 10^9 nanoseconds in a second.) The entries in Table 1.1 record only the deviation from 24,800 nanoseconds. The table entry 28 is short for the original 0.000024828. The entry −44 stands for 24,756 nanoseconds.

Measurement

instrument

A detailed answer to the question "What variable is being measured?" requires a description of the **instrument** used to make the measurement. Then we must judge whether the variable measured is appropriate for our purpose. This judgment often requires expert knowledge of the particular field of study. For example, Newcomb invented a novel and complicated apparatus to measure the passage time of light. We accept the judgment of physicists that this instrument is appropriate for its intended task and more accurate than earlier instruments.

Newcomb measured a clearly defined and easily understood variable, the time light takes to travel a fixed distance. Questions about measurement are often harder to answer in the social and behavioral sciences than in the physical sciences. We can agree that a tape measure is an appropriate way to measure a person's height. But how shall we measure her intelligence? Questionnaires and interview forms as well as tape measures are measuring instruments. A psychologist wishing to measure "general intelligence" might use the Wechsler Adult Intelligence Scale (WAIS). The WAIS is a standard "IQ test" that asks subjects to solve a large number of problems. The appropriateness of this instrument as a measure of intelligence is not accepted by all psychologists, many of whom doubt that there is such a thing as "general intelligence." Because questions about measurement usually require knowledge of the particular field of study, we will say little about them.

rate

You should nonetheless always ask yourself if a particular variable really does measure what you want it to. Often, for example, the **rate** at which something occurs is a more meaningful measure than a simple **count** of occurrences.

EXAMPLE 1.3

The government's Fatal Accident Reporting System says that 20,818 passenger car occupants were killed in crashes in 1999, but only 2472 motorcycle riders were killed. Does this mean that motorcycles are safer than cars? Not at all—there are many more cars than motorcycles, and they are driven many more miles.

A better measure of the dangers of driving is a *rate*, the number of deaths divided by the number of miles driven. In 1999, passenger cars drove 1,566,979,000,000 miles in the United States. Because this number is so large, it is usual to give the death rate per 100 million miles driven. The 1999 fatality rate for passenger cars is

$$\frac{\text{traffic deaths}}{\text{hundreds of millions of miles driven}} = \frac{20{,}818}{15{,}669.79}$$
$$= 1.3$$

Motorcycles drove only 10,584,000,000 miles in 1999. There were 23.4 deaths per 100 million miles. Motorcycles are, as we surmise, much more dangerous than cars.[4]

Variation

Look again at Newcomb's observations in Table 1.1. They are not all the same. Why should this be? After all, each observation records the travel time of light over the same path, measured by a skilled observer using the same apparatus each time. Newcomb knew that careful measurements almost always vary. The environment of every measurement is slightly different. The apparatus changes a bit with the temperature, the density of the atmosphere changes from day to day, and so on. Newcomb did his best to eliminate the sources of variation that he could anticipate, but even the most careful experiments produce variable results. Newcomb took many measurements rather than just one because the average of 66 passage times should be less variable than the result of a single measurement. The average does not depend on the specific conditions at just one time.

Putting ourselves in Newcomb's place, we might rush to compute the average passage time, convert this time to a new and better estimate of the speed of light, and publish the result. The temptation to do a routine calculation and announce an answer arises whenever data have been collected to answer a specific question. Because computers are very good at routine calculations, succumbing to temptation is the easy road. The first step toward statistical sophistication is to resist the temptation to calculate without thinking. Since variation is surely present in any set of data, we must first examine the nature of the variation.

The distribution of a quantitative variable describes the pattern of variation of its values. Just as pie charts and bar graphs display the distribution of categorical variables, distributions of quantitative variables are best displayed graphically.

Stemplots

A *stemplot* (also called a stem-and-leaf plot) gives a quick picture of the shape of a distribution while including the actual numerical values in the graph. Stemplots work best for small numbers of observations that are all greater than 0.

> **Stemplot**
>
> To make a **stemplot:**
>
> 1. Separate each observation into a **stem** consisting of all but the final (rightmost) digit and a **leaf,** the final digit. Stems may have as many digits as needed, but each leaf contains only a single digit.
> 2. Write the stems in a vertical column with the smallest at the top, and draw a vertical line at the right of this column.
> 3. Write each leaf in the row to the right of its stem, in increasing order out from the stem.

EXAMPLE 1.4

Here are the numbers of home runs that Babe Ruth hit in each of his 15 years with the New York Yankees, 1920 to 1934:

54	59	35	41	46	25	47	60	54	46	49	46	41	34	22

To make a stemplot for these data, we use the first digits as stems and the second digits as leaves. Thus 54, for example, appears as a leaf 4 on the stem 5. Figure 1.2 shows the steps in making the plot. We see from the completed stemplot that Ruth hit 54 home runs in two different years.

```
2 |               2 | 52            2 | 25
3 |               3 | 54            3 | 45
4 |               4 | 1676961       4 | 1166679
5 |               5 | 494           5 | 449
6 |               6 | 0             6 | 0
   (a)               (b)               (c)
```

FIGURE 1.2 Making a stemplot of the data in Example 1.4. (a) Write the stems. (b) Go through the data writing each leaf on the proper stem. (c) Arrange the leaves on each stem in order out from the stem.

back-to-back stemplot

When you wish to compare two related distributions, a **back-to-back stemplot** with common stems is useful. The leaves on each side are ordered out from the common stem. The leading contemporary power hitter is Mark McGwire of the St. Louis Cardinals, who retired at the end of the 2001 season.[5] Here is a back-to-back stemplot comparing the home run data for Ruth and McGwire.

Ruth		McGwire
	0	9 9
	1	
5 2	2	2 9
5 4	3	2 2 3 9 9
9 7 6 6 6 1 1	4	2 9
9 4 4	5	2 8
0	6	5
	7	0

McGwire matches up well with the legendary Ruth. The two low observations are explained by an injury in 1993 and a players' strike in 1994. In fact, McGwire was injured in 2000 and 2001 as well. He hit 32 and 29 home runs in half seasons in those years, respectable full-season totals for most players.

splitting stems
rounding

Stemplots do not work well for large data sets, where each stem must hold a large number of leaves. Fortunately, there are several modifications of the basic stemplot that are helpful when plotting the distribution of a moderate number of observations. You can increase the number of stems in a plot by **splitting each stem** into two: one with leaves 0 to 4 and the other with leaves 5 through 9. When the observed values have many digits, it is often best to **round** the numbers to just a few digits before making a stemplot. You must use your judgment in deciding whether to split stems and whether to round, though statistical software will often make these choices for you. Remember that the purpose of a stemplot is to display the shape of a distribution. If a stemplot has fewer than about five stems, you should usually split the stems unless there are few observations. If many stems have no leaves or only one leaf, rounding may help. Here is an example that makes use of both of these modifications.

EXAMPLE 1.5

A marketing consultant observed 50 consecutive shoppers at a supermarket. One variable of interest was how much each shopper spent in the store. Here are the data (in dollars), arranged in increasing order:

3.11	8.88	9.26	10.81	12.69	13.78	15.23	15.62	17.00	17.39
18.36	18.43	19.27	19.50	19.54	20.16	20.59	22.22	23.04	24.47
24.58	25.13	26.24	26.26	27.65	28.06	28.08	28.38	32.03	34.98
36.37	38.64	39.16	41.02	42.97	44.08	44.67	45.40	46.69	48.65
50.39	52.75	54.80	59.07	61.22	70.32	82.70	85.76	86.37	93.34

To make a stemplot of this distribution, we first round the purchases to the nearest dollar. We can then use tens of dollars as our stems and dollars as leaves. This gives us the single-digit leaves that a stemplot requires. For example, we round 12.69 to

(a)

0	3 9 9
1	1 3 4 5 6 7 7 8 8 9 9
2	0 0 1 2 3 4 5 5 6 6 8 8 8 8
3	2 5 6 9 9
4	1 3 4 5 5 7 9
5	0 3 5 9
6	1
7	0
8	3 6 6
9	3

(b)

0	3
0	9 9
1	1 3 4
1	5 6 7 7 8 8 9 9
2	0 0 1 2 3 4
2	5 5 6 6 8 8 8 8
3	2
3	5 6 9 9
4	1 3 4
4	5 5 7 9
5	0 3
5	5 9
6	1
6	
7	0
7	
8	3
8	6 6
9	3

FIGURE 1.3 Stemplots of the amounts spent by 50 supermarket shoppers (in dollars), with and without splitting the stems.

13 and write it as a leaf of 3 on the 1 stem. Figure 1.3(a) displays the stemplot. For comparison, we split each stem to make the stemplot in Figure 1.3(b). Both graphs show the distribution shape, but the split stems provide more detail.

Examining distributions

Making a statistical graph is not an end in itself. The purpose of the graph is to help us understand the data. After you make a graph, always ask, "What do I see?" Once you have displayed a distribution, you can see its important features as follows.

> **Examining a Distribution**
>
> In any graph of data, look for the **overall pattern** and for striking **deviations** from that pattern.
>
> You can describe the overall pattern of a distribution by its **shape, center,** and **spread.**
>
> An important kind of deviation is an **outlier,** an individual value that falls outside the overall pattern.

We will learn how to describe center and spread numerically in Section 1.2. For now, we can describe the center of a distribution by its *midpoint*, the value with roughly half the observations taking smaller values and half taking larger values. We can describe the spread of a distribution by giving the range between the *smallest and largest values*. To see the shape of a distribution more clearly, turn the stemplot on its side so that the larger values lie to the right. Some things to look for in describing shape are:

modes
unimodal

- Does the distribution have one or several major peaks, called **modes?** A distribution with one major peak is called **unimodal.**

symmetric
skewed

- Is it approximately symmetric or is it skewed in one direction? A distribution is **symmetric** if the values smaller and larger than its midpoint are mirror images of each other. It is **skewed to the right** if the right tail (larger values) is much longer than the left tail (smaller values).

EXAMPLE 1.6

The distribution of Babe Ruth's home run counts (Figure 1.2) is symmetric and unimodal. Note that when describing data we do not insist on exact symmetry. Our goal is an approximate overall description, not mathematical exactness. The midpoint is 46 home runs (count up to the eighth of the 15 values), and the range is from 22 to 60. There are no outliers. In particular, Ruth's record 60 home runs in one season does not stand outside his overall pattern.

EXAMPLE 1.7

Mark McGwire's 70 home run season is also not an outlier. There are, however, two low outliers in McGwire's career, two seasons in which he hit only 9 home runs. You should search for an explanation for any outlier. Sometimes outliers point to errors made in recording the data. In other cases, the outlying observation may be caused by equipment failure or other unusual circumstances. In McGwire's case, the low outliers are not complete seasons: he played only 27 games in 1993 because of an injury and only 47 games in 1994 due to a players' strike. McGwire's home run counts, like many distributions with only a few observations, have an irregular shape. We might describe the distribution, omitting the outliers, as skewed to the right with midpoint around 40 and range 22 to 70.

EXAMPLE 1.8

The distribution of supermarket spending (Figure 1.3) is skewed to the right. There are many moderate values and a few quite large dollar amounts. The direction of the long tail (to the right) determines the direction of the skewness. Because the largest amounts do not fall outside the overall skewed pattern, we do not call them outliers. The midpoint of the distribution (count up 25 entries in the stemplot) is \$28. The range is \$3 to \$93. The shape revealed by the stemplot will interest the store management. If the big spenders can be studied in more detail, steps to attract them may be profitable. Finally, this distribution is unimodal. Although there are several minor bulges in the stemplot, there is only one major peak, containing the 22 shoppers who spent between \$15 and \$28.

The right-skewed shape in Figure 1.3 is common among distributions of money amounts. There are many moderately priced houses, for example, but the few very expensive mansions give the distribution of house prices a strong right skew.

Histograms

Stemplots display the actual values of the observations. This feature makes stemplots awkward for large data sets. Moreover, the picture presented by a stemplot divides the observations into groups (stems) determined by the number system rather than by judgment. Histograms do not have these limitations. A **histogram** breaks the range of values of a variable into intervals and displays only the count or percent of the observations that fall into each interval. You can choose any convenient number of intervals, but you should always choose intervals of equal width. Histograms are slower to construct by hand than stemplots and do not display the actual values observed. For these reasons we prefer stemplots for small data sets. The construction of a histogram is best shown by example. Any statistical software package will of course make a histogram for you.

histogram

EXAMPLE 1.9

One of the most striking findings of the 2000 census was the growth of the Hispanic population in the United States. Table 1.2 presents the percent of adults (age 18 and over) in each of the 50 states who identified themselves in the 2000 census as "Spanish/Hispanic/Latino."[6] To make a histogram of this distribution, proceed as follows:

1. Divide the range of the data into classes of equal width. The data in Table 1.2 range from 0.6 to 38.7, so we choose as our classes

$$0.0 < \text{percent Hispanic} \le 5.0$$
$$5.0 < \text{percent Hispanic} \le 10.0$$
$$\vdots$$
$$35.0 < \text{percent Hispanic} \le 40.0$$

TABLE 1.2 Percent of Hispanics in the adult population, by state (2000)

State	Percent	State	Percent	State	Percent
Alabama	1.5	Louisiana	2.4	Ohio	1.6
Alaska	3.6	Maine	0.6	Oklahoma	4.3
Arizona	21.3	Maryland	4.0	Oregon	6.5
Arkansas	2.8	Massachusetts	5.6	Pennsylvania	2.6
California	28.1	Michigan	2.7	Rhode Island	7.0
Colorado	14.9	Minnesota	2.4	South Carolina	2.2
Connecticut	8.0	Mississippi	1.3	South Dakota	1.2
Delaware	4.0	Missouri	1.8	Tennessee	2.0
Florida	16.1	Montana	1.6	Texas	28.6
Georgia	5.0	Nebraska	4.5	Utah	8.1
Hawaii	5.7	Nevada	16.7	Vermont	0.8
Idaho	6.4	New Hampshire	1.4	Virginia	4.2
Illinois	10.7	New Jersey	12.3	Washington	6.0
Indiana	3.1	New Mexico	38.7	West Virginia	0.6
Iowa	2.3	New York	13.8	Wisconsin	2.9
Kansas	5.8	North Carolina	4.3	Wyoming	5.5
Kentucky	1.3	North Dakota	1.0		

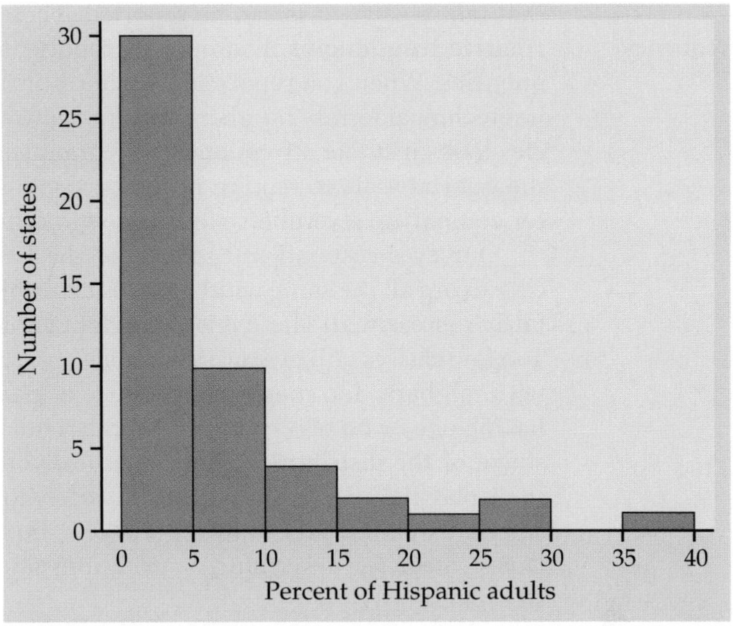

FIGURE 1.4 Histogram of the percent of each state's adult residents who identified themselves as Hispanic in the 2000 census.

Be sure to specify the classes precisely so that each individual falls into exactly one class. A state with 5.0% Hispanic residents would fall into the first class, but 5.1% falls into the second.

frequency

2. Count the number of individuals in each class. These counts are called **frequencies,** and a table of frequencies for all classes is a **frequency table.**

Class	Count	Percent	Class	Count	Percent
0.1 to 5.0	30	60.0	20.1 to 25.0	1	2.0
5.1 to 10.0	10	20.0	25.1 to 30.0	2	4.0
10.1 to 15.0	4	8.0	30.1 to 35.0	0	0.0
15.1 to 20.0	2	4.0	35.1 to 40.0	1	2.0

3. Draw the histogram. First mark the scale for the variable whose distribution you are displaying on the horizontal axis. That's the percent of adults who are Hispanic. The scale runs from 0 to 40 because that is the span of the classes we chose. The vertical axis contains the scale of counts. Each bar represents a class. The base of the bar covers the class, and the bar height is the class count. There is no horizontal space between the bars unless a class is empty, so that its bar has height zero. Figure 1.4 is our histogram.

Large sets of data are often reported in the form of frequency tables when it is not practical to publish the individual observations. In addition to the frequency (count) for each class, the fraction or percent of the observations relative frequency

that fall in each class can be reported. These fractions are sometimes called **relative frequencies.** Example 1.9 reports both frequencies and relative frequencies. When you report statistical results, however, it is clearer to use the nontechnical terms "number" and "percent." A histogram of relative frequencies looks just like a frequency histogram such as Figure 1.4. Simply relabel the vertical scale to read in percents. Use histograms of relative frequencies for comparing several distributions with different numbers of observations.

Our eyes respond to the *area* of the bars in a histogram. Because the classes are all the same width, area is determined by height and all classes are fairly represented. There is no one right choice of the classes in a histogram. Too few classes will give a "skyscraper" graph, with all values in a few classes with tall bars. Too many will produce a "pancake" graph, with most classes having one or no observations. Neither choice will give a good picture of the shape of the distribution. You must use your judgment in choosing classes to display the shape. Statistical software will choose the classes for you. The software's choice is usually a good one, but you can change it if you want. Our strategy for describing distributions applies to histograms as well as to stemplots.

EXAMPLE 1.10

What does the histogram in Figure 1.4 tell us? **Shape:** The distribution is *skewed to the right* with a *single peak* at the left. Most states have no more than 10% Hispanics, but some states have much higher percents. **Center:** About half the states have fewer than 4% Hispanics among their adult residents and half have more. So the midpoint of the distribution is close to 4%. **Spread:** The spread is from about 0% to 40%, but only one state falls above 30%.

Outliers: The one state that stands out is New Mexico, with 38.7% Hispanics. Other states (Arizona, California, Texas) with high percents are part of the long right tail of the distribution, but New Mexico is 10% higher than any other state. Outliers often point to the special nature of some observations. New Mexico is heavily Hispanic by history and location.

Although histograms resemble bar graphs, their details and uses are distinct. A histogram shows the distribution of frequencies or relative frequencies among the values of a single variable. A bar graph compares the size of different items. The horizontal axis of a bar graph need not have any measurement

Exploring the Web

The U.S. government offers a river of information online. Want to know the percent of each state's residents who are Hispanic? Go to the Census Bureau, `www.census.gov`, and look at the results of the 2000 census. What about fatal traffic accidents? Visit the home of the Fatal Accident Reporting System at the National Highway Traffic Safety Administration, `www.nhtsa.dot.gov`. Doctoral degrees earned by women? The National Center for Education Statistics has the facts at `nces.ed.gov`. Best of all, there is one gateway to the more than 70 federal Web sites: FedStats, at `www.fedstats.gov`.

scale but simply identifies the items being compared. Draw bar graphs with blank space between the bars to separate the items being compared. Draw histograms with no space, to indicate that all values of the variable are covered. Some software not primarily intended for statistics, such as spreadsheet programs, will draw histograms as if they were bar graphs, with space between the bars. You can tell the software to eliminate the space in order to produce a proper histogram.

Dealing with outliers

What of Simon Newcomb and the speed of light? Scientists know that careful measurements will vary, but they hope to see a symmetric unimodal distribution. If there is no systematic bias in the measurements, the best estimate of the true value of the measured quantity is the center of the distribution. The histogram in Figure 1.5 shows that Newcomb's data do have a symmetric unimodal distribution—but there are two low outliers that stand outside this pattern. What should he do?

The handling of outliers is a matter for judgment. Outliers can point to the unusual nature of some observations, such as Mark McGwire's incomplete seasons in Example 1.7. Newcomb's data, however, are just repeated measurements of the same quantity, so the two outliers (-44 and -2 in Table 1.1) were disturbing. Outliers in research studies can result from equipment failure or errors in recording the data. If equipment failure or some other abnormal condition caused the outlier, we can simply delete it. We may be able to correct recording errors by looking at the original data record. When no cause is found, the decision is difficult. The outlier may be evidence of an extraordinary occurrence or just a mistake that went unnoticed. Newcomb decided to drop the worse outlier (-44) and to keep the other. He based his estimate of

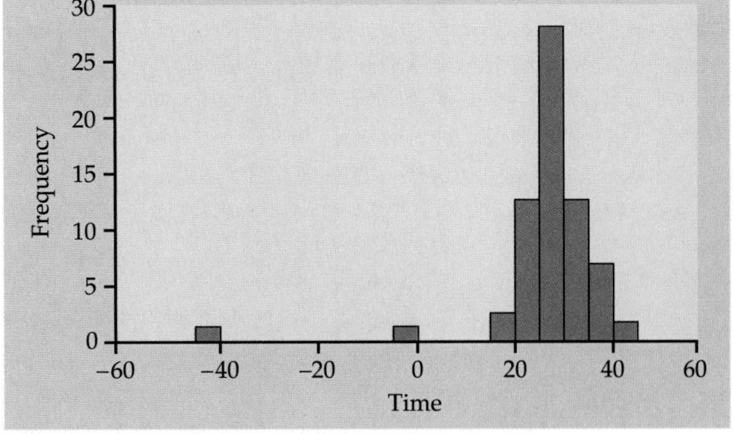

FIGURE 1.5 Histogram of Simon Newcomb's 66 measurements of the passage time of light, from Table 1.1.

the speed of light on the average (the mean, in the language of Section 1.2) of his observations. The mean of all 66 observations is 26.21. The mean of the 65 observations that he kept is 27.29. The strong effect of the single value −44 on the mean is one motivation for discarding it when we are interested in the center of the distribution as a whole.

Time plots

We can gain more insight into Newcomb's data from a different kind of graph. When data represent similar observations made over time, it is wise to plot them against either time or against the order in which the observations were taken. Figure 1.6 plots Newcomb's passage times for light in the order that they were recorded. There is some suggestion in this plot that the variability (the vertical spread in the plot) is decreasing over time. In particular, both outlying observations were made early on. Perhaps Newcomb became more adept at using his equipment as he gained experience. Learning effects like this are quite common. If we allow Newcomb 20 observations for learning, the mean of the remaining 46 measurements is 28.15. The best modern measurements suggest that the "true value" for the passage time in Newcomb's experiment is 33.02. Allowing for learning does move the average result closer to the true value.

> **Time Plot**
>
> A **time plot** of a variable plots each observation against the time at which it was measured. Always put time on the horizontal scale of your plot and the variable you are measuring on the vertical scale. Connecting the data points by lines helps emphasize any change over time.

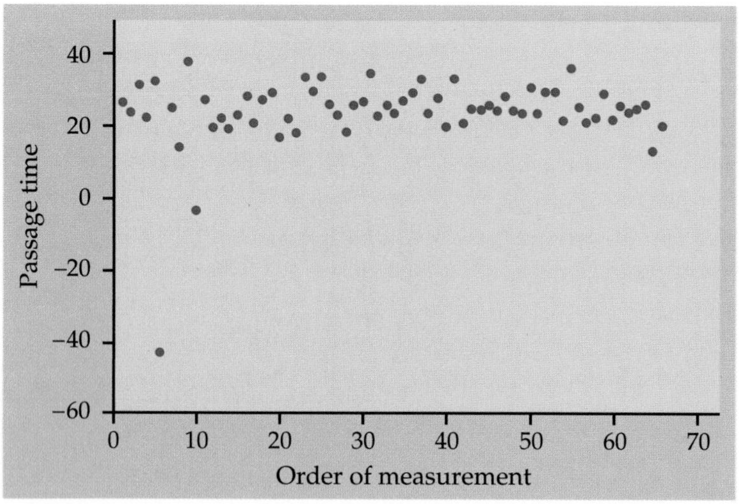

FIGURE 1.6 Plot of Newcomb's measurements against the time order in which they were made.

Whenever data are collected over time, it is a good idea to plot the observations in time order. Summaries of the distribution of a variable that ignore time order, such as stemplots and histograms, can be misleading when there is systematic change over time.

time series

Many interesting data sets are **time series,** measurements of a variable taken at regular intervals over time. Government economic and social data are often published as time series. Some examples are the monthly unemployment rate and the quarterly gross domestic product. Weather records, the demand for electricity, and measurements on the items produced by a manufacturing process are other examples of time series.

Time plots can reveal the main features of a time series. As in our examination of distributions, we look first for overall patterns and then for striking deviations from those patterns. Here are some types of overall patterns to look for in a time series.

Seasonal Variation and Trend

A pattern in a time series that repeats itself at known regular intervals of time is called **seasonal variation.**

A **trend** in a time series is a persistent, long-term rise or fall.

EXAMPLE 1.11

Figure 1.7 is a time plot of the average retail price of gasoline over the period 1988 to April 2001.[7] The plot shows the upward spike in prices due to the 1990 Iraqi invasion of Kuwait, the drop in 1998 when an economic crisis in Asia reduced demand for fuel,

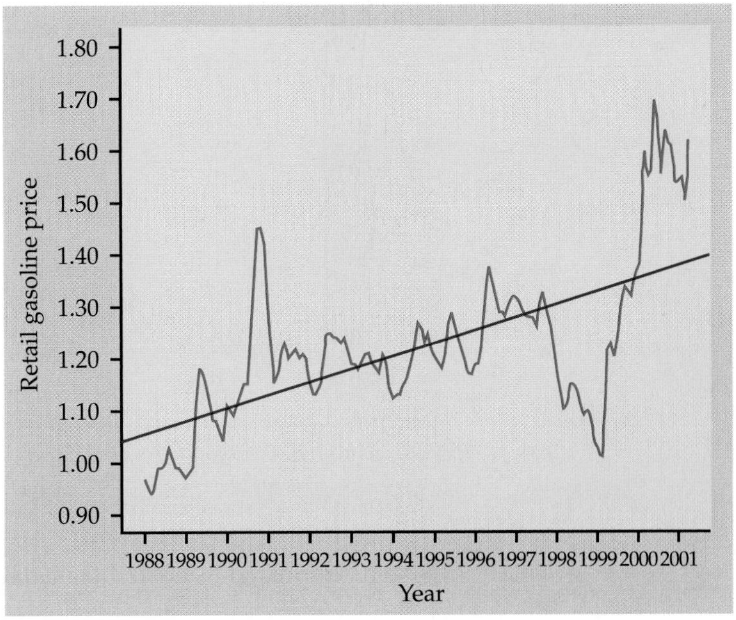

FIGURE 1.7 Time plot of average retail price of gasoline, 1988 to April 2001. The line shows the increasing trend.

and the rapid price increase in 2000 when demand recovered and OPEC reduced production. These deviations are so large that overall patterns are hard to see.

There is nonetheless a clear *trend* of increasing price. Using statistical software, we have drawn a line on the time plot in Figure 1.7 that represents the trend. Superimposed on this trend is *seasonal variation*, a regular rise and fall that recurs each year. Americans drive more in the summer vacation season, so the price of gasoline rises each spring, then drops in the fall as demand goes down.

seasonally adjusted

Because many economic time series show strong seasonal variation, government agencies often adjust for this variation before releasing economic data. The data are then said to be **seasonally adjusted.** Seasonal adjustment helps avoid misinterpretation. A rise in the unemployment rate from December to January, for example, does not mean that the economy is slipping. Unemployment almost always rises in January as temporary holiday help is laid off and outdoor employment in the north drops because of bad weather. The seasonally adjusted unemployment rate reports an increase only if unemployment rises more than normal from December to January.

Beyond the basics ▸ decomposing time series*

Statistical software can help us examine a time series by "decomposing" the data into systematic patterns such as trends and seasonal variation and the *residuals* that remain after we remove these patterns. Figure 1.8 shows the combined trend and seasonal variation for the time series of gasoline prices

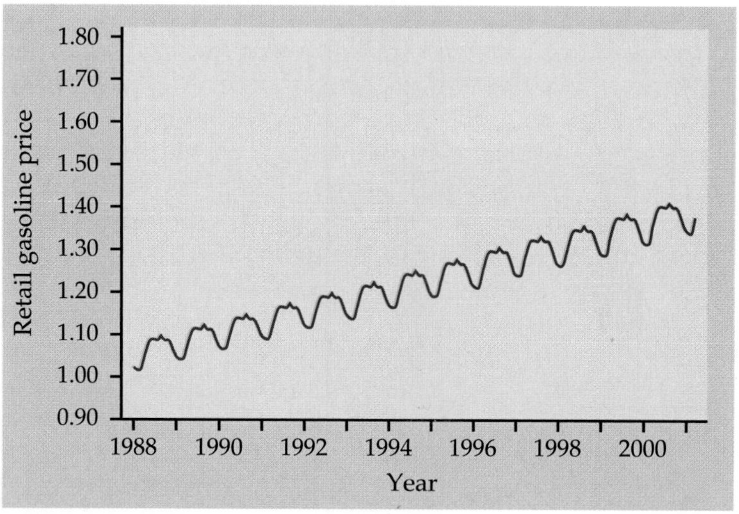

FIGURE 1.8 The estimated trend and seasonal variation in gasoline prices.

*The "Beyond the Basics" sections briefly discuss supplementary topics. Your software may make some of these available to you. For example, the results plotted in Figures 1.8 and 1.9 come from the Minitab statistical software.

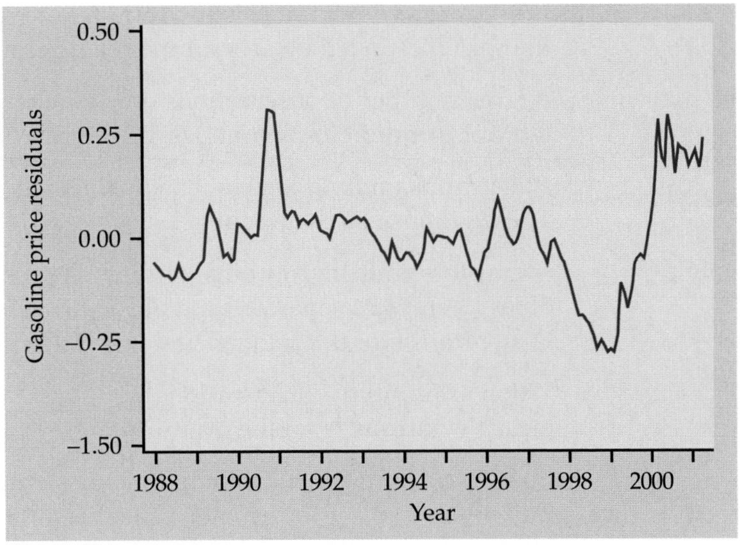

FIGURE 1.9 The erratic fluctuations that remain when both the trend and the seasonal variation are removed from gasoline prices.

in Figure 1.7. The seasonal variation represents an average of the seasonal pattern for all the years in the original data, automatically extracted by software. Figure 1.8 superimposes this average seasonal pattern on the trend. In terms of our strategy for looking at graphs, Figure 1.8 displays the overall pattern of the time plot. This pattern is well hidden by the erratic fluctuations in the data, so decomposing the time series is helpful.

Can we use the overall pattern in Figure 1.8 to predict next year's gasoline prices? Software also subtracts the trend and the seasonal variation from the original data, giving the residuals in Figure 1.9. This is a graph of the deviations of gasoline prices from the overall pattern of Figure 1.8. Figure 1.9 shows a great deal of erratic fluctuation, as well as the major deviations in 1990, 1998, and 2000 that we pointed out in Example 1.11. It is clear that trend and seasonal variation do not allow us to predict gasoline prices at all accurately.

SUMMARY

A data set contains information on a collection of **individuals.** Individuals may be people, animals, or things. The data for one individual make up a **case.** For each individual, the data give values for one or more **variables.** A variable describes some characteristic of an individual, such as a person's height, gender, or salary.

Some variables are **categorical** and others are **quantitative.** A categorical variable places each individual into a category, such as male or female. A quantitative variable has numerical values that measure some characteristic of each individual, such as height in centimeters or annual salary in dollars.

Exploratory data analysis uses graphs and numerical summaries to describe the variables in a data set and the relations among them.

A large number of observations on a single variable can be summarized in a table of **frequencies** (counts) or **relative frequencies** (percents or fractions).

Bar graphs and **pie charts** display the distributions of categorical variables. These graphs use the frequencies or relative frequencies of the categories.

Stemplots and **histograms** display the distributions of quantitative variables. Stemplots separate each observation into a **stem** and a one-digit **leaf.** Histograms plot the frequencies or relative frequencies of classes of values.

When examining a distribution, look for **shape, center,** and **spread** and for clear **deviations** from the overall shape.

Some distributions have simple shapes, such as **symmetric** and **skewed.** The number of **modes** (major peaks) is another aspect of overall shape. Not all distributions have a simple overall shape, especially when there are few observations.

Outliers are observations that lie outside the overall pattern of a distribution. Always look for outliers and try to explain them.

When observations on a variable are taken over time, make a **time plot** that graphs time horizontally and the values of the variable vertically. A time plot can reveal **trends** or other changes over time.

SECTION 1.1 EXERCISES

1.1 Data from a medical study contain values of many variables for each of the people who were the subjects of the study. Which of the following variables are categorical and which are quantitative?

(a) Gender (female or male)

(b) Age (years)

(c) Race (Asian, black, white, or other)

(d) Smoker (yes or no)

(e) Systolic blood pressure (millimeters of mercury)

(f) Level of calcium in the blood (micrograms per milliliter)

1.2 Here are the first lines of a professor's data set at the end of a statistics course:

NAME	MAJOR	POINTS	GRADE
ADVANI, SURA	COMM	397	B
BARTON, DAVID	HIST	323	C
BROWN, ANNETTE	BIOL	446	A
CHIU, SUN	PSYC	405	B
CORTEZ, MARIA	PSYC	461	A

What are the individuals and the variables in these data? Which variables are categorical and which are quantitative?

1.3 Here is a small part of a data set that describes mutual funds available to the public:

Fund	Category	Net assets (millions of $)	Year-to-date return	Largest holding
⋮				
Fidelity Low-Priced Stock	Small value	6,189	4.56%	Dallas Semiconductor
Price International Stock	International stock	9,745	−0.45%	Vodafone
Vanguard 500 Index	Large blend	89,394	3.45%	General Electric
⋮				

What individuals does this data set describe? In addition to the fund's name, how many variables does the data set contain? Which of these variables are categorical and which are quantitative?

1.4 You want to measure the amount of "leisure time" that college students enjoy. Write a brief discussion of two issues:

(a) How will you define "leisure time"?

(b) Once you have defined leisure time, how will you measure Sally's leisure time this week?

1.5 You want to measure the "physical fitness" of college students. Describe several variables you might use to measure fitness. What instrument or instruments does each measurement require?

1.6 Popular magazines rank colleges and universities on their "academic quality" in serving undergraduate students. Describe five variables that you would like to see measured for each college if you were choosing where to study. Give reasons for each of your choices.

1.7 Congress wants the medical establishment to show that progress is being made in fighting cancer. Some variables that might be used are:[8]

(a) Total deaths from cancer. These have risen over time, from 331,000 in 1970 to 505,000 in 1980 to 552,000 in 2000.

(b) The percent of all Americans who die from cancer. The percent of deaths due to cancer has also risen steadily, from 17.2% in 1970 to 20.9% in 1980 to 23.0% in 2000.

(c) The percent of cancer patients who survive for five years from the time the disease is discovered. These rates are rising slowly. For whites, the five-year survival rate was 50.9% in the 1974 to 1979 period and 61.5% from 1989 to 1996.

Discuss the usefulness of each of these variables as a measure of the effectiveness of cancer treatment. In particular, explain why both (a) and (b) could increase even if treatment is getting more effective, and why (c) could increase even if treatment is getting less effective.

1.8 Is driving becoming more dangerous? Traffic deaths declined for years, bottoming out at 39,250 killed in 1992, then began to increase again. In 1999, 41,611 people died in traffic accidents. But more people drove more miles in 1999 than in 1992. In fact, the government says that motor vehicles traveled 2247 billion miles in 1992 and 2691 billion miles in 1999. Reports on transportation use deaths per 100 million miles as a measure of risk. Compare the rates for 1992 and 1999. What do you conclude?

1.9 The National Highway Traffic Safety Administration says that an average of 11 children die each year in school bus accidents, and an average of 600 school-age children die each year in auto accidents during school hours. These numbers suggest that riding the bus is safer than driving to school with a parent. The *counts* aren't fully convincing, however. What *rates* would you like to know to compare the safety of bus and private auto?

1.10 You are writing an article for a consumer magazine based on a survey of the magazine's readers on the reliability of their household appliances. Of 13,376 readers who reported owning Brand A dishwashers, 2942 required a service call during the past year. Only 192 service calls were reported by the 480 readers who owned Brand B dishwashers. Describe an appropriate variable to measure the reliability of a make of dishwasher, and compute the values of this variable for Brand A and for Brand B.

1.11 In 1997, there were 12,298,000 undergraduate students in U.S. colleges. According to the U.S. Department of Education, there were 127,000 American Indian or Alaskan Native students, 737,000 Asian or Pacific Islander, 1,380,000 non-Hispanic black, 1,108,000 Hispanic, and 8,682,000 non-Hispanic white students. In addition, 265,000 foreign undergraduates were enrolled in U.S. colleges.[9]

roundoff error

(a) Each number, including the total, is rounded to the nearest thousand. Separate rounding may cause **roundoff errors,** so that the sum of the counts does not equal the total given. Are roundoff errors present in these data?

(b) Present the data in a graph.

1.12 The number of deaths among persons aged 15 to 24 years in the United States in 1999 due to the eight leading causes of death for this age group were: accidents, 13,602; homicide, 4989; suicide, 3885; cancer, 1724; heart disease, 1048; congenital defects, 430; respiratory disease, 208; AIDS, 197.[10]

(a) Make a bar graph to display these data.

(b) What additional information do you need to make a pie chart?

1.13 According to the 2000 census, there are 105.5 million households in the United States. A household consists of people living together in the same residence, regardless of their relationship to each other. Of these, 71.8 million were "family households" in which at least one other person was related to the householder by blood, marriage, or adoption. The family households include 54.5 million headed by a married couple and 17.3 million other families (for example, a single parent with children). The other 33.7 million households are "nonfamily households." Of these, 27.2 million contain a person living alone, 5.5 million are unmarried couples living together, and 1 million consist of other unrelated people living together.[11] Be creative: make a bar graph that displays these facts, including the distinction between family and nonfamily households.

1.14 Here are the percents of doctoral degrees in each of several subjects that were earned by women in 1997–98: psychology, 67.5%; education, 63.2%; life sciences, 42.5%; business, 31.4%; physical sciences, 25.2%; engineering, 12.2%.[12]

(a) Explain clearly why we cannot use a pie chart to display these data, even if we knew the percent female for every academic subject.

(b) Make a bar graph of the data. (Comparisons are easier if you order the bars by height, which is the order in which we give the percents. A bar graph ordered from tallest to shortest bar is sometimes called a **Pareto chart,** after the Italian economist who recommended this procedure.)

Pareto chart

1.15 Here is a stemplot of the percents of residents aged 25 to 34 in each of the 50 states. The stems are whole percents and the leaves are tenths of a percent.

```
10 | 9
11 | 0
12 | 1 3 4 4 6 7 7 8 8 9
13 | 0 0 1 2 4 5 5 5 6 6 7 8 9 9 9 9
14 | 1 1 2 2 2 3 4 4 4 4 5 7 8 9
15 | 2 4 4 7 8 9 9 9
```

(a) Montana and Wyoming have the smallest percents of young adults, perhaps because they lack job opportunities. What are the percents for these two states?

(b) Ignoring Montana and Wyoming, describe the shape, center, and spread of this distribution.

1.16 Make another stemplot of the percent of residents aged 25 to 34 in each of the 50 states by splitting the stems in the plot from the previous exercise. Which plot do you prefer? Why?

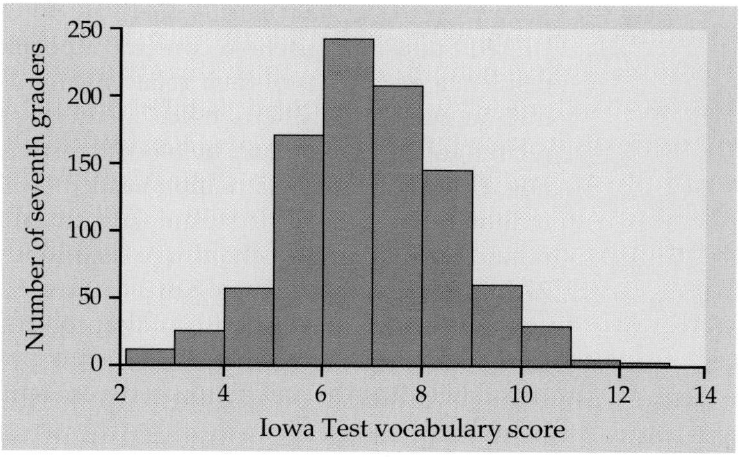

FIGURE 1.10 Histogram of the Iowa Test of Basic Skills vocabulary scores of seventh-grade students in Gary, Indiana, for Exercise 1.17.

1.17 Figure 1.10 displays the scores of all 947 seventh-grade students in the public schools of Gary, Indiana, on the vocabulary part of the Iowa Test of Basic Skills.[13] Give a brief description of the overall pattern (shape, center, spread) of this distribution.

1.18 Figure 1.11 is a histogram of the lengths of words used in Shakespeare's plays. Because there are so many words in the plays, we use a relative

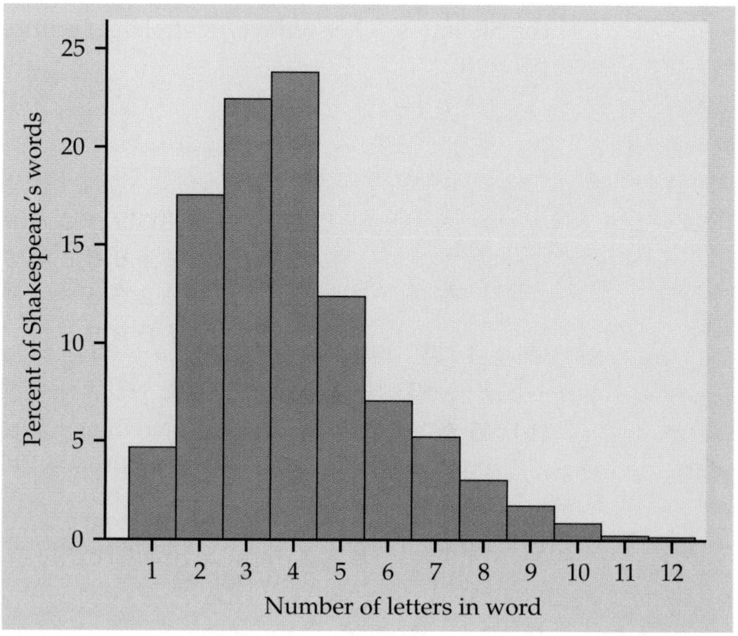

FIGURE 1.11 Histogram of lengths of words used in Shakespeare's plays, for Exercise 1.18.

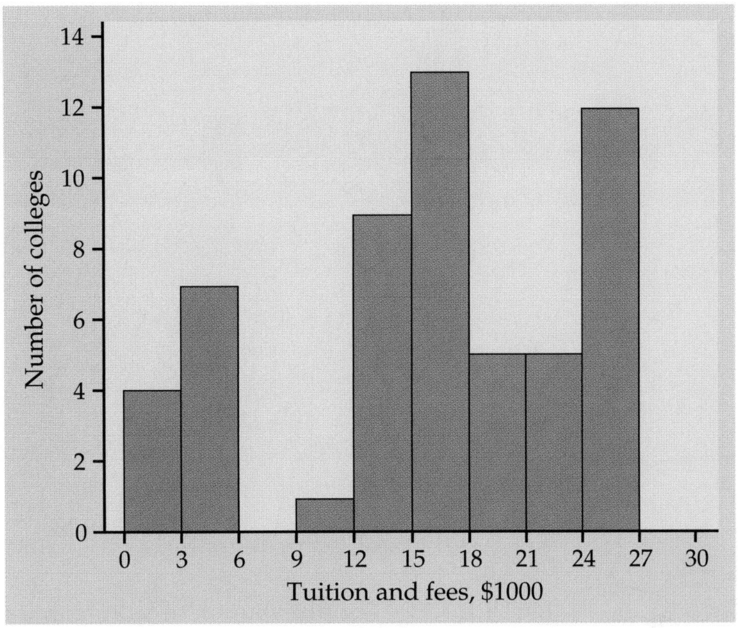

FIGURE 1.12 Histogram of the tuition and fees charged by all four-year colleges in Massachusetts, for Exercise 1.19.

frequency histogram. What is the overall shape of this distribution? What does this shape say about word lengths in Shakespeare? Do you expect other authors to have word length distributions of the same general shape? Why?

1.19 Jeanna plans to attend college in her home state of Massachusetts. In the College Board's *Annual Survey of Colleges*, she finds data on college tuition and fees for the 2000–2001 academic year. Figure 1.12 displays the costs for all 56 four-year colleges in Massachusetts (omitting art schools and other special colleges). For state schools, we used the in-state tuition. What is the most important aspect of the overall pattern of this distribution? Why do you think this pattern appears?

1.20 Outliers are sometimes the most interesting feature of a distribution. Figure 1.13 displays the distribution of batting averages for all 167 American League baseball players who batted at least 200 times in the 1980 season. The outlier is the .390 batting average of George Brett, the highest batting average in the major leagues since Ted Williams hit .406 in 1941. (See Exercise 1.87 on page 86 for a comparison of Brett and Williams.) Is the overall shape (ignoring the outlier) roughly symmetric or clearly skewed? What is the approximate midpoint of American League batting averages? What is the range if we ignore the outlier?

1.21 How much oil the wells in a given field will ultimately produce is key information in deciding whether to drill more wells. Here are the estimated

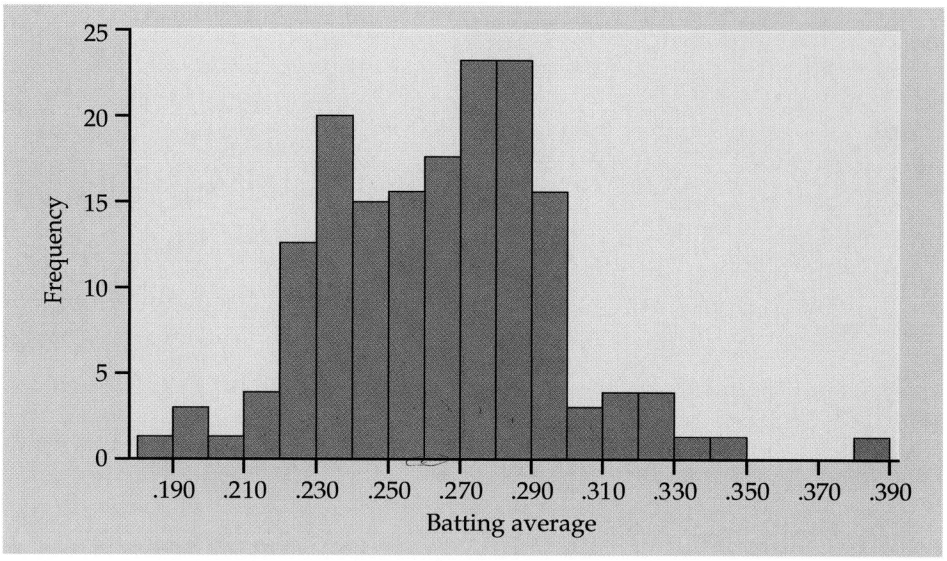

FIGURE 1.13 The distribution of batting averages of American League players in 1980, for Exercise 1.20.

total amounts of oil recovered from 64 wells in the Devonian Richmond Dolomite area of the Michigan basin, in thousands of barrels:[14]

21.71	53.2	46.4	42.7	50.4	97.7	103.1	51.9
43.4	69.5	156.5	34.6	37.9	12.9	2.5	31.4
79.5	26.9	18.5	14.7	32.9	196	24.9	118.2
82.2	35.1	47.6	54.2	63.1	69.8	57.4	65.6
56.4	49.4	44.9	34.6	92.2	37.0	58.8	21.3
36.6	64.9	14.8	17.6	29.1	61.4	38.6	32.5
12.0	28.3	204.9	44.5	10.3	37.7	33.7	81.1
12.1	20.1	30.5	7.1	10.1	18.0	3.0	2.0

Graph the distribution and describe its main features.

1.22 Sketch a histogram for a distribution that is skewed to the left. Suppose that you and your friends emptied your pockets of coins and recorded the year marked on each coin. The distribution of dates would be skewed to the left. Explain why.

1.23 The Survey of Study Habits and Attitudes (SSHA) is a psychological test that evaluates college students' motivation, study habits, and attitudes toward school. A selective private college gives the SSHA to a sample of 18 of its incoming first-year college women. Their scores are

154	109	137	115	152	140	154	178	101
103	126	126	137	165	165	129	200	148

The college also administers the test to a sample of 20 first-year college men. Their scores are

108	140	114	91	180	115	126	92	169	146
109	132	75	88	113	151	70	115	187	104

(a) Make a back-to-back stemplot of the men's and women's scores. The overall shapes of the distributions are indistinct, as often happens when only a few observations are available. Are there any outliers?

(b) Compare the midpoints and the ranges of the two distributions. What is the most noticeable contrast between the female and male scores?

1.24 There is some evidence that increasing the amount of calcium in the diet can lower blood pressure. In a medical experiment one group of men was given a daily calcium supplement, while a control group received a placebo (a dummy pill). The seated systolic blood pressure of all the men was measured before the treatments began and again after 12 weeks. The blood pressure distributions in the two groups should have been similar at the beginning of the experiment. Here are the initial blood pressure readings for the two groups:

Calcium group

107	110	123	129	112	111	107	112	136	102

Placebo group

123	109	112	102	98	114	119	112	110	117	130

Make a back-to-back stemplot of these data. Does your plot show any major differences in the two groups before the treatments began? In particular, are the centers of the two blood pressure distributions close together?

1.25 Do women study more than men? We asked the students in a large first-year college class how many minutes they studied on a typical week night. Here are the responses of random samples of 30 women and 30 men from the class:

Women					Men				
180	120	180	360	240	90	120	30	90	200
120	180	120	240	170	90	45	30	120	75
150	120	180	180	150	150	120	60	240	300
200	150	180	150	180	240	60	120	60	30
120	60	120	180	180	30	230	120	95	150
90	240	180	115	120	0	200	120	120	180

(a) Examine the data. Why are you not surprised that most responses are multiples of 10 minutes? We eliminated one student who claimed to study 30,000 minutes per night. Are there any other responses you consider suspicious?

(b) Make a back-to-back stemplot of these data. Report the approximate midpoints of both groups. Does it appear that women study more than men (or at least claim that they do)?

1.26 The Degree of Reading Power (DRP) test is often used to measure the reading ability of children. Here are the DRP scores of 44 third-grade students, measured during research on ways to improve reading performance:[15]

40	26	39	14	42	18	25	43	46	27	19
47	19	26	35	34	15	44	40	38	31	46
52	25	35	35	33	29	34	41	49	28	52
47	35	48	22	33	41	51	27	14	54	45

Make a stemplot of these data. Then make a histogram. Which display do you prefer, and why? Describe the main features of the distribution.

1.27 In 1798 the English scientist Henry Cavendish measured the density of the earth by careful work with a torsion balance. The variable recorded was the density of the earth as a multiple of the density of water. Here are Cavendish's 29 measurements:[16]

5.50	5.61	4.88	5.07	5.26	5.55	5.36	5.29	5.58	5.65
5.57	5.53	5.62	5.29	5.44	5.34	5.79	5.10	5.27	5.39
5.42	5.47	5.63	5.34	5.46	5.30	5.75	5.68	5.85	

Present these measurements graphically by either a stemplot or a histogram and explain the reason for your choice. Then briefly discuss the main features of the distribution. In particular, what is your estimate of the density of the earth based on these measurements?

1.28 Table 1.3 gives the number of medical doctors per 100,000 people in each state.[17]

(a) Why is the number of doctors per 100,000 people a better measure of the availability of health care than a simple count of the number of doctors in a state?

(b) Make a graph to display the distribution of doctors per 100,000 people. Write a brief description of the distribution. Are there any outliers? If so, can you explain them?

TABLE 1.3 Medical doctors per 100,000 population, by state (1998)

State	Doctors	State	Doctors	State	Doctors
Alabama	198	Louisiana	246	Ohio	235
Alaska	167	Maine	223	Oklahoma	169
Arizona	202	Maryland	374	Oregon	225
Arkansas	190	Massachusetts	412	Pennsylvania	291
California	247	Michigan	224	Rhode Island	338
Colorado	238	Minnesota	249	South Carolina	207
Connecticut	354	Mississippi	163	South Dakota	184
Delaware	234	Missouri	230	Tennessee	246
Florida	238	Montana	190	Texas	203
Georgia	211	Nebraska	218	Utah	200
Hawaii	265	Nevada	173	Vermont	305
Idaho	154	New Hampshire	237	Virginia	241
Illinois	260	New Jersey	295	Washington	235
Indiana	195	New Mexico	212	West Virginia	215
Iowa	173	New York	387	Wisconsin	227
Kansas	203	North Carolina	232	Wyoming	171
Kentucky	209	North Dakota	222	D.C.	737

TABLE 1.4 Monthly percent returns on Philip Morris stock (July 1990 to May 1997)

−5.7	1.2	4.1	3.2	7.3	7.5	18.6	3.7	−1.8	2.4
−6.5	6.7	9.4	−2.0	−2.8	−3.4	19.2	−4.8	0.5	−0.6
2.8	−0.5	−4.5	8.7	2.7	4.1	−10.3	4.8	−2.3	−3.1
−10.2	−3.7	−26.6	7.2	−2.9	−2.3	3.5	−4.6	17.2	4.2
0.5	8.3	−7.1	−8.4	7.7	−9.6	6.0	6.8	10.9	1.6
0.2	−2.4	−2.4	3.9	1.7	9.0	3.6	7.6	3.2	−3.7
4.2	13.2	0.9	4.2	4.0	2.8	6.7	−10.4	2.7	10.3
5.7	0.6	−14.2	1.3	2.9	11.8	10.6	5.2	13.8	−14.7
3.5	11.7	1.3							

1.29 Table 1.4 gives the monthly percent returns on Philip Morris stock for the period from July 1990 to May 1997. (The return on an investment consists of the change in its price plus any cash payments made, given here as a percent of its price at the start of each month.)

(a) Make either a histogram or a stemplot of these data. How did you decide which graph to make?

(b) There is one clear outlier. What is the value of this observation? (It is explained by news of action against smoking, which depressed this tobacco company stock.) Describe the shape, center, and spread of the data after you omit the outlier.

TABLE 1.5 Survival times (days) of guinea pigs in a medical experiment

43	45	53	56	56	57	58	66	67	73
74	79	80	80	81	81	81	82	83	83
84	88	89	91	91	92	92	97	99	99
100	100	101	102	102	102	103	104	107	108
109	113	114	118	121	123	126	128	137	138
139	144	145	147	156	162	174	178	179	184
191	198	211	214	243	249	329	380	403	511
522	598								

(c) The data appear in time order reading from left to right across each row in turn, beginning with the −5.7% return in July 1990. Make a time plot of the data. This was a period of increasing action against smoking, so we might expect a trend toward lower returns. But it was also a period in which stocks in general rose sharply, which would produce an increasing trend. What does your time plot show?

1.30 Table 1.5 gives the survival times in days of 72 guinea pigs after they were injected with tubercle bacilli in a medical experiment.[18] Make a suitable graph and describe the shape, center, and spread of the distribution of survival times. Are there any outliers?

1.31 Table 1.6 presents data on 78 seventh-grade students in a rural midwestern school.[19] The researcher was interested in the relationship between the students' "self-concept" and their academic performance. The data we give here include each student's grade point average (GPA), score on a standard IQ test, and gender, taken from school records. Gender is coded as F for female and M for male. The students are identified only by an observation number (OBS). The missing OBS numbers show that some students dropped out of the study. The final variable is each student's score on the Piers-Harris Children's Self-Concept Scale, a psychological test administered by the researcher.

(a) How many variables does this data set contain? Which are categorical variables and which are quantitative variables?

(b) Make a stemplot of the distribution of GPA, after rounding to the nearest tenth of a point.

(c) Describe the shape, center, and spread of the GPA distribution. Identify any suspected outliers from the overall pattern.

(d) Make a back-to-back stemplot of the rounded GPAs for female and male students. Write a brief comparison of the two distributions.

1.32 Make a graph of the distribution of IQ scores for the seventh-grade students in Table 1.6. Describe the shape, center, and spread of the distribution, as well as any outliers. IQ scores are usually said to be centered at 100. Is the midpoint for these students close to 100, clearly above, or clearly below?

TABLE 1.6 Educational data for 78 seventh-grade students

OBS	GPA	IQ	Gender	Self-concept	OBS	GPA	IQ	Gender	Self-concept
001	7.940	111	M	67	043	10.760	123	M	64
002	8.292	107	M	43	044	9.763	124	M	58
003	4.643	100	M	52	045	9.410	126	M	70
004	7.470	107	M	66	046	9.167	116	M	72
005	8.882	114	F	58	047	9.348	127	M	70
006	7.585	115	M	51	048	8.167	119	M	47
007	7.650	111	M	71	050	3.647	97	M	52
008	2.412	97	M	51	051	3.408	86	F	46
009	6.000	100	F	49	052	3.936	102	M	66
010	8.833	112	M	51	053	7.167	110	M	67
011	7.470	104	F	35	054	7.647	120	M	63
012	5.528	89	F	54	055	0.530	103	M	53
013	7.167	104	M	54	056	6.173	115	M	67
014	7.571	102	F	64	057	7.295	93	M	61
015	4.700	91	F	56	058	7.295	72	F	54
016	8.167	114	F	69	059	8.938	111	F	60
017	7.822	114	F	55	060	7.882	103	F	60
018	7.598	103	F	65	061	8.353	123	M	63
019	4.000	106	M	40	062	5.062	79	M	30
020	6.231	105	F	66	063	8.175	119	M	54
021	7.643	113	M	55	064	8.235	110	M	66
022	1.760	109	M	20	065	7.588	110	M	44
024	6.419	108	F	56	068	7.647	107	M	49
026	9.648	113	M	68	069	5.237	74	F	44
027	10.700	130	F	69	071	7.825	105	M	67
028	10.580	128	M	70	072	7.333	112	F	64
029	9.429	128	M	80	074	9.167	105	M	73
030	8.000	118	M	53	076	7.996	110	M	59
031	9.585	113	M	65	077	8.714	107	F	37
032	9.571	120	F	67	078	7.833	103	F	63
033	8.998	132	F	62	079	4.885	77	M	36
034	8.333	111	F	39	080	7.998	98	F	64
035	8.175	124	M	71	083	3.820	90	M	42
036	8.000	127	M	59	084	5.936	96	F	28
037	9.333	128	F	60	085	9.000	112	F	60
038	9.500	136	M	64	086	9.500	112	F	70
039	9.167	106	M	71	087	6.057	114	M	51
040	10.140	118	F	72	088	6.057	93	F	21
041	9.999	119	F	54	089	6.938	106	M	56

1.33 Based on a suitable graph, briefly describe the distribution of self-concept scores for the students in Table 1.6. Be sure to identify any suspected outliers.

1.34 The distribution of the ages of a nation's population has a strong influence on economic and social conditions. The following table shows the age distribution of U.S. residents in 1950 and 2050, in millions of people. The 1950 data come from that year's census, while the 2050 data are projections made by the Census Bureau.

Age group	1950	2050
Under 10 years	29.3	53.3
10 to 19 years	21.8	53.2
20 to 29 years	24.0	51.2
30 to 39 years	22.8	50.5
40 to 49 years	19.3	47.5
50 to 59 years	15.5	44.8
60 to 69 years	11.0	40.7
70 to 79 years	5.5	30.9
80 to 89 years	1.6	21.7
90 to 99 years	0.1	8.8
100 to 109 years	—	1.1
Total	151.1	403.7

(a) Because the total population in 2050 is much larger than the 1950 population, comparing percents (relative frequencies) in each age group is clearer than comparing counts. Make a table of the percent of the total population in each age group for both 1950 and 2050.

(b) Make a relative frequency histogram of the 1950 age distribution. Describe the main features of the distribution. In particular, look at the percent of children relative to the rest of the population.

(c) Make a relative frequency histogram of the projected age distribution for the year 2050. Use the same scales as in (b) for easy comparison. What are the most important changes in the U.S. age distribution projected for the century between 1950 and 2050?

1.35 **(Optional)** Sometimes you want to make a histogram from data that are already grouped into classes of unequal width. A report on the recent graduates of a large state university includes the following relative frequency table of the first-year salaries of last year's graduates. Salaries are in \$1000 units, and it is understood that each class includes its left endpoint but not its right endpoint—for example, a salary of exactly \$20,000 belongs in the second class.

Salary	15–20	20–25	25–30	30–35	35–40	40–50	50–60	60–80
Percent	7	14	29	23	13	9	4	1

The last three classes are wider than the others. An accurate histogram must take this into account. If the base of each bar in the histogram covers a class and the height is the percent of graduates with salaries in that class, the areas of the three rightmost bars will overstate the percent who have salaries in these classes. To make a correct histogram, the area of each bar must be proportional to the percent in that class. Most classes are \$5000 wide. A class *twice* as wide (\$10,000) should have a bar *half* as tall as the percent in that class. This keeps the area proportional to the percent. How should you treat the height of the bar for a class \$20,000 wide? Make a correct histogram with the heights of the bars for the last three classes adjusted so that the areas of the bars reflect the percent in each class.

1.36 "Major hurricanes account for just over 20% of the tropical storms and hurricanes that strike the United States but cause more than 80% of the damage." So say investigators who have shown that major hurricanes (with sustained wind speeds at least 50 meters per second) are tied to ocean temperature and other variables. These variables change slowly, so the high level of hurricane activity that began in 1995 "is likely to persist for an additional 10 to 40 years." This is bad news for people with beach houses on the Atlantic coast. Table 1.7 gives the counts of major hurricanes for each year between 1944 and 2000.[20]

(a) What is the average number of major hurricanes per year during the period 1944 to 2000?

TABLE 1.7 Major hurricanes (sustained winds in excess of 50 meters per second), 1944 to 2000

Year	Count	Year	Count	Year	Count	Year	Count	Year	Count
1944	3	1956	2	1968	0	1980	2	1992	1
1945	2	1957	2	1969	3	1981	3	1993	1
1946	1	1958	4	1970	2	1982	1	1994	0
1947	2	1959	2	1971	1	1983	1	1995	5
1948	4	1960	2	1972	0	1984	1	1996	6
1949	3	1961	6	1973	1	1985	3	1997	1
1950	7	1962	0	1974	2	1986	0	1998	3
1951	2	1963	2	1975	3	1987	1	1999	5
1952	3	1964	5	1976	2	1988	3	2000	3
1953	3	1965	1	1977	1	1989	2		
1954	2	1966	3	1978	2	1990	1		
1955	5	1967	1	1979	2	1991	2		

(b) Make a time plot of the count of major hurricanes. Draw a line across your plot at the average number of hurricanes per year. This helps divide the plot into three periods. Describe the pattern you see.

1.37 Treasury bills are short-term borrowing by the U.S. government. They are important in financial theory because the interest rate for Treasury bills is a "risk-free rate" that says what return investors can get while taking (almost) no risk. More risky investments should in theory offer higher returns in the long run. Here are the annual returns on Treasury bills from 1970 to 2000:[21]

Year	Rate	Year	Rate	Year	Rate	Year	Rate
1970	6.52	1978	7.19	1986	6.16	1994	3.91
1971	4.39	1979	10.38	1987	5.47	1995	5.60
1972	3.84	1980	11.26	1988	6.36	1996	5.20
1973	6.93	1981	14.72	1989	8.38	1997	5.25
1974	8.01	1982	10.53	1990	7.84	1998	4.85
1975	5.80	1983	8.80	1991	5.60	1999	4.69
1976	5.08	1984	9.84	1992	3.50	2000	5.69
1977	5.13	1985	7.72	1993	2.90		

(a) Make a time plot of the returns paid by Treasury bills in these years.

cycles

(b) Interest rates, like many economic variables, show **cycles,** clear but irregular up-and-down movements. In which years did the interest rate cycle reach temporary peaks?

(c) A time plot may show a consistent trend underneath cycles. When did interest rates reach their overall peak during these years? Has there been a general trend downward since that year?

1.38 Time series data often display the effects of changes in policy. Here are data on motor vehicle deaths in the United States. As in Example 1.3, we look at the death rate per 100 million miles driven.

Year	Rate	Year	Rate	Year	Rate	Year	Rate
1960	5.1	1970	4.7	1980	3.3	1990	2.1
1962	5.1	1972	4.3	1982	2.8	1992	1.7
1964	5.4	1974	3.5	1984	2.6	1994	1.7
1966	5.5	1976	3.2	1986	2.5	1996	1.7
1968	5.2	1978	3.3	1988	2.3	1998	1.6

(a) Make a time plot of these death rates. During these years, safety requirements for motor vehicles became stricter and interstate highways replaced older roads. How does the pattern of your plot reflect these changes?

(b) In 1974 the national speed limit was lowered to 55 miles per hour in an attempt to conserve gasoline after the 1973 Arab-Israeli War. In the

mid-1980s most states raised speed limits on interstate highways to 65 miles per hour. Some said that the lower speed limit saved lives. Is the effect of lower speed limits between 1974 and the mid-1980s visible in your plot?

(c) Does it make sense to make a histogram of these 20 death rates? Explain your answer.

1.39 The impression that a time plot gives depends on the scales you use on the two axes. If you stretch the vertical axis and compress the time axis, change appears to be more rapid. Compressing the vertical axis and stretching the time axis make change appear slower. Make two more time plots of the data in Exercise 1.38, one that makes motor vehicle death rates appear to decrease very rapidly and one that shows only a slow decrease. The moral of this exercise is: pay close attention to the scales when you look at a time plot.

1.40 The following table gives the times (in minutes, rounded to the nearest minute) for the winning man in the Boston Marathon in the years 1959 to 2001:

Year	Time	Year	Time	Year	Time	Year	Time
1959	143	1970	131	1981	129	1992	128
1960	141	1971	139	1982	129	1993	130
1961	144	1972	136	1983	129	1994	127
1962	144	1973	136	1984	131	1995	129
1963	139	1974	134	1985	134	1996	129
1964	140	1975	130	1986	128	1997	131
1965	137	1976	140	1987	132	1998	128
1966	137	1977	135	1988	129	1999	130
1967	136	1978	130	1989	129	2000	130
1968	142	1979	129	1990	128	2001	130
1969	134	1980	132	1991	131		

Women were allowed to enter the race in 1972. Here are the times for the winning woman from 1972 to 2001:

Year	Time	Year	Time	Year	Time	Year	Time
1972	190	1980	154	1988	145	1996	147
1973	186	1981	147	1989	144	1997	146
1974	167	1982	150	1990	145	1998	143
1975	162	1983	143	1991	144	1999	143
1976	167	1984	149	1992	144	2000	146
1977	168	1985	154	1993	145	2001	144
1978	165	1986	145	1994	142		
1979	155	1987	146	1995	145		

Make time plots of the men's and women's winning times on the same graph for easy comparison. Give a brief description of the pattern of Boston Marathon winning times over these years. Have times stopped improving in recent years? If so, when did improvement end for men and for women?

1.2 Describing Distributions with Numbers

Interested in an exotic car? Worried that it may use too much gas? The Environmental Protection Agency lists most such vehicles in its "minicompact" or "two-seater" categories. Table 1.8 gives the city and highway gas mileage for all cars in these groups.[22] (The mileages are for the basic engine and transmission combination for each car.) We want to compare minicompacts with two-seaters and city mileage with highway mileage. We can begin with graphs, but numerical summaries make the comparisons more specific.

A brief description of a distribution should include its *shape* and numbers describing its *center* and *spread*. We describe the shape of a distribution based on inspection of a histogram or a stemplot. Now we will learn specific ways to use numbers to measure the center and spread of a distribution. We can calculate these numerical measures for any quantitative variable. But to interpret measures of center and spread, and to choose among the several measures

TABLE 1.8 Fuel economy (miles per gallon) for model year 2001 cars

Minicompact cars			Two-seater cars		
Model	City	Highway	Model	City	Highway
Audi TT Coupe	22	31	Acura NSX	17	24
BMW 325CI Convertible	19	27	Audi TT Roadster	22	30
BMW 330CI Convertible	20	28	BMW Z3 Coupe	21	28
BMW M3 Convertible	16	23	BMW Z3 Roadster	20	27
Jaguar XK8 Convertible	17	24	BMW Z8	13	21
Jaguar XKR Convertible	16	22	Chevrolet Corvette	18	26
Mercedes-Benz CLK320	20	28	Dodge Viper	11	21
Mercedes-Benz CLK430	18	24	Ferrari Modena	11	16
Mitsubishi Eclipse	22	30	Ferrari Maranello	8	13
Porsche 911 Carrera	17	25	Honda Insight	61	68
Porsche 911 Turbo	15	22	Honda S2000	20	26
			Lamborghini Diablo	10	13
			Mazda Miata	22	28
			Mercedes-Benz SL500	16	23
			Mercedes-Benz SL600	13	19
			Mercedes-Benz SLK320	21	27
			Plymouth Prowler	17	23
			Porsche Boxster	19	27
			Toyota MR2	25	30

we will learn, you must think about the shape of the distribution and the meaning of the data. The numbers, like graphs, are aids to understanding, not "the answer" in themselves.

Measuring center: the mean

Numerical description of a distribution begins with a measure of its center or average. The two common measures of center are the *mean* and the *median*. The mean is the "average value" and the median is the "middle value." These are two different ideas for "center," and the two measures behave differently. We need precise recipes for the mean and the median.

The Mean $\bar{x}$

To find the **mean** $\bar{x}$ of a set of observations, add their values and divide by the number of observations. If the n observations are $x_1, x_2, \ldots, x_n$, their mean is

$$\bar{x} = \frac{x_1 + x_2 + \cdots + x_n}{n}$$

or, in more compact notation,

$$\bar{x} = \frac{1}{n}\sum x_i$$

The $\sum$ (capital Greek sigma) in the formula for the mean is short for "add them all up." The bar over the x indicates the mean of all the x-values. Pronounce the mean $\bar{x}$ as "x-bar." This notation is so common that writers who are discussing data use $\bar{x}$, $\bar{y}$, etc. without additional explanation. The subscripts on the observations x_i are just a way of keeping the n observations distinct. They do not necessarily indicate order or any other special facts about the data.

EXAMPLE 1.12

The mean highway mileage for the 19 two-seaters in Table 1.8 is

$$\begin{aligned} \bar{x} &= \frac{x_1 + x_2 + \cdots + x_n}{n} \\ &= \frac{24 + 30 + 28 + \cdots + 30}{19} \\ &= \frac{490}{19} = 25.8 \text{ miles per gallon} \end{aligned}$$

In practice, you can key the data into your calculator and hit the $\bar{x}$ key.

The data for Example 1.12 contain an outlier: the Honda Insight is a hybrid gas-electric car that doesn't belong in the same category as the 18 gasoline-powered two-seater cars. If we exclude the Insight, the mean highway mileage drops to 23.4 MPG. The single outlier adds 2.4 MPG to the mean highway mileage. This illustrates an important weakness of the mean as a

measure of center: it is sensitive to the influence of a few extreme observations. These may be outliers, but a skewed distribution that has no outliers will also pull the mean toward its long tail. Because the mean cannot resist the influence of extreme observations, we say that it is not a **resistant measure** of center. A measure that is resistant does more than limit the influence of outliers. Its value does not respond strongly to changes in a few observations, no matter how large those changes may be. The mean fails this requirement because we can make the mean as large as we wish by making a large enough increase in just one observation.

resistant measure

Measuring center: the median

We used the midpoint of a distribution as an informal measure of center in the previous section. The *median* is the formal version of the midpoint, with a specific rule for calculation.

The Median M

The **median M** is the midpoint of a distribution, the number such that half the observations are smaller and the other half are larger. To find the median of a distribution:

1. Arrange all observations in order of size, from smallest to largest.
2. If the number of observations n is odd, the median M is the center observation in the ordered list. Find the location of the median by counting $(n+1)/2$ observations up from the bottom of the list.
3. If the number of observations n is even, the median M is the mean of the two center observations in the ordered list. The location of the median is again $(n+1)/2$ from the bottom of the list.

Note that the formula $(n+1)/2$ does *not* give the median, just the location of the median in the ordered list. Medians require little arithmetic, so they are easy to find by hand for small sets of data. Arranging even a moderate number of observations in order is tedious, however, so that finding the median by hand for larger sets of data is unpleasant. Even simple calculators have an $\bar{x}$ button, but you will need computer software or a graphing calculator to automate finding the median.

EXAMPLE 1.13

To find the median highway mileage for 2001 model two-seater cars, arrange the data in increasing order:

13	13	16	19	21	21	23	23	24	**26**
26	27	27	27	28	28	30	30	68	

The median is the bold 26, the 10th observation in the ordered list. You can find the median by eye—there are 9 observations to the left and 9 to the right. Or you can use the recipe $(n+1)/2 = 20/2 = 10$ to locate the median in the list.

What happens if we drop the Honda Insight? The remaining 18 cars have highway mileages

13	13	16	19	21	21	23	23	**24**	**26**	26	27	27
27	28	28	30	30								

Because the number of observations $n = 18$ is even, there is no center observation. There is a center pair—the bold 24 and 26 have 8 observations to their left and 8 to their right. The median M is the mean of the center pair:

$$M = \frac{24 + 26}{2} = \frac{50}{2} = 25$$

The recipe $(n + 1)/2 = 19/2 = 9.5$ for the position of the median in the list says that the median is at location "nine and one-half," that is, halfway between the 9th and 10th observations.

You see that the median is more resistant than the mean. Removing the Honda Insight reduces the median highway mileage by just 1 MPG. And if we mistakenly enter the Insight's mileage as 680 rather than 68, the median remains 25. The very high value is simply one observation to the right of center. The *Mean and Median* applet available on the Web site `www.whfreeman.com/ips` is an excellent way to compare the resistance of M and $\overline{x}$. See Exercises 1.51 and 1.52 for use of this applet.

Mean versus median

The median and mean are the most common measures of the center of a distribution. The mean and median of a symmetric distribution are close together. If the distribution is exactly symmetric, the mean and median are exactly the same. In a skewed distribution, the mean is farther out in the long tail than is the median. For example, the distribution of house prices is strongly skewed to the right. There are many moderately priced houses and a few very expensive mansions. The few expensive houses pull the mean up but do not affect the median. The mean price of existing houses sold in 2000 was \$176,200, but the median price for these same houses was only \$139,000. Reports about house prices, incomes, and other strongly skewed distributions usually give the median ("middle value") rather than the mean ("arithmetic average value"). However, if you are a tax assessor interested in the total value of houses in your area, use the mean. The total is the mean times the number of houses, but it has no connection with the median. The mean and median measure center in different ways, and both are useful.

We now have two general strategies for dealing with outliers and other irregularities in data. The first strategy asks us to detect outliers and investigate their causes. We can then correct outliers if they are wrongly recorded, delete them for good reason, or otherwise give them individual attention. The second strategy is to use resistant methods, so that outliers have little influence over our conclusions. In general, we prefer to examine data carefully and consider any irregularity in the light of the individual situation. This will sometimes lead to a decision to employ resistant methods.

EXAMPLE 1.14

Newcomb's data are supposed to be repeated measurements of the same quantity, but they contain unexplained outliers. Even though we can't explain the outlying values, we should not give them much influence over our estimate of the velocity of light. We can delete the outliers from a computation of the mean, or we can report the median. If the two outliers are omitted, the mean of the remaining 64 measurements is $\bar{x} = 27.75$. The median of 66 observations has position $(66 + 1)/2 = 33.5$ in the ordered list of passage times from Table 1.1. Its value is $M = 27$. Either of these values is preferable to the overall mean ($\bar{x} = 26.21$) as a statement of Newcomb's value for the travel time of light.

Measuring spread: the quartiles

A measure of center alone can be misleading. Two nations with the same median family income are very different if one has extremes of wealth and poverty and the other has little variation among families. A drug with the correct mean concentration of active ingredient is dangerous if some batches are much too high and others much too low. We are interested in the *spread* or *variability* of incomes and drug potencies as well as their centers. **The simplest useful numerical description of a distribution consists of both a measure of center and a measure of spread.**

percentile

We can describe the spread or variability of a distribution by giving several percentiles. The ***p*th percentile** of a distribution is the value such that p percent of the observations fall at or below it. The median is just the 50th percentile, so the use of percentiles to report spread is particularly appropriate when the median is our measure of center. The most commonly used percentiles other than the median are the *quartiles*. The first quartile is the 25th percentile, and the third quartile is the 75th percentile. (The second quartile is the median itself.) To calculate a percentile, arrange the observations in increasing order and count up the required percent from the bottom of the list. Our definition of percentiles is a bit inexact, because there is not always a value with exactly p percent of the data at or below it. We will be content to take the nearest observation for most percentiles, but the quartiles are important enough to require an exact recipe.

The Quartiles Q_1 and Q_3

To calculate the quartiles:

1. Arrange the observations in increasing order and locate the median M in the ordered list of observations.
2. The **first quartile Q_1** is the median of the observations whose position in the ordered list is to the left of the location of the overall median.
3. The **third quartile Q_3** is the median of the observations whose position in the ordered list is to the right of the location of the overall median.

EXAMPLE 1.15

The highway mileages of the 18 gasoline-powered two-seater cars, arranged in increasing order, are

13	13	16	19	21	21	23	23	24	|	26	26	27	27
27	28	28	30	30									

The median is midway between the center pair of observations. We have marked its position in the list by |. The first quartile is the median of the 9 observations to the left of the position of the median. It is the 5th of these, $Q_1 = 21$. Similarly, the third quartile is the median of the 9 observations to the right of the |. Check that $Q_3 = 27$.

When there is an odd number of observations, the median is the unique center observation, and the rule for finding the quartiles excludes this center value. The highway mileages of the 11 minicompact cars in Table 1.8 are (in order)

22	22	23	24	24	**25**	27	28	28	30	31

The median is the bold 25. The first quartile is the median of the 5 observations falling to the left of this point in the list, $Q_1 = 23$. Similarly, $Q_3 = 28$.

We find other percentiles more informally. For example, we take the 90th percentile of the 18 two-seater mileages to be the 16th in the ordered list, because $0.90 \times 18 = 16.2$, which we round to 16. The 90th percentile is therefore 28 MPG.

EXAMPLE 1.16

Statistical software often provides several numerical measures in response to a single command. Figure 1.14 displays such output from the Minitab and SPSS software for the shopping data from Example 1.5. Both give us the number of observations, the mean and median, and the quartiles, as well as other measures. Minitab has been set to round its output to two decimal places. You can check that the values for the quartiles differ slightly from those given by our rule. For example, the first quartile is \$19.27 by hand and \$19.06 by software. Software often uses more elaborate rules that aren't suited for hand calculation. Moreover, not all software uses the same rule. For example, the Microsoft Excel spreadsheet gives the first quartile as 19.3275. These differences are too small to affect conclusions based on the data. Just use the values that your software gives you.

The five-number summary and boxplots

In Section 1.1, we used the smallest and largest observations to indicate the spread of a distribution. These single observations tell us little about the distribution as a whole, but they give information about the tails of the distribution that is missing if we know only Q_1, M, and Q_3. To get a quick summary of both center and spread, combine all five numbers.

SPSS

N	**Valid**	50
	Missing	0
Mean		34.7022
Median		27.8550
Std. Deviation		21.6974
Range		90.23
Minimum		3.11
Maximum		93.34
Percentiles	25	19.0600
	50	27.8550
	75	45.7225

Minitab

```
Descriptive  Statistics

Variable         N      Mean     Median     TrMean      StDev    SEMean
spending        50     34.70      27.85      32.92      21.70      3.07

Variable       Min       Max         Q1         Q3
spending      3.11     93.34      19.06      45.72
```

FIGURE 1.14 Numerical descriptions of the unrounded shopping data from the SPSS and Minitab software.

The Five-Number Summary

The **five-number summary** of a set of observations consists of the smallest observation, the first quartile, the median, the third quartile, and the largest observation, written in order from smallest to largest. In symbols, the five-number summary is

$$\text{Minimum} \quad Q_1 \quad M \quad Q_3 \quad \text{Maximum}$$

These five numbers offer a reasonably complete description of center and spread. The five-number summaries for highway gas mileages are

13	21	25	27	30

for two-seaters and

22	23	25	28	31

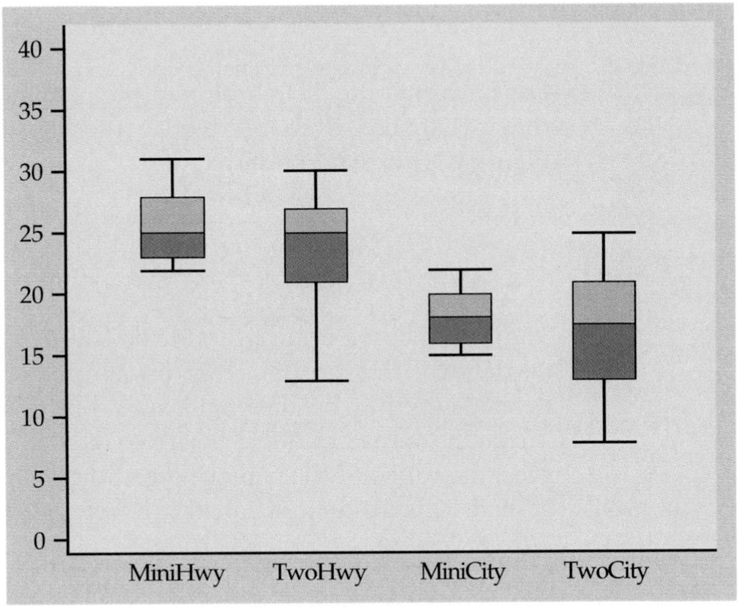

FIGURE 1.15 Boxplots of the highway and city gas mileages for cars classified as two-seaters and as minicompacts by the Environmental Protection Agency.

for minicompacts. The median describes the center of the distribution; the quartiles show the spread of the center half of the data; the minimum and maximum show the full spread of the data. The five-number summary leads to another visual representation of a distribution, the *boxplot*. Figure 1.15 shows boxplots for both city and highway gas mileages for our two groups of cars.

Boxplot

A **boxplot** is a graph of the five-number summary.

- A central box spans the quartiles Q_1 and Q_3.
- A line in the box marks the median M.
- Lines extend from the box out to the smallest and largest observations.

Because boxplots show less detail than histograms or stemplots, they are best used for side-by-side comparison of more than one distribution, as in Figure 1.15. When you look at a boxplot, first locate the median, which marks the center of the distribution. Then look at the spread. The quartiles show the spread of the middle half of the data, and the extremes (the smallest and largest observations) show the spread of the entire data set. We see at once that city mileages are lower than highway mileages. The median gas mileages for the two groups of cars are very close together, but the two-seaters are more variable. In particular, the two-seater group contains some cars with quite low gas mileages.

The 1.5 × *IQR* criterion for suspected outliers

Look again at the distribution of the percent of each state's adult population that is Hispanic, given in Table 1.2 (page 14) and displayed in the histogram in Figure 1.4. You can check that the five-number summary for this distribution is

0.6	2.0	4.1	7.0	38.7

The largest observation (New Mexico, 38.7% Hispanic) is an outlier. How shall we describe the spread of this distribution? The range of all the observations depends entirely on the smallest and largest observations and does not describe the spread of the majority of the data. The distance between the quartiles, the range of the center half of the data, is a more resistant measure of spread. This distance is called the *interquartile range*.

The Interquartile Range *IQR*

The **interquartile range *IQR*** is the distance between the first and third quartiles,

$$IQR = Q_3 - Q_1$$

For our data on Hispanics in the states, $IQR = 7.0 - 2.0 = 5.0$. The quartiles and the *IQR* are not affected by changes in either tail of the distribution. They are therefore resistant, because changes in a few data points have no further effect once these points move outside the quartiles. You should be aware that no single numerical measure of spread, such as *IQR*, is very useful for describing skewed distributions. The two sides of a skewed distribution have different spreads. We can often detect skewness from the five-number summary by comparing how far the first quartile and the minimum are from the median (left tail) with how far the third quartile and the maximum (right tail) are from the median. The interquartile range is mainly used as the basis of a rule of thumb for identifying suspected outliers.

The 1.5 × *IQR* Criterion for Outliers

Call an observation a suspected outlier if it falls more than $1.5 \times IQR$ above the third quartile or below the first quartile.

EXAMPLE 1.17

For the Hispanics data,

$$1.5 \times IQR = 1.5 \times 5 = 7.5$$

Any values below $2.0 - 7.5 = -5.5$ or above $7.0 + 7.5 = 14.5$ are flagged as possible outliers. There are no low outliers, but 7 states are flagged as possible high outliers.

This distribution is strongly skewed and has a quite compact middle half (small *IQR*). The histogram in Figure 1.4 (page 15) suggests that only New Mexico is truly an outlier in the sense of deviating from the overall pattern of the distribution. The other 6 states are just part of the long right tail. You see that the $1.5 \times IQR$ rule does not remove the need to look at the distribution and use judgment. It is useful mainly when large volumes of data must be scanned automatically.

modified boxplot

We can modify boxplots to plot suspected outliers individually. In a **modified boxplot,** the lines extend out from the central box only to the smallest and largest observations that are not suspected outliers. Observations more than $1.5 \times IQR$ outside the box are plotted as individual points. Figure 1.16 is a modified boxplot of the data from Table 1.2.

The modified boxplot in Figure 1.16, and especially the histogram in Figure 1.4, tell us much more about the distribution than the five-number summary or other numerical measures. The routine methods of statistics compute numerical measures and draw conclusions based on their values. These methods are very useful, and we will study them carefully in later chapters. But they cannot be applied blindly, by feeding data to a computer program, because **statistical measures and methods based on them are**

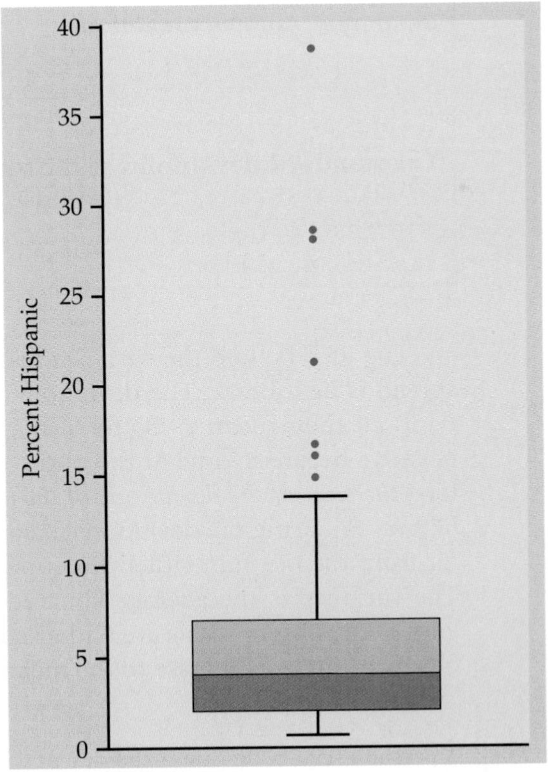

FIGURE 1.16 Modified boxplot of the percent of adults in each state who identified themselves as Hispanic in the 2000 census.

generally meaningful only for distributions of sufficiently regular shape. This principle will become clearer as we progress, but it is good to be aware at the beginning that quickly resorting to fancy calculations is the mark of a statistical amateur. Look, think, and choose your calculations selectively.

Measuring spread: the standard deviation

The five-number summary is not the most common numerical description of a distribution. That distinction belongs to the combination of the mean to measure center and the *standard deviation* to measure spread. The standard deviation measures spread by looking at how far the observations are from their mean.

The Standard Deviation s

The **variance s^2** of a set of observations is the average of the squares of the deviations of the observations from their mean. In symbols, the variance of n observations $x_1, x_2, \ldots, x_n$ is

$$s^2 = \frac{(x_1 - \overline{x})^2 + (x_2 - \overline{x})^2 + \cdots + (x_n - \overline{x})^2}{n - 1}$$

or, in more compact notation,

$$s^2 = \frac{1}{n-1}\sum(x_i - \overline{x})^2$$

The **standard deviation s** is the square root of the variance s^2:

$$s = \sqrt{\frac{1}{n-1}\sum(x_i - \overline{x})^2}$$

The idea behind the variance and the standard deviation as measures of spread is as follows: The deviations $x_i - \overline{x}$ display the spread of the values x_i about their mean $\overline{x}$. Some of these deviations will be positive and some negative because some of the observations fall on each side of the mean. In fact, *the sum of the deviations of the observations from their mean will always be zero*. Squaring the deviations makes them all positive, so that observations far from the mean in either direction have large positive squared deviations. The variance is the average squared deviation. Therefore, s^2 and s will be large if the observations are widely spread about their mean, and small if the observations are all close to the mean.

EXAMPLE 1.18

A person's metabolic rate is the rate at which the body consumes energy. Metabolic rate is important in studies of weight gain, dieting, and exercise. Here are the metabolic rates of 7 men who took part in a study of dieting. (The units are calories per 24 hours. These are the same calories used to describe the energy content of foods.)

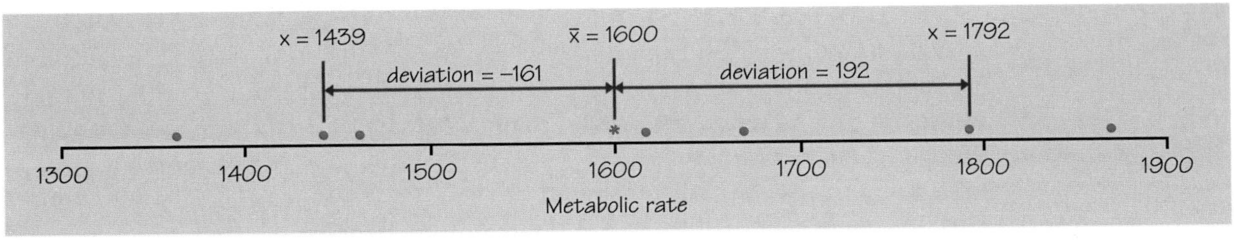

FIGURE 1.17 Metabolic rates for seven men, with the mean (*) and the deviations of two observations from the mean.

1792	1666	1362	1614	1460	1867	1439

Enter these data into your calculator or software and verify that

$$\bar{x} = 1600 \text{ calories} \quad s = 189.24 \text{ calories}$$

Figure 1.17 plots these data as dots on the calorie scale, with their mean marked by an asterisk (*). The arrows mark two of the deviations from the mean. If you were calculating s by hand, you would find the first deviation as

$$x_1 - \bar{x} = 1792 - 1600 = 192$$

Exercise 1.62 asks you to calculate the seven deviations, square them, and find s^2 and s directly from the deviations. Working one or two short examples by hand helps you understand how the standard deviation is obtained. In practice you will always use either software or a calculator that will find s from keyed-in data.

The idea of the variance is straightforward: it is the average of the squares of the deviations of the observations from their mean. The details we have just presented, however, raise some questions.

Why do we square the deviations? Why not just average the distances of the observations from their mean? There are two reasons, neither of them obvious. First, the sum of the squared deviations of any set of observations from their mean is the smallest that the sum of squared deviations from any number can possibly be. This is not true of the unsquared distances. So squared deviations point to the mean as center in a way that distances do not. Second, the standard deviation turns out to be the natural measure of spread for a particularly important class of symmetric unimodal distributions, the *normal distributions*. We will meet the normal distributions in the next section. We commented earlier that the usefulness of many statistical procedures is tied to distributions of particular shapes. This is distinctly true of the standard deviation.

Why do we emphasize the standard deviation rather than the variance? One reason is that s, not s^2, is the natural measure of spread for normal distributions. There is also a more general reason to prefer s to s^2. Because the variance involves squaring the deviations, it does not have the same unit of measurement as the original observations. The variance of the metabolic rates, for example, is measured in squared calories. Taking the square root remedies this. The standard deviation s measures spread about the mean in the original scale.

Why do we average by dividing by $n - 1$ rather than n in calculating the variance? Because the sum of the deviations is always zero, the last deviation can be found once we know the other $n - 1$. So we are not averaging n unrelated numbers. Only $n - 1$ of the squared deviations can vary freely, and we average by dividing the total by $n - 1$. The number $n - 1$ is called the **degrees of freedom** of the variance or standard deviation. Many calculators offer a choice between dividing by n and dividing by $n - 1$, so be sure to use $n - 1$.

degrees of freedom

Properties of the standard deviation

Here are the basic properties of the standard deviation s as a measure of spread.

Properties of the Standard Deviation

- s measures spread about the mean and should be used only when the mean is chosen as the measure of center.
- $s = 0$ only when there is *no spread*. This happens only when all observations have the same value. Otherwise, $s > 0$. As the observations become more spread out about their mean, s gets larger.
- s, like the mean $\bar{x}$, is not resistant. A few outliers can make s very large.

The use of squared deviations renders s even more sensitive than $\bar{x}$ to a few extreme observations. For example, dropping the Honda Insight from our list of two-seater cars reduces the mean highway mileage from 25.8 MPG to 23.4 MPG. It cuts the standard deviation more than half, from 11.4 MPG with the Insight to 5.3 MPG without it. Distributions with outliers and strongly skewed distributions have large standard deviations. The number s does not give much helpful information about such distributions.

Choosing measures of center and spread

How do we choose between the five-number summary and $\bar{x}$ and s to describe the center and spread of a distribution? Because the two sides of a strongly skewed distribution have different spreads, no single number such as s describes the spread well. The five-number summary, with its two quartiles and two extremes, does a better job.

Choosing a Summary

The five-number summary is usually better than the mean and standard deviation for describing a skewed distribution or a distribution with strong outliers. Use $\bar{x}$ and s only for reasonably symmetric distributions that are free of outliers.

EXAMPLE 1.19

A central principle in the study of investments is that taking bigger risks is rewarded by higher returns, at least on the average over long periods of time. It is usual in finance to measure risk by the standard deviation of returns, on the grounds that investments whose returns show a large spread from year to year are less predictable and therefore more risky than those whose returns have small spread. Compare, for example, the approximate mean and standard deviation of the annual percent returns on American common stocks and U.S. Treasury bills over the period from 1950 to 2000:

Investment	Mean return	Standard deviation
Common stocks	13.3%	17.1%
Treasury bills	5.2%	2.9%

Stocks are risky. They went up more than 13% per year on the average during this period, but they dropped almost 28% in the worst year. The large standard deviation reflects the fact that stocks have produced both large gains and large losses. When you buy a Treasury bill, on the other hand, you are lending money to the government for one year. You know that the government will pay you back with interest. That is much less risky than buying stocks, so (on the average) you get a smaller return.

Are $\bar{x}$ and s good summaries for distributions of investment returns? Figure 1.18 displays stemplots of the annual returns for both investments. (Because stock returns are so much more spread out, a back-to-back stemplot does not work well. The stems in the stock stemplot are tens of percents; the stems for bills are percents. The lowest returns are -28% for stocks and 0.9% for bills.) You see that returns on Treasury bills have a right-skewed distribution. Convention in the financial world calls for $\bar{x}$ and s because some parts of investment theory use them. For describing this right-skewed distribution, however, the five-number summary would be more informative.

Remember that a graph gives the best overall picture of a distribution. Numerical measures of center and spread report specific facts about a distribution, but they do not describe its entire shape. Numerical summaries do not disclose the presence of multiple modes or gaps, for example. **Always plot your data.**

Changing the unit of measurement

The same variable can be recorded in different units of measurement. Americans commonly record distances in miles and temperatures in degrees Fahrenheit, while the rest of the world measures distances in kilometers and temperatures in degrees Celsius. Fortunately, it is easy to convert numerical descriptions of a distribution from one unit of measurement to another. This is true because a change in the measurement unit is a *linear transformation* of the measurements.

−2	8
−1	9 1 1 0 0 0
−0	9 6 4 3
0	0 0 0 1 2 3 8 9 9
1	1 3 3 4 4 6 6 6 7 8
2	0 1 1 2 3 4 4 4 5 7 7 9 9
3	1 1 3 4 6 7
4	5
5	0

(a)

0	9
1	255668
2	15779
3	01155899
4	24778
5	1122256678
6	24569
7	278
8	048
9	8
10	45
11	3
12	
13	
14	7

(b)

FIGURE 1.18 Stemplots of annual returns for stocks and Treasury bills, 1950 to 2000. (a) Stock returns, in whole percents. (b) Treasury bill returns, in percents and tenths of a percent.

Linear Transformations

A **linear transformation** changes the original variable x into the new variable x_{new} given by an equation of the form

$$x_{\text{new}} = a + bx$$

Adding the constant a shifts all values of x upward or downward by the same amount. In particular, such a shift changes the origin (zero point) of the variable. Multiplying by the positive constant b changes the size of the unit of measurement.

EXAMPLE 1.20

(a) If a distance x is measured in kilometers, the same distance in miles is

$$x_{\text{new}} = 0.62x$$

For example, a 10-kilometer race covers 6.2 miles. This transformation changes the units without changing the origin—a distance of 0 kilometers is the same as a distance of 0 miles.

(b) A temperature x measured in degrees Fahrenheit must be reexpressed in degrees Celsius to be easily understood by the rest of the world. The transformation is

$$x_{\text{new}} = \frac{5}{9}(x - 32) = -\frac{160}{9} + \frac{5}{9}x$$

Thus, the high of 95° F on a hot American summer day translates into 35° C. In this case

$$a = -\frac{160}{9} \quad \text{and} \quad b = \frac{5}{9}$$

This linear transformation changes both the unit size and the origin of the measurements. The origin in the Celsius scale (0° C, the temperature at which water freezes) is 32° in the Fahrenheit scale.

Linear transformations do not change the shape of a distribution. If measurements on a variable x have a right-skewed distribution, any new variable x_{new} obtained by a linear transformation $x_{\text{new}} = a + bx$ (for $b > 0$) will also have a right-skewed distribution. If the distribution of x is symmetric and unimodal, the distribution of x_{new} remains symmetric and unimodal.

Although a linear transformation preserves the basic shape of a distribution, the center and spread will change. Because linear changes of measurement scale are common, we must be aware of their effect on numerical descriptive measures of center and spread. Fortunately, the changes follow a simple pattern.

EXAMPLE 1.21

Mary and John both measure the weights of the same five newly hatched pythons. Mary measures in ounces and John uses grams. There are 28.35 grams in an ounce, so each of John's measurements is 28.35 times as large as Mary's measurement of the same python. Here are their results:

Python	1	2	3	4	5
Mary	1.13 oz	1.02 oz	1.23 oz	1.06 oz	1.16 oz
John	32 g	29 g	35 g	30 g	33 g

These two sets of numbers measure the same five weights in different units. Their means represent the same average weight in different units, so John's mean (in grams) must be 28.35 times as large as Mary's mean (in ounces). The spread of John's numbers is likewise 28.35 times the spread of Mary's, so John's standard deviation (again in grams) is 28.35 times Mary's standard deviation (in ounces). You can verify these facts (up to roundoff error) with your calculator.

Here is a summary of such common sense relations in more formal language.

Effect of a Linear Transformation

To see the effect of a linear transformation on measures of center and spread, apply these rules:

- Multiplying each observation by a positive number b multiplies both measures of center (mean and median) and measures of spread (interquartile range and standard deviation) by b.
- Adding the same number a (either positive or negative) to each observation adds a to measures of center and to quartiles and other percentiles but does not change measures of spread.

The measures of spread IQR and s do not change when we add the same number a to all of the observations because adding a constant changes the location of the distribution but leaves the spread unaltered. You can find the effect of a linear transformation $x_{\text{new}} = a + bx$ by combining these rules. For example, if x has mean $\bar{x}$, the transformed variable x_{new} has mean $a + b\bar{x}$.

SUMMARY

A numerical summary of a distribution should report its **center** and its **spread** or **variability.**

The **mean $\bar{x}$** and the **median M** describe the center of a distribution in different ways. The mean is the arithmetic average of the observations, and the median is their midpoint.

When you use the median to describe the center of the distribution, describe its spread by giving the **quartiles.** The **first quartile Q_1** has one-fourth of the observations below it, and the **third quartile Q_3** has three-fourths of the observations below it.

The **interquartile range** is the difference between the quartiles. It is the spread of the center half of the data. The **1.5 × IQR criterion** flags observations more than 1.5 × IQR beyond the quartiles as possible outliers.

The **five-number summary** consisting of the median, the quartiles, and the smallest and largest individual observations provides a quick overall description of a distribution. The median describes the center, and the quartiles and extremes show the spread.

Boxplots based on the five-number summary are useful for comparing several distributions. The box spans the quartiles and shows the spread of the central half of the distribution. The median is marked within the box. Lines extend from the box to the extremes and show the full spread of the data. In a **modified boxplot,** points identified by the 1.5 × IQR criterion are plotted individually.

The **variance s^2** and especially its square root, the **standard deviation s**, are common measures of spread about the mean as center. The standard deviation s is zero when there is no spread and gets larger as the spread increases.

A **resistant measure** of any aspect of a distribution is relatively unaffected by changes in the numerical value of a small proportion of the total number of observations, no matter how large these changes are. The median and quartiles are resistant, but the mean and the standard deviation are not.

The mean and standard deviation are good descriptions for symmetric distributions without outliers. They are most useful for the normal distributions introduced in the next section. The five-number summary is a better exploratory summary for skewed distributions.

Linear transformations have the form $x_{new} = a + bx$. A linear transformation changes the origin if $a \neq 0$ and changes the size of the unit of measurement if $b > 0$. Linear transformations do not change the overall shape of a distribution. A linear transformation multiplies a measure of spread by b and changes a percentile or measure of center m into $a + bm$.

Numerical measures of particular aspects of a distribution, such as center and spread, do not report the entire shape of most distributions. In some cases, particularly distributions with multiple peaks and gaps, these measures may not be very informative.

SECTION 1.2 EXERCISES

1.41 Figure 1.12 (page 27) is a histogram of the tuition and fees charged by all 56 four-year colleges in the state of Massachusetts. Here are those charges (in dollars), arranged in increasing order:

2,508	2,830	2,883	2,914	3,018	3,044	3,357	4,129	4,222	4,255
5,212	11,894	12,520	13,170	13,600	13,660	13,800	14,062	14,645	14,720
14,800	15,500	15,504	15,618	15,830	16,412	16,450	16,775	17,270	17,300
17,320	17,410	17,500	17,820	18,320	18,910	19,590	20,014	20,890	23,262
23,485	23,520	23,584	23,815	24,154	24,450	24,790	24,850	25,044	25,128
25,360	25,714	26,050	26,080	26,125	26,241				

Find the five-number summary and make a boxplot. What distinctive feature of the histogram do these summaries miss? Remember that numerical summaries are not a substitute for looking at the data.

The INDIVIDUALS data set, described in the Data Appendix, contains the income and level of education of 55,899 people between the ages of 25 and 65. The boxplots in Figure 1.19 compare the distributions of income for people with six levels of education. This figure is a variation on the boxplot idea: because large data sets often contain very extreme observations, the lines extend from the central box only to the 5th and 95th percentiles. Exercises 1.42 to 1.44 concern these data.

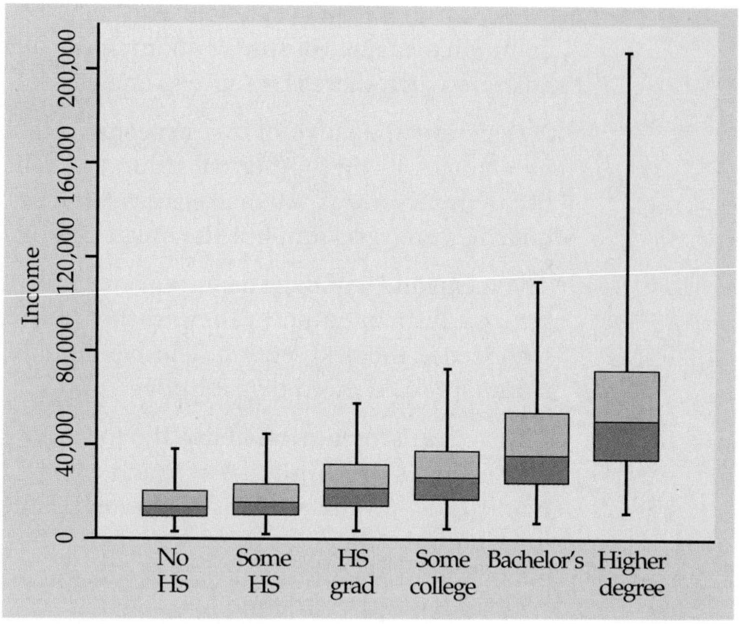

FIGURE 1.19 Boxplots comparing the distributions of income for people aged 25 to 65 years with six different levels of education. The lines extend from the quartiles to the 5th and 95th percentiles.

1.42 The data include 10,991 people whose highest level of education is a bachelor's degree.

(a) What is the position of the median in the ordered list of incomes (1 to 10,991)? From the plot, about what is the median income of people with a bachelor's degree?

(b) What is the position of the first and third quartiles in the ordered list of incomes for these people? About what are the numerical values of Q_1 and Q_3?

1.43 About what are the positions of the 5th and 95th percentiles in the ordered list of incomes of people with a bachelor's degree? Incomes outside this range do not appear in the boxplot. About what are the numerical values of the 5th and 95th percentiles of income? (For comparison, the largest income among these 10,991 people was $419,304. That one person made this much tells us less about the group than does the 95th percentile.)

1.44 Write a brief description of how the distribution of income changes with the highest level of education reached. Be sure to discuss center, spread, and skewness. Give some specifics read from the graph to back up your statements.

1.45 How much do users pay for Internet service? Here are the monthly fees (in dollars) paid by a random sample of 50 users of commercial Internet service providers in August 2000:[23]

20	40	22	22	21	21	20	10	20	20
20	13	18	50	20	18	15	8	22	25
22	10	20	22	22	21	15	23	30	12
9	20	40	22	29	19	15	20	20	20
20	15	19	21	14	22	21	35	20	22

(a) Make a stemplot of these data. Briefly describe the pattern you see. About how much do you think America Online and its larger competitors were charging in August 2000?

(b) Which observations are suspected outliers by the $1.5 \times IQR$ criterion? Which observations would you call outliers based on the stemplot?

1.46 Stemplots help you find the five-number summary because they arrange the observations in increasing order. Exercise 1.15 (page 25) includes a stemplot of the percent of residents aged 25 to 34 in each of the 50 states.

(a) Find the five-number summary of this distribution.

(b) Does the $1.5 \times IQR$ criterion flag Montana and Wyoming as suspected outliers?

(c) How much does the median change if you omit Montana and Wyoming?

1.47 Look at the histogram of lengths of words in Shakespeare's plays, Figure 1.11 (page 26). The heights of the bars tell us what percent of words have each length. What is the median length of words used by Shakespeare? Similarly, what are the quartiles? Give the five-number summary for Shakespeare's word lengths.

1.48 Exercise 1.21 (page 27) gives data on the total oil recovered from 64 wells. Your graph in that exercise shows that the distribution is strongly right-skewed.

(a) Find the mean and median of the amounts recovered. Explain how the relationship between the mean and the median reflects the shape of the distribution.

(b) Give the five-number summary and explain briefly how it reflects the shape of the distribution.

1.49 Here are the scores on the Survey of Study Habits and Attitudes (SSHA) for 18 first-year college women:

154	109	137	115	152	140	154	178	101
103	126	126	137	165	165	129	200	148

and for 20 first-year college men:

108	140	114	91	180	115	126	92	169	146
109	132	75	88	113	151	70	115	187	104

(a) Make a back-to-back stemplot of these data, or use your result from Exercise 1.23.

(b) Find the mean $\bar{x}$ and the median M for both sets of SSHA scores. What feature of each distribution explains the fact that $\bar{x} > M$?

(c) Find the five-number summaries for both sets of SSHA scores. Your plot in (a) suggests that there is an outlier among the women's scores. Does the $1.5 \times IQR$ criterion flag this observation? Make side-by-side modified boxplots for the two distributions.

(d) Use your results to write a brief comparison of the two groups. Do women as a group score higher than men? Which of your descriptions (stemplots, boxplots, numerical measures) show this? Which group of scores is more spread out when we ignore outliers? Which of your descriptions shows this most clearly?

1.50 The SSHA data for women given in the previous exercise contain one high outlier. Calculate the mean $\bar{x}$ and the median M for these data with and without the outlier. How does removing the outlier affect $\bar{x}$? How does it affect M? Your results illustrate the greater resistance of the median.

1.51 The *Mean and Median* applet allows you to place observations on a line and see their mean and median visually. Place two observations on the line. Why does only one arrow appear?

1.52 In the *Mean and Median* applet, place three observations on the line, two close together near the center of the line, and one somewhat to the right of these two.

(a) Pull the single rightmost observation out to the right. (Place the cursor on the point, hold down a mouse button, and drag the point.) How does the mean behave? How does the median behave? Explain briefly why each measure acts as it does.

(b) Now drag the rightmost point to the left as far as you can. What happens to the mean? What happens to the median as you drag this point past the other two (watch carefully)?

1.53 Table 1.3 (page 31) gives the number of medical doctors per 100,000 people in each state. Your graph of the distribution in Exercise 1.28 shows that the District of Columbia (D.C.) is a high outlier. Because D.C. is a city rather than a state, we will omit it here.

(a) Calculate both the five-number summary and $\bar{x}$ and s for the number of doctors per 100,000 people in the 50 states. Based on your graph, which description do you prefer?

(b) What facts about the distribution can you see in the graph that the numerical summaries don't reveal? Remember that measures of center and spread are not complete descriptions of a distribution.

1.54 Here are the percents of the popular vote won by the successful candidate in each U.S. presidential election from 1948 to 2000:

Year	1948	1952	1956	1960	1964	1968	1972
Percent	49.6	55.1	57.4	49.7	61.1	43.4	60.7
Year	1976	1980	1984	1988	1992	1996	2000
Percent	50.1	50.7	58.8	53.9	43.2	49.2	47.9

(a) Make a graph to display the distribution of winners' percents. What are the main features of this distribution?

(b) What is the median percent of the vote won by the successful candidate in presidential elections?

(c) Call an election a landslide if the winner's percent falls at or above the third quartile. Which elections were landslides?

Table 1.9 shows Consumer Reports *magazine's laboratory test results for calories and milligrams of sodium (mostly due to salt) in a number of major brands of hot dogs.*[24] *There are three types: all beef, "meat" (mainly pork and beef, but government regulations allow up to 15% poultry meat), and poultry. Exercises 1.55 to 1.57 analyze these data.*

1.55 Find the five-number summaries of the calorie content of the three types of hot dogs. Then use the 1.5 × *IQR* criterion to check for suspected outliers.

TABLE 1.9 Calories and sodium in three types of hot dogs

Beef hot dogs		Meat hot dogs		Poultry hot dogs	
Calories	Sodium	Calories	Sodium	Calories	Sodium
186	495	173	458	129	430
181	477	191	506	132	375
176	425	182	473	102	396
149	322	190	545	106	383
184	482	172	496	94	387
190	587	147	360	102	542
158	370	146	387	87	359
139	322	139	386	99	357
175	479	175	507	170	528
148	375	136	393	113	513
152	330	179	405	135	426
111	300	153	372	142	513
141	386	107	144	86	358
153	401	195	511	143	581
190	645	135	405	152	588
157	440	140	428	146	522
131	317	138	339	144	545
149	319				
135	298				
132	253				

Make modified boxplots to compare the three distributions. Write a brief discussion of your findings.

1.56 Make a stemplot of the calorie content of the 17 brands of meat hot dogs. What is the most important feature of the overall pattern of the distribution? Are there any outliers? Note that the five-number summary misses the big feature of this distribution. Routine numerical summaries are never a substitute for looking at the data.

1.57 Use graphs and numerical summaries to compare the sodium content of the three types of hot dogs. Write a summary of your findings suitable for readers who know no statistics. Can we hold down our sodium intake by buying poultry hot dogs?

1.58 Exercise 1.25 (page 29) presented data on the nightly study time claimed by first-year college men and women. The most common methods for formal comparison of two groups use $\bar{x}$ and s to summarize the data. We wonder if this is appropriate here. Look at your back-to-back stemplot from Exercise 1.25, or make one now if you have not done so.

(a) What kinds of distributions are best summarized by $\bar{x}$ and s? It isn't easy to decide whether small data sets with irregular distributions fit the criteria. We will learn a better tool for making this decision in the next section.

(b) Each set of study times appears to contain a high outlier. Are these points flagged as suspicious by the $1.5 \times IQR$ criterion? How much does removing the outlier change $\bar{x}$ and s for each group? The presence of outliers makes us reluctant to use the mean and standard deviation for these data unless we remove the outliers on the grounds that these students were exaggerating.

1.59 Last year a small accounting firm paid each of its five clerks \$25,000, two junior accountants \$60,000 each, and the firm's owner \$255,000. What is the mean salary paid at this firm? How many of the employees earn less than the mean? What is the median salary?

1.60 In computing the median income of any group, some federal agencies omit all members of the group who had no income. Give an example to show that the reported median income of a group can go down even though the group becomes economically better off. Is this also true of the mean income?

1.61 This year, the firm in Exercise 1.59 gives no raises to the clerks and junior accountants, while the owner's take increases to \$455,000. How does this change affect the mean? How does it affect the median?

1.62 Calculate the mean and standard deviation of the metabolic rates in Example 1.18 (page 48), showing each step in detail. First find the mean $\bar{x}$ by summing the 7 observations and dividing by 7. Then find each of the deviations $x_i - \bar{x}$ and their squares. Check that the deviations have sum 0. Calculate the variance as an average of the squared deviations (remember to divide by $n - 1$). Finally, obtain s as the square root of the variance.

1.63 Give an example of a small set of data for which the mean is larger than the third quartile.

1.64 Create a set of 5 positive numbers (repeats allowed) that have median 10 and mean 7. What thought process did you use to create your numbers?

1.65 This is a standard deviation contest. You must choose four numbers from the whole numbers 0 to 10, with repeats allowed.

(a) Choose four numbers that have the smallest possible standard deviation.

(b) Choose four numbers that have the largest possible standard deviation.

(c) Is more than one choice possible in either (a) or (b)? Explain.

1.66 Use the definition of the mean $\overline{x}$ to show that the sum of the deviations $x_i - \overline{x}$ of the observations from their mean is always zero. This is one reason why the variance and standard deviation use squared deviations.

1.67 This exercise requires a calculator with a standard deviation button, or statistical software on a computer. The observations

$$10,001 \quad 10,002 \quad 10,003$$

have mean $\overline{x} = 10,002$ and standard deviation $s = 1$. Adding a 0 in the center of each number, the next set becomes

$$100,001 \quad 100,002 \quad 100,003$$

The standard deviation remains $s = 1$ as more 0s are added. Use your calculator or computer to calculate the standard deviation of these numbers, adding extra 0s until you get an incorrect answer. How soon did you go wrong? This demonstrates that calculators and computers cannot handle an arbitrary number of digits correctly.

1.68 We saw in Example 1.19 that it is usual in the study of investments to use the mean and standard deviation to summarize and compare investment returns. Table 1.4 (page 31) gives the monthly returns on one company's stock for 83 consecutive months.

(a) Find the mean monthly return and the standard deviation of the returns. If you invested $100 in this stock at the beginning of a month and got the mean return, how much would you have at the end of the month?

(b) The distribution can be described as "symmetric and unimodal, with one low outlier." If you invested $100 in this stock at the beginning of the worst month in the data (the outlier), how much would you have at the end of the month? Find the mean and standard deviation again, this time leaving out the low outlier. How much did this one observation affect the summary measures? Would leaving out this one observation change the median? The quartiles? How do you know, without actual calculation? (Returns over longer periods of time, or returns on portfolios containing several investments, tend to follow a normal distribution more closely than these monthly returns do. So use of the mean and standard deviation is better justified for such data.)

1.69 Table 1.5 (page 32) gives the survival times of 72 guinea pigs in a medical study. Survival times—whether of cancer patients after treatment or of car batteries in everyday use—are almost always right-skewed. Make a graph to verify that this is true of these survival times. Then give a numerical summary that is appropriate for such data. Explain in simple language, to someone who knows no statistics, what your summary tells us about the guinea pigs.

1.70 Find the 10th and 90th percentiles of the distribution of doctors per 100,000 population in the states, from Table 1.3 (page 31). Which states are in the top 10%? In the bottom 10%?

quintiles

deciles

1.71 Find the **quintiles** (the 20th, 40th, 60th, and 80th percentiles) of the IQ scores in Table 1.6 (page 33). For quite large sets of data, the quintiles or the **deciles** (10th, 20th, 30th, etc. percentiles) give a more detailed summary than the quartiles.

1.72 In each of the following settings, give the values of a and b for the linear transformation $x_{new} = a + bx$ that expresses the change in units of measurement.

(a) Change a speed x measured in miles per hour into the metric system value x_{new} in kilometers per hour. (A kilometer is 0.62 mile.) What is 65 miles per hour in metric units?

(b) You are writing a report on the power of car engines. Your sources use horsepower x. Reexpress power in watts x_{new}. (One horsepower is 746 watts.) What is the power in watts of a 140-horsepower engine?

1.73 In each of the following settings, give the values of a and b for the linear transformation $x_{new} = a + bx$ that expresses the change in units of measurement.

(a) You want to restate water temperature x in a swimming pool, measured in degrees Fahrenheit, as the difference x_{new} between x and the "normal" body temperature of 98.6 degrees.

(b) The recommended daily allowance (RDA) for vitamin C was recently increased to 120 milligrams. You measure milligrams of vitamin C in foods and want to convert your results to percent of RDA.

1.74 Henry Cavendish (see Exercise 1.27 on page 30) used $\bar{x}$ to summarize his 29 measurements of the density of the earth.

(a) Find $\bar{x}$ and s for his data.

(b) Cavendish recorded the density of the earth as a multiple of the density of water. The density of water is almost exactly 1 gram per cubic centimeter, so his measurements have these units. In American units, the density of water is 62.43 pounds per cubic foot. This is the weight of a cube of water measuring 1 foot (that is, 30.48 cm) on each side. Express Cavendish's first result for the earth (5.50 g/cm^3) in pounds per cubic foot. Then find $\bar{x}$ and s in pounds per cubic foot.

1.75 A change of units that multiplies each unit by b, such as the change $x_{new} = 2.54x$ from inches x to centimeters x_{new}, multiplies our usual measures of spread by b. This is true of *IQR* and the standard deviation. What happens to the variance when we change units in this way?

trimmed mean **1.76** The **trimmed mean** is a measure of center that is more resistant than the mean but uses more of the available information than the median. To compute the 10% trimmed mean, discard the highest 10% and the lowest 10% of the observations and compute the mean of the remaining 80%. Trimming eliminates the effect of a small number of outliers. Compute the 10% trimmed mean of the guinea pig survival time data in Table 1.5 (page 32). Then compute the 20% trimmed mean. Compare the values of these measures with the median and the ordinary untrimmed mean.

1.3 The Normal Distributions

We now have a kit of graphical and numerical tools for describing distributions. What is more, we have a clear strategy for exploring data on a single quantitative variable:

1. Always plot your data: make a graph, usually a stemplot or a histogram.
2. Look for the overall pattern and for striking deviations such as outliers.
3. Calculate an appropriate numerical summary to briefly describe center and spread.

Here is one more step to add to this strategy:

4. Sometimes the overall pattern of a large number of observations is so regular that we can describe it by a smooth curve.

Figure 1.20 is a histogram of the scores of all 947 seventh-grade students in Gary, Indiana, on the vocabulary part of the Iowa Test of Basic Skills. Scores of many students on this national test have a very regular distribution. The histogram is symmetric, and both tails fall off quite smoothly from a single center peak. There are no large gaps or obvious outliers. The smooth curve drawn through the tops of the histogram bars in Figure 1.20 is a good description of the overall pattern of the data. The curve is a **mathematical model** for the distribution. A mathematical model is an idealized description. It gives a compact picture of the overall pattern of the data but ignores minor irregularities as well as any outliers.

mathematical model

The use of mathematical models is a common and powerful tool in statistics, as in other sciences. Suppose, for example, that we are watching a caterpillar crawl forward. Once each minute we measure the distance it has moved. When we plot the observed values against time, we see that the points fall close to a straight line. It would be natural to draw a straight line on our graph as a compact description of these data. The straight line, with its equation of the form $y = a + bt$ (where y is distance and t is time), is a mathematical model

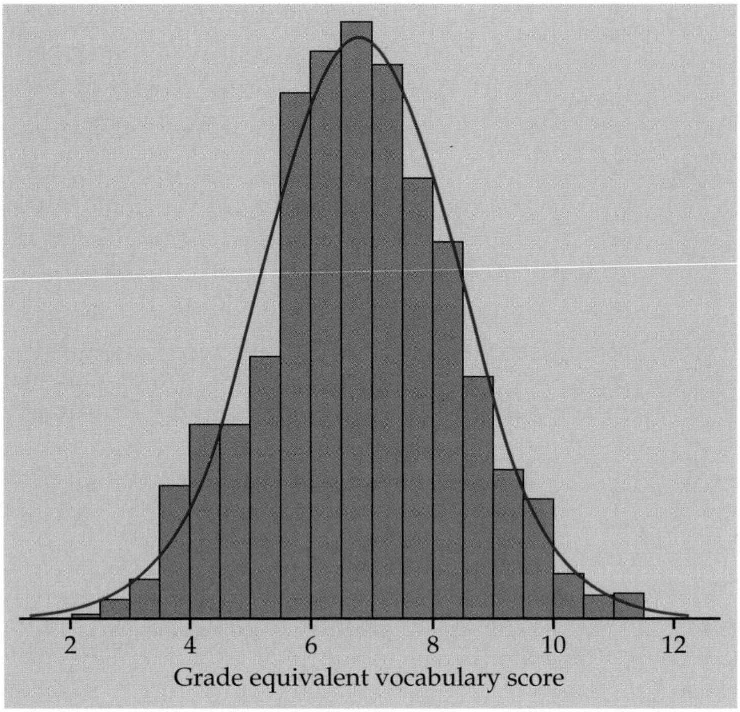

FIGURE 1.20 Histogram of the Iowa Test vocabulary scores for Gary, Indiana, seventh graders, showing the approximation of the distribution by a normal curve.

of the caterpillar's progress. It is an idealized description because the points on the graph do not fall *exactly* on the line. We want to give a similarly compact description of the distribution of a quantitative variable, to replace stemplots and histograms. Just as a straight line is one of many types of curves that can be used to describe a plot of distance traveled against time, there are many types of mathematical distributions that can be used to describe a set of single-variable data. This section will concentrate on one type, the normal distributions.

Density curves

We will see that it is easier to work with the smooth curve in Figure 1.20 than with the histogram. The reason is that the histogram depends on our choice of classes, while with a little care we can use a curve that does not depend on any choices we make. Here's the idea.

EXAMPLE 1.22

In a histogram, the *areas* of the bars represent either counts or proportions of the observations. Figure 1.21(a) is a copy of Figure 1.20 with the leftmost bars shaded. The area of the shaded bars in Figure 1.21(a) represents the students with vocabulary scores 6.0 or lower. There are 287 such students, who make up the proportion $287/947 = 0.303$ of all Gary seventh graders.

In Figure 1.21(b), the area under the smooth curve to the left of 6.0 is shaded. Adjust the scale so that the total area under the curve is exactly 1. Areas under the curve

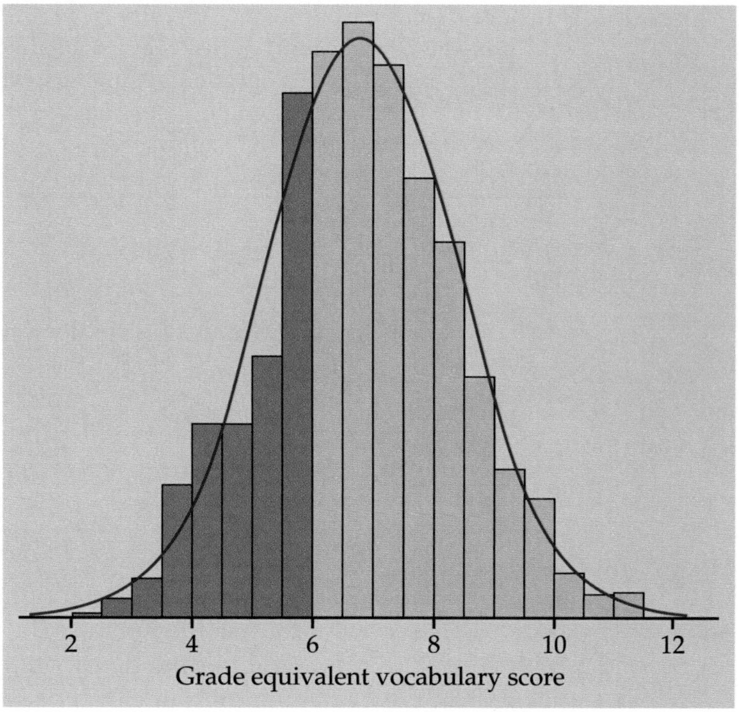

FIGURE 1.21(a) The relative frequency of scores less than or equal to 6.0 from the histogram is 0.303.

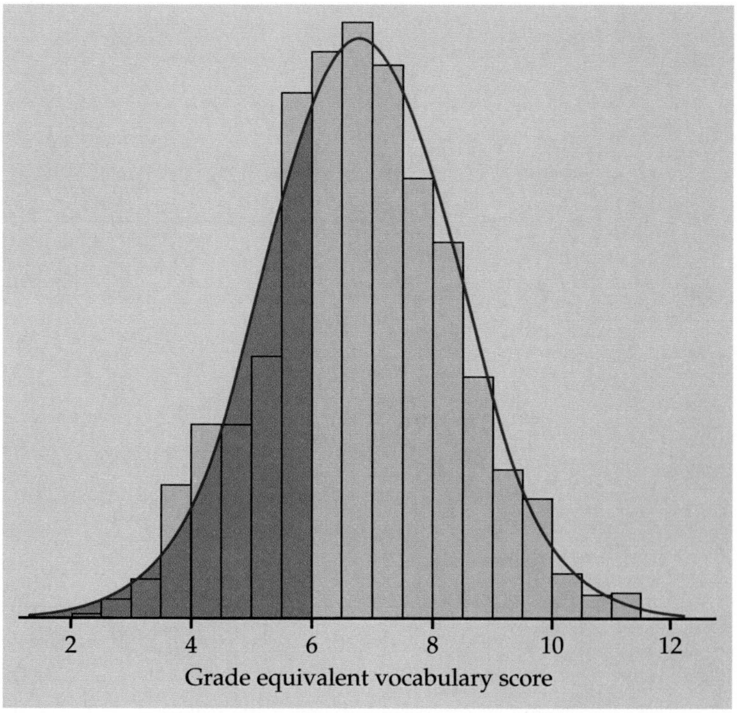

FIGURE 1.21(b) The relative frequency of scores less than or equal to 6.0 from the density curve is 0.293.

then represent proportions of the observations. That is, *area* = *relative frequency*. The curve is then a *density curve*. The shaded area under the density curve in Figure 1.21(b) represents the proportion of students with score 6.0 or lower. This area is 0.293, only 0.010 away from the histogram result. You can see that areas under the density curve give quite good approximations of areas given by the histogram.

Density Curve

A **density curve** is a curve that

- is always on or above the horizontal axis and
- has area exactly 1 underneath it.

A density curve describes the overall pattern of a distribution. The area under the curve and above any range of values is the relative frequency of all observations that fall in that range.

The density curve in Figures 1.20 and 1.21 is a *normal curve*. Density curves, like distributions, come in many shapes. Figure 1.22 shows two density curves, a symmetric normal density curve and a right-skewed curve. A density curve of the appropriate shape is often an adequate description of the overall pattern of a distribution. Outliers, which are deviations from the overall pattern, are not described by the curve.

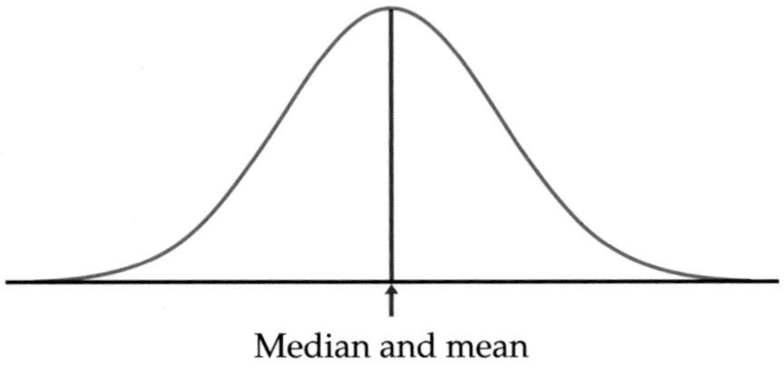

FIGURE 1.22(a) A symmetric density curve with its mean and median marked.

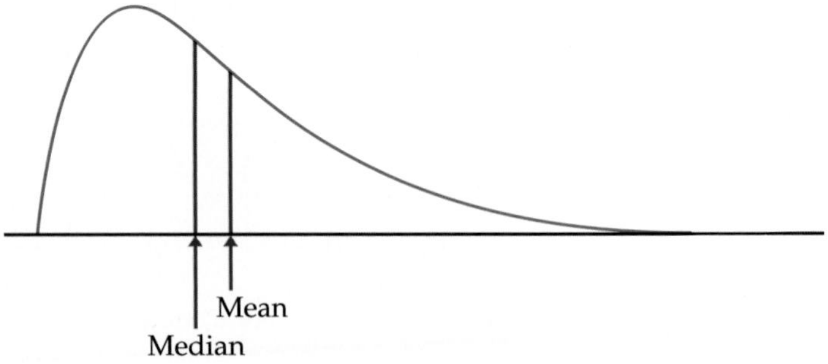

FIGURE 1.22(b) A right-skewed density curve with its mean and median marked.

Measuring center and spread for density curves

Our measures of center and spread apply to density curves as well as to actual sets of observations, but only some of these measures are easily seen from the curve. A **mode** of a distribution described by a density curve is a peak point of the curve, the location where the curve is highest. Because areas under a density curve represent proportions of the observations, the **median** is the point with half the total area on each side. You can roughly locate the **quartiles** by dividing the area under the curve into quarters as accurately as possible by eye. The *IQR* is then the distance between the first and third quartiles. There are mathematical ways of calculating areas under curves. These allow us to locate the median and quartiles exactly on any density curve.

What about the mean and standard deviation? The mean of a set of observations is their arithmetic average. If we think of the observations as weights strung out along a thin rod, the mean is the point at which the rod would balance. This fact is also true of density curves. The mean is the point at which the curve would balance if it were made out of solid material. Figure 1.23 illustrates this interpretation of the mean. We have marked the mean and median on the density curves in Figure 1.22. A symmetric curve, such as the normal curve in Figure 1.22(a), balances at its center of symmetry. Half the area under a symmetric curve lies on either side of its center, so this is also the median. For a right-skewed curve, such as that shown in Figure 1.22(b), the small area in the long right tail tips the curve more than the same area near the center. The mean (the balance point) therefore lies to the right of the median. It is hard to locate the balance point by eye on a skewed curve. There are mathematical ways of calculating the mean for any density curve, so we are able to mark the mean as well as the median in Figure 1.22(b). The standard deviation can also be calculated mathematically, but it can't be located by eye on most density curves.

> **Median and Mean of a Density Curve**
>
> The **median** of a density curve is the equal-areas point, the point that divides the area under the curve in half.
>
> The **mean** of a density curve is the balance point, at which the curve would balance if made of solid material.
>
> The median and mean are the same for a symmetric density curve. They both lie at the center of the curve. The mean of a skewed curve is pulled away from the median in the direction of the long tail.

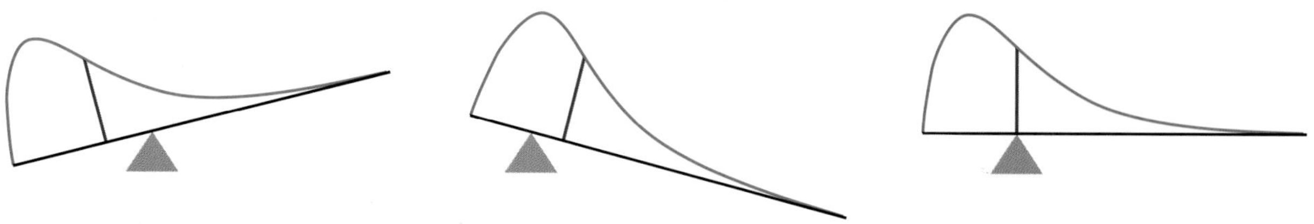

FIGURE 1.23 The mean of a density curve is the point at which it would balance.

A density curve is an idealized mathematical model for a distribution of data. For example, the symmetric density curve in Figures 1.20 and 1.21 is exactly symmetric, but the histogram of vocabulary scores is only approximately symmetric. We therefore need to distinguish between the mean and standard deviation of the density curve and the numbers $\bar{x}$ and s computed from the actual observations. The usual notation for the mean of an idealized distribution is μ (the Greek letter mu). We write the standard deviation of a density curve as σ (the Greek letter sigma).

mean μ

standard deviation σ

Normal distributions

normal curves

One particularly important class of density curves has already appeared in Figures 1.20 and 1.22(a). These density curves are symmetric, unimodal, and bell-shaped. They are called **normal curves,** and they describe *normal distributions*. All normal distributions have the same overall shape. The exact density curve for a particular normal distribution is specified by giving its mean μ and its standard deviation σ. The mean is located at the center of the symmetric curve and is the same as the median. Changing μ without changing σ moves the normal curve along the horizontal axis without changing its spread. The standard deviation σ controls the spread of a normal curve. Figure 1.24 shows two normal curves with different values of σ. The curve with the larger standard deviation is more spread out.

The standard deviation σ is the natural measure of spread for normal distributions. Not only do μ and σ completely determine the shape of a normal curve, but we can locate σ by eye on the curve. Here's how. As we move out in either direction from the center μ, the curve changes from falling ever more steeply

to falling ever less steeply

The points at which this change of curvature takes place are located at distance σ on either side of the mean μ. You can feel the change as you run a pencil along a normal curve, and so find the standard deviation. Remember that μ

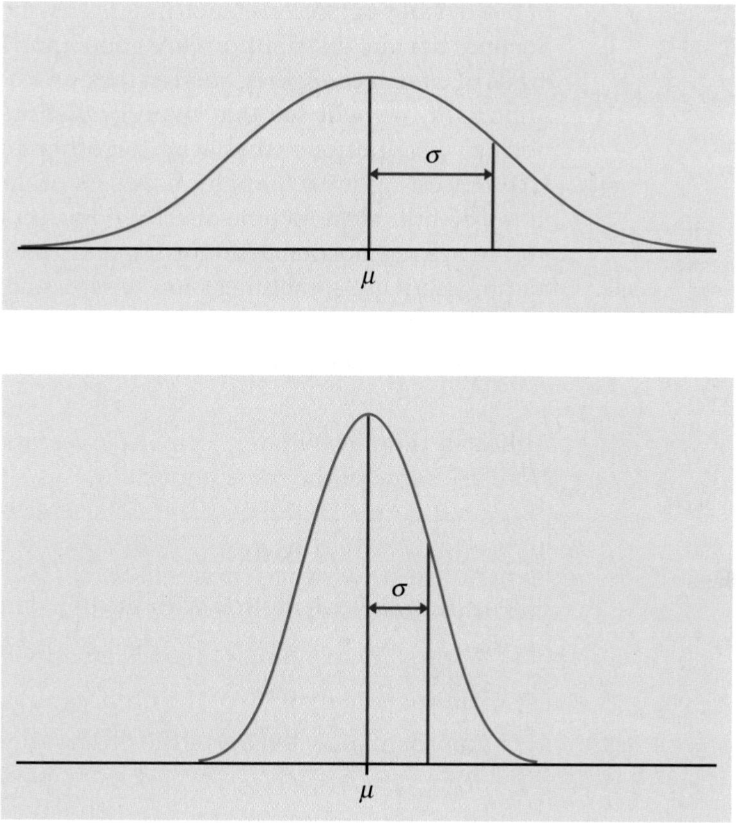

FIGURE 1.24 Two normal curves, showing the mean μ and standard deviation σ.

and σ alone do not specify the shape of most distributions, and that the shape of density curves in general does not reveal σ. These are special properties of normal distributions.

There are other symmetric bell-shaped density curves that are not normal. The normal density curves are specified by a particular equation. The height of the density curve at any point x is given by

$$\frac{1}{\sigma\sqrt{2\pi}}e^{-\frac{1}{2}\left(\frac{x-\mu}{\sigma}\right)^2}$$

We will not make direct use of this fact, although it is the basis of mathematical work with normal distributions. Note that the equation of the curve is completely determined by the mean μ and the standard deviation σ.

Why are the normal distributions important in statistics? Here are three reasons. First, normal distributions are good descriptions for some distributions of *real data*. Distributions that are often close to normal include scores on tests taken by many people (such as SAT exams and many psychological tests), repeated careful measurements of the same quantity, and characteristics of biological populations (such as lengths of baby pythons and yields of corn). Second, normal distributions are good approximations to the results of many kinds of *chance outcomes*, such as tossing a coin many times. Third, and most important, we will see that many *statistical inference* procedures based on normal distributions work well for other roughly symmetric distributions. HOWEVER ... even though many sets of data follow a normal distribution, many do not. Most income distributions, for example, are skewed to the right and so are not normal. Nonnormal data, like nonnormal people, not only are common but are sometimes more interesting than their normal counterparts.

The 68–95–99.7 rule

Although there are many normal curves, they all have common properties. Here are some of the most important.

The 68–95–99.7 Rule

In the normal distribution with mean μ and standard deviation σ:

- Approximately **68%** of the observations fall within σ of the mean μ.
- Approximately **95%** of the observations fall within 2σ of μ.
- Approximately **99.7%** of the observations fall within 3σ of μ.

Figure 1.25 illustrates the 68–95–99.7 rule. By remembering these three numbers, you can think about normal distributions without constantly making detailed calculations.

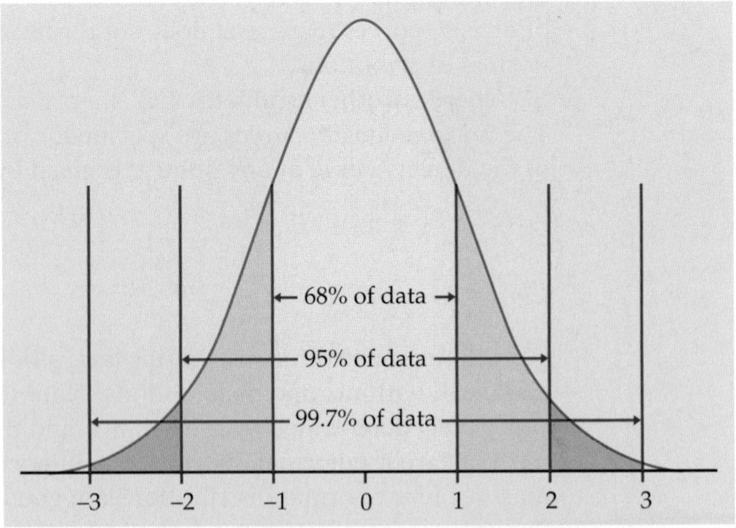

FIGURE 1.25 The 68–95–99.7 rule for normal distributions.

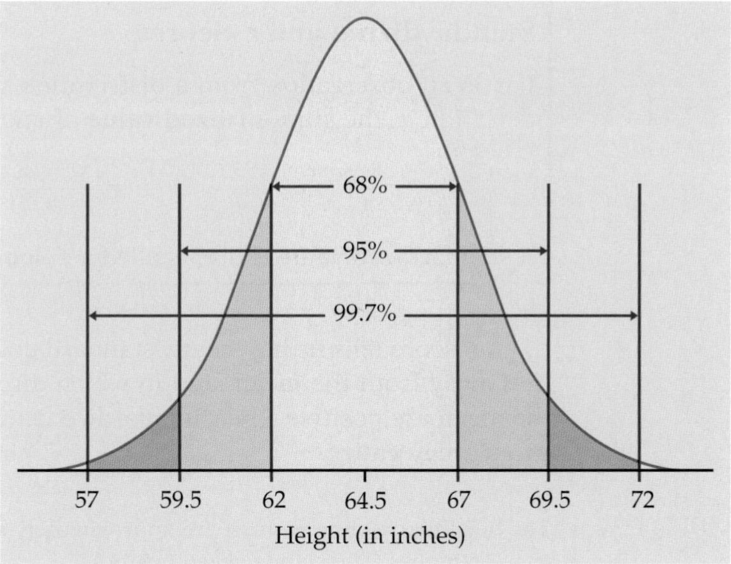

FIGURE 1.26 The 68–95–99.7 rule applied to the heights of young women.

EXAMPLE 1.23

The distribution of heights of young women aged 18 to 24 is approximately normal with mean $\mu = 64.5$ inches and standard deviation $\sigma = 2.5$ inches. Figure 1.26 shows what the 68–95–99.7 rule says about this distribution.

Two standard deviations is 5 inches for this distribution. The 95 part of the 68–95–99.7 rule says that the middle 95% of young women are between $64.5 - 5$ and $64.5 + 5$ inches tall, that is, between 59.5 inches and 69.5 inches. This fact is exactly true for an exactly normal distribution. It is approximately true for the heights of young women because the distribution of heights is approximately normal.

The other 5% of young women have heights outside the range from 59.5 to 69.5 inches. Because the normal distributions are symmetric, half of these women are on the tall side. So the tallest 2.5% of young women are taller than 69.5 inches.

$N(\mu, \sigma)$

Because we will mention normal distributions often, a short notation is helpful. We abbreviate the normal distribution with mean μ and standard deviation σ as $N(\mu, \sigma)$. For example, the distribution of young women's heights is $N(64.5, 2.5)$.

Standardizing observations

As the 68–95–99.7 rule suggests, all normal distributions share many properties. In fact, all normal distributions are the same if we measure in units of size σ about the mean μ as center. Changing to these units is called *standardizing*. To standardize a value, subtract the mean of the distribution and then divide by the standard deviation.

> **Standardizing and z-Scores**
>
> If x is an observation from a distribution that has mean μ and standard deviation σ, the **standardized value** of x is
>
> $$z = \frac{x - \mu}{\sigma}$$
>
> A standardized value is often called a **z-score.**

A z-score tells us how many standard deviations the original observation falls away from the mean, and in which direction. Observations larger than the mean are positive when standardized, and observations smaller than the mean are negative.

EXAMPLE 1.24

The heights of young women are approximately normal with $\mu = 64.5$ inches and $\sigma = 2.5$ inches. The standardized height is

$$z = \frac{\text{height} - 64.5}{2.5}$$

A woman's standardized height is the number of standard deviations by which her height differs from the mean height of all young women. A woman 68 inches tall, for example, has standardized height

$$z = \frac{68 - 64.5}{2.5} = 1.4$$

or 1.4 standard deviations above the mean. Similarly, a woman 5 feet (60 inches) tall has standardized height

$$z = \frac{60 - 64.5}{2.5} = -1.8$$

or 1.8 standard deviations less than the mean height.

We need a way to write variables, such as "height" in Example 1.24, that follow a theoretical distribution such as a normal distribution. We use capital letters near the end of the alphabet for such variables. If X is the height of a young woman, we can then shorten "the height of a young woman is less than 68 inches" to "$X < 68$." We will use lowercase x to stand for any specific value of the variable X.

We often standardize observations from symmetric distributions to express them in a common scale. We might, for example, compare the height of two children of different ages by calculating their z-scores. The standardized heights tell us where each child stands in the distribution for his or her age group.

Standardizing is a linear transformation that transforms the data into the standard scale of z-scores. We know that a linear transformation does not change the shape of a distribution, and that the mean and standard deviation change in a simple manner. In fact, *any variable obtained from a normal variable by a linear transformation remains normal.* Of course, the mean and standard deviation will change in the usual way. If X has the normal distribution with mean μ and standard deviation σ, then $X_{new} = a + bX$ for a positive b has the normal distribution with mean $a + b\mu$ and standard deviation $b\sigma$. In particular, *the standardized values for any distribution always have mean 0 and standard deviation 1*. If the original distribution was normal, the standardized values have the normal distribution with mean 0 and standard deviation 1.

The standard normal distribution

If the variable we standardize has a normal distribution, standardizing does more than give a common scale. It makes all normal distributions into a single distribution, and this distribution is still normal. Standardizing a variable that has any normal distribution produces a new variable that has the *standard normal distribution*.

The Standard Normal Distribution

The **standard normal distribution** is the normal distribution $N(0, 1)$ with mean 0 and standard deviation 1.

If a variable X has any normal distribution $N(\mu, \sigma)$ with mean μ and standard deviation σ, then the standardized variable

$$Z = \frac{X - \mu}{\sigma}$$

has the standard normal distribution.

Table A in the back of the book gives areas under the standard normal curve. Table A also appears on the inside front cover. You can use Table A to do normal calculations, although software or a statistical calculator is usually more efficient.

The Standard Normal Table

Table A is a table of areas under the standard normal curve. The table entry for each value z is the area under the curve to the left of z.

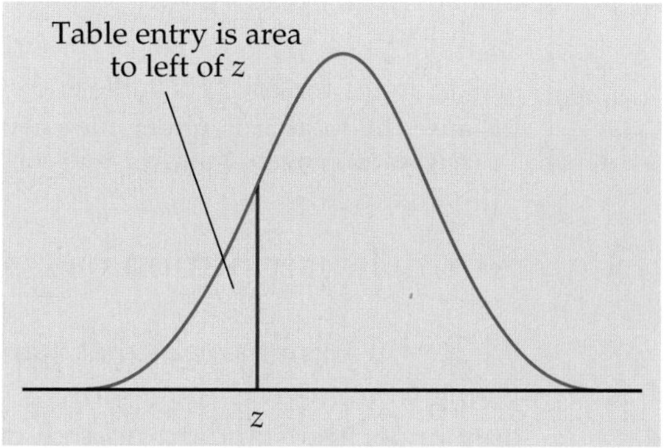

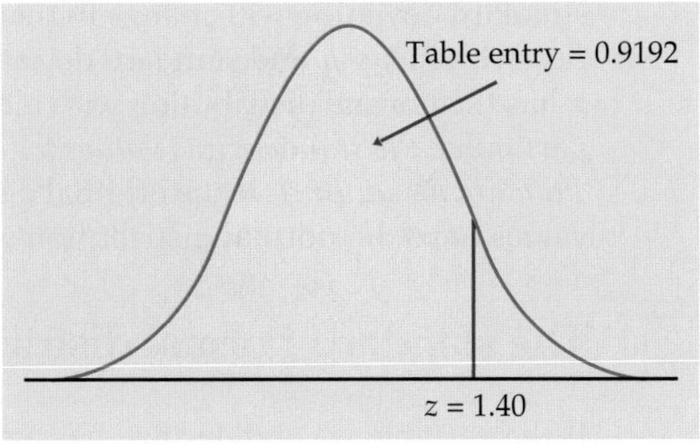

FIGURE 1.27(a) The area under a standard normal curve to the left of the point $z = 1.40$ is 0.9192. Table A gives areas under the standard normal curve.

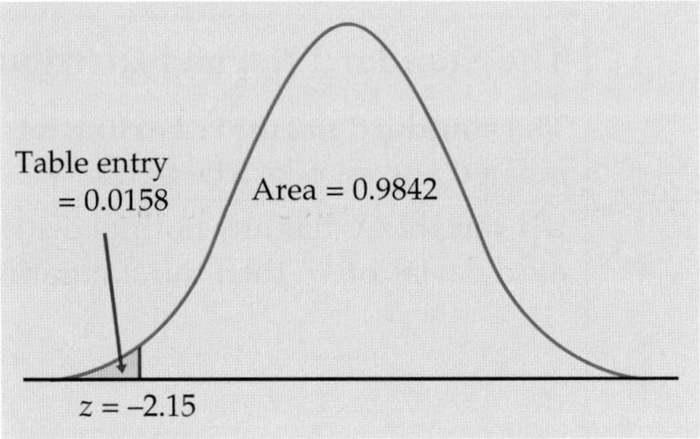

FIGURE 1.27(b) Areas under the standard normal curve to the right and left of $z = -2.15$. Table A gives the area to the left.

EXAMPLE 1.25

Problem: What proportion of observations on a standard normal variable Z take values less than 1.4?

Solution: To find the area to the left of 1.40, locate 1.4 in the left-hand column of Table A, then locate the remaining digit 0 as .00 in the top row. The entry opposite 1.4 and under .00 is 0.9192. This is the area we seek. Figure 1.27(a) illustrates this area.

Problem: Find the proportion of observations from the standard normal distribution that are greater than -2.15.

Solution: Enter Table A under $z = -2.15$. That is, find -2.1 in the left-hand column and .05 in the top row. The table entry is 0.0158. This is the area to the *left* of -2.15. Because the total area under the curve is 1, the area lying to the *right* of -2.15 is $1 - 0.0158 = 0.9842$. Figure 1.27(b) shows these areas.

Normal distribution calculations

We saw in Example 1.24 that standardizing the height $x = 68$ inches for a young woman gives the z-score $z = 1.4$. Example 1.25 then showed that the area under the standard normal curve to the left of $z = 1.4$ is 0.9192. This is the relative frequency of z-scores less than 1.4. But the z-score is less than 1.4 whenever the height is less than 68 inches. So the proportion of all young women who are less than 68 inches tall is 0.9192, or about 92%. We can find relative frequencies for *any* normal distribution by standardizing and using Table A. Of course, software or a statistical calculator often automates the calculation. Here is an example.

EXAMPLE 1.26

The National Collegiate Athletic Association (NCAA) requires Division I athletes to score at least 820 on the combined mathematics and verbal parts of the SAT exam in order to compete in their first college year. (Higher scores are required for students with poor high school grades.) In 2000, the scores of the more than one million students taking the SATs were approximately normal with mean 1019 and standard deviation 209. What percent of all students had SAT scores of at least 820?

1. *State the problem*. Call the SAT score of a randomly chosen student X. The variable X has the $N(1019, 209)$ distribution. We want the proportion of students with $X > 820$.
2. *Standardize*. Subtract the mean, then divide by the standard deviation, to transform the problem about X into a problem about a standard normal Z:

$$X > 820$$

$$\frac{X - 1019}{209} > \frac{820 - 1019}{209}$$

$$Z > -0.95$$

Figure 1.28 shows the standard normal curve with the area of interest shaded.

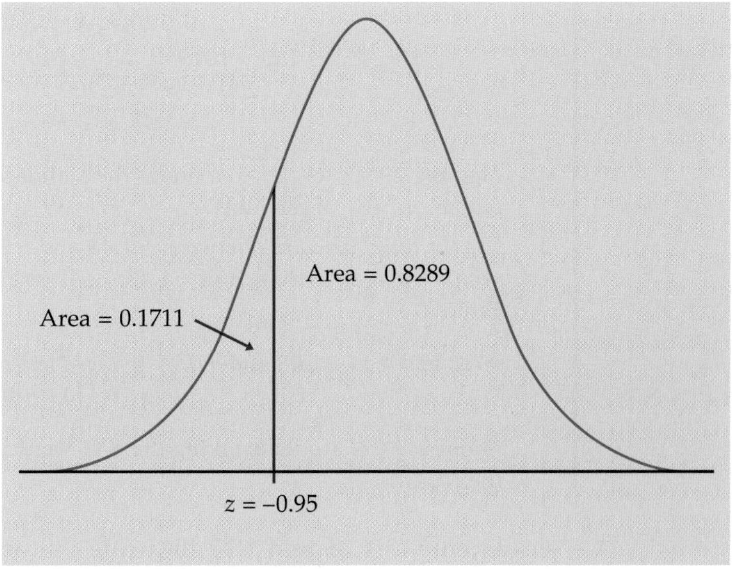

FIGURE 1.28 Areas under the standard normal curve for Example 1.26.

3. *Use the table*. From Table A, we see that the proportion of observations less than -0.95 is 0.1711. About 17% of students who take the SAT score less than 820. The area to the right of -0.95 is therefore $1 - 0.1711 = 0.8289$. This is about 0.83, or 83%. About 83% of all students taking the SAT would be academically qualified for Division I college athletics.

There is *no* area under a smooth curve and exactly over the point 820. Consequently, the area to the right of 820 (the relative frequency of $X > 820$) is the same as the area at or to the right of this point (the relative frequency of $X \geq 820$). The actual data may contain a student who scored exactly 820 on the SAT. The fact that it is not possible for a normal variable to have $X = 820$ exactly is a consequence of the idealized smoothing of normal models for data.

Sometimes we encounter a value of z more extreme than those appearing in Table A. For example, the area to the left of $z = -4$ is not given directly in the table. The z-values in Table A leave only area 0.0002 in each tail unaccounted for. For practical purposes, we can act as if there is zero area outside the range of Table A.

The key to using either software or Table A to do a normal calculation is to sketch the area you want, then match that area with the area that the table or software gives you. Here is another example.

EXAMPLE 1.27

The NCAA considers a student a "partial qualifier" eligible to practice and receive an athletic scholarship, but not to compete, if the combined SAT score is at least 720. What percent of all students who take the SAT would be partial qualifiers?

1. *State the problem*. We want the proportion of SAT scores in the interval $720 \leq X < 820$.

2. *Standardize*.

$$720 \leq X < 820$$
$$\frac{720 - 1019}{209} \leq \frac{X - 1019}{209} < \frac{820 - 1019}{209}$$
$$-1.43 \leq Z < -0.95$$

Figure 1.29 shows the area under the standard normal curve. This picture guides our use of the table.

3. *Use the table*. The area between -1.43 and -0.95 is the area to the left of -0.95 *minus* the area to the left of -1.43. Look at Figure 1.29 to visualize this. From Table A,

$$\begin{aligned}\text{area between } -1.43 \text{ and } -0.95 &= (\text{area left of } -0.95) - (\text{area left of } -1.43)\\ &= 0.1711 - 0.0764 = 0.0947\end{aligned}$$

About 9.5% of students taking the SAT would be partial qualifiers in the eyes of the NCAA.

Examples 1.26 and 1.27 illustrate the use of Table A to find the relative frequency of a given event, such as "SAT score between 720 and 820." We may instead want to find the observed value corresponding to a given relative

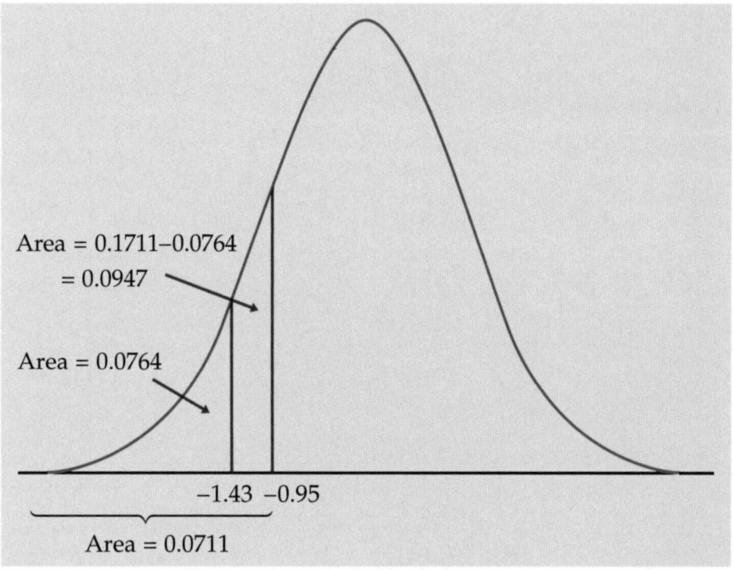

FIGURE 1.29 Areas under the standard normal curve for Example 1.27.

frequency. To do this, we use Table A backward, finding the desired relative frequency in the body of the table and then reading the corresponding z from the left column and top row.

EXAMPLE 1.28

Scores on the SAT verbal test in recent years follow approximately the $N(505, 110)$ distribution. How high must a student score in order to place in the top 10% of all students taking the SAT?

1. *State the problem*. We want to find the SAT score x with area 0.1 to its *right* under the normal curve with mean $\mu = 505$ and standard deviation $\sigma = 110$. That's the same as finding the SAT score x with area 0.9 to its *left*. Figure 1.30 poses the question in graphical form. Because Table A gives the areas to the left of z-values, always state the problem in terms of the area to the left of x.
2. *Use the table*. Look in the body of Table A for the entry closest to 0.9. It is 0.8997. This is the entry corresponding to $z = 1.28$. So $z = 1.28$ is the standardized value with area 0.9 to its left.
3. *Unstandardize* to transform the solution from the z back to the original x scale. We know that the standardized value of the unknown x is $z = 1.28$. So x itself satisfies

$$\frac{x - 505}{110} = 1.28$$

Solving this equation for x gives

$$x = 505 + (1.28)(110) = 645.8$$

This equation should make sense: it finds the x that lies 1.28 standard deviations above the mean on this particular normal curve. That is the

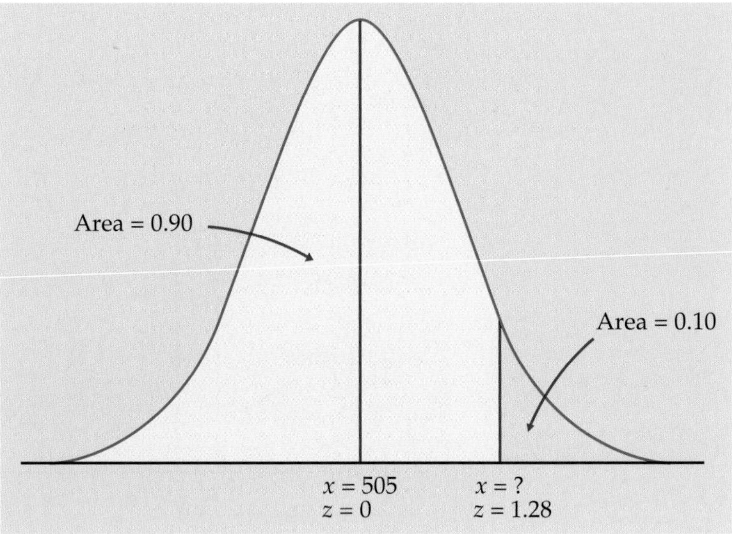

FIGURE 1.30 Locating the point on a normal curve with area 0.10 to its right.

"unstandardized" meaning of $z = 1.28$. We see that a student must score at least 646 to place in the highest 10%. The general rule for unstandardizing a z-score is

$$x = \mu + z\sigma$$

Normal quantile plots

The normal distributions provide good models for some distributions of real data. Examples include the Gary vocabulary scores and Newcomb's measurements of the passage time of light (after dropping the outliers). The distributions of some other common variables are usually skewed and therefore distinctly nonnormal. Examples include economic variables such as personal income and gross sales of business firms, the survival times of cancer patients after treatment, and the service lifetime of mechanical or electronic components. While experience can suggest whether or not a normal model is plausible in a particular case, it is risky to assume that a distribution is normal without actually inspecting the data.

The decision to describe a distribution by a normal model may determine the later steps in our analysis of the data. Both relative frequency calculations and statistical inference based on such calculations follow from the choice of a model. How can we judge whether data are approximately normal?

A histogram or stemplot can reveal distinctly nonnormal features of a distribution, such as outliers (Newcomb's histogram, Figure 1.5 on page 17), pronounced skewness (the supermarket purchase stemplot, Figure 1.3 on page 12), or gaps and clusters (the histogram of Massachusetts college tuitions in Figure 1.12, page 27). If the stemplot or histogram appears roughly symmetric and unimodal, however, we need a more sensitive way to judge the adequacy of a normal model. The most useful tool for assessing normality is another graph, the **normal quantile plot.***

normal quantile plot

Here is the idea of a simple version of a normal quantile plot. It is not feasible to make normal quantile plots by hand, but software makes them for us, using more sophisticated versions of this basic idea.

1. Arrange the observed data values from smallest to largest. Record what percentile of the data each value occupies. For example, the smallest observation in a set of 20 is at the 5% point, the second smallest is at the 10% point, and so on.
2. Do normal distribution calculations to find the z-scores at these same percentiles. For example, $z = -1.645$ is the 5% point of the standard normal distribution, and $z = -1.282$ is the 10% point.
3. Plot each data point x against the corresponding z. If the data distribution is close to standard normal, the plotted points will lie close to the 45-degree line $x = z$. If the data distribution is close to any normal distribution, the plotted points will lie close to some straight line.

Any normal distribution produces a straight line on the plot because standardizing turns any normal distribution into a standard normal. Standardizing is a linear transformation that can change the slope and intercept of the line in our plot but cannot turn a line into a curved pattern.

Use of Normal Quantile Plots

If the points on a normal quantile plot lie close to a straight line, the plot indicates that the data are normal. Systematic deviations from a straight line indicate a nonnormal distribution. Outliers appear as points that are far away from the overall pattern of the plot.

Figures 1.31 to 1.34 are normal quantile plots for data we have met earlier. The data x are plotted vertically against the corresponding standard normal z-score plotted horizontally. The z-score scale extends from -3 to 3 because almost all of a standard normal curve lies between these values. These figures show how normal quantile plots behave.

EXAMPLE 1.29

Figure 1.31 is a normal quantile plot of Newcomb's passage time data. Most of the points lie close to a straight line, indicating that a normal model fits well. The two outliers deviate from the line and show how the plot responds to low outliers. These two points are below the line formed by the rest of the data—they are farther out

*Some software calls these graphs *normal probability plots*. There is a technical distinction between the two types of graphs, but the terms are often used loosely.

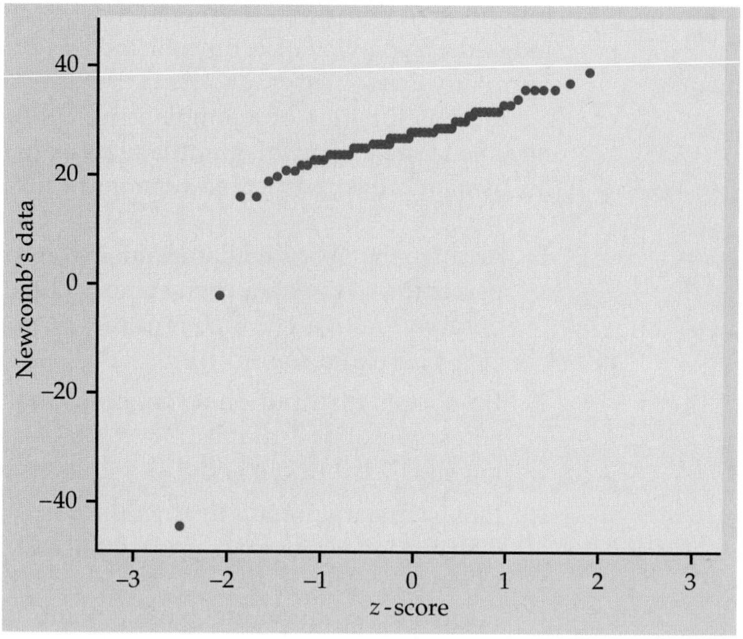

FIGURE 1.31 Normal quantile plot for Newcomb's 66 measurements of the passage time of light. There are two low outliers.

in the low direction than we expect normal data to be. Compare the histogram of these data (Figure 1.5, page 17).

Figure 1.32 is a normal quantile plot of Newcomb's measurements of the passage time of light with the two outliers omitted. The effect of omitting the outliers is to magnify the plot of the remaining data. As we saw from Figure 1.31, a normal distribution fits quite well. The only important deviation from normality is the "stair-step" appearance caused by numerous short horizontal runs of points. Each run represents repeated observations having the same value—there are six measurements at 27 and seven at 28, for example. This phenomenon is called **granularity.** It is caused by the limited precision of the measurements. The granularity would disappear if Newcomb had been able to measure one more decimal place to spread out the "steps." This minor granularity does not prevent us from adopting a normal distribution as a model.

granularity

EXAMPLE 1.30

Figure 1.33 is a normal quantile plot of the supermarket spending data. The stemplots in Figure 1.3 (page 12) show that the distribution is right-skewed. To see the right skewness in the normal quantile plot, draw a line through the leftmost points, which correspond to the smaller observations. The larger observations fall systematically above this line. That is, the right-of-center observations have larger values than in a normal distribution. *In a right-skewed distribution, the largest observations fall distinctly above a line drawn through the main body of points.* Similarly, left skewness is evident when the smallest observations fall below the line. Unlike Figure 1.31, there are no individual outliers.

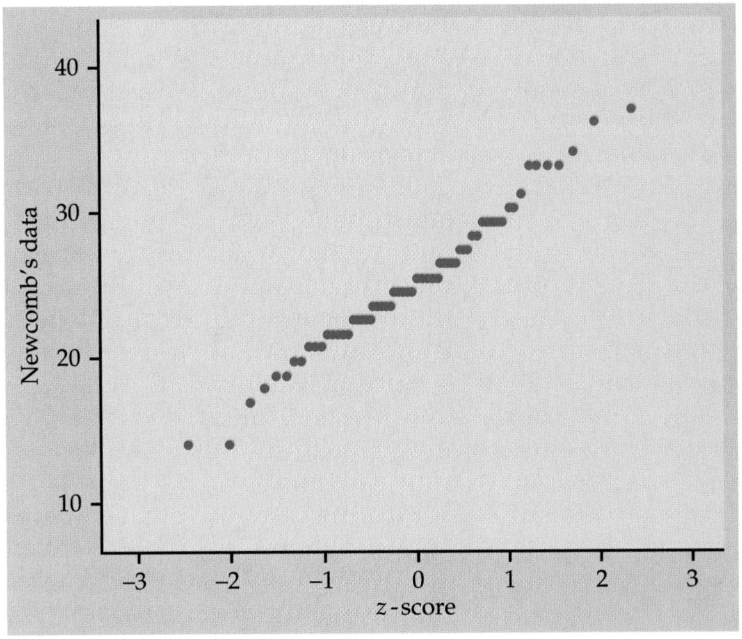

FIGURE 1.32 Newcomb's data without the two outliers. The data are close to normal except for slight granularity.

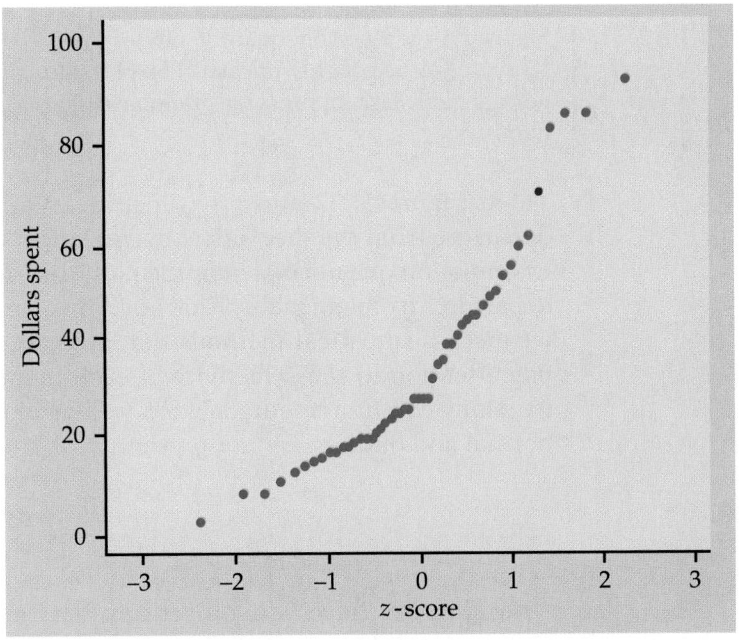

FIGURE 1.33 Normal quantile plot of the supermarket spending data. The pattern bends up at the right, showing right skewness.

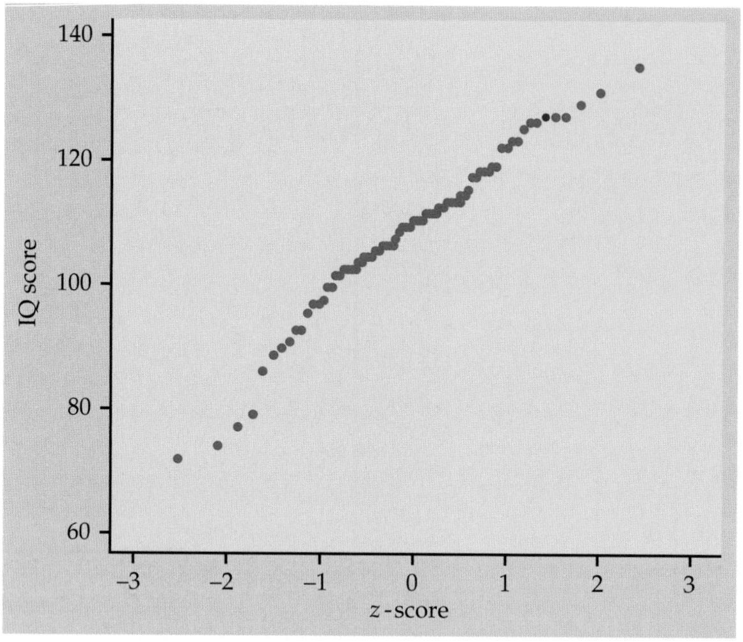

FIGURE 1.34 Normal quantile plot of the IQ scores of 78 seventh-grade students. The straight pattern shows that this distribution is close to normal.

EXAMPLE 1.31

Figure 1.34 is a normal quantile plot of the 78 IQ scores from Table 1.6. We expect IQ scores to be roughly normal. The plot is roughly straight, showing good fit to a normal model except for some granularity due to the fact that all IQ scores are whole numbers.

As Figures 1.32 and 1.34 illustrate, real data almost always show some departure from the theoretical normal model. It is important to confine your examination of a normal quantile plot to searching for shapes that show clear departures from normality. Don't overreact to minor wiggles in the plot. When we discuss statistical methods that are based on the normal model, we will pay attention to the sensitivity of each method to departures from normality. Many common methods work well as long as the data are approximately normal and outliers are not present.

Beyond the basics ▸ density estimation

A density curve gives a compact summary of the overall shape of a distribution. Figure 1.20 (page 64) shows a normal density curve that provides a good summary of the distribution of Gary vocabulary scores. Many distributions do not have the normal shape. There are other families of density curves that are used as mathematical models for various distribution shapes.

density estimation

Modern software offers a more flexible option, **density estimation.** A density estimator does not start with any specific shape, such as the normal shape.

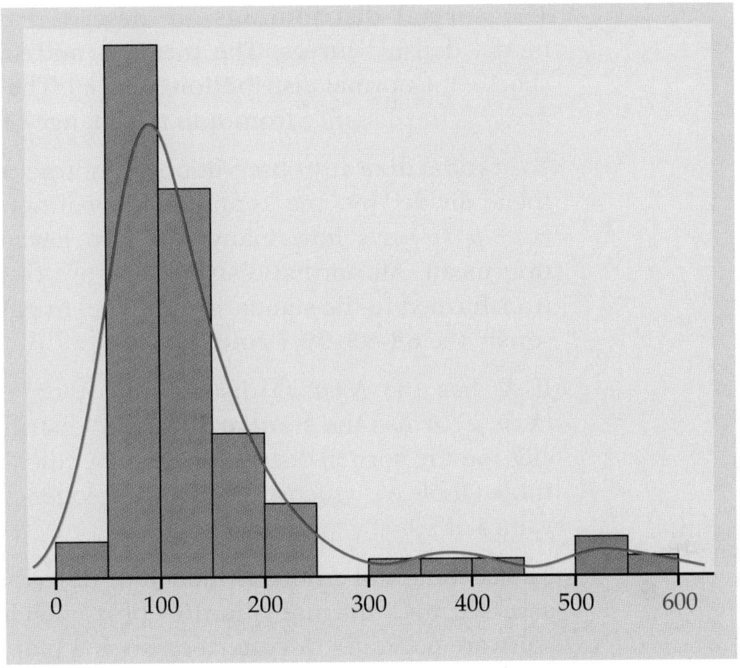

FIGURE 1.35 A density curve for the guinea pig survival data, calculated from the data by density estimation software.

It looks at the data and draws a density curve that describes the overall shape of the data. Figure 1.35 shows a histogram of the strongly right-skewed guinea pig survival data from Table 1.5. A density estimator produced the density curve drawn over the histogram. You can see that this curve does catch the overall pattern: the strong right skewness, the major peak near 100 days, and the two smaller clusters in the right tail. You can also see that the density estimator works blindly: in order to get a smooth curve, it starts below 0, even though survival times less than 0 days are not possible. Nonetheless, density estimation offers a quick way to picture the overall shapes of distributions. It is another useful tool for exploratory data analysis.[25]

SUMMARY

The overall pattern of a distribution can often be described compactly by a **density curve.** A density curve has total area 1 underneath it. Areas under a density curve give **relative frequencies** for the distribution.

The **mean** μ (balance point), the **median** (equal-areas point), and the **quartiles** can be approximately located by eye on a density curve. The **standard deviation** σ cannot be located by eye on most density curves. The mean and median are equal for symmetric density curves, but the mean of a skewed curve is located farther toward the long tail than is the median.

The **normal distributions** are described by bell-shaped, symmetric, unimodal density curves. The mean μ and standard deviation σ completely specify the normal distribution $N(\mu, \sigma)$. The mean is the center of symmetry, and σ is the distance from μ to the change-of-curvature points on either side.

To **standardize** any observation x, subtract the mean of the distribution and then divide by the standard deviation. The resulting ***z*-score** $z = (x - \mu)/\sigma$ says how many standard deviations x lies from the distribution mean. All normal distributions are the same when measurements are transformed to the standardized scale. In particular, all normal distributions satisfy the **68–95–99.7 rule.**

If X has the $N(\mu, \sigma)$ distribution, then the standardized variable $Z = (X - \mu)/\sigma$ has the **standard normal** distribution $N(0, 1)$. Relative frequencies for any normal distribution can be calculated from the **standard normal table** (Table A), which gives relative frequencies for the events $Z < z$ for many values of z.

The adequacy of a normal model for describing a distribution of data is best assessed by a **normal quantile plot,** which is available in most statistical software packages. A pattern on such a plot that deviates substantially from a straight line indicates that the data are not normal.

SECTION 1.3 EXERCISES

1.77 **(a)** Sketch a density curve that is symmetric but has a shape different from that of the normal curves.

(b) Sketch a density curve that is strongly skewed to the left.

1.78 If you ask a computer to generate "random numbers" between 0 and 1, you will get observations from a **uniform distribution.** Figure 1.36 graphs the density curve for a uniform distribution.

uniform distribution

(a) Check that the area under this curve is 1.

(b) What proportion of the outcomes lie between 0.1 and 0.9? (Sketch the density curve, shade the area that represents the proportion, then find that area.)

1.79 Many random number generators allow users to specify the range of the random numbers to be produced. Suppose that you specify that the

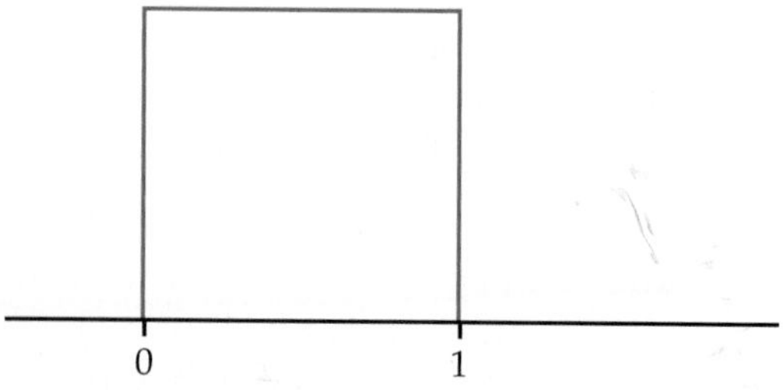

FIGURE 1.36 The density curve of a uniform distribution, for Exercise 1.78.

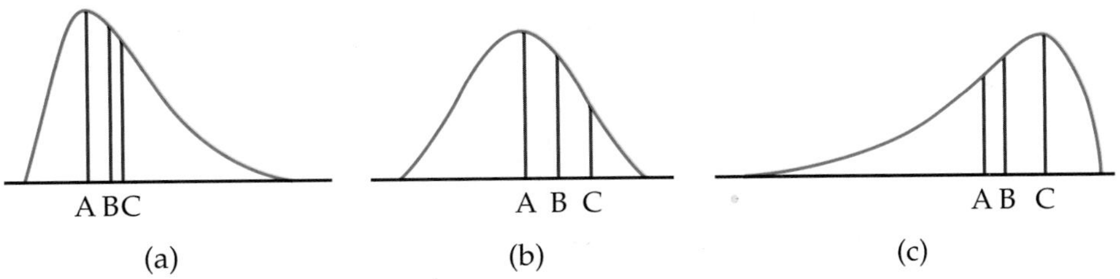

FIGURE 1.37 Three density curves, for Exercise 1.81.

outcomes are to be distributed uniformly between 0 and 2. Then the density curve of the outcomes has constant height between 0 and 2, and height 0 elsewhere.

(a) What is the height of the density curve between 0 and 2? Draw a graph of the density curve.

(b) Use your graph from (a) and the fact that areas under the curve are relative frequencies of outcomes to find the proportion of outcomes that are less than 1.

(c) Find the proportion of outcomes that lie between 0.5 and 1.3.

1.80 What are the mean and the median of the uniform distribution in Figure 1.36? What are the quartiles?

1.81 Figure 1.37 displays three density curves, each with three points marked on it. At which of these points on each curve do the mean and the median fall?

1.82 The length of human pregnancies from conception to birth varies according to a distribution that is approximately normal with mean 266 days and standard deviation 16 days. Use the 68–95–99.7 rule to answer the following questions.

(a) Between what values do the lengths of the middle 95% of all pregnancies fall?

(b) How short are the shortest 2.5% of all pregnancies? How long do the longest 2.5% last?

1.83 Bigger animals tend to carry their young longer before birth. The length of horse pregnancies from conception to birth varies according to a roughly normal distribution with mean 336 days and standard deviation 3 days. Use the 68–95–99.7 rule to answer the following questions.

(a) Almost all (99.7%) of horse pregnancies fall in what range of lengths?

(b) What percent of horse pregnancies are longer than 339 days?

applet

1.84 The 68–95–99.7 rule for normal distributions is a useful approximation. You can use the *Normal Curve* applet on the Web site for this book to see how accurate the rule is. Drag one flag across the other so that the applet shows the area under the curve between the two flags.

(a) Place the flags one standard deviation on either side of the mean. What is the area between these two values? What does the 68–95–99.7 rule say this area is?

(b) Repeat for locations two and three standard deviations on either side of the mean. Again compare the 68–95–99.7 rule with the area given by the applet.

1.85 The normal quantile plot in Figure 1.34 shows that the IQ scores of the 78 seventh-grade students reported in Table 1.6 are approximately normal. How well do these scores satisfy the 68–95–99.7 rule? To find out, calculate the mean $\bar{x}$ and standard deviation s of the scores. Then calculate the percent of the 78 scores that fall between $\bar{x} - s$ and $\bar{x} + s$ and compare your result with 68%. Do the same for the intervals covering two and three standard deviations on either side of the mean. (The 68–95–99.7 rule is exact for any theoretical normal distribution. It will hold only approximately for actual data.)

1.86 Eleanor scores 680 on the mathematics part of the SAT examination. The distribution of SAT scores in a reference population is normal with mean 500 and standard deviation 100. Gerald takes the ACT mathematics test and scores 27. ACT scores are normally distributed with mean 18 and standard deviation 6. Find the z-scores for both students. Assuming that both tests measure the same kind of ability, who has the higher score?

1.87 Three landmarks of baseball achievement are Ty Cobb's batting average of .420 in 1911, Ted Williams's .406 in 1941, and George Brett's .390 in 1980. These batting averages cannot be compared directly because the distribution of major league batting averages has changed over the decades. The distributions are quite symmetric and (except for outliers such as Cobb, Williams, and Brett) reasonably normal. While the mean batting average has been held roughly constant by rule changes and the balance between hitting and pitching, the standard deviation has dropped over time. Here are the facts:[26]

Decade	Mean	Std. dev.
1910s	.266	.0371
1940s	.267	.0326
1970s	.261	.0317

Compute the standardized batting averages for Cobb, Williams, and Brett to compare how far each stood above his peers.

1.88 Using either Table A or your calculator or software, find the proportion of observations from a standard normal distribution that satisfies each of the following statements. In each case, sketch a standard normal curve and shade the area under the curve that is the answer to the question.

(a) $Z < 2.85$

(b) $Z > 2.85$

(c) $Z > -1.66$

(d) $-1.66 < Z < 2.85$

1.89 Using either Table A or your calculator or software, find the relative frequency of each of the following events in a standard normal distribution. In each case, sketch a standard normal curve with the area representing the relative frequency shaded.

(a) $Z \leq -2.25$

(b) $Z \geq -2.25$

(c) $Z > 1.77$

(d) $-2.25 < Z < 1.77$

1.90 Find the value z of a standard normal variable Z that satisfies each of the following conditions. (If you use Table A, report the value of z that comes closest to satisfying the condition.) In each case, sketch a standard normal curve with your value of z marked on the axis.

(a) The point z with 25% of the observations falling below it.

(b) The point z with 40% of the observations falling above it.

1.91 The variable Z has a standard normal distribution.

(a) Find the number z such that the event $Z < z$ has relative frequency 0.8.

(b) Find the number z such that the event $Z > z$ has relative frequency 0.35.

1.92 In 2000, the scores of men on the math part of the SAT approximately followed a normal distribution with mean 533 and standard deviation 115.

(a) What proportion of men scored above 500?

(b) What proportion scored between 400 and 600?

1.93 Too much cholesterol in the blood increases the risk of heart disease. Young women are generally less afflicted with high cholesterol than other groups. The cholesterol levels for women aged 20 to 34 follow an approximately normal distribution with mean 185 milligrams per deciliter (mg/dL) and standard deviation 39 mg/dL.[27]

(a) Cholesterol levels above 240 mg/dL demand medical attention. What percent of young women have levels above 240 mg/dL?

(b) Levels above 200 mg/dL are considered borderline high. What percent of young women have blood cholesterol between 200 and 240 mg/dL?

1.94 It is possible to score higher than 800 on the SAT, but scores above 800 are reported as 800. (That is, a student can get a reported score of 800 without a perfect paper.) In 2000, the scores of men on the math part of the SAT approximately followed a normal distribution with mean 533 and standard deviation 115. What percent of scores were above 800 (and so reported as 800)?

1.95 Middle-aged men are more susceptible to high cholesterol than the young women of Exercise 1.93. The blood cholesterol levels of men aged 55 to 64 are approximately normal with mean 222 mg/dL and standard deviation

37 mg/dL. What percent of these men have high cholesterol (levels above 240 mg/dL)? What percent have borderline high cholesterol (between 200 and 240 mg/dL)?

1.96 Osteoporosis is a condition in which the bones become brittle due to loss of minerals. To diagnose osteoporosis, an elaborate apparatus measures bone mineral density (BMD). BMD is usually reported in standardized form. The standardization is based on a population of healthy young adults. The World Health Organization (WHO) criterion for osteoporosis is a BMD 2.5 standard deviations below the mean for young adults. BMD measurements in a population of people similar in age and sex roughly follow a normal distribution.

(a) What percent of healthy young adults have osteoporosis by the WHO criterion?

(b) Women aged 70 to 79 are of course not young adults. The mean BMD in this age is about -2 on the standard scale for young adults. Suppose that the standard deviation is the same as for young adults. What percent of this older population has osteoporosis?

1.97 Changing the mean of a normal distribution by a moderate amount can greatly change the percent of observations in the tails. Suppose that a college is looking for applicants with SAT math scores 750 and above.

(a) In 2000, the scores of men on the math SAT followed a normal distribution with mean 533 and standard deviation 115. What percent of men scored 750 or better?

(b) Women's scores that year had a normal distribution with mean 498 and standard deviation 109. What percent of women scored 750 or better? You see that the percent of men above 750 is almost three times the percent of women with such high scores.

1.98 The yearly rate of return on stock indexes (which combine many individual stocks) is approximately normal. Between 1950 and 2000, U.S. common stocks had a mean yearly return of about 13%, with a standard deviation of about 17%. Take this normal distribution to be the distribution of yearly returns over a long period.

(a) In what range do the middle 95% of all yearly returns lie?

(b) The market is down for the year if the return is less than zero. In what percent of years is the market down?

(c) In what percent of years does the index gain 25% or more?

1.99 The length of human pregnancies from conception to birth varies according to a distribution that is approximately normal with mean 266 days and standard deviation 16 days.

(a) What percent of pregnancies last less than 240 days (that's about 8 months)?

(b) What percent of pregnancies last between 240 and 270 days (roughly between 8 months and 9 months)?

(c) How long do the longest 20% of pregnancies last?

1.100 Some companies "grade on a bell curve" to compare the performance of their managers and professional workers. This forces the use of some low performance ratings, so that not all workers are graded "above average." Until the threat of lawsuits forced a change, Ford Motor Company's "performance management process" assigned 10% A grades, 80% B grades, and 10% C grades to the company's 18,000 managers.[28] It isn't clear that the "bell curve" of ratings is really a normal distribution. Nonetheless, suppose that Ford's performance scores are normally distributed. One year, managers with scores less than 25 received C's and those with scores above 475 received A's. What are the mean and standard deviation of the scores?

1.101 We have met the Survey of Study Habits and Attitudes (SSHA) as a common psychological instrument to evaluate the attitudes of students. The SSHA is used for subjects from seventh grade through college. Different groups have different distributions. To prepare to use the SSHA to evaluate future teachers, researchers gave the test to 238 college juniors majoring in elementary education. Their scores were roughly normal with mean 114 and standard deviation 30. Take this as the distribution of SSHA scores in the population of future elementary school teachers.

A study of Native American education students in Canada found that this relatively disadvantaged group had mean SSHA score 99.[29] What percentile of the overall distribution is this?

1.102 Scores on the Wechsler Adult Intelligence Scale for the 20 to 34 age group are approximately normally distributed with mean 110 and standard deviation 25. Scores for the 60 to 64 age group are approximately normally distributed with mean 90 and standard deviation 25.

Sarah, who is 30, scores 135 on this test. Sarah's mother, who is 60, also takes the test and scores 120. Who scored higher relative to her age group, Sarah or her mother? Who has the higher absolute level of the variable measured by the test? At what percentile of their age groups are Sarah and her mother? (That is, what percent of the age group has lower scores?)

1.103 How high a score on the SSHA test of Exercise 1.101 must an elementary education student obtain to be among the highest-scoring 30% of the population? What scores make up the lowest 30%?

1.104 The scores of a reference population on the Wechsler Intelligence Scale for Children (WISC) are normally distributed with $\mu = 100$ and $\sigma = 15$.

(a) What percent of this population have WISC scores below 100?

(b) Below 80?

(c) Above 140?

(d) Between 100 and 120?

1.105 The distribution of scores on the WISC is described in the previous exercise. What score will place a child in the top 5% of the population? In the top 1%?

1.106 The median of any normal distribution is the same as its mean. We can use normal calculations to find the quartiles and related descriptive measures for normal distributions.

(a) What is the area under the standard normal curve to the left of the first quartile? Use this to find the value of the first quartile for a standard normal distribution. Find the third quartile similarly.

(b) Your work in (a) gives the z-scores for the quartiles of any normal distribution. Scores on the Wechsler Intelligence Scale for Children (WISC) are normally distributed with mean 100 and standard deviation 15. What are the quartiles of WISC scores?

(c) What is the value of the *IQR* for the standard normal distribution?

(d) What percent of the observations in the standard normal distribution are suspected outliers according to the $1.5 \times IQR$ criterion? (This percent is the same for any normal distribution.)

1.107 The lower and upper deciles of any distribution are the points that mark off the lowest 10% and the highest 10%. On a density curve, these are the points with area 0.1 and 0.9 to their left under the curve.

(a) What are the lower and upper deciles of the standard normal distribution?

(b) The length of human pregnancies is approximately normal with mean 266 days and standard deviation 16 days. What are the lower and upper deciles of this distribution?

1.108 Figure 1.38 is a normal quantile plot of the DRP scores from Exercise 1.26 (page 30). Are these scores approximately normally distributed? Discuss any major deviations from normality that appear in the plot.

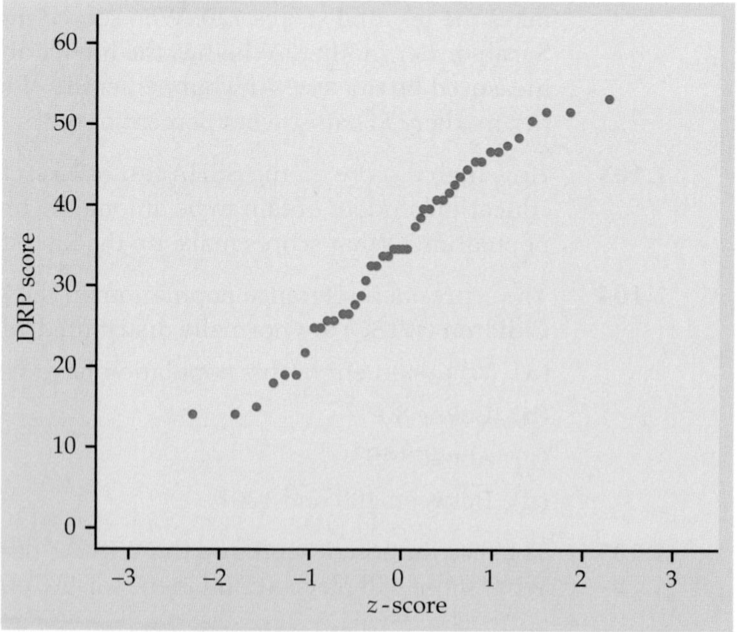

FIGURE 1.38 Normal quantile plot of DRP scores, for Exercise 1.108.

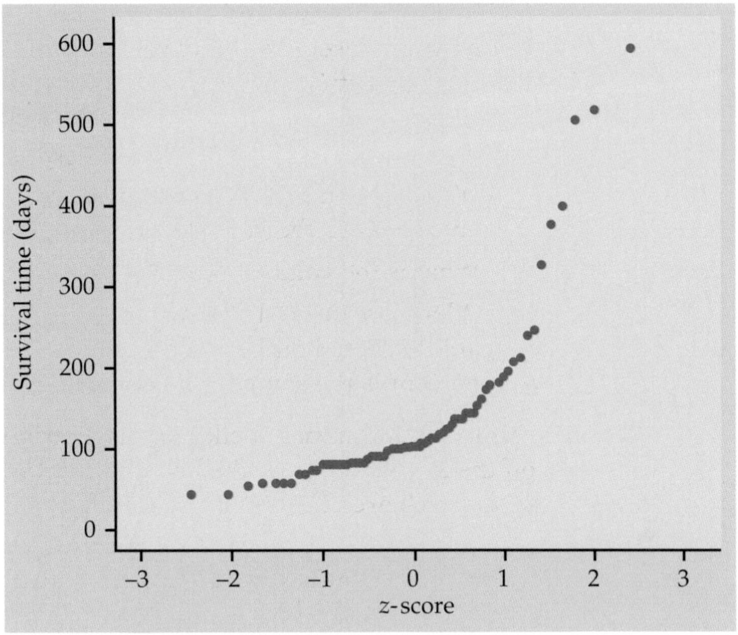

FIGURE 1.39 Normal quantile plot of guinea pig survival times, for Exercise 1.109.

1.109 Figure 1.39 is a normal quantile plot of the survival times of the guinea pigs in a medical experiment, from Table 1.5. Explain carefully how the strong right skewness of this distribution is seen in the plot.

1.110 The distance between two mounting holes is important to the performance of an electrical meter. The manufacturer measures this distance regularly for quality control purposes, recording the data as thousandths of an inch more than 0.600 inches. For example, 0.644 is recorded as 44. Figure 1.40 is a normal quantile plot of the distances for the last 90 electrical meters measured.[30] Is the overall shape of the distribution approximately normal? Why does the plot have a "stair-step" appearance?

The remaining exercises for this section require the use of software that will make normal quantile plots.

1.111 We expect repeated careful measurements of the same quantity to be approximately normal. Make a normal quantile plot for Cavendish's measurements in Exercise 1.27 (page 30). Are the data approximately normal? If not, describe any clear deviations from normality.

1.112 The distribution of Internet access costs in Exercise 1.45 (page 56) has a compact center with a long tail on either side. Make a normal quantile plot of these data. Explain carefully why the pattern of this plot is typical of a "long-tailed" distribution.

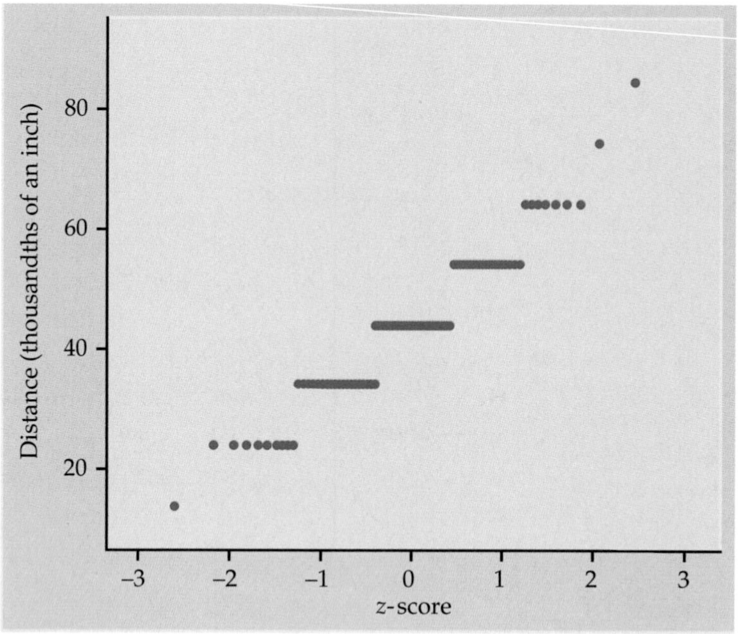

FIGURE 1.40 Normal quantile plot of distance between mounting holes, for Exercise 1.110.

1.113 Table 1.6 (page 33) reports the scores of 78 seventh-grade students on the Piers-Harris Children's Self-Concept Scale. Give a careful description of the distribution of self-concept scores using both graphs and numbers. Are there outliers? Is the distribution roughly normal if we ignore the outliers?

1.114 Table 1.6 (page 33) gives the genders and Piers-Harris self-concept scores for 78 seventh-grade students. Gender is coded as F for females and M for males. Compare the distributions of scores for female and male students, using both graphs and numbers. Report your findings. Some people think that female students in mixed classes have lower self-concept than male students. Do you see any evidence that this is true for these students?

1.115 Use software to generate 100 observations from the standard normal distribution. Make a histogram of these observations. How does the shape of the histogram compare with a normal density curve? Make a normal quantile plot of the data. Does the plot suggest any important deviations from normality? (Repeating this exercise several times is a good way to become familiar with how normal quantile plots look when data actually are close to normal.)

1.116 Use software to generate 100 observations from the uniform distribution described in Exercise 1.78. Make a histogram of these observations. How does the histogram compare with the density curve in Figure 1.36? Make a normal quantile plot of your data. According to this plot, how does the uniform distribution deviate from normality?

CHAPTER 1 EXERCISES

1.117 Product preference depends in part on the age, income, and gender of the consumer. A market researcher selects a large sample of potential car buyers. For each consumer, she records gender, age, household income, and automobile preference. Which of these variables are categorical and which are quantitative?

1.118 What type of graph or graphs would you plan to make in a study of each of the following issues?

(a) What makes of cars do students drive? How old are their cars?

(b) How many hours per week do students study? How does the number of study hours change during a semester?

(c) Which radio stations are most popular with students?

(d) When many students measure the concentration of the same solution for a chemistry course laboratory assignment, do their measurements follow a normal distribution?

1.119 Are the rich different? Figure 1.41 shows histograms of the number of people in a household for households in the top 5% of incomes (left) and for the other 95% of households (right).[31]

(a) These are relative frequency histograms, in which the vertical scales show the proportion of observations in a class rather than the count of observations. Why are these preferable to frequency histograms to display these data?

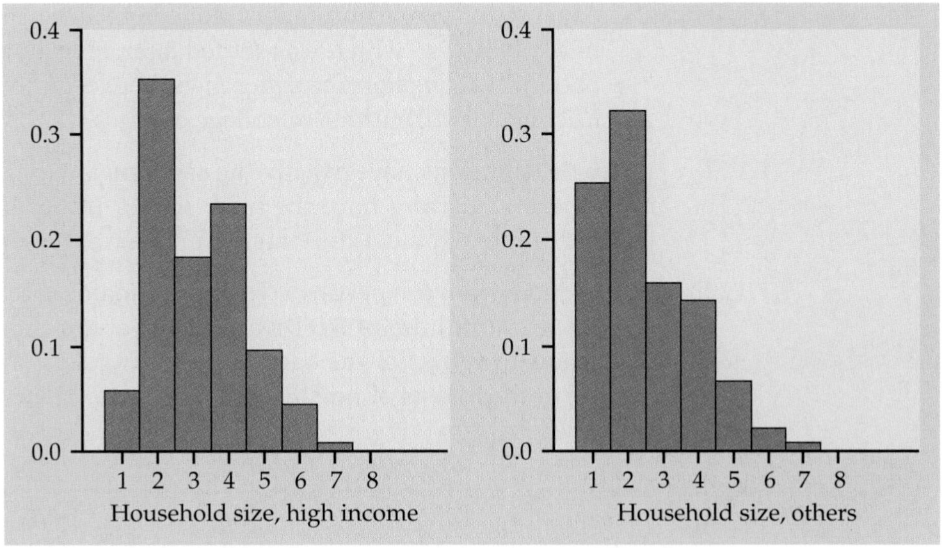

FIGURE 1.41 The distributions of household size for households in the top 5% of income (left) and for other households (right), for Exercise 1.119.

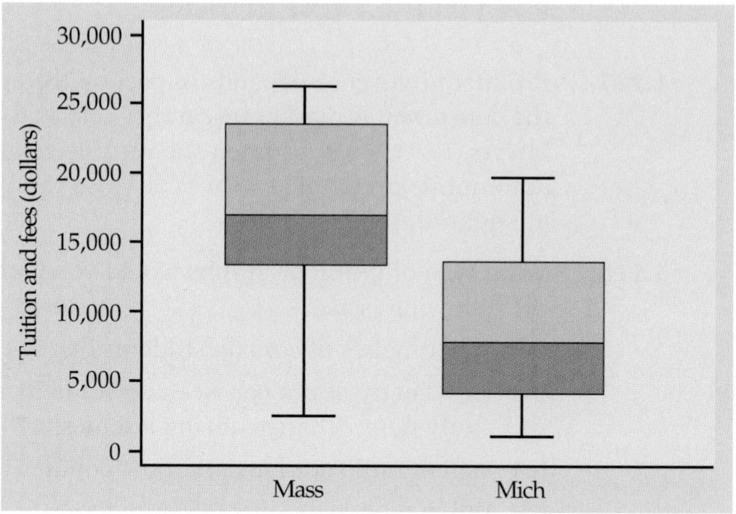

FIGURE 1.42 Boxplots comparing the tuition and fees charged by four-year colleges in Massachusetts and Michigan, for Exercise 1.120.

(b) What are the most important ways in which the sizes of high-income households differ from other households?

1.120 Figure 1.42 displays side-by-side boxplots comparing the 2000–2001 tuition and fees charged by all four-year colleges in Massachusetts and in Michigan. For state schools, we used the in-state tuition.

(a) Use the boxplots to write a brief comparison of the two distributions.

(b) Eastern states had many private colleges before state-supported higher education became common. Public higher education is more prevalent in the Midwest, which was settled later. How does this historical pattern help explain the differences between the Massachusetts and Michigan distributions of college charges?

1.121 If two distributions have exactly the same mean and standard deviation, must their histograms have the same shape? If they have the same five-number summary, must their histograms have the same shape? Explain.

1.122 By-products from the pesticide DDT were major threats to the survival of birds of prey until use of DDT was banned at the end of 1972. Can time plots show the effect of the ban? Here are two sets of data for bald eagles nesting in the forests of northwestern Ontario.[32] The first data set gives the mean number of young per breeding area:

Year	1966	1967	1968	1969	1970	1971	1972	1973
Young	1.26	0.73	0.89	0.84	0.54	0.60	0.54	0.78
Year	1974	1975	1976	1977	1978	1979	1980	1981
Young	0.46	0.77	0.86	0.96	0.82	0.98	0.93	1.12

The second set of data are measurements of the chemical DDE, the by-product of DDT that most threatens birds of prey, from bald eagle eggs in the same area of Canada. These are in parts per million (ppm). There are often several measurements per year.

Year	1967	1967	1968	1971	1971	1972	1976	1976	1976	1976
DDE	44	95	121	125	95	87	13.3	16.4	50.4	59.8
Year	1976	1977	1977	1980	1980	1980	1981	1981	1981	
DDE	56.4	0.6	23.8	16.6	14.5	24.0	7.8	48.2	53.4	

Make time plots of eagle young and of mean DDE concentration in eggs. How does the effect of banning DDT appear in your plots?

1.123 Our direct comparison of Mark McGwire and Babe Ruth (page 11) ignores the historical context. Here are the number of home runs by the major league leader for each year in baseball history, 1876 to 2001, in order from left to right. Make a time plot. (Be sure to add the scale of years.)

5	3	4	9	6	7	7	10	27	11	11	17	14	20	14	16	13	19
18	17	13	11	15	25	12	16	16	13	10	9	12	10	12	9	10	21
14	19	19	24	12	12	**11**	**29**	**54**	**59**	42	**41**	**46**	39	**47**	**60**	**54**	**46**
56	**46**	58	48	49	36	49	46	58	35	43	37	36	34	33	28	44	51
40	54	47	42	37	47	49	51	52	44	47	46	41	61	49	45	49	52
49	44	44	49	45	48	40	44	36	38	38	52	46	48	48	31	39	40
43	40	40	**49**	42	47	51	44	43	46	43	50	**52**	**58**	**70**	**65**	50	73

(a) Describe the effect of World War II (1942 to 1945 seasons).

(b) Ruth led in the 11 years in boldface between 1918 and 1931. McGwire led in the 5 boldface years between 1987 and 1999. Briefly compare the achievements of Ruth and McGwire in the context of their times.

1.124 The single-season home run record is now held by Barry Bonds of the San Francisco Giants, who hit 73 in 2001. Here are Bonds's home run totals from 1986 (his first year) to 2001:

16	25	24	19	33	25	34	46	37	33	42	40	37	34	49	73

Bonds's record year is a high outlier. How do his career mean and median number of home runs change when we drop the record 73? What general fact about the mean and median does your result illustrate?

1.125 The Bureau of Justice Statistics says that in 1999, 51% of homicides were committed with handguns, 14% with other firearms, 13% with knives, and 6% with blunt objects. Make a graph to display these data. Do you need an "other methods" category?

1.126 The Internal Revenue Service reports that in 1998 about 124 million individual income tax returns showed adjusted gross income (AGI) greater than 0. The mean and median AGI on these tax returns were \$25,491 and \$44,186. Which of these numbers is the mean and which is the median? How do you know?

1.127 Raw scores on behavioral tests are often transformed for easier comparison. A test of reading ability has mean 75 and standard deviation 10 when given to third graders. Sixth graders have mean score 82 and standard deviation 11 on the same test. To provide separate "norms" for each grade, we want scores in each grade to have mean 100 and standard deviation 20.

(a) What linear transformation will change third-grade scores x into new scores $x_{\text{new}} = a + bx$ that have the desired mean and standard deviation? (Use $b > 0$ to preserve the order of the scores.)

(b) Do the same for the sixth-grade scores.

(c) David is a third-grade student who scores 78 on the test. Find David's transformed score. Nancy is a sixth-grade student who scores 78. What is her transformed score? Who scores higher within his or her grade?

(d) Suppose that the distribution of scores in each grade is normal. Then both sets of transformed scores have the $N(100, 20)$ distribution. What percent of third graders have scores less than 78? What percent of sixth graders have scores less than 78?

1.128 The Chapin Social Insight Test evaluates how accurately the subject appraises other people. In the reference population used to develop the test, scores are approximately normally distributed with mean 25 and standard deviation 5. The range of possible scores is 0 to 41.

(a) What proportion of the population has scores below 20 on the Chapin test?

(b) What proportion has scores below 10?

(c) How high a score must you have in order to be in the top quarter of the population in social insight?

1.129 Exercise 1.25 (page 29) presents data on the nightly study time claimed by first-year college men and women. We noted in Exercise 1.58 that each group contains an outlier. Make a normal quantile plot for each group, omitting the outlier. How does the fact that most students estimated study time in multiples of 10 minutes affect the appearance of the plots? Are the distributions roughly normal in other respects?

1.130 The Chapin Social Insight Test described in Exercise 1.128 has a mean of 25 and a standard deviation of 5. You want to rescale the test using a linear

TABLE 1.10 Position and weight (pounds) for a major college football team

QB	235	QB	220	QB	215	QB	228	K	175	K	205
K	220	K	185	RB	200	RB	188	RB	190	RB	215
RB	190	RB	225	RB	225	RB	240	RB	237	R	205
R	185	R	185	R	190	R	195	R	201	R	190
R	195	R	180	OL	291	OL	280	OL	300	OL	320
OL	325	OL	285	OL	305	OL	305	OL	290	OL	310
OL	315	OL	285	OL	290	OL	325	OL	310	OL	256
OL	305	OL	300	DB	170	DB	207	DB	185	DB	175
DB	180	DB	190	DB	210	DB	200	DB	180	DB	195
DB	185	DB	170	DB	180	DB	190	DB	190	LB	220
LB	212	LB	233	LB	190	LB	195	LB	215	LB	220
LB	240	LB	230	LB	220	LB	225	LB	200	DL	260
DL	245	DL	264	DL	254	DL	215	DL	250	DL	295
DL	240	DL	275	DL	255	DL	285	DL	245	DL	270
DL	250	DL	250	TE	255	TE	245	TE	260	TE	245

transformation so that the mean is 100 and the standard deviation is 20. Let x denote the score in the original scale and x_{new} be the transformed score.

(a) Find the linear transformation required. That is, find the values of a and b in the equation $x_{new} = a + bx$.

(b) Give the rescaled score for someone who scores 30 in the original scale.

(c) What are the quartiles of the rescaled scores?

1.131 The Florida State University Seminoles have been among the more successful teams in college football. Table 1.10 gives the weights in pounds and positions of the players on the 2000–2001 football team, which was defeated in the national title game by the University of Oklahoma.[33] The positions are quarterback (QB), running back (RB), offensive line (OL), receiver (R), tight end (TE), kicker (K), defensive back (DB), linebacker (LB), and defensive line (DL).

(a) Make side-by-side modified boxplots of the weights for running backs, receivers, offensive linemen, defensive linemen, linebackers, and defensive backs.

(b) Briefly compare the weight distributions. Which position has the heaviest players overall? Which has the lightest?

(c) Are any individual players outliers within their position?

1.132 Most statistical software packages have routines for generating values of variables having specified distributions. Use your statistical software to generate 25 observations from the $N(20, 5)$ distribution. Compute the mean and standard deviation $\bar{x}$ and s of the 25 values you obtain. How close are $\bar{x}$ and s to the μ and σ of the distribution from which the observations were drawn?

Repeat 20 times the process of generating 25 observations from the $N(20, 5)$ distribution and recording $\bar{x}$ and s. Make a stemplot of the 20 values of $\bar{x}$ and another stemplot of the 20 values of s. Make normal quantile plots of both sets of data. Briefly describe each of these distributions. Are they symmetric or skewed? Are they roughly normal? Where are their centers? (The distributions of measures like $\bar{x}$ and s when repeated sets of observations are made from the same theoretical distribution will be very important in later chapters.)

1.133 Table 1.11 shows the salaries paid to the members of the New York Yankees baseball team as of opening day of the 2001 season. Display this distribution with a graph and describe its main features. Find the mean and median salary and explain how the pattern of the distribution explains the relationship between these two measures of center. Find the standard deviation and the quartiles. Do you prefer the five-number summary or $\bar{x}$ and s as a quick description of this distribution?

1.134 The American Housing Survey provides data on all housing units in the United States—houses, apartments, mobile homes, and so on. Here are the years in which a random sample of 100 housing units were built. The survey does not produce exact dates for years before 1990. Years before 1920 are

TABLE 1.11 2001 salaries for the New York Yankees baseball team

Player	Salary	Player	Salary
Derek Jeter	\$12,600,000	Joe Oliver	\$1,100,000
Bernie Williams	12,357,143	Henry Rodriguez	850,000
Roger Clemens	10,300,000	Alfonso Soriano	630,000
Mike Mussina	10,000,000	Luis Sojo	500,000
Mariano Rivera	9,150,000	Brian Boehringer	350,000
David Justice	7,000,000	Shane Spencer	320,000
Andy Pettitte	7,000,000	Todd Williams	320,000
Paul O'Neill	6,500,000	Carlos Almanzar	270,000
Chuck Knoblauch	6,000,000	Clay Bellinger	230,000
Tino Martinez	6,000,000	Darrell Einertson	206,000
Scott Brosius	5,250,000	Randy Choate	204,750
Jorge Posada	4,050,000	Michael Coleman	204,000
Mike Stanton	2,450,000	D'Angelo Jimenez	200,000
Orlando Hernandez	2,050,000	Christian Parker	200,000
Allen Watson	1,700,000	Scott Seabol	200,000
Ramiro Mendoza	1,600,000		

given as 1919. Dates between 1920 and 1970 are given in ten-year blocks, so that a unit built in 1956 appears as 1950. Dates between 1970 and 1990 are given in five-year blocks, so that 1987 appears as 1985.[34]

1960	1920	1991	1919	1985	1985	1975	1980	1975	1985
1930	1993	1985	1975	1970	1970	1975	1980	1940	1940
1980	1919	1980	1950	1940	1950	1993	1985	1975	1960
1919	1950	1960	1975	1950	1919	1920	1985	1970	1975
1930	1975	1960	1920	1940	1950	1985	1990	1950	1970
1985	1920	1950	1980	1975	1950	1950	1919	1919	1985
1985	1991	1980	1960	1940	1960	1930	1998	1994	1960
1919	1975	1919	1950	1975	1930	1919	1970	1920	1930
1950	1975	1970	1985	1919	1960	1930	1980	1960	1950
1996	1940	1950	1998	1930	1919	1930	1950	1950	1920

(a) Make a histogram of these dates, using classes 10 years wide beginning with 1910 to 1919. The first class will contain all housing units built before 1920. In which decades after 1920 were most housing units that still exist built?

(b) Give the five-number summary of these data. Write a brief warning on how to interpret your results. For example, what does the fact that the median is 1960 tell us about the age of American housing?

1.135 At the time the salaries in Table 1.11 were announced, one dollar was worth 1.72 Swiss francs. Answer these questions without doing any calculations in addition to those you did in Exercise 1.133.

(a) What transformation converts a salary in dollars into the same salary in Swiss francs?

(b) What are the mean, median, and quartiles of the distribution in francs?

(c) What are the standard deviation and interquartile range of the distribution in francs?

1.136 You are planning a sample survey of households in California. You decide to select households separately within each county and to choose more households from the more populous counties. To aid in the planning, Table 1.12 gives the populations of California counties from the 2000 census. Examine the distribution of county populations both graphically and numerically, using whatever tools are most suitable. Write a brief description of the main features of this distribution. Sample surveys often select households from all of the most populous counties but from only some of the less populous. How would you divide California counties into three groups according to population, with the intent of including all of the

TABLE 1.12 Population of California counties, 2000 census

County	Population	County	Population	County	Population
Alameda	1,443,741	Marin	247,289	San Mateo	707,161
Alpine	1,208	Mariposa	17,130	Santa Barbara	399,347
Amador	35,100	Mendocino	86,265	Santa Clara	1,682,585
Butte	203,171	Merced	210,554	Santa Cruz	255,602
Calaveras	40,554	Modoc	9,449	Shasta	163,256
Colusa	18,804	Mono	12,853	Sierra	3,555
Contra Costa	948,816	Monterey	401,762	Siskiyou	44,301
Del Norte	27,507	Napa	124,279	Solano	394,542
El Dorado	156,299	Nevada	92,033	Sonoma	458,614
Fresno	799,407	Orange	2,846,289	Stanislaus	446,997
Glenn	26,453	Placer	248,399	Sutter	78,930
Humboldt	126,518	Plumas	20,824	Tehama	56,039
Imperial	142,361	Riverside	1,545,387	Trinity	13,022
Inyo	17,945	Sacramento	1,223,499	Tulare	368,021
Kern	661,645	San Benito	53,234	Tuolumne	54,501
Kings	129,461	San Bernardino	1,709,434	Ventura	753,197
Lake	58,309	San Diego	2,813,833	Yolo	168,660
Lassen	33,828	San Francisco	776,733	Yuba	60,219
Los Angeles	9,519,338	San Joaquin	563,598		
Madera	123,109	San Luis Obispo	246,681		

first group, half of the second, and a smaller fraction of the third in your survey?

1.137 Each March, the Bureau of Labor Statistics collects detailed information about more than 50,000 randomly selected households. The INDIVIDUALS data set contains data on 55,899 people from the March 2000 survey. The Data Appendix describes this data set in detail. Give a brief description of the distribution of incomes for these people, using graphs and numbers to report your findings. Because this is a very large randomly selected sample, your results give a good description of individual incomes for all Americans aged 25 to 65 who work outside of agriculture.

1.138 Continue your study of the INDIVIDUALS data set. Do a statistical analysis to compare the incomes of people whose main work experience is (1) in the private sector, (2) in government, and (3) self-employed. Use graphs and numerical descriptions to report your findings.

1.139 The CHEESE data set described in the Data Appendix records measurements on 30 specimens of Australian cheddar cheese. Investigate the distributions of the variables H2S and LACTIC using graphical and numerical summaries of your choice. Write a short description of the notable features for each distribution.

1.140 The CSDATA data set described in the Data Appendix contains information on 234 computer science students. We are interested in comparing the SAT mathematics scores and grade point averages of female students with those of male students. Make two sets of side-by-side boxplots to carry out these comparisons. Write a brief discussion of the male-female comparisons. Then make normal quantile plots of grade point averages and SAT math scores separately for men and women. Which of the four distributions are approximately normal?

Prelude: will stocks go up or down?

Predicting the future course of the stock market could make you rich. No wonder lots of people and lots of computers pore over market data looking for patterns. Some popular patterns are a bit bizarre. The "Super Bowl Indicator" says that the football Super Bowl, played in January, predicts how stocks will behave each year. The current National Football League (NFL) was formed by merging the original NFL with the American Football League (AFL). The indicator claims that stocks go up in years when a team from the old NFL wins and down when an AFL team wins.

The Super Bowl Indicator was right in 28 of 31 years between the first Super Bowl in 1967 and 1997. Sounds impressive. But stocks went down only 6 times in those 31 years, so just predicting "up" every year would have been right 25 times. Original NFL teams won most Super Bowls for good reasons: there are 17 of them against only 11 old AFL teams, and the NFL was the established and stronger league. So "NFL wins" was pretty much the same as "up every year." The indicator is now fading fast: the great 1990s boom in stocks roared on unhindered by AFL wins in 1998 and 1999, while stocks fell after NFL wins in 2000 and 2001.

The Super Bowl predictor is simpleminded. There are statistical methods to predict one variable from others that go well beyond just counting ups and downs. *Regression*, which we will meet in this chapter, is the starting point for these methods. But, alas, even the fanciest statistical tools can't predict stock prices.

CHAPTER 2

Laure Cuvillier, monotype, *Human Body in Motion*

Looking at Data—Relationships

Introduction

A medical study finds that short women are more likely to have heart attacks than women of average height, while tall women have the fewest heart attacks. An insurance group reports that heavier cars have fewer accident deaths per 10,000 vehicles registered than do lighter cars. These and many other statistical studies look at the relationship between two variables. To understand such a relationship, we must often examine other variables as well. To conclude that shorter women have higher risk from heart attacks, for example, the researchers had to eliminate the effect of other variables such as weight and exercise habits. Our topic in this chapter is relationships between variables. One of our main themes is that the relationship between two variables can be strongly influenced by other variables that are lurking in the background.

To study the relationship between two variables, we measure both variables on the same individuals. If we measure both the height and the weight of each of a large group of people, we know which height goes with each weight. These data allow us to study the connection between height and weight. A list of the heights and a separate list of the weights, two sets of single-variable data, do not show the connection between the two variables. In fact, taller people also tend to be heavier. And people who smoke more cigarettes per day tend not to live as long as those who smoke fewer. We say that pairs of variables such as height and weight or smoking and life expectancy are *associated*.

Association between Variables

Two variables measured on the same individuals are **associated** if some values of one variable tend to occur more often with some values of the second variable than with other values of that variable.

Statistical associations are overall tendencies, not ironclad rules. They allow individual exceptions. Although smokers on the average die earlier than nonsmokers, some people live to 90 while smoking three packs a day.

Examining relationships

When you examine the relationship between two or more variables, first ask the preliminary questions that are familiar from Chapter 1:

- What *individuals* do the data describe?
- What *variables* are present? How are they measured?
- Which variables are *quantitative* and which are *categorical*?

A medical study, for example, may record each subject's sex (female, male) and smoking status (current smoker, former smoker, never smoked) along with quantitative variables such as weight and blood pressure. We may be interested in possible associations between two quantitative variables (such as a person's weight and blood pressure), between a categorical and a quantitative variable (such as sex and blood pressure), or between two categorical variables (such as sex and smoking status).

When you examine the relationship between two variables, a new question becomes important:

- Is your purpose simply to explore the nature of the relationship, or do you hope to show that one of the variables can explain variation in the other? That is, are some of the variables *response variables* and others *explanatory variables*?

Response Variable, Explanatory Variable

A **response variable** measures an outcome of a study. An **explanatory variable** explains or causes changes in the response variables.

It is easiest to identify explanatory and response variables when we actually set values of one variable in order to see how it affects another variable.

EXAMPLE 2.1

Alcohol has many effects on the body. One effect is a drop in body temperature. To study this effect, researchers give several different amounts of alcohol to mice, then measure the change in each mouse's body temperature in the 15 minutes after taking the alcohol. Amount of alcohol is the explanatory variable, and change in body temperature is the response variable.

When you don't set the values of either variable but just observe both variables, there may or may not be explanatory and response variables. Whether there are depends on how you plan to use the data.

EXAMPLE 2.2

A political scientist looks at the Democrats' share of the popular vote for president in each state in the two most recent presidential elections. She does not wish to explain one year's data by the other's, but rather to find a pattern that may shed light on political conditions. The political scientist has two related variables, but she does not use the explanatory-response distinction.

A political consultant looks at the same data with an eye to including the previous election result in a state as one of several explanatory variables in a model to predict how the state will vote in the next election. Now the earlier election result is an explanatory variable and the later result is a response variable.

In many studies, the goal is to show that changes in one or more explanatory variables actually *cause* changes in a response variable. But not all explanatory-response relationships involve direct causation. The SAT scores of high school students may help predict the students' future college grades, but high SAT scores certainly don't cause high college grades.

independent variable

dependent variable

Some of the statistical techniques in this chapter require us to distinguish explanatory from response variables; others make no use of this distinction. You will often see explanatory variables called **independent variables** and response variables called **dependent variables.** The idea behind this language is that response variables depend on explanatory variables. Because the words "independent" and "dependent" have other meanings in statistics that are unrelated to the explanatory-response distinction, we prefer to avoid those words.

Most statistical studies examine data on more than one variable. Fortunately, statistical analysis of several-variable data builds on the tools used for examining individual variables. The principles that guide our work also remain the same:

- Start with graphical display, then add numerical summaries.
- Look for overall patterns and deviations from those patterns.
- When the overall pattern is quite regular, use a compact mathematical model to describe it.

2.1 Scatterplots

Relationships between two quantitative variables are best displayed graphically. The most useful graph for this purpose is a *scatterplot*.

Scatterplot

A **scatterplot** shows the relationship between two quantitative variables measured on the same individuals. The values of one variable appear on the horizontal axis, and the values of the other variable appear on the vertical axis. Each individual in the data appears as the point in the plot fixed by the values of both variables for that individual.

Always plot the explanatory variable, if there is one, on the horizontal axis (the x axis) of a scatterplot. As a reminder, we usually call the explanatory variable x and the response variable y. If there is no explanatory-response distinction, either variable can go on the horizontal axis.

EXAMPLE 2.3

Over a million American students take the SAT college entrance examination each year. We sometimes see the states "rated" by the average SAT scores of their high school seniors. For example, North Dakota students average 609 on the math part of the SAT, which looks better than the 513 average of Massachusetts students. This makes little sense, because average SAT score is largely explained by what percent of a state's students take the SAT. The scatterplot in Figure 2.1 allows us to see how the mean SAT mathematics score in each state is related to the percent of that state's high school seniors who take the SAT.[1]

Each point on the plot represents a single individual—that is, a single state. Because we think that the percent taking the exam influences mean score, percent taking is the explanatory variable and we plot it horizontally. For example, 19% of

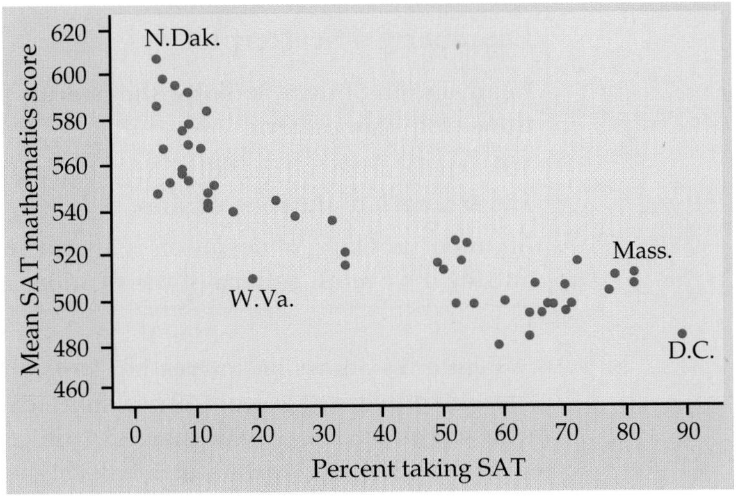

FIGURE 2.1 State mean SAT mathematics scores plotted against the percent of high school seniors in each state who take the SAT exams.

West Virginia high school seniors take the SAT, and their mean math score is 511. West Virginia appears as the point (19, 511) in the scatterplot, above 19 on the x axis and to the right of 511 on the y axis.

We see at once that state average SAT score is closely related to the percent of students who take the SAT. North Dakota has a high mean score, but only 4% of that state's seniors take the SAT. In Massachusetts, on the other hand, 78% of seniors take the exam. In addition to these states, we have labeled two states that may be *outliers* in Figure 2.1. The District of Columbia is a city rather than a state; 89% of its seniors take the SAT. West Virginia appears to have a lower mean score than we would expect in a state where only 19% take the exam.

Interpreting scatterplots

To look more closely at a scatterplot such as Figure 2.1, apply the strategies of exploratory analysis learned in Chapter 1.

Exploring the Web

The Web site of the College Board, `www.collegeboard.com`, appears at first glance to be of interest mainly to students applying to college and planning to take the SAT. The site in fact contains a mine of information about these exams, but it is deeply buried. Search the site for the Research Library. The URL recently was `www.collegeboard.org/toc/html/tocresearch000.html`, but this may change. Here you can find data on state and national performance on the latest SAT, such as the data that lie behind Figure 2.1. The online "Research Reports" offered here concern such issues as the effect of coaching on SAT scores and using SAT scores and high school grades to predict college GPA (Example 2.22, page 166).

Examining a Scatterplot

In any graph of data, look for the **overall pattern** and for striking **deviations** from that pattern.

You can describe the overall pattern of a scatterplot by the **form, direction,** and **strength** of the relationship.

An important kind of deviation is an **outlier,** an individual value that falls outside the overall pattern of the relationship.

clusters

Figure 2.1 shows an interesting *form*: there are two distinct **clusters** of states. In one cluster, at least 49% of high school seniors take the SAT, and the mean scores are low. Fewer than 35% of seniors in states in the other cluster take the SAT, and these states have higher mean scores. There are in fact two common college entrance examinations, the SAT and the ACT. Each state tends to prefer one or the other. In ACT states (the left cluster in Figure 2.1) most students who take the SAT are applying to selective out-of-state colleges. This select group performs well. In SAT states (the right cluster), many seniors take the SAT, and this broader group has a lower mean score.

The relationship in Figure 2.1 also has a clear *direction*: states in which a higher percent of students take the SAT tend to have lower mean scores. This is true both between the clusters and within each cluster. This is a *negative association* between the two variables.

Positive Association, Negative Association

Two variables are **positively associated** when above-average values of one tend to accompany above-average values of the other and below-average values also tend to occur together.

Two variables are **negatively associated** when above-average values of one accompany below-average values of the other, and vice versa.

linear relationship

When a scatterplot shows distinct clusters, it is often useful to describe the overall pattern separately within each cluster. The *form* of the relationship in the ACT states is roughly **linear.** That is, the points roughly follow a straight line. The *strength* of a relationship in a scatterplot is determined by how closely the points follow a clear form. The linear pattern among the ACT states is moderately strong because the points show only modest scatter about the straight-line pattern. In summary, the ACT states in Figure 2.1 show a moderately strong negative linear relationship. The cluster of SAT states shows a much weaker relationship between percent taking the SAT and mean SAT mathematics score.

Adding categorical variables to scatterplots

The Census Bureau groups the states into four broad regions, named Midwest, Northeast, South, and West. We might ask about regional patterns in

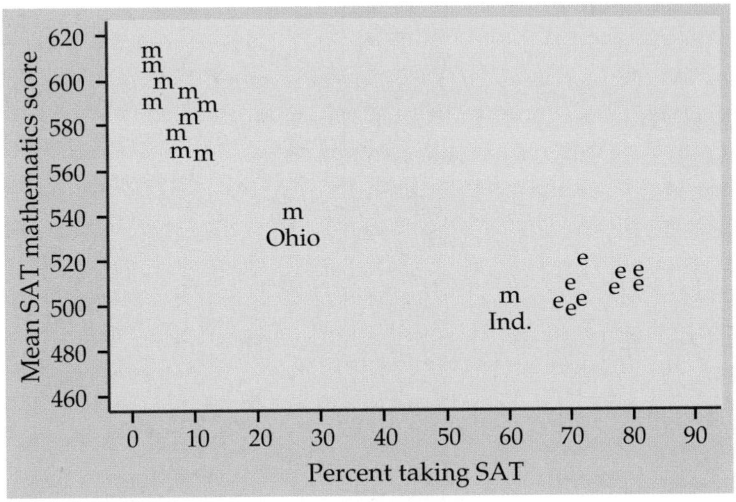

FIGURE 2.2 State mean SAT mathematics scores and percent taking the SAT for the northeastern states (plot symbol "e") and the midwestern states (plot symbol "m").

SAT exam scores. Figure 2.2 repeats part of Figure 2.1, with an important difference. We have plotted only the Northeast and Midwest groups of states, using the plot symbol "e" for the northeastern states and the symbol "m" for the midwestern states.

The regional comparison is striking. The nine northeastern states are all SAT states—in fact, at least 68% of high school graduates in each of these states take the SAT. The twelve midwestern states are mostly ACT states. In ten of these states, the percent taking the SAT is between 4% and 12%. One midwestern state is clearly an outlier within the region. Indiana is an SAT state (60% take the SAT) that falls close to the northeastern cluster. Ohio, where 26% take the SAT, also lies outside the midwestern cluster.

In dividing the states into regions, we introduced a third variable into the scatterplot. "Region" is a categorical variable that has four values, although we plotted data from only two of the four regions. The two regions are displayed by the two different plotting symbols.[2]

Categorical Variables in Scatterplots

To add a categorical variable to a scatterplot, use a different plot color or symbol for each category.

More examples of scatterplots

Experience in examining scatterplots is the foundation for more detailed study of relationships among quantitative variables. Here is an example with a stronger pattern than that in Figure 2.1.

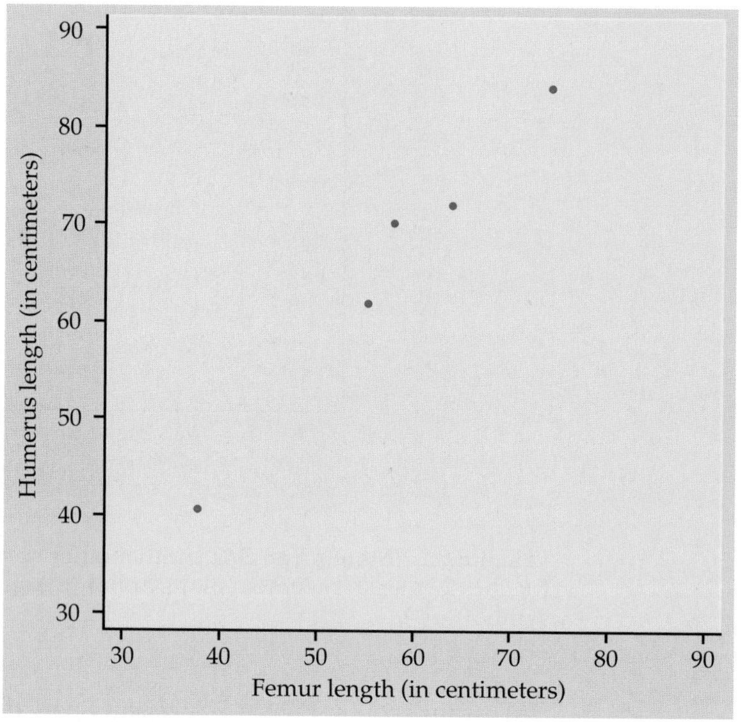

FIGURE 2.3 The lengths of two bones in the five surviving fossil specimens of the extinct beast *Archaeopteryx*.

EXAMPLE 2.4

Archaeopteryx is an extinct beast having feathers like a bird but teeth and a long bony tail like a reptile. Only six fossil specimens are known. Because these specimens differ greatly in size, some scientists think they are different species rather than individuals from the same species. Here are data on the lengths in centimeters of the femur (a leg bone) and the humerus (a bone in the upper arm) for the five specimens that preserve both bones:[3]

Femur	38	56	59	64	74
Humerus	41	63	70	72	84

Because there is no explanatory-response distinction, we can put either measurement on the x axis of a scatterplot. The plot appears in Figure 2.3.

The plot shows a strong positive linear association. The form is linear and the association is strong because the points lie close to a straight line. The association is positive because as the length of one bone increases, so does the length of the other bone. These data suggest that all five specimens belong to the same species and differ in size because some are younger than others. We expect that a specimen from a different species would have a different relationship between the lengths of the two bones, so that it would appear as an outlier.

Some scatterplots appear quite different from the cloud of points in Figure 2.1 and the strong linear pattern in Figure 2.3. This is true, for example, in experiments in which measurements of a response variable are taken at a few selected levels of the explanatory variable. The following example illustrates the use of scatterplots in this setting.

EXAMPLE 2.5

How much corn per acre should a farmer plant to obtain the highest yield? Too few plants will produce a low yield. On the other hand, if there are too many plants, they will compete with each other for moisture and nutrients, and yields will fall. Table 2.1 shows the results of several years of field experiments in Nebraska.[4] Each entry is the mean yield of four small plots planted at the same rate per acre. All plots were irrigated, fertilized, and cultivated identically. The yield of each plot should therefore depend only on the planting rate—and of course on the uncontrolled aspects of each growing season, such as temperature and wind. The experiment lasted several years in order to avoid misleading conclusions due to the peculiarities of a single growing season.

TABLE 2.1 Corn yields (bushels per acre) in an agricultural experiment

Plants per acre	1956	1958	1959	1960	Mean
12,000	150.1	113.0	118.4	142.6	131.0
16,000	166.9	120.7	135.2	149.8	143.2
20,000	165.3	130.1	139.6	149.9	146.2
24,000		134.7	138.4	156.1	143.1
28,000			119.0	150.5	134.8
Mean	160.8	124.6	130.1	149.8	

The scatterplot in Figure 2.4 displays the results of this experiment. Because planting rate is the explanatory variable, we plot it horizontally as the x variable. The yield is the response variable y. The vertical spread of points over each planting rate shows the year-to-year variation in yield. We show the overall pattern by plotting the mean yield for each planting rate (averaged over all years). These means are marked

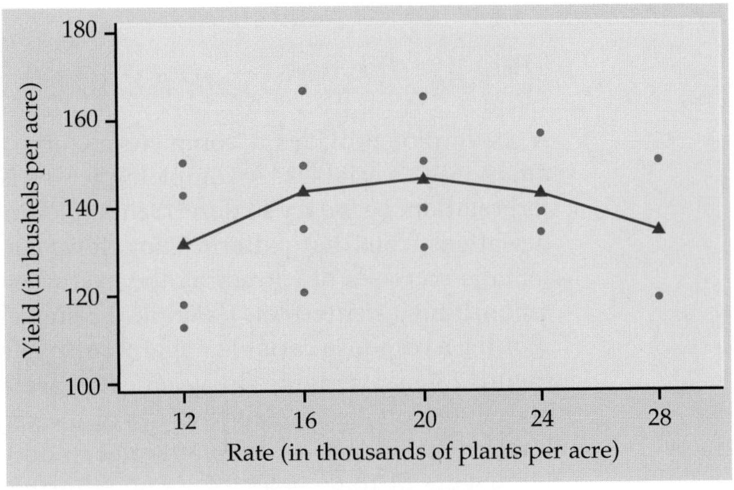

FIGURE 2.4 Yield of corn plotted against planting rate, from an agricultural experiment.

by triangles and joined by line segments. As expected, the form of the relationship is not linear. Yields first increase with the planting rate, then decrease when too many plants are crowded in. Because there is no consistent direction, we cannot describe the association as either positive or negative. It appears that the best choice is about 20,000 plants per acre.

Plotting the means as in Figure 2.4 helps us see the overall pattern of results when responses are measured at a few fixed levels of an explanatory variable such as planting rate. The scatterplot also displays deviations from the pattern by showing how much variability lies behind each mean.

In this case, however, we must regard the means with caution. As the gaps in Table 2.1 show, the agronomists used only the three lowest planting rates in all four years. The means therefore cover different spans of time. In particular, 1956 was a year of very high yields, so the means that include 1956 are biased upward relative to those that do not. A closer look at Table 2.1 shows that 24,000 plants per acre was superior to 20,000 in two of the three years in which both rates were planted. The agronomists in fact concluded that yields continue to increase somewhat up to 24,000 plants per acre.

Incomplete data, as in Example 2.5, are common in practice. The data would be more convincing if all five planting rates had been used in all four years—but it was only after the first results were in that the researchers realized that higher planting rates might give better results. Incomplete data often complicate a statistical analysis.

Example 2.5 illustrates a very important fact: **the relationship between two variables often cannot be fully understood without knowledge about other variables.** Figure 2.4 plots corn yield against planting rate, but the experiment spanned several growing seasons that differed from each other. If the agronomists had not carefully controlled many other variables (moisture, fertilizer, etc.), these variables would have confused the situation completely. You should be cautious in drawing conclusions from a strong relationship appearing in a scatterplot until you understand what other variables may be lurking in the background.

Beyond the basics ▸ scatterplot smoothers

A scatterplot provides a complete picture of the relationship between two quantitative variables. A complete picture is often too detailed for easy interpretation, so we try to describe the plot in terms of an overall pattern and deviations from that pattern. Though we can often do this by eye, more systematic methods of extracting the overall pattern are helpful. This is called **smoothing** a scatterplot. Example 2.5 suggests how to proceed when we are plotting a response variable y against an explanatory variable x. We smoothed Figure 2.4 by averaging the y-values separately for each x-value. Though not all scatterplots have many y-values at the same value of x, as did Figure 2.4, modern software provides scatterplot smoothers that form an overall pattern by looking at the y-values for points in the neighborhood of each x-value. Smoothers use *resistant* calculations, so they are not affected by outliers in the plot.

smoothing

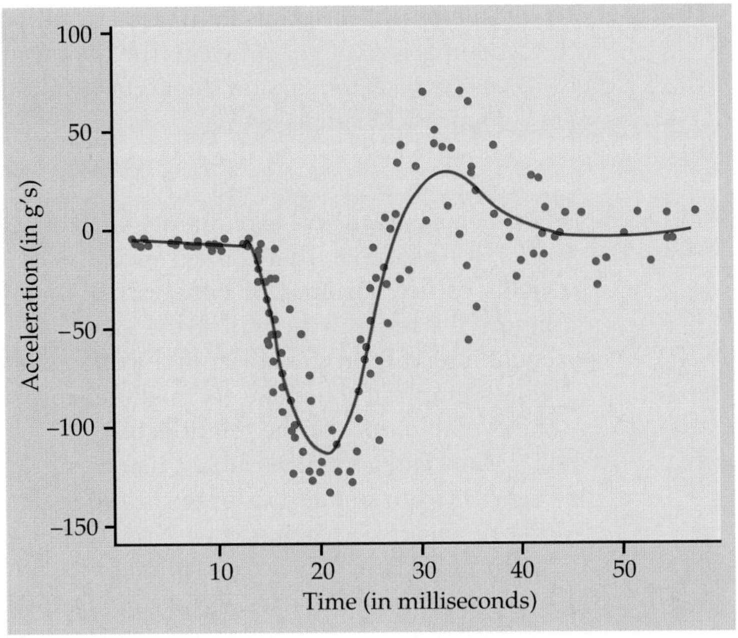

FIGURE 2.5 Time plot of the acceleration of the head of a crash test dummy as a motorcycle hits a wall, with the overall pattern calculated by a scatterplot smoother.

EXAMPLE 2.6

Crash a motorcycle into a wall. The rider, fortunately, is a dummy with an instrument to measure acceleration (change of velocity) mounted in its head. Figure 2.5 plots the acceleration of the dummy's head against time in milliseconds.[5] Acceleration is measured in g's, or multiples of the acceleration due to gravity at the earth's surface. The motorcycle approaches the wall at a constant speed (acceleration near 0). As it hits, the dummy's head snaps forward and decelerates violently (negative acceleration reaching more than 100 g's), then snaps back again (up to 75 g's) and wobbles a bit before coming to rest.

The scatterplot has a clear overall pattern, but it does not obey a simple form such as linear. Moreover, the strength of the pattern varies, from quite strong at the left of the plot to weaker (much more scatter) at the right. A scatterplot smoother deals with this complexity quite effectively and draws a line on the plot to represent the overall pattern.

Categorical explanatory variables

Scatterplots display the association between two quantitative variables. To display a relationship between a categorical explanatory variable and a quantitative response variable, make a side-by-side comparison of the distributions of the response for each category. We have already met some tools for such comparisons:

- A back-to-back stemplot compares two distributions. See the comparison of home run counts (the quantitative response) hit by Babe Ruth and Mark McGwire (the categories) on page 11.

- Side-by-side boxplots compare any number of distributions. See the comparison of gas mileage (the quantitative response) for minicompact and two-seater cars on the highway and in the city (four categories) in Figure 1.15 on page 45.

You can sometimes use a type of scatterplot to display the association between a categorical explanatory variable and a quantitative response. Suppose, for example, that the agronomists of Example 2.5 had compared the yields of five varieties of corn rather than five planting rates. The plot in Figure 2.4 remains helpful if the varieties A, B, C, D, and E are marked at equal intervals on the horizontal axis in place of the planting rate. In particular, a graph of the mean (or median) response for each category will show the overall nature of the relationship.

Many categorical variables, like corn variety or blood type, have no natural order from smallest to largest. In such situations we cannot speak of a positive or negative association with the response variable. If the mean responses in our plot increase as we go from left to right, we could make them decrease by writing the categories in the opposite order. The plot simply presents a side-by-side comparison of several distributions. The categorical variable labels the distributions. Some categorical variables do have a least-to-most order, however. We can then speak of the direction of the association between the categorical explanatory variable and the quantitative response. Here is an example.

EXAMPLE 2.7

We suspect that there is a positive association between how much education a person has and his or her income. The Census Bureau publishes relevant data, but "educational attainment" appears as a categorical variable with values:[6]

1 = did not reach high school
2 = some high school, but no high school diploma

up to

6 = postgraduate degree (master's, professional, or doctorate)

The categories 1 to 6 are ordered from least education to most education. The numerical values are just the codes used to identify education category in the data file. The side-by-side boxplots in Figure 2.6 show the distributions of personal income in March 2000 for people between the ages of 25 and 65 years at each level of education. We arrange the boxplots in order of increasing education to show the association. The lines in the plots extend out to the 5th and 95th percentiles of the distributions rather than to the single smallest and largest values. This variation on the boxplot idea is common when we have very many observations—55,899 people in this example.

Figure 2.6 makes clear the strong positive association between education and earnings. The 5th percentile is low in all groups, but the median and quartiles increase with more education. The first quartile in the college-educated groups exceeds the third quartile of the incomes of people who did not attend high school. Most strikingly, the 95th percentile shoots up in the college-educated groups. These groups contain many people with high incomes.

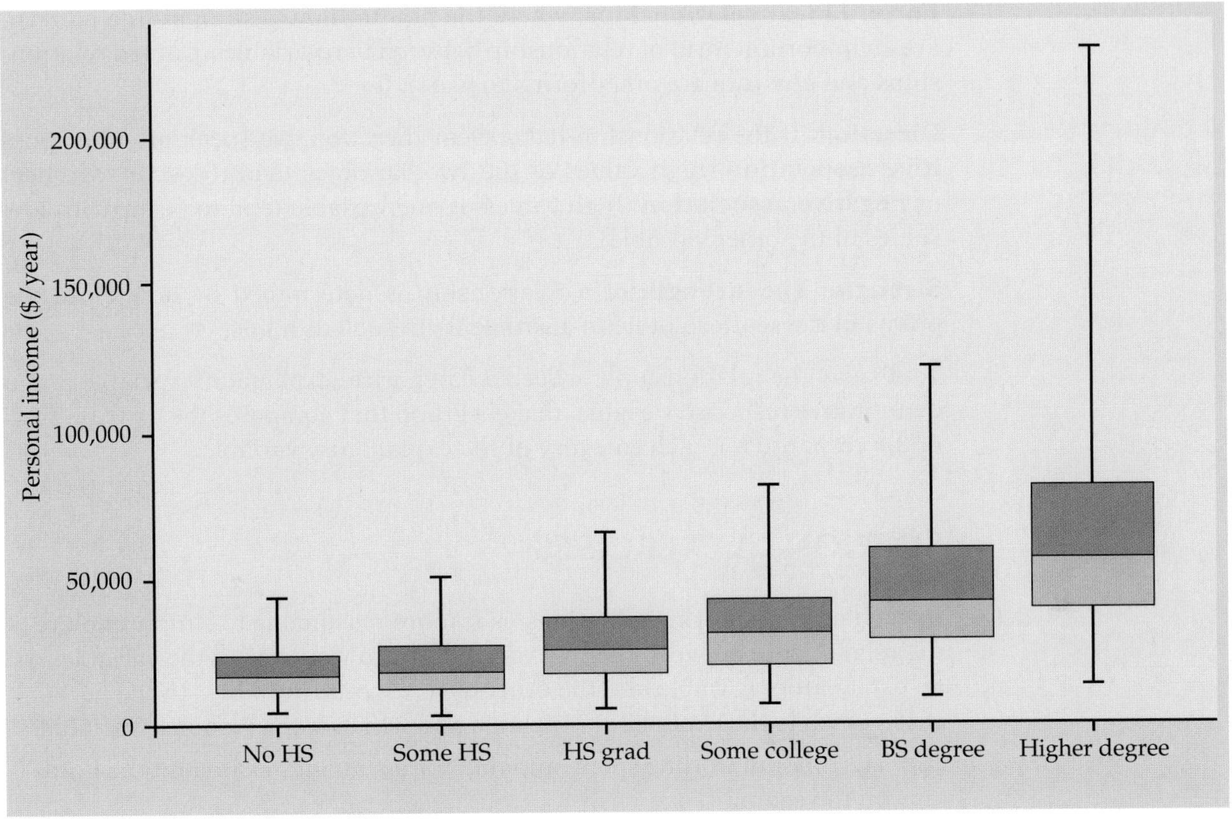

FIGURE 2.6 Boxplots for the incomes of people aged 25 to 65, by highest education level attained. The lines reach from the 5th percentile of incomes to the 95th percentile.

SUMMARY

To study relationships between variables, we must measure the variables on the same group of individuals.

If we think that a variable x may explain or even cause changes in another variable y, we call x an **explanatory variable** and y a **response variable.**

A **scatterplot** displays the relationship between two quantitative variables. Mark values of one variable on the horizontal axis (x axis) and values of the other variable on the vertical axis (y axis). Plot each individual's data as a point on the graph.

Always plot the explanatory variable, if there is one, on the x axis of a scatterplot. Plot the response variable on the y axis.

Plot points with different colors or symbols to see the effect of a categorical variable in a scatterplot.

In examining a scatterplot, look for an overall pattern showing the **form, direction,** and **strength** of the relationship, and then for **outliers** or other deviations from this pattern.

Form: Linear relationships, where the points show a straight-line pattern, are an important form of relationship between two variables. Curved relationships and **clusters** are other forms to watch for.

Direction: If the relationship has a clear direction, we speak of either **positive association** (high values of the two variables tend to occur together) or **negative association** (high values of one variable tend to occur with low values of the other variable).

Strength: The **strength** of a relationship is determined by how close the points in the scatterplot lie to a simple form such as a line.

To display the relationship between a categorical explanatory variable and a quantitative response variable, make a graph that compares the distributions of the response for each category of the explanatory variable.

SECTION 2.1 EXERCISES

2.1 In each of the following situations, is it more reasonable to simply explore the relationship between the two variables or to view one of the variables as an explanatory variable and the other as a response variable? In the latter case, which is the explanatory variable and which is the response variable?

(a) The amount of time spent studying for a statistics exam and the grade on the exam.

(b) The weight in kilograms and height in centimeters of a person.

(c) Inches of rain in the growing season and the yield of corn in bushels per acre.

(d) A student's scores on the SAT math exam and the SAT verbal exam.

(e) A family's income and the years of education their eldest child completes.

2.2 How well does a child's height at age 6 predict height at age 16? To find out, measure the heights of a large group of children at age 6, wait until they reach age 16, then measure their heights again. What are the explanatory and response variables here? Are these variables categorical or quantitative?

2.3 One measure of a state's prosperity is the median income of its households. Another measure is the mean personal income per person in the state. Figure 2.7 is a scatterplot of these two variables, both measured in thousands of dollars. Because both variables have the same units, the plot is square with the same scales on both axes.

(a) Explain why you expect a positive association between these variables. Also explain why you expect household income to be generally higher than income per person.

(b) Nonetheless, the mean income per person in a state can be higher than the median household income. In fact, the District of Columbia has median income \$33,433 per household and mean income \$38,429 per person. Explain why this can happen.

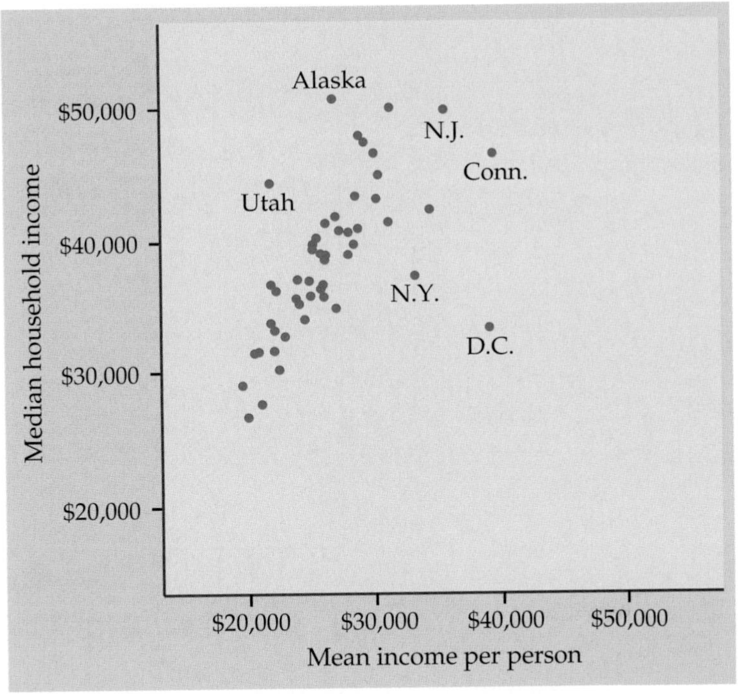

FIGURE 2.7 Median household income and mean income per person in the states, for Exercise 2.3.

(c) Describe the overall pattern of the plot, ignoring outliers.

(d) We have labeled some interesting states on the graph. New York has low median household income relative to its mean individual income. In Utah, median household income is high relative to mean individual income. What facts about a state could make median household income unusually high or low relative to mean individual income?

2.4 When interest rates are high, investors may shun stocks because they can get high returns with less risk. Figure 2.8 plots the annual return on U.S. common stocks for the years 1950 to 2000 against the returns on Treasury bills for the same years.[7] (The interest rate paid by Treasury bills is a measure of how high interest rates were that year.)

(a) Approximately what were the highest and lowest annual percent returns on stocks during this period? What were the highest and lowest returns for Treasury bills?

(b) Describe the pattern you see. Are high interest rates bad for stocks? Is the relationship between interest rates and stock returns strong or weak?

2.5 Often the percent of an animal species in the wild that survive to breed again is lower following a successful breeding season. This is part of nature's self-regulation, tending to keep population size stable. A study of merlins (small falcons) in northern Sweden observed the number of

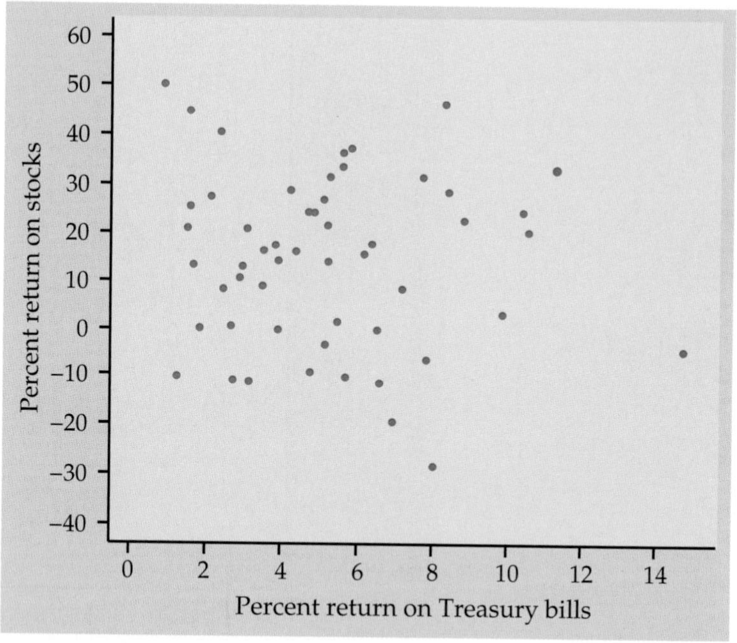

FIGURE 2.8 Percent returns on Treasury bills and common stocks for the years 1950 to 2000, for Exercise 2.4.

breeding pairs in an isolated area and the percent of males (banded for identification) who returned the next breeding season. Here are data for nine years:[8]

Breeding pairs	Percent of males returning
28	82
29	83, 70, 61
30	69
32	58
33	43
38	50, 47

(a) Why is the response variable the *percent* of males that return rather than the *number* of males that return?

(b) Make a scatterplot. Describe the pattern. Do the data support the theory that a smaller percent of birds survive following a successful breeding season?

2.6 Table 1.8 (page 38) gives the city and highway gas mileages for minicompact and two-seater cars. We expect a positive association between the city and highway mileages of a group of vehicles. We have already seen

TABLE 2.2 Wine consumption and heart attacks

Country	Alcohol from wine	Heart disease deaths	Country	Alcohol from wine	Heart disease deaths
Australia	2.5	211	Netherlands	1.8	167
Austria	3.9	167	New Zealand	1.9	266
Belgium	2.9	131	Norway	0.8	227
Canada	2.4	191	Spain	6.5	86
Denmark	2.9	220	Sweden	1.6	207
Finland	0.8	297	Switzerland	5.8	115
France	9.1	71	United Kingdom	1.3	285
Iceland	0.8	211	United States	1.2	199
Ireland	0.7	300	West Germany	2.7	172
Italy	7.9	107			

that the Honda Insight is a different type of car, so omit it as you work with these data.

(a) Make a scatterplot that shows the relationship between city and highway mileage, using city mileage as the explanatory variable. Use different plotting symbols for the two types of cars.

(b) Interpret the plot. Is there a positive association? Is the form of the plot roughly linear? Is the form of the relationship similar for the two types of car? What is the most important difference between the two types?

2.7 There is some evidence that drinking moderate amounts of wine helps prevent heart attacks. Table 2.2 gives data on yearly wine consumption (liters of alcohol from drinking wine, per person) and yearly deaths from heart disease (deaths per 100,000 people) in 19 developed nations.[9]

(a) Make a scatterplot that shows how national wine consumption helps explain heart disease death rates.

(b) Describe the form of the relationship. Is there a linear pattern? How strong is the relationship?

(c) Is the direction of the association positive or negative? Explain in simple language what this says about wine and heart disease. Do you think these data give good evidence that drinking wine *causes* a reduction in heart disease deaths? Why?

2.8 Metatarsus adductus (call it MA) is a turning in of the front part of the foot that is common in adolescents and usually corrects itself. Hallux abducto valgus (call it HAV) is a deformation of the big toe that is not common in youth and often requires surgery. Perhaps the severity of MA can help predict the severity of HAV. Table 2.3 gives data on 38 consecutive patients who came to a medical center for HAV surgery.[10] Using X-rays, doctors measured the angle of deformity for both MA and HAV. They speculated that there is a positive association—more serious MA is associated with more serious HAV.

TABLE 2.3 Two measurements of foot deformities

HAV angle	MA angle	HAV angle	MA angle	HAV angle	MA angle
28	18	21	15	16	10
32	16	17	16	30	12
25	22	16	10	30	10
34	17	21	7	20	10
38	33	23	11	50	12
26	10	14	15	25	25
25	18	32	12	26	30
18	13	25	16	28	22
30	19	21	16	31	24
26	10	22	18	38	20
28	17	20	10	32	37
13	14	18	15	21	23
20	20	26	16		

(a) Make a scatterplot of the data in Table 2.3. (Which is the explanatory variable?)

(b) Describe the form, direction, and strength of the relationship between MA angle and HAV angle. Are there any clear outliers in your graph?

(c) Do you think the data confirm the doctors' speculation?

2.9 The National Assessment of Educational Progress (NAEP) assesses what students know in several subject areas based on large representative samples. Table 2.4 reports some findings of the NAEP year 2000 Mathematics Assessment for fourth graders in the 40 states that participated. For each state we give the mean NAEP math score (out of 500) and also the percent of students who were at least "proficient" in the sense of being able to use math skills to solve real-world problems.[11] Nationally, about 25% of students are "proficient" by NAEP's standards. We expect that average performance and percent of proficient performers will be strongly related.

(a) Make a scatterplot, using mean NAEP score as the explanatory variable. Notice that there are several pairs of states with identical values. Use a different symbol for points that represent two states.

(b) Describe the form, direction, and strength of the relationship.

(c) Circle your home state's point in the scatterplot. Although there are no clear outliers, there are some points that you may consider interesting, perhaps because they are on the edge of the pattern. Choose one such point: which state is this, and in what way is it interesting?

TABLE 2.4 State performance on the NAEP Mathematics Assessment of fourth graders

State	Mean NAEP score	Percent proficient	Percent poverty	State	Mean NAEP score	Percent proficient	Percent poverty
Alabama	218	14	21.8	Missouri	229	24	14.4
Arizona	219	17	23.6	Montana	230	25	21.2
Arkansas	217	14	13.1	Nebraska	226	24	14.8
California	214	15	22.3	Nevada	220	16	12.8
Connecticut	234	32	13.4	New Mexico	214	12	23.5
Georgia	220	18	24.7	New York	227	22	28.9
Hawaii	216	14	14.5	North Carolina	232	28	21.3
Idaho	227	21	17.4	North Dakota	231	25	17.2
Illinois	225	22	12.2	Ohio	231	26	16.0
Indiana	234	31	12.6	Oklahoma	225	17	19.9
Iowa	233	28	14.2	Oregon	227	24	19.4
Kansas	232	30	13.3	Rhode Island	225	23	20.5
Kentucky	221	17	16.7	South Carolina	220	18	17.6
Louisiana	218	14	29.8	Tennessee	220	13	14.5
Maine	231	24	12.0	Texas	233	27	20.1
Maryland	222	22	8.1	Utah	227	24	11.8
Massachusetts	235	33	15.0	Vermont	232	30	12.2
Michigan	231	29	14.8	Virginia	230	25	7.9
Minnesota	235	34	12.6	West Virginia	225	18	25.7
Mississippi	211	9	19.3	Wyoming	229	25	13.0

2.10 How does the fuel consumption of a car change as its speed increases? Here are data for a British Ford Escort. Speed is measured in kilometers per hour, and fuel consumption is measured in liters of gasoline used per 100 kilometers traveled.[12]

Speed (km/h)	Fuel used (liters/100 km)	Speed (km/h)	Fuel used (liter/100 km)
10	21.00	90	7.57
20	13.00	100	8.27
30	10.00	110	9.03
40	8.00	120	9.87
50	7.00	130	10.79
60	5.90	140	11.77
70	6.30	150	12.83
80	6.95		

explanitory

(a) Make a scatterplot. (Which variable should go on the x axis?)

(b) Describe the form of the relationship. In what way is it not linear? Explain why the form of the relationship makes sense.

(c) It does not make sense to describe the variables as either positively associated or negatively associated. Why not?

(d) Is the relationship reasonably strong or quite weak? Explain your answer.

2.11 Metabolic rate, the rate at which the body consumes energy, is important in studies of weight gain, dieting, and exercise. The table below gives data on the lean body mass and resting metabolic rate for 12 women and 7 men who are subjects in a study of dieting. Lean body mass, given in kilograms, is a person's weight leaving out all fat. Metabolic rate is measured in calories burned per 24 hours, the same calories used to describe the energy content of foods. The researchers believe that lean body mass is an important influence on metabolic rate.

Subject	Sex	Mass	Rate	Subject	Sex	Mass	Rate
1	M	62.0	1792	11	F	40.3	1189
2	M	62.9	1666	12	F	33.1	913
3	F	36.1	995	13	M	51.9	1460
4	F	54.6	1425	14	F	42.4	1124
5	F	48.5	1396	15	F	34.5	1052
6	F	42.0	1418	16	F	51.1	1347
7	M	47.4	1362	17	F	41.2	1204
8	F	50.6	1502	18	M	51.9	1867
9	F	42.0	1256	19	M	46.9	1439
10	M	48.7	1614				

TABLE 2.5 Growth of icicles over time

Run 8903				Run 8905			
Time (min)	Length (cm)	Time (min)	Length (cm)	Time (min)	Length (cm)	Time (min)	Length (cm)
10	0.6	130	18.1	10	0.3	130	10.4
20	1.8	140	19.9	20	0.6	140	11.0
30	2.9	150	21.0	30	1.0	150	11.9
40	4.0	160	23.4	40	1.3	160	12.7
50	5.0	170	24.7	50	3.2	170	13.9
60	6.1	180	27.8	60	4.0	180	14.6
70	7.9			70	5.3	190	15.8
80	10.1			80	6.0	200	16.2
90	10.9			90	6.9	210	17.9
100	12.7			100	7.8	220	18.8
110	14.4			110	8.3	230	19.9
120	16.6			120	9.6	240	21.1

(a) Make a scatterplot of the data, using different symbols or colors for men and women.

(b) Is the association between these variables positive or negative? What is the form of the relationship? How strong is the relationship? Does the pattern of the relationship differ for women and men? How do the male subjects as a group differ from the female subjects as a group?

2.12 How fast do icicles grow? Japanese researchers measured the growth of icicles in a cold chamber under various conditions of temperature, wind, and water flow.[13] Table 2.5 contains data produced under two sets of conditions. In both cases, there was no wind and the temperature was set at $-11°$ C. Water flowed over the icicle at a higher rate (29.6 milligrams per second) in run 8905 and at a slower rate (11.9 mg/s) in run 8903.

(a) Make a scatterplot of the length of the icicle in centimeters versus time in minutes, using separate symbols for the two runs.

(b) Write a careful explanation of what your plots show about the growth of icicles.

2.13 Table 2.6 shows the progress of world record times (in seconds) for the 10,000-meter run.[14] Concentrate first on the women's world record. Make a scatterplot with year as the explanatory variable. Describe the pattern of improvement over time that your plot displays.

2.14 Table 1.6 (page 33) reports data on 78 seventh-grade students. We wonder if IQ score helps explain grade point average (GPA). Make a scatterplot and describe the form, direction, and strength of the relationship between IQ and GPA. There are some unusual observations, in particular a student with IQ above 100 but very low GPA and another student with very low IQ but a respectable GPA. Write the observation numbers for these students next to the points on your plot. What are their genders?

TABLE 2.6 World record times for the 10,000-meter run

Men				Women	
Record year	Time (seconds)	Record year	Time (seconds)	Record year	Time (seconds)
1912	1880.8	1962	1698.2	1967	2286.4
1921	1840.2	1963	1695.6	1970	2130.5
1924	1835.4	1965	1659.3	1975	2100.4
1924	1823.2	1972	1658.4	1975	2041.4
1924	1806.2	1973	1650.8	1977	1995.1
1937	1805.6	1977	1650.5	1979	1972.5
1938	1802.0	1978	1642.4	1981	1950.8
1939	1792.6	1984	1633.8	1981	1937.2
1944	1775.4	1989	1628.2	1982	1895.3
1949	1768.2	1993	1627.9	1983	1895.0
1949	1767.2	1993	1618.4	1983	1887.6
1949	1761.2	1994	1612.2	1984	1873.8
1950	1742.6	1995	1603.5	1985	1859.4
1953	1741.6	1996	1598.1	1986	1813.7
1954	1734.2	1997	1591.3	1993	1771.8
1956	1722.8	1997	1587.8		
1956	1710.4	1998	1582.7		
1960	1698.8				

2.15 Table 2.6 shows the progress of world record times (in seconds) for the 10,000-meter run for both men and women.

(a) Make a scatterplot of world record time against year, using separate symbols for men and women. Describe the pattern for each sex. Then compare the progress of men and women.

(b) Women began running this long distance later than men, so we might expect their improvement to be more rapid. Moreover, it is often said that men have little advantage over women in distance running as opposed to sprints, where muscular strength plays a greater role. Do the data appear to support these claims?

2.16 To demonstrate the effect of nematodes (microscopic worms) on plant growth, a botanist introduces different numbers of nematodes into 16 planting pots. He then transplants a tomato seedling into each pot. Here are data on the increase in height of the seedlings (in centimeters) 14 days after planting:[15]

Nematodes	Seedling growth			
0	10.8	9.1	13.5	9.2
1,000	11.1	11.1	8.2	11.3
5,000	5.4	4.6	7.4	5.0
10,000	5.8	5.3	3.2	7.5

(a) Make a scatterplot of the response variable (growth) against the explanatory variable (nematode count). Then compute the mean growth for each group of seedlings, plot the means against the nematode counts, and connect these four points with line segments.

(b) Briefly describe the conclusions about the effects of nematodes on plant growth that these data suggest.

2.17 Fidelity Investments, like other large mutual-fund companies, offers many "sector funds" that concentrate their investments in narrow segments of the stock market. These funds often rise or fall by much more than the market as a whole. We can group them by broader market sector to compare returns. Here are percent total returns for 23 Fidelity "Select Portfolios" funds for the year 2000, a year in which stocks declined and technology shares were especially hard-hit.[16]

Market sector	Fund returns (percent)						
Consumer	−9.3	29.8	−24.4	−23.1	−11.3		
Financial services	18.3	28.1	28.5	50.2	53.3		
Technology	−1.6	−30.4	−28.8	−17.5	−20.2	−32.3	−37.4
Utilities and natural resources	31.8	50.3	−18.1	71.3	30.4	−13.5	

(a) Make a plot of total return against market sector (space the four market sectors equally on the horizontal axis). Compute the mean return for each sector, add the means to your plot, and connect the means with line segments.

(b) Based on the data, which market sectors were good places to invest in 2000? Hindsight is wonderful.

(c) Does it make sense to speak of a positive or negative association between market sector and total return?

2.18 The presence of harmful insects in farm fields is detected by erecting boards covered with a sticky material and examining the insects trapped on the boards. Some colors are more attractive to insects than others. In an experiment aimed at determining the best color for attracting cereal leaf beetles, six boards of each of four colors were placed in a field of oats in July. The table below gives data on the number of cereal leaf beetles trapped.[17]

Board color	Insects trapped					
Lemon yellow	45	59	48	46	38	47
White	21	12	14	17	13	17
Green	37	32	15	25	39	41
Blue	16	11	20	21	14	7

(a) Make a plot of the counts of insects trapped against board color (space the four colors equally on the horizontal axis). Compute the mean count for each color, add the means to your plot, and connect the means with line segments.

(b) Based on the data, state your conclusions about the attractiveness of these colors to the beetles.

(c) Does it make sense to speak of a positive or negative association between board color and insect count?

2.2 Correlation

A scatterplot displays the form, direction, and strength of the relationship between two quantitative variables. Linear (straight-line) relations are particularly important because a straight line is a simple pattern that is quite common. We say a linear relationship is strong if the points lie close to a straight line, and weak if they are widely scattered about a line. Our eyes are not good judges of how strong a relationship is. The two scatterplots in Figure 2.9

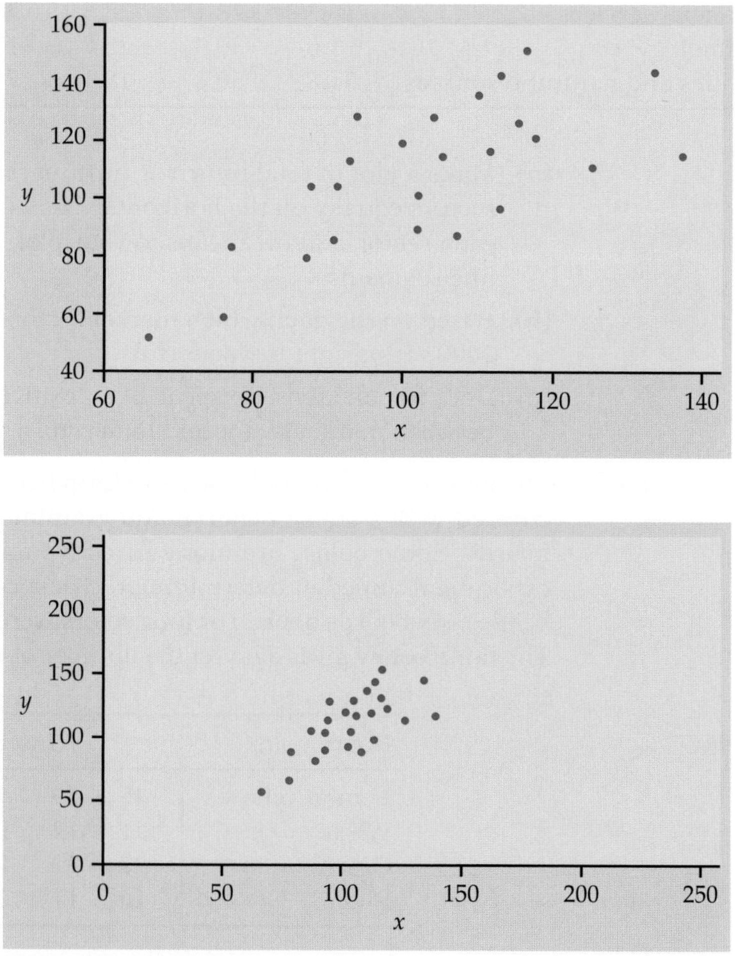

FIGURE 2.9 Two scatterplots of the same data; the linear pattern in the lower plot appears stronger because of the surrounding white space.

depict exactly the same data, but the lower plot is drawn smaller in a large field. The lower plot seems to show a stronger relationship. Our eyes can be fooled by changing the plotting scales or the amount of white space around the cloud of points in a scatterplot.[18] We need to follow our strategy for data analysis by using a numerical measure to supplement the graph. *Correlation* is the measure we use.*

The correlation r

We have data on variables x and y for n individuals. Think, for example, of measuring height and weight for n people. Then x_1 and y_1 are your height and your weight, x_2 and y_2 are my height and my weight, and so on. For the ith individual, height x_i goes with weight y_i. Here is the definition of correlation.

Correlation

The **correlation** measures the direction and strength of the linear relationship between two quantitative variables. Correlation is usually written as r.

Suppose that we have data on variables x and y for n individuals. The means and standard deviations of the two variables are $\overline{x}$ and s_x for the x-values, and $\overline{y}$ and s_y for the y-values. The correlation r between x and y is

$$r = \frac{1}{n-1}\sum\left(\frac{x_i - \overline{x}}{s_x}\right)\left(\frac{y_i - \overline{y}}{s_y}\right)$$

As always, the summation sign $\sum$ means "add these terms for all the individuals." The formula for the correlation r is a bit complex. It helps us see what correlation is but is not convenient for actually calculating r. In practice you should use software or a calculator that finds r from keyed-in values of two variables x and y. Exercise 2.19 asks you to calculate a correlation step-by-step from the definition to solidify its meaning.

The formula for r begins by standardizing the observations. Suppose, for example, that x is height in centimeters and y is weight in kilograms and that we have height and weight measurements for n people. Then $\overline{x}$ and s_x are the mean and standard deviation of the n heights, both in centimeters. The value

$$\frac{x_i - \overline{x}}{s_x}$$

is the standardized height of the ith person, familiar from Chapter 1. The standardized height says how many standard deviations above or below the mean a person's height lies. Standardized values have no units—in this

*Correlation was introduced by the English gentleman-scientist Francis Galton (1822–1911) in 1888. Galton called r the "index of co-relation" and applied it to measurements such as the forearm lengths and the heights of a group of people.

example, they are no longer measured in centimeters. Standardize the weights also. The correlation r is an average of the products of the standardized height and the standardized weight for the n people.

Properties of correlation

The formula for correlation helps us see that r is positive when there is a positive association between the variables. Height and weight, for example, have a positive association. People who are above average in height tend to also be above average in weight. Both the standardized height and the standardized weight for such a person are positive. People who are below average in height tend also to have below-average weight. Then both standardized height and standardized weight are negative. In both cases, the products in the formula for r are mostly positive and so r is positive. In the same way, we can see that r is negative when the association between x and y is negative. More detailed study of the formula gives more detailed properties of r. Here is what you need to know in order to interpret correlation:

- Correlation makes no use of the distinction between explanatory and response variables. It makes no difference which variable you call x and which you call y in calculating the correlation.
- Correlation requires that both variables be quantitative, so that it makes sense to do the arithmetic indicated by the formula for r. We cannot calculate a correlation between the incomes of a group of people and what city they live in, because city is a categorical variable.
- Because r uses the standardized values of the observations, r does not change when we change the units of measurement of x, y, or both. Measuring height in inches rather than centimeters and weight in pounds rather than kilograms does not change the correlation between height and weight. The correlation r itself has no unit of measurement; it is just a number.
- Positive r indicates positive association between the variables, and negative r indicates negative association.
- The correlation r is always a number between -1 and 1. Values of r near 0 indicate a very weak linear relationship. The strength of the relationship increases as r moves away from 0 toward either -1 or 1. Values of r close to -1 or 1 indicate that the points lie close to a straight line. The extreme values $r = -1$ and $r = 1$ occur only when the points in a scatterplot lie exactly along a straight line.
- Correlation measures the strength of only the linear relationship between two variables. Correlation does not describe curved relationships between variables, no matter how strong they are.
- Like the mean and standard deviation, the correlation is not resistant: r is strongly affected by a few outlying observations. Use r with caution when outliers appear in the scatterplot.

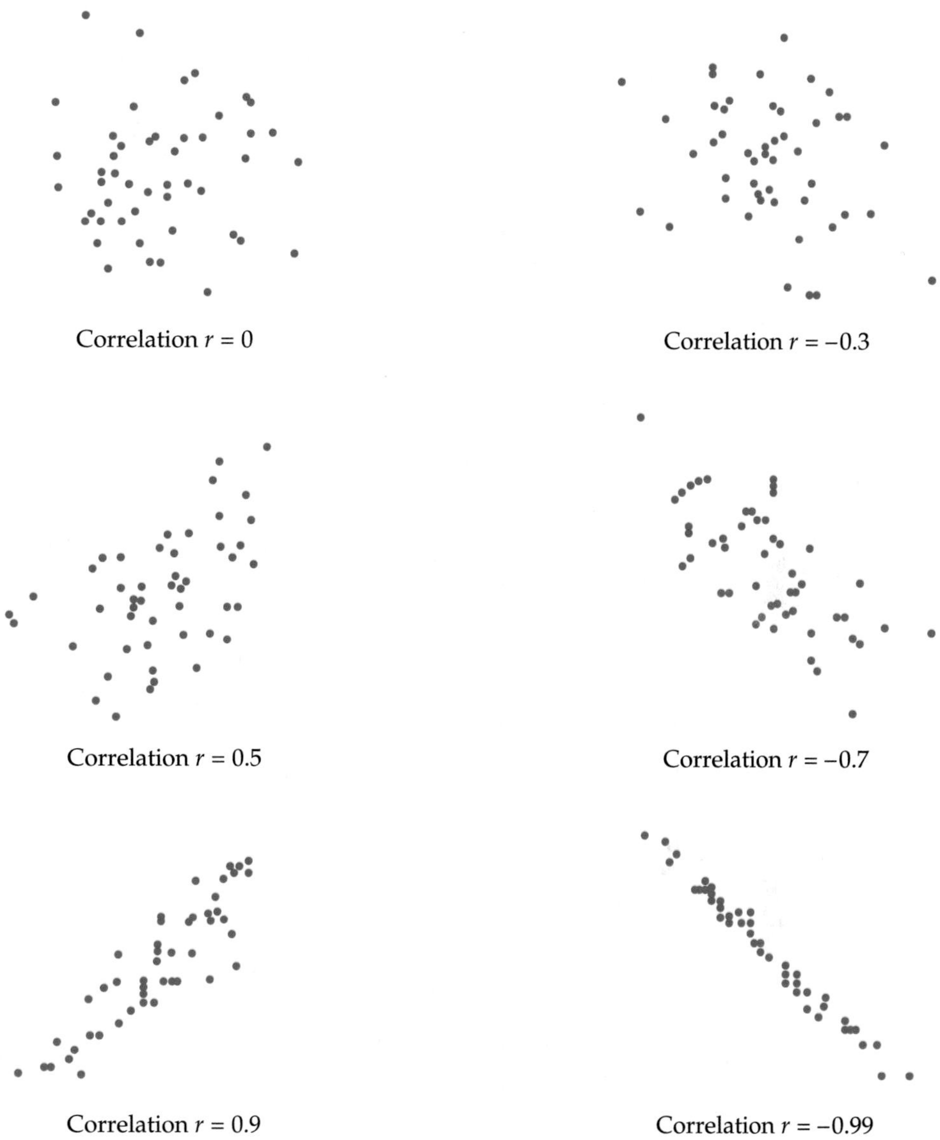

FIGURE 2.10 How the correlation r measures the direction and strength of linear association.

The scatterplots in Figure 2.10 illustrate how values of r closer to 1 or -1 correspond to stronger linear relationships. To make the essential meaning of r clear, the standard deviations of both variables in these plots are equal and the horizontal and vertical scales are the same. In general, it is not so easy to guess the value of r from the appearance of a scatterplot. Remember that changing the plotting scales in a scatterplot may mislead our eyes, but it does not change the standardized values of the variables and therefore cannot change the correlation. To explore how extreme observations can influence r, use the *Correlation and Regression* applet available on the Web site `www.whfreeman.com/ips`.

EXAMPLE 2.8

The real data we have examined also illustrate the behavior of correlation.

Figure 2.1 (page 107), despite the clusters, shows a moderately strong negative linear association between the percent of a state's high school seniors who take the SAT exam and their mean SAT math score. The correlation is $r = -0.864$.

Figure 2.8 (page 118) shows a very weak relationship between returns on Treasury bills and on common stocks. We expect a small negative r, and calculation gives $r = -0.098$.

The correlation between time and acceleration for the motorcycle crash data graphed in Figure 2.5 (page 113) is $r = 0.296$. Because the relationship is not at all linear, r provides no useful information. Always plot your data before calculating common statistical measures such as correlation.

Finally, remember that **correlation is not a complete description of two-variable data,** even when the relationship between the variables is linear. You should give the means and standard deviations of both x and y along with the correlation. (Because the formula for correlation uses the means and standard deviations, these measures are the proper choices to accompany a correlation.) Conclusions based on correlations alone may require rethinking in the light of a more complete description of the data.

EXAMPLE 2.9

Competitive divers are scored on their form by a panel of judges who use a scale from 1 to 10. The subjective nature of the scoring often results in controversy. We have the scores awarded by two judges, Ivan and George, on a large number of dives. How well do they agree? We do some calculation and find that the correlation between their scores is $r = 0.9$. But the mean of Ivan's scores is 3 points lower than George's mean.

These facts do not contradict each other. They are simply different kinds of information. The mean scores show that Ivan awards much lower scores than George. But because Ivan gives *every* dive a score about 3 points lower than George, the correlation remains high. Adding the same number to all values of either x or y does not change the correlation. If Ivan and George both rate several divers, the contest is fairly scored because Ivan and George agree on which dives are better than others. The high r shows their agreement. But if Ivan scores one diver and George another, we must add 3 points to Ivan's scores to arrive at a fair comparison.

SUMMARY

The **correlation *r*** measures the direction and strength of the linear association between two quantitative variables x and y. Although you can calculate a correlation for any scatterplot, r measures only linear relationships.

Correlation indicates the direction of a linear relationship by its sign: $r > 0$ for a positive association and $r < 0$ for a negative association.

Correlation always satisfies $-1 \leq r \leq 1$ and indicates the strength of a relationship by how close it is to -1 or 1. Perfect correlation, $r = \pm 1$, occurs only when the points lie exactly on a straight line.

Correlation ignores the distinction between explanatory and response variables. The value of r is not affected by changes in the unit of measurement of either variable. Correlation is not resistant, so outliers can greatly change the value of r.

SECTION 2.2 EXERCISES

2.19 Example 2.4 (page 110) gives the lengths of two bones in five fossil specimens of the extinct beast *Archaeopteryx*. The scatterplot in Figure 2.3 shows a strong positive linear relationship.

(a) Find the correlation r step-by-step. That is, find the mean and standard deviation of the femur lengths and of the humerus lengths. Then find the five standardized values for each variable and use the formula for r.

(b) Now enter these data into your calculator or software and use the correlation function to find r. Check that you get the same result as in (a).

2.20 Exercise 2.11 (page 122) gives data on the lean body mass and metabolic rate for 12 women and 7 men.

(a) If you did not do Exercise 2.11, make a scatterplot. Use different symbols or colors for women and men. Do you think the correlation will be about the same for men and women or quite different for the two groups? Why?

(b) Find r for women alone and also for men alone.

(c) Calculate the mean body mass for the women and for the men. Does the fact that the men are heavier than the women on the average influence the correlations? If so, in what way?

(d) Lean body mass was measured in kilograms. How would the correlations change if we measured body mass in pounds? (There are about 2.2 pounds in a kilogram.)

2.21 A student wonders if tall women tend to date taller men than do short women. She measures herself, her dormitory roommate, and the women in the adjoining rooms; then she measures the next man each woman dates. Here are the data (heights in inches):

Women (x)	66	64	66	65	70	65
Men (y)	72	68	70	68	71	65

(a) Make a scatterplot of these data. Based on the scatterplot, do you expect the correlation to be positive or negative? Near ± 1 or not?

(b) Find the correlation r between the heights of the men and women.

(c) How would r change if all the men were 6 inches shorter than the heights given in the table? Does the correlation tell us whether women tend to date men taller than themselves?

(d) If heights were measured in centimeters rather than inches, how would the correlation change? (There are 2.54 centimeters in an inch.)

(e) If every woman dated a man exactly 3 inches taller than herself, what would be the correlation between male and female heights?

2.22 Make a scatterplot of the following data:

x	1	2	3	4	10	10
y	1	3	3	5	1	11

Use your calculator to show that the correlation is about 0.5. What feature of the data is responsible for reducing the correlation to this value despite a strong straight-line association between x and y in most of the observations?

2.23 Exercise 2.10 (page 122) gives data on gas mileage against speed for a small car. Make a scatterplot if you have not already done so, then find the correlation r. Explain why r is close to zero despite a strong relationship between speed and gas used.

2.24 Changing the units of measurement can dramatically alter the appearance of a scatterplot. Return to the fossil data from Example 2.4:

Femur	38	56	59	64	74
Humerus	41	63	70	72	84

These measurements are in centimeters. Suppose a deranged scientist measured the femur in meters and the humerus in millimeters. The data would then be

Femur	0.38	0.56	0.59	0.64	0.74
Humerus	410	630	700	720	840

(a) Draw an x axis extending from 0 to 75 and a y axis extending from 0 to 850. Plot the original data on these axes. Then plot the new data on the same axes in a different color. The two plots look very different.

(b) Nonetheless, the correlation is exactly the same for the two sets of measurements. Why do you know that this is true without doing any calculations? Find the two correlations to verify that they are the same.

2.25 Table 1.8 (page 38) gives the city and highway gas mileages for 19 two-seater cars, including the Honda Insight gas-electric hybrid car.

(a) Make a scatterplot of highway mileage y against city mileage x for all 19 cars. There is a strong positive linear association. The Insight lies far from the other points. Does the Insight extend the linear pattern of the other cars, or is it far from the line they form?

(b) Find the correlation between city and highway mileages both without and with the Insight. Based on your answer to (a), explain why r changes in this direction when you add the Insight.

2.26 The *Correlation and Regression* applet at `www.whfreeman.com/ips` allows you to create a scatterplot and to move points by dragging with the mouse. Click to create a group of 10 points in the lower-left corner of the scatterplot with a strong straight-line pattern (correlation about 0.9).

(a) Add one point at the upper right that is in line with the first 10. How does the correlation change?

(b) Drag this last point down until it is opposite the group of 10 points. How small can you make the correlation? Can you make the correlation negative? You see that a single outlier can greatly strengthen or weaken a correlation. Always plot your data to check for outlying points.

2.27 Many mutual funds compare their performance with that of a benchmark, an index of the returns on all securities of the kind the fund buys. The Vanguard International Growth Fund, for example, takes as its benchmark the Morgan Stanley EAFE (Europe, Australasia, Far East) index of overseas stock market performance. Here are the percent returns for the fund and for the EAFE from 1982 (the first full year of the fund's existence) to 2000:[19]

Year	Fund	EAFE	Year	Fund	EAFE
1982	5.27	−0.86	1992	−5.79	−11.85
1983	43.08	24.61	1993	44.74	32.94
1984	−1.02	7.86	1994	0.76	8.06
1985	56.94	56.72	1995	14.89	11.55
1986	56.71	69.94	1996	14.65	6.36
1987	12.48	24.93	1997	4.12	2.06
1988	11.61	28.59	1998	16.93	20.33
1989	24.76	10.80	1999	26.34	27.30
1990	−12.05	−23.20	2000	−8.60	−13.96
1991	4.74	12.50			

Make a scatterplot suitable for predicting fund returns from EAFE returns. Is there a clear straight-line pattern? How strong is this pattern? (Give a numerical measure.) Are there any extreme outliers from the straight-line pattern?

2.28 A mutual-fund company's newsletter says, "A well-diversified portfolio includes assets with low correlations." The newsletter includes a table of correlations between the annual returns on various classes of investments. For example, the correlation between municipal bonds and large-cap stocks is 0.50 and the correlation between municipal bonds and small-cap stocks is 0.21.

(a) Rachel invests heavily in municipal bonds. She wants to diversify by adding an investment whose returns do not closely follow the returns on her bonds. Should she choose large-cap stocks or small-cap stocks for this purpose? Explain your answer.

(b) If Rachel wants an investment that tends to increase when the return on her bonds drops, what kind of correlation should she look for?

2.29 Consider again the correlation r between the speed of a car and its gas consumption, from the data in Exercise 2.10 (page 122).

(a) Transform the data so that speed is measured in miles per hour and fuel consumption in gallons per mile. (There are 1.609 kilometers in a mile and 3.785 liters in a gallon.) Make a scatterplot and find the correlation for both the original and the transformed data. How did the change of units affect your results?

(b) Now express fuel consumption in miles per gallon. (So each value is $1/x$ if x is gallons per mile.) Again make a scatterplot and find the correlation. How did this change of units affect your results?

(*Lesson:* The effects of a linear transformation of the form $x_{\text{new}} = a + bx$ are simple. The effects of a nonlinear transformation are more complex.)

2.30 Table 1.6 (page 33) reports data on 78 seventh-grade students. Some people think that self-concept is related to school performance. Make a scatterplot of school GPA against self-concept score. Describe the overall pattern that you see. Does it make sense to use the correlation as a summary? If so, find r and interpret its value in words.

2.31 Suppose that women always married men 2 years older than themselves. Draw a scatterplot of the ages of 5 married couples, with the wife's age as the explanatory variable. What is the correlation r for your data? Why?

2.32 Investment reports now often include correlations. Following a table of correlations among mutual funds, a report adds, "Two funds can have perfect correlation, yet different levels of risk. For example, Fund A and Fund B may be perfectly correlated, yet Fund A moves 20% whenever Fund B moves 10%." Write a brief explanation, for someone who knows no statistics, of how this can happen. Include a sketch to illustrate your explanation.

2.33 A college newspaper interviews a psychologist about student ratings of the teaching of faculty members. The psychologist says, "The evidence indicates that the correlation between the research productivity and teaching rating of faculty members is close to zero." The paper reports this as "Professor McDaniel said that good researchers tend to be poor teachers, and vice versa." Explain why the paper's report is wrong. Write a statement in plain language (don't use the word "correlation") to explain the psychologist's meaning.

2.34 Each of the following statements contains a blunder. Explain in each case what is wrong.

(a) "There is a high correlation between the gender of American workers and their income."

(b) "We found a high correlation ($r = 1.09$) between students' ratings of faculty teaching and ratings made by other faculty members."

(c) "The correlation between planting rate and yield of corn was found to be $r = 0.23$ bushel."

2.3 Least-Squares Regression

Correlation measures the direction and strength of the linear (straight-line) relationship between two quantitative variables. If a scatterplot shows a linear relationship, we would like to summarize this overall pattern by drawing a line on the scatterplot. A *regression line* summarizes the relationship between two variables, but only in a specific setting: when one of the variables helps explain or predict the other. That is, regression describes a relationship between an explanatory variable and a response variable.*

Regression Line

A **regression line** is a straight line that describes how a response variable y changes as an explanatory variable x changes. We often use a regression line to **predict** the value of y for a given value of x. Regression, unlike correlation, requires that we have an explanatory variable and a response variable.

EXAMPLE 2.10

How do children grow? The pattern of growth varies from child to child, so we can best understand the general pattern by following the average height of a number of children. Table 2.7 presents the mean heights of a group of children in Kalama, an Egyptian village that was the site of a study of nutrition in developing countries.[20] The data were obtained by measuring the heights of 161 children from the village each month from 18 to 29 months of age.

Figure 2.11 is a scatterplot of the data in Table 2.7. Age is the explanatory variable, which we plot on the x axis. The plot shows a strong positive linear association with no outliers. The correlation is $r = 0.994$, close to the $r = 1$ of points that lie exactly on a line. A line drawn through the points will describe these data very well.

*The term "regression" and the general methods for studying relationships now included under this term were introduced by Francis Galton. Galton was engaged in the study of heredity. One of his observations was that the children of tall parents tended to be taller than average but not as tall as their parents. This "regression toward mediocrity" gave these statistical methods their name.

TABLE 2.7 Mean height of Kalama children	
Age x in months	Height y in centimeters
18	76.1
19	77.0
20	78.1
21	78.2
22	78.8
23	79.7
24	79.9
25	81.1
26	81.2
27	81.8
28	82.8
29	83.5

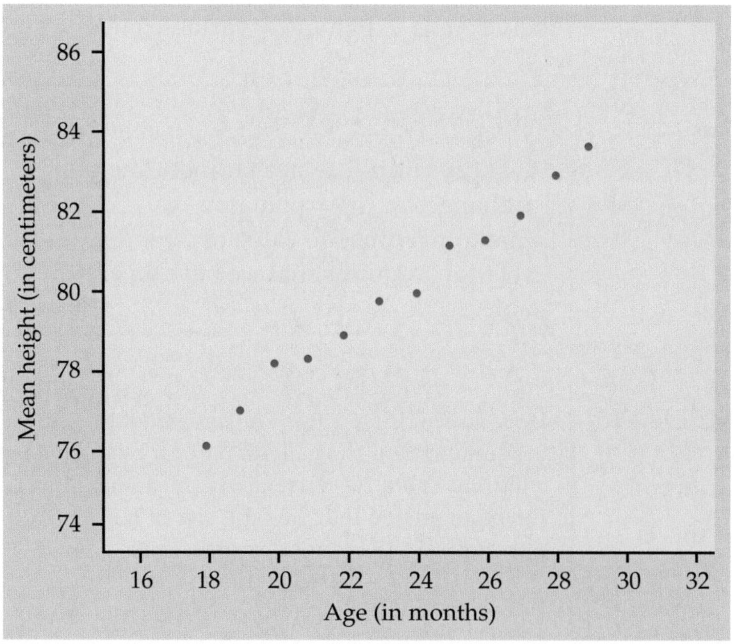

FIGURE 2.11 Mean height of children in Kalama, Egypt, plotted against age from 18 to 29 months, from Table 2.7.

Fitting a line to data

fitting a line

When a scatterplot displays a linear pattern, we can describe the overall pattern by drawing a straight line through the points. Of course, no straight line passes exactly through all of the points. **Fitting a line** to data means drawing a line that comes as close as possible to the points. The equation of a line fitted to the data gives a compact description of the dependence of the response variable y on the explanatory variable x. It is a mathematical model for the straight-line relationship.

Straight Lines

Suppose that y is a response variable (plotted on the vertical axis) and x is an explanatory variable (plotted on the horizontal axis). A straight line relating y to x has an equation of the form

$$y = a + bx$$

In this equation, b is the **slope,** the amount by which y changes when x increases by one unit. The number a is the **intercept,** the value of y when $x = 0$.

Any straight line describing the Kalama data has the form

$$\text{height} = a + (b \times \text{age})$$

In Figure 2.12 we have drawn the regression line with the equation

$$\text{height} = 64.93 + (0.635 \times \text{age})$$

The figure shows that this line fits the data well. The slope $b = 0.635$ tells us that the height of Kalama children increases by about 0.6 centimeter for each month of age. The slope b of a line $y = a + bx$ is the *rate of change* in the

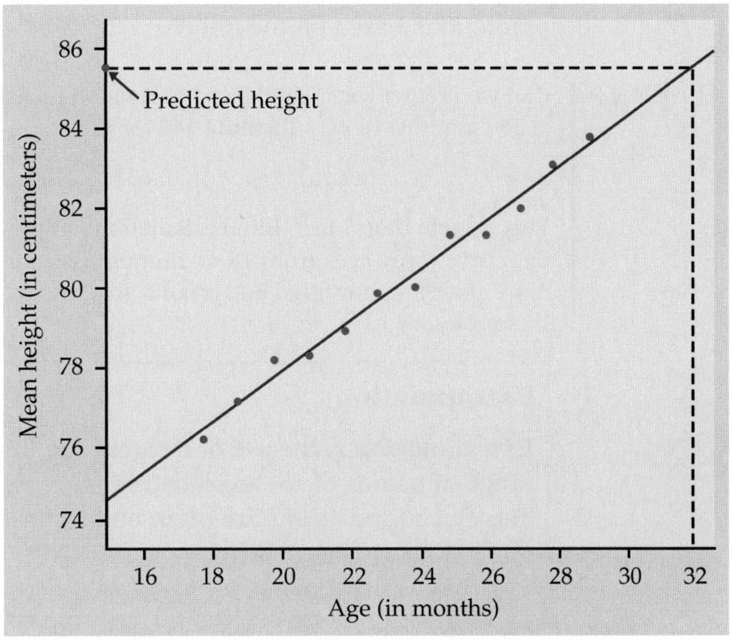

FIGURE 2.12 The regression line fitted to the Kalama data and used to predict height at age 32 months.

response y as the explanatory variable x changes. The slope of a regression line is an important numerical description of the relationship between the two variables. The intercept, $a = 64.93$ centimeters, would be the mean height at birth (age = 0) if the straight-line pattern of growth were true starting at birth. Children don't grow at a fixed rate from birth, so the intercept a is not important in our situation except as part of the equation of the line.

Prediction

prediction

We can use a regression line to **predict** the response y for a specific value of the explanatory variable x.

EXAMPLE 2.11

Based on the strong linear pattern, we want to predict the mean height of Kalama children at 32 months of age. To use our fitted line to predict height, go "up and over" on the graph in Figure 2.12. From age 32 months on the x axis, go up to the fitted line and over to the y axis. It appears that the predicted height is a bit more than 85 centimeters.

If we have the equation of the line, it is faster and more accurate to substitute 32 for the age in the equation. Our predicted height is

$$\text{height} = 64.93 + (0.635 \times 32) = 85.25 \text{ centimeters}$$

The accuracy of predictions from a regression line depends on how much scatter about the line the data show. In the Kalama example, the data points are all very close to the line, so we are confident that our prediction is accurate. If the data show a linear pattern with considerable spread, we may use a regression line to summarize the pattern but we will put less confidence in predictions based on the line.

EXAMPLE 2.12

Can we predict the mean height of Kalama residents at age 20 years? Twenty years is 240 months, so we substitute 240 for the age. The prediction is

$$\text{height} = 64.93 + (0.635 \times 240) = 217.33 \text{ centimeters}$$

This is more than 7 feet. Blind calculation has produced an unreasonable result. Our data cover only ages from 18 to 29 months. As people grow older, they gain height more slowly, so our fitted line is not a good model at ages far removed from the data that produced it.

Extrapolation

Extrapolation is the use of a regression line for prediction far outside the range of values of the explanatory variable x that you used to obtain the line. Such predictions are often not accurate.

Notice that age 0 (birth) lies well outside the range of the data in Table 2.7. Taking the intercept 64.93 centimeters (over 25 inches) as an estimate of height at birth is therefore another example of extrapolation. The linear growth pattern does not begin at birth, and this estimate of height at birth is not accurate. In addition, we know that the pattern we discovered in Figure 2.12 does not hold for all children everywhere. We cannot use data from Kalama to predict the height of children in Cairo, much less Los Angeles. Statistical methods such as fitting regression lines cannot go beyond the limitations of the data that feed them. Do not let calculation override common sense.

Least-squares regression

Different people might draw different lines by eye on a scatterplot. This is especially true when the points are widely scattered. We need a way to draw a regression line that doesn't depend on our guess as to where the line should go. No line will pass exactly through all the points, but we want one that is as close as possible. We will use the line to predict y from x, so we want a line that is as close as possible to the points in the *vertical* direction. That's because the prediction errors we make are errors in y, which is the vertical direction in the scatterplot. If we predict 85.25 centimeters for the mean height at age 32 months and the actual mean turns out to be 84 centimeters, our error is

$$\begin{aligned} \text{error} &= \text{observed height} - \text{predicted height} \\ &= 84 - 85.25 = -1.25 \text{ centimeters} \end{aligned}$$

We want a regression line that makes these prediction errors as small as possible. Figure 2.13 illustrates the idea. For clarity, the plot shows only three of the points from Figure 2.12, along with the line, on an expanded scale. The line passes below two of the points and above one of them. The vertical

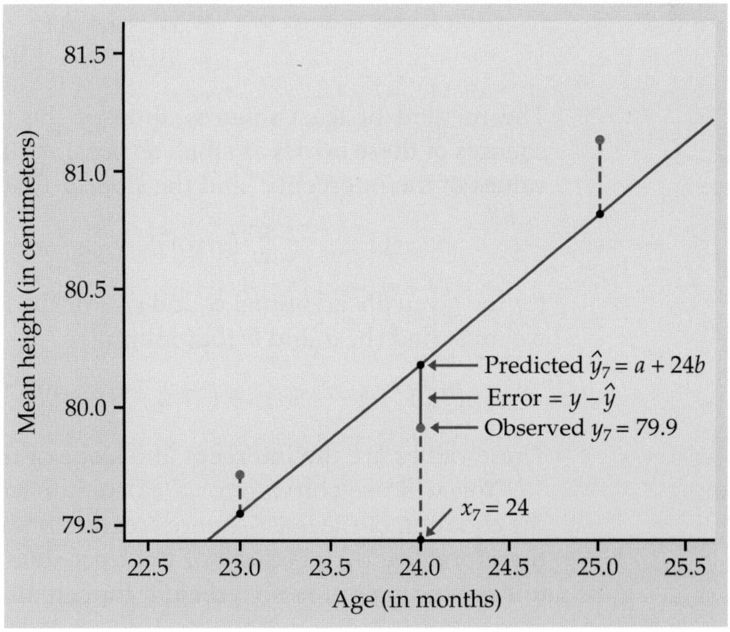

FIGURE 2.13 The least-squares idea: make the errors in predicting y as small as possible by minimizing the sum of their squares.

distances of the data points from the line appear as vertical line segments. A "good" regression line makes these distances as small as possible. There are many ways to make "as small as possible" precise. The most common is the *least-squares* idea. The line in Figures 2.12 and 2.13 is in fact the least-squares regression line for predicting mean height from age.*

Least-Squares Regression Line

The **least-squares regression line of y on x** is the line that makes the sum of the squares of the vertical distances of the data points from the line as small as possible.

Here is the least-squares idea expressed as a mathematical problem. We represent n observations on two variables x and y as

$$(x_1, y_1),\ (x_2, y_2), \ldots, (x_n, y_n)$$

If we draw a line $y = a + bx$ through the scatterplot of these observations, the line predicts the value of y corresponding to x_i as $\hat{y}_i = a + bx_i$. We write $\hat{y}$ (read "y hat") in the equation of a regression line to emphasize that the line gives a *predicted* response $\hat{y}$ for any x. The predicted response will usually not be exactly the same as the actually *observed* response y. For example, the seventh observation in Table 2.7 is $x_7 = 24$, $y_7 = 79.9$. Figure 2.13 shows the prediction error for this point, which is

$$\begin{aligned} \text{error} &= \text{observed } y - \text{predicted } y \\ &= y_7 - \hat{y}_7 \\ &= y_7 - a - bx_7 \\ &= 79.9 - a - 24b \end{aligned}$$

The method of least-squares chooses the line that makes the sum of the squares of these errors as small as possible. To find this line, we must find the values of the intercept a and the slope b that minimize

$$\sum (\text{error})^2 = \sum (y_i - a - bx_i)^2$$

for the given observations x_i and y_i. For the Kalama growth data in Table 2.7, we must find the a and b that minimize

$$(76.1 - a - 18b)^2 + (77.0 - a - 19b)^2 + \cdots + (83.5 - a - 29b)^2$$

These values are the intercept and slope of the least-squares line.

You will use software or a calculator with a regression function to find the equation of the least-squares regression line from data on x and y. We will therefore give the equation of the least-squares line in a form that helps our understanding but is not efficient for calculation.

*The method of least squares was first published by the French mathematician Adrien-Marie Legendre (1752–1833) in 1805. It at once became the standard method for combining observations in astronomy and surveying and remains one of the most-used statistical tools.

Equation of the Least-Squares Regression Line

We have data on an explanatory variable x and a response variable y for n individuals. The means and standard deviations of the sample data are $\bar{x}$ and s_x for x and $\bar{y}$ and s_y for y, and the correlation between x and y is r. The equation of the least-squares regression line of y on x is

$$\hat{y} = a + bx$$

with **slope**

$$b = r\frac{s_y}{s_x}$$

and **intercept**

$$a = \bar{y} - b\bar{x}$$

EXAMPLE 2.13

Verify from the data in Table 2.7 that the mean and standard deviation of the 12 ages are

$$\bar{x} = 23.5 \text{ months} \quad \text{and} \quad s_x = 3.606 \text{ months}$$

The mean and standard deviation of the 12 heights are

$$\bar{y} = 79.85 \text{ cm} \quad \text{and} \quad s_y = 2.302 \text{ cm}$$

The correlation between height and age is $r = 0.9944$. The least-squares regression line of height y on age x therefore has slope

$$\begin{aligned} b &= r\frac{s_y}{s_x} = 0.9944\frac{2.302}{3.606} \\ &= 0.6348 \text{ cm per month} \end{aligned}$$

and intercept

$$\begin{aligned} a &= \bar{y} - b\bar{x} = 79.85 - (0.6348)(23.5) \\ &= 64.932 \text{ cm} \end{aligned}$$

The equation of the least-squares line is

$$\hat{y} = 64.932 + 0.6348x$$

Interpreting the regression line

The slope $b = 0.6348$ centimeters per month in Example 2.13 is the rate of change in the mean height as age increases. The units "centimeters per month" come from the units of y (centimeters) and x (months). Although the correlation does not change when we change the units of measurement, the equation of the least-squares line does change. The slope in centimeters per month is 2.54 times as large as the slope in inches per month, because there are 2.54 centimeters in an inch.

The expression $b = rs_y/s_x$ for the slope says that along the regression line, **a change of one standard deviation in x corresponds to a change of r standard deviations in y.** When the variables are perfectly correlated ($r = 1$ or $r = -1$), the change in the predicted response $\hat{y}$ is the same (in standard deviation units) as the change in x. Otherwise, because $-1 \leq r \leq 1$, the change in $\hat{y}$ is less than the change in x. As the correlation grows less strong, the prediction $\hat{y}$ moves less in response to changes in x.

The least-squares regression line always passes through the point $(\bar{x}, \bar{y})$ on the graph of y against x. Check that when you substitute $\bar{x} = 23.5$ into the equation of the regression line in Example 2.13, the result is $\hat{y} = 79.8498$, equal to the mean of y except for roundoff error. So the least-squares regression line of y on x is the line with slope rs_y/s_x that passes through the point $(\bar{x}, \bar{y})$. We can describe regression entirely in terms of the basic descriptive measures $\bar{x}$, s_x, $\bar{y}$, s_y, and r. If both x and y are standardized variables, so that their means are 0 and their standard deviations are 1, then the regression line has slope r and passes through the origin.

Figure 2.14 displays the basic regression output for the Kalama data from a graphing calculator and two statistical software packages. You can find the

TI-83

```
LinReg
 y=a+bx
 a=64.92832168
 b=.634965035
 r²=.9887639371
 r=.9943660981
```

SPSS

Model Summary

Model	R	R Square	Adjusted R Square	Std. Error of the Estimate
1	.994[a]	.989	.988	.2560

a. Predictors: (Constant), AGE

Coefficients[a]

Model		Unstandardized Coefficients		Standardized Coefficients		
		B	Std. Error	Beta	t	Sig.
1	(Constant)	64.928	.508		127.709	.000
	AGE	.635	.021	.994	29.665	.000

a. Dependent Variable: HEIGHT

Minitab

```
The regression equation is
height = 64.9 + 0.635 age

Predictor        Coef      Stdev    t-ratio        p
Constant      64.9283     0.5084     127.71    0.000
height        0.63497    0.02140      29.66    0.000

s = 0.2560        R-sq = 98.9%       R-sq(adj) = 98.8%
```

FIGURE 2.14 Regression results for the Kalama data, from a graphing calculator and two statistical software packages.

slope and intercept of the least-squares line, calculated to more decimal places than we need, in all these outputs. The software also provides information that we do not yet need. Part of the art of using software is to ignore the extra information that is almost always present. We will make use of other parts of the output in Chapter 10.

Correlation and regression

Least-squares regression looks at the distances of the data points from the line only in the y direction. So the two variables x and y play different roles in regression.

EXAMPLE 2.14

Figure 2.15 is a scatterplot of data that played a central role in the discovery that the universe is expanding. They are the distances from Earth of 24 spiral galaxies and the speed at which these galaxies are moving away from us, reported by the astronomer Edwin Hubble in 1929.[21] There is a positive linear relationship, $r = 0.7842$. More distant galaxies are moving away more rapidly. Astronomers believe that there is in fact a perfect linear relationship, and that the scatter is caused by imperfect measurements.

The two lines on the plot are the two least-squares regression lines. The regression line of velocity on distance is solid. The regression line of distance on velocity is dashed. *Although there is only one correlation between velocity and distance, regression of velocity on distance and regression of distance on velocity give different lines.* In the regression setting you must choose which variable is explanatory.

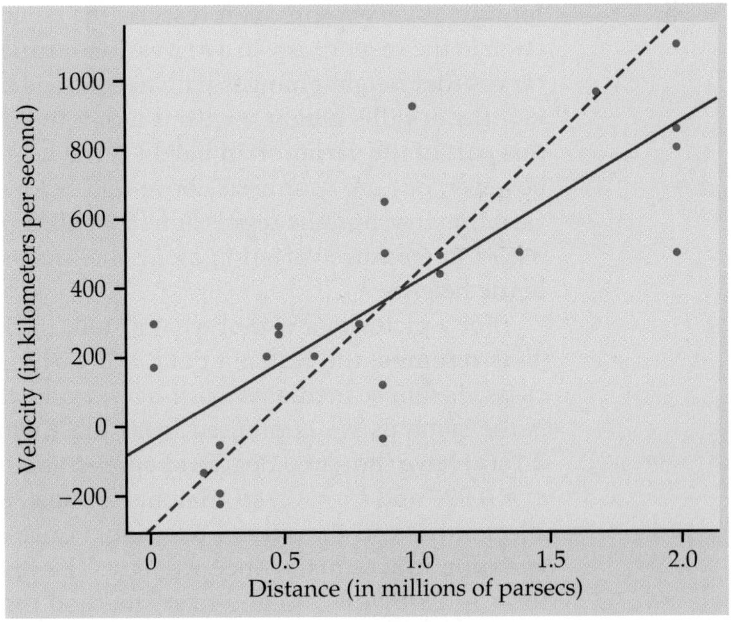

FIGURE 2.15 Hubble's data on the velocity and distance of 24 galaxies. The lines are least-squares regression lines of velocity on distance (solid) and of distance on velocity (dashed).

Although the correlation r ignores the distinction between explanatory and response variables, there is a close connection between correlation and regression. We saw that the slope of the least-squares line involves r. Another connection between correlation and regression is even more important. In fact, the numerical value of r as a measure of the strength of a linear relationship is best interpreted by thinking about regression. Here is the fact we need.

r^2 in Regression

The **square of the correlation, r^2**, is the fraction of the variation in the values of y that is explained by the least-squares regression of y on x.

The correlation between the mean height and age of Kalama children is $r = 0.9944$. Because $r^2 = 0.9888$, the straight-line relationship between height and age explains 98.88% (almost all) of the variation in heights in Table 2.7. When you report a regression, give r^2 as a measure of how successfully the regression explains the response. All the software outputs in Figure 2.14 include r^2, either in decimal form or as a percent. When you see a correlation, square it to get a better feel for the strength of the association. Perfect correlation ($r = -1$ or $r = 1$) means the points lie exactly on a line. Then $r^2 = 1$ and all of the variation in one variable is accounted for by the linear relationship with the other variable. If $r = -0.7$ or $r = 0.7$, $r^2 = 0.49$ and about half the variation is accounted for by the linear relationship. In the r^2 scale, correlation ± 0.7 is about halfway between 0 and ± 1.

The use of r^2 to describe the success of regression in explaining the response y is very common. It rests on the fact that there are two sources of variation in the responses y in a regression setting. One reason the Kalama heights vary is that height changes with age: as age x increases from 18 months to 29 months, it pulls height y with it along the regression line. The line explains this part of the variation in heights. The heights do not lie exactly on the line, however, but are scattered above and below it. This is the second source of variation in y, and the regression line tells us nothing about how large it is. We use r^2 to measure variation along the line as a fraction of the total variation in the heights.

For a pictorial grasp of what r^2 tells us, look at Figure 2.16. Both scatterplots resemble the Kalama data, but with many more observations. In both cases, height y increases with age x, and the least-squares regression line is the same as we computed from the Kalama data. The heights in Figure 2.16(a) have the same degree of spread about the line as the Kalama data, so $r = 0.994$ and $r^2 = 0.989$. Heights for age 28 months are higher than for age 20 months, and the spread of heights once we fix age is rather small. Little variation in y remains once we fix x. The figure shows that a prediction of y for $x = 28$ by the "up-and-over" method that takes account of the variation in y for a fixed x pins down height quite closely.

The heights in Figure 2.16(b) have the same overall linear pattern expressed by the regression line, but there is more scatter about the fitted line.

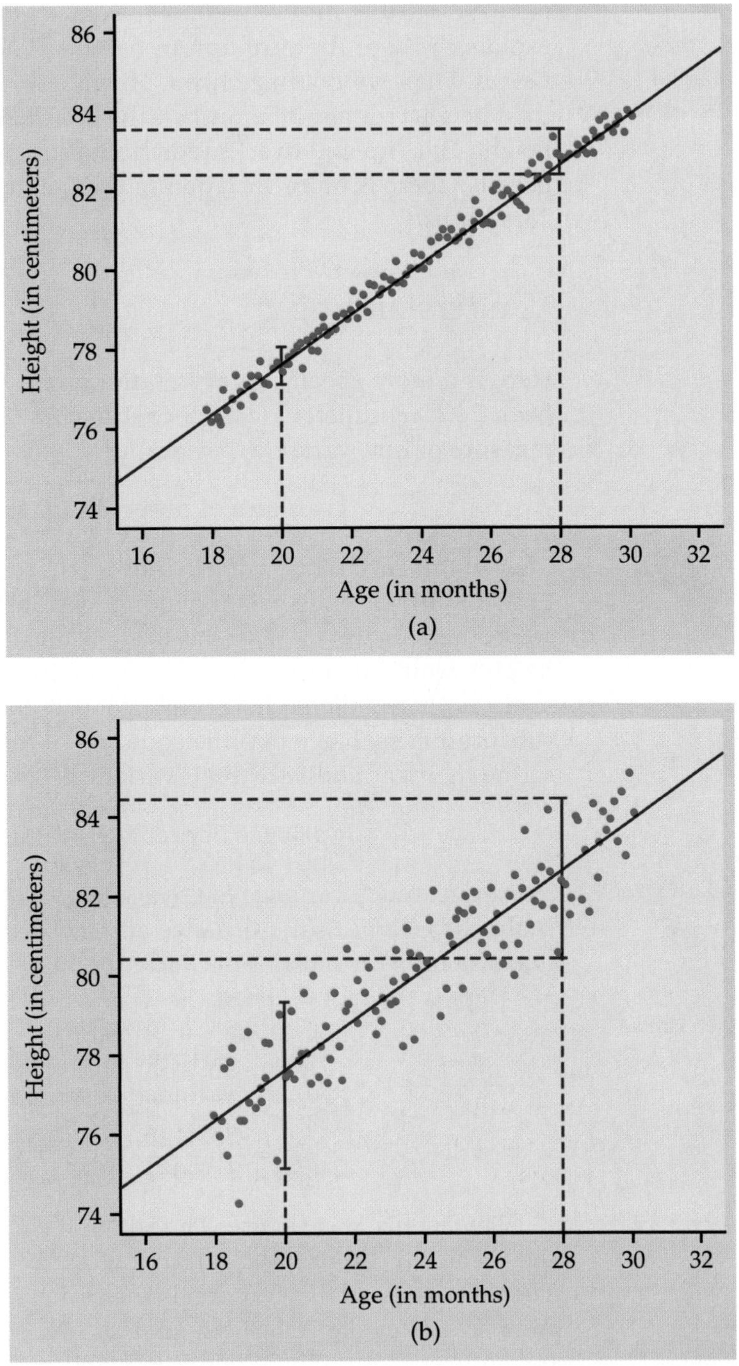

FIGURE 2.16 Explained versus unexplained variation. In (a), almost all of the variation in height is explained by the linear relationship between height and age. The remaining variation (the spread of heights when $x = 28$ months, for example) is small. In (b), the linear relationship explains a smaller part of the variation in height. The remaining variation (illustrated again for $x = 28$) is larger.

Here, $r = 0.921$ and $r^2 = 0.849$. That is, the link between height and age explains 85% of the variation in heights. The remaining 15% is variation in height when we hold age fixed. Heights for age 28 months are still greater than heights for age 20 months—the line still explains part of the variation in height. But "up-and-over" prediction of height y when $x = 28$ is less precise because there is more variation in the heights for individuals with ages near 28 months.

*Understanding r^2

Here is a more specific interpretation of r^2. The heights y in Table 2.7 range from 76.1 centimeters to 83.5 centimeters. The variance of these heights, a measure of how variable they are, is

$$\text{variance of observed values } y = 5.30091$$

Most of this variability is due to the fact that as age x increases from 18 months to 29 months it pulls height y along with it. If the only variability in the observed heights were due to the straight-line dependence of height on age, the heights would lie exactly on the regression line. That is, they would be the same as the predicted heights $\hat{y}$. We can compute the predicted heights by substituting each age into the equation of the least-squares line and then describe their variability by their variance. The result is

$$\text{variance of predicted values } \hat{y} = 5.24135$$

There isn't much variance left over—the variance of points exactly on the line is almost as large as the variance of the observed heights. This shows that the line accounts for most of the variation in y, but what does it have to do with r^2? Here is the fact we need:

$$\begin{aligned} r^2 &= \frac{\text{variance of predicted values } \hat{y}}{\text{variance of observed values } y} \\ &= \frac{5.24135}{5.30091} = 0.9888 \end{aligned}$$

This fact is always true. The squared correlation gives us the variance of the predicted responses as a fraction of the variance of the actual responses. The numerator of this fraction is the variance the heights would have if there were no scatter about the least-squares line, so that all the variation in y comes from the straight-line tie between y and x. The denominator is the variance the heights actually do have, including both the tie to x and scatter about the fitted line. This gives an exact meaning for "fraction of variation explained" as an interpretation of r^2.

These connections with correlation are special properties of least-squares regression. They are not true for other methods of fitting a line to data. One

*This explanation is optional reading.

reason that least squares is the most common method for fitting a regression line to data is that it has many convenient special properties.

SUMMARY

A **regression line** is straight line that describes how a response variable y changes as an explanatory variable x changes.

The most common method of fitting a line to a scatterplot is least squares. The **least-squares regression line** is the straight line $\hat{y} = a + bx$ that minimizes the sum of the squares of the vertical distances of the observed y-values from the line.

You can use a regression line to **predict** the value of y for any value of x by substituting this x into the equation of the line. **Extrapolation** beyond the range of x-values spanned by the data is risky.

The **slope** b of a regression line $\hat{y} = a + bx$ is the rate at which the predicted response $\hat{y}$ changes along the line as the explanatory variable x changes. Specifically, b is the change in $\hat{y}$ when x increases by 1.

The **intercept** a of a regression line $\hat{y} = a + bx$ is the predicted response $\hat{y}$ when the explanatory variable $x = 0$. This prediction is of no statistical use unless x can actually take values near 0.

The least-squares regression line of y on x is the line with slope $b = rs_y/s_x$ and intercept $a = \bar{y} - b\bar{x}$. This line always passes through the point $(\bar{x}, \bar{y})$.

Correlation and regression are closely connected. The correlation r is the slope of the least-squares regression line when we measure both x and y in standardized units. The square of the correlation r^2 is the fraction of the variance of one variable that is explained by least-squares regression on the other variable.

SECTION 2.3 EXERCISES

2.35 (Review of straight lines) Fred keeps his savings in his mattress. He begins with \$500 from his mother and adds \$100 each year. His total savings y after x years are given by the equation

$$y = 500 + 100x$$

(a) Draw a graph of this equation. (*Hint:* Choose two values of x, such as 0 and 10. Find the corresponding values of y from the equation. Plot these two points on graph paper and draw the straight line joining them.)

(b) After 20 years, how much will Fred have in his mattress?

(c) If Fred adds \$200 instead of \$100 each year to his initial \$500, what is the equation that describes his savings after x years?

2.36 (Review of straight lines) Sound travels at a speed of 1500 meters per second in sea water. You dive into the sea from your yacht. Give an equation for the distance y at which a shark can hear your splash in terms of the number of seconds x since you hit the water.

2.37 (Review of straight lines) During the period after birth, a male white rat gains 40 grams (g) per week. (This rat is unusually regular in his growth, but 40 g per week is a realistic rate.)

(a) If the rat weighed 100 g at birth, give an equation for his weight after x weeks. What is the slope of this line?

(b) Draw a graph of this line between birth and 10 weeks of age.

(c) Would you be willing to use this line to predict the rat's weight at age 2 years? Do the prediction and think about the reasonableness of the result. (There are 454 g in a pound. To help you assess the result, note that a large cat weighs about 10 pounds.)

2.38 (Review of straight lines) A cellular telephone company offers two plans. Plan A charges \$20 a month for up to 75 minutes of air time and \$0.45 per minute above 75 minutes. Plan B charges \$30 a month for up to 250 minutes and \$0.40 per minute above 250 minutes.

(a) Draw a graph of the Plan A charge against minutes used from 0 to 250 minutes.

(b) How many minutes a month must the user talk in order for Plan B to be less expensive than Plan A?

2.39 Researchers studying acid rain measured the acidity of precipitation in an isolated wilderness area in Colorado for 150 consecutive weeks. The acidity of a solution is measured by pH, with lower pH values indicating that the solution is more acid. The acid rain researchers observed a linear pattern over time. They reported that the least-squares line

$$\text{pH} = 5.43 - (0.0053 \times \text{weeks})$$

fit the data well.[22]

(a) Draw a graph of this line. Note that the linear change is decreasing rather than increasing.

(b) According to the fitted line, what was the pH at the beginning of the study (weeks = 1)? At the end (weeks = 150)?

(c) What is the slope of the fitted line? Explain clearly what this slope says about the change in the pH of the precipitation in this wilderness area.

2.40 Keeping water supplies clean requires regular measurement of levels of pollutants. The measurements are indirect—a typical analysis involves forming a dye by a chemical reaction with the dissolved pollutant, then passing light through the solution and measuring its "absorbance." To calibrate such measurements, the laboratory measures known standard solutions and uses regression to relate absorbance to pollutant concentration. This is usually done every day. Here is one series of data on the absorbance for different levels of nitrates. Nitrates are measured in milligrams per liter of water.[23]

Nitrates	50	50	100	200	400	800	1200	1600	2000	2000
Absorbance	7.0	7.5	12.8	24.0	47.0	93.0	138.0	183.0	230.0	226.0

(a) Chemical theory says that these data should lie on a straight line. If the correlation is not at least 0.997, something went wrong and the calibration procedure is repeated. Plot the data and find the correlation. Must the calibration be done again?

(b) What is the equation of the least-squares line for predicting absorbance from concentration? If the lab analyzed a specimen with 500 milligrams of nitrates per liter, what do you expect the absorbance to be? Based on your plot and the correlation, do you expect your predicted absorbance to be very accurate?

2.41 Table 2.5 (page 123) gives data on the growth of icicles at two rates of water flow. You examined these data in Exercise 2.12. Use least-squares regression to estimate the rate (centimeters per minute) at which icicles grow at these two flow rates.

2.42 Your scatterplot in Exercise 2.8 (page 119) suggests that the severity of the mild foot deformity called MA can help predict the severity of the more serious deformity called HAV. Table 2.3 (page 120) gives data for 38 young patients.

(a) Find the equation of the least-squares regression line for predicting HAV angle from MA angle. Add this line to the scatterplot you made in Exercise 2.8.

(b) A new patient has MA angle 25 degrees. What do you predict this patient's HAV angle to be?

(c) Does knowing MA angle allow doctors to predict HAV angle accurately? Explain your answer from the scatterplot, then calculate a numerical measure to support your finding.

2.43 The scatterplot in Figure 2.8 (page 118) suggests that returns on common stocks may be somewhat lower in years with high interest rates. Here is part of the output from a spreadsheet program for the regression of stock returns on the Treasury bill returns for the same years:

```
Multiple R       -0.0975
R Square          0.0095

                 Coefficients     Standard Error   t Stat
Intercept        16.2352          4.9603           3.2731
Tbill return     -0.5727          0.8347          -0.6860
```

(a) What is the equation of the least-squares line? Use this line to predict the percent return on stocks in a year when Treasury bills return 5%.

(b) Explain what the slope of the regression line tells us. Does the slope confirm that high interest rates are in general bad for stocks?

(c) If you knew the return on Treasury bills for next year, do you think you could predict the return on stocks quite accurately? Use both the scatterplot in Figure 2.8 and the regression output to justify your answer.

2.44 How do the incomes of working-age people change with age? Because many older women have been out of the labor force for much of their lives, we look only at men between the ages of 25 and 65. Because education strongly influences income, we look only at men who have a bachelor's degree but no higher degree. A government sample survey tells us the age in years and income in dollars of a random sample of 5712 such men. (This is part of the INDIVIDUALS data set, described in the Data Appendix.) Figure 2.17 is a scatterplot of these data. Here is software output for regressing income on age. The line in the scatterplot is the least-squares regression line from this output.

	Coefficients	Standard Error	t Stat	P-value
Intercept	24874.3745	2637.4198	9.4313	5.75E-21
AGE	892.1135	61.7639	14.4439	1.79E-46

(a) The scatterplot in Figure 2.17 has a distinctive form. Why do the points fall into vertical stacks?

(b) Give some reasons why older men in this population might earn more than younger men. Give some reasons why younger men might earn more than older men. What do the data show about the relationship between age and income in the sample? Is the relationship very strong?

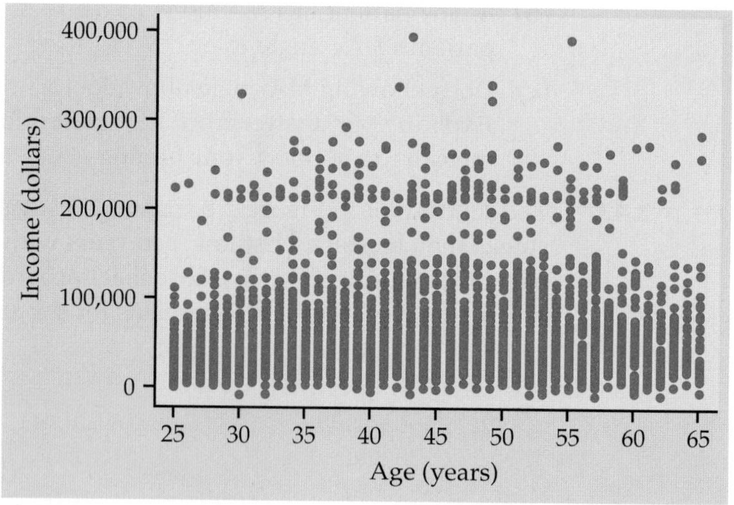

FIGURE 2.17 Scatterplot of income in dollars against age in years for 5712 men aged 25 to 65. All have a bachelor's degree as their highest level of education. See Exercise 2.44.

(c) What is the equation of the least-squares line for predicting income from age? What specifically does the slope of this line tell us?

2.45 Table 2.6 (page 124) shows the progress of world record times (in seconds) for the 10,000-meter run for both men and women. You made a scatterplot comparing the progress over time of men's and women's records in Exercise 2.15.

(a) The plot of running time in seconds against year is very linear for both men and women. Find the least-squares regression lines for predicting record time from year separately for the two sexes.

(b) Explain in simple language what the slopes of these two lines tell us about the progress of men and women in the 10,000-meter run.

(c) If the linear patterns continue, about when will the women's world record be the same as the men's record? (You can give a rough answer from the scatterplot, or you can use the fitted lines and do some algebra. Your result is a reminder that extrapolation is risky.)

2.46 Investors ask about the relationship between returns on investments in the United States and on investments overseas. Table 2.8 gives the total returns on U.S. and overseas common stocks over a 30-year period. (The total return is change in price plus any dividends paid, converted into U.S. dollars. Both returns are averages over many individual stocks.)[24]

(a) Make a scatterplot suitable for predicting overseas returns from U.S. returns.

(b) Find the correlation and r^2. Describe the relationship between U.S. and overseas returns in words, using r and r^2 to make your description more precise.

TABLE 2.8 Annual total return on overseas and U.S. stocks

Year	Overseas % return	U.S. % return	Year	Overseas % return	U.S. % return
1971	29.6	14.6	1986	69.4	18.6
1972	36.3	18.9	1987	24.6	5.1
1973	−14.9	−14.8	1988	28.5	16.8
1974	−23.2	−26.4	1989	10.6	31.5
1975	35.4	37.2	1990	−23.0	−3.1
1976	2.5	23.6	1991	12.8	30.4
1977	18.1	−7.4	1992	−12.1	7.6
1978	32.6	6.4	1993	32.9	10.1
1979	4.8	18.2	1994	6.2	1.3
1980	22.6	32.3	1995	11.2	37.6
1981	−2.3	−5.0	1996	6.4	23.0
1982	−1.9	21.5	1997	2.1	33.4
1983	23.7	22.4	1998	20.3	28.6
1984	7.4	6.1	1999	27.2	21.0
1985	56.2	31.6	2000	−14.0	−9.1

(c) Find the least-squares regression line of overseas returns on U.S. returns. Draw the line on the scatterplot. What are the predicted return $\hat{y}$ and the observed return y for 1993?

(d) Are you confident that predictions using the regression line will be quite accurate? Why?

2.47 Find the mean and standard deviation of the times and icicle lengths for the data on run 8903 in Table 2.5 (page 123). Find the correlation between the two variables. Use these five numbers to find the equation of the regression line for predicting length from time. Verify that your result agrees with that in Exercise 2.41. Use the same five numbers to find the equation of the regression line for predicting the time an icicle has been growing from its length. What units does the slope of each of these lines have?

2.48 Compute the mean and the standard deviation of the metabolic rates and mean body masses in Exercise 2.11 (page 122) and the correlation between these two variables. Use these values to find the slope of the regression line of metabolic rate on lean body mass. Also find the slope of the regression line of lean body mass on metabolic rate. What are the units for each of the two slopes?

2.49 The mean height of American women in their early twenties is about 64.5 inches and the standard deviation is about 2.5 inches. The mean height of men the same age is about 68.5 inches, with standard deviation about 2.7 inches. If the correlation between the heights of husbands and wives is about $r = 0.5$, what is the equation of the regression line of the husband's height on the wife's height in young couples? Draw a graph of this regression line. Predict the height of the husband of a woman who is 67 inches tall.

2.50 Use the equation for the least-squares regression line to show that this line always passes through the point $(\bar{x}, \bar{y})$. That is, set $x = \bar{x}$ and show that the line predicts that $y = \bar{y}$.

2.51 The data for run 8903 in Table 2.5 (page 123) describe how the length y in centimeters of an icicle increases over time x. Time is measured in minutes.

(a) What are the numerical values and units of measurement for each of $\bar{x}$, s_x, $\bar{y}$, s_y, and the correlation r between x and y?

(b) There are 2.54 centimeters in an inch. If we measure length y in inches rather than in centimeters, what are the new values of $\bar{y}$, s_y, and the correlation r?

(c) If we measure length y in inches rather than in centimeters, what is the new value of the slope b of the least-squares line for predicting length from time?

2.52 In Professor Friedman's economics course the correlation between the students' total scores before the final examination and their final examination scores is $r = 0.6$. The pre-exam totals for all students in the course have mean 280 and standard deviation 30. The final exam scores have mean 75 and standard deviation 8. Professor Friedman has lost Julie's final exam but knows that her total before the exam was 300. He decides to predict her final exam score from her pre-exam total.

(a) What is the slope of the least-squares regression line of final exam scores on pre-exam total scores in this course? What is the intercept?

(b) Use the regression line to predict Julie's final exam score.

(c) Julie doesn't think this method accurately predicts how well she did on the final exam. Calculate r^2 and use the value you get to argue that her actual score could have been much higher or much lower than the predicted value.

2.53 A study of class attendance and grades among first-year students at a state university showed that in general students who attended a higher percent of their classes earned higher grades. Class attendance explained 16% of the variation in grade index among the students. What is the numerical value of the correlation between percent of classes attended and grade index?

2.54 The number of people living on American farms has declined steadily during the past century. Here are data on the farm population (millions of persons) from 1935 to 1980:

Year	1935	1940	1945	1950	1955	1960	1965	1970	1975	1980
Population	32.1	30.5	24.4	23.0	19.1	15.6	12.4	9.7	8.9	7.2

(a) Make a scatterplot of these data and find the least-squares regression line of farm population on year.

(b) According to the regression line, how much did the farm population decline each year on the average during this period? What percent of the observed variation in farm population is accounted for by linear change over time?

(c) Use the regression equation to predict the number of people living on farms in 1990. Is this result reasonable? Why?

2.55 Literacy is an important contributor to the economic and social development of a country. The data set ex02_055.dat, found on the Web site `www.whfreeman.com/ips`, gives the percents of men and women at least 15 years old who were illiterate in 138 developing nations in the year 2000.[25] Use software to analyze these data.

(a) Make a scatterplot of female illiteracy versus male illiteracy. Because schooling for women often lags behind schooling for men, we take the percent of illiterate males as our explanatory variable. Describe the form, direction, and strength of the relationship, giving a numerical measure of the strength.

(b) Find the equation of the least-squares regression line for predicting female illiteracy from male illiteracy and draw this line on your plot.

(c) In which countries is female illiteracy farthest above the predicted level? In which countries is female illiteracy farthest below the predicted level? Which countries have the highest rates of illiteracy for both men and women?

2.56 Table 1.6 (page 33) reports data on 78 seventh-grade students. We want to know how well each of IQ score and self-concept score predicts GPA using least-squares regression. We also want to know which of these explanatory variables predicts GPA better. Give numerical measures that answer these questions, and explain your answers.

2.57 You have data on an explanatory variable x and a response variable y, and have found the least-squares regression line of y on x. Add one more data point, with x equal to the mean $\bar{x}$ for the existing points and y greater than the mean $\bar{y}$ for the existing points. Show that the new least-squares line is parallel to the existing line. (To do this, you must show that the slope does not change when you add the new point. Start with the fact that the slope is $b = rs_y/s_x$ and substitute the definition of the correlation r.)

2.4 Cautions about Regression and Correlation

Correlation and regression are among the most common statistical tools. They are used in more elaborate form to study relationships among many variables, a situation in which we cannot grasp the essentials by studying a single scatterplot. We need a firm grasp of the use and limitations of these tools, both now and as a foundation for more advanced statistics.

Residuals

A regression line is a mathematical model for the overall pattern of a linear relationship between an explanatory variable and a response variable. Deviations from the overall pattern are also important. In the regression setting, we see deviations by looking at the scatter of the data points about the regression line. The vertical distances from the points to the least- squares regression line are as small as possible in the sense that they have the smallest possible sum of squares. Because they represent "left-over" variation in the response after fitting the regression line, these distances are called *residuals*.

Residuals

A **residual** is the difference between an observed value of the response variable and the value predicted by the regression line. That is,

$$\text{residual} = \text{observed } y - \text{predicted } y$$
$$= y - \hat{y}$$

EXAMPLE 2.15

According to Table 2.7, the observed mean height of Kalama children at 24 months was 79.9 cm. That is, the observed y is 79.9 when $x = 24$. The least-squares regression line predicts the height to be

$$\hat{y} = 64.93 + (0.635 \times 24) = 80.17 \text{ cm}$$

The residual at 24 months is therefore

$$\begin{aligned} \text{residual} &= \text{observed } y - \text{predicted } y \\ &= y - \hat{y} \\ &= 79.9 - 80.17 = -0.27 \text{ cm} \end{aligned}$$

The 12 data points used in fitting the least-squares line produce 12 residuals. They are

−0.26	0.01	0.47	−0.06	−0.10	0.17
−0.27	0.30	−0.24	−0.27	0.09	0.16

Most regression software will calculate and store these residuals for you.

Because the residuals show how far the data fall from our regression line, examining the residuals helps assess how well the line describes the data. Although residuals can be calculated from any model fitted to the data, the residuals from the least-squares line have a special property: **the mean of the least-squares residuals is always zero.** Figure 2.18(a) is a scatterplot of the Kalama data with the regression line fitted. Compare this with the *residual plot* in Figure 2.18(b).

Residual Plots

A **residual plot** is a scatterplot of the regression residuals against the explanatory variable. Residual plots help us assess the fit of a regression line.

In Figure 2.18(b), we plot each residual from Example 2.15 against the corresponding value of the explanatory variable x. Because the mean of the residuals is always zero, the horizontal line at zero helps orient us. This line (residual = 0) corresponds to the fitted line in Figure 2.18(a). The residual plot magnifies the deviations from the line to make patterns easier to see. If the regression line catches the overall pattern of the data, there should be *no pattern* in the residuals. The residuals in Figure 2.18(b) have the irregular scatter that is typical of data that do not deviate from the model in any systematic way.

Figure 2.19 (page 157) shows the overall patterns of some typical residual plots in simplified form. The residuals are plotted in the vertical direction against the corresponding values of the explanatory variable x in the horizontal direction. If all is well, the pattern of a residual plot will be an unstructured horizontal band centered at zero (the mean of the residuals) and symmetric about zero, as in Figure 2.19(a). A curved pattern like the one in Figure 2.19(b) indicates that the relationship between y and x is curved rather than linear. A straight line is not a good description of such a relationship. A fan-shaped pattern like the one in Figure 2.19(c) shows that the variation of y about the line increases as x increases. Predictions of y will therefore be more precise for smaller values of x, where y shows less spread about the line.

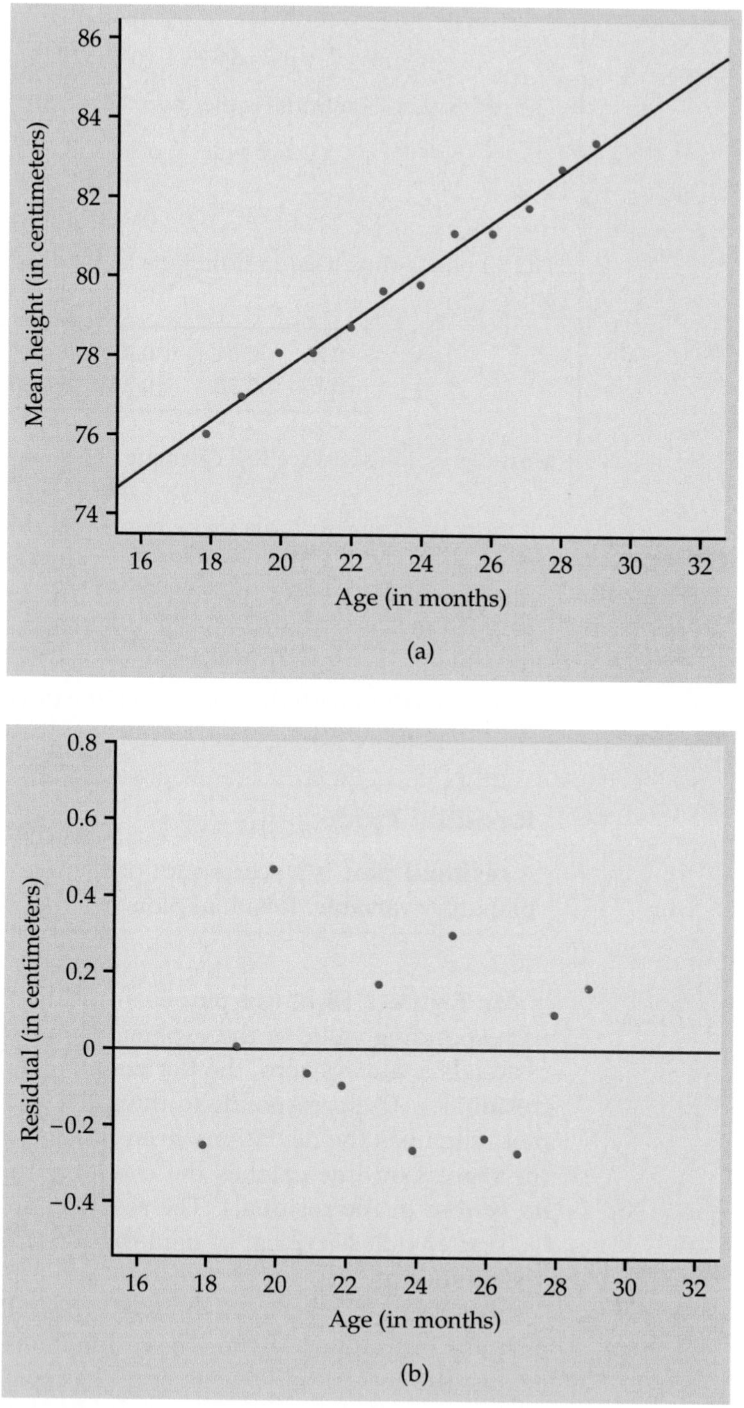

FIGURE 2.18 (a) The Kalama growth data with the least-squares line. (b) Plot of the residuals from the regression in (a) against the explanatory variable.

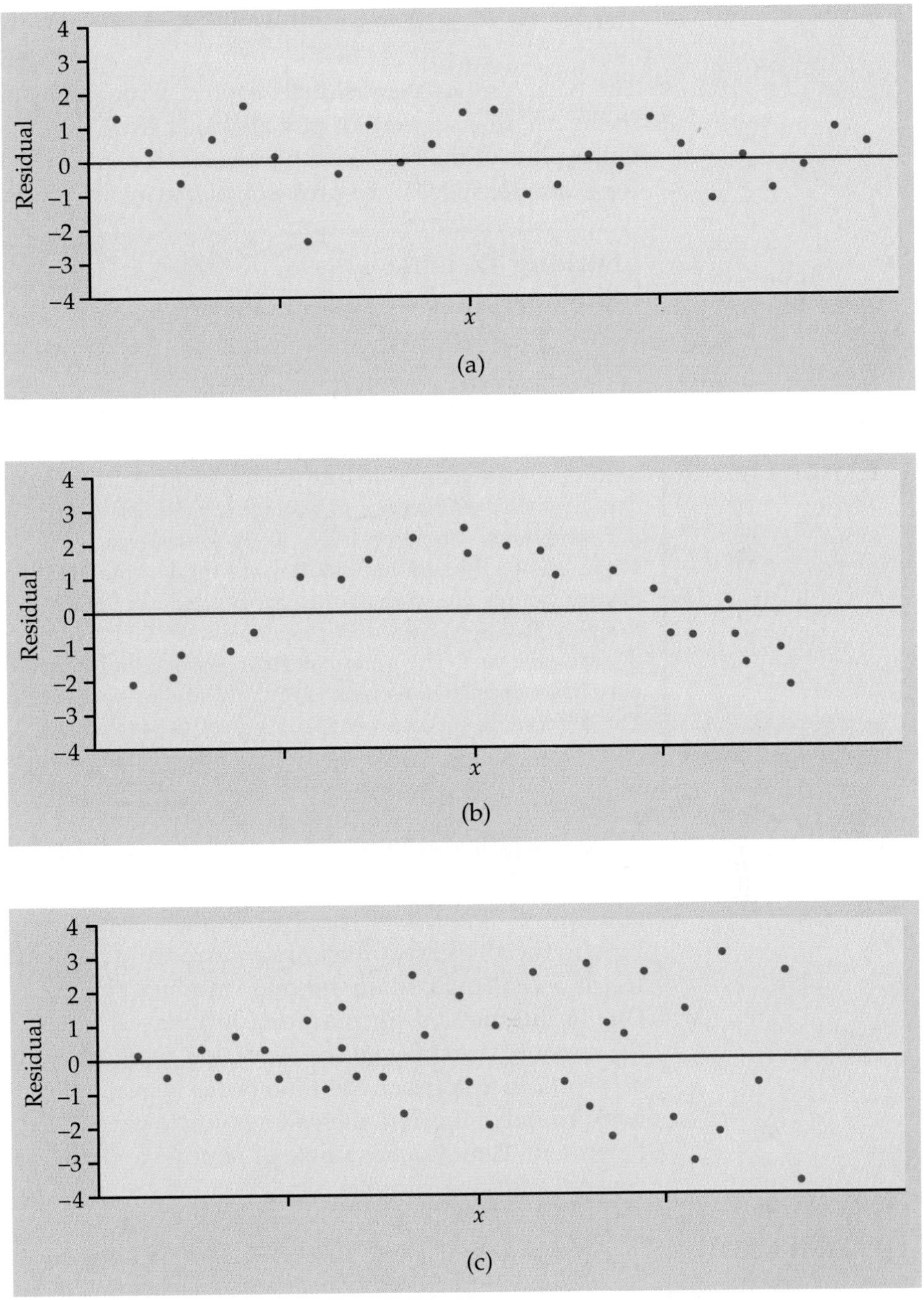

FIGURE 2.19 Simplified patterns in plots of least-squares residuals.

Lurking variables

The basic plot of the residuals against x magnifies patterns that we can see in the original scatterplot of y against x if we add the fitted line to that plot. Other plots of the residuals, however, can reveal new information. They may, for example, suggest the presence of *lurking variables*.

> **Lurking Variable**
>
> A **lurking variable** is a variable that is not among the explanatory or response variables in a study and yet may influence the interpretation of relationships among those variables.

EXAMPLE 2.16

Studies of the relationship between treatment of heart disease and the patients' gender show that women are in general treated less aggressively than men with similar symptoms. Women are less likely to undergo bypass operations, for example. Might this be discrimination? Beware the lurking variable: although half of heart disease victims are women, they are on the average much older than male victims. Roughly 36,000 women under age 65 and 85,000 men under age 65 die from heart disease each year. The American Heart Association says, "Risk of death due to coronary heart disease in women is roughly similar to that of men 10 years younger."[26] The differences in treatment may reflect the fact that women patients are on the average older and more likely to have other health problems that make aggressive treatment risky.

Never forget that the relationship between two variables can be strongly influenced by other variables that are lurking in the background. Lurking variables can dramatically change the conclusions of a regression study. Because lurking variables are often unrecognized and unmeasured, detecting their effect is a challenge. Many lurking variables change systematically over time. One useful method for detecting lurking variables is therefore to *plot both the response variable and the regression residuals against the time order of the observations* whenever the time order is available. An understanding of the background of the data then allows you to guess what lurking variables might be present. Here is an example of plotting and interpreting residuals that uncovered a lurking variable.

EXAMPLE 2.17

The mathematics department of a large state university must plan the number of sections and instructors required for its elementary courses. The department hopes that the number of students in these courses can be predicted from the number of first-year students, which is known before the new students actually choose courses. The table below contains data for several years.[27] The explanatory variable x is the number of first-year students. The response variable y is the number of students who enroll in elementary mathematics courses.

Year	1993	1994	1995	1996	1997	1998	1999	2000
x	4595	4827	4427	4258	3995	4330	4265	4351
y	7364	7547	7099	6894	6572	7156	7232	7450

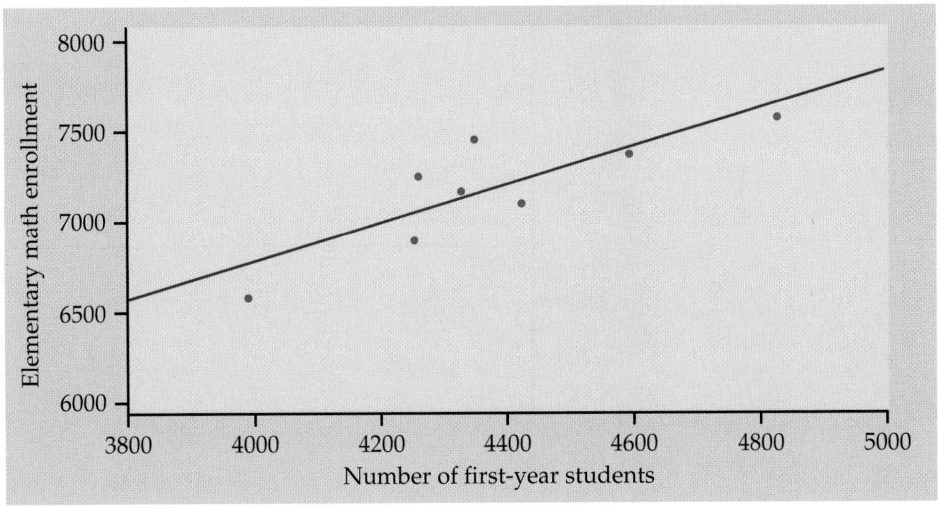

FIGURE 2.20 The regression of elementary mathematics enrollment on number of first-year students at a large university.

A scatterplot (Figure 2.20) shows a reasonably linear pattern with a cluster of points near the center. We use regression software to obtain the equation of the least-squares regression line:

$$\hat{y} = 2492.69 + 1.0663x$$

The software also tells us that $r^2 = 0.694$. That is, linear dependence on x explains about 70% of the variation in y. The line appears to fit reasonably well.

A plot of the residuals against x (Figure 2.21) magnifies the vertical deviations of the points from the line. We can see that a somewhat different line

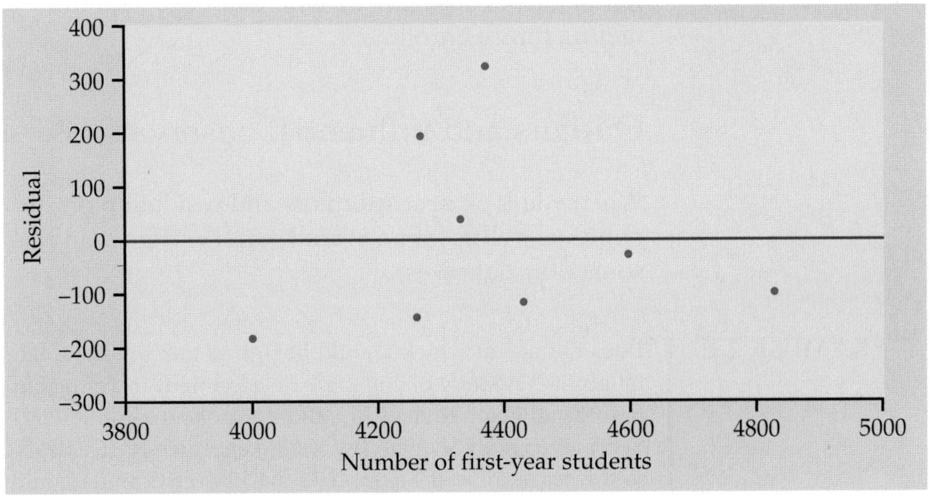

FIGURE 2.21 Residual plot for the regression of mathematics enrollment on number of first-year students.

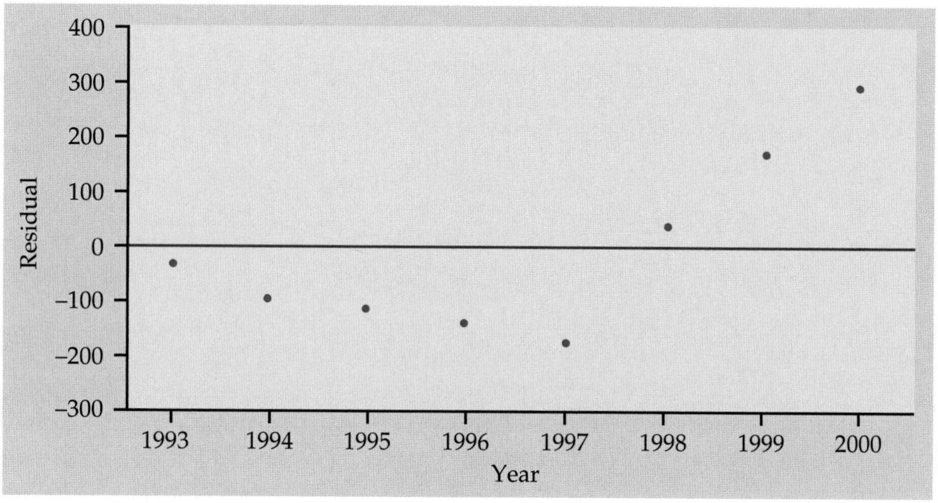

FIGURE 2.22 Plot of the residuals against year for the regression of mathematics enrollment on number of first-year students.

would fit the five lower points well. The three points above the line represent a different relation between the number of first-year students x and mathematics enrollments y.

A second plot of the residuals clarifies the situation. Figure 2.22 is a plot of the residuals against year. We now see that the five negative residuals are from the years 1993 to 1997, and the three positive residuals represent the years 1998 to 2000. This plot suggests that a change took place between 1997 and 1998 that caused a higher proportion of students to take mathematics courses beginning in 1998. In fact, one of the schools in the university changed its program to require that entering students take another mathematics course. This change is the lurking variable that explains the pattern we observed. The mathematics department should not use data from years before 1998 for predicting future enrollment.

Outliers and influential observations

When you look at scatterplots and residual plots, look for striking individual points as well as for an overall pattern. Here is an example of data that contain some unusual cases.

EXAMPLE 2.18

Does the age at which a child begins to talk predict a later score on a test of mental ability? A study of cognitive development in young children recorded the age in months at which each of 21 children spoke their first word and their Gesell Adaptive Score, the result of an ability test taken much later. The data appear in Table 2.9 and in the scatterplot of Figure 2.23.[28] Children 3 and 13, and also Children 16 and 21, have identical values of both variables. We use a different plotting symbol to show that one point stands for two individuals.

TABLE 2.9 Age (in months) at first word and Gesell score

Child	Age	Score	Child	Age	Score
1	15	95	12	9	96
2	26	71	13	10	83
3	10	83	14	11	84
4	9	91	15	11	102
5	15	102	16	10	100
6	20	87	17	12	105
7	18	93	18	42	57
8	11	100	19	17	121
9	8	104	20	11	86
10	20	94	21	10	100
11	7	113			

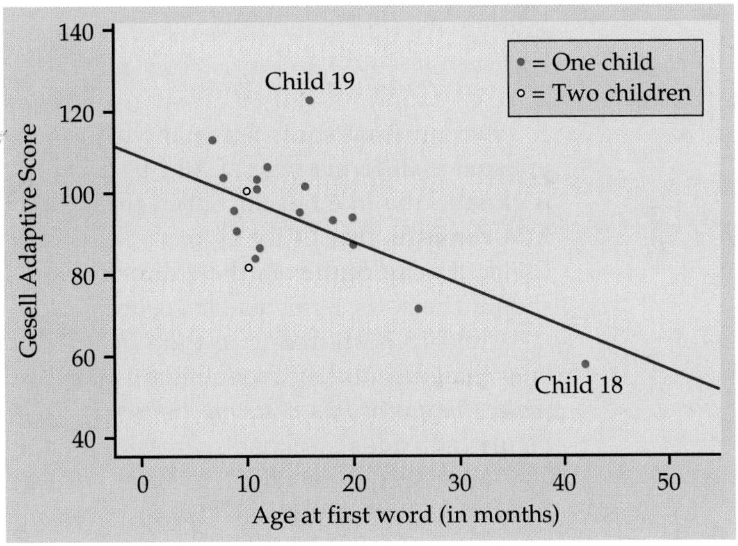

FIGURE 2.23 Regression of Gesell Adaptive Score on age at first word for 21 children, from Table 2.9.

The plot shows a negative linear relationship. That is, children who begin to speak later tend to have lower test scores than early talkers. The line on the plot is the least-squares regression line for predicting Gesell score from age at first word. Its equation is

$$\hat{y} = 109.87 - 1.127x$$

The relationship is not strong, but the proportion of variation in Gesell score explained by regression on age at first word, $r^2 = 0.41$, is quite respectable for a study of this kind. We will see that this result is misleading.

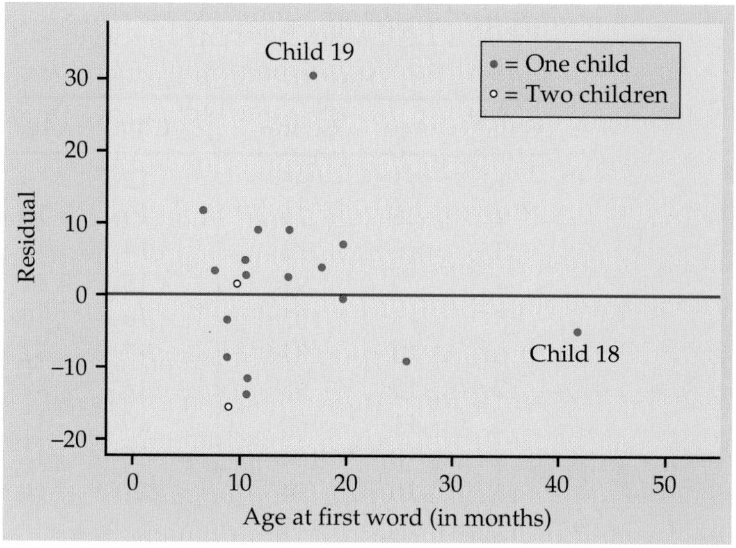

FIGURE 2.24 Residual plot for the regression of Gesell score on age at first word. Child 19 is an outlier in y. Child 18 is an influential case that does not have a large residual.

Two unusual cases are marked in Figure 2.23. Children 18 and 19 are unusual in different ways. Child 19 lies far from the regression line. Child 18 is close to the line but far out in the x direction. The residual plot in Figure 2.24 confirms that Child 19 has a large residual and that Child 18 does not. Child 19 is an outlier in the y direction, with a Gesell score so high that we should check for a mistake in recording it. In fact, the score is correct.

Child 18 is an outlier in the x direction. This child began to speak much later than any of the other children. *Because of its extreme position on the age scale, this point has a strong influence on the position of the regression line.* Figure 2.25 adds a second regression line, calculated after leaving out Child 18. You can see that this one point moves the line quite a bit. Least-squares lines make the sum of squares of the vertical distances to the points as small as possible. A point that is extreme in the x direction with no other points near it pulls the line toward itself. We call such points *influential.*

Outliers and Influential Observations in Regression

An **outlier** is an observation that lies outside the overall pattern of the other observations. Points that are outliers in the y direction of a scatterplot have large regression residuals, but other outliers need not have large residuals.

An observation is **influential** for a statistical calculation if removing it would markedly change the result of the calculation. Points that are outliers in the x direction of a scatterplot are often influential for the least-squares regression line.

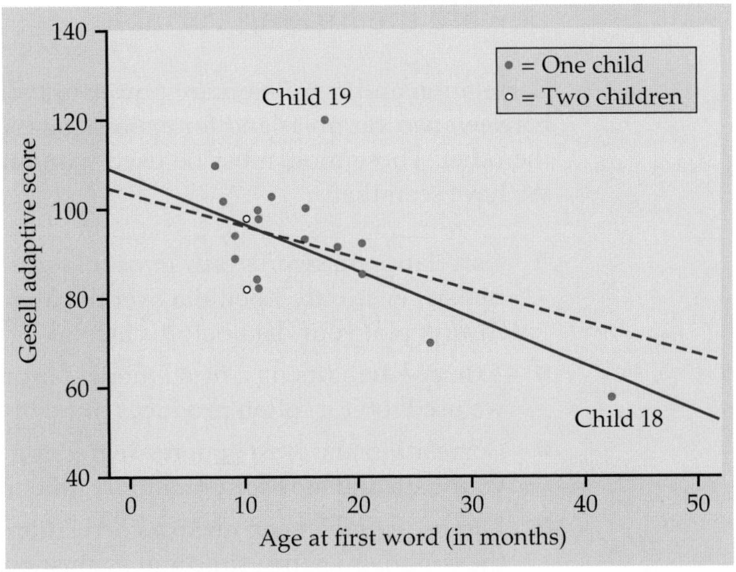

FIGURE 2.25 Least-squares regression of Gesell score on age at first word, with and without an influential observation. The dashed line was calculated leaving out Child 18.

We did not need the distinction between outliers and influential observations in Chapter 1. A single large salary that pulls up the mean salary $\overline{x}$ for a group of workers is an outlier because it lies far above the other salaries. It is also influential, because the mean changes when it is removed. In the regression setting, however, not all outliers are influential. The least-squares regression line is most likely to be heavily influenced by observations that are outliers in the x direction. Because influential observations draw the regression line toward themselves, we cannot spot them by looking for large residuals. The basic scatterplot of y against x will alert you to observations that are extreme in x and may therefore be influential. The surest way to verify that a point is influential is to find the regression line both with and without the suspect point, as in Figure 2.25. If the line moves more than a small amount when the point is deleted, the point is influential. The *Correlation and Regression* applet available at `www.whfreeman.com/ips` allows you to move points and watch how they influence the least-squares line.

applet

EXAMPLE 2.19

The strong influence of Child 18 makes the original regression of Gesell score on age at first word misleading. The original data have $r^2 = 0.41$. That is, the age at which a child begins to talk explains 41% of the variation on a later test of mental ability. This relationship is strong enough to be interesting to parents. If we leave out Child 18, r^2 drops to only 11%. The apparent strength of the association was largely due to a single influential observation.

What should the child development researcher do? She must decide whether Child 18 is so slow to speak that this individual should not be allowed to influence the analysis. If she excludes Child 18, much of the evidence for a connection between the age at which a child begins to talk and later ability score vanishes. If she keeps Child 18, she needs data on other children who were also slow to begin talking, so that the conclusion no longer depends so heavily on just one child.

Beware the lurking variable

Correlation and regression are powerful tools for measuring the association between two variables and for expressing the dependence of one variable on the other. These tools must be used with an awareness of their limitations. We have seen that:

- Correlation measures *only linear association*, and fitting a straight line makes sense only when the overall pattern of the relationship is linear. Always plot your data before calculating.
- *Extrapolation* (using a fitted model far outside the range of the data that we used to fit it) often produces unreliable predictions.
- Correlation and least-squares regression are *not resistant*. Always plot your data and look for potentially influential points.
- *Lurking variables* can make a correlation or regression misleading. Plot the residuals against time and against other variables that may influence the relationship between x and y.

Here are two more examples that illustrate how lurking variables can influence the relationship between an explanatory variable and a response variable.

EXAMPLE 2.20

Measure the number of television sets per person x and the average life expectancy y for the world's nations. There is a high positive correlation—nations with many TV sets have higher life expectancies. Could we lengthen the lives of people in Rwanda by shipping them TV sets? No. Rich nations have more TV sets than poor nations. Rich nations also have longer life expectancies because they offer better nutrition, clean water, and better health care. There is no cause-and-effect tie between TV sets and length of life.

Correlations such as that in Example 2.20 are sometimes called "nonsense correlations." The correlation is real. What is nonsense is the conclusion that changing one of the variables *causes* changes in the other. The question of causation is important enough to merit separate treatment in Section 2.5. For now, just remember that an association between two variables x and y can reflect many types of relationship among x, y, and one or more lurking variables.

Association Does Not Imply Causation

An association between an explanatory variable x and a response variable y, even if it is very strong, is not by itself good evidence that changes in x actually cause changes in y.

Lurking variables sometimes create "nonsense correlations" between x and y, as in Example 2.20. They can also hide a true relationship between x and y, as the following example illustrates.

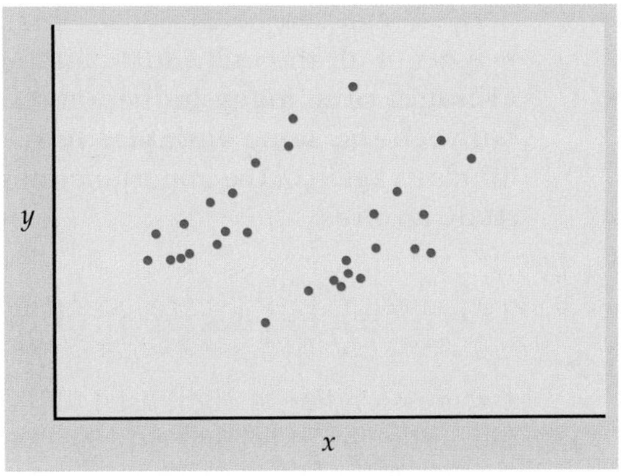

FIGURE 2.26 This scatterplot has a low r even though there is a strong correlation within each of the two clusters.

EXAMPLE 2.21

A study of housing conditions and health in the city of Hull, England, measured a large number of variables for each of the wards into which the city is divided. Two of the variables were an index x of overcrowding and an index y of the lack of indoor toilets. Because x and y are both measures of inadequate housing, we expect a high correlation. Yet the correlation was only $r = 0.08$. How can this be? Investigation disclosed that some poor wards were dominated by public housing. These wards had high values of x but low values of y because public housing always includes indoor toilets. Other poor wards lacked public housing, and in these wards high values of x were accompanied by high values of y. Because the relationship between x and y differed in the two types of wards, analyzing all wards together obscured the nature of the relationship.[29]

Figure 2.26 shows in simplified form how groups formed by a categorical lurking variable, as in the housing example, can make the correlation r misleading. The groups appear as clusters of points in the scatterplot. There is a strong relationship between x and y within each of the clusters. In fact, $r = 0.85$ and $r = 0.91$ in the two clusters. However, because similar values of x correspond to quite different values of y in the two clusters, x alone is of little value for predicting y. The correlation for all points displayed is therefore low: $r = 0.14$. This example is another reminder to plot the data rather than simply calculate numerical measures such as the correlation.

Beware correlations based on averaged data

Many regression or correlation studies work with averages or other measures that combine information from many individuals. We saw, for example, a very strong linear relationship ($r = 0.994$) between age and the mean height of children in the Egyptian village of Kalama.

Averaging over many children smooths out the variation in height among children the same age, resulting in a very high correlation between mean height and age. Because children of each age vary in height, the correlation between age and height would be much smaller if we used the data on the heights of all the individual children at each age. **A correlation based on averages over many individuals is usually higher than the correlation between the same variables based on data for individuals.** This fact reminds us again of the importance of noting exactly what variables a statistical study involves.

The restricted-range problem

A regression line is often used to *predict* the response y to a given value x of the explanatory variable. Successful prediction does not require a cause-and-effect relationship. If both x and y respond to the same underlying unmeasured variables, x may help us predict y even though x has no direct influence on y. For example, the scores of SAT exams taken in high school help predict college grades. There is no cause-and-effect tie between SAT scores and college grades. Rather, both reflect a student's ability and knowledge.

How well do SAT scores, perhaps with the help of high school grades, predict college GPA? We can use the correlation r and its square to get a rough answer. There is, however, a subtle difficulty.

EXAMPLE 2.22

Combining several studies for students graduating from college since 1980, the College Board reports these correlations between explanatory variables and the overall GPA of college students:

SAT math and verbal	High school grades	SAT plus grades
$r = 0.36$	$r = 0.42$	$r = 0.52$

Because $0.52^2 = 0.27$, we see that SAT scores plus students' high school records explain about 27% of the variation in GPA among college students.

The subtle problem? Colleges differ greatly in the range of students they attract. Almost all students at Princeton have high SAT scores and did well in high school. At Generic State College, most students are in the middle range of SAT scores and high school performance. Both sets of students receive the full spread of grades. We suspect that if Princeton admitted weaker students they would get lower grades, and that the typical Princeton student would get very high grades at Generic State. This is the *restricted-range problem:* the data do not contain information on the full range of both explanatory and response variables. **When data suffer from restricted range, r and r^2 are lower than they would be if the full range could be observed.**

Thus $r = 0.52$ understates the actual ability of SAT scores and high school grades to predict college GPA. One investigator found 21 colleges that enrolled the full range of high school graduates. Sure enough, at these colleges, $r = 0.65$.[30]

Did you notice that two of the correlations in Example 2.22 involve more than one explanatory variable? It is common to use several explanatory variables together to predict a response. This is called *multiple regression*. Each

r in the example is a *multiple correlation coefficient*, whose square is the proportion of variation in the response explained by the multiple regression. Chapter 11 introduces multiple regression.

Beyond the basics ▸ data mining

Chapters 1 and 2 of this book are devoted to the important aspect of statistics called *exploratory data analysis* (EDA). We use graphs and numerical summaries to examine data, searching for patterns and paying attention to striking deviations from the patterns we find. In discussing regression, we advanced to using the pattern we find (in this case, a linear pattern) for prediction.

Suppose now that we have a truly enormous data base, such as all purchases recorded by the cash register scanners of a national retail chain during the past week. Surely this trove of data contains patterns that might guide business decisions. If we could see clearly the types of activewear preferred in large California cities and compare the preferences of small Midwest cities—right now, not at the end of the season—we might improve profits in both parts of the country by matching stock with demand. This sounds much like EDA, and indeed it is. There is, however, a saying in computer science that a big enough difference of scale amounts to a difference of kind. Exploring really large data bases in the hope of finding useful patterns is called **data mining.** Here are some distinctive features of data mining:

data mining

- When you have 100 gigabytes of data, even straightforward calculations and graphics become impossibly time-consuming. So efficient algorithms are very important.
- The structure of the data base and the process of storing the data (the fashionable term is *data warehousing*), perhaps by unifying data scattered across many departments of a large corporation, require careful consideration.
- Data mining requires automated tools that work based on only vague queries by the user. The process is too complex to do step-by-step as we have done in EDA.

All of these features point to the need for sophisticated computer science as a basis for data mining. Indeed, data mining is often thought of as a part of computer science. Yet many statistical ideas and tools—mostly tools for dealing with multidimensional data, not the sort of thing that appears in a first statistics course—are very helpful. Like many modern developments, data mining crosses the boundaries of traditional fields of study.

Do remember that the perils we associate with blind use of correlation and regression are yet more perilous in data mining, where the fog of an immense data base prevents clear vision. Extrapolation, ignoring lurking variables, and confusing association with causation are traps for the unwary data miner.

SUMMARY

You can examine the fit of a regression line by plotting the **residuals,** which are the differences between the observed and predicted values of y. Be on the lookout for points with unusually large residuals and also for nonlinear patterns and uneven variation about the line.

Evidence of the effects of **lurking variables** on the relationship between x and y may be provided by plots of y and the residuals against the time order of the observations.

Also look for **influential observations,** individual points that substantially change the regression line. Influential observations are often outliers in the x direction, but they need not have large residuals.

Correlation and regression must be **interpreted with caution.** Plot the data to be sure that the relationship is roughly linear and to detect outliers and influential observations.

Lurking variables may explain the relationship between the explanatory and response variables. Correlation and regression can be misleading if you ignore important lurking variables.

We cannot conclude that there is a cause-and-effect relationship between two variables just because they are strongly associated. **High correlation does not imply causation.**

A correlation based on averages is usually higher than if we used data for individuals. A correlation based on data with a **restricted range** is often lower than would be the case if we could observe the full range of the variables.

SECTION 2.4 EXERCISES

2.58 Exercise 2.10 (page 122) gives data on the fuel consumption y of a car at various speeds x. Fuel consumption is measured in liters of gasoline per 100 kilometers driven, and speed is measured in kilometers per hour. Software tells us that the equation of the least-squares regression line is

$$\hat{y} = 11.058 - 0.01466x$$

The residuals, in the same order as the observations, are

10.09	2.24	−0.62	−2.47	−3.33	−4.28	−3.73	−2.94
−2.17	−1.32	−0.42	0.57	1.64	2.76	3.97	

(a) Make a scatterplot of the observations and draw the regression line on your plot.

(b) Would you use the regression line to predict y from x? Explain your answer.

(c) Check that the residuals have sum zero (up to roundoff error).

(d) Make a plot of the residuals against the values of x. Draw a horizontal line at height zero on your plot. Notice that the residuals show the same pattern about this line as the data points show about the regression line in the scatterplot in (a).

2.59 A study of nutrition in developing countries collected data from the Egyptian village of Nahya. Here are the mean weights (in kilograms) for 170 infants in Nahya who were weighed each month during their first year of life:[31]

Age (months)	1	2	3	4	5	6	7	8	9	10	11	12
Weight (kg)	4.3	5.1	5.7	6.3	6.8	7.1	7.2	7.2	7.2	7.2	7.5	7.8

(a) Plot weight against time.

(b) A hasty user of statistics enters the data into software and computes the least-squares line without plotting the data. The result is

```
The regression equation is
weight = 4.88 + 0.267 age
```

Plot this line on your graph. Is it an acceptable summary of the overall pattern of growth? Remember that you can calculate the least-squares line for *any* set of two-variable data. It's up to you to decide if it makes sense to fit a line.

(c) Fortunately, the software also prints out the residuals from the least-squares line. In order of age along the rows, they are

−0.85	−0.31	0.02	0.35	0.58	0.62
0.45	0.18	−0.08	−0.35	−0.32	−0.28

Verify that the residuals have sum 0 (except for roundoff error). Plot the residuals against age and add a horizontal line at 0. Describe carefully the pattern that you see.

2.60 Table 1.9 (page 59) gives the calories and sodium content for each of 17 brands of meat hot dogs. "Eat Slim Veal Hot Dogs," with just 107 calories, is a low outlier in the distribution of calories.

(a) Make a scatterplot of sodium content y against calories x. Describe the main features of the relationship. Is "Eat Slim Veal Hot Dogs" an outlier in this plot? Circle its point.

(b) Calculate two least-squares regression lines, one using all of the observations and the other omitting "Eat Slim." Draw both lines on your plot. Does a comparison of the two regression lines show that the "Eat Slim" brand is influential? Explain your answer.

(c) A new brand of meat hot dog (not made with veal) has 150 calories per frank. How many milligrams of sodium do you estimate that one of these hot dogs contains?

2.61 Here are the golf scores of 11 members of a women's golf team in two rounds of tournament play:

Player	1	2	3	4	5	6	7	8	9	10	11
Round 1	89	90	87	95	86	81	105	83	88	91	79
Round 2	94	85	89	89	81	76	89	87	91	88	80

(a) Plot the data with the Round 1 scores on the x axis and the Round 2 scores on the y axis. There is a generally linear pattern except for one potentially influential observation. Circle this observation on your graph.

(b) Here are the equations of two least-squares lines. One of them is calculated from all 11 data points and the other omits the influential observation.

$$\hat{y} = 20.49 + 0.754x$$
$$\hat{y} = 50.01 + 0.410x$$

Draw both lines on your scatterplot. Which line omits the influential observation? How do you know this?

2.62 Gas chromatography is a technique used to detect very small amounts of a substance, for example, a contaminant in drinking water. Laboratories use regression to calibrate such techniques. The data below show the results of five measurements for each of four amounts of the substance being investigated.[32] The explanatory variable x is the amount of substance in the specimen, measured in nanograms (ng), units of 10^{-9} gram. The response variable y is the reading from the gas chromatograph.

Amount (ng)	Response				
0.25	6.55	7.98	6.54	6.37	7.96
1.00	29.7	30.0	30.1	29.5	29.1
5.00	211	204	212	213	205
20.00	929	905	922	928	919

(a) Make a scatterplot of these data. The relationship appears to be approximately linear, but the wide variation in the response values makes it hard to see detail in this graph.

(b) Compute the least-squares regression line of y on x, and plot this line on your graph.

(c) Now compute the residuals and make a plot of the residuals against x. It is much easier to see deviations from linearity in the residual plot. Describe carefully the pattern displayed by the residuals.

2.63 Table 2.4 (page 121) gives information on states' performance on the National Assessment of Educational Progress (NAEP) year 2000 Mathematics Assessment. We saw in Exercise 2.9 that the two measures of performance are closely related. (In fact, the correlation is about $r = 0.95$.) Table 2.4 also gives the percent of children aged 5 to 17 years in each state who lived in households with incomes below the federal poverty level in 1998. We expect that poverty among children will be related to NAEP performance.

(a) Make a scatterplot suitable for predicting mean NAEP score from poverty percent. Describe the relationship, using correlation as a measure to complement your verbal description.

(b) Use software to find the least-squares regression line for predicting mean NAEP score from poverty rate, and the residuals from this line. We might call states with large positive residuals overachievers, because their fourth graders do better than the state poverty rate would lead us to guess. Similarly, states with large negative residuals might be called underachievers. What are the three states with the largest positive residuals and the three states with the largest negative residuals?

2.64 We might expect states with more poverty to have fewer doctors. Table 2.10 gives data on the percent of each state's residents living below the poverty line and on the number of M.D.'s per 100,000 residents in each state.[33]

(a) Make a scatterplot and calculate a regression line suitable for predicting M.D.'s per 100,000 from poverty rate. Draw the line on your plot. Surprise: the slope is positive, so poverty and M.D.'s go up together.

(b) The District of Columbia is an outlier, with both very many M.D.'s and a high poverty rate. (D.C. is a city rather than a state.) Circle the point for D.C. on your plot and explain why this point may strongly influence the least-squares line.

(c) Calculate the regression line for the 50 states, omitting D.C. Add the new line to your scatterplot. Was this point highly influential? Does the number of doctors now go down with increasing poverty, as we initially expected?

2.65 Runners are concerned about their form when racing. One measure of form is the stride rate, the number of steps taken per second. As running speed increases, the stride rate should also increase. In a study of 21 of the best American female runners, researchers measured the stride rate for different

TABLE 2.10 Poverty and M.D.'s in the states

State	Poverty percent	M.D.'s per 100,000	State	Poverty percent	M.D.'s per 100,000
Alabama	15.1	198	Montana	15.9	190
Alaska	8.6	167	Nebraska	11.0	218
Arizona	15.2	202	Nevada	11.0	173
Arkansas	16.4	190	New Hampshire	8.9	237
California	15.3	247	New Jersey	8.5	295
Colorado	8.6	238	New Mexico	20.8	212
Connecticut	8.4	354	New York	15.7	387
Delaware	10.1	234	North Carolina	13.0	232
D.C.	19.7	737	North Dakota	13.9	222
Florida	13.3	238	Ohio	11.4	235
Georgia	13.7	211	Oklahoma	13.5	169
Hawaii	11.9	265	Oregon	13.1	225
Idaho	13.9	154	Pennsylvania	10.6	291
Illinois	10.4	260	Rhode Island	11.4	338
Indiana	8.3	195	South Carolina	12.8	207
Iowa	8.7	173	South Dakota	11.7	184
Kansas	10.5	203	Tennessee	13.2	246
Kentucky	13.8	209	Texas	15.6	203
Louisiana	18.2	246	Utah	7.9	200
Maine	10.4	223	Vermont	9.6	305
Maryland	7.6	374	Virginia	9.8	241
Massachusetts	10.9	412	Washington	9.2	235
Michigan	10.3	224	West Virginia	16.7	215
Minnesota	9.1	249	Wisconsin	8.5	227
Mississippi	16.8	163	Wyoming	11.9	171
Missouri	11.1	230			

speeds. The following table gives the speeds (in feet per second) and the mean stride rates for these runners:[34]

Speed	15.86	16.88	17.50	18.62	19.97	21.06	22.11
Stride rate	3.05	3.12	3.17	3.25	3.36	3.46	3.55

(a) Plot the data with speed on the x axis and stride rate on the y axis. Does a straight line adequately describe these data?

(b) Find the equation of the regression line of stride rate on speed. Draw this line on your plot.

(c) For each of the speeds given, obtain the predicted value and the residual. Verify that the residuals add to 0.

(d) Plot the residuals against speed. Describe the pattern. What does the plot indicate about the adequacy of the linear fit? Are there any potentially influential observations? Can you plot the residuals against the time at which the observations were made?

2.66 Exercise 2.46 (page 151) examined the relationship between returns on U.S. and overseas stocks. Return to the scatterplot and regression line for predicting overseas returns from U.S. returns.

(a) Circle the point that has the largest residual (either positive or negative). What year is this? Redo the regression without this point and add the new regression line to your plot. Was this observation very influential?

(b) Whenever we regress two variables that both change over time, we should plot the residuals against time as a check for time-related lurking variables. Make this plot for the stock returns data. Are there any suspicious patterns in the residuals?

applet

2.67 The *Correlation and Regression* applet at `www.whfreeman.com/ips` allows you to create a scatterplot and to move points by dragging with the mouse. Create a group of 10 points in the lower-left corner of the scatterplot with a strong straight-line pattern (correlation at least 0.9). Click the "Show least-squares line" box to display the regression line.

(a) Add one point at the upper right that is far from the other 10 points but exactly on the regression line. Why does this outlier have no effect on the line even though it changes the correlation?

(b) Now drag this last point down until it is opposite the group of 10 points. You see that one end of the least-squares line chases this single point, while the other end remains near the middle of the original group of 10. What makes the last point so influential?

2.68 Table 2.11 presents four sets of data prepared by the statistician Frank Anscombe to illustrate the dangers of calculating without first plotting the data.[35]

(a) Without making scatterplots, find the correlation and the least-squares regression line for all four data sets. What do you notice? Use the regression line to predict y for $x = 10$.

(b) Make a scatterplot for each of the data sets and add the regression line to each plot.

(c) In which of the four cases would you be willing to use the regression line to describe the dependence of y on x? Explain your answer in each case.

2.69 If you need medical care, should you go to a hospital that handles many cases like yours? Figure 2.27 presents some data for heart attacks. The figure plots mortality rate (the proportion of patients who died) against the number of heart attack patients treated for a large number of hospitals in

TABLE 2.11 Four data sets for exploring correlation and regression

Data Set A

x	10	8	13	9	11	14	6	4	12	7	5
y	8.04	6.95	7.58	8.81	8.33	9.96	7.24	4.26	10.84	4.82	5.68

Data Set B

x	10	8	13	9	11	14	6	4	12	7	5
y	9.14	8.14	8.74	8.77	9.26	8.10	6.13	3.10	9.13	7.26	4.74

Data Set C

x	10	8	13	9	11	14	6	4	12	7	5
y	7.46	6.77	12.74	7.11	7.81	8.84	6.08	5.39	8.15	6.42	5.73

Data Set D

x	8	8	8	8	8	8	8	8	8	8	19
y	6.58	5.76	7.71	8.84	8.47	7.04	5.25	5.56	7.91	6.89	12.50

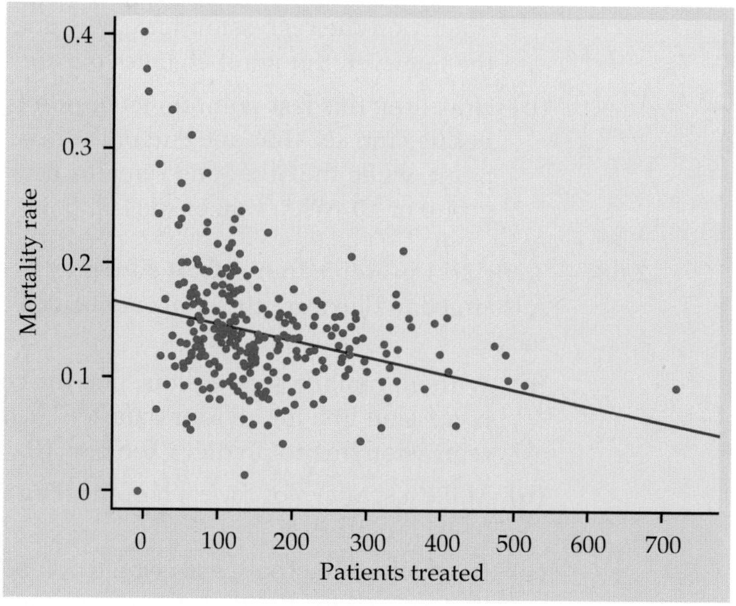

FIGURE 2.27 Mortality of heart attack patients and number of heart attack cases treated for a large group of hospitals, for Exercise 2.69.

a recent year. The line on the plot is the least-squares regression line for predicting mortality from number of patients.

(a) Do the plot and regression generally support the thesis that mortality is lower at hospitals that treat more heart attacks? Is the relationship very strong?

(b) In what way is the pattern of the plot nonlinear? Does the nonlinearity strengthen or weaken the conclusion that heart attack patients should avoid hospitals that treat few heart attacks? Why?

2.70 Exercise 2.46 (page 151) examined the relationship between returns on U.S. and overseas stocks. Investors also want to know what typical returns are and how much year-to-year variability (called *volatility* in finance) there is. Regression and correlation don't answer questions about center and spread.

(a) Find the five-number summaries for both U.S. and overseas returns, and make side-by-side boxplots to compare the two distributions.

(b) Were returns generally higher in the United States or overseas during this period? Explain your answer.

(c) Were returns more volatile (more variable) in the United States or overseas during this period? Explain your answer.

2.71 Your company gives a test of cognitive ability to job applicants before deciding whom to hire. Your boss has asked you to use company records to see if this test really helps predict the performance ratings of employees. Explain carefully to your boss why the restricted-range problem may make it difficult to see a strong relationship between test scores and performance ratings.

2.72 The Standard & Poor's 500-stock index is an average of the price of 500 stocks. There is a moderately strong correlation (roughly $r = 0.6$) between how much this index changes in January and how much it changes during the entire year. If we looked instead at data on all 500 individual stocks, we would find a quite different correlation. Would the correlation be higher or lower? Why?

2.73 Airborne particles such as dust and smoke are an important part of air pollution. To measure particulate pollution, a vacuum motor draws air through a filter for 24 hours. Weigh the filter at the beginning and end of the period. The weight gained is a measure of the concentration of particles in the air. A study of air pollution made measurements every 6 days with identical instruments in the center of a small city and at a rural location 10 miles southwest of the city. Because the prevailing winds blow from the west, we suspect that the rural readings will be generally lower than the city readings, but that the city readings can be predicted from the rural readings. Table 2.12 gives readings taken every 6 days over a 7-month period. The entry NA means that the reading for that date is not available, usually because of equipment failure.[36]

(a) We hope to use the rural particulate level to predict the city level on the same day. Make a graph to examine the relationship. Does the graph suggest that using the least-squares regression line for prediction will give approximately correct results over the range of values appearing in the data? Calculate the least-squares line for predicting city pollution from rural pollution. What percent of the observed variation in city pollution levels does this straight-line relationship account for?

TABLE 2.12 Particulate levels (grams) in two nearby locations

Day	1	2	3	4	5	6	7	8	9	10	11	12
Rural	NA	67	42	33	46	NA	43	54	NA	NA	NA	NA
City	39	68	42	34	48	82	45	NA	NA	60	57	NA

Day	13	14	15	16	17	18	19	20	21	22	23	24
Rural	38	88	108	57	70	42	43	39	NA	52	48	56
City	39	NA	123	59	71	41	42	38	NA	57	50	58

Day	25	26	27	28	29	30	31	32	33	34	35	36
Rural	44	51	21	74	48	84	51	43	45	41	47	35
City	45	69	23	72	49	86	51	42	46	NA	44	42

(b) Find the residuals from your least-squares fit. Plot the residuals both against x and against the time order of the observations, and comment on the results.

(c) Which observation appears to be the most influential? Circle this observation on your plot. Is it the observation with the largest residual?

(d) On the fourteenth date in the series, the rural reading was 88 and the city reading was not available. What do you estimate the city reading to be for that date?

(e) Make a normal quantile plot of the residuals. (Make a stemplot or histogram if your software does not make normal quantile plots.) Is the distribution of the residuals approximately normal?

2.74 A multimedia statistics learning system includes a test of skill in using the computer's mouse. The software displays a circle at a random location on the computer screen. The subject tries to click in the circle with the mouse as quickly as possible. A new circle appears as soon as the subject clicks the old one. Table 2.13 gives data for one subject's trials, 20 with each hand. Distance is the distance from the cursor location to the center of the new circle, in units whose actual size depends on the size of the screen. Time is the time required to click in the new circle, in milliseconds.[37]

(a) We suspect that time depends on distance. Make a scatterplot of time against distance, using separate symbols for each hand.

(b) Describe the pattern. How can you tell that the subject is right-handed?

(c) Find the regression line of time on distance separately for each hand. Draw these lines on your plot. Which regression does a better job of predicting time from distance? Give numerical measures that describe the success of the two regressions.

(d) It is possible that the subject got better in later trials due to learning. It is also possible that he got worse due to fatigue. Plot the residuals from

TABLE 2.13 Reaction times in a computer game

Time	Distance	Hand	Time	Distance	Hand
115	190.70	right	240	190.70	left
96	138.52	right	190	138.52	left
110	165.08	right	170	165.08	left
100	126.19	right	125	126.19	left
111	163.19	right	315	163.19	left
101	305.66	right	240	305.66	left
111	176.15	right	141	176.15	left
106	162.78	right	210	162.78	left
96	147.87	right	200	147.87	left
96	271.46	right	401	271.46	left
95	40.25	right	320	40.25	left
96	24.76	right	113	24.76	left
96	104.80	right	176	104.80	left
106	136.80	right	211	136.80	left
100	308.60	right	238	308.60	left
113	279.80	right	316	279.80	left
123	125.51	right	176	125.51	left
111	329.80	right	173	329.80	left
95	51.66	right	210	51.66	left
108	201.95	right	170	201.95	left

each regression against the time order of the trials (down the columns in Table 2.13). Is either of these systematic effects of time visible in the data?

2.75 There is a strong positive correlation between years of education and income for economists employed by business firms. (In particular, economists with doctorates earn more than economists with only a bachelor's degree.) There is also a strong positive correlation between years of education and income for economists employed by colleges and universities. But when all economists are considered, there is a *negative* correlation between education and income. The explanation for this is that business pays high salaries and employs mostly economists with bachelor's degrees, while colleges pay lower salaries and employ mostly economists with doctorates. Sketch a scatterplot with two groups of cases (business and academic) that illustrates how a strong positive correlation within each group and a negative overall correlation can occur together. (*Hint:* Begin by studying Figure 2.26.)

2.76 Exercise 2.65 gives data on the mean stride rate of a group of 21 elite female runners at various running speeds. Find the correlation between speed and stride rate. Would you expect this correlation to increase or decrease if we had data on the individual stride rates of all 21 runners at each speed? Why?

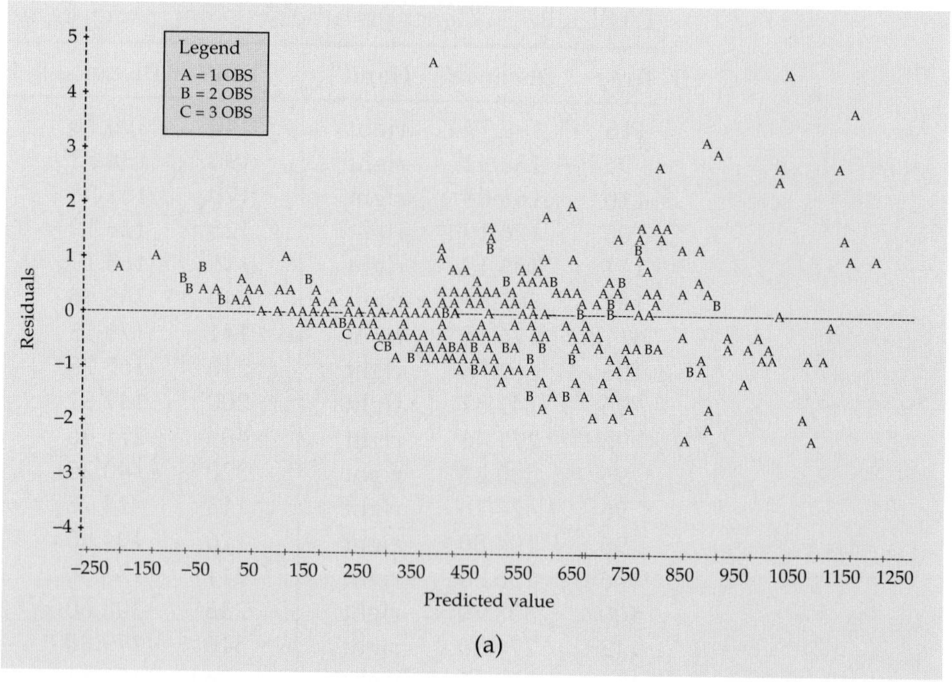

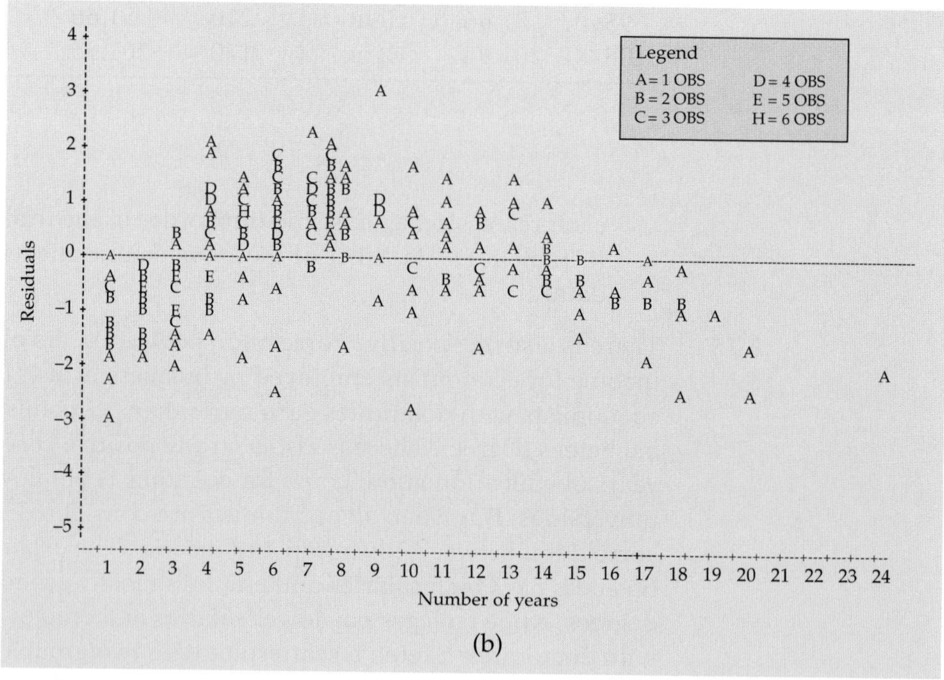

FIGURE 2.28 Two residual plots for the regression of baseball players' salaries on their performance, for Exercise 2.77.

2.77 Are baseball players paid according to their performance? To study this question, a statistician analyzed the salaries of over 260 major league hitters along with such explanatory variables as career batting average, career home runs per time at bat, and years in the major leagues.[38] This is a *multiple regression* with several explanatory variables. More detail on multiple regression appears in Chapter 11, but the fit of the model is assessed just as we have done in this chapter, by calculating and plotting the residuals:

$$\text{residual} = \text{observed } y - \text{predicted } y$$

(a) Figure 2.28(a) is a plot of the residuals against the predicted salary. When points are too close together to plot separately, the plot uses letters of the alphabet to show how many points there are at each position. Describe the pattern that appears on this residual plot. Will the regression model predict high or low salaries more precisely?

(b) After studying the residuals in more detail, the statistician decided to predict the logarithm of salary rather than the salary itself. One reason was that while salaries are not normally distributed (the distribution is skewed to the right), their logarithms are nearly normal. When the response variable is the logarithm of salary, a plot of the residuals against the predicted value is satisfactory—it looks like Figure 2.19(a). Figure 2.28(b) is a plot of the residuals against the number of years the player has been in the major leagues. Describe the pattern that you see. Will the model overestimate or underestimate the salaries of players who are new to the majors? Of players who have been in the major leagues about 8 years? Of players with more than 15 years in the majors?

2.5 The Question of Causation

In many studies of the relationship between two variables, the goal is to establish that changes in the explanatory variable *cause* changes in the response variable. Even when a strong association is present, the conclusion that this association is due to a causal link between the variables is often elusive. What ties between two variables (and others lurking in the background) can explain an observed association? What constitutes good evidence for causation? We begin our consideration of these questions with a set of examples. In each case, there is a clear association between an explanatory variable x and a response variable y. Moreover, the association is positive whenever the direction makes sense.

EXAMPLE 2.23

Here are some examples of observed association between x and y:

1. x = mother's body mass index
y = daughter's body mass index

2. x = amount of the artificial sweetener saccharin in a rat's diet
y = count of tumors in the rat's bladder

3. x = a student's SAT score as a high school senior
   y = the student's first-year college grade point average

4. x = monthly flow of money into stock mutual funds
   y = monthly rate of return for the stock market

5. x = whether a person regularly attends religious services
   y = how long the person lives

6. x = the number of years of education a worker has
   y = the worker's income

Explaining association: causation

Figure 2.29 shows in outline form how a variety of underlying links between variables can explain association. The dashed double-arrow line represents an observed association between the variables x and y. Some associations are explained by a direct cause-and-effect link between these variables. The first diagram in Figure 2.29 shows "x causes y" by a solid arrow running from x to y.

EXAMPLE 2.24

Items 1 and 2 in Example 2.23 are examples of direct causation. Thinking about these examples, however, shows that "causation" is not a simple idea.

1. A study of Mexican American girls aged 9 to 12 years recorded body mass index (BMI), a measure of weight relative to height, for both the girls and their mothers. People with high BMI are overweight or obese. The study also measured hours of television, minutes of physical activity, and intake of several kinds of food. The strongest correlation ($r = 0.506$) was between the BMI of daughters and the BMI of their mothers.[39]

   Body type is in part determined by heredity. Daughters inherit half their genes from their mothers. There is therefore a direct causal link between the BMI of mothers and daughters. Yet the mothers' BMIs explain only 25.6% (that's r^2 again) of the variation among the daughters' BMIs. Other factors,

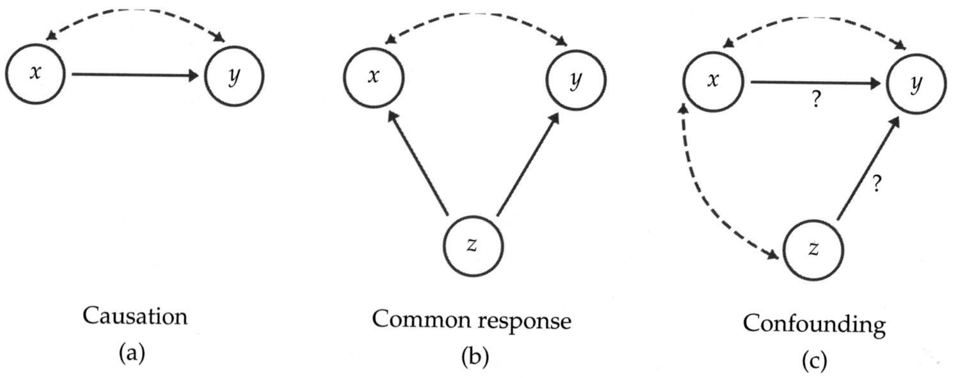

FIGURE 2.29 Some explanations for an observed association. The dashed double-arrow lines show an association. The solid arrows show a cause-and-effect link. The variable x is explanatory, y is a response variable, and z is a lurking variable.

such as diet and exercise, also influence BMI. **Even when direct causation is present, it is rarely a complete explanation of an association between two variables.**

2. The best evidence for causation comes from experiments that actually change x while holding all other factors fixed. If y changes, we have good reason to think that x caused the change in y. Experiments show conclusively that large amounts of saccharin in the diet cause bladder tumors in rats. Should we avoid saccharin as a replacement for sugar in food? Rats are not people. Although we can't experiment with people, studies of people who consume different amounts of saccharin show little association between saccharin and bladder tumors.[40] **Even well-established causal relations may not generalize to other settings.**

Explaining association: common response

common response

"Beware the lurking variable" is good advice when thinking about an association between two variables. The second diagram in Figure 2.29 illustrates **common response.** The observed association between the variables x and y is explained by a lurking variable z. Both x and y change in response to changes in z. This common response creates an association even though there may be no direct causal link between x and y.

EXAMPLE 2.25

The third and fourth items in Example 2.23 illustrate how common response can create an association.

3. Students who are smart and who have learned a lot tend to have both high SAT scores and high college grades. The positive correlation is explained by this common response to students' ability and knowledge.
4. There is a strong positive correlation between how much money individuals add to mutual funds each month and how well the stock market does the same month. Is the new money driving the market up? The correlation may be explained in part by common response to underlying investor sentiment: when optimism reigns, individuals send money to funds and large institutions also invest more. The institutions would drive up prices even if individuals did nothing. In addition, what causation there is may operate in the other direction: when the market is doing well, individuals rush to add money to their mutual funds.[41]

Explaining association: confounding

We noted in Example 2.24 that inheritance no doubt explains part of the association between the body mass indexes (BMIs) of daughters and their mothers. Can we use r or r^2 to say how much inheritance contributes to the daughters' BMIs? No. It may well be that mothers who are overweight also set an example of little exercise, poor eating habits, and lots of television. Their daughters pick up these habits to some extent, so the influence of heredity is mixed up with influences from the girls' environment. We call this mixing of influences *confounding*.

Confounding

Two variables are **confounded** when their effects on a response variable cannot be distinguished from each other. The confounded variables may be either explanatory variables or lurking variables.

When many variables interact with each other, confounding of several variables often prevents us from drawing conclusions about causation. The third diagram in Figure 2.29 illustrates confounding. Both the explanatory variable x and the lurking variable z may influence the response variable y. Because x is confounded with z, we cannot distinguish the influence of x from the influence of z. We cannot say how strong the direct effect of x on y is. In fact, it can be hard to say if x influences y at all.

EXAMPLE 2.26

The last two associations in Example 2.23 (Items 5 and 6) are explained in part by confounding.

5. Many studies have found that people who are active in their religion live longer than nonreligious people. But people who attend church or mosque or synagogue also take better care of themselves than nonattenders. They are less likely to smoke, more likely to exercise, and less likely to be overweight. The effects of these good habits are confounded with the direct effects of attending religious services.

6. It is likely that more education is a cause of higher income—many highly paid professions require advanced education. However, confounding is also present. People who have high ability and come from prosperous homes are more likely to get many years of education than people who are less able or poorer. Of course, people who start out able and rich are more likely to have high earnings even without much education. We can't say how much of the higher income of well-educated people is actually caused by their education.

Many observed associations are at least partly explained by lurking variables. Both common response and confounding involve the influence of a lurking variable (or variables) z on the response variable y. The distinction between these two types of relationship is less important than the common element, the influence of lurking variables. The most important lesson of these examples is one we have already emphasized: **even a very strong association between two variables is not by itself good evidence that there is a cause-and-effect link between the variables.**

Establishing causation

How can a direct causal link between x and y be established? The best method—indeed, the only fully compelling method—of establishing causation is to conduct a carefully designed experiment in which the effects of possible lurking variables are controlled. Much of Chapter 3 is devoted to the art of designing convincing experiments.

Many of the sharpest disputes in which statistics plays a role involve questions of causation that cannot be settled by experiment. Does gun control reduce violent crime? Does living near power lines cause cancer? Has increased free trade helped to increase the gap between the incomes of more educated and less educated American workers? All of these questions have become public issues. All concern associations among variables. And all have this in common: they try to pinpoint cause and effect in a setting involving complex relations among many interacting variables. Common response and confounding, along with the number of potential lurking variables, make observed associations misleading. Experiments are not possible for ethical or practical reasons. We can't assign some people to live near power lines or compare the same nation with and without free-trade agreements.

EXAMPLE 2.27

Electric currents generate magnetic fields. So living with electricity exposes people to magnetic fields. Living near power lines increases exposure to these fields. Really strong fields can disturb living cells in laboratory studies. Some people claim the weaker fields we experience if we live near power lines cause leukemia in children.

It isn't ethical to do experiments that expose children to magnetic fields. It's hard to compare cancer rates among children who happen to live in more and less exposed locations, because leukemia is rare and locations vary in many ways other than magnetic fields. We must rely on studies that compare children who have leukemia with children who don't.

A careful study of the effect of magnetic fields on children took five years and cost \$5 million. The researchers compared 638 children who had leukemia and 620 who did not. They went into the homes and actually measured the magnetic fields in the children's bedrooms, in other rooms, and at the front door. They recorded facts about nearby power lines for the family home and also for the mother's residence when she was pregnant. Result: no evidence of more than a chance connection between magnetic fields and childhood leukemia.[42]

"No evidence" that magnetic fields are connected with childhood leukemia doesn't prove that there is no risk. It says only that a careful study could not find any risk that stands out from the play of chance that distributes leukemia cases across the landscape. Critics continue to argue that the study failed to measure some lurking variables, or that the children studied don't fairly represent all children. Nonetheless, a carefully designed study comparing children with and without leukemia is a great advance over haphazard and sometimes emotional counting of cancer cases.

EXAMPLE 2.28

Despite the difficulties, it is sometimes possible to build a strong case for causation in the absence of experiments. The evidence that smoking causes lung cancer is about as strong as nonexperimental evidence can be.

Doctors had long observed that most lung cancer patients were smokers. Comparison of smokers and similar nonsmokers showed a very strong association between smoking and death from lung cancer. Could the association be due to common response? Might there be, for example, a genetic factor that predisposes people both to nicotine addiction and to lung cancer? Smoking and lung cancer would then be positively associated even if smoking had no direct effect on the lungs. Or perhaps confounding is to blame. It might be that smokers live unhealthy lives in other ways (diet, alcohol, lack of exercise) and that some other habit confounded with smoking is a cause of lung cancer. How were these objections overcome?

Exploring the Web

The Web abounds in information and misinformation about health issues. If you are interested in data and statistical studies, the place to start is the site of the Centers for Disease Control and Prevention, `www.cdc.gov`. More than you want to know about smoking and tobacco, for example, appears at `www.cdc.gov/tobacco/data.htm`. Tobacco use, advertising, taxation, surveys of teenagers—it's all here.

Let's answer this question in general terms: What are the criteria for establishing causation when we cannot do an experiment?

- *The association is strong*. The association between smoking and lung cancer is very strong.
- *The association is consistent*. Many studies of different kinds of people in many countries link smoking to lung cancer. That reduces the chance that a lurking variable specific to one group or one study explains the association.
- *Higher doses are associated with stronger responses*. People who smoke more cigarettes per day or who smoke over a longer period get lung cancer more often. People who stop smoking reduce their risk.
- *The alleged cause precedes the effect in time*. Lung cancer develops after years of smoking. The number of men dying of lung cancer rose as smoking became more common, with a lag of about 30 years. Lung cancer kills more men than any other form of cancer. Lung cancer was rare among women until women began to smoke. Lung cancer in women rose along with smoking, again with a lag of about 30 years, and has now passed breast cancer as the leading cause of cancer death among women.
- *The alleged cause is plausible*. Experiments with animals show that tars from cigarette smoke do cause cancer.

Medical authorities do not hesitate to say that smoking causes lung cancer. The U.S. Surgeon General states that cigarette smoking is "the largest avoidable cause of death and disability in the United States."[43] The evidence for causation is overwhelming—but it is not as strong as the evidence provided by well-designed experiments.

SUMMARY

Some observed associations between two variables are due to a **cause-and-effect** relationship between these variables, but others are explained by **lurking variables.**

The effect of lurking variables can operate through **common response** if changes in both the explanatory and response variables are caused by changes in lurking variables. **Confounding** of two variables (either explanatory or lurking variables) means that we cannot distinguish their effects on the response variable.

That an association is due to causation is best established by an **experiment** that changes the explanatory variable while controlling other influences on the response.

In the absence of experimental evidence, be cautious in accepting claims of causation. Good evidence of causation requires a strong association that appears consistently in many studies, a clear explanation for the alleged causal link, and careful examination of possible lurking variables.

SECTION 2.5 EXERCISES

2.78 An article entitled "The Health and Wealth of Nations" says: "The positive correlation between health and income per capita is one of the best-known relations in international development. This correlation is commonly thought to reflect a causal link running from income to health. . . . Recently, however, another intriguing possibility has emerged: that the health-income correlation is partly explained by a causal link running the other way—from health to income."[44]

Explain how higher income in a nation can cause better health. Then explain how better health can cause higher income. There is no simple way to determine the direction of the link.

2.79 In 1998, the Kalamazoo (Michigan) Symphony advertised a "Mozart for Minors" program with this statement: "Question: Which students scored 51 points higher in verbal skills and 39 points higher in math? Answer: Students who had experience in music."[45] What do you think of the claim that "experience in music" causes higher test scores?

2.80 A study finds that high school students who take the SAT, enroll in an SAT coaching course, and then take the SAT a second time raise their SAT mathematics scores from a mean of 521 to a mean of 561.[46] What factors other than "taking the course causes higher scores" might explain this improvement?

2.81 A news report says that minority children who take algebra and geometry in high school succeed in college about as well as whites. We should hesitate to conclude that requiring more math in high school would cause more minority students to succeed in college. Why? What lurking variables might influence both the choice of high school courses and college ambitions, creating a relationship between these variables? Use a diagram like those in Figure 2.29 to present your suggestions.

2.82 The National Halothane Study was a major investigation of the safety of the anesthetics used in surgery. Records of over 850,000 operations performed in 34 major hospitals showed the following death rates y for four common anesthetics x:[47]

Anesthetic	A	B	C	D
Death rate	1.7%	1.7%	3.4%	1.9%

Do these data prove that Anesthetic C is causing more deaths than the others? Use a diagram like those in Figure 2.29 to describe other explanations.

2.83 A study showed that women who work in the production of computer chips have abnormally high numbers of miscarriages. The union claimed that exposure to chemicals used in production caused the miscarriages. Another possible explanation is that these workers spend most of their work time standing up. Illustrate these relationships in a diagram like those in Figure 2.29.

2.84 A study shows that there is a positive correlation between the size of a hospital (measured by its number of beds x) and the median number of days y that patients remain in the hospital. Does this mean that you can shorten a hospital stay by choosing a small hospital? Use a diagram like those in Figure 2.29 to explain the association.

2.85 Children who watch many hours of television get lower grades in school on the average than those who watch less TV. Explain clearly why this fact does not show that watching TV *causes* poor grades. In particular, suggest some other variables that may be confounded with heavy TV viewing and may contribute to poor grades.

2.86 People who use artificial sweeteners in place of sugar tend to be heavier than people who use sugar. Does this mean that artificial sweeteners cause weight gain? Give a more plausible explanation for this association.

2.87 Members of a high school language club believe that study of a foreign language improves a student's command of English. From school records, they obtain the scores on an English achievement test given to all seniors. The mean score of seniors who had studied a foreign language for at least two years is much higher than the mean score of seniors who studied no foreign language. The club's advisor says that these data are not good evidence that language study strengthens English skills. Identify the explanatory and response variables in this study. Then explain what lurking variable prevents the conclusion that language study improves students' English scores.

2.88 People who do well tend to feel good about themselves. Perhaps helping people feel good about themselves will help them do better in their jobs and in life. Raising self-esteem became for a time a goal in many schools and companies. Can you think of explanations for the association between high self-esteem and good performance other than "Self-esteem causes better work"?

2.89 Return to the study of the safety of anesthetics from Exercise 2.82. For ethical reasons, surgeons are unwilling to do an experiment in which similar patients are given different anesthetics. However, you have very detailed records for the 850,000 operations that were investigated in the study cited in Exercise 2.82. What kinds of information would you seek in these records in order to establish that Anesthetic C does or does not cause extra deaths among patients?

2.6 Transforming Relationships*

How is the weight of an animal's brain related to the weight of its body? Figure 2.30 is a scatterplot of brain weight against body weight for 96 species of mammals.[48] The line is the least-squares regression line for predicting brain weight from body weight. The outliers are interesting. We might say that dolphins and humans are smart, hippos are dumb, and African elephants are just big. That's because dolphins and humans have larger brains than

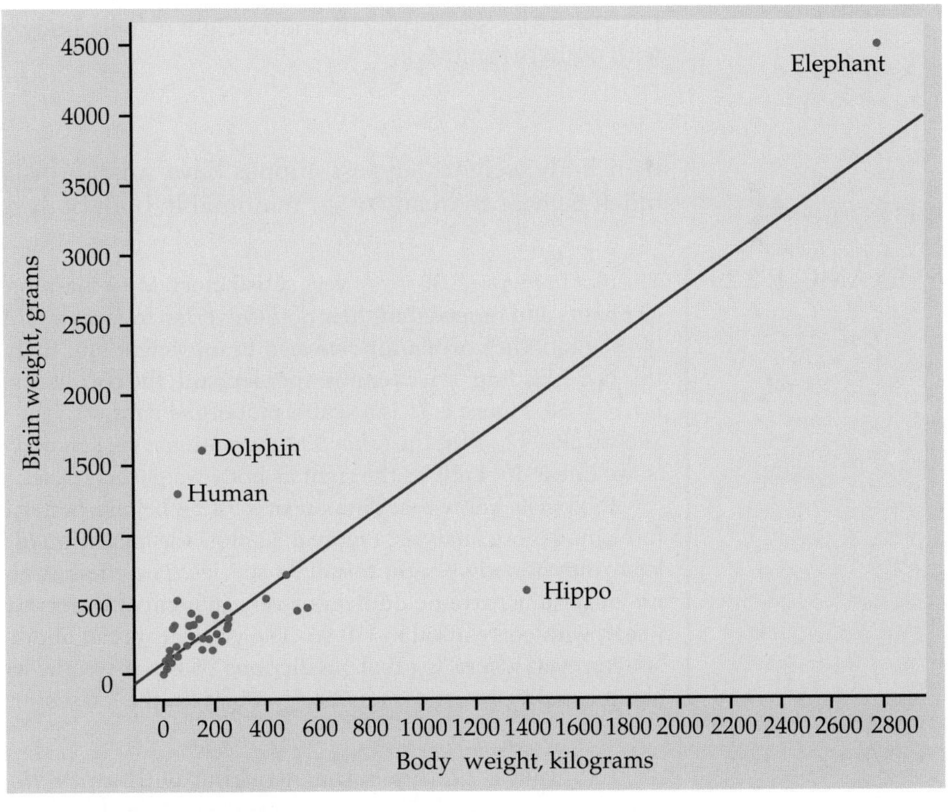

FIGURE 2.30 Scatterplot of brain weight against body weight for 96 species of mammals.

*This section can be omitted without loss of continuity.

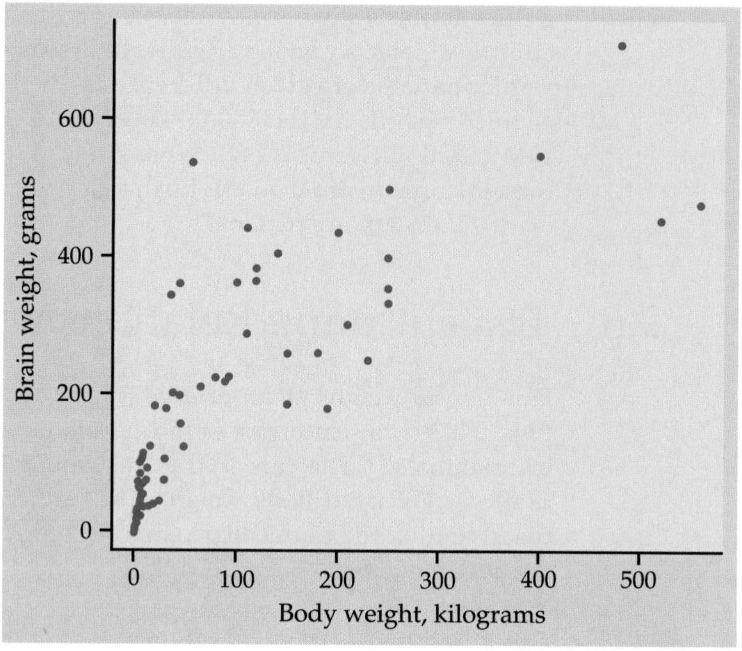

FIGURE 2.31 Scatterplot of brain weight against body weight for mammals, with outliers removed.

their body weights suggest, hippos have smaller brains, and the elephant is much heavier than any other mammal in both body and brain.

EXAMPLE 2.29

The plot in Figure 2.30 is not very satisfactory. Most mammals are so small relative to elephants and hippos that their points overlap to form a blob in the lower-left corner of the plot. The correlation between brain weight and body weight is $r = 0.86$, but this is misleading. If we remove the elephant, the correlation for the other 95 species is $r = 0.50$. Figure 2.31 is a scatterplot of the data with the four outliers removed to allow a closer look at the other 92 observations. We can now see that the relationship is not linear. It bends to the right as body weight increases.

Biologists know that data on sizes often behave better if we take *logarithms* before doing more analysis. Figure 2.32 plots the logarithm of brain weight against the logarithm of body weight for all 96 species. The effect is almost magical. There are no longer any extreme outliers or very influential observations. The pattern is very linear, with correlation $r = 0.96$. The vertical spread about the least-squares line is similar everywhere, so that predictions of brain weight from body weight will be about equally precise for any body weight (in the log scale).

transforming
reexpressing

Example 2.29 shows that working with a *function* of our original measurements can greatly simplify statistical analysis. Applying a function such as the logarithm or square root to a quantitative variable is called **transforming** or **reexpressing** the data. We will see in this section that understanding how simple functions work helps us choose and use transformations. Because we may want to transform either the explanatory variable x or the response

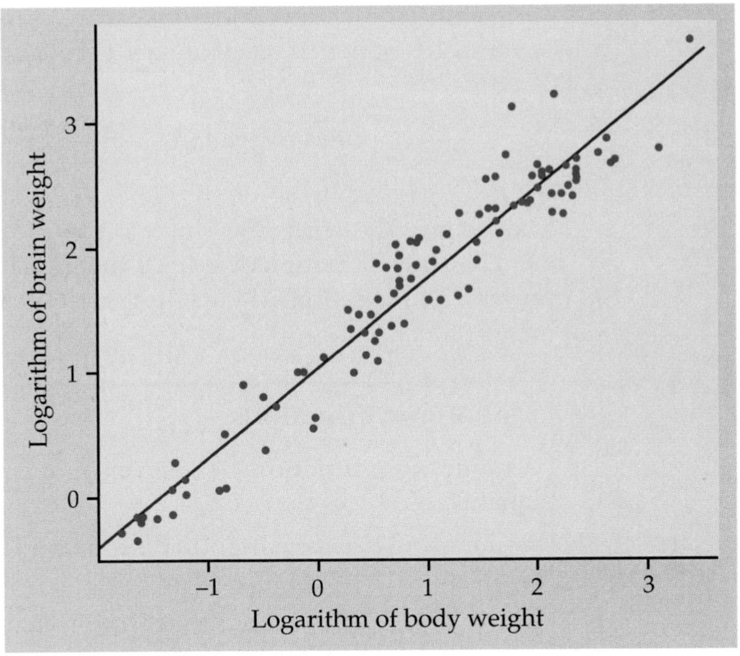

FIGURE 2.32 Scatterplot of the logarithm of brain weight against the logarithm of body weight for 96 species of mammals.

variable y in a scatterplot, or both, we will call the variable t when talking about transforming in general.

First steps in transforming

Transforming data amounts to changing the scale of measurement that was used when the data were collected. We can choose to measure temperature in degrees Fahrenheit or in degrees Celsius, distance in miles or in kilometers. These changes of units are *linear transformations*, discussed on pages 51 to 54. **Linear transformations cannot straighten a curved relationship between two variables.** To do that, we resort to functions that are not linear. The logarithm, applied in Example 2.29, is a nonlinear function. Here are some others.

- How shall we measure the size of a sphere or of such roughly spherical objects as grains of sand or bubbles in a liquid? The size of a sphere can be expressed in terms of the diameter t, in terms of surface area (proportional to t^2), or in terms of volume (proportional to t^3). Any one of these *powers* of the diameter may be natural in a particular application.
- We commonly measure the fuel consumption of a car in miles per gallon—that is, how many miles the car travels on 1 gallon of fuel. Engineers prefer to measure in gallons per mile—that is, how many

gallons of fuel the car needs to travel 1 mile. This is a *reciprocal* transformation. A car that gets 25 miles per gallon uses

$$\frac{1}{\text{miles per gallon}} = \frac{1}{25} = 0.04 \text{ gallons per mile}$$

The reciprocal is a *negative power* $1/t = t^{-1}$.

The transformations we have mentioned—linear, positive and negative powers, and logarithms—are those used in most statistical problems. They are all *monotonic*.

Monotonic Functions

A **monotonic function** $f(t)$ moves in one direction as its argument t increases.

A **monotonic increasing** function preserves the order of data. That is, if $a > b$, then $f(a) > f(b)$.

A **monotonic decreasing** function reverses the order of data. That is, if $a > b$, then $f(a) < f(b)$.

The graph of a linear function is a straight line. The graph of a monotonic increasing function is increasing everywhere. A monotonic decreasing function has a graph that is decreasing everywhere. A function can be monotonic over some range of t without being everywhere monotonic. For example, the square function t^2 is monotonic increasing for $t \geq 0$. If the range of t includes both positive and negative values, the square is not monotonic—it decreases as t increases for negative values of t and increases as t increases for positive values.

Figure 2.33 compares three monotonic increasing functions and three monotonic decreasing functions for positive values of the argument t. Many variables take only 0 or positive values, so we are particularly interested in how functions behave for positive values of t. The increasing functions are

Linear	$a + bt$, slope $b > 0$
Square	t^2
Logarithm	$\log t$

The decreasing functions in the lower panel of Figure 2.33 are

Linear	$a + bt$, slope $b < 0$
Reciprocal square root	$1/\sqrt{t}$, or $t^{-1/2}$
Reciprocal	$1/t$, or t^{-1}

Nonlinear monotonic transformations change data enough to alter the shape of distributions and the form of relations between two variables, yet are simple enough to preserve order and allow recovery of the original data. We will concentrate on powers and logarithms. The even-numbered powers t^2, t^4, and so on are monotonic increasing for $t \geq 0$, but not when t can take both negative and positive values. The logarithm is not even defined unless

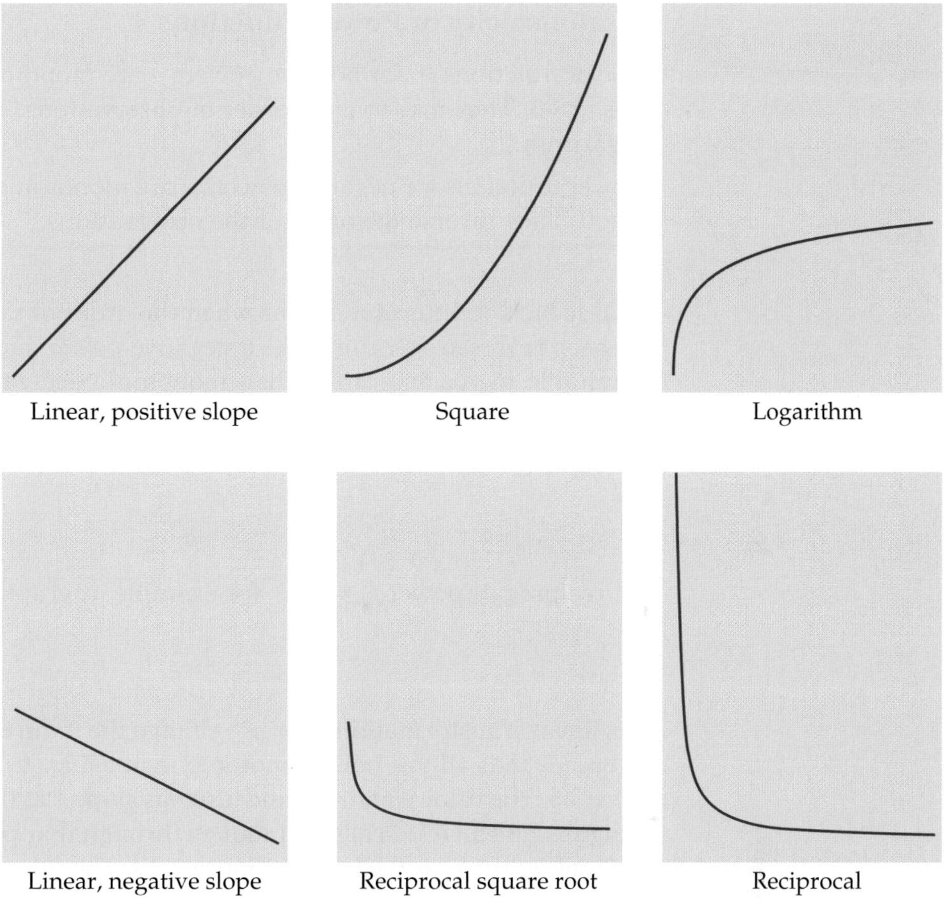

FIGURE 2.33 Monotonic functions, increasing (top panel) and decreasing (bottom panel).

$t > 0$. Our strategy for transforming data is therefore as follows:

1. If the variable to be transformed takes values that are 0 or negative, first apply a linear transformation to make the values all positive. Often we just add a constant to all the observations.
2. Then choose a power or logarithmic transformation that simplifies the data, for example, one that approximately straightens a scatterplot.

The ladder of power transformations

Though simple in algebraic form and easy to compute with a calculator, the power and logarithm functions are varied in their behavior. It is natural to think of powers such as

$$\ldots, t^{-1}, t^{-1/2}, t^{1/2}, t, t^2, \ldots$$

as a hierarchy or ladder. Some facts about this ladder will help us choose transformations. In all cases, we look only at positive values of the argument t.

Monotonicity of Power Functions

Power functions t^p for positive powers p are monotonic increasing for values $t > 0$. They preserve the order of observations. This is also true of the logarithm.

Power functions for negative powers p are monotonic decreasing for values $t > 0$. They reverse the order of the observations.

It is hard to interpret graphs when the order of the original observations has been reversed. We can make a negative power such as the reciprocal $1/t$ monotonic increasing rather than monotonic decreasing by using $-1/t$ instead. Figure 2.34 takes this idea a step farther. This graph compares the ladder of power functions in the form

$$\frac{t^p - 1}{p}$$

The reciprocal (power $p = -1$), for example, is graphed as

$$\frac{1/t - 1}{-1} = 1 - \frac{1}{t}$$

This linear transformation does not change the nature of the power functions t^p, except that all are now monotonic increasing. It is chosen so that every power has the value 0 at $t = 1$ and also has slope 1 at that point. So the graphs in Figure 2.34 all touch at $t = 1$ and go through that point at the same slope.

Look at the $p = 0$ graph in Figure 2.34. The 0th power t^0 is just the constant 1, which is not very useful. The $p = 0$ entry in the figure is not constant. In fact, it is the logarithm, $\log t$. That is, **the logarithm fits into the ladder of power transformations at $p = 0$.**[49]

Figure 2.34 displays another key fact about these functions. The graph of a linear function (power $p = 1$) is a straight line. Powers greater than 1 give graphs that bend upward. That is, the transformed variable grows ever faster as t gets larger. Powers less than 1 give graphs that bend downward. The transformed values continue to grow with t, but at a rate that decreases as t increases. What is more, the sharpness of the bend increases as we move away from $p = 1$ in either direction.

Concavity of Power Functions

Power transformations t^p for powers p greater than 1 are *concave up*; that is, they have the shape . These transformations push out the right tail of a distribution and pull in the left tail. This effect gets stronger as the power p moves up away from 1.

Power transformations t^p for powers p less than 1 (and the logarithm for $p = 0$) are *concave down*; that is, they have the shape . These transformations pull in the right tail of a distribution and push out the left tail. This effect gets stronger as the power p moves down away from 1.

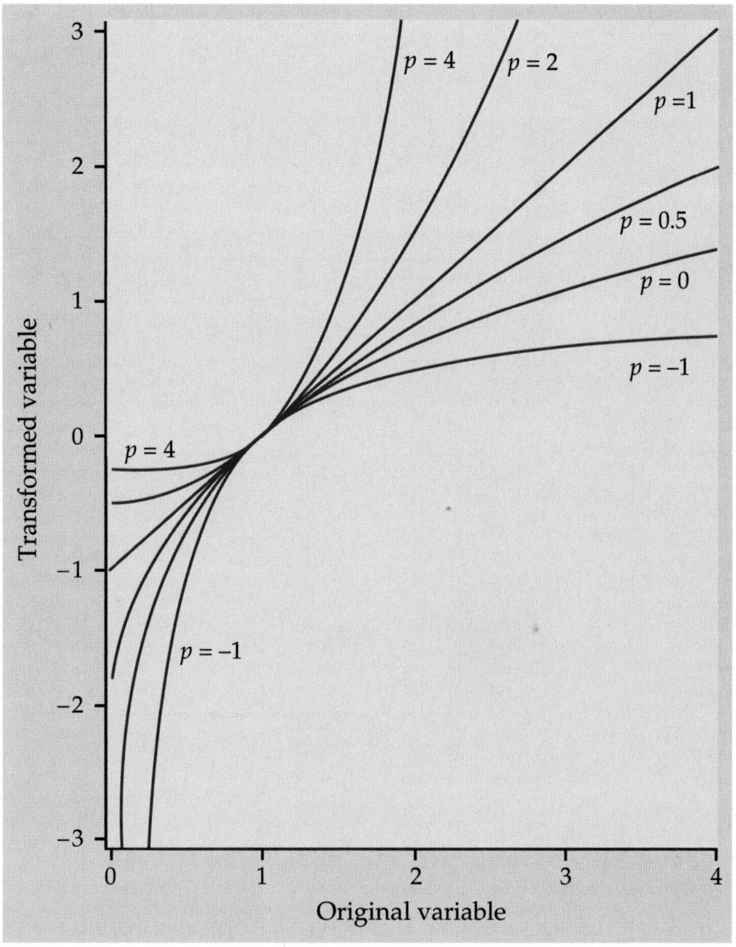

FIGURE 2.34 The ladder of power and logarithm transformations. For easier comparison, the functions have been scaled so that all meet at $t = 1$ and have slope 1 at that point.

EXAMPLE 2.30

Figure 2.35(a) is a scatterplot of data from the World Bank.[50] The individuals are all the world's nations for which data are available. The explanatory variable x is a measure of how rich a country is: the gross domestic product (GDP) per person. GDP is the total value of the goods and services produced in a country, converted into dollars. The response variable y is life expectancy at birth.

Life expectancy increases in richer nations, but only up to a point. The pattern in Figure 2.35(a) at first rises rapidly as GDP increases but then levels out. Three African nations (Botswana, Gabon, and Namibia) are outliers with much lower life expectancy than the overall pattern suggests. Can we straighten the overall pattern by transforming?

Life expectancy does not have a large range, but we can see that the distribution of GDP is right-skewed and very spread out. So GDP is a better candidate for transformation. We want to pull in the long right tail, so we try transformations with $p < 1$. Figures 2.35(b), (c), and (d) show the results of three transformations of GDP. The r-value in each figure is the correlation when the three outliers are omitted.

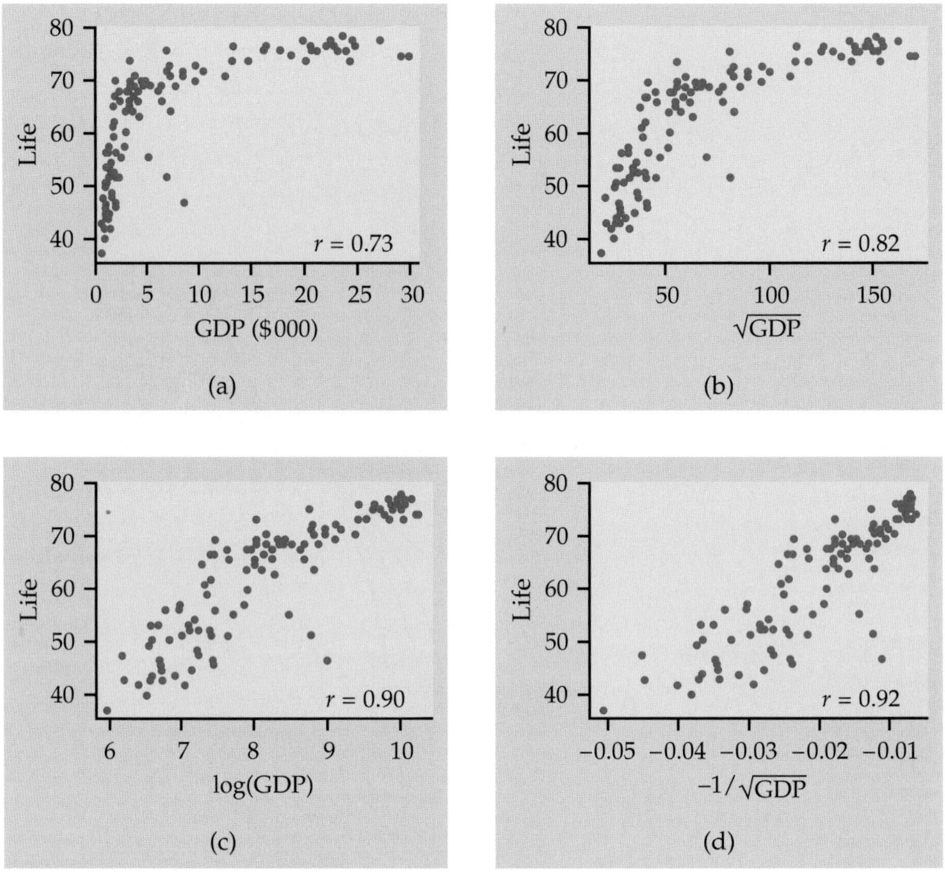

FIGURE 2.35 The ladder of transformations at work. The data are life expectancy and gross domestic product (GDP) for 115 nations. Panel (a) displays the original data. Panels (b), (c), and (d) transform GDP, moving down the ladder away from linear functions.

The square root $\sqrt{x}$, with $p = 1/2$, reduces the curvature of the scatterplot, but not enough. The logarithm $\log x$ ($p = 0$) straightens the pattern more, but it still bends to the right. The reciprocal square root $1/\sqrt{x}$, with $p = -1/2$, gives a pattern that is quite straight except for the outliers. To avoid reversing the order of the observations, we actually used $-1/\sqrt{x}$.

Example 2.30 shows the ladder of powers at work. As we move down the ladder from linear transformations (power $p = 1$), the scatterplot gets straighter. Moving farther down the ladder, to the reciprocal $1/x = x^{-1}$, begins to bend the plot in the other direction. But this "try it and see" approach isn't very satisfactory. That life expectancy depends linearly on $1/\sqrt{\text{GDP}}$ does not increase our understanding of the relationship between the health and wealth of nations. We don't recommend just pushing buttons on your calculator to try to straighten a scatterplot.

It is much more satisfactory to begin with a theory or mathematical model that we expect to describe a relationship. The transformation needed to make the relationship linear is then a consequence of the model. One of the most common models is *exponential growth*.

Exponential growth

A variable grows linearly over time if it *adds* a fixed increment in each equal time period. Exponential growth occurs when a variable is *multiplied* by a fixed number in each time period. To grasp the effect of multiplicative growth, consider a population of bacteria in which each bacterium splits into two each hour. Beginning with a single bacterium, we have 2 after one hour, 4 at the end of two hours, 8 after three hours, then 16, 32, 64, 128, and so on. These first few numbers are deceiving. After 1 day of doubling each hour, there are 2^{24} (16,777,216) bacteria in the population. That number then doubles the next hour! Try successive multiplications by 2 on your calculator to see for yourself the very rapid increase after a slow start. Figure 2.36 shows the growth of the bacteria population over 24 hours. For the first 15 hours, the population is too small to rise visibly above the zero level on the graph. It is characteristic of exponential growth that the increase appears slow for a long period, then seems to explode.

Linear versus Exponential Growth

Linear growth increases by a fixed *amount* in each equal time period. **Exponential growth** increases by a fixed *percentage* of the previous total.

Populations of living things—like bacteria—tend to grow exponentially if not restrained by outside limits such as lack of food or space. More pleasantly, money also displays exponential growth when returns to an investment are compounded. Compounding means that last period's income earns income this period.

EXAMPLE 2.31

A dollar invested at an annual rate of 6% turns into $1.06 in a year. The original dollar remains and has earned $0.06 in interest. That is, 6% annual interest means that any amount on deposit for the entire year is multiplied by 1.06. If the $1.06 remains invested for a second year, the new amount is therefore 1.06×1.06, or 1.06^2. That is only $1.12, but this in turn is multiplied by 1.06 during the third year, and so on. After x years, the dollar has become 1.06^x dollars.

If the Native Americans who sold Manhattan Island for $24 in 1626 had deposited the $24 in a savings account at 6% annual interest, they would now have almost $80 billion. Our savings accounts don't make us billionaires, because we don't stay around long enough. A century of growth at 6% per year turns $24 into $8143. That's 1.06^{100} times $24. By 1826, two centuries after the sale, the account would hold a bit over $2.7 million. Only after a patient 302 years do we finally reach $1 billion. That's real money, but 302 years is a long time.

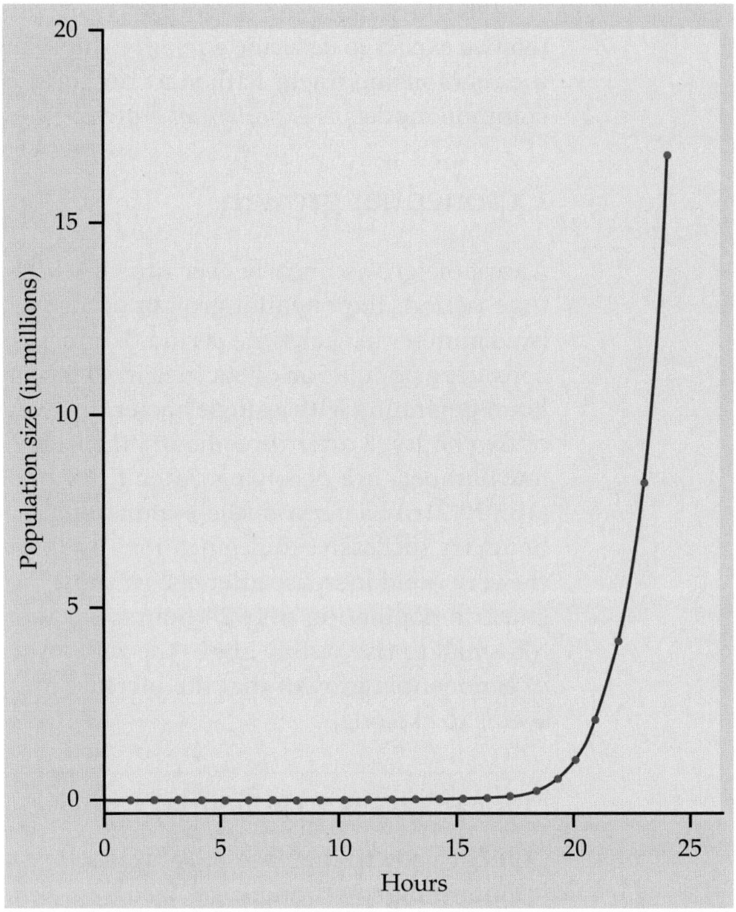

FIGURE 2.36 Growth of a bacteria population that doubles each hour.

exponential growth model

The count of bacteria after x hours is 2^x. The value of \$24 invested for x years at 6% interest is 24×1.06^x. Both are examples of the **exponential growth model** $y = a \times b^x$ for different constants a and b. In this model, the response y is multiplied by b in each time period.

EXAMPLE 2.32

Does the exponential growth model sometimes describe real data that don't arise from any obvious process of multiplying by a fixed number over and over again? Let's look at the world's production of oil. Petroleum is the most important source of energy for the developed nations and has been the cause of economic dislocation and even war. Table 2.14 and Figure 2.37 show the growth of annual world crude oil production, measured in millions of barrels per year.[51] There is an increasing trend, but the overall pattern is not linear. Oil production has increased much faster than linear growth. The pattern of growth follows a smooth curve until 1973, when a Mideast war touched off a vast price increase and a change in the previous pattern of production. In fact, the curve of oil production from 1880 to 1973 looks a lot like an exponential curve. Is this exponential growth?

TABLE 2.14 Annual world crude oil production, 1880–1998 (millions of barrels)

Year	Mbbl.	Year	Mbbl.	Year	Mbbl.
1880	30	1945	2,595	1976	20,188
1890	77	1950	3,803	1978	21,922
1900	149	1955	5,626	1980	21,722
1905	215	1960	7,674	1982	19,411
1910	328	1962	8,882	1984	19,837
1915	432	1964	10,310	1986	20,246
1920	689	1966	12,016	1988	21,338
1925	1,069	1968	14,104	1990	22,100
1930	1,412	1970	16,690	1992	22,028
1935	1,655	1972	18,584	1994	22,234
1940	2,150	1974	20,389	1996	23,380
				1998	24,441

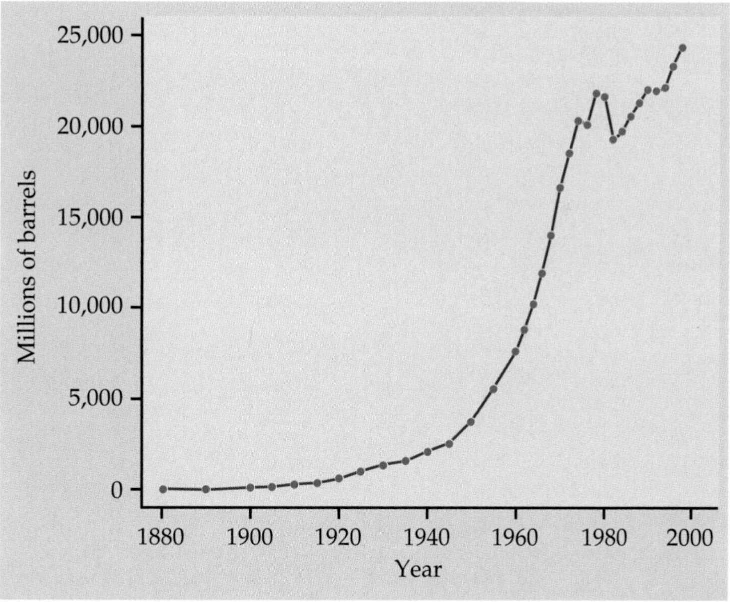

FIGURE 2.37 World production of crude oil from 1880 to 1998. This time plot has a regular pattern prior to 1973.

The logarithm transformation

The shape of the growth curve for oil production in Figure 2.37 does look like the exponential curve in Figure 2.36, but our eyes are not very good at comparing curves of roughly similar shape. We need a better way to check whether growth is exponential. Our eyes are quite good at judging whether or not points lie along a straight line. So we will apply a mathematical transformation that changes exponential growth into linear growth—and patterns of growth that are not exponential into something other than linear.

The necessary transformation is carried out by taking the *logarithm* of the data points. Use a calculator with a `log` button to compute logarithms. Better yet, most statistical software will calculate the logarithms of all the values of a variable in response to a single command. The essential property of the logarithm for our purposes is that it straightens an exponential growth curve. **If a variable grows exponentially, its logarithm grows linearly.**

EXAMPLE 2.33

Figure 2.38(a) plots the logarithms of the bacteria counts in Figure 2.36. Sure enough, exact exponential growth turns into an exact straight line when we plot the logarithms. After 15 hours, for example, the population contains 2^{15} (32,768)

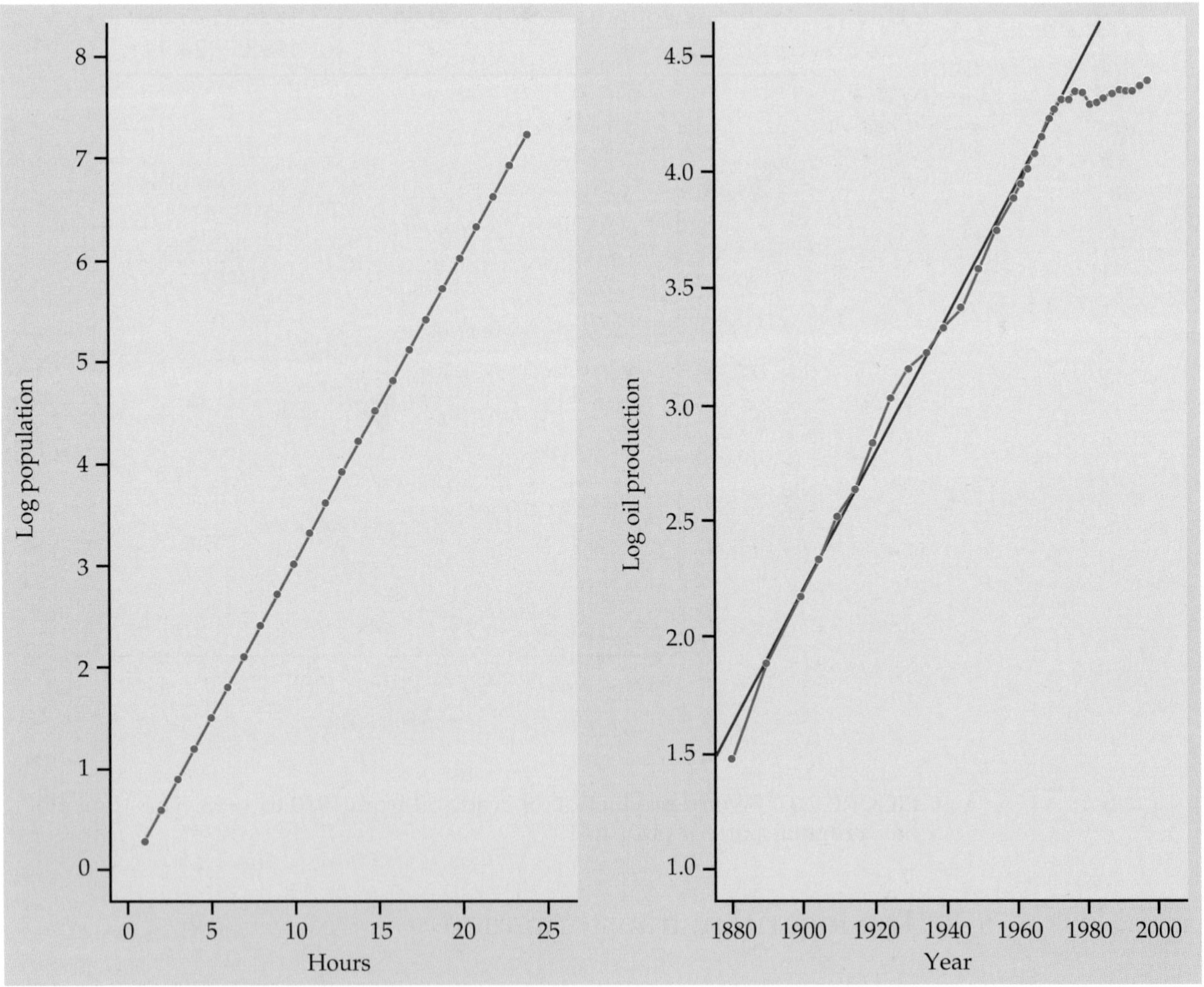

FIGURE 2.38 Taking logarithms turns exponential growth into linear growth. (a) Exact exponential growth: bacteria doubling each hour. (b) Oil production from 1880 to 1998. Growth was approximately exponential until 1973.

bacteria. The logarithm of 32,768 is 4.515, and this point appears above the 15-hour mark in Figure 2.38(a).

Next, take the logarithms of world oil production from 1880 to 1998 and plot them against time. Figure 2.38(b) is the result. It looks straight until 1974! The regression line based on the years from 1880 to 1972, added to the plot, describes the long-term trend. The increase in oil production was indeed close to exponential over most of the twentieth century. Perhaps this should have been a warning: in the real world, nothing can grow exponentially for a long period.

Figure 2.39 plots the residuals of the actual logarithms for the years between 1880 and 1972 from the regression line. The residual plot gives us a detailed picture of oil production in these years. Oil production increased more rapidly than the long-term rate in the years between 1900 and the beginning of the Depression in 1929. Production grew more slowly not only during the Depression but also during World War II. That is, the turning points in the residual plot occur in 1925 (the last point before 1929) and 1945 (the end of World War II). Only after the war did oil production return to an above-average growth rate, which lasted until 1973. The production of oil was of course increasing during the entire period up to 1973—the runs of declining residuals show periods of slower growth, not an actual drop in oil production.

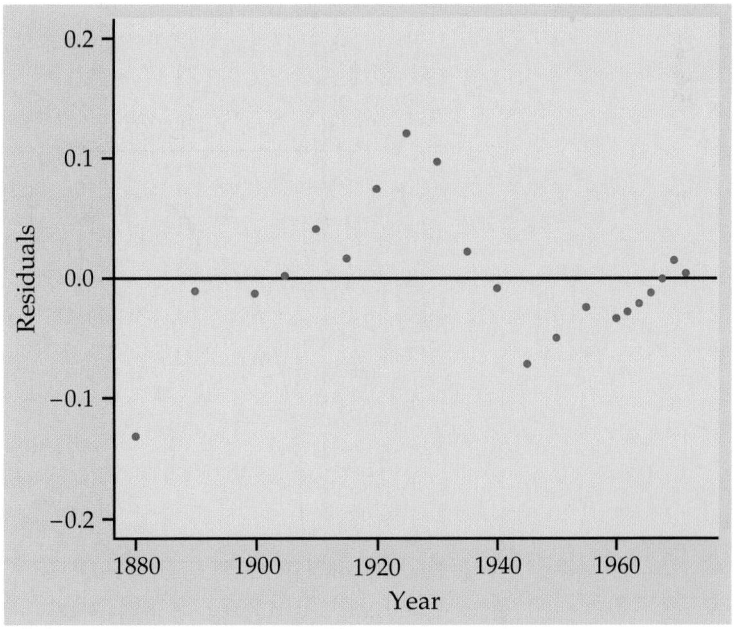

FIGURE 2.39 Residuals from the least-squares regression line of the logarithms of oil production against year, ending in 1972. The residual plot magnifies deviations from the least-squares line and helps us see details of the history of oil production.

Prediction in the exponential growth model

Regression is often used for prediction. When we fit a least-squares regression line, we find the predicted response $\hat{y}$ for any value of the explanatory variable x by substituting our x-value into the equation of the line. In the case of exponential growth, the logarithms rather than the actual responses follow a linear pattern. To do prediction, we "undo" the logarithm transformation to return to the original units of measurement. The same idea works for any monotonic transformation. There is always exactly one original value behind any transformed value, so we can always go back to our original scale.

EXAMPLE 2.34

Figure 2.40 shows the idea for the logarithm transformation. The logarithm of $x = 2$ is $\log 2 = 0.3010$. The graph shows that we can go "up and over" from $x = 2$ to $\log x = 0.3010$. Because the logarithm function is monotonic, $x = 2$ is the *only* number whose logarithm is 0.3010. So if we are told that $\log x = 0.3010$, we can go "over and down" to recover $x = 2$.

In the form of equations, any number x can be obtained from its common (base 10) logarithm $\log x$ by

$$x = 10^{\log x}$$

If we are told that $\log x = 0.3010$, then

$$x = 10^{\log x} = 10^{0.3010} = 1.9999$$

That's $x = 2$ up to the error caused by rounding $\log x$ to four places.

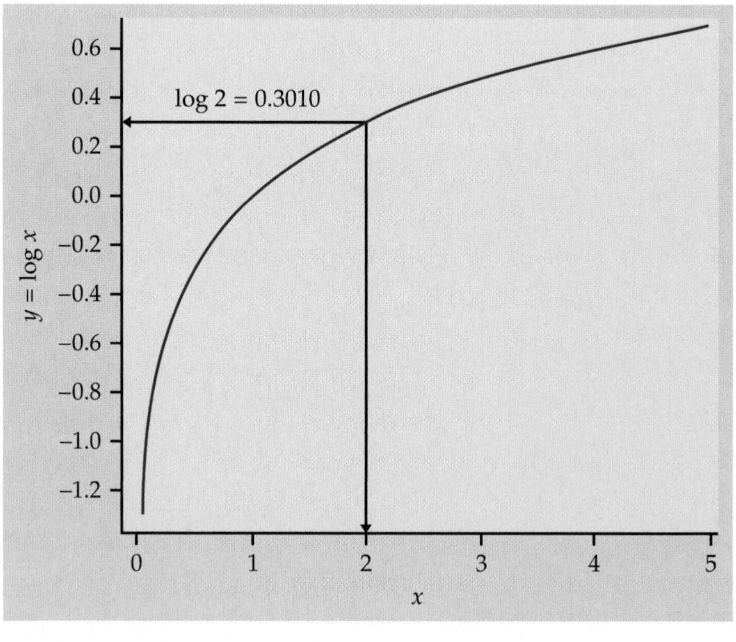

FIGURE 2.40 Any monotonic function has an inverse function that undoes the transformation and returns to the original data. This graph illustrates the process of inverting the logarithm transformation.

The regression line for predicting the logarithm of oil production from year x is

$$\log \hat{y} = -52.7 + 0.0289x$$

This line was fitted to data up to 1972. For 1977, we would predict that

$$\begin{aligned}\log \hat{y} &= -52.7 + (0.0289 \times 1977) \\ &= 4.4353\end{aligned}$$

The predicted production in the original units (millions of barrels of oil) is therefore

$$\hat{y} = 10^{4.4353} = 27,246$$

The actual production in 1977 was 21,787 million barrels. The prediction is much too high because the exponential growth model fails to fit post-1973 oil production.

Power law models

When you visit a pizza parlor, you order a pizza by its diameter, say 10 inches, 12 inches, or 14 inches. But the amount you get to eat depends on the *area* of the pizza. The area of a circle is π times the square of its radius. So the area of a round pizza with diameter x is

$$\text{area} = \pi r^2 = \pi(x/2)^2 = \pi(x^2/4) = (\pi/4)x^2$$

power law model

This is a **power law model** of the form

$$y = a \times x^p$$

When we are dealing with things of the same general form, whether circles or fish or people, we expect area to go up with the square of a dimension such as diameter or height. Volume should go up with the cube of a linear dimension. That is, geometry tells us to expect power laws in some settings.

Biologists have found that many characteristics of living things are described quite closely by power laws. There are more mice than elephants, and more flies than mice—the abundance of species follows a power law with body weight as the explanatory variable. So do pulse rate, length of life, the number of eggs a bird lays, and so on. Sometimes the powers can be predicted from geometry, but sometimes they are mysterious. Why, for example, does the rate at which animals use energy go up as the 3/4 power of their body weight? Biologists call this relationship *Kleiber's law*. It has been found to work all the way from bacteria to whales. The search goes on for some physical or geometrical explanation for why life follows power laws. There is as yet no general explanation, but power laws are a good place to start in simplifying relationships for living things.

Exponential growth models become linear when we apply the logarithm transformation to the response variable y. **Power law models become linear**

when we apply the logarithm transformation to both variables. Here are the details:

1. The power law model is

$$y = a \times x^p$$

2. Take the logarithm of both sides of this equation. You see that

$$\log y = \log a + p \log x$$

That is, taking the logarithm of both variables straightens the scatterplot of y against x.

3. Look carefully: The *power* p in the power law becomes the *slope* of the straight line that links $\log y$ to $\log x$.

Prediction in power law models

If taking the logarithms of both variables makes a scatterplot linear, a power law is a reasonable model for the original data. We can even roughly estimate what power p the law involves by regressing $\log y$ on $\log x$ and using the slope of the regression line as an estimate of the power. Remember that the slope is only an estimate of the p in an underlying power model. The greater the scatter of the points in the scatterplot about the fitted line, the smaller our confidence that this estimate is accurate.

EXAMPLE 2.35

The magical success of the logarithm transformation in Example 2.29 would not surprise a biologist. We suspect that a power law governs this relationship. Least-squares regression for the scatterplot in Figure 2.32 gives the line

$$\log \hat{y} = 1.01 + 0.72 \times \log x$$

for predicting the logarithm of brain weight from the logarithm of body weight. To undo the logarithm transformation, remember that for common logarithms with base 10, $y = 10^{\log y}$. We see that

$$\begin{aligned} \hat{y} &= 10^{1.01 + 0.72 \log x} \\ &= 10^{1.01} \times 10^{0.72 \log x} \\ &= 10.2 \times (10^{\log x})^{0.72} \end{aligned}$$

Because $10^{\log x} = x$, the estimated power model connecting predicted brain weight $\hat{y}$ with body weight x for mammals is

$$\hat{y} = 10.2 \times x^{0.72}$$

Based on footprints and some other sketchy evidence, some people think that a large apelike animal, called Sasquatch or Bigfoot, lives in the Pacific Northwest. His weight is estimated to be about 280 pounds, or 127 kilograms. How big is Bigfoot's brain? Based on the power law estimated from data on other mammals, we predict

$$\begin{aligned} \hat{y} &= 10.2 \times 127^{0.72} \\ &= 10.2 \times 32.7 \\ &= 333.7 \text{ grams} \end{aligned}$$

For comparison, gorillas have an average body weight of about 140 kilograms and an average brain weight of about 406 grams. Of course, Bigfoot may have a larger brain than his weight predicts—after all, he has avoided being captured, shot, or videotaped for many years.

SUMMARY

Nonlinear relationships between two quantitative variables can sometimes be changed into linear relationships by **transforming** one or both of the variables.

The most common transformations belong to the family of **power transformations** t^p. The logarithm $\log t$ fits into the power family at position $p = 0$.

When the variable being transformed takes only positive values, the power transformations are all **monotonic.** This implies that there is an inverse transformation that returns to the original data from the transformed values. The effect of the power transformations on data becomes stronger as we move away from linear transformations ($p = 1$) in either direction.

Transformation is particularly effective when there is reason to think that the data are governed by some mathematical model. The **exponential growth model** $y = ab^x$ becomes linear when we plot $\log y$ against x. The **power law model** $y = ax^p$ becomes linear when we plot $\log y$ against $\log x$.

We can fit exponential growth and power models to data by finding the least-squares regression line for the transformed data, then doing the inverse transformation.

SECTION 2.6 EXERCISES

2.90 Which of these transformations are monotonic increasing? Monotonic decreasing? Not monotonic? Give an equation for each transformation.

(a) You transform height in inches to height in centimeters.

(b) You transform typing speed in words per minute into seconds needed to type a word.

(c) You transform the diameter of a coin to its circumference.

(d) A composer insists that her new piece of music should take exactly 5 minutes to play. You time several performances, then transform the time in minutes into squared error, the square of the difference between 5 minutes and the actual time.

2.91 Suppose that t is an angle, measured in degrees between 0° and 180°. On what part of this range is the function $\sin t$ monotonic increasing? Monotonic decreasing?

2.92 Gordon Moore, one of the founders of Intel Corporation, predicted in 1965 that the number of transistors on an integrated circuit chip would double every 18 months. This is "Moore's law," one way to measure the revolution in computing. Here are data on the dates and number of transistors for Intel microprocessors:[52]

Processor	Date	Transistors	Processor	Date	Transistors
4004	1971	2,250	486 DX	1989	1,180,000
8008	1972	2,500	Pentium	1993	3,100,000
8080	1974	5,000	Pentium II	1997	7,500,000
8086	1978	29,000	Pentium III	1999	24,000,000
286	1982	120,000	Pentium 4	2000	42,000,000
386	1985	275,000			

(a) Explain why Moore's law says that the number of transistors grows exponentially over time.

(b) Make a plot suitable to check for exponential growth. Does it appear that the number of transistors on a chip has in fact grown approximately exponentially?

2.93 A handbook on safe handling of radioactive isotopes gives the following information on the decay of the isotope iodine 125, commonly used in medical and biological experiments:

Days	Activity	Days	Activity
0	1.000	120	0.250
20	0.794	140	0.198
40	0.630	160	0.157
60	0.500	180	0.125
80	0.397	200	0.099
100	0.315	220	0.079

(a) Make plots to decide whether a linear model, an exponential growth or decay model, or a power model best fits these data.

(b) Use regression on transformed data to fit your chosen model. What equation would you use to predict activity after x days?

(c) Do you think that the handbook reports actual laboratory data or data calculated from a theoretical model? Why?

2.94 (**Optional**) Return to Moore's law, described in Exercise 2.92.

(a) Find the least-squares regression line for predicting the logarithm of the number of transistors on a chip from the date. Before calculating your line, subtract 1971 from all the dates so that 1971 becomes year 0, 1972 is year 1, and so on.

(b) Suppose that Moore's law is exactly correct. That is, the number of transistors is 2,250 in year 0 (1971) and doubles every 18 months

(1.5 years) thereafter. Write the model for predicting transistors in year x after 1971. What is the equation of the line that, according to your model, connects the logarithm of transistors with x? Explain why a comparison of this line with your regression line from (a) shows that, while transistor counts have grown exponentially, they have grown a bit more slowly than Moore's law predicts.

2.95 Here are data for 12 perch caught in a lake in Finland:[53]

Weight (grams)	Length (cm)	Width (cm)	Weight (grams)	Length (cm)	Width (cm)
5.9	8.8	1.4	300.0	28.7	5.1
100.0	19.2	3.3	300.0	30.1	4.6
110.0	22.5	3.6	685.0	39.0	6.9
120.0	23.5	3.5	650.0	41.4	6.0
150.0	24.0	3.6	820.0	42.5	6.6
145.0	25.5	3.8	1000.0	46.6	7.6

(a) Make a scatterplot of weight against length. Describe the pattern you see.

(b) How do you expect the weight of animals of the same species to change as their length increases? Make a transformation of weight that should straighten the plot if your expectation is correct. Plot the transformed weights against length. Is the plot now roughly linear?

2.96 Plot the widths of the 12 perch in the previous problem against their lengths. What is the pattern of the plot? Explain why we should expect this pattern.

2.97 (Exact exponential growth) The common intestinal bacterium *E. coli* is one of the fastest-growing bacteria. Under ideal conditions, the number of *E. coli* in a colony doubles about every 15 minutes until restrained by lack of resources. Starting from a single bacterium, how many *E. coli* will there be in 1 hour? In 5 hours?

2.98 (Exact exponential growth) A paper in a scholarly journal once claimed (I am not making this up), "Every year since 1950, the number of American children gunned down has doubled."[54] To see that this is silly, suppose that in 1950 just 1 child was "gunned down" and suppose that the paper's claim is exactly right.

(a) Make a table of the number of children killed in each of the next 10 years, 1951 to 1960.

(b) Plot the number of deaths against the year and connect the points with a smooth curve. This is an exponential curve.

(c) The paper appeared in 1995, 45 years after 1950. How many children were killed in 1995, according to the paper?

(d) Take the logarithm of each of your counts from (a). Plot these logarithms against the year. You should get a straight line.

(e) From your graph in (d) find the approximate values of the slope b and the intercept a for the line. Use the equation $y = a + bx$ to predict the logarithm of the count for the 45th year. Check your result by taking the logarithm of the count you found in (c).

2.99 (Exact exponential growth) Maria is given a savings bond at birth. The bond is initially worth $500 and earns interest at 7.5% each year. This means that the value is multiplied by 1.075 each year.

(a) Find the value of the bond at the end of 1 year, 2 years, and so on up to 10 years.

(b) Plot the value y against years x. Connect the points with a smooth curve. This is an exponential curve.

(c) Take the logarithm of each of the values y that you found in (a). Plot the logarithm $\log y$ against years x. You should obtain a straight line.

2.100 (Exact exponential growth) Fred and Alice were born the same year, and each began life with $500. Fred added $100 each year, but earned no interest. Alice added nothing, but earned interest at 7.5% annually. After 25 years, Fred and Alice are getting married. Who has more money?

2.101 The following table gives the resident population of the United States from 1790 to 2000, in millions of persons:

Date	Pop.	Date	Pop.	Date	Pop.	Date	Pop.
1790	3.9	1850	23.2	1910	92.0	1970	203.3
1800	5.3	1860	31.4	1920	105.7	1980	226.5
1810	7.2	1870	39.8	1930	122.8	1990	248.7
1820	9.6	1880	50.2	1940	131.7	2000	281.4
1830	12.9	1890	62.9	1950	151.3		
1840	17.1	1900	76.0	1960	179.3		

(a) Plot population against time. The growth of the American population appears roughly exponential.

(b) Plot the logarithms of population against time. The pattern of growth is now clear. An expert says that "the population of the United States increased exponentially from 1790 to about 1880. After 1880 growth was still approximately exponential, but at a slower rate." Explain how this description is obtained from the graph.

2.102 Do mold colonies grow exponentially? In an investigation of the growth of molds, biologists inoculated flasks containing a growth medium with equal amounts of spores of the mold *Aspergillus nidulans*. They measured the size of a colony by analyzing how much remains of a radioactive tracer substance that is consumed by the mold as it grows. Each size measurement requires destroying that colony, so that the data below refer to 30 separate colonies. To smooth the pattern, we take the mean size of the three colonies measured at each time.[55]

Hours	Colony sizes			Mean
0	1.25	1.60	0.85	1.23
3	1.18	1.05	1.32	1.18
6	0.80	1.01	1.02	0.94
9	1.28	1.46	2.37	1.70
12	2.12	2.09	2.17	2.13
15	4.18	3.94	3.85	3.99
18	9.95	7.42	9.68	9.02
21	16.36	13.66	12.78	14.27
24	25.01	36.82	39.83	33.89
36	138.34	116.84	111.60	122.26

(a) Graph the mean colony size against time. Then graph the logarithm of the mean colony size against time.

(b) On the basis of data like these, microbiologists divide the growth of mold colonies into three phases that follow each other in time. Exponential growth occurs during only one of these phases. Briefly describe the three phases, making specific reference to the graphs to support your description.

(c) The exponential growth phase for these data lasts from about 6 hours to about 24 hours. Find the least-squares regression line of the logarithms of mean size on hours for only the data between 6 and 24 hours. Use this line to predict the size of a colony 10 hours after inoculation. (The line predicts the logarithm. You must obtain the size from its logarithm.)

2.103 It is easy to measure the "diameter at breast height" of a tree. It's hard to measure the total "aboveground biomass" of a tree, because to do this you must cut and weigh the tree. The biomass is important for studies of ecology, so ecologists commonly estimate it using a power law. Combining data on 378 trees in tropical rain forests gives this relationship between biomass y measured in kilograms and diameter x measured in centimeters:[56]

$$\log_e y = -2.00 + 2.42 \log_e x$$

Note that the investigators chose to use *natural logarithms*, with base $e = 2.71828$, rather than common logarithms with base 10.

(a) Translate the line given into a power model. Use the fact that for natural logarithms,

$$y = e^{\log_e y}$$

(b) Estimate the biomass of a tropical tree 30 centimeters in diameter.

TABLE 2.15 Body weight and life span for several species of mammals

Species	Weight (kg)	Life span (years)
Baboon	32	20
Beaver	25	5
Cat, domestic	2.5	12
Chimpanzee	45	20
Dog	8.5	12
Elephant	2800	35
Goat, domestic	30	8
Gorilla	140	20
Grizzly bear	250	25
Guinea pig	1	4
Hippopotamus	1400	41
Horse	480	20
Lion	180	15
Mouse, house	0.024	3
Pig, domestic	190	10
Red fox	6	7
Sheep, domestic	30	12

2.104 Find the correlation between the logarithm of mean size and hours for the data between 6 and 24 hours in Exercise 2.102. Make a scatterplot of the logarithms of the individual size measurements against hours for this same period and find the correlation. Why do we expect the second r to be smaller? Is it in fact smaller?

2.105 Table 2.15 gives the average weight and average life span in captivity for several species of mammals. Some writers on power laws in biology claim that life span depends on body weight according to a power law with power $p = 0.2$. Fit a power law model to these data (using logarithms). Does this small set of data appear to follow a power law with power close to 0.2? Use your fitted model to predict the average life span for humans (average weight 65 kilograms). Humans are an exception to the rule.

2.106 Bigger people are generally stronger than smaller people, though there's a lot of individual variation. Let's find a theoretical model. Body weight increases as the cube of height. The strength of a muscle increases with its cross-sectional area, which we expect to go up as the square of height. Put these together: What power law should describe how muscle strength increases with weight?

2.107 Let's apply your result from the previous problem. Graph the power law relation between strength and body weight for weights from (say) 1 to 1000. (Constants in the power law just reflect the units of measurement used, so we can ignore them.) Use the graph to explain why a person 1 million times as heavy as an ant can't lift a million times as much as an ant can lift.

2.108 Physiologists say that the resting heart rate of humans is related to our body weight by a power law. Specifically, average heart rate y (beats per minute) is found from body weight x (kilograms) by[57]

$$y = 241 \times x^{-1/4}$$

Let's try to make sense of this. Kleiber's law says that energy use in animals, including humans, increases as the 3/4 power of body weight. But the weight of human hearts and lungs and the volume of blood in the body are directly proportional to body weight. Given these facts, you should not be surprised that heart rate is proportional to the $-1/4$ power of body weight. Why not?

2.109 Galileo studied motion by rolling balls down ramps. Newton later showed how Galileo's data fit his general laws of motion. Imagine that you are Galileo, without Newton's laws to guide you. He rolled a ball down a ramp at different heights above the floor and measured the horizontal distance the ball traveled before it hit the floor. Here are Galileo's data when he placed a horizontal shelf at the end of the ramp so that the ball is moving horizontally when it starts to fall. (We won't try to describe the obscure seventeenth-century units Galileo used to measure distance.)[58]

Distance	Height
1500	1000
1340	828
1328	800
1172	600
800	300

Plot distance y against height x. The pattern is very regular, as befits data described by a physical law. We want to find distance as a function of height. That is, we want to transform x to straighten the graph.

(a) Think before you calculate: Will powers x^p for $p < 1$ or $p > 1$ tend to straighten the graph. Why?

(b) Move along the ladder of transformations in the direction you have chosen until the graph is nearly straight. What transformation do you suggest?

2.110 Table 2.16 gives data on the mean number of seeds produced in a year by several common tree species and the mean weight (in milligrams) of the seeds produced. (Some species appear twice because their seeds were counted in two locations.) We might expect that trees with heavy seeds produce fewer of them, but what is the form of the relationship?[59]

TABLE 2.16 Count and weight of seeds produced by common tree species

Tree species	Seed count	Seed weight (mg)	Tree species	Seed count	Seed weight (mg)
Paper birch	27,239	0.6	American beech	463	247
Yellow birch	12,158	1.6	American beech	1,892	247
White spruce	7,202	2.0	Black oak	93	1,851
Engelmann spruce	3,671	3.3	Scarlet oak	525	1,930
Red spruce	5,051	3.4	Red oak	411	2,475
Tulip tree	13,509	9.1	Red oak	253	2,475
Ponderosa pine	2,667	37.7	Pignut hickory	40	3,423
White fir	5,196	40.0	White oak	184	3,669
Sugar maple	1,751	48.0	Chestnut oak	107	4,535
Sugar pine	1,159	216.0			

(a) Make a scatterplot showing how the weight of tree seeds helps explain how many seeds the tree produces. Describe the form, direction, and strength of the relationship.

(b) If a power law holds for this relationship, the logarithms of the original data will display a linear pattern. Use your calculator or software to obtain the logarithms of both the seed weights and the seed counts in Table 2.16. Make a new scatterplot using these new variables. Now what are the form, direction, and strength of the relationship?

2.111 Long ago, Thomas Malthus (1766–1834) claimed that population grows exponentially while food supplies grow only linearly. Result: starvation holds down population. Table 2.17 gives data on the population of the world, as estimated by the U.S. Census Bureau. Do a careful analysis, using our study of world petroleum production as a model, and write a brief discussion of the pattern of increase in the world's population since 1950. Has population grown exponentially since 1950? Has population growth slowed in recent years?

TABLE 2.17 Midyear world population, billions

Year	Population	Year	Population	Year	Population
1950	2.555	1968	3.558	1986	4.938
1952	2.635	1970	3.708	1988	5.110
1954	2.728	1972	3.862	1990	5.284
1956	2.833	1974	4.015	1992	5.450
1958	2.945	1976	4.160	1994	5.611
1960	3.039	1978	4.305	1996	5.769
1962	3.136	1980	4.457	1998	5.925
1964	3.277	1982	4.613	2000	6.080
1966	3.416	1984	4.774	2002	6.234

CHAPTER 2 EXERCISES

2.112 Here is one way in which nature regulates the size of animal populations: high population density attracts predators, who remove a higher proportion of the population than when the density of the prey is low. One study looked at kelp perch and their common predator, the kelp bass. The researcher set up four large circular pens on sandy ocean bottom in southern California. He chose young perch at random from a large group and placed 10, 20, 40, and 60 perch in the four pens. Then he dropped the nets protecting the pens, allowing bass to swarm in, and counted the perch left after 2 hours. Here are data on the percent of perch eaten in four repeats of this setup:[60]

Perch	Proportion killed			
10	0.0	0.1	0.3	0.3
20	0.2	0.3	0.3	0.6
40	0.075	0.3	0.6	0.725
60	0.517	0.55	0.7	0.817

(a) Why is the response variable the *proportion* of perch in a pen who are killed rather than the *count* of perch killed?

(b) Make a graph to examine how proportion killed changes with the number of perch in the pen. The relationship is roughly linear. Give a numerical measure of its strength.

(c) Do the data support the principle that "more prey attract more predators, who drive down the number of prey"?

2.113 Endangered animal species often live in isolated patches of habitat. If the population in a patch varies a lot (due to weather, for example), the species is more likely to disappear from that patch in a bad year. Here is a general question: Is there less variation in population size when a patch of habitat has more diverse vegetation? If so, maintaining habitat diversity can help protect endangered species.

A researcher measured the variation over time in the population of a cricket species in 45 habitat patches. He also measured the diversity of each patch.[61] He reported his results by giving the least-squares equation

$$\text{population variation} = 84.4 - 0.13 \times \text{diversity}$$

along with the fact that $r^2 = 0.34$. Do these results support the idea that more diversity goes with less variation in population size? Is the relationship very strong or only moderately strong?

Studies of disease often ask people about their diet in years past in order to discover links between diet and disease. How well do people remember their past diet? Can we predict actual past diet as well or better from what subjects eat now as from their memory of past habits? Data on actual past diet are available for 91 people who were asked about their diet when they were 18 years old and again when they were 30. Researchers asked them at about age 55 to describe their eating habits at ages 18 and 30 and also their current diet. The study report says:

> The first study aim, to determine how accurately this group of participants remembered past consumption, was addressed by correlations between recalled and historical consumption in each time period. To evaluate the second study aim, that is, whether recalled intake or current intake more accurately predicts historical intake of food groups at age 30 years, we performed regression analysis.[62]

Exercises 2.114 to 2.116 ask you to interpret the results of this paper to a group of people who know no statistics.

2.114 Explain in nontechnical language what "correlation" means, why correlation suits the first aim of the study, what "regression" means, and why regression fits the second study aim. Be sure to point out the distinction between correlation and regression.

2.115 The study looked at the correlations between actual intake of many foods at age 18 and the intake the subjects now remember for age 18. The median correlation was $r = 0.217$. The authors say, "We conclude that memory of food intake in the distant past is fair to poor." Explain to your audience why $r = 0.217$ points to this conclusion.

2.116 The authors used regression to predict the intake of a number of foods at age 30 from current intake of those foods and from what the subjects now remember about their intake at age 30. They conclude that "recalled intake more accurately predicted historical intake at age 30 years than did current diet." As evidence, they present r^2-values for the regressions. Explain to your audience why comparing r^2-values is one way to compare how well different explanatory variables predict a response.

2.117 To study the energy savings due to adding solar heating panels to a house, researchers measured the natural gas consumption of the house for more than a year, then installed solar panels and observed the natural gas consumption for almost two years. The explanatory variable x is degree-days per day during the several weeks covered by each observation, and the response variable y is gas consumption (in hundreds of cubic feet) per day during the same period. Figure 2.41 plots y against x, with separate symbols for observations taken before and after the installation of the solar panels.[63] The least-squares regression lines were computed separately for the before and after data and are drawn on the plot. The regression lines are:

$$\text{Before: } \hat{y} = 1.089 + 0.189x$$
$$\text{After: } \hat{y} = 0.853 + 0.157x$$

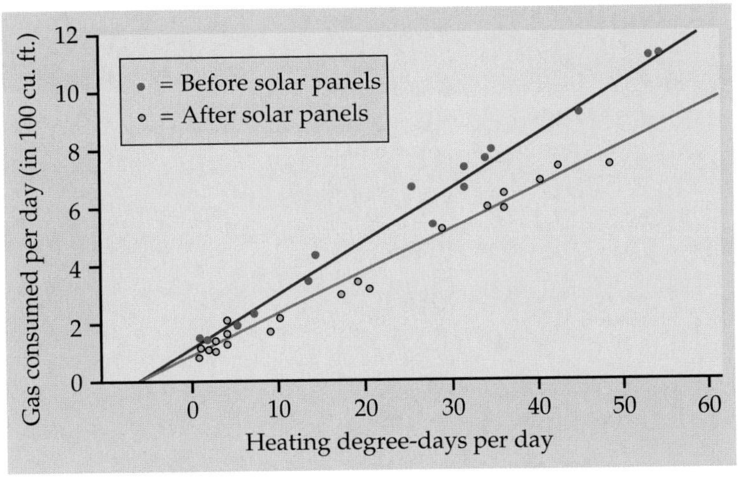

FIGURE 2.41 The regression of residential natural gas consumption on heating degree-days before and after installation of solar heating panels, for Exercise 2.117.

(a) Does the scatterplot suggest that a straight line is an appropriate description of the relationship between degree-days and natural gas consumption? Do any individual observations appear to have large residuals or to be highly influential?

(b) About how much additional natural gas was consumed per day for each additional degree-day before the panels were added? After the panels were added?

(c) The daily average temperature during January in this location is about 30°, which corresponds to 35 degree-days per day. Use the regression lines to predict daily gas usage for a day with 35 degree-days before and after installation of the panels.

(d) The Energy Information Agency predicted that natural gas would cost consumers about \$0.96 per hundred cubic feet in the winter of 2000–2001. At this rate, how much money do the solar panels save in the 31 days of a typical January?

2.118 We have seen that investment theory uses the standard deviation of returns to describe the volatility or risk of an investment. To describe how the risk of a specific security is related to that of the market as a whole, we use least-squares regression. Figure 2.42 plots the monthly percent total return y on Philip Morris common stock against the monthly return x on the Standard & Poor's 500-stock index, which represents the market, for the period between August 1990 and June 1997. The one clear outlier turns out not to be very influential. Here are the basic descriptive measures:

$$\bar{x} = 1.304 \quad s_x = 3.392 \quad r = 0.5251$$
$$\bar{y} = 1.878 \quad s_y = 7.554$$

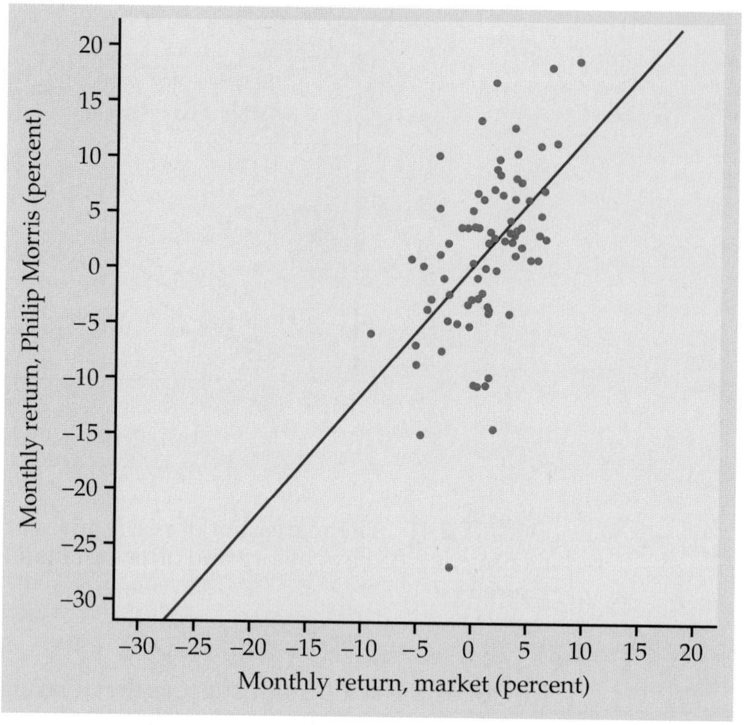

FIGURE 2.42 Monthly returns on an individual stock plotted against the returns for the stock market as a whole, with the least-squares line. See Exercise 2.118.

(a) Find the equation of the least-squares line from this information. What percent of the volatility in Philip Morris stock is explained by the linear relationship with the market as a whole?

(b) Explain carefully what the slope of the line tells us about how Philip Morris stock responds to changes in the market. This slope is called "beta" in investment theory.

(c) Returns on most individual stocks have a positive correlation with returns on the entire market. That is, when the market goes up, an individual stock tends to also go up. Explain why an investor should prefer stocks with beta > 1 when the market is rising and stocks with beta < 1 when the market is falling.

2.119 Lamb's-quarter is a common weed that interferes with the growth of corn. An agriculture researcher planted corn at the same rate in 16 small plots of ground, then weeded the plots by hand to allow a fixed number of lamb's-quarter plants to grow in each meter of corn row. No other weeds were allowed to grow. Here are the yields of corn (bushels per acre) in each of the plots:[64]

Weeds per meter	Corn yield	Weeds per meter	Corn yield	Weeds per meter	Corn yield	Weeds per meter	Corn yield
0	166.7	1	166.2	3	158.6	9	162.8
0	172.2	1	157.3	3	176.4	9	142.4
0	165.0	1	166.7	3	153.1	9	162.8
0	176.9	1	161.1	3	156.0	9	162.4

(a) What are the explanatory and response variables in this experiment?

(b) Make side-by-side stemplots of the yields, after rounding to the nearest bushel. What do you conclude about the effect of this weed on corn yield?

(c) Make a scatterplot of corn yield against weeds per meter. Find the least-squares regression line and add it to your plot. The advantage of regression over the side-by-side comparison in (b) is that we can use the fitted model to draw conclusions for counts of weeds other than the ones the researcher actually used. What does the slope of the fitted line tell us about the effect of lamb's-quarter on corn yield?

(d) Predict the yield for corn grown under these conditions with 6 lamb's-quarter plants per meter of row.

2.120 The table below gives data on the relationship between running speed (feet per second) and stride rate (steps taken per second) for elite female runners:[65]

Speed	15.86	16.88	17.50	18.62	19.97	21.06	22.11
Stride rate	3.05	3.12	3.17	3.25	3.36	3.46	3.55

Here are the corresponding data from the same source for male runners:

Speed	15.86	16.88	17.50	18.62	19.97	21.06	22.11
Stride rate	2.92	2.98	3.03	3.11	3.22	3.31	3.41

(a) Plot the data for both groups on one graph using different symbols to distinguish between the points for females and those for males.

(b) Suppose now that the data came to you without identification as to gender. Compute the least-squares line from all of the data and plot it on your graph.

(c) Compute the residuals from this line for each observation. Make a plot of the residuals against speed. How does the fact that the data come from two distinct groups show up in the residual plot?

2.121 Wood scientists are interested in replacing solid-wood building material by less expensive products made from wood flakes. They collected the following data to examine the relationship between the length (in inches) and the strength (in pounds per square inch) of beams made from wood flakes.[66]

Length	5	6	7	8	9	10	11	12	13	14
Strength	446	371	334	296	249	254	244	246	239	234

(a) Make a scatterplot that shows how the length of a beam affects its strength.

(b) Describe the overall pattern of the plot. Are there any outliers?

(c) Fit a least-squares line to the entire set of data. Graph the line on your scatterplot. Does a straight line adequately describe these data?

(d) The scatterplot suggests that the relation between length and strength can be described by *two* straight lines, one for lengths less than 9 inches and another for lengths greater than 9 inches. Fit least-squares lines to these two subsets of the data, and draw the lines on your plot. Do they describe the data adequately? What question would you now ask the wood experts?

2.122 The *Correlation and Regression* applet at www.whfreeman.com/ips allows you to create a scatterplot and to move points by dragging with the mouse. You are going to make scatterplots with 10 points that have correlation close to 0.7. The lesson is that many patterns can have the same correlation. Always plot your data before you trust a correlation.

(a) Stop after adding the first two points. What is the value of the correlation? Why does it have this value?

(b) Make a lower-left to upper-right pattern of 10 points with correlation about $r = 0.7$. (You can drag points up or down to adjust r after you have 10 points.) Make a rough sketch of your scatterplot.

(c) Make another scatterplot with 9 points in a vertical stack at the left of the plot. Add one point far to the right and move it until the correlation is close to 0.7. Make a rough sketch of your scatterplot.

(d) Make yet another scatterplot with 10 points in a curved pattern that starts at the lower left, rises to the right, then falls again at the far right. Adjust the points up or down until you have a quite smooth curve with correlation close to 0.7. Make a rough sketch of this scatterplot also.

2.123 In the previous exercise, you created three scatterplots having correlation about $r = 0.7$ between the horizontal variable x and the vertical variable y. Correlation $r = 0.7$ is considered reasonably strong in many areas of scientific work. Because there is a reasonably strong correlation, we might use a regression line to predict y from x. Re-create your three plots after clicking "Show least-squares line." In which of your three scatterplots does it make sense to use a straight line for prediction?

2.124 One reason to invest abroad is that markets in different countries don't move in step. When American stocks go down, foreign stocks may go up. So an investor who holds both bears less risk. That's the theory. Now we read: "The correlation between changes in American and European share prices has risen from 0.4 in the mid-1990s to 0.8 in 2000."[67] Explain to an investor who knows no statistics why this fact reduces the protection provided by buying European stocks.

2.125 The same article that claims that the correlation between changes in stock prices in Europe and the United States was 0.8 in 2000 goes on to say: "Crudely, that means that movements on Wall Street can explain 80% of price movements in Europe." Is this true? What is the correct percent explained if $r = 0.8$?

2.126 Someone says, "There is a strong positive correlation between the number of firefighters at a fire and the amount of damage the fire does. So sending lots of firefighters just causes more damage." Explain why this reasoning is wrong.

2.127 The following table gives the U.S. resident population of voting age and the votes cast for president, both in thousands, for presidential elections between 1960 and 2000:

Year	Population	Votes	Year	Population	Votes
1960	109,672	68,838	1984	173,995	92,653
1964	114,090	70,645	1988	181,956	91,595
1968	120,285	73,212	1992	189,524	104,425
1972	140,777	77,719	1996	196,511	96,456
1976	152,308	81,556	2000	209,128	105,363
1980	163,945	86,515			

(a) For each year compute the percent of people who voted. Make a time plot of the percent who voted. Describe the change over time in participation in presidential elections.

(b) Before proposing political explanations for this change, we should examine possible lurking variables. The minimum voting age in presidential elections dropped from 21 to 18 years in 1970. Use this fact to propose a partial explanation for the trend you saw in (a).

2.128 We expect that students who do well on the midterm exam in a course will usually also do well on the final exam. Gary Smith of Pomona College looked at the exam scores of all 346 students who took his statistics class over a 10-year period.[68] The least-squares line for predicting final exam score from midterm exam score was $\hat{y} = 46.6 + 0.41x$.

Octavio scores 10 points above the class mean on the midterm. How many points above the class mean do you predict that he will score on the final? (*Hint:* Use the fact that the least-squares line passes through the point

$(\bar{x}, \bar{y})$ and the fact that Octavio's midterm score is $\bar{x} + 10$. This is an example of the phenomenon that gave "regression" its name: students who do well on the midterm will on the average do less well on the final, but still above the class mean.)

2.129 (**Optional**) Expose marine bacteria to X-rays for time periods from 1 to 15 minutes. Here are the number of surviving bacteria (in hundreds) on a culture plate after each exposure time:[69]

Time *t*	Count *y*	Time *t*	Count *y*
1	355	9	56
2	211	10	38
3	197	11	36
4	166	12	32
5	142	13	21
6	106	14	19
7	104	15	15
8	60		

Theory suggests an exponential growth or decay model. Do the data appear to conform to this theory?

2.130 (**Optional**) Electronic fund transfers, from bank automatic teller machines and the use of debit cards by consumers, have grown rapidly in the United States. Here are data on the number of such transfers (in millions):[70]

Year	EFT	Year	EFT	Year	EFT
1985	3,579	1991	6,642	1996	11,780
1987	4,108	1992	7,537	1997	12,580
1988	4,581	1993	8,135	1998	13,160
1989	5,274	1994	9,078	1999	13,316
1990	5,942	1995	10,464		

Write a clear account of the pattern of growth of electronic transfers over time, supporting your description with plots and calculations as needed. Has the pattern changed in the most recent years?

2.131 (**Optional**) What is the relationship between the mature size of a fish species and the size of the fish when they begin to reproduce? To answer this question, scientists collected data on 85 species of fish from many published studies. The data set `ex02_131.dat`, available on the Web site `www.whfreeman.com/ips`, contains data on two variables for these 85 species.[71] The first variable is the "asymptotic body length," the length (in centimeters) to which female fish will eventually grow. The second variable is the length (also in centimeters) at which 50% of females first reproduce.

Explore the relationship between these variables carefully. Does any simple model describe the relationship well?

2.132 Can we predict college grade point average from SAT scores and high school grades? The CSDATA data set described in the Data Appendix contains information on this issue for a large group of computer science students. We will look only at SAT mathematics scores as a predictor of later college GPA, using the variables SATM and GPA from CSDATA. Make a scatterplot, obtain r and r^2, and draw on your plot the least-squares regression line for predicting GPA from SATM. Then write a brief discussion of the ability of SATM alone to predict GPA. (In Chapter 11 we will see how combining several explanatory variables improves our ability to predict.)

Prelude: how to produce data badly

In Rapides Parish, Louisiana, only one company is allowed to provide ambulance service. The local paper, the *Town Talk*, asked readers to call in to offer their opinion on whether the company should keep its monopoly. Call-in polls are generally automated: call one telephone number to vote "Yes" and call another number to vote "No." Telephone companies often charge callers to these numbers.

The *Town Talk* got 3763 calls, which suggests unusual interest in ambulance service. Investigation showed that 638 calls came from the ambulance company office or from the homes of its executives. Many more no doubt came from lower-level employees. "We've got employees who are concerned about this situation, their job stability and their families and maybe called more than they should have," said a company vice-president. Other sources said employees were told to, as they say in Chicago, "vote early and often."[1]

Sampling is one common type of data production: we gather information from a sample of people in order to draw conclusions about a larger group of people. We will see that a sample size of 3763 is quite respectable—if the sample is properly designed. Inviting people to call (and call, and call . . .) isn't proper sample design. In this chapter we will learn how the pros take a sample, and why it beats the *Town Talk* method.

CHAPTER 3

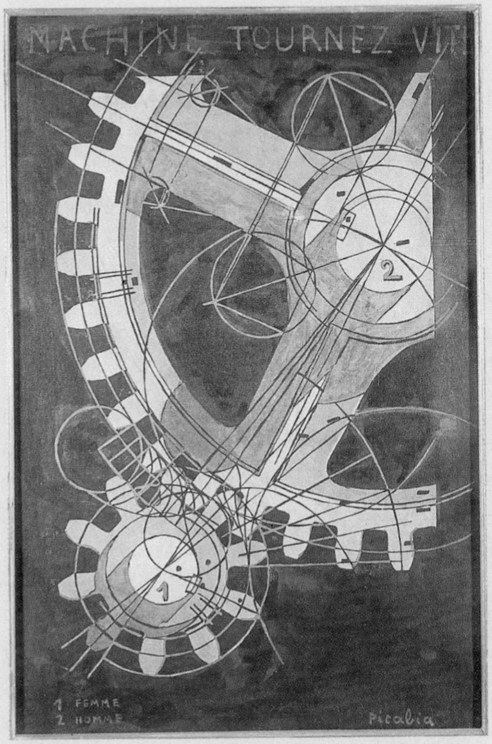

Francis Picabia, *Machines Turn Quickly*.

Producing Data

Introduction

Numerical data are raw material for the growth of knowledge. Statistics is a tool that helps data produce knowledge rather than confusion. As such, it must be concerned with producing data as well as with interpreting already available data.

Chapters 1 and 2 explored the art of *data analysis*. They showed how to use graphs and numbers to uncover the nature of a set of data. We can distinguish two different attitudes toward data. We may ask, "What do we see in the data?" Or we may ask, "What answer do the data give to this specific question?" This distinction becomes important as we turn to producing data, because trustworthy answers to specific questions often require data produced for exactly that purpose.

exploratory data analysis

"What do we see?" is the attitude of **exploratory data analysis.** Sometimes we meet data that seem interesting and that deserve careful exploration. We do not know what the data may show, and we may have no specific questions in mind before we begin our examination. We have explored the relationship between SAT scores in the states and the percent of students who take the SAT, looked for explanations of obesity in girls, and examined the relationship between education and income. Exploratory analysis relies heavily on plotting the data. It seeks patterns that suggest interesting conclusions or questions for further study. However, exploratory analysis alone can rarely provide convincing evidence for its conclusions, because striking patterns in data can arise from many sources.

We may also use data to provide clear answers to specific questions such as "What is the speed of light?" or "How many plants per acre give the highest yield of corn?" or "Does the age at which a child starts to talk predict a later test score?" Data that answer such questions are as much the product of intelligent effort as hybrid tomatoes and DVD players. This chapter is devoted to developing the skills needed to produce trustworthy data and to judge the quality of data produced by others. The techniques for producing data discussed in this chapter involve no formulas or graphs, but they are among the most important ideas in statistics.

statistical inference

Statistical techniques for producing data open the door to a further advance in data analysis, **formal statistical inference,** which answers specific questions with a known degree of confidence. Statistical inference uses the descriptive tools we have already met, in combination with new kinds of reasoning. It is primarily numerical rather than graphical. The later chapters of this book are devoted to inference. We will see that careful design of data production is the most important prerequisite for trustworthy inference.

3.1 First Steps

The first questions you must answer when you plan to produce data are familiar: "What individuals shall I study?" "What variables shall I measure?" Answering these questions gives a clear statement of your interests. We will take for granted in this chapter that all the variables we work with are clearly defined and satisfactorily measured.

Statistics begins to play an essential role when the focus turns from measurements on individuals (whether ball bearings, rats, or people) to arrangements for collecting data from many individuals. These arrangements or patterns for producing data are called **designs.** Some questions that a design must address are: How many individuals shall we collect data from? How shall we select the individuals to be studied? If, as in many experiments, several groups of individuals are to receive different treatments, how shall we form the groups? If there is no systematic design for producing data, we may be misled by haphazard or incomplete data or by confounding.

designs

Where to find data: the library and the Internet

You want data on a question of interest to you. What are the social and economic backgrounds of college undergraduates? What causes of death are most common among young people? Or you are writing a paper on homelessness and wonder how many people are homeless. How will you get relevant data?

It is tempting to simply base our conclusions on our own experience, making no use of systematic data. We think (without really thinking) that the students at our college are typical. Or we recall an unusual incident that sticks in our memory exactly because it is unusual. We remember an airplane crash that killed several hundred people. We may ignore the fact that data on all flights show that flying is much safer than driving. We are relying on *anecdotes* rather than data.

Anecdotal Evidence

Anecdotal evidence is based on haphazardly selected individual cases, which often come to our attention because they are striking in some way. These cases need not be representative of any larger group of cases.

A better tactic is to head for the library or the Internet. There we find abundant data, not gathered specifically to answer our questions but available for our use. The annual *Statistical Abstract of the United States* is available both in print and online. It contains data that shed light on both college students and causes of death. You can learn, for example, that more than 33% of college students are employed full-time and that accidents are the leading cause of death among men aged 25 to 44 but rank behind cancer in killing women in that age group. The *Statistical Abstract* has nothing on the homeless, however.

More detailed and more recent data are often available on the Internet, though locating them can be time-consuming. Government statistical offices are the primary source for demographic, economic, and social data. The United States does not have a central government statistical office. More than 70 different federal agencies collect data. Fortunately, you can reach most

The Web sites of government statistical offices are prime sources of data. Here is the home page of the Census Bureau's American Community Survey.

of them through the government's handy FedStats site (`www.fedstats.gov`). If you select the National Center for Education Statistics, you can search the more than 400 tables of *The Digest of Education Statistics* to satisfy your curiosity about undergraduates. Many nations do have a single national statistical office, such as Statistics Canada (`www.statcan.ca`) or Mexico's INEGI (`www.inegi.gob.mx`).

EXAMPLE 3.1

You seek data on homeless people for a paper you are writing. Your Internet harvest will change, probably from week to week, but here is an example of search results. Searching government sites through FedStats discloses that the 2000 census sent enumerators to emergency shelters and other areas known to be occupied by homeless people. The homeless are included in census counts, but there is as yet no separate count available. At the Department of Housing and Urban Development, you find a detailed study of the people who use homeless assistance programs, based on a national sample of such programs.[2] Your report will profit from this study, but it makes no attempt to count the homeless. It was much faster to use the keyword search facilities at these agencies than to spend an afternoon in the library.

One of the best sources of data on public opinion and social trends is the General Social Survey (GSS) operated by the University of Chicago's National Opinion Research Center. A search of the GSS archives `www.icpsr.umich.edu/GSS` shows that in 2001 the GSS asked if "you or members of your family" gave money to "the homeless or street people" during the past year. But it appears that since 1991 the GSS has not asked its respondents if they themselves have been homeless—and the GSS interviews only people who live in households.

Various advocacy groups do offer counts. The most careful of these is the National Coalition for the Homeless (`www.nationalhomeless.org`). A fact sheet at this site explains why homelessness is hard to measure and collects several different estimates of the number of homeless people, with references and comments. These estimates are not all equally trustworthy, but the data there will strengthen your paper.

The library and the Internet are sources of *available data*.

> **Available Data**
>
> **Available data** are data that were produced in the past for some other purpose but that may help answer a present question.

Available data are the only data used in most student reports. Because producing new data is expensive, we all use available data whenever possible. However, the clearest answers to present questions often require data produced to answer those specific questions. Statistical designs for producing data rely on either *sampling* or *experiments*.

Sampling

sample
population

The General Social Survey interviews about 3000 adult residents of the United States every second year. That is, the GSS selects a **sample** of adults to represent the larger **population** of all adults living in the United States. The idea of *sampling* is to study a part in order to gain information about the whole. Data are often produced by sampling a population of people or things. Opinion polls, for example, report the views of the entire country based on interviews with a sample of about 1000 to 1500 people. Government reports on employment and unemployment are produced from a monthly sample of about 55,000 households. The quality of manufactured items is monitored by inspecting small samples each hour or each shift.

census

In all of these examples, the expense of examining every item in the population makes sampling a practical necessity. Timeliness is another reason for preferring a sample to a **census,** which is an attempt to contact every individual in the entire population. We want information on current unemployment and public opinion next week, not next year. Moreover, a carefully conducted sample is often more accurate than a census. Accountants, for example, sample a firm's inventory to verify the accuracy of the records. Attempting to count every last item in the warehouse would be not only expensive but inaccurate. Bored people do not count carefully.

If conclusions based on a sample are to be valid for the entire population, a sound design for selecting the sample is required. Sampling designs are the topic of Section 3.3.

Experiments

A sample survey collects information about a population by selecting and measuring a sample from the population. The goal is a picture of the population, disturbed as little as possible by the act of gathering information. Sample surveys are one kind of *observational study*.

Observation versus Experiment

An **observational study** observes individuals and measures variables of interest but does not attempt to influence the responses.

An **experiment** deliberately imposes some treatment on individuals in order to observe their responses.

An observational study, even one based on a statistical sample, is a poor way to gauge the effect of an intervention. To see the response to a change, we must actually impose the change. When our goal is to understand cause and effect, experiments are the only source of fully convincing data.

EXAMPLE 3.2

A large study of child care enrolled 1364 children in 1991 and has now followed them for more than a decade. In 2001, the researchers announced that "children who spend most of their time in child care are three times as likely to exhibit behavioral problems in kindergarten as those who are cared for primarily by their mothers."[3]

This is an observational study. Parents made all child care decisions and the study did not attempt to influence them. We can't conclude that child care *causes* behavioral problems. Perhaps employed parents who use child care are under stress and the children react to their parents' stress. Perhaps single parents are more likely to use child care. Perhaps parents are more likely to place in child care children who already have behavior problems. We can imagine an experiment that would remove these difficulties. From a large group of young children, choose some to be placed in child care and others to remain at home. This is an experiment because the treatment (child care or not) is imposed on the children. Of course, this particular experiment is neither practical nor ethical.

In Example 3.2, the effect of child care on behavior is *confounded* with (mixed up with) the characteristics of families who use child care. Observational studies of the effect of one variable on another often fail because the explanatory variable is confounded with lurking variables. Because experiments allow us to pin down the effects of specific variables of interest to us, they are the preferred method for gaining knowledge in science, medicine, and industry.

Experiments, like samples, require careful design. Because of the unique importance of experiments as a method for producing data, we begin the discussion of statistical designs for data collection in Section 3.2 with the principles underlying the design of experiments.

SUMMARY

Data analysis is sometimes **exploratory** in nature. Exploratory analysis asks what the data tell us about the variables and their relations to each other. The conclusions of an exploratory analysis may not generalize beyond the specific data studied.

Statistical inference produces answers to specific questions, along with a statement of how confident we can be that the answer is correct. The conclusions of statistical inference are usually intended to apply beyond the individuals actually studied. Successful statistical inference usually requires **production of data** intended to answer the specific questions posed.

Anecdotal evidence based on a few individual cases is rarely trustworthy. **Available data** collected for other purposes, such as the data produced by government statistical offices, are often helpful.

We can produce data intended to answer specific questions by sampling or experimentation. **Sampling** selects a part of a population of interest to represent the whole. **Experiments** are distinguished from **observational studies** such as sample surveys by the active imposition of some treatment on the subjects of the experiment.

SECTION 3.1 EXERCISES

3.1 Yvette is a young banker. She and all her friends carry cell phones and use them heavily. Last year, two of Yvette's acquaintances developed brain tumors. Yvette wonders if the tumors are related to use of cell phones. Explain briefly why the experience of Yvette's friends does not provide good evidence that cell phones cause brain tumors.

3.2 When the discussion turns to the pros and cons of wearing automobile seat belts, Herman always brings up the case of a friend who survived an accident because he was not wearing a seat belt. The friend was thrown out of the car and landed on a grassy bank, suffering only minor injuries, while the car burst into flames and was destroyed. Explain briefly why this anecdote does not provide good evidence that it is safer not to wear seat belts.

3.3 One study of cell phones and the risk of brain cancer looked at a group of 469 people who have brain cancer. The investigators matched each cancer patient with a person of the same sex, age, and race who did not have brain cancer, then asked about use of cell phones.[4] Result: "Our data suggest that use of handheld cellular telephones is not associated with risk of brain cancer." Is this an observational study or an experiment? Why? What are the explanatory and response variables?

3.4 There is strong public support for "term limits" that restrict the number of terms that legislators can serve. One possible explanation for this support is that voters are dissatisfied with the performance of Congress and other legislative bodies. A political scientist asks a sample of voters if they support term limits for members of Congress and also asks several questions that gauge their satisfaction with Congress. He finds no relationship between approval of Congress and support for term limits. Is this an observational study or an experiment? Why? What are the explanatory and response variables?

3.5 An educational software company wants to compare the effectiveness of its computer animation for teaching cell biology with that of a textbook presentation. The company tests the biological knowledge of each of a group of first-year college students, then divides them into two groups. One group uses the animation, and the other studies the text. The company retests all the students and compares the increase in understanding of cell biology in the two groups. Is this an experiment? Why or why not? What are the explanatory and response variables?

3.6 Many studies have found that people who drink alcohol in moderation have lower risk of heart attacks than either nondrinkers or heavy drinkers. Does alcohol consumption also improve survival after a heart attack? One study followed 1913 people who were hospitalized after severe heart attacks. In the year before their heart attack, 47% of these people did not drink, 36% drank moderately, and 17% drank heavily. After four years, fewer of the moderate drinkers had died.[5] Is this an observational study or an experiment? Why? What are the explanatory and response variables?

3.7 Some people believe that exercise raises the body's metabolic rate for as long as 12 to 24 hours, enabling us to continue to burn off fat after our workout has ended. In a study of this effect, subjects walked briskly on a treadmill for several hours. Their metabolic rates were measured before, immediately after, and 12 hours after the exercise. The study was criticized because eating raises the metabolic rate, and no record was kept of what the subjects ate after exercising. Was this study an experiment? Why or why not? What are the explanatory and response variables?

3.8 The National Halothane Study was a major investigation of the safety of anesthetics used in surgery. Records of over 850,000 operations performed in 34 major hospitals showed the following death rates for four common anesthetics:[6]

Anesthetic	A	B	C	D
Death rate	1.7%	1.7%	3.4%	1.9%

There is a clear association between the anesthetic used and the death rate of patients. Anesthetic C appears dangerous.

(a) Explain why we call the National Halothane Study an observational study rather than an experiment, even though it compared the results of using different anesthetics in actual surgery.

(b) When the study looked at other variables that are confounded with a doctor's choice of anesthetic, it found that Anesthetic C was not causing extra deaths. Suggest several variables that are mixed up with what anesthetic a patient receives.

3.2 Design of Experiments

A study is an experiment when we actually do something to people, animals, or objects in order to observe the response. Here is the basic vocabulary of experiments.

> **Experimental Units, Subjects, Treatment**
>
> The individuals on which the experiment is done are the **experimental units.** When the units are human beings, they are called **subjects.** A specific experimental condition applied to the units is called a **treatment.**

Because the purpose of an experiment is to reveal the response of one variable to changes in other variables, the distinction between explanatory and response variables is important. The explanatory variables in an experiment are often called **factors.** Many experiments study the joint effects of several factors. In such an experiment, each treatment is formed by combining a specific value (often called a **level**) of each of the factors.

factors

level of a factor

EXAMPLE 3.3

Does regularly taking aspirin help protect people against heart attacks? The Physicians' Health Study was a medical experiment that helped answer this question. In fact, the Physicians' Health Study looked at the effects of two drugs: aspirin and beta carotene. The body converts beta carotene into vitamin A, which may help prevent some forms of cancer. The *subjects* were 21,996 male physicians. There were two *factors*, each having two levels: aspirin (yes or no) and beta carotene (yes or no). Combinations of the levels of these factors form the four *treatments* shown in Figure 3.1. One-fourth of the subjects were assigned to each of these treatments.

On odd-numbered days, the subjects took a white tablet that contained either aspirin or a **placebo,** a dummy pill that looked and tasted like the aspirin but had no active ingredient. On even-numbered days, they took a red capsule containing either beta carotene or a placebo. There were several *response variables*—the study looked

placebo

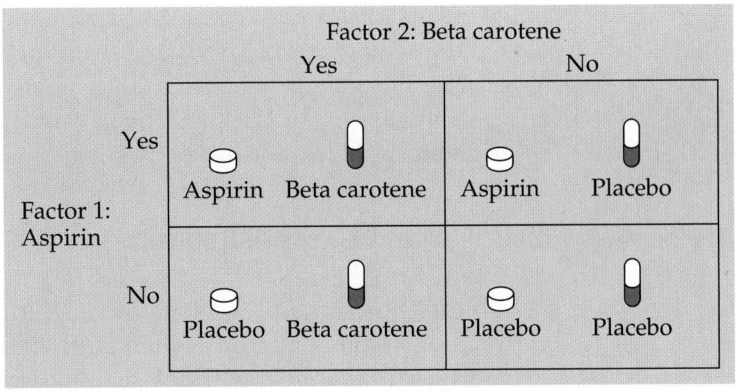

FIGURE 3.1 The treatments in the Physicians' Health Study, for Example 3.3.

for heart attacks, several kinds of cancer, and other medical outcomes. After several years, 239 of the placebo group but only 139 of the aspirin group had suffered heart attacks. This difference is large enough to give good evidence that taking aspirin does reduce heart attacks.[7] It did not appear, however, that beta carotene had any effect.

EXAMPLE 3.4

Does studying a foreign language in high school increase students' verbal ability in English? Julie obtains lists of all seniors in her high school who did and did not study a foreign language. Then she compares their scores on a standard test of English reading and grammar given to all seniors. The average score of the students who studied a foreign language is much higher than the average score of those who did not.

This observational study gives no evidence that studying another language builds skill in English. Students decide for themselves whether or not to elect a foreign language. Those who choose to study a language are mostly students who are already better at English than most students who avoid foreign languages. The difference in average test scores just shows that students who choose to study a language differ (on the average) from those who do not. We can't say whether studying languages *causes* this difference.

Examples 3.3 and 3.4 illustrate the big advantage of experiments over observational studies. **In principle, experiments can give good evidence for causation.** All the doctors in the Physicians' Health Study took a pill every other day, and all got the same schedule of checkups and information. The only difference was the content of the pill. When one group had many fewer heart attacks, we conclude that it was the content of the pill that made the difference. Julie's observational study—a census of all seniors in her high school—does a good job of describing differences between seniors who have studied foreign languages and those who have not. But she can say nothing about cause and effect.

Another advantage of experiments is that they allow us to study the specific factors we are interested in, while controlling the effects of lurking variables. The subjects in the Physicians' Health Study were all middle-aged male doctors and all followed the same schedule of medical checkups. These similarities reduce variation among the subjects and make any effects of aspirin or beta carotene easier to see. Experiments also allow us to study the combined effects of several factors. The interaction of several factors can produce effects that could not be predicted from looking at the effects of each factor alone. The Physicians' Health Study tells us that aspirin helps prevent heart attacks, at least in middle-aged men, and that beta carotene taken with the aspirin neither helps nor hinders aspirin's protective powers.

Comparative experiments

Laboratory experiments in science and engineering often have a simple design with only a single treatment, which is applied to all of the experimental units. The design of such an experiment can be outlined as

Treatment $\longrightarrow$ **Observe response**

For example, we may subject a beam to a load (treatment) and measure its deflection (observation). We rely on the controlled environment of the laboratory to protect us from lurking variables. When experiments are conducted in the field or with living subjects, such simple designs often yield invalid data. That is, we cannot tell whether the response was due to the treatment or to lurking variables. Another medical example will show what can go wrong.

EXAMPLE 3.5

"Gastric freezing" is a clever treatment for ulcers in the upper intestine. The patient swallows a deflated balloon with tubes attached, then a refrigerated liquid is pumped through the balloon for an hour. The idea is that cooling the stomach will reduce its production of acid and so relieve ulcers. An experiment reported in the *Journal of the American Medical Association* showed that gastric freezing did reduce acid production and relieve ulcer pain. The treatment was safe and easy and was widely used for several years. The design of the experiment was

Gastric freezing $\longrightarrow$ **Observe pain relief**

placebo effect

The gastric freezing experiment was poorly designed. The patients' response may have been due to the **placebo effect.** A placebo is a dummy treatment. Many patients respond favorably to any treatment, even a placebo. This may be due to trust in the doctor and expectations of a cure, or simply to the fact that medical conditions often improve without treatment. The response to a dummy treatment is the placebo effect.

A later experiment divided ulcer patients into two groups. One group was treated by gastric freezing as before. The other group received a placebo treatment in which the liquid in the balloon was at body temperature rather than freezing. The results: 34% of the 82 patients in the treatment group improved, but so did 38% of the 78 patients in the placebo group. This and other properly designed experiments showed that gastric freezing was no better than a placebo, and its use was abandoned.[8]

control group

The first gastric freezing experiment gave misleading results because the effects of the explanatory variable were *confounded* with (mixed up with) the placebo effect. We can defeat confounding by *comparing* two groups of patients, as in the second gastric freezing experiment. The placebo effect and other lurking variables now operate on both groups. The only difference between the groups is the actual effect of gastric freezing. The group of patients who received a sham treatment is called a **control group,** because it enables us to control the effects of outside variables on the outcome. Control is the first basic principle of statistical design of experiments. Comparison of several treatments in the same environment is the simplest form of control.

Without control, experimental results in medicine and the behavioral sciences can be dominated by such influences as the details of the experimental arrangement, the selection of subjects, and the placebo effect. The result is often *bias*.

Bias

The design of a study is **biased** if it systematically favors certain outcomes.

An uncontrolled study of a new medical therapy, for example, is biased in favor of finding the treatment effective because of the placebo effect. It should not surprise you to learn that uncontrolled studies in medicine give new therapies a much higher success rate than proper comparative experiments. Well-designed experiments, like the Physicians' Health Study and the second gastric freezing study, usually compare several treatments.

Randomization

The design of an experiment first describes the response variable or variables, the factors (explanatory variables), and the layout of the treatments, with comparison as the leading principle. Figure 3.1 illustrates this aspect of the design of the Physicians' Health Study. The second aspect of design is the rule used to assign the experimental units to the treatments. Comparison of the effects of several treatments is valid only when all treatments are applied to similar groups of experimental units. If one corn variety is planted on more fertile ground, or if one cancer drug is given to more seriously ill patients, comparisons among treatments are meaningless. Systematic differences among the groups of experimental units in a comparative experiment cause bias. How can we assign experimental units to treatments in a way that is fair to all of the treatments?

Experimenters often attempt to match groups by elaborate balancing acts. Medical researchers, for example, try to match the patients in a "new drug" experimental group and a "standard drug" control group by age, sex, physical condition, smoker or not, and so on. Matching is helpful but not adequate—there are too many lurking variables that might affect the outcome. The experimenter is unable to measure some of these variables and will not think of others until after the experiment. Some important variables, such as how advanced a cancer patient's disease is, are so subjective that an experimenter might bias the study by, for example, assigning more advanced cancer cases to a promising new treatment in the unconscious hope that it will help them.

The statistician's remedy is to rely on chance to make an assignment that does not depend on any characteristic of the experimental units and that does not rely on the judgment of the experimenter in any way. The use of chance can be combined with matching, but the simplest design creates groups by chance alone. Here is an example.

EXAMPLE 3.6

A food company assesses the nutritional quality of a new "instant breakfast" product by feeding it to newly weaned male white rats and measuring their weight gain over a 28-day period. A control group of rats receives a standard diet for comparison. This nutrition experiment has a single factor (the diet) with two levels. The researchers use 30 rats for the experiment and so must divide them into two groups of 15. To do this in a completely unbiased fashion, they put the cage numbers of the 30 rats in a hat, mix them up, and draw 15. These rats form the experimental group and the remaining 15 make up the control group. Figure 3.2 outlines the design of this experiment.

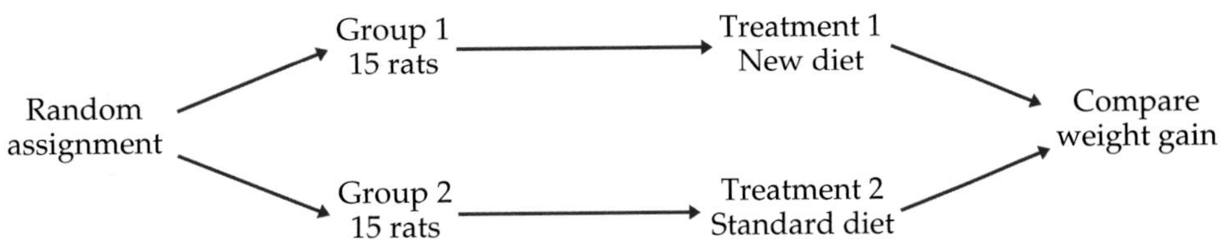

FIGURE 3.2 Outline of a randomized comparative experiment, for Example 3.6.

randomization

The use of chance to divide experimental units into groups is called **randomization.*** The design in Figure 3.2 combines comparison and randomization to arrive at the simplest randomized comparative design. This "flowchart" outline presents all the essentials: randomization, the sizes of the groups and which treatment they receive, and the response variable. There are, as we will see later, statistical reasons for generally using treatment groups about equal in size.

Randomized comparative experiments

The logic behind the randomized comparative design in Figure 3.2 is as follows:

- Randomization produces groups of rats that should be similar in all respects before the treatments are applied.
- Comparative design ensures that influences other than the diets operate equally on both groups.
- Therefore, differences in average weight gain must be due either to the diets or to the play of chance in the random assignment of rats to the two diets.

That "either-or" deserves more thought. We cannot say that *any* difference in the average weight gains of rats fed the two diets must be caused by a difference between the diets. There would be some difference even if both groups received the same diet, because the natural variability among rats means that some grow faster than others. Chance assigns the faster-growing rats to one group or the other, and this creates a chance difference between the groups. We would not trust an experiment with just one rat in each group, for example. The results would depend too much on which group got lucky and received the faster-growing rat. If we assign many rats to each diet, however, the effects of chance will average out and there will be little difference in the average weight gains in the two groups unless the diets themselves cause a

*The use of randomization in experimental design was pioneered by R. A. Fisher (1890–1962). Fisher worked with agricultural field experiments and genetics at the Rothamsted Experiment Station in England. He invented statistical design of experiments and formal statistical methods for analyzing experimental data and developed them to an advanced state.

difference. "Use enough experimental units to reduce chance variation" is the third big idea of statistical design of experiments.

Principles of Experimental Design

The basic principles of statistical design of experiments are

1. Control the effects of lurking variables on the response, most simply by comparing two or more treatments.
2. Randomize—use impersonal chance to assign experimental units to treatments.
3. Replicate each treatment on many units to reduce chance variation in the results.

We hope to see a difference in the responses so large that it is unlikely to happen just because of chance variation. We can use the laws of probability, which give a mathematical description of chance behavior, to learn if the treatment effects are larger than we would expect to see if only chance were operating. If they are, we call them *statistically significant*.

Statistical Significance

An observed effect so large that it would rarely occur by chance is called **statistically significant.**

You will often see the phrase "statistically significant" in reports of investigations in many fields of study. It tells you that the investigators found good evidence for the effect they were seeking. The Physicians' Health Study, for example, reported statistically significant evidence that aspirin reduces the number of heart attacks compared with a placebo.

Exploring the Web

Randomized comparative experiments are so important in medicine that the *Journal of the American Medical Association* now regularly includes the words "a randomized controlled trial" in the title of articles to alert readers. The Web sites of major medical journals offer an endless supply of both experiments and observational studies:

- *British Medical Journal:* `www.bmj.com`
- *Journal of the American Medical Association:* `jama.ama-assn.org`
- *New England Journal of Medicine:* `www.nejm.org`

We include the smaller *British Medical Journal* because it makes the full text of its articles available free online. Most journals offer only summaries and some selected full articles free of charge.

How to randomize

The idea of randomization is to assign subjects to treatments by drawing names from a hat. In practice, experimenters use software to carry out randomization. Most statistical software will choose 15 out of a list of 30 at random, for example. The list can contain the cage numbers of 30 rats or the names of 30 human subjects. The 15 chosen form one group, and the 15 that remain form the second group. The companion Web site for this text, www.whfreeman.com/ips, contains a *Simple Random Sample* applet that will choose treatment groups at random for you. Exercises 3.24 to 3.27 illustrate the use of this applet.

applet

We can randomize without software by using a *table of random digits*. Thinking about random digits helps understand randomization even if you will use software in practice. Table B at the back of the book and on the back endpaper is a table of random digits.

Random Digits

A **table of random digits** is a list of the digits 0, 1, 2, 3, 4, 5, 6, 7, 8, 9 that has the following properties:

1. The digit in any position in the list has the same chance of being any one of 0, 1, 2, 3, 4, 5, 6, 7, 8, 9.
2. The digits in different positions are independent in the sense that the value of one has no influence on the value of any other.

You can think of Table B as the result of asking an assistant (or a computer) to mix the digits 0 to 9 in a hat, draw one, then replace the digit drawn, mix again, draw a second digit, and so on. The assistant's mixing and drawing saves us the work of mixing and drawing when we need to randomize. Table B begins with the digits 19223950340575628713. To make the table easier to read, the digits appear in groups of five and in numbered rows. The groups and rows have no meaning—the table is just a long list of digits having the properties 1 and 2 described above.

Our goal is to use random digits for experimental randomization. We need the following facts about random digits, which are consequences of the basic properties 1 and 2:

- Any *pair* of random digits has the same chance of being any of the 100 possible pairs: 00, 01, 02, . . . , 98, 99.
- Any *triple* of random digits has the same chance of being any of the 1000 possible triples: 000, 001, 002, . . . , 998, 999.
- . . . and so on for groups of four or more random digits.

EXAMPLE 3.7 | In the nutrition experiment of Example 3.6, we must divide 30 rats at random into two groups of 15 rats each.

Step 1. Label: Give each rat a numerical label, using as few digits as possible. Two digits are needed to label 30 rats, so we use labels

01, 02, 03, . . . , 29, 30

It is also correct to use labels 00 to 29 or some other choice of 30 two-digit labels.

Step 2. Table: Start anywhere in Table B and read two-digit groups. Suppose we begin at line 130, which is

69051	64817	87174	09517	84534	06489	87201	97245

The first 10 two-digit groups in this line are

69	05	16	48	17	87	17	40	95	17

Each of these two-digit groups is a label. The labels 00 and 31 to 99 are not used in this example, so we ignore them. The first 15 labels between 01 and 30 that we encounter in the table choose rats for the experimental group. Of the first 10 labels in line 130, we ignore five because they are too high (over 30). The others are 05, 16, 17, 17, and 17. The rats labeled 05, 16, and 17 go into the experimental group. Ignore the second and third 17s because that rat is already in the group. Run your finger across line 130 (and continue to line 131 if needed) until you have chosen 15 rats. They are the rats labeled

05, 16, 17, 20, 19, 04, 25, 29, 18, 07, 13, 02, 23, 27, 21

These rats form the experimental group. The remaining 15 are the control group.

As Example 3.7 illustrates, randomization requires two steps: assign labels to the experimental units and then use Table B to select labels at random. Be sure that all labels are the same length so that all have the same chance to be chosen. Use the shortest possible labels—one digit for 9 or fewer individuals, two digits for 10 to 100 individuals, and so on. Don't try to scramble the labels as you assign them. Table B will do the required randomizing, so assign labels in any convenient manner, such as in alphabetical order for human subjects. You can read digits from Table B in any order—along a row, down a column, and so on—because the table has no order. As an easy standard practice, we recommend reading along rows.

EXAMPLE 3.8

Many utility companies have programs to encourage their customers to conserve energy. An electric company is considering placing electronic meters in households to show what the cost would be if the electricity use at that moment continued for a month. Will meters reduce electricity use? Would cheaper methods work almost as well? The company decides to design an experiment.

One cheaper approach is to give customers a chart and information about monitoring their electricity use. The experiment compares these two approaches (meter, chart) with each other and also with a control group of customers who receive no

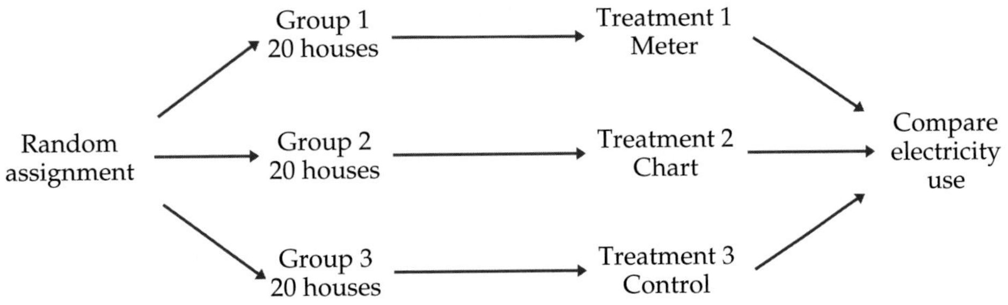

FIGURE 3.3 Outline of a completely randomized design comparing three treatments, for Example 3.8.

help in monitoring electricity use. The response variable is total electricity used in a year. The company finds 60 single-family residences in the same city willing to participate, so it assigns 20 residences at random to each of the three treatments. The outline of the design appears in Figure 3.3.

To carry out the random assignment, label the 60 houses 01 to 60. Then enter Table B and read two-digit groups until you have selected 20 houses to receive the meters. Continue in Table B to select 20 more to receive charts. The remaining 20 form the control group. The process is simple but tedious.

completely randomized design

When all experimental units are allocated at random among all treatments, the experimental design is **completely randomized.** The designs in Figures 3.2 and 3.3 are both completely randomized. Completely randomized designs can compare any number of treatments. The treatments can be formed by more than one factor. The Physicians' Health Study had two factors, which combine to form the four treatments shown in Figure 3.1. The study used a completely randomized design that assigned 5499 of the 21,996 subjects to each of the four treatments.

Cautions about experimentation

double-blind

The logic of a randomized comparative experiment depends on our ability to treat all the experimental units identically in every way except for the actual treatments being compared. Good experiments therefore require careful attention to details. For example, the subjects in both the Physicians' Health Study (Example 3.3) and the second gastric freezing experiment (Example 3.5) all got the same medical attention over the several years the studies continued. Moreover, these studies were **double-blind**—neither the subjects themselves nor the medical personnel who worked with them knew which treatment any subject had received. The double-blind method avoids unconscious bias by, for example, a doctor who doesn't think that "just a placebo" can benefit a patient.

lack of realism

The most serious potential weakness of experiments is **lack of realism.** The subjects or treatments or setting of an experiment may not realistically duplicate the conditions we really want to study. Here are some examples.

EXAMPLE 3.9

A study compares two television advertisements by showing TV programs to student subjects. The students know it's "just an experiment." We can't be sure that the results apply to everyday television viewers. Many behavioral science experiments use as subjects students who know they are subjects in an experiment. That's not a realistic setting.

An industrial experiment uses a small-scale pilot production process to find the choices of catalyst concentration and temperature that maximize yield. These may not be the best choices for the operation of a full-scale plant.

Lack of realism can limit our ability to apply the conclusions of an experiment to the settings of greatest interest. Most experimenters want to generalize their conclusions to some setting wider than that of the actual experiment. Statistical analysis of the original experiment cannot tell us how far the results will generalize. Nonetheless, the randomized comparative experiment, because of its ability to give convincing evidence for causation, is one of the most important ideas in statistics.

Matched pairs designs

Completely randomized designs are the simplest statistical designs for experiments. They illustrate clearly the principles of control, randomization, and replication. However, completely randomized designs are often inferior to more elaborate statistical designs. In particular, matching the subjects in various ways can produce more precise results than simple randomization.

EXAMPLE 3.10

Are cereal leaf beetles more strongly attracted by the color yellow or by the color green? Agriculture researchers want to know, because they detect the presence of the pests in farm fields by mounting sticky boards to trap insects that land on them. The board color should attract beetles as strongly as possible. We must design an experiment to compare yellow and green by mounting boards on poles in a large field of oats.

The experimental units are locations within the field far enough apart to represent independent observations. We erect a pole at each location to hold the boards. We might employ a completely randomized design in which we randomly select half the poles to receive a yellow board while the remaining poles receive green. The locations vary widely in the number of beetles present. For example, the alfalfa that borders the oats on one side is a natural host of the beetles, so locations near the alfalfa will have extra beetles. This variation among experimental units can hide the systematic effect of the board color.

matched pairs design

It is more efficient to use a **matched pairs design** in which we mount boards of both colors on each pole. The observations (numbers of beetles trapped) are matched in pairs from the same poles. We compare the number of trapped beetles on a yellow board with the number trapped by the green board on the same pole. Because the boards are mounted one above the other, we select the color of the top board at random. Just toss a coin for each board—if the coin falls heads, the yellow board is mounted above the green board.

Matched pairs designs compare just two treatments. We choose *blocks* of two units that are as closely matched as possible. In Example 3.10, two boards on the same pole form a block. We assign one of the treatments to each unit by tossing a coin or reading odd and even digits from Table B. Alternatively, each block in a matched pairs design may consist of just one subject, who gets both treatments one after the other. Each subject serves as his or her own control. The *order* of the treatments can influence the subject's response, so we randomize the order for each subject, again by a coin toss.

Block designs

The matched pairs design of Example 3.10 uses the principles of comparison of treatments, randomization, and replication on several experimental units. However, the randomization is not complete (all locations randomly assigned to treatment groups) but restricted to assigning the order of the boards at each location. The matched pairs design reduces the effect of variation among locations in the field by comparing the pair of boards at each location. Matched pairs are an example of *block designs*.

Block Design

A **block** is a group of experimental units or subjects that are known before the experiment to be similar in some way that is expected to affect the response to the treatments. In a **block design,** the random assignment of units to treatments is carried out separately within each block.

Block designs can have blocks of any size. A block design combines the idea of creating equivalent treatment groups by matching with the principle of forming treatment groups at random. Blocks are another form of *control*. They control the effects of some outside variables by bringing those variables into the experiment to form the blocks. Here are some typical examples of block designs.

EXAMPLE 3.11

The progress of a type of cancer differs in women and men. A clinical experiment to compare three therapies for this cancer therefore treats sex as a blocking variable. Two separate randomizations are done, one assigning the female subjects to the treatments and the other assigning the male subjects. Figure 3.4 outlines the design of this experiment. Note that there is no randomization involved in making up the blocks. They are groups of subjects who differ in some way (sex in this case) that is apparent before the experiment begins.

EXAMPLE 3.12

The soil type and fertility of farmland differ by location. Because of this, a test of the effect of tillage type (two types) and pesticide application (three application schedules) on soybean yields uses small fields as blocks. Each block is divided into six plots, and the six treatments are randomly assigned to plots separately within each block.

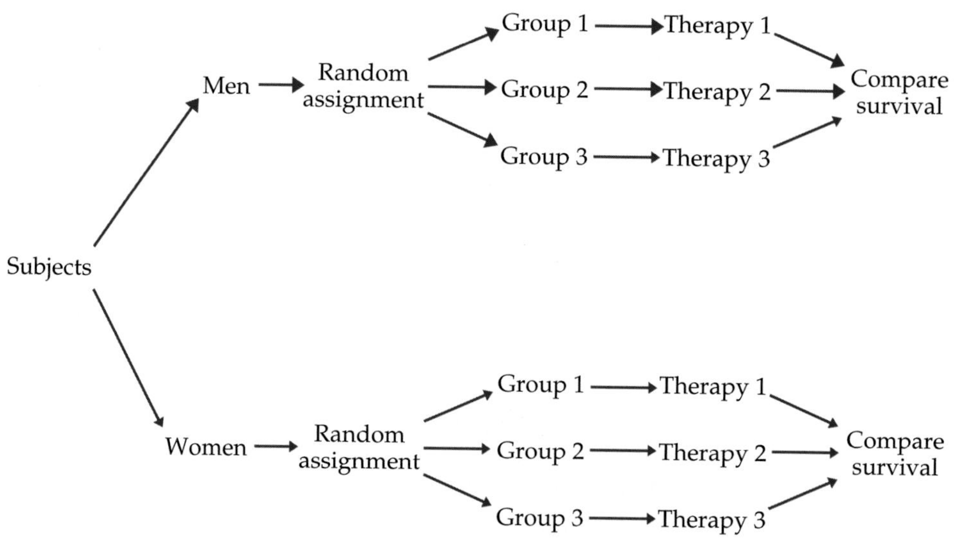

FIGURE 3.4 Outline of a block design. The blocks consist of male and female subjects. The treatments are three therapies for cancer.

EXAMPLE 3.13

A social policy experiment will assess the effect on family income of several proposed new welfare systems and compare them with the present welfare system. Because the income of a family under any welfare system is strongly related to its present income, the families who agree to participate are divided into blocks of similar income levels. The families in each block are then allocated at random among the welfare systems.

Blocks allow us to draw separate conclusions about each block, for example, about men and women in the cancer study in Example 3.11. Blocking also allows more precise overall conclusions, because the systematic differences between men and women can be removed when we study the overall effects of the three therapies. The idea of blocking is an important additional principle of statistical design of experiments. A wise experimenter will form blocks based on the most important unavoidable sources of variability among the experimental units. Randomization will then average out the effects of the remaining variation and allow an unbiased comparison of the treatments.

SUMMARY

In an experiment, one or more **treatments** are imposed on the **experimental units** or **subjects.** Each treatment is a combination of **levels** of the explanatory variables, which we call **factors.**

The **design** of an experiment refers to the choice of treatments and the manner in which the experimental units or subjects are assigned to the treatments.

The basic principles of statistical design of experiments are **control, randomization,** and **replication.**

The simplest form of control is **comparison.** Experiments should compare two or more treatments in order to prevent **confounding** the effect of a treatment with other influences, such as lurking variables.

Randomization uses chance to assign subjects to the treatments. Randomization creates treatment groups that are similar (except for chance variation) before the treatments are applied. Randomization and comparison together prevent **bias,** or systematic favoritism, in experiments.

You can carry out randomization by giving numerical labels to the experimental units and using a **table of random digits** to choose treatment groups.

Replication of the treatments on many units reduces the role of chance variation and makes the experiment more sensitive to differences among the treatments.

Good experiments require attention to detail as well as good statistical design. Many behavioral and medical experiments are **double-blind. Lack of realism** in an experiment can prevent us from generalizing its results.

In addition to comparison, a second form of control is to restrict randomization by forming **blocks** of experimental units that are similar in some way that is important to the response. Randomization is then carried out separately within each block.

Matched pairs are a common form of blocking for comparing just two treatments. In some matched pairs designs, each subject receives both treatments in a random order. In others, the subjects are matched in pairs as closely as possible, and one subject in each pair receives each treatment.

SECTION 3.2 EXERCISES

For each of the experimental situations described in Exercises 3.9 to 3.11, identify the experimental units or subjects, the factors, the treatments, and the response variables.

3.9 Ability to grow in shade may help pines found in the dry forests of Arizona resist drought. How well do these pines grow in shade? Investigators planted pine seedlings in a greenhouse in either full light or light reduced to 5% of normal by shade cloth. At the end of the study, they dried the young trees and weighed them.

3.10 A manufacturer of food products uses package liners that are sealed at the top by applying heated jaws after the package is filled. The customer peels the sealed pieces apart to open the package. What effect does the temperature of the jaws have on the force required to peel the liner? To answer this question, the engineers prepare 20 pairs of pieces of package liner. They seal five pairs at each of 250° F, 275° F, 300° F, and 325° F. Then they measure the peel strength of each seal.

3.11 How can we reduce the rate of refusals in telephone surveys? Most people who answer at all listen to the interviewer's introductory remarks and then decide whether to continue. One study made telephone calls to randomly selected households to ask opinions about the next election. In some calls, the interviewer gave her name, in others she identified the university she was representing, and in still others she identified both herself and the university. For each type of call, the interviewer either did or did not offer to send a copy of the final survey results to the person interviewed. Do these differences in the introduction affect whether the interview is completed?

3.12 People who eat lots of fruits and vegetables have lower rates of colon cancer than those who eat little of these foods. Fruits and vegetables are rich in "antioxidants" such as vitamins A, C, and E. Will taking antioxidants help prevent colon cancer? A clinical trial studied this question with 864 people who were at risk of colon cancer. The subjects were divided into four groups: daily beta carotene, daily vitamins C and E, all three vitamins every day, and daily placebo. After four years, the researchers were surprised to find no significant difference in colon cancer among the groups.[9]

(a) What are the explanatory and response variables in this experiment?

(b) Outline the design of the experiment. Use your judgment in choosing the group sizes.

(c) Assign labels to the 864 subjects and use Table B starting at line 118 to choose the first 5 subjects for the beta carotene group.

(d) The study was double-blind. What does this mean?

(e) What does "no significant difference" mean in describing the outcome of the study?

(f) Suggest some lurking variables that could explain why people who eat lots of fruits and vegetables have lower rates of colon cancer. The experiment suggests that these variables, rather than the antioxidants, may be responsible for the observed benefits of fruits and vegetables.

3.13 Doctors identify "chronic tension-type headaches" as headaches that occur almost daily for at least six months. Can antidepressant medications or stress management training reduce the number and severity of these headaches? Are both together more effective than either alone? Investigators compared four treatments: antidepressant alone, placebo alone, antidepressant plus stress management, and placebo plus stress management. Outline the design of the experiment. The headache sufferers named below have agreed to participate in the study. Use the applet or Table B at line 130 to randomly assign the subjects to the treatments.

Acosta	Duncan	Han	Liang	Padilla	Valasco
Asihiro	Durr	Howard	Maldonado	Plochman	Vaughn
Bennett	Edwards	Hruska	Marsden	Rosen	Wei
Bikalis	Farouk	Imrani	Montoya	Solomon	Wilder
Chen	Fleming	James	O'Brian	Trujillo	Willis
Clemente	George	Kaplan	Ogle	Tullock	Zhang

3.14 Surgery patients are often cold because the operating room is kept cool and the body's temperature regulation is disturbed by anesthetics. Will warming patients to maintain normal body temperature reduce infections after surgery? In one experiment, patients undergoing colon surgery received intravenous fluids from a warming machine and were covered with a blanket through which air circulated. For some patients, the fluid and the air were warmed; for others, they were not. The patients received identical treatment in all other respects.[10]

(a) Outline the design of a randomized comparative experiment for this study.

(b) The following subjects have given consent to participate in this study. Do the random assignment required by your design. (If you use Table B, begin at line 121.)

Abbott	Decker	Gutierrez	Lucero	Rosen
Adamson	Devlin	Howard	Masters	Sugiwara
Afifi	Engel	Hwang	McNeill	Thompson
Brown	Fluharty	Iselin	Morse	Travers
Cansico	Garcia	Janle	Ng	Turing
Chen	Gerson	Kaplan	Quinones	Ullmann
Cordoba	Green	Kim	Rivera	Williams
Curzakis	Gupta	Lattimore	Roberts	Wong

3.15 Outline the design of the package liner experiment of Exercise 3.10. Label the pairs of liner pieces 01 to 20 and carry out the randomization that your design calls for. (If you use Table B, start at line 120.)

3.16 Here are some questions about the study of heating surgery patients in Exercise 3.14.

(a) To simplify the setup of the study, we might warm the fluids and air blanket for one operating team and not for another doing the same kind of surgery. Why might this design result in bias?

(b) The operating team did not know whether fluids and air blanket were heated, nor did the doctors who followed the patients after surgery. What is this practice called? Why was it used here?

3.17 "Bee pollen is effective for combating fatigue, depression, cancer, and colon disorders." So says a Web site that offers the pollen for sale. We wonder if bee pollen really does prevent colon disorders. Here are two ways to study this question. Explain why the second design will produce more trustworthy data.

(1) Find 200 women who take bee pollen regularly. Match each with a woman of the same age, race, and occupation who does not take bee pollen. Follow both groups for 5 years.

(2) Find 400 women who do not have colon disorders. Assign 200 to take bee pollen capsules and the other 200 to take placebo capsules that are identical in appearance. Follow both groups for 5 years.

3.18 Example 3.8 (page 236) describes an experiment to learn whether providing households with electronic meters or charts will reduce their electricity consumption. An executive of the electric company objects to including a control group. He says, "It would be simpler to just compare electricity use last year [before the meter or chart was provided] with consumption in the same period this year. If households use less electricity this year, the indicator or chart must be working." Explain clearly why this design is inferior to that in Example 3.8.

3.19 Puzzling but true: stocks tend to go down on Mondays. There is no convincing explanation for this fact. A recent study looked at this "Monday effect" in more detail, using data on the daily returns of stocks on several U.S. exchanges over a 30-year period. Here are some of the findings:

To summarize, our results indicate that the well-known Monday effect is caused largely by the Mondays of the last two weeks of the month. The mean Monday return of the first three weeks of the month is, in general, not significantly different from zero and is generally significantly higher than the mean Monday return of the last two weeks. Our finding seems to make it more difficult to explain the Monday effect.[11]

A friend thinks that "significantly" in this article has its plain English meaning, roughly "I think this is important." Explain in simple language what "significantly higher" and "not significantly different from zero" actually tell us here.

3.20 Health care providers are giving more attention to relieving the pain of cancer patients. An article in the journal *Cancer* surveyed a number of studies and concluded that controlled-release morphine tablets, which release the pain killer gradually over time, are more effective than giving standard morphine when the patient needs it.[12] The "methods" section of the article begins: "Only those published studies that were controlled (i.e., randomized, double blind, and comparative), repeated-dose studies with CR morphine tablets in cancer pain patients were considered for this review." Explain the terms in parentheses to someone who knows nothing about medical trials.

3.21 Does the herb Saint-John's-wort relieve major depression? Here are some excerpts from the report of a study of this issue.[13] The study concluded that the herb is no more effective than a placebo.

(a) "Design: Randomized, double-blind, placebo-controlled clinical trial . . . ." Explain the meaning of each of the terms in this description.

(b) "Participants . . . were randomly assigned to receive either Saint-John's-wort extract ($n = 98$) or placebo ($n = 102$) . . . . The primary outcome measure was the rate of change in the Hamilton Rating Scale for Depression over the treatment period." Based on this information, use a diagram to outline the design of this clinical trial.

3.22 Once a person has been convicted of drunk driving, one purpose of court-mandated treatment or punishment is to prevent future offenses of the same kind. Suggest three different treatments that a court might require. Then outline the design of an experiment to compare their effectiveness. Be sure to specify the response variables you will measure.

3.23 You want to determine the best color for attracting cereal leaf beetles to boards on which they will be trapped. You will compare four colors: blue, green, white, and yellow. The response variable is the count of beetles trapped. You will mount one board on each of 16 poles evenly spaced in a square field, with four poles in each of four rows. Sketch the field with the locations of the 16 poles. Outline the design of a completely randomized experiment to compare the colors. Randomly assign colors to the poles, and mark on your sketch the color assigned to each pole. (If you use Table B, start at line 115.)

3.24 You can use the *Simple Random Sample* applet to choose a treatment group at random once you have labeled the subjects. Example 3.7 (page 235) uses Table B to choose 15 rats from a group of 30 for the treatment group in a nutrition experiment. Use the applet to choose the 15 rats for the treatment group. Which rats did you choose? The remaining 15 rats make up the control group.

3.25 The *Simple Random Sample* applet allows you to randomly assign experimental units to more than two groups without difficulty. Example 3.8 (page 236) describes a randomized comparative experiment in which 60 houses are randomly assigned to three groups of 20.

(a) Use the applet to randomly choose 20 out of 60 houses to form the first group. Which houses are in this group?

(b) The "Population hopper" now contains the 40 houses that were not chosen, in scrambled order. Click "Sample" again to choose 20 of these remaining houses to make up the second group. Which houses were chosen?

(c) The 20 houses remaining in the "Population hopper" form the third group. Which houses are these?

3.26 Suppose that the 15 even-numbered rats among the 30 rats available in the setting of Exercise 3.24 are (unknown to the experimenters) a fast-growing variety. We hope that these rats will be roughly equally distributed between the two groups. Use the *Simple Random Sample* applet to choose 10 samples of size 15 from the 30 rats. (Be sure to click "Reset" after each sample.) Record the counts of even-numbered rats in each of your 10 samples. You see that there is considerable chance variation but no systematic bias in favor of one or the other group in assigning the fast-growing rats. Larger samples from larger populations will on the average do a better job of making the two groups equivalent.

3.27 You plan to compare two methods of teaching math skills to seventh-grade students. Your subjects are the 78 students described in Table 1.6 (page 33). You will choose 39 students at random to be taught by Method A. The remaining 39 will be taught using Method B. Because IQ may affect learning, you want the two groups to have about the same mean IQ. Randomization is supposed to make the groups similar.

(a) Use the *Simple Random Sample* applet to choose 39 of the 78 IQ scores at random. This is the same as choosing 39 of the students to form Group A. What are the scores of the 39 Group A students?

(b) What is the mean IQ of the 39 students chosen? What is the mean IQ of the 39 not chosen, who make up Group B? Do the two groups have similar mean IQs? (Randomization will *usually* produce two groups with similar means. There is, however, no guarantee: your randomization might, just by chance, produce unbalanced groups. The chance of this goes down as the size of the groups increases.)

3.28 A mathematics education researcher is studying where in high school mathematics texts it is most effective to insert questions. She wants to know whether it is better to present questions as motivation before the text passage or as review after the passage. The result may depend on the type of question asked: simple fact, computation, or word problem.

(a) This experiment has two factors. What are they? How many treatments do all combinations of levels of these factors form? List the treatments.

(b) Because it is disruptive to assign high school students at random to the treatment groups, the researcher will assign two classes of the same grade level to each treatment. The response variable is score on a mathematics test taken by all the students in these classes. Outline the design of the experiment. Carry out the random assignment required.

3.29 Is the right hand generally stronger than the left in right-handed people? You can crudely measure hand strength by placing a bathroom scale on a shelf with the end protruding, then squeezing the scale between the thumb below and the four fingers above it. The reading of the scale shows the force exerted. Describe the design of a matched pairs experiment to compare the strength of the right and left hands, using 15 right-handed people as subjects. Use a coin to do the required randomization.

3.30 The concentration of carbon dioxide (CO_2) in the atmosphere is increasing rapidly due to our use of fossil fuels. Because plants use CO_2 to fuel photosynthesis, more CO_2 may cause trees and other plants to grow faster. An elaborate apparatus allows researchers to pipe extra CO_2 to a 30-meter circle of forest. We want to compare the growth in base area of trees in treated and untreated areas to see if extra CO_2 does in fact increase growth. We can afford to treat three circular areas.[14]

(a) Describe the design of a completely randomized experiment using 6 well-separated 30-meter circular areas in a pine forest. Sketch the circles and carry out the randomization your design calls for.

(b) Areas within the forest may differ in soil fertility. Describe a matched pairs design using three pairs of circles that will reduce the extra variation due to different fertility. Sketch the circles and carry out the randomization your design calls for.

3.31 Return to the experiment of Exercise 3.23 on the attractiveness of several colors to cereal leaf beetles. The researchers decide to use two oat fields in different locations and to space eight poles equally within each field. Outline a randomized block design using the fields as blocks. Then use Table B, beginning at line 105, to carry out the random assignment of colors to poles. Report your results by means of a sketch of the two fields with the color at each pole noted.

3.32 Twenty overweight females have agreed to participate in a study of the effectiveness of four reducing regimens, A, B, C, and D. The researcher first calculates how overweight each subject is by comparing the subject's actual weight with her "ideal" weight. The subjects and their excess weights in pounds are

Birnbaum	35	Hernandez	25	Moses	25	Smith	29
Brown	34	Jackson	33	Nevesky	39	Stall	33
Brunk	30	Kendall	28	Obrach	30	Tran	35
Dixon	34	Loren	32	Rodriguez	30	Wilansky	42
Festinger	24	Mann	28	Santiago	27	Williams	22

The response variable is the weight lost after eight weeks of treatment. Because the initial amount overweight will influence the response variable, a block design is appropriate.

(a) Arrange the subjects in order of increasing excess weight. Form five blocks by grouping the four least overweight, then the next four, and so on.

(b) Use Table B to do the required random assignment of subjects to the four reducing regimens separately within each block. Be sure to explain exactly how you used the table.

3.33 Will people spend less on health care if their health insurance requires them to pay some part of the cost themselves? An experiment on this issue asked if the percent of medical costs that are paid by health insurance has an effect either on the amount of medical care that people use or on their health. The treatments were four insurance plans. Each plan paid all medical costs above a ceiling. Below the ceiling, the plans paid 100%, 75%, 50%, or 0% of costs incurred.

(a) Outline the design of a randomized comparative experiment suitable for this study.

(b) Describe briefly the practical and ethical difficulties that might arise in such an experiment.

3.34 An ultramarathon, as you might guess, is a foot race longer than the 26.2 miles of a marathon. Runners commonly develop respiratory infections after an ultramarathon. Will taking 600 milligrams of vitamin C daily reduce these infections? Researchers randomly assigned ultramarathon runners to receive either vitamin C or a placebo. Separately, they also randomly assigned these treatments to a group of nonrunners the same age as the runners. All subjects were watched for 14 days after the big race to see if infections developed.[15]

(a) What is the name for this experimental design?

(b) Use a diagram to outline the design.

(c) The report of the study said:

Sixty-eight percent of the runners in the placebo group reported the development of symptoms of upper respiratory tract infection after the race; this was significantly more than that reported by the vitamin C-supplemented group (33%).

Explain to someone who knows no statistics why "significantly more" means there is good reason to think that vitamin C works.

3.35 Table B is a table of random digits. Software can also generate random digits for use in random selection. Which of the following statements are true of a table of random digits, and which are false?

(a) There are exactly four 0s in each row of 40 digits.

(b) Each pair of digits has chance 1/100 of being 00.

(c) The digits 0000 can never appear as a group, because this pattern is not random.

3.3 Sampling Design

A political scientist wants to know what percent of the voting age population consider themselves conservatives. An automaker hires a market research firm to learn what percent of adults aged 18 to 35 recall seeing television advertisements for a new sport utility vehicle. Government economists inquire about average household income. In all these cases, we want to gather information about a large group of individuals. We will not, as in an experiment, impose a treatment in order to observe the response. Time, cost, and inconvenience forbid contacting every individual. In such cases, we gather information about only part of the group in order to draw conclusions about the whole.

> **Population and Sample**
>
> The entire group of individuals that we want information about is called the **population.**
>
> A **sample** is a part of the population that we actually examine in order to gather information.

Notice that "population" is defined in terms of our desire for knowledge. If we wish to draw conclusions about all U.S. college students, that group is our population even if only local students are available for questioning. The sample design

sample is the part from which we draw conclusions about the whole. The **design** of a sample refers to the method used to choose the sample from the population. Poor sample designs can produce misleading conclusions, as the following examples illustrate.

EXAMPLE 3.14

A mill produces large coils of thin steel for use in manufacturing home appliances. The quality engineer wants to submit a sample of 5-centimeter squares to detailed laboratory examination. She asks a technician to cut a sample of 10 such squares. Wanting to provide "good" pieces of steel, the technician carefully avoids the visible defects in the coil material when cutting the sample. The laboratory results are wonderful but the customers complain about the material they are receiving.

EXAMPLE 3.15

Television news programs like to conduct call-in polls of public opinion. The program announces a question and asks viewers to call one telephone number to respond "Yes" and another for "No." Telephone companies charge for these calls. The ABC network program *Nightline* once asked whether the United Nations should continue to have its headquarters in the United States. More than 186,000 callers responded, and 67% said "No."

People who spend the time and money to respond to call-in polls are not representative of the entire adult population. In fact, they tend to be the same people who call radio talk shows. People who feel strongly, especially those with strong negative opinions, are more likely to call. It is not surprising that a properly designed sample showed that 72% of adults want the UN to stay.[16]

In both Examples 3.14 and 3.15, the sample was selected in a manner that guaranteed that it would not be representative of the entire population. These sampling schemes display *bias*, or systematic error, in favoring some parts of the population over others. *Nightline* used a *voluntary response sample*, a particularly common form of biased sample.

Voluntary Response Sample

A **voluntary response sample** consists of people who choose themselves by responding to a general appeal. Voluntary response samples are biased because people with strong opinions, especially negative opinions, are most likely to respond.

The remedy for bias in choosing a sample is to allow impersonal chance to do the choosing, so that there is neither favoritism by the sampler (as in Example 3.14) nor voluntary response (as in Example 3.15). Random selection of a sample eliminates bias by giving all individuals an equal chance to be chosen, just as randomization eliminates bias in assigning experimental subjects.

Simple random samples

The simplest sampling design amounts to placing names in a hat (the population) and drawing out a handful (the sample). This is simple random sampling.

Simple Random Sample

A **simple random sample (SRS)** of size n consists of n individuals from the population chosen in such a way that every set of n individuals has an equal chance to be the sample actually selected.

Each treatment group in a completely randomized experimental design is an SRS drawn from the available experimental units. We select an SRS by labeling all the individuals in the population and using a table of random digits to select a sample of the desired size, just as in experimental randomization. Notice that an SRS not only gives each individual an equal chance to be chosen (thus avoiding bias in the choice) but gives every possible sample an equal chance to be chosen. There are other random sampling designs that give each individual, but not each sample, an equal chance. One such design, systematic random sampling, is described in Exercise 3.46.

EXAMPLE 3.16

An academic department wishes to choose a three-member advisory committee at random from the members of the department. To choose an SRS of size 3 from the 28 faculty listed below, first label the members of the population as shown.

00	Abbott	07	Goodwin	14	Pillotte	21	Theobald
01	Cicirelli	08	Haglund	15	Raman	22	Vader
02	Cuellar	09	Johnson	16	Riemann	23	Wang
03	Dunsmore	10	Keegan	17	Rodriguez	24	Wieczorek
04	Engle	11	Luo	18	Rowe	25	Williams
05	Fitzpatrick	12	Martinez	19	Salazar	26	Wilson
06	Garcia	13	Nguyen	20	Stone	27	Wong

Now enter Table B, and read two-digit groups until you have chosen three committee members. If you enter at line 140, the committee consists of Martinez (12), Nguyen (13), and Engle (04). Most statistical software will select an SRS for you, eliminating the need for Table B. The *Simple Random Sample* applet at www.whfreeman.com/ips will also automate this task.

Stratified samples

The general framework for designs that use chance to choose a sample is a *probability sample*.

Probability Sample

A **probability sample** is a sample chosen by chance. We must know what samples are possible and what chance, or probability, each possible sample has.

Some probability sampling designs (such as an SRS) give each member of the population an *equal* chance to be selected. This may not be true in more elaborate sampling designs. In every case, however, the use of chance to select the sample is the essential principle of statistical sampling.

Designs for sampling from large populations spread out over a wide area are usually more complex than an SRS. For example, it is common to sample important groups within the population separately, then combine these samples. This is the idea of a *stratified sample*.

> **Stratified Random Sample**
>
> To select a **stratified random sample,** first divide the population into groups of similar individuals, called **strata.** Then choose a separate SRS in each stratum and combine these SRSs to form the full sample.

Choose the strata based on facts known before the sample is taken. For example, a population of election districts might be divided into urban, suburban, and rural strata. A stratified design can produce more exact information than an SRS of the same size by taking advantage of the fact that individuals in the same stratum are similar to one another. If all individuals in each stratum are identical, for example, just one individual from each stratum is enough to completely describe the population. Strata for sampling are similar to blocks in experiments. We have two names because the idea of grouping similar units before randomizing arose separately in sampling and in experiments.

EXAMPLE 3.17

A radio station that broadcasts a piece of music owes a royalty to the composer. The organization of composers (called ASCAP) collects these royalties for all its members by charging stations a license fee for the right to play members' songs. ASCAP has four million songs in its catalog and collects $435 million in fees each year. How should ASCAP distribute this income among its members? By sampling: ASCAP tapes about 60,000 hours from the 53 million hours of local radio programs across the country each year.

Radio stations are stratified by type of community (metropolitan, rural), geographic location (New England, Pacific, etc.), and the size of the license fee paid to ASCAP, which reflects the size of the audience. In all, there are 432 strata. Tapes are made at random hours for randomly selected members of each stratum. The tapes are reviewed by experts who can recognize almost every piece of music ever written, and the composers are then paid according to their popularity.[17]

Multistage samples

Another common means of restricting random selection is to choose the sample in stages. This is usual practice for national samples of households or people. For example, data on employment and unemployment are gathered by the government's Current Population Survey, which conducts interviews in about 55,000 households each month. It is not practical to maintain a list of all U.S. households from which to select an SRS. Moreover, the cost of sending interviewers to the widely scattered households in an SRS would be multistage sample

too high. The Current Population Survey therefore uses a **multistage sampling design.** The final sample consists of clusters of nearby households that an interviewer can easily visit. Most opinion polls and other national samples are also multistage, though interviewing in most national samples today is done by telephone rather than in person, eliminating the economic need for clustering. The Current Population Survey sampling design is roughly as follows:[18]

Stage 1. Divide the United States into 2007 geographical areas called Primary Sampling Units, or PSUs. Select a sample of 754 PSUs. This sample includes the 428 PSUs with the largest population and a stratified sample of 328 of the others.

Stage 2. Divide each PSU selected into smaller areas called "blocks." Stratify the blocks using ethnic and other information and take a stratified sample of the blocks in each PSU.

Stage 3. Sort the housing units in each block into clusters of four nearby units. Interview the households in a random sample of these clusters.

Analysis of data from sampling designs more complex than an SRS takes us beyond basic statistics. But the SRS is the building block of more elaborate designs, and analysis of other designs differs more in complexity of detail than in fundamental concepts.

Cautions about sample surveys

Random selection eliminates bias in the choice of a sample from a list of the population. Sample surveys of large human populations, however, require much more than a good sampling design.[19] To begin, we need an accurate and complete list of the population. Because such a list is rarely available, most samples suffer from some degree of *undercoverage*. A sample survey of households, for example, will miss not only homeless people but prison

Exploring the Web

Here are the Web sites of some major sample surveys:

- Current Population Survey: `www.bls.gov/cps`
- Gallup Poll: `www.gallup.com/poll`
- General Social Survey: `www.norc.org`
- Nielsen Media Research: `www.nielsenmedia.com`

Look at Gallup's site for plain-language explanations of sample surveys and lots of recent opinion poll results. The Nielsen people do the ratings of TV programs. The CPS and GSS sites give all the details about these two important surveys.

inmates and students in dormitories. An opinion poll conducted by telephone will miss the 6% of American households without residential phones. The results of national sample surveys therefore have some bias if the people not covered—who most often are poor people—differ from the rest of the population.

A more serious source of bias in most sample surveys is *nonresponse*, which occurs when a selected individual cannot be contacted or refuses to cooperate. Nonresponse to sample surveys often reaches 50% or more, even with careful planning and several callbacks. Because nonresponse is higher in urban areas, most sample surveys substitute other people in the same area to avoid favoring rural areas in the final sample. If the people contacted differ from those who are rarely at home or who refuse to answer questions, some bias remains.

Undercoverage and Nonresponse

Undercoverage occurs when some groups in the population are left out of the process of choosing the sample.

Nonresponse occurs when an individual chosen for the sample can't be contacted or does not cooperate.

EXAMPLE 3.18

How bad is nonresponse? The Current Population Survey (CPS) has the lowest nonresponse rate of any poll we know: only about 6% or 7% of the households in the CPS sample don't respond. People are more likely to respond to a government survey such as the CPS, and the CPS contacts its sample in person before doing later interviews by phone.

The General Social Survey (Example 3.1) is the nation's most important social science survey. The GSS also contacts its sample in person, and it is run by a university. Despite these advantages, its most recent survey had a 30% rate of nonresponse.

What about polls done by the media and by market research and opinion polling firms? We don't know their rates of nonresponse, because they won't say. That itself is a bad sign. The Pew Research Center imitated a careful telephone survey and published the results: out of 2879 households called, 1658 were never at home, refused, or would not finish the interview. That's a nonresponse rate of 58%.[20]

response bias

The behavior of the respondent or of the interviewer can cause **response bias** in sample results. Respondents may lie, especially if asked about illegal or unpopular behavior. The sample then underestimates the occurrence of such behavior in the population. An interviewer whose attitude suggests that some answers are more desirable than others will get these answers more often. The race or sex of the interviewer can influence responses to questions about race relations or attitudes toward feminism. Answers to questions that ask respondents to recall past events are often inaccurate because of faulty memory. For example, many people "telescope" events in the past, bringing them forward in memory to more recent time periods. "Have you visited a dentist in the last 6 months?" will often elicit a "Yes" from someone who last visited a dentist 8 months ago.[21] Careful training of interviewers and careful supervision to avoid variation among the interviewers can greatly reduce response bias. Good interviewing technique is another aspect of a well-done sample survey.

wording of questions

The **wording of questions** is the most important influence on the answers given to a sample survey. Confusing or leading questions can introduce strong bias, and even minor changes in wording can change a survey's outcome. Here are some examples.

EXAMPLE 3.19

How do Americans feel about government help for the poor? Only 13% think we are spending too much on "assistance to the poor," but 44% think we are spending too much on "welfare." How do the Scots feel about the movement to become independent from England? Well, 51% would vote for "independence for Scotland," but only 34% support "an independent Scotland separate from the United Kingdom." And 46% of Americans have high confidence in "colleges," but only 29% feel that way about "professors."[22]

It seems that "assistance to the poor" and "independence" are nice, hopeful words. "Welfare" and "separate" are negative words. We won't comment on "professors."

Never trust the results of a sample survey until you have read the exact questions posed. The sampling design, the amount of nonresponse, and the date of the survey are also important. Good statistical design is a part, but only a part, of a trustworthy survey.

SUMMARY

A sample survey selects a **sample** from the **population** of all individuals about which we desire information. We base conclusions about the population on data about the sample.

The **design** of a sample refers to the method used to select the sample from the population. **Probability sampling designs** use impersonal chance to select a sample.

The basic probability sample is a **simple random sample (SRS).** An SRS gives every possible sample of a given size the same chance to be chosen.

Choose an SRS by labeling the members of the population and using a **table of random digits** to select the sample. Software can automate this process.

To choose a **stratified random sample,** divide the population into **strata,** groups of individuals that are similar in some way that is important to the response. Then choose a separate SRS from each stratum and combine them to form the full sample.

Multistage samples select successively smaller groups within the population in stages, resulting in a sample consisting of clusters of individuals. Each stage may employ an SRS, a stratified sample, or another type of sample.

Failure to use probability sampling often results in **bias,** or systematic errors in the way the sample represents the population. **Voluntary response** samples, in which the respondents choose themselves, are particularly prone to large bias.

In human populations, even probability samples can suffer from bias due to **undercoverage** or **nonresponse,** from **response bias** due to the behavior of the interviewer or the respondent, or from misleading results due to **poorly worded questions.**

SECTION 3.3 EXERCISES

3.36 A political scientist wants to know how college students feel about the Social Security system. She obtains a list of the 3456 undergraduates at her college and mails a questionnaire to 250 students selected at random. Only 104 questionnaires are returned. What is the population in this study? What is the sample from which information was actually obtained? What is the rate (percent) of nonresponse?

3.37 Different types of writing can sometimes be distinguished by the lengths of the words used. A student interested in this fact wants to study the lengths of words used by Tom Clancy in his novels. She opens a Clancy novel at random and records the lengths of each of the first 250 words on the page. What is the population in this study? What is the sample? What is the variable measured?

3.38 For each of the following sampling situations, identify the population as exactly as possible. That is, say what kind of individuals the population consists of and say exactly which individuals fall in the population. If the information given is not complete, complete the description of the population in a reasonable way.

(a) An opinion poll contacts 1161 adults and asks them, "Which political party do you think has better ideas for leading the country in the twenty-first century?"

(b) A furniture maker buys hardwood in large lots. The supplier is supposed to dry the wood before shipping—wood that is not dry won't hold its size and shape. The furniture maker chooses 5 pieces of wood from each lot and tests their moisture content. If any piece exceeds 12% moisture content, the entire lot is sent back.

(c) The American Community Survey will contact 3 million households, including some in every county in the United States. This new Census Bureau survey will ask each household questions about their housing, economic, and social status.

3.39 A newspaper article about an opinion poll says that "43% of Americans approve of the president's overall job performance." Toward the end of the article, you read: "The poll is based on telephone interviews with 1210 adults from around the United States, excluding Alaska and Hawaii." What variable did this poll measure? What population do you think the newspaper wants information about? What was the sample? Are there any sources of bias in the sampling method used?

3.40 You are planning a report on apartment living in a college town. You decide to select 5 apartment complexes at random for in-depth interviews with residents. Use Table B, starting at line 117, to select a simple random sample of 5 of the following apartment complexes.

Ashley Oaks	Country View	Mayfair Village
Bay Pointe	Country Villa	Nobb Hill
Beau Jardin	Crestview	Pemberly Courts
Bluffs	Del-Lynn	Peppermill
Brandon Place	Fairington	Pheasant Run
Briarwood	Fairway Knolls	Richfield
Brownstone	Fowler	Sagamore Ridge
Burberry	Franklin Park	Salem Courthouse
Cambridge	Georgetown	Village Manor
Chauncey Village	Greenacres	Waterford Court
Country Squire	Lahr House	Williamsburg

3.41 A firm wants to understand the attitudes of its minority managers toward its system for assessing management performance. Below is a list of all the firm's managers who are members of minority groups. Use Table B at line 139 to choose 6 to be interviewed in detail about the performance appraisal system.

Agarwal	Dewald	Huang	Puri
Alfonseca	Fleming	Kim	Richards
Baxter	Fonseca	Lujan	Rodriguez
Bowman	Gates	Mourning	Santiago
Brown	Goel	Nunez	Shen
Cortez	Gomez	Peters	Vega
Cross	Hernandez	Pliego	Watanabe

applet

3.42 After you have labeled the individuals in a population, the *Simple Random Sample* applet automates the task of choosing an SRS. Use the applet to choose the sample in Exercise 3.40.

applet

3.43 You want to choose an SRS of 25 of a city's 440 voting precincts for special voting-fraud surveillance on election day. How will you label the 440 precincts? Choose the SRS, and list the precincts you selected. (Use the *Simple Random Sample* applet. If you use Table B, enter at line 117 and select only the first 5 precincts in the sample.)

3.44 The Census Bureau divides the entire country into "census tracts" that contain about 4000 people. Each tract is in turn divided into small "blocks,"

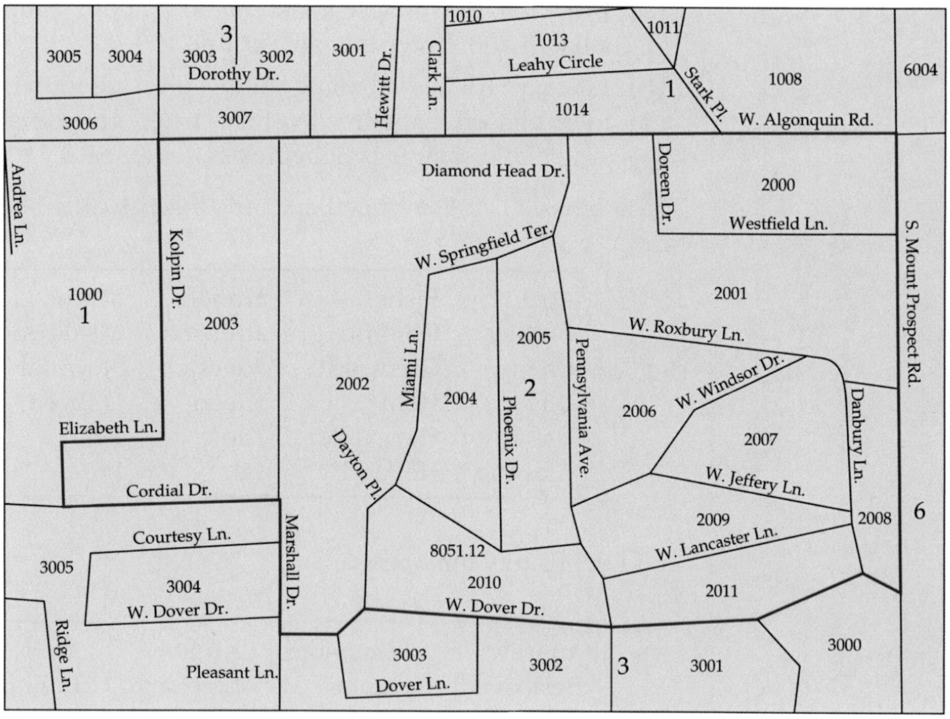

FIGURE 3.5 Census blocks in Cook County, Illinois, for Exercise 3.44. The outlined area is a block group.

which in urban areas are bounded by local streets. An SRS of blocks from a census tract is often the next-to-last stage in a multistage sample. Figure 3.5 shows part of census tract 8051.12, in Cook County, Illinois, west of Chicago. The 44 blocks in this tract are divided into three "block groups." Group 1 contains 6 blocks numbered 1000 to 1005; group 2 (outlined in Figure 3.5) contains 12 blocks numbered 2000 to 2011; group 3 contains 26 blocks numbered 3000 to 3025. Use Table B beginning at line 125 to choose an SRS of 5 of the 44 blocks in this census tract. Explain carefully how you labeled the blocks.

3.45 In using Table B repeatedly to choose samples or do experimental randomization, you should not always begin at the same place, such as line 101. Why not?

systematic random sample

3.46 **Systematic random samples** are often used to choose a sample of apartments in a large building or dwelling units in a block at the last stage of a multistage sample. An example will illustrate the idea of a systematic sample. Suppose that we must choose 4 addresses out of 100. Because $100/4 = 25$, we can think of the list as four lists of 25 addresses. Choose 1 of the first 25 at random, using Table B. The sample contains this address and the addresses 25, 50, and 75 places down the list from it. If 13 is chosen, for example, then the systematic random sample consists of the addresses numbered 13, 38, 63, and 88.

(a) Use Table B to choose a systematic random sample of 5 addresses from a list of 200. Enter the table at line 120.

(b) Like an SRS, a systematic sample gives all individuals the same chance to be chosen. Explain why this is true, then explain carefully why a systematic sample is nonetheless *not* an SRS.

3.47 A club has 30 student members and 10 faculty members. The students are

Abel	Fisher	Huber	Moran	Reinmann
Carson	Golomb	Jimenez	Moskowitz	Santos
Chen	Griswold	Jones	Neyman	Shaw
David	Hein	Kiefer	O'Brien	Thompson
Deming	Hernandez	Klotz	Pearl	Utts
Elashoff	Holland	Liu	Potter	Vlasic

and the faculty members are

Andrews	Fernandez	Kim	Moore	Rabinowitz
Besicovitch	Gupta	Lightman	Phillips	Yang

The club can send 4 students and 2 faculty members to a convention and decides to choose those who will go by random selection. Use Table B to choose a stratified random sample of 4 students and 2 faculty members.

3.48 A university has 2000 male and 500 female faculty members. The equal opportunity employment officer wants to poll the opinions of a random sample of faculty members. In order to give adequate attention to female faculty opinion, he decides to choose a stratified random sample of 200 males and 200 females. He has alphabetized lists of female and male faculty members. Explain how you would assign labels and use random digits to choose the desired sample. Enter Table B at line 122 and give the labels of the first 5 females and the first 5 males in the sample.

3.49 Accountants use stratified samples during audits to verify a company's records of such things as accounts receivable. The stratification is based on the dollar amount of the item and often includes 100% sampling of the largest items. One company reports 5000 accounts receivable. Of these, 100 are in amounts over \$50,000; 500 are in amounts between \$1000 and \$50,000; and the remaining 4400 are in amounts under \$1000. Using these groups as strata, you decide to verify all of the largest accounts and to sample 5% of the midsize accounts and 1% of the small accounts. How would you label the two strata from which you will sample? Use Table B, starting at line 115, to select the first 5 accounts from each of these strata.

3.50 At a party there are 30 students over age 21 and 20 students under age 21. You choose at random 3 of those over 21 and separately choose at random 2 of those under 21 to interview about attitudes toward alcohol. You have given every student at the party the same chance to be interviewed: what is that chance? Why is your sample not an SRS?

3.51 The list of individuals from which a sample is actually selected is called the **sampling frame.** Ideally, the frame should list every individual in the population, but in practice this is often difficult. A frame that leaves out part of the population is a common source of undercoverage.

sampling frame

(a) Suppose that a sample of households in a community is selected at random from the telephone directory. What households are omitted from this frame? What types of people do you think are likely to live in these households? These people will probably be underrepresented in the sample.

(b) It is usual in telephone surveys to use random digit dialing equipment that selects the last four digits of a telephone number at random after being given the exchange (the first three digits). Which of the households that you mentioned in your answer to (a) will be included in the sampling frame by random digit dialing?

3.52 A common form of nonresponse in telephone surveys is "ring-no-answer." That is, a call is made to an active number but no one answers. The Italian National Statistical Institute looked at nonresponse to a government survey of households in Italy during the periods January 1 to Easter and July 1 to August 31. All calls were made between 7 and 10 p.m., but 21.4% gave "ring-no-answer" in one period versus 41.5% "ring-no-answer" in the other period.[23] Which period do you think had the higher rate of no answers? Why? Explain why a high rate of nonresponse makes sample results less reliable.

3.53 The Excite Poll can be found online at `news.excite.com`. The question appears on the screen, and you simply click buttons to vote "Yes," "No," or "Not sure." On January 25, 2000, the question was "Should female athletes be paid the same as men for the work they do?" In all, 13,147 (44%) said "Yes," another 15,182 (50%) said "No," and the remaining 1448 said "Not sure."

(a) What is the sample size for this poll?

(b) That's a much larger sample than standard sample surveys. In spite of this, we can't trust the result to give good information about any clearly defined population. Why?

(c) More men than women use the Web. How might this fact affect the poll results?

3.54 During the 2000 presidential campaign, the candidates debated what to do with the large government surplus. The Pew Research Center asked two questions of random samples of adults. Both questions stated that Social Security would be "fixed." Here are the uses suggested for the remaining surplus:

Should the money be used for a tax cut, or should it be used to fund new government programs?

Should the money be used for a tax cut, or should it be spent on programs for education, the environment, health care, crime-fighting and military defense?

One of these questions drew 60% favoring a tax cut; the other, only 22%. Which wording pulls respondents toward a tax cut? Why?

3.55 Comment on each of the following as a potential sample survey question. Is the question clear? Is it slanted toward a desired response?

(a) "Some cell phone users have developed brain cancer. Should all cell phones come with a warning label explaining the danger of using cell phones?"

(b) "Do you agree that a national system of health insurance should be favored because it would provide health insurance for everyone and would reduce administrative costs?"

(c) "In view of escalating environmental degradation and incipient resource depletion, would you favor economic incentives for recycling of resource-intensive consumer goods?"

3.56 Write your own examples of bad sample survey questions.

(a) Write a biased question designed to get one answer rather than another.

(b) Write a question that is confusing, so that it is hard to answer.

3.57 Spain's Centro de Investigaciones Sociológicos carried out a sample survey on the economic attitudes of Spaniards.[24] Of the 2496 adults interviewed, 72% agreed that "Employees with higher performance must get higher pay." On the other hand, 71% agreed that "Everything a society produces should be distributed among its members as equally as possible and there should be no major differences." Use these conflicting results as an example in a short explanation of why opinion polls often fail to reveal public attitudes clearly.

3.4 Toward Statistical Inference

A market research firm interviews a random sample of 2500 adults. Result: 66% find shopping for clothes frustrating and time-consuming. That's the truth about the 2500 people in the sample. What is the truth about the almost 210 million American adults who make up the population? Because the sample was chosen at random, it's reasonable to think that these 2500 people represent the entire population fairly well. So the market researchers turn the *fact* that 66% of the *sample* find shopping frustrating into an *estimate* that about 66% of *all adults* feel this way. That's a basic move in statistics: use a fact about a sample to estimate the truth about the whole population. We call this **statistical inference** because we infer conclusions about the wider population from data on selected individuals. To think about inference, we must keep straight whether a number describes a sample or a population. Here is the vocabulary we use.

statistical inference

Parameters and Statistics

A **parameter** is a number that describes the **population.** A parameter is a fixed number, but in practice we do not know its value.

A **statistic** is a number that describes a **sample.** The value of a statistic is known when we have taken a sample, but it can change from sample to sample. We often use a statistic to estimate an unknown parameter.

EXAMPLE 3.20

Are attitudes toward shopping changing? Sample surveys show that fewer people enjoy shopping than in the past. A survey by the market research firm Yankelovich Clancy Shulman asked a nationwide random sample of 2500 adults if they agreed or disagreed that "I like buying new clothes, but shopping is often frustrating and time-consuming." Of the respondents, 1650, or 66%, said they agreed.[25] The proportion of the sample who agree is

$$\hat{p} = \frac{1650}{2500} = 0.66 = 66\%$$

The number $\hat{p} = 0.66$ is a *statistic*. The corresponding *parameter* is the proportion (call it p) of all adult U.S. residents who would have said "Agree" if asked the same question. We don't know the value of the parameter p, so we use the statistic $\hat{p}$ to estimate it.

Sampling variability

sampling variability

If Yankelovich took a second random sample of 2500 adults, the new sample would have different people in it. It is almost certain that there would not be exactly 1650 positive responses. That is, the value of the statistic $\hat{p}$ will vary from sample to sample. This basic fact is called **sampling variability:** the value of a statistic varies in repeated random sampling. Could it happen that one random sample finds that 66% of adults find clothes shopping frustrating and a second random sample finds that only 42% feel this way? Random samples eliminate *bias* from the act of choosing a sample, but they can still be wrong because of the *variability* that results when we choose at random. If the variation when we take repeat samples from the same population is too great, we can't trust the results of any one sample.

We are saved by the second great advantage of random samples. The first advantage is that choosing at random eliminates favoritism. That is, random sampling attacks bias. The second advantage is that if we take lots of random samples of the same size from the same population, the variation from sample to sample will follow a predictable pattern. **All of statistical inference is based on one idea: to see how trustworthy a procedure is, ask what would happen if we repeated it many times.**

To understand why sampling variability is not fatal, we ask, "What would happen if we took many samples?" Here's how to answer that question:

- Take a large number of samples from the same population.
- Calculate the sample proportion $\hat{p}$ for each sample.

- Make a histogram of the values of $\hat{p}$.
- Examine the distribution displayed in the histogram for shape, center, and spread, as well as outliers or other deviations.

In practice it is too expensive to take many samples from a large population such as all adult U.S. residents. But we can imitate many samples by using random digits. Using random digits from a table or computer software to imitate chance behavior is called **simulation.**

simulation

EXAMPLE 3.21

We will simulate drawing simple random samples (SRSs) of size 100 from the population of all adult U.S. residents. Suppose that in fact 60% of the population find clothes shopping time-consuming and frustrating. Then the true value of the parameter we want to estimate is $p = 0.6$. (Of course, we would not sample in practice if we already knew that $p = 0.6$. We are sampling here to learn how sampling behaves.)

We can imitate the population by a table of random digits, with each entry standing for a person. Six of the ten digits (say 0 to 5) stand for people who find shopping frustrating. The remaining four digits, 6 to 9, stand for those who do not. Because all digits in a random number table are equally likely, this assignment produces a population proportion of frustrated shoppers equal to $p = 0.6$. We then imitate an SRS of 100 people from the population by taking 100 consecutive digits from Table B. The statistic $\hat{p}$ is the proportion of 0s to 5s in the sample.

The first 100 entries in Table B contain 63 digits between 0 and 5, so $\hat{p} = 63/100 = 0.63$. A second SRS based on the second 100 entries in Table B gives a different result, $\hat{p} = 0.56$. The two sample results are different, and neither is equal to the true population value $p = 0.6$. That's sampling variability.

Sampling distributions

Simulation is a powerful tool for studying chance. Now that we see how simulation works, it is faster to abandon Table B and to use a computer programmed to produce random digits.

EXAMPLE 3.22

Figure 3.6 illustrates the process of choosing many samples and finding the sample proportion $\hat{p}$ for each one. Follow the flow of the figure from the population at the

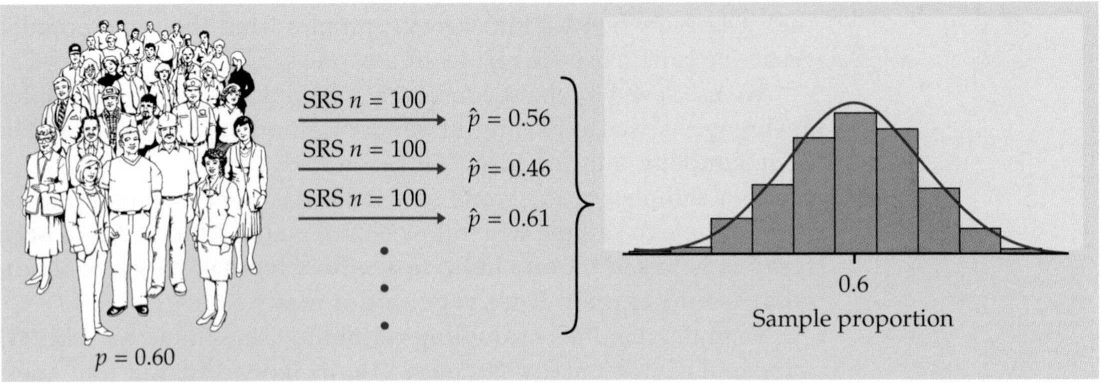

FIGURE 3.6 The results of many SRSs have a regular pattern. Here, we draw 1000 SRSs of size 100 from the same population. The population proportion is $p = 0.60$. The histogram shows the distribution of the 1000 sample proportions $\hat{p}$.

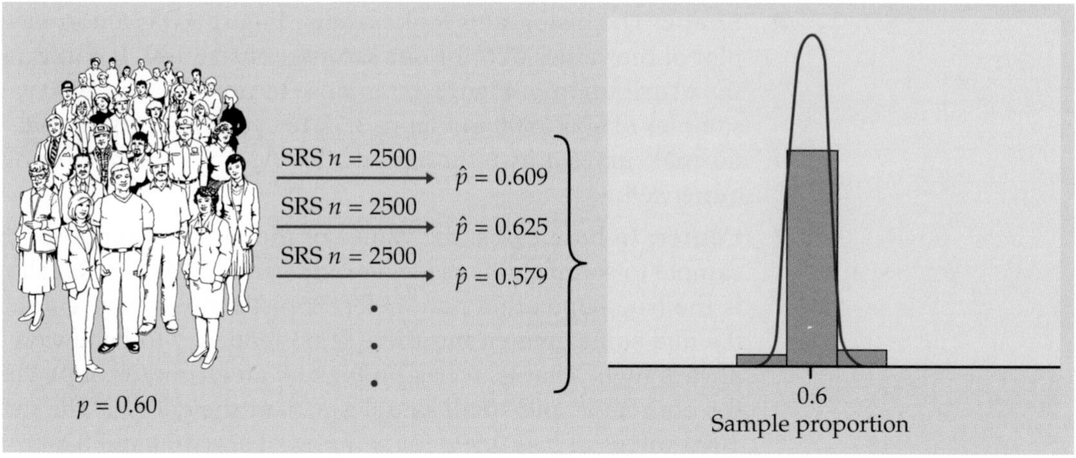

FIGURE 3.7 The distribution of sample proportions $\hat{p}$ for 1000 SRSs of size 2500 drawn from the same population as in Figure 3.6. The two histograms have the same scale. The statistic from the larger sample is less variable.

left, to choosing an SRS and finding the $\hat{p}$ for this sample, to collecting together the $\hat{p}$'s from many samples. The histogram at the right of the figure shows the distribution of the values of $\hat{p}$ from 1000 separate SRSs of size 100 drawn from a population with $p = 0.6$.

Of course, Yankelovich interviewed 2500 people, not just 100. Figure 3.7 is parallel to Figure 3.6. It shows the process of choosing 1000 SRSs, each of size 2500, from a population in which the true proportion is $p = 0.6$. The 1000 values of $\hat{p}$ from these samples form the histogram at the right of the figure. Figures 3.6 and 3.7 are drawn on the same scale. Comparing them shows what happens when we increase the size of our samples from 100 to 2500. These histograms display the *sampling distribution* of the statistic $\hat{p}$ for two sample sizes.

Sampling Distribution

The **sampling distribution** of a statistic is the distribution of values taken by the statistic in all possible samples of the same size from the same population.

Strictly speaking, the sampling distribution is the ideal pattern that would emerge if we looked at all possible samples of size 100 from our population. A distribution obtained from a fixed number of trials, like the 1000 trials in Figures 3.6 and 3.7, is only an approximation to the sampling distribution. We will see that probability theory, the mathematics of chance behavior, can sometimes describe sampling distributions exactly. The interpretation of a sampling distribution is the same, however, whether we obtain it by simulation or by the mathematics of probability.

We can use the tools of data analysis to describe any distribution. Let's apply those tools to Figures 3.6 and 3.7.

- **Shape:** The histograms look normal. Figure 3.8 is a normal quantile plot of the values of $\hat{p}$ for our samples of size 100. It confirms that the distribution in Figure 3.6 is close to normal. The 1000 values for samples of size 2500 in Figure 3.7 are even closer to normal. The normal curves drawn through the histograms describe the overall shape quite well.
- **Center:** In both cases, the values of the sample proportion $\hat{p}$ vary from sample to sample, but the values are centered at 0.6. Recall that $p = 0.6$ is the true population parameter. Some samples have a $\hat{p}$ less than 0.6 and some greater, but there is no tendency to be always low or always high. That is, $\hat{p}$ has no **bias** as an estimator of p. This is true for both large and small samples. (Want the details? The mean of the 1000 values of $\hat{p}$ is 0.598 for samples of size 100 and 0.6002 for samples of size 2500. The median value of $\hat{p}$ is exactly 0.6 for samples of both sizes.)
- **Spread:** The values of $\hat{p}$ from samples of size 2500 are much less spread out than the values from samples of size 100. In fact, the standard deviations are 0.051 for Figure 3.6 and 0.0097, or about 0.01, for Figure 3.7.

Although these results describe just two sets of simulations, they reflect facts that are true whenever we use random sampling.

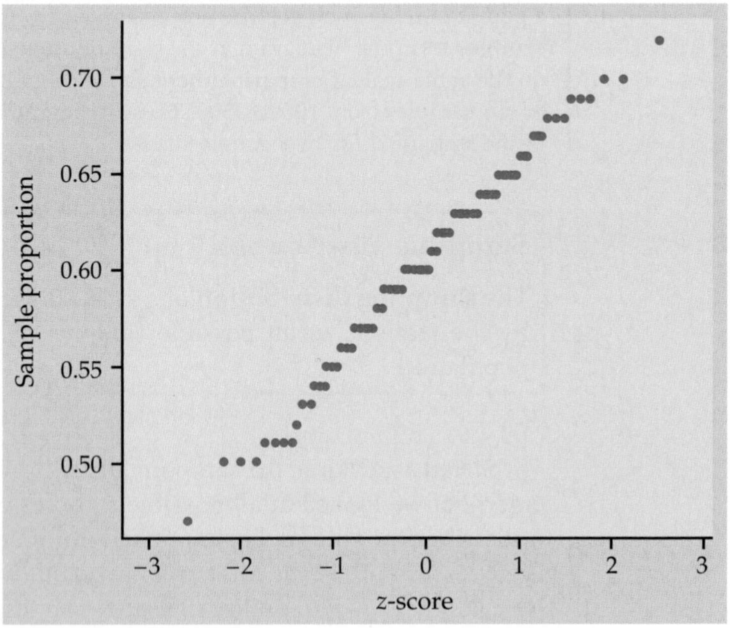

FIGURE 3.8 Normal quantile plot of the sample proportions in Figure 3.6. The distribution is close to normal except for some granularity due to the fact that sample proportions from a sample of size 100 can take only values that are multiples of 0.01. Because a plot of 1000 points is hard to read, this plot presents only every 10th value.

Bias and variability

Our simulations show that a sample of size 2500 will almost always give an estimate $\hat{p}$ that is close to the truth about the population. Figure 3.7 illustrates this fact for just one value of the population proportion, but it is true for any population. Samples of size 100, on the other hand, might give an estimate of 50% or 70% when the truth is 60%.

Thinking about Figures 3.6 and 3.7 helps us restate the idea of bias when we use a statistic like $\hat{p}$ to estimate a parameter like p. It also reminds us that variability matters as much as bias.

Bias and Variability

Bias concerns the center of the sampling distribution. A statistic used to estimate a parameter is **unbiased** if the mean of its sampling distribution is equal to the true value of the parameter being estimated.

The **variability of a statistic** is described by the spread of its sampling distribution. This spread is determined by the sampling design and the sample size n. Statistics from larger samples have smaller spreads.

We can think of the true value of the population parameter as the bull's-eye on a target, and of the sample statistic as an arrow fired at the bull's-eye. Bias and variability describe what happens when an archer fires many arrows at the target. *Bias* means that the aim is off, and the arrows land consistently off the bull's-eye in the same direction. The sample values do not center about the population value. Large *variability* means that repeated shots are widely scattered on the target. Repeated samples do not give similar results but differ widely among themselves. Figure 3.9 shows this target illustration of the two types of error.

Notice that small variability (repeated shots are close together) can accompany large bias (the arrows are consistently away from the bull's-eye in one direction). And small bias (the arrows center on the bull's-eye) can accompany large variability (repeated shots are widely scattered). A good sampling scheme, like a good archer, must have both small bias and small variability. Here's how we do this.

Managing Bias and Variability

To reduce bias, use random sampling. When we start with a list of the entire population, simple random sampling produces unbiased estimates—the values of a statistic computed from an SRS neither consistently overestimate nor consistently underestimate the value of the population parameter.

To reduce the variability of a statistic from an SRS, use a larger sample. You can make the variability as small as you want by taking a large enough sample.

In practice, Yankelovich takes only one sample. We don't know how close to the truth an estimate from this one sample is because we don't know what

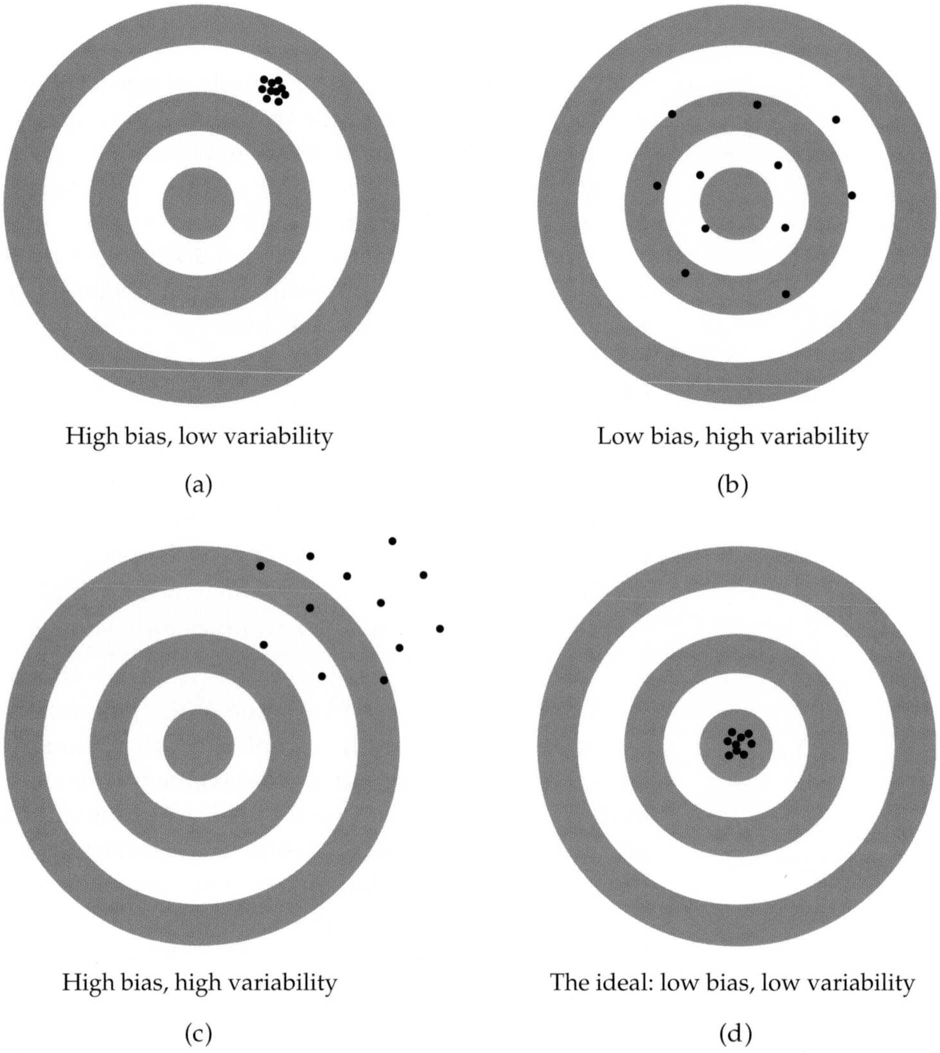

FIGURE 3.9 Bias and variability in shooting arrows at a target. Bias means the archer systematically misses in the same direction. Variability means that the arrows are scattered.

the truth about the population is. But *large random samples almost always give an estimate that is close to the truth*. Looking at the pattern of many samples shows that we can trust the result of one sample. The Current Population Survey's sample of 55,000 households estimates the national unemployment rate very accurately. Of course, only probability samples carry this guarantee. *Nightline*'s voluntary response sample (Example 3.15) is worthless even though 186,000 people called in. Using a probability sampling design and taking care to deal with practical difficulties reduce bias in a sample. The size of the sample then determines how close to the population truth the sample result is likely to fall. Results from a sample survey usually come with a *margin of error* that sets bounds on the size of the likely error. How to do this is part of the detail of statistical inference. We will describe the reasoning in Chapter 6.

Sampling from large populations

Yankelovich's sample of 2500 adults is only about 1 out of every 85,000 adults in the United States. Does it matter whether we sample 1-in-100 individuals in the population or 1-in-85,000?

> **Population Size Doesn't Matter**
>
> The variability of a statistic from a random sample does not depend on the size of the population, as long as the population is at least 100 times larger than the sample.

Why does the size of the population have little influence on the behavior of statistics from random samples? To see why this is plausible, imagine sampling harvested corn by thrusting a scoop into a lot of corn kernels. The scoop doesn't know whether it is surrounded by a bag of corn or by an entire truckload. As long as the corn is well mixed (so that the scoop selects a random sample), the variability of the result depends only on the size of the scoop.

The fact that the variability of sample results is controlled by the size of the sample has important consequences for sampling design. An SRS of size 2500 from the more than 210 million adult residents of the United States gives results as precise as an SRS of size 2500 from the 665,000 adult inhabitants of San Francisco. This is good news for designers of national samples but bad news for those who want accurate information about the citizens of San Francisco. If both use an SRS, both must use the same size sample to obtain equally trustworthy results.

Why randomize?

Why randomize? The act of randomizing guarantees that the results of analyzing our data are subject to the laws of probability. The behavior of statistics is described by a sampling distribution. The form of the distribution is known, and in many cases is approximately normal. Often the center of the distribution lies at the true parameter value, so that the notion that randomization eliminates bias is made more precise. The spread of the distribution describes the variability of the statistic and can be made as small as we wish by choosing a large enough sample. In a randomized experiment, we can reduce variability by choosing larger groups of subjects for each treatment.

These facts are at the heart of formal statistical inference. Later chapters will have much to say in more technical language about sampling distributions and the way statistical conclusions are based on them. What any user of statistics must understand is that all the technical talk has its basis in a simple question: What would happen if the sample or the experiment were repeated many times? The reasoning applies not only to an SRS but also to the complex sampling designs actually used by opinion polls and other national sample surveys. The same conclusions hold as well for randomized experimental designs. The details vary with the design but the basic facts are true whenever randomization is used to produce data.

Remember that proper statistical design is not the only aspect of a good sample or experiment. The sampling distribution shows only how a statistic varies due to the operation of chance in randomization. It reveals nothing about possible bias due to undercoverage or nonresponse in a sample, or to lack of realism in an experiment. The true distance of a statistic from the parameter it is estimating can be much larger than the sampling distribution suggests. What is worse, there is no way to say how large the added error is. The real world is less orderly than statistics textbooks imply.

Beyond the basics ▸ capture-recapture sampling

Sockeye salmon return to reproduce in the river where they were hatched four years earlier. How many salmon survived natural perils and heavy fishing to make it back this year? How many mountain sheep are there in Colorado? Are migratory songbird populations in North America decreasing or holding their own? These questions concern the size of animal populations. Biologists address them with a special kind of repeated sampling, called *capture-recapture sampling*.

EXAMPLE 3.23

You are interested in the number of least flycatchers migrating along a major route in the north-central United States. You set up "mist nets" that capture the birds but do not harm them. The birds caught in the net are fitted with a small aluminum leg band and released. Last year you banded and released 200 least flycatchers. This year you repeat the process. Your net catches 120 least flycatchers, 12 of which have tags from last year's catch.

The proportion of your second sample that have bands should estimate the proportion in the entire population that are banded. So if N is the unknown number of least flycatchers, we should have approximately

$$\text{proportion banded in sample} = \text{proportion banded in population}$$

$$\frac{12}{120} = \frac{200}{N}$$

Solve for N to estimate that the total number of flycatchers migrating while your net was up this year is approximately

$$N = 200 \times \frac{120}{12} = 2000$$

The capture-recapture idea extends the use of a sample proportion to estimate a population proportion. The idea works well if both samples are SRSs from the population and the population remains unchanged between samples. In practice, complications arise because, for example, some of the birds tagged last year died before this year's migration. Variations on capture-recapture samples are widely used in wildlife studies and are now finding other applications. One way to estimate the census undercount in a district is to consider the census as "capturing and marking" the households that respond. Census workers then visit the district, take an SRS of households, and see how many of those counted by the census show up in the sample. Capture-recapture estimates the total count of households in the district. As with estimating wildlife populations, there are many practical pitfalls. Our final word is as before: the real world is less orderly than statistics textbooks imply.

SUMMARY

A number that describes a population is a **parameter.** A number that can be computed from the data is a **statistic.** The purpose of sampling or experimentation is usually to use statistics to make statements about unknown parameters.

A statistic from a probability sample or randomized experiment has a **sampling distribution** that describes how the statistic varies in repeated data production. The sampling distribution answers the question "What would happen if we repeated the sample or experiment many times?" Formal statistical inference is based on the sampling distributions of statistics.

A statistic as an estimator of a parameter may suffer from **bias** or from high **variability.** Bias means that the center of the sampling distribution is not equal to the true value of the parameter. The variability of the statistic is described by the spread of its sampling distribution.

Properly chosen statistics from randomized data production designs have no bias resulting from the way the sample is selected or the way the experimental units are assigned to treatments. We can reduce the variability of the statistic by increasing the size of the sample or the size of the experimental groups.

SECTION 3.4 EXERCISES

State whether each boldface number in Exercises 3.58 to 3.61 is a parameter or a statistic.

3.58 Voter registration records show that **68%** of all voters in Indianapolis are registered as Republicans. To test a random digit dialing device, you use the device to call 150 randomly chosen residential telephones in Indianapolis. Of the registered voters contacted, **73%** are registered Republicans.

3.59 A carload lot of ball bearings has a mean diameter of **2.503** centimeters (cm). This is within the specifications for acceptance of the lot by the purchaser. The inspector happens to inspect 100 bearings from the lot with a mean diameter of **2.515** cm. This is outside the specified limits, so the lot is mistakenly rejected.

3.60 A telemarketing firm in Los Angeles uses a device that dials residential telephone numbers in that city at random. Of the first 100 numbers dialed, **43** are unlisted. This is not surprising, because **52%** of all Los Angeles residential phones are unlisted.

3.61 The Carolina Abecedarian Project investigated the effect of high-quality preschool programs on children from poor families. Children were randomly assigned to two groups. One group participated in a year-round

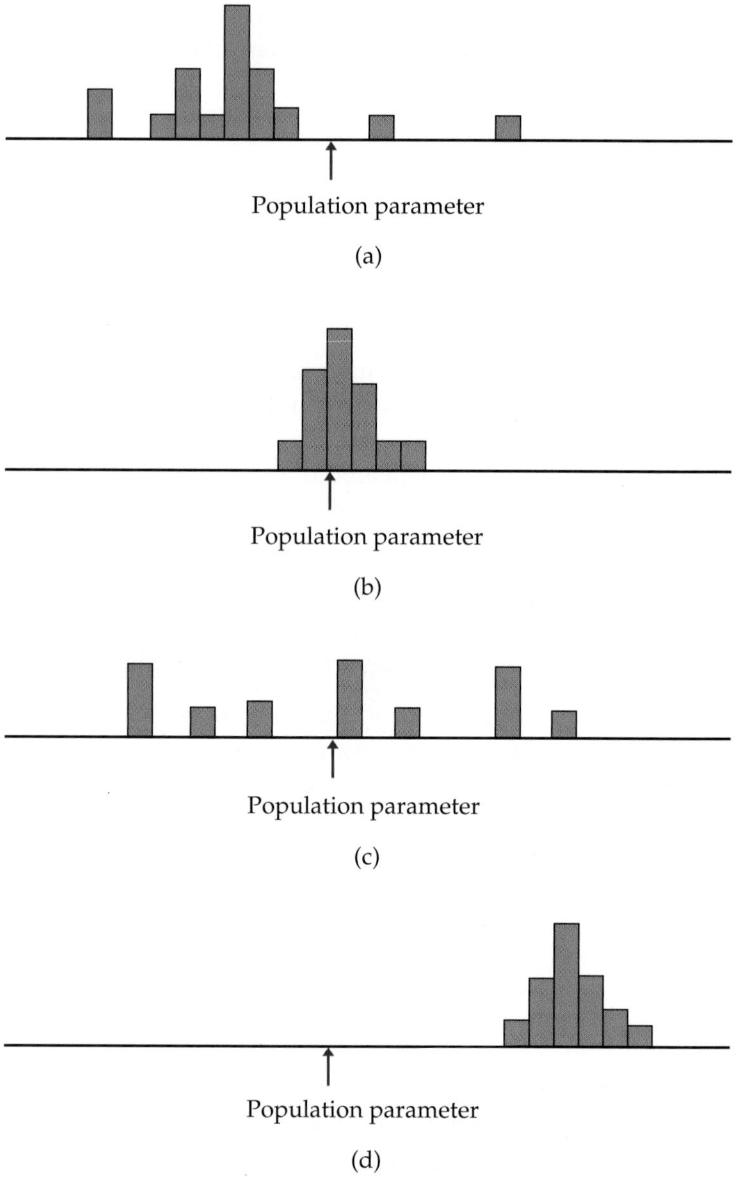

FIGURE 3.10 Which of these sampling distributions displays high or low bias and high or low variability? (For Exercise 3.62.)

preschool program from age three months. The control group received social services but no preschool. At age 21, **35%** of the treatment group and **14%** of the control group were attending a four-year college or had already graduated from college.

3.62 Figure 3.10 shows histograms of four sampling distributions of statistics intended to estimate the same parameter. Label each distribution relative to the others as high or low bias and as high or low variability.

3.63 Just before a presidential election, a national opinion polling firm increases the size of its weekly sample from the usual 1500 people to 4000 people. Why do you think the firm does this?

3.64 Each March, the Bureau of Labor Statistics carries out an Annual Demographic Supplement to its monthly Current Population Survey. The March 2000 survey recorded the incomes of 55,899 people between the ages of 25 and 65 who had worked but whose main work was not in agriculture. We will treat these 55,899 people as a population. As is usually the case, the distribution of incomes in this population is strongly skewed to the right. To estimate the mean income in this population, we can select an SRS and use the sample mean $\bar{x}$ to estimate the unknown population mean. How will the sample mean behave when we take many samples?

We used software to choose 1000 SRSs of size 25 and another 1000 SRSs of size 100. Figure 3.11 shows histograms of the two sets of 1000 sample means, using the same classes and drawn to the same scale for easy comparison.

(a) About what is the range (from smallest to largest) of the sample means for samples of size 25? For samples of size 100? What general fact about sampling distributions does this comparison illustrate?

(b) Based on the sample means for samples of size 100, about what is the value of the mean income for this entire population?

3.65 An agency of the federal government plans to take an SRS of residents in each state to estimate the proportion of owners of real estate in each state's population. The populations of the states range from about 494,000 people in Wyoming to more than 34 million in California.

(a) Will the variability of the sample proportion vary from state to state if an SRS of size 2000 is taken in each state? Explain your answer.

(b) Will the variability of the sample proportion change from state to state if an SRS of 1/10 of 1% (0.001) of the state's population is taken in each state? Explain your answer.

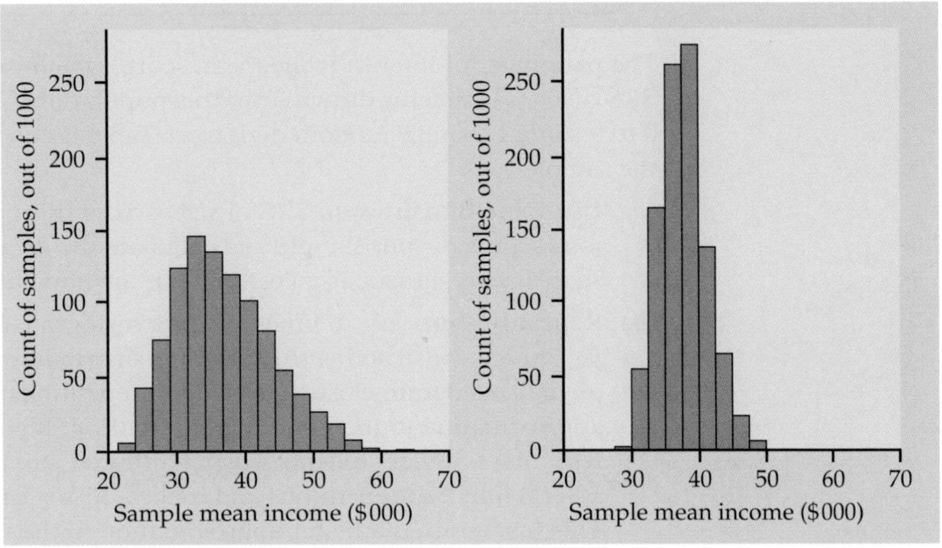

FIGURE 3.11 The distribution of sample means $\bar{x}$ for 1000 SRSs of size 25 (left) and of size 100 (right) from the same population, for Exercise 3.64.

3.66 The Ministry of Health in the Canadian province of Ontario wants to know whether the national health care system is achieving its goals in the province. The ministry conducted the Ontario Health Survey, which interviewed a probability sample of 61,239 adults who live in Ontario.[26]

(a) What is the population for this sample survey? What is the sample?

(b) The survey found that 76% of males and 86% of females in the sample had visited a general practitioner at least once in the past year. Do you think these estimates are close to the truth about the entire population? Why?

3.67 Coin tossing can illustrate the idea of a sampling distribution. The population is all outcomes (heads or tails) we would get if we tossed a coin forever. The parameter p is the proportion of heads in this population. We suspect that p is close to 0.5; that is, we think the coin will show about one-half heads in the long run. The sample is the outcomes of 20 tosses, and the statistic $\hat{p}$ is the proportion of heads in these 20 tosses.

(a) Toss a coin 20 times and record the value of $\hat{p}$.

(b) Repeat this sampling process 10 times. Make a stemplot of the 10 values of $\hat{p}$. Is the center of this distribution close to 0.5? (Ten repetitions give only a crude approximation to the sampling distribution. If possible, pool your work with that of other students to obtain several hundred repetitions and make a histogram of the values of $\hat{p}$.)

3.68 We can construct a sampling distribution by hand in the case of a very small sample from a very small population. The population contains 10 students. Here are their scores on an exam:

Student	0	1	2	3	4	5	6	7	8	9
Score	82	62	80	58	72	73	65	66	74	62

The parameter of interest is the mean score, which is 69.4. The sample is an SRS of $n = 4$ students drawn from this population. The students are labeled 0 to 9 so that a single random digit from Table B chooses one student for the sample.

(a) Use Table B to draw an SRS of size 4 from this population. Write the four scores in your sample and calculate the mean $\bar{x}$ of the sample scores. This statistic is an estimate of the population parameter.

(b) Repeat this process 10 times. Make a histogram of the 10 values of $\bar{x}$. You are constructing the sampling distribution of $\bar{x}$. Is the center of your histogram close to 69.4? (Ten repetitions give only a crude approximation to the sampling distribution. If possible, pool your work with that of other students—using different parts of Table B—to obtain several hundred repetitions and make a histogram of the values of $\bar{x}$. This histogram is a better approximation to the sampling distribution.)

3.69 An entomologist samples a field for egg masses of a harmful insect by placing a yard-square frame at random locations and examining the ground within the frame carefully. She wishes to estimate the proportion of square yards in which egg masses are present. Suppose that in a large field egg masses are present in 20% of all possible yard-square areas—that is, $p = 0.2$ in this population.

(a) Use Table B to simulate the presence or absence of egg masses in each square yard of an SRS of 10 square yards from the field. Be sure to explain clearly which digits you used to represent the presence and the absence of egg masses. What proportion of your 10 sample areas had egg masses?

(b) Repeat (a) with different lines from Table B, until you have simulated the results of 20 SRSs of size 10. What proportion of the square yards in each of your 20 samples had egg masses? Make a stemplot from these 20 values to display the sampling distribution of $\hat{p}$ in this case. What is the mean of this distribution? What is its shape?

3.70 The *What Is Probability?* applet can speed up simulations like that described in Example 3.21. This applet simulates tossing a coin many times, but choosing a random sample from a very large population is the same in principle.

You have a population in which 60% of the individuals find clothes shopping frustrating. You want to take many samples from this population to observe how the sample proportion who find shopping frustrating varies from sample to sample. Set the "Probability of heads" in the applet to 0.6 and the number of tosses to 25. This simulates an SRS of size 25 from this population. Each head in the 25 tosses is a person who finds shopping frustrating. By alternating between "Toss" and "Reset" you can take many samples quickly.

(a) Take 50 samples, recording the number of heads in each sample. Make a histogram of the 50 sample proportions.

(b) Another population contains only 20% who approve of legal gambling. Take 50 samples of size 25 from this population, record the number in each sample who approve, and make a histogram of the 50 sample proportions. How do the centers of your two histograms reflect the differing truths about the two populations?

3.71 The *Simple Random Sample* applet can illustrate the idea of a sampling distribution. Form a population labeled 1 to 100. We will choose an SRS of 10 of these numbers. That is, in this exercise, the numbers themselves are the population, not just labels for 100 individuals. The mean of the whole numbers 1 to 100 is 50.5. This is the population mean.

(a) Use the applet to choose an SRS of size 10. Which 10 numbers were chosen? What is their mean? This is the sample mean $\bar{x}$.

(b) Although the population and its mean 50.5 remain fixed, the sample mean changes as we take more samples. Take another SRS of size 10. (Use the "Reset" button to return to the original population before taking the second sample.) What are the 10 numbers in your sample? What is their mean? This is another value of $\bar{x}$.

(c) Take 8 more SRSs from this same population and record their means. You now have 10 values of the sample mean $\overline{x}$ from 10 SRSs of the same size from the same population. Make a histogram of the 10 values and mark the population mean 50.5 on the horizontal axis. Are your 10 sample values roughly centered at the population value? (If you kept going forever, your $\overline{x}$-values would form the sampling distribution of the sample mean; the population mean would indeed be the center of this distribution.)

3.72 The CSDATA data file contains the college grade point averages (GPAs) of all 224 students in a university entering class who planned to major in computer science. This is our population. Statistical software can take repeated samples to illustrate sampling variability.

(a) Using software, describe this population with a histogram and with numerical summaries. In particular, what is the mean GPA in the population?

(b) Choose an SRS of 20 members from this population. Make a histogram of the GPAs in the sample and find their mean. Briefly compare the distributions of GPA in the sample and in the population.

(c) Repeat the process of choosing an SRS of size 20 four more times (five in all). Record the five histograms of your sample GPAs. Does it seem reasonable to you from this small trial that an SRS will usually produce a sample that is generally representative of the population?

3.73 Continue the previous exercise, using software to illustrate the idea of a sampling distribution.

(a) Choose 20 more SRSs of size 20 in addition to the 5 you have already chosen. Don't make histograms of these latest samples—just record the mean GPA for each sample. Make a histogram of the 25 sample means. This histogram is a rough approximation to the sampling distribution of the mean.

(b) One sign of bias would be that the distribution of the sample means was systematically on one side of the true population mean. Mark the population mean GPA on your histogram of the 25 sample means. Is there a clear bias?

(c) Find the mean and standard deviation of your 25 sample means. We expect that the mean will be close to the true mean of the population. Is it? We also expect that the standard deviation of the sampling distribution will be smaller than the standard deviation of the population. Is it?

CHAPTER 3 EXERCISES

3.74 A large observational study used records from Canada's national health care system to compare the effectiveness of two ways to treat prostate disease. The two treatments are traditional surgery and a new method that does not require surgery. The records described many patients whose doctors had chosen one or the other method. The study found that patients treated by the new method were significantly more likely to die within 8 years.[27]

(a) Further study of the data showed that this conclusion was wrong. The extra deaths among patients who received the new treatment could be explained by lurking variables. What lurking variables might be confounded with a doctor's choice of surgical or nonsurgical treatment?

(b) You have 300 prostate patients who are willing to serve as subjects in an experiment to compare the two methods. Use a diagram to outline the design of a randomized comparative experiment.

3.75 A researcher studying the effect of price promotions on consumers' expectations makes up two different histories of the store price of a hypothetical brand of laundry detergent for the past year. Students in a marketing course view one or the other price history on a computer. Some students see a steady price, while others see regular promotions that temporarily cut the price. Then the students are asked what price they would expect to pay for the detergent. Is this study an experiment? Why? What are the explanatory and response variables?

3.76 What is the best way to answer each of the questions below: an experiment, a sample survey, or an observational study that is not a sample survey? Explain your choices.

(a) Are people generally satisfied with how things are going in the country right now?

(b) Do college students learn basic accounting better in a classroom or using an online course?

(c) How long do your teachers wait on the average after they ask their class a question?

3.77 Give an example of a question about college students, their behavior, or their opinions that would best be answered by

(a) a sample survey.

(b) an observational study that is not a sample survey.

(c) an experiment.

3.78 Do consumers prefer the taste of a cheeseburger from McDonald's or from Wendy's in a blind test in which neither burger is identified? Describe briefly the design of a matched pairs experiment to investigate this question. How will you use randomization?

3.79 How does the time it takes a bicycle rider to travel 100 meters depend on which gear is used and how steep the course is? It may be, for example, that higher gears are faster on the level but lower gears are faster on steep inclines. Discuss the design of a two-factor experiment to investigate this issue, using one bicycle with three gears and one rider. How will you use randomization?

3.80 The previous two exercises illustrate the use of statistically designed experiments to answer questions that arise in everyday life. Select a question of interest to you that an experiment might answer and carefully discuss the design of an appropriate experiment.

3.81 We often see players on the sidelines of a football game inhaling oxygen. Their coaches think this will speed their recovery. We might measure recovery from intense exercise as follows: Have a football player run 100 yards three times in quick succession. Then allow three minutes to rest before running 100 yards again. Time the final run. Because players vary greatly in speed, you plan a matched pairs experiment using 25 football players as subjects. Discuss the design of such an experiment to investigate the effect of inhaling oxygen during the rest period.

3.82 It is possible to use a computer to make telephone calls over the Internet. How will the cost affect the behavior of users of this service? You will offer the service to all 200 rooms in a college dormitory. Some rooms will pay a flat rate. Others will pay higher rates at peak periods and very low rates off-peak. You are interested in the amount and time of use and in the effect on the congestion of the network. Outline the design of an experiment to study the effect of rate structure. Use Table B, starting at line 125, to assign the first 5 rooms to the flat-rate group.

3.83 The National Institutes of Mental Health (NIMH) wants to know whether intense education about the risks of AIDS will help change the behavior of people who now report sexual activities that put them at risk of infection. NIMH investigators screened 38,893 people to identify 3706 suitable subjects. The subjects were assigned to a control group (1855 people) or an intervention group (1851 people). The control group attended a one-hour AIDS education session; the intervention group attended seven single-sex discussion sessions, each lasting 90 to 120 minutes. After 12 months, 64% of the intervention group and 52% of the control group said they used condoms. (None of the subjects used condoms regularly before the study began.)[28]

(a) Because none of the subjects used condoms when the study started, we might just offer the intervention sessions and find that 64% used condoms 12 months after the sessions. Explain why this greatly overstates the effectiveness of the intervention.

(b) Outline the design of this experiment.

(c) You must randomly assign 3706 subjects. How would you label them? Use line 119 of Table B to choose the first 5 subjects for the intervention group.

3.84 What are the most important goals of schoolchildren? Do girls and boys have different goals? Are goals different in urban, suburban, and rural areas? To find out, researchers wanted to ask children in the fourth, fifth, and sixth grades this question:

What would you most like to do at school?

A. *Make good grades.*
B. *Be good at sports.*
C. *Be popular.*

Because most children live in heavily populated urban and suburban areas, an SRS might contain few rural children. Moreover, it is too expensive to choose children at random—we must start by choosing schools rather than children. Describe a suitable sample design for this study, using the ideas of stratified and multistage samples. Explain the reasoning behind your choice.[29]

3.85 You want to investigate the attitudes of students at your school about the faculty's commitment to teaching. The student government will pay the costs of contacting about 500 students.

(a) Specify the exact population for your study; for example, will you include part-time students?

(b) Describe your sample design. Will you use a stratified sample?

(c) Briefly discuss the practical difficulties that you anticipate; for example, how will you contact the students in your sample?

3.86 A drug manufacturer is studying how a new drug behaves in patients. Investigators compare 2 doses, 5 milligrams (mg) and 10 mg. The drug can be administered by injection, by a skin patch, or by intravenous drip. Concentration in the blood after 30 minutes (the response variable) may depend both on the dose and on the method of administration.

(a) Make a sketch that describes the treatments formed by combining dosage and method. Then use a diagram to outline a completely randomized design for this two-factor experiment.

(b) "How many subjects?" is a tough issue. We will explain the basic ideas in Chapter 6. What can you say now about the advantage of using larger groups of subjects?

3.87 Few people want to eat discolored french fries. Potatoes are kept refrigerated before being cut for french fries to prevent spoiling and preserve flavor. But immediate processing of cold potatoes causes discoloring due to complex chemical reactions. The potatoes must therefore be brought to room temperature before processing. Design an experiment in which tasters will rate the color and flavor of french fries prepared from several groups of potatoes. The potatoes will be fresh picked or stored for a month at room temperature or stored for a month refrigerated. They will then be sliced and cooked either immediately or after an hour at room temperature.

(a) What are the factors and their levels, the treatments, and the response variables?

(b) Describe and outline the design of this experiment.

(c) It is efficient to have each taster rate fries from all treatments. How will you use randomization in presenting fries to the tasters?

3.88 The drug that is the subject of the experiment in Exercise 3.86 may behave differently in men and women. How would you modify your experimental design to take this into account?

3.89 A psychologist is interested in the effect of room temperature on the performance of tasks requiring manual dexterity. She chooses temperatures of 70° F and 90° F as treatments. The response variable is the number of correct insertions, during a 30-minute period, in an elaborate peg-and-hole apparatus that requires the use of both hands simultaneously. Each subject is trained on the apparatus and then asked to make as many insertions as possible in 30 minutes of continuous effort.

(a) Outline a completely randomized design to compare dexterity at 70° and 90°. Twenty subjects are available.

(b) Because individuals differ greatly in dexterity, the wide variation in individual scores may hide the systematic effect of temperature unless there are many subjects in each group. Describe in detail the design of a matched pairs experiment in which each subject serves as his or her own control.

3.90 The requirement that human subjects give their informed consent to participate in an experiment can greatly reduce the number of available subjects. For example, a study of new teaching methods asks the consent of parents for their children to be taught by either a new method or the standard method. Many parents do not return the forms, so their children must continue to follow the standard curriculum. Why is it not correct to consider these children as part of the control group along with children who are randomly assigned to the standard method?

3.91 Advice columnist Ann Landers once asked her female readers whether they would be content with affectionate treatment by men, with no sex ever. Over 90,000 women wrote in, with 72% answering "Yes." Many of the letters described unfeeling treatment by men. Explain why this sample is certainly biased. What is the likely direction of the bias? That is, is that 72% probably higher or lower than the truth about the population of all adult women?

3.92 Statistical software allows you to simulate the results of, say, 100 SRSs of size n drawn from a population in which proportion p would say "Yes" to a certain question. We will see in Chapter 5 that "binomial" is the key word to look for in the software menus.

Draw 100 samples of size $n = 50$ from populations with $p = 0.1$, $p = 0.3$, and $p = 0.5$. Make a stemplot of the 100 values of $\hat{p}$ obtained in each simulation. Compare your three stemplots. Do they show about the same variability? How does changing the parameter p affect the sampling distribution? If your software permits, make a normal quantile plot of the $\hat{p}$-values from the population with $p = 0.5$. Is the sampling distribution approximately normal?

3.93 Draw 100 samples each of sizes $n = 50$, $n = 200$, and $n = 800$ from a population with $p = 0.6$. Prepare a frequency histogram of the $\hat{p}$-values for each simulation, using the same horizontal and vertical scales so that the three graphs can be compared easily. How does increasing the size of an SRS affect the sampling distribution of $\hat{p}$?

PART 2

Probability and Inference

Prelude: getting through tollbooths faster

We all know and hate the tollbooth lines at bridges, tunnels, and toll roads. Statistics can't reduce the number of vehicles trying to get through, but it can help them get through faster.

Suppose that we must design the toll plaza for a new tunnel. Traffic in both directions will pay at the same location. Drivers may use cash, credit cards, or an electronic pay-while-moving system. How many toll booths do we need? Should each booth accept all types of payment, or should they specialize? What will happen in tourist season, when fewer drivers use the local electronic payment system and more pay cash?

Surprisingly, systems such as this can be described by the same mathematics that describes tossing coins and rolling dice. Like the fall of coins and dice, the arrival of vehicles at a toll booth is *random:* we can't say whether the next driver will command a car or a semi, or whether she will use cash or credit, but we know the *probabilities* of all combinations as well as probabilities for the time between vehicles and how long each type of vehicle takes to pay. By combining these probabilities into mathematical models, we can study the system on a computer before it is built. Are lines often too long at rush hour in tourist season? What happens if we add a booth? What happens if we make that booth "cash only"? What happens if we make the approach longer so drivers have more time to see the shortest line?

Probability theory, which begins with tossing coins and rolling dice, ends by describing much in the world. Probability, focused on its service to statistics, is our topic in this chapter.

CHAPTER 4

Wassily Kandinsky, *Yellow, Red, and Blue.*

Probability—The Study of Randomness

Introduction

The reasoning of statistical inference rests on asking, "How often would this method give a correct answer if I used it very many times?" When we produce data by random sampling or randomized comparative experiments, the laws of probability answer the question "What would happen if we did this many times?"

4.1 Randomness

Toss a coin, or choose an SRS. The result can't be predicted in advance, because the result will vary when you toss the coin or choose the sample repeatedly. But there is still a regular pattern in the results, a pattern that emerges clearly only after many repetitions. This remarkable fact is the basis for the idea of probability.

EXAMPLE 4.1

When you toss a coin, there are only two possible outcomes, heads or tails. Figure 4.1 shows the results of tossing a coin 5000 times twice. For each number of tosses from 1 to 5000, we have plotted the proportion of those tosses that gave a head. Trial A (red line) begins tail, head, tail, tail. You can see that the proportion of heads for Trial A starts at 0 on the first toss, rises to 0.5 when the second toss gives a head, then falls to 0.33 and 0.25 as we get two more tails. Trial B (green line), on the other hand, starts with five straight heads, so the proportion of heads is 1 until the sixth toss.

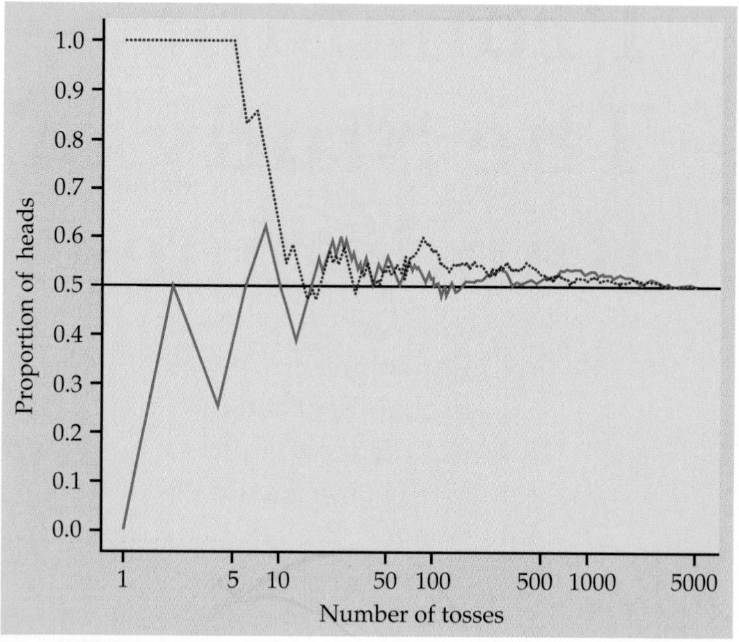

FIGURE 4.1 The proportion of tosses of a coin that give a head changes as we make more tosses. Eventually, however, the proportion approaches 0.5, the probability of a head. Here are the results of two trials of 5000 tosses each.

The proportion of tosses that produce heads is quite variable at first. Trial A starts low and Trial B starts high. As we make more and more tosses, however, the proportion of heads for both trials gets close to 0.5 and stays there. If we made yet a third trial at tossing the coin a great many times, the proportion of heads would again settle down to 0.5 in the long run. We say that 0.5 is the *probability* of a head. The probability 0.5 appears as a horizontal line on the graph.

applet

The *What Is Probability?* applet available on the Web site for this book, www.whfreeman.com/ips, animates Figure 4.1. It allows you to choose the probability of a head and simulate any number of tosses of a coin with that probability. Experience shows that the proportion of heads gradually settles down close to the probability. Equally important, it also shows that the proportion in a small or moderate number of tosses can be far from the probability. Probability describes *only* what happens in the long run.

The language of probability

"Random" in statistics is not a synonym for "haphazard" but a description of a kind of order that emerges only in the long run. We often encounter the unpredictable side of randomness in our everyday experience, but we rarely see enough repetitions of the same random phenomenon to observe the long-term regularity that probability describes. You can see that regularity emerging in Figure 4.1. In the very long run, the proportion of tosses that give a head is 0.5. This is the intuitive idea of probability. Probability 0.5 means "occurs half the time in a very large number of trials."

We might suspect that a coin has probability 0.5 of coming up heads just because the coin has two sides. As Exercises 4.1 and 4.2 illustrate, such suspicions are not always correct. The idea of probability is empirical. That is, it is based on observation rather than theorizing. Probability describes what happens in very many trials, and we must actually observe many trials to pin down a probability. In the case of tossing a coin, some diligent people have in fact made thousands of tosses.

EXAMPLE 4.2

The French naturalist Count Buffon (1707–1788) tossed a coin 4040 times. Result: 2048 heads, or proportion 2048/4040 = 0.5069 for heads.

Around 1900, the English statistician Karl Pearson heroically tossed a coin 24,000 times. Result: 12,012 heads, a proportion of 0.5005.

While imprisoned by the Germans during World War II, the South African statistician John Kerrich tossed a coin 10,000 times. Result: 5067 heads, proportion of heads 0.5067.

Randomness and Probability

We call a phenomenon **random** if individual outcomes are uncertain but there is nonetheless a regular distribution of outcomes in a large number of repetitions.

The **probability** of any outcome of a random phenomenon is the proportion of times the outcome would occur in a very long series of repetitions. That is, probability is long-term relative frequency.

Thinking about randomness

That some things are random is an observed fact about the world. The outcome of a coin toss, the time between emissions of particles by a radioactive source, and the sexes of the next litter of lab rats are all random. So is the outcome of a random sample or a randomized experiment. Probability theory is the branch of mathematics that describes random behavior. Of course, we can never observe a probability exactly. We could always continue tossing the coin, for example. Mathematical probability is an idealization based on imagining what would happen in an indefinitely long series of trials.

The best way to understand randomness is to observe random behavior—not only the long-run regularity but the unpredictable results of short runs. You can do this with physical devices, as in Exercises 4.1 to 4.4, but computer simulations (imitations) of random behavior allow faster exploration. Exercises 4.6 to 4.9 suggest some simulations of random behavior. As you explore randomness, remember:

independence

- You must have a long series of **independent** trials. That is, the outcome of one trial must not influence the outcome of any other. Imagine a crooked gambling house where the operator of a roulette wheel can stop it where she chooses—she can prevent the proportion of "red" from settling down to a fixed number. These trials are not independent.
- The idea of probability is empirical. Computer simulations start with given probabilities and imitate random behavior, but we can estimate a real-world probability only by actually observing many trials.
- Nonetheless, computer simulations are very useful because we need long runs of trials. In situations such as coin tossing, the proportion of an outcome often requires several hundred trials to settle down to the probability of that outcome. The kinds of physical random devices suggested in the exercises are too slow for this. Short runs give only rough estimates of a probability.

The uses of probability

Probability theory originated in the study of games of chance. Tossing dice, dealing shuffled cards, and spinning a roulette wheel are examples of deliberate randomization that are similar to random sampling. Although games of chance are ancient, they were not studied by mathematicians until the sixteenth and seventeenth centuries. It is only a mild simplification to say that probability as a branch of mathematics arose when seventeenth-century French gamblers asked the mathematicians Blaise Pascal and Pierre de Fermat for help. Gambling is still with us, in casinos and state lotteries. We will make use of games of chance as simple examples that illustrate the principles of probability.

Careful measurements in astronomy and surveying led to further advances in probability in the eighteenth and nineteenth centuries because the results of repeated measurements are random and can be described by distributions much like those arising from random sampling. Similar distributions appear in data on human life span (mortality tables) and in data on lengths or weights in a population of skulls, leaves, or cockroaches.[1] Now, we employ the mathematics of probability to describe the flow of traffic through a highway system, the Internet, or a computer processor; the genetic makeup of individuals or populations; the energy states of subatomic particles; the spread of epidemics or rumors; and the rate of return on risky investments. Although we are interested in probability because of its usefulness in statistics, the mathematics of chance is important in many fields of study.

SUMMARY

A **random phenomenon** has outcomes that we cannot predict but that nonetheless have a regular distribution in very many repetitions.

The **probability** of an event is the proportion of times the event occurs in many repeated trials of a random phenomenon.

SECTION 4.1 EXERCISES

4.1 Hold a penny upright on its edge under your forefinger on a hard surface, then snap it with your other forefinger so that it spins for some time before falling. Based on 50 spins, what is the probability of heads?

4.2 You may feel that it is obvious that the probability of a head in tossing a coin is about 1/2 because the coin has two faces. Such opinions are not always correct. The previous exercise asked you to spin a penny rather than toss it—that changes the probability of a head. Now try another variation. Stand a nickel on edge on a hard, flat surface. Pound the surface with your hand so that the nickel falls over. What is the probability that it falls with heads upward? Make at least 50 trials to estimate the probability of a head.

4.3 Toss a thumbtack on a hard surface 100 times. How many times did the thumbtack land with the point up? What is the approximate probability of landing point up?

4.4 In the game of Heads or Tails, Betty and Bob toss a coin four times. Betty wins a dollar from Bob for each head and pays Bob a dollar for each tail—that is, she wins or loses the difference between the number of heads and the number of tails. For example, if there are one head and three tails, Betty loses \$2. You can check that Betty's possible outcomes are

$$\{-4, -2, 0, 2, 4\}$$

Assign probabilities to these outcomes by playing the game 20 times and using the proportions of the outcomes as estimates of the probabilities. If possible, combine your trials with those of other students to obtain long-run proportions that are closer to the probabilities.

4.5 You read in a book on poker that the probability of being dealt three of a kind in a five-card poker hand is 1/50. Explain in simple language what this means.

applet

4.6 The idea of probability is that the *proportion* of heads in many tosses of a balanced coin eventually gets close to 0.5. But does the actual *count* of heads get close to one-half the number of tosses? Let's find out. Set the "Probability of heads" in the *What Is Probability?* applet to 0.5 and the number of tosses to 40. You can extend the number of tosses by clicking "Toss" again to get 40 more. Don't click "Reset" during this exercise.

(a) After 40 tosses, what is the proportion of heads? What is the count of heads? What is the difference between the count of heads and 20 (one-half the number of tosses)?

(b) Keep going to 120 tosses. Again record the proportion and count of heads and the difference between the count and 60 (half the number of tosses).

(c) Keep going. Stop at 240 tosses and again at 480 tosses to record the same facts. Although it may take a long time, the laws of probability say that the proportion of heads will always get close to 0.5 and also that the difference between the count of heads and half the number of tosses will always grow without limit.

applet

4.7 The basketball player Shaquille O'Neal makes about half of his free throws over an entire season. Use the *What Is Probability?* applet or software to simulate 100 free throws shot independently by a player who has probability 0.5 of making each shot. (In most software, the key phrase to look for is "Bernoulli trials." This is the technical term for independent trials with "Yes/No" outcomes. Our outcomes here are "Hit" and "Miss.")

(a) What percent of the 100 shots did he hit?

(b) Examine the sequence of hits and misses. How long was the longest run of shots made? Of shots missed? (Sequences of random outcomes often show runs longer than our intuition thinks likely.)

applet

4.8 A recent opinion poll showed that about 73% of married women agree that their husbands do at least their fair share of household chores. Suppose that this is exactly true. Choosing a married woman at random then has probability 0.73 of getting one who agrees that her husband does his share. You can use the *What Is Probability?* applet or software to simulate choosing many women independently. (In most software, the key phrase to look for is "Bernoulli trials." This is the technical term for independent trials with "Yes/No" outcomes. Our outcomes here are "Agree" and "Disagree.")

(a) Simulate drawing 20 women, then 80 women, then 320 women. What proportion agree in each case? We expect (but because of chance variation we can't be sure) that the proportion will be closer to 0.73 in longer runs of trials.

(b) Simulate drawing 20 women 10 times and record the percents in each trial who agree. Then simulate drawing 320 women 10 times and again record the 10 percents. Which set of 10 results is less variable? We expect the results of larger samples to be more predictable (less variable) than the results of smaller samples. That is "long-run regularity" showing itself.

4.9 Continue the exploration begun in Exercise 4.8. Software allows you to simulate many independent "Yes/No" trials more quickly if all you want to save is the count of "Yes" outcomes. The keyword "Binomial" simulates n independent Bernoulli trials, each with probability p of a "Yes," and records just the count of "Yes" outcomes.

(a) Simulate 100 draws of 20 women from the population in Exercise 4.8. Record the number who say "Agree" on each draw. What is the approximate probability that out of 20 women drawn at random at least 14 agree?

(b) Convert the counts who agree into percents of the 20 women in each trial who agree. Make a histogram of these 100 percents. Describe the shape, center, and spread of this distribution.

(c) Now simulate drawing 320 women. Do this 100 times and record the percent who agree on each of the 100 draws. Make a histogram of the percents and describe the shape, center, and spread of the distribution.

(d) In what ways are the distributions in parts (b) and (c) alike? In what ways do they differ? (Because regularity emerges in the long run, we expect the results of drawing 320 women to be less variable than the results of drawing 20 women.)

4.2 Probability Models

Earlier chapters gave mathematical models for linear relationships (in the form of the equation of a line) and for some distributions of data (in the form of normal density curves). Now we must give a mathematical description or model for randomness. To see how to proceed, think first about a very simple random phenomenon, tossing a coin once. When we toss a coin, we cannot know the outcome in advance. What do we know? We are willing to say that the outcome will be either heads or tails. Because the coin appears to be balanced, we believe that each of these outcomes has probability 1/2. This description of coin tossing has two parts:

- A list of possible outcomes
- A probability for each outcome

Such a description is the basis for all probability models. We will begin by describing the outcomes of a random phenomenon and then learn how to assign probabilities to the outcomes.

Sample spaces

A probability model first tells us what outcomes are possible.

> **Sample Space**
>
> The **sample space S** of a random phenomenon is the set of all possible outcomes.

The name "sample space" is natural in random sampling, where each possible outcome is a sample and the sample space contains all possible samples. To specify S, we must state what constitutes an individual outcome and then state which outcomes can occur. We often have some freedom in defining the sample space, so the choice of S is a matter of convenience as well as correctness. The idea of a sample space, and the freedom we may have in specifying it, are best illustrated by examples.

EXAMPLE 4.3

Toss a coin. There are only two possible outcomes, and the sample space is

$$S = \{\text{heads, tails}\}$$

or more briefly, $S = \{\text{H, T}\}$.

EXAMPLE 4.4

Let your pencil point fall blindly into Table B of random digits and record the value of the digit it lands on. The possible outcomes are

$$S = \{0, 1, 2, 3, 4, 5, 6, 7, 8, 9\}$$

EXAMPLE 4.5

Toss a coin four times and record the results. That's a bit vague. To be exact, record the results of each of the four tosses in order. A typical outcome is then HTTH. Counting shows that there are 16 possible outcomes. The sample space S is the set of all 16 strings of four H's and T's.

Suppose that our only interest is the number of heads in four tosses. Now we can be exact in a simpler fashion. The random phenomenon is to toss a coin four times and count the number of heads. The sample space contains only five outcomes:

$$S = \{0, 1, 2, 3, 4\}$$

This example illustrates the importance of carefully specifying what constitutes an individual outcome.

Although these examples seem remote from the practice of statistics, the connection is surprisingly close. Suppose that in conducting an opinion poll you select four people at random from a large population and ask each if he or she favors reducing federal spending on low-interest student loans. The answers are "Yes" or "No." The possible outcomes—the sample space—are exactly as in Example 4.3 if we replace heads by "Yes" and tails by "No." Similarly, the possible outcomes of an SRS of 1500 people are the same in principle as the possible outcomes of tossing a coin 1500 times. One of the great advantages of mathematics is that the essential features of quite different phenomena can be described by the same mathematical model.

EXAMPLE 4.6

Many computing systems have a function that will generate a random number between 0 and 1. The sample space is

$$S = \{\text{all numbers between 0 and 1}\}$$

This S is a mathematical idealization. Any specific random number generator produces numbers with some limited number of decimal places so that, strictly speaking, not all numbers between 0 and 1 are possible outcomes. The entire interval from 0 to 1 is easier to think about. It also has the advantage of being a suitable sample space for different computers that produce random numbers with different numbers of significant digits.

Intuitive probability

A sample space S lists the possible outcomes of a random phenomenon. To complete a mathematical model for the random phenomenon, we must also give the probabilities with which these outcomes occur.

The true long-term relative frequency of any outcome—say, "exactly 2 heads in four tosses of a coin"—can be found only empirically, and then only approximately. How then can we describe probability mathematically? Rather than immediately attempting to give "correct" probabilities, let's confront the easier task of laying down rules that any assignment of probabilities must satisfy.* We need to assign probabilities not only to single outcomes but also to sets of outcomes.

Event

An **event** is an outcome or a set of outcomes of a random phenomenon. That is, an event is a subset of the sample space.

EXAMPLE 4.7

Take the sample space S for four tosses of a coin to be the 16 possible outcomes in the form HTHH. Then "exactly 2 heads" is an event. Call this event A. The event A expressed as a set of outcomes is

$$A = \{\text{HHTT, HTHT, HTTH, THHT, THTH, TTHH}\}$$

In a probability model, events have probabilities. What properties must any assignment of probabilities to events have? Here are some basic facts about any probability model. These facts follow from the idea of probability as "the long-run proportion of repetitions on which an event occurs."

1. **Any probability is a number between 0 and 1.** Any proportion is a number between 0 and 1, so any probability is also a number between 0

*Probability is an old branch of mathematics, but it was only in 1933 that the Russian mathematician A. N. Kolmogorov (1903–1987) showed that the entire theory could be derived by starting with a few basic rules.

and 1. An event with probability 0 never occurs, and an event with probability 1 occurs on every trial. An event with probability 0.5 occurs in half the trials in the long run.

2. **All possible outcomes together must have probability 1.** Because some outcome must occur on every trial, the sum of the probabilities for all possible outcomes must be exactly 1.
3. **The probability that an event does not occur is 1 minus the probability that the event does occur.** If an event occurs in (say) 70% of all trials, it fails to occur in the other 30%. The probability that an event occurs and the probability that it does not occur always add to 100%, or 1.
4. **If two events have no outcomes in common, the probability that one or the other occurs is the sum of their individual probabilities.** If one event occurs in 40% of all trials, a different event occurs in 25% of all trials, and the two can never occur together, then one or the other occurs on 65% of all trials because 40% + 25% = 65%.

Probability rules

Formal probability uses mathematical notation to state Facts 1 to 4 more concisely. We use capital letters near the beginning of the alphabet to denote events. If A is any event, we write its probability as $P(A)$. Here are our probability facts in formal language. As you apply these rules, remember that they are just another form of intuitively true facts about long-run proportions.

Probability Rules

Rule 1. The probability $P(A)$ of any event A satisfies $0 \leq P(A) \leq 1$.

Rule 2. If S is the sample space in a probability model, then $P(S) = 1$.

Rule 3. The **complement** of any event A is the event that A does not occur, written as A^c. The **complement rule** states that

$$P(A^c) = 1 - P(A)$$

Rule 4. Two events A and B are **disjoint** if they have no outcomes in common and so can never occur simultaneously. If A and B are disjoint,

$$P(A \text{ or } B) = P(A) + P(B)$$

This is the **addition rule for disjoint events.**

You may find it helpful to draw a picture to remind yourself of the meaning of complements and disjoint events. A picture like Figure 4.2 that shows the sample space S as a rectangular area and events as areas within S is called a **Venn diagram.** The events A and B in Figure 4.2 are disjoint because they do not overlap. As Figure 4.3 shows, the complement A^c contains exactly the outcomes that are not in A.

Venn diagram

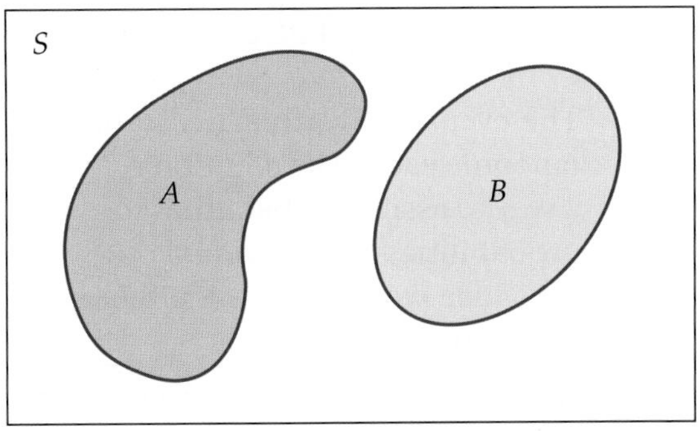

FIGURE 4.2 Venn diagram showing disjoint events A and B.

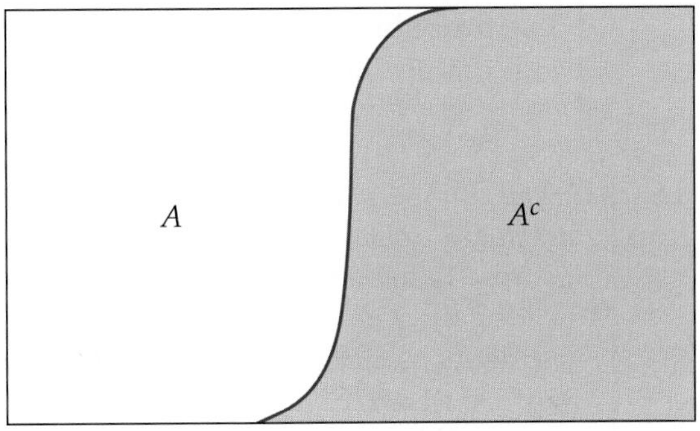

FIGURE 4.3 Venn diagram showing the complement A^c of an event A.

EXAMPLE 4.8

Draw a woman aged 25 to 34 years old at random and record her marital status. "At random" means that we give every such woman the same chance to be the one we choose. That is, we choose an SRS of size 1. The probability of any marital status is just the proportion of all women aged 25 to 34 who have that status—if we drew many women, this is the proportion we would get. Here is the probability model:

Marital status	Never married	Married	Widowed	Divorced
Probability	0.298	0.622	0.005	0.075

Each probability is between 0 and 1. The probabilities add to 1, because these outcomes together make up the sample space S.

The probability that the woman we draw is not married is, by the complement rule,

$$\begin{aligned} P(\text{not married}) &= 1 - P(\text{married}) \\ &= 1 - 0.622 = 0.378 \end{aligned}$$

That is, if 62.2% are married, then the remaining 37.8% are not married.

"Never married" and "divorced" are disjoint events, because no woman can be both never married and divorced. So the addition rule says that

$$\begin{aligned} P(\text{never married or divorced}) &= P(\text{never married}) + P(\text{divorced}) \\ &= 0.298 + 0.075 = 0.373 \end{aligned}$$

That is, 37.3% of women in this age group are either never married or divorced.

Assigning probabilities: finite number of outcomes

The events "never married" and "divorced" in Example 4.8 contain just one outcome each. The addition rule applies to individual outcomes and provides a way to assign probabilities to events with more than one outcome: start with probabilities for individual outcomes and add to get probabilities for events. This idea works well when there are only a finite (fixed and limited) number of outcomes.

Probabilities in a Finite Sample Space

Assign a probability to each individual outcome. These probabilities must be numbers between 0 and 1 and must have sum 1.

The probability of any event is the sum of the probabilities of the outcomes making up the event.

EXAMPLE 4.9

Benford's Law

Faked numbers in tax returns, payment records, invoices, expense account claims, and many other settings often display patterns that aren't present in legitimate records. Some patterns, like too many round numbers, are obvious and easily avoided by a clever crook. Others are more subtle. It is a striking fact that the first digits of numbers in legitimate records often follow a distribution known as **Benford's Law.** Here it is (note that a first digit can't be 0):[2]

First digit	1	2	3	4	5	6	7	8	9
Probability	0.301	0.176	0.125	0.097	0.079	0.067	0.058	0.051	0.046

Benford's Law usually applies to the first digits of the sizes of similar quantities, such as invoices, expense account claims, and county populations. Investigators can detect fraud by comparing the first digits in records such as invoices paid by a business with these probabilities.

Consider the events

$$A = \{\text{first digit is 1}\}$$
$$B = \{\text{first digit is 6 or greater}\}$$

From the table of probabilities,

$$P(A) = P(1) = 0.301$$
$$P(B) = P(6) + P(7) + P(8) + P(9)$$
$$= 0.067 + 0.058 + 0.051 + 0.046 = 0.222$$

Note that $P(B)$ is not the same as the probability that a first digit is strictly greater than 6. The probability $P(6)$ that a first digit is 6 is included in "6 or greater" but not in "greater than 6."

The probability that a first digit is anything other than 1 is, by the complement rule,

$$P(A^c) = 1 - P(A)$$
$$= 1 - 0.301 = 0.699$$

The events A and B are disjoint, so the probability that a first digit is either 1 or 6 or greater is, by the addition rule,

$$\begin{aligned} P(A \text{ or } B) &= P(A) + P(B) \\ &= 0.301 + 0.222 = 0.523 \end{aligned}$$

Be careful to apply the addition rule only to disjoint events. Check that the probability of the event C that a first digit is odd is

$$P(C) = P(1) + P(3) + P(5) + P(7) + P(9) = 0.609$$

The probability

$$P(B \text{ or } C) = P(1) + P(3) + P(5) + P(6) + P(7) + P(8) + P(9) = 0.727$$

is *not* the sum of $P(B)$ and $P(C)$, because events B and C are not disjoint. Outcomes 7 and 9 are common to both events.

Assigning probabilities: equally likely outcomes

Assigning correct probabilities to individual outcomes often requires long observation of the random phenomenon. In some circumstances, however, we are willing to assume that individual outcomes are equally likely because of some balance in the phenomenon. Ordinary coins have a physical balance that should make heads and tails equally likely, for example, and the table of random digits comes from a deliberate randomization.

EXAMPLE 4.10

You might think that first digits are distributed "at random" among the digits 1 to 9. The 9 possible outcomes would then be equally likely. The sample space for a single digit is

$$S = \{1, 2, 3, 4, 5, 6, 7, 8, 9\}$$

Because the total probability must be 1, the probability of each of the 9 outcomes must be $1/9$. That is, the assignment of probabilities to outcomes is

First digit	1	2	3	4	5	6	7	8	9
Probability	1/9	1/9	1/9	1/9	1/9	1/9	1/9	1/9	1/9

The probability of the event B that a randomly chosen first digit is 6 or greater is

$$\begin{aligned} P(B) &= P(6) + P(7) + P(8) + P(9) \\ &= \frac{1}{9} + \frac{1}{9} + \frac{1}{9} + \frac{1}{9} = \frac{4}{9} = 0.444 \end{aligned}$$

Compare this with the Benford's Law probability in Example 4.9. A crook who fakes data by using "random" digits will end up with too many first digits 6 or greater and too few 1s and 2s.

In Example 4.10 all outcomes have the same probability. Because there are 9 equally likely outcomes, each must have probability 1/9. Because exactly

4 of the 9 equally likely outcomes are 6 or greater, the probability of this event is 4/9. In the special situation where all outcomes are equally likely, we have a simple rule for assigning probabilities to events.

Equally Likely Outcomes

If a random phenomenon has k possible outcomes, all equally likely, then each individual outcome has probability $1/k$. The probability of any event A is

$$\begin{aligned} P(A) &= \frac{\text{count of outcomes in } A}{\text{count of outcomes in } S} \\ &= \frac{\text{count of outcomes in } A}{k} \end{aligned}$$

Most random phenomena do not have equally likely outcomes, so the general rule for finite sample spaces is more important than the special rule for equally likely outcomes.

Independence and the multiplication rule

Rule 4, the addition rule for disjoint events, describes the probability that *one or the other* of two events A and B will occur in the special situation when A and B cannot occur together because they are disjoint. Our final rule describes the probability that *both* events A and B occur, again only in a special situation. More general rules appear in Section 4.5, but in our study of statistics we will need only the rules that apply to special situations.

Suppose that you toss a balanced coin twice. You are counting heads, so two events of interest are

$$\begin{aligned} A &= \{\text{first toss is a head}\} \\ B &= \{\text{second toss is a head}\} \end{aligned}$$

The events A and B are not disjoint. They occur together whenever both tosses give heads. We want to compute the probability of the event $\{A \text{ and } B\}$ that *both* tosses are heads. The Venn diagram in Figure 4.4 illustrates the event $\{A \text{ and } B\}$ as the overlapping area that is common to both A and B.

The coin tossing of Buffon, Pearson, and Kerrich described at the beginning of this chapter makes us willing to assign probability $1/2$ to a head when we toss a coin. So

$$\begin{aligned} P(A) &= 0.5 \\ P(B) &= 0.5 \end{aligned}$$

What is $P(A \text{ and } B)$? Our common sense says that it is $1/4$. The first coin will give a head half the time and then the second will give a head on half of those

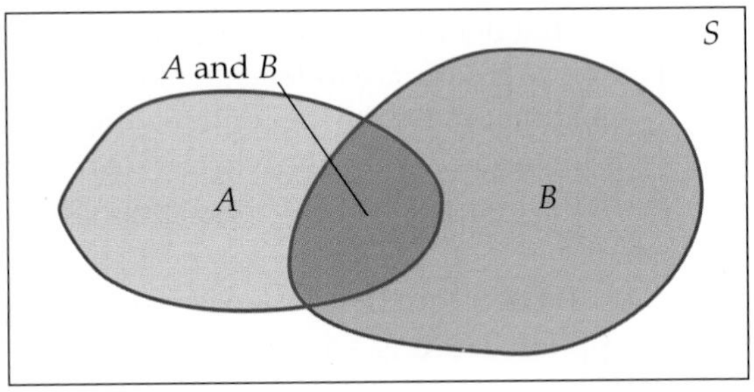

FIGURE 4.4 Venn diagram showing the event {A and B}. This event consists of outcomes common to A and B.

trials, so both coins will give heads on $1/2 \times 1/2 = 1/4$ of all trials in the long run. This reasoning assumes that the second coin still has probability $1/2$ of a head after the first has given a head. This is true—we can verify it by tossing two coins many times and observing the proportion of heads on the second toss after the first toss has produced a head. We say that the events "head on the first toss" and "head on the second toss" are *independent*. Here is our final probability rule.

The Multiplication Rule for Independent Events

Rule 5. Two events A and B are **independent** if knowing that one occurs does not change the probability that the other occurs. If A and B are independent,

$$P(A \text{ and } B) = P(A)P(B)$$

This is the **multiplication rule for independent events.**

Our definition of independence is rather informal. We will make this informal idea precise in Section 4.5. In practice, though, we rarely need a precise definition of independence, because independence is usually *assumed* as part of a probability model when we want to describe random phenomena that seem to be physically unrelated to each other.

EXAMPLE 4.11

Because a coin has no memory and most coin-tossers cannot influence the fall of the coin, it is safe to assume that successive coin tosses are independent. For a balanced coin this means that after we see the outcome of the first toss, we still assign probability $1/2$ to heads on the second toss.

On the other hand, the colors of successive cards dealt from the same deck are not independent. A standard 52-card deck contains 26 red and 26 black cards. For the first card dealt from a shuffled deck, the probability of a red card is $26/52 = 0.50$ because the 52 possible cards are equally likely. Once we see that the first card is red, we know that there are only 25 reds among the remaining 51 cards. The probability that the second card is red is therefore only $25/51 = 0.49$. Knowing the outcome of the first deal changes the probabilities for the second.

If a doctor measures your blood pressure twice, it is reasonable to assume that the two results are independent because the first result does not influence the instrument that makes the second reading. But if you take an IQ test or other mental test twice in succession, the two test scores are not independent. The learning that occurs on the first attempt influences your second attempt.

When independence is part of a probability model, the multiplication rule applies. Here is an example.

EXAMPLE 4.12

Gregor Mendel used garden peas in some of the experiments that revealed that inheritance operates randomly. The seed color of Mendel's peas can be either green or yellow. Two parent plants are "crossed" (one pollinates the other) to produce seeds. Each parent plant carries two genes for seed color, and each of these genes has probability $1/2$ of being passed to a seed. The two genes that the seed receives, one from each parent, determine its color. The parents contribute their genes independently of each other.

Suppose that both parents carry the G and the Y genes. The seed will be green if both parents contribute a G gene; otherwise it will be yellow. If M is the event that the male contributes a G gene and F is the event that the female contributes a G gene, then the probability of a green seed is

$$\begin{aligned} P(M \text{ and } F) &= P(M)P(F) \\ &= (0.5)(0.5) = 0.25 \end{aligned}$$

In the long run, $1/4$ of all seeds produced by crossing these plants will be green.

The multiplication rule $P(A \text{ and } B) = P(A)P(B)$ holds if A and B are independent but not otherwise. The addition rule $P(A \text{ or } B) = P(A)+P(B)$ holds if A and B are disjoint but not otherwise. Resist the temptation to use these simple formulas when the circumstances that justify them are not present. You must also be certain not to confuse disjointness and independence. If A and B are disjoint, then the fact that A occurs tells us that B cannot occur—look again at Figure 4.2. So disjoint events are not independent. Unlike disjointness or complements, independence cannot be pictured by a Venn diagram, because it involves the probabilities of the events rather than just the outcomes that make up the events.

Applying the probability rules

If two events A and B are independent, then their complements A^c and B^c are also independent and A^c is independent of B. Suppose, for example, that 75% of all registered voters in a suburban district are Republicans. If an opinion poll interviews two voters chosen independently, the probability that the first is a Republican and the second is not a Republican is $(0.75)(0.25) = 0.1875$. The multiplication rule also extends to collections of more than two events, provided that all are independent. Independence of events A, B, and C means that no information about any one or any two can change the probability of the remaining events. The formal definition is a bit messy. Fortunately, independence is usually assumed in setting up a probability model. We can then use the multiplication rule freely, as in this example.

EXAMPLE 4.13

The first successful transatlantic telegraph cable was laid in 1866. The first telephone cable across the Atlantic did not appear until 1956—the barrier was designing "repeaters," amplifiers needed to boost the signal, that could operate for years on the sea bottom. This first cable had 52 repeaters. The last copper cable, laid in 1983 and retired in 1994, had 662 repeaters. The first fiber optic cable was laid in 1988 and has 109 repeaters. There are now more than 400,000 miles of undersea cable, with more being laid every year to handle the flood of Internet traffic.

Repeaters in undersea cables must be very reliable. To see why, suppose that each repeater has probability 0.999 of functioning without failure for 25 years. Repeaters fail independently of each other. (This assumption means that there are no "common causes" such as earthquakes that would affect several repeaters at once.) Denote by A_i the event that the ith repeater operates successfully for 25 years.

The probability that two repeaters both last 25 years is

$$\begin{aligned} P(A_1 \text{ and } A_2) &= P(A_1)P(A_2) \\ &= 0.999 \times 0.999 = 0.998 \end{aligned}$$

For a cable with 10 repeaters the probability of no failures in 25 years is

$$\begin{aligned} P(A_1 \text{ and } A_2 \text{ and } \ldots \text{ and } A_{10}) &= P(A_1)P(A_2)\cdots P(A_{10}) \\ &= 0.999 \times 0.999 \times \cdots \times 0.999 \\ &= 0.999^{10} = 0.990 \end{aligned}$$

Cables with 2 or 10 repeaters would be quite reliable. Unfortunately, the last copper transatlantic cable had 662 repeaters. The probability that all 662 work for 25 years is

$$P(A_1 \text{ and } A_2 \text{ and } \ldots \text{ and } A_{662}) = 0.999^{662} = 0.516$$

This cable will fail to reach its 25-year design life about half the time if each repeater is 99.9% reliable over that period. The multiplication rule for probabilities shows that repeaters must be much more than 99.9% reliable.

By combining the rules we have learned, we can compute probabilities for rather complex events. Here is an example.

EXAMPLE 4.14

Screening large numbers of blood samples for HIV, the virus that causes AIDS, uses an enzyme immunoassay (EIA) test that detects antibodies to the virus. Samples that test positive are retested using a more accurate "Western blots" test. Applied to people who have no HIV antibodies, EIA has probability about 0.006 of producing a false positive (that is, of falsely indicating that antibodies are present). If the 140 employees of a medical clinic are tested and all 140 are free of HIV antibodies, what is the probability that at least one false positive will occur?

It is reasonable to assume as part of the probability model that the test results for different individuals are independent. The probability that the test is positive for a single person is 0.006, so the probability of a negative result is $1 - 0.006 = 0.994$ by the complement rule. The probability of at least one false positive among the

140 people tested is therefore

$$\begin{aligned} P(\text{at least one positive}) &= 1 - P(\text{no positives}) \\ &= 1 - P(140 \text{ negatives}) \\ &= 1 - 0.994^{140} \\ &= 1 - 0.431 = 0.569 \end{aligned}$$

The probability is greater than 1/2 that at least one of the 140 people will test positive for HIV, even though no one has the virus.

SUMMARY

A **probability model** for a random phenomenon consists of a sample space S and an assignment of probabilities P.

The **sample space** S is the set of all possible outcomes of the random phenomenon. Sets of outcomes are called **events.** P assigns a number $P(A)$ to an event A as its probability.

The **complement** A^c of an event A consists of exactly the outcomes that are not in A. Events A and B are **disjoint** if they have no outcomes in common. Events A and B are **independent** if knowing that one event occurs does not change the probability we would assign to the other event.

Any assignment of probability must obey the rules that state the basic properties of probability:

1. $0 \leq P(A) \leq 1$ for any event A.
2. $P(S) = 1$.
3. **Complement rule:** For any event A, $P(A^c) = 1 - P(A)$.
4. **Addition rule:** If events A and B are **disjoint,** then $P(A \text{ or } B) = P(A) + P(B)$.
5. **Multiplication rule:** If events A and B are **independent,** then $P(A \text{ and } B) = P(A)P(B)$.

SECTION 4.2 EXERCISES

4.10 A randomly chosen subject arrives for a study of exercise and fitness. Describe a sample space for each of the following. (In some cases, you may have some freedom in your choice of S.)

(a) The subject is either female or male.

(b) After 10 minutes on an exercise bicycle, you ask the subject to rate his or her effort on the Rate of Perceived Exertion (RPE) scale. RPE ranges in whole-number steps from 6 (no exertion at all) to 20 (maximal exertion).

(c) You measure VO2, the maximum volume of oxygen consumed per minute during exercise. VO2 is generally between 2.5 liters per minute and 6 l/min.

(d) You measure the maximum heart rate (beats per minute).

4.11 Choose a student at random from a large statistics class. Describe a sample space *S* for each of the following. (In some cases you may have some freedom in specifying *S*.)

(a) Ask how much time the student spent studying during the past 24 hours.

(b) Ask how much money in coins (not bills) the student is carrying.

(c) Record the student's letter grade at the end of the course.

(d) Ask whether the student did or did not take a math class in each of the two previous years of school.

4.12 The percent return on U.S. common stocks in the next year is random. Figure 2.8 (page 118) reports historical data for the years 1950 to 2000. Give a reasonable sample space for the possible returns next year. Explain how you chose this *S*.

4.13 All human blood can be "ABO-typed" as one of O, A, B, or AB, but the distribution of the types varies a bit among groups of people. Here is the distribution of blood types for a randomly chosen person in the United States:

Blood type	O	A	B	AB
U.S. probability	0.45	0.40	0.11	?

(a) What is the probability of type AB blood in the United States?

(b) Maria has type B blood. She can safely receive blood transfusions from people with blood types O and B. What is the probability that a randomly chosen American can donate blood to Maria?

4.14 The distribution of blood types in China differs from the U.S. distribution given in the previous exercise:

Blood type	O	A	B	AB
China probability	0.35	0.27	0.26	0.12

Choose an American and a Chinese at random, independently of each other. What is the probability that both have type O blood? What is the probability that both have the same blood type?

4.15 In each of the following situations, state whether or not the given assignment of probabilities to individual outcomes is legitimate, that is, satisfies the rules of probability. If not, give specific reasons for your answer.

(a) When a coin is spun, $P(\mathrm{H}) = 0.55$ and $P(\mathrm{T}) = 0.45$.

(b) When two coins are tossed, $P(\mathrm{HH}) = 0.4$, $P(\mathrm{HT}) = 0.4$, $P(\mathrm{TH}) = 0.4$, and $P(\mathrm{TT}) = 0.4$.

(c) When a die is rolled, the number of spots on the up-face has $P(1) = 1/2$, $P(4) = 1/6$, $P(5) = 1/6$, and $P(6) = 1/6$.

4.16 Wabash Red, when asked to predict the Big Ten Conference men's basketball champion, follows the modern practice of giving probabilistic predictions. He says, "Michigan State has probability 0.3 of winning. Michigan, Minnesota, Northwestern, and Penn State have no chance. That leaves 6 teams. Iowa, Illinois, and Purdue all have the same probability of winning. Indiana, Ohio State, and Wisconsin also have the same probability, but that probability is one-half that of the first 3." What probability does Red give to each of the 11 teams?

4.17 Chose a student in grades 9 to 12 at random and ask if he or she is studying a language other than English. Here is the distribution of results:

Language	Spanish	French	German	All others	None
Probability	0.26	0.09	0.03	0.03	0.59

(a) Explain why this is a legitimate probability model.

(b) What is the probability that a randomly chosen student is studying a language other than English?

(c) What is the probability that a randomly chosen student is studying French, German, or Spanish?

4.18 Choose an acre of land in Canada at random. The probability is 0.35 that it is forest and 0.03 that it is pasture.

(a) What is the probability that the acre chosen is not forested?

(b) What is the probability that it is either forest or pasture?

(c) What is the probability that a randomly chosen acre in Canada is something other than forest or pasture?

4.19 Choose a new car or light truck at random and note its color. Here are the probabilities of the most popular colors for vehicles made in North America in 2000:[3]

Color	Silver	White	Black	Dark green	Dark blue	Medium red
Probability	0.176	0.172	0.113	0.089	0.088	0.067

(a) What is the probability that the vehicle you choose has any color other than the six listed?

(b) What is the probability that a randomly chosen vehicle is either silver or white?

(c) Choose two vehicles at random. What is the probability that both are silver or white?

4.20 The 2000 census allowed each person to choose from a long list of races. That is, in the eyes of the Census Bureau, you belong to whatever race you say you belong to. "Hispanic/Latino" is a separate category; Hispanics may be of any race. If we choose a resident of the United States at random, the 2000 census gives these probabilities:

	Hispanic	Not Hispanic
Asian	0.000	0.036
Black	0.003	0.121
White	0.060	0.691
Other	0.062	0.027

Let A be the event that a randomly chosen American is Hispanic, and let B be the event that the person chosen is white.

(a) Verify that the table gives a legitimate assignment of probabilities.

(b) What is $P(A)$?

(c) Describe B^c in words and find $P(B^c)$ by the complement rule.

(d) Express "the person chosen is a non-Hispanic white" in terms of events A and B. What is the probability of this event?

4.21 Exercise 4.20 assigns probabilities for the ethnic background of a randomly chosen resident of the United States. Let A be the event that the person chosen is Hispanic, and let B be the event that he or she is white. Are events A and B independent? How do you know?

4.22 Human blood is typed as O, A, B, or AB and also as Rh-positive or Rh-negative. ABO type and Rh-factor type are independent because they are governed by different genes. In the American population, 84% of people are Rh-positive. Use the information about ABO type in Exercise 4.13 to give the probability distribution of blood type (ABO and Rh) for a randomly chosen person.

4.23 A roulette wheel has 38 slots, numbered 0, 00, and 1 to 36. The slots 0 and 00 are colored green, 18 of the others are red, and 18 are black. The dealer spins the wheel and at the same time rolls a small ball along the wheel in the opposite direction. The wheel is carefully balanced so that the ball is equally likely to land in any slot when the wheel slows. Gamblers can bet on various combinations of numbers and colors.

(a) What is the probability that the ball will land in any one slot?

(b) If you bet on "red," you win if the ball lands in a red slot. What is the probability of winning?

(c) The slot numbers are laid out on a board on which gamblers place their bets. One column of numbers on the board contains all multiples of 3, that is, 3, 6, 9, . . . , 36. You place a "column bet" that wins if any of these numbers comes up. What is your probability of winning?

4.24 A company that offers courses to prepare would-be MBA students for the GMAT examination has the following information about its customers: 20% are currently undergraduate students in business; 15% are undergraduate students in other fields of study; 60% are college graduates who are currently employed; and 5% are college graduates who are not employed.

(a) Is this a legitimate assignment of probabilities to customer backgrounds? Why?

(b) What percent of customers are currently undergraduates?

4.25 Psychologists sometimes use tetrahedral dice to study our intuition about chance behavior. A tetrahedron is a pyramid-like solid with four identical faces, each a triangle with all sides equal in length. (See the sketch below.) Label the four faces of a tetrahedral die with 1, 2, 3, and 4 spots. Give a probability model for rolling such a die and recording the number of spots on the down-face. Explain why you think your model is at least close to correct.

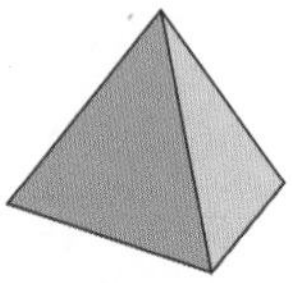

Tetrahedron

4.26 The "pick four" games in many state lotteries announce a four-digit winning number each day. The winning number is essentially a four-digit group from a table of random digits. You win if your choice matches the winning digits. Suppose your chosen number is 5974.

(a) What is the probability that your number matches the winning number exactly?

(b) What is the probability that your number matches the digits in the winning number *in any order*?

4.27 The personal identification numbers (PINs) for automatic teller machines usually consist of four digits. You notice that most of your PINs have at least one 0, and you wonder if the issuers use lots of 0s to make the numbers easy to remember. Suppose that PINs are assigned at random, so that all four-digit numbers are equally likely.

(a) How many possible PINs are there?

(b) What is the probability that a PIN assigned at random has at least one 0?

4.28 Tetrahedral dice are described in Exercise 4.25. Give a probability model for rolling two such dice independently. That is, write down all possible outcomes and give a probability to each. What is the probability that the sum of the down-faces is 5?

4.29 Abby, Deborah, Mei-Ling, Sam, and Roberto work in a firm's public relations office. Their employer must choose two of them to attend a conference in Paris. To avoid unfairness, the choice will be made by drawing two names from a hat. (This is an SRS of size 2.)

(a) Write down all possible choices of two of the five names. This is the sample space.

(b) The random drawing makes all choices equally likely. What is the probability of each choice?

(c) What is the probability that Mei-Ling is chosen?

(d) What is the probability that neither of the two men (Sam and Roberto) is chosen?

4.30 People with type O-negative blood are universal donors. That is, any patient can receive a transfusion of O-negative blood. Only 7% of the American population have O-negative blood. If 10 people appear at random to give blood, what is the probability that at least one of them is a universal donor?

4.31 A general can plan a campaign to fight one major battle or three small battles. He believes that he has probability 0.6 of winning the large battle and probability 0.8 of winning each of the small battles. Victories or defeats in the small battles are independent. The general must win either the large battle or all three small battles to win the campaign. Which strategy should he choose?

4.32 Most sample surveys use random-digit dialing equipment to call residential telephone numbers at random. The telephone polling firm Zogby International reports that the probability that a call reaches a live person is 0.2.[4] Calls are independent.

(a) A polling firm places 5 calls. What is the probability that none of them reaches a person?

(b) When calls are made to New York City, the probability of reaching a person is only 0.08. What is the probability that none of 5 calls made to New York City reaches a person?

4.33 Government data show that 27% of employed people have at least 4 years of college and that 14% of employed people work as laborers or operators of machines or vehicles. Can you conclude that because $(0.27)(0.14) = 0.038$ about 3.8% of employed people are college-educated laborers or operators? Explain your answer.

4.34 A string of holiday lights contains 20 lights. The lights are wired in series, so that if any light fails the whole string will go dark. Each light has probability 0.02 of failing during a 3-year period. The lights fail independently of each other. What is the probability that the string of lights will remain bright for 3 years?

4.35 The "random walk" theory of securities prices holds that price movements in disjoint time periods are independent of each other. Suppose that we record only whether the price is up or down each year, and that the probability that our portfolio rises in price in any one year is 0.65. (This probability is approximately correct for a portfolio containing equal dollar amounts of all common stocks listed on the New York Stock Exchange.)

(a) What is the probability that our portfolio goes up for 3 consecutive years?

(b) If you know that the portfolio has risen in price 2 years in a row, what probability do you assign to the event that it will go down next year?

(c) What is the probability that the portfolio's value moves in the same direction in both of the next 2 years?

Mendelian inheritance. *Some traits of plants and animals depend on inheritance of a single gene. This is called Mendelian inheritance, after Gregor Mendel (1822–1884). Exercises 4.36 to 4.39 are based on the following information about Mendelian inheritance of blood type.*

Each of us has an ABO blood type, which describes whether two characteristics called A and B are present. Every human being has two blood type alleles (gene forms), one inherited from our mother and one from our father. Each of these alleles can be A, B, or O. Which two we inherit determines our blood type. Here is a table that shows what our blood type is for each combination of two alleles:

Alleles inherited	Blood type
A and A	A
A and B	AB
A and O	A
B and B	B
B and O	B
O and O	O

We inherit each of a parent's two alleles with probability 0.5. We inherit independently from our mother and father.

4.36 Hannah and Jacob both have alleles A and B.

(a) What blood types can their children have?

(b) What is the probability that their next child has each of these blood types?

4.37 Nancy and David both have alleles B and O.

(a) What blood types can their children have?

(b) What is the probability that their next child has each of these blood types?

4.38 Jennifer has alleles A and O. José has alleles A and B. They have two children. What is the probability that both children have blood type A? What is the probability that both children have the same blood type?

4.39 Jasmine has alleles A and O. Joshua has alleles B and O.

(a) What is the probability that a child of these parents has blood type O?

(b) If Jasmine and Joshua have three children, what is the probability that all three have blood type O? What is the probability that the first child has blood type O and the next two do not?

4.3 Random Variables

Sample spaces need not consist of numbers. When we toss a coin four times, we can record the outcome as a string of heads and tails, such as HTTH. In statistics, however, we are most often interested in numerical outcomes such as the count of heads in the four tosses. It is convenient to use a shorthand notation: Let X be the number of heads. If our outcome is HTTH, then $X = 2$. If the next outcome is TTTH, the value of X changes to $X = 1$. The possible values of X are 0, 1, 2, 3, and 4. Tossing a coin four times will give X one of these possible values. Tossing four more times will give X another and probably different value. We call X a *random variable* because its values vary when the coin tossing is repeated.

Random Variable

A **random variable** is a variable whose value is a numerical outcome of a random phenomenon.

We usually denote random variables by capital letters near the end of the alphabet, such as X or Y. Of course, the random variables of greatest interest to us are outcomes such as the mean $\overline{x}$ of a random sample, for which we will keep the familiar notation.[5] As we progress from general rules of probability toward statistical inference, we will concentrate on random variables. When a random variable X describes a random phenomenon, the sample space S just lists the possible values of the random variable. We usually do not mention S separately. There remains the second part of any probability model, the assignment of probabilities to events. In this section, we will learn two ways of assigning probabilities to the values of a random variable. The two types of probability models that result will dominate our application of probability to statistical inference.

Discrete random variables

We have learned several rules of probability but only one method of assigning probabilities: state the probabilities of the individual outcomes and assign probabilities to events by summing over the outcomes. The outcome probabilities must be between 0 and 1 and have sum 1. When the outcomes are numerical, they are values of a random variable. We will now attach a name to random variables having probability assigned in this way.[6]

Discrete Random Variable

A **discrete random variable** X has a finite number of possible values. The **probability distribution** of X lists the values and their probabilities:

Value of X	x_1	x_2	x_3	$\cdots$	x_k
Probability	p_1	p_2	p_3	$\cdots$	p_k

The probabilities p_i must satisfy two requirements:

1. Every probability p_i is a number between 0 and 1.
2. $p_1 + p_2 + \cdots + p_k = 1$.

Find the probability of any event by adding the probabilities p_i of the particular values x_i that make up the event.

EXAMPLE 4.15

North Carolina State University posts the grade distributions for its courses online.[7] Students in Accounting 210 in the spring 2001 semester received 18% A's, 32% B's, 34% C's, 9% D's, and 7% F's. Choose an Accounting 210 student at random. To "choose at random" means to give every student the same chance to be chosen. The student's grade on a four-point scale (with A = 4) is a random variable X.

The value of X changes when we repeatedly choose students at random, but it is always one of 0, 1, 2, 3, or 4. Here is the distribution of X:

Value of X	0	1	2	3	4
Probability	0.07	0.09	0.34	0.32	0.18

The probability that the student got a B or better is the sum of the probabilities of an A and a B:

$$\begin{aligned} P(\text{grade is 3 or 4}) &= P(X = 3) + P(X = 4) \\ &= 0.32 + 0.18 = 0.5 \end{aligned}$$

probability histogram

We can use histograms to show probability distributions as well as distributions of data. Figure 4.5 displays **probability histograms** that compare the probability model for random digits (Example 4.10) with the model given by Benford's Law (Example 4.9). The height of each bar shows the probability of the outcome at its base. Because the heights are probabilities, they add to 1. As usual, all the bars in a histogram have the same width. So the areas also display the assignment of probability to outcomes. Think of these histograms as idealized pictures of the results of very many trials. The histograms make it easy to quickly compare the two distributions.

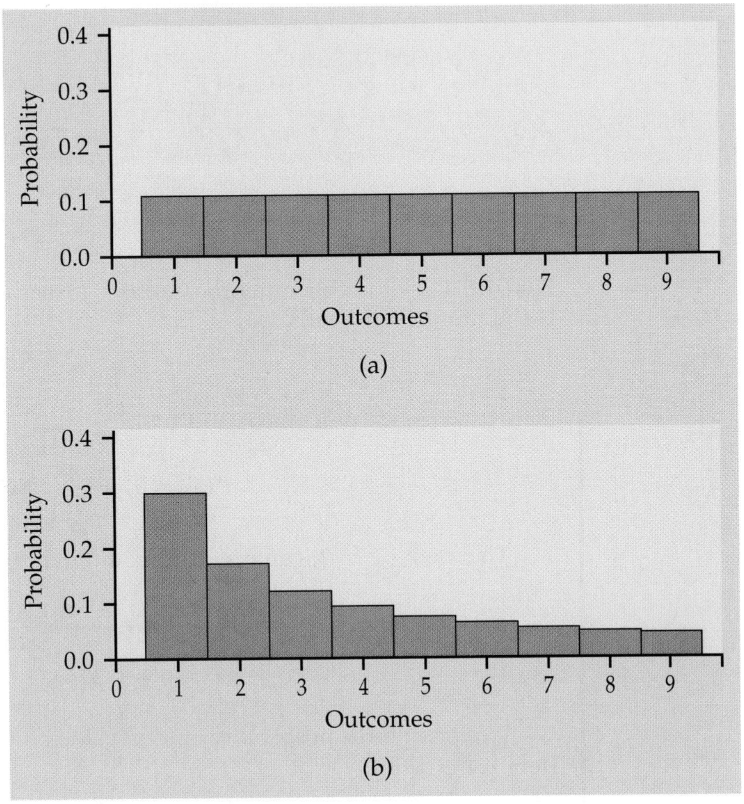

FIGURE 4.5 Probability histograms for (a) random digits 1 to 9 and (b) Benford's Law. The height of each bar shows the probability assigned to a single outcome.

EXAMPLE 4.16

What is the probability distribution of the discrete random variable X that counts the number of heads in four tosses of a coin? We can derive this distribution if we make two reasonable assumptions:

- The coin is balanced, so each toss is equally likely to give H or T.
- The coin has no memory, so tosses are independent.

The outcome of four tosses is a sequence of heads and tails such as HTTH. There are 16 possible outcomes in all. Figure 4.6 lists these outcomes along with the value of X for each outcome. The multiplication rule for independent events tells us that, for example,

$$P(\text{HTTH}) = \frac{1}{2} \times \frac{1}{2} \times \frac{1}{2} \times \frac{1}{2} = \frac{1}{16}$$

Each of the 16 possible outcomes similarly has probability 1/16. That is, these outcomes are equally likely.

The number of heads X has possible values 0, 1, 2, 3, and 4. These values are *not* equally likely. As Figure 4.6 shows, there is only one way that $X = 0$ can occur:

		HTTH		
		HTHT		
	HTTT	THTH	HHHT	
	THTT	HHTT	HHTH	
	TTHT	THHT	HTHH	
TTTT	TTTH	TTHH	THHH	HHHH
$X = 0$	$X = 1$	$X = 2$	$X = 3$	$X = 4$

FIGURE 4.6 Possible outcomes in four tosses of a coin. The random variable X is the number of heads.

namely, when the outcome is TTTT. So

$$P(X = 0) = \frac{1}{16} = 0.0625$$

The event $\{X = 2\}$ can occur in six different ways, so that

$$P(X = 2) = \frac{\text{count of ways } X = 2 \text{ can occur}}{16} = \frac{6}{16} = 0.375$$

We can find the probability of each value of X from Figure 4.6 in the same way. Here is the result:

Value of X	0	1	2	3	4
Probability	0.0625	0.25	0.375	0.25	0.0625

Figure 4.7 is a probability histogram for this distribution. The probability distribution is exactly symmetric. It is an idealization of the relative frequency distribution

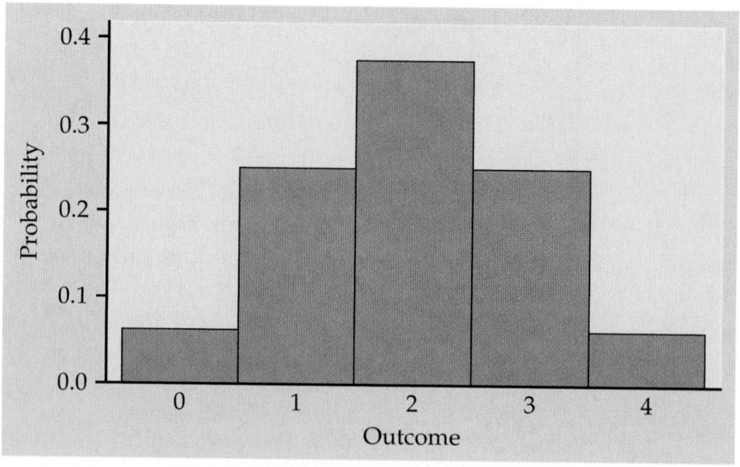

FIGURE 4.7 Probability histogram for the number of heads in four tosses of a coin.

of the number of heads after many tosses of four coins, which would be nearly symmetric but is unlikely to be exactly symmetric.

Any event involving the number of heads observed can be expressed in terms of X, and its probability can be found from the distribution of X. For example, the probability of tossing at least two heads is

$$P(X \geq 2) = 0.375 + 0.25 + 0.0625 = 0.6875$$

The probability of at least one head is most simply found by use of the complement rule:

$$\begin{aligned} P(X \geq 1) &= 1 - P(X = 0) \\ &= 1 - 0.0625 = 0.9375 \end{aligned}$$

Recall that tossing a coin n times is similar to choosing an SRS of size n from a large population and asking a yes-or-no question. We will extend the results of Example 4.16 when we return to sampling distributions in the next chapter.

Continuous random variables

When we use the table of random digits to select a digit between 0 and 9, the result is a discrete random variable. The probability model assigns probability 1/10 to each of the 10 possible outcomes. Suppose that we want to choose a number at random between 0 and 1, allowing *any* number between 0 and 1 as the outcome. Software random number generators will do this. You can visualize such a random number by thinking of a spinner (Figure 4.8) that turns freely on its axis and slowly comes to a stop. The pointer can come to rest anywhere on a circle that is marked from 0 to 1. The sample space is now an entire interval of numbers:

$$S = \{\text{all numbers } x \text{ such that } 0 \leq x \leq 1\}$$

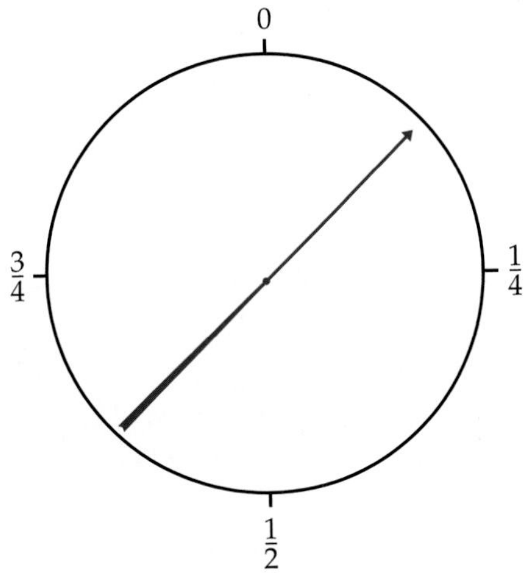

FIGURE 4.8 A spinner that generates a random number between 0 and 1.

How can we assign probabilities to events such as $\{0.3 \leq x \leq 0.7\}$? As in the case of selecting a random digit, we would like all possible outcomes to be equally likely. But we cannot assign probabilities to each individual value of x and then sum, because there are infinitely many possible values. Instead we use a new way of assigning probabilities directly to events—as *areas under a density curve*. Any density curve has area exactly 1 underneath it, corresponding to total probability 1.

EXAMPLE 4.17

uniform distribution

The random number generator will spread its output uniformly across the entire interval from 0 to 1 as we allow it to generate a long sequence of numbers. The results of many trials are represented by the density curve of a **uniform distribution** (Figure 4.9). This density curve has height 1 over the interval from 0 to 1. The area under the density curve is 1, and the probability of any event is the area under the density curve and above the event in question.

As Figure 4.9(a) illustrates, the probability that the random number generator produces a number X between 0.3 and 0.7 is

$$P(0.3 \leq X \leq 0.7) = 0.4$$

because the area under the density curve and above the interval from 0.3 to 0.7 is 0.4. The height of the density curve is 1, and the area of a rectangle is the product of height and length, so the probability of any interval of outcomes is just the length of the interval.

Similarly,

$$\begin{aligned} P(X \leq 0.5) &= 0.5 \\ P(X > 0.8) &= 0.2 \\ P(X \leq 0.5 \text{ or } X > 0.8) &= 0.7 \end{aligned}$$

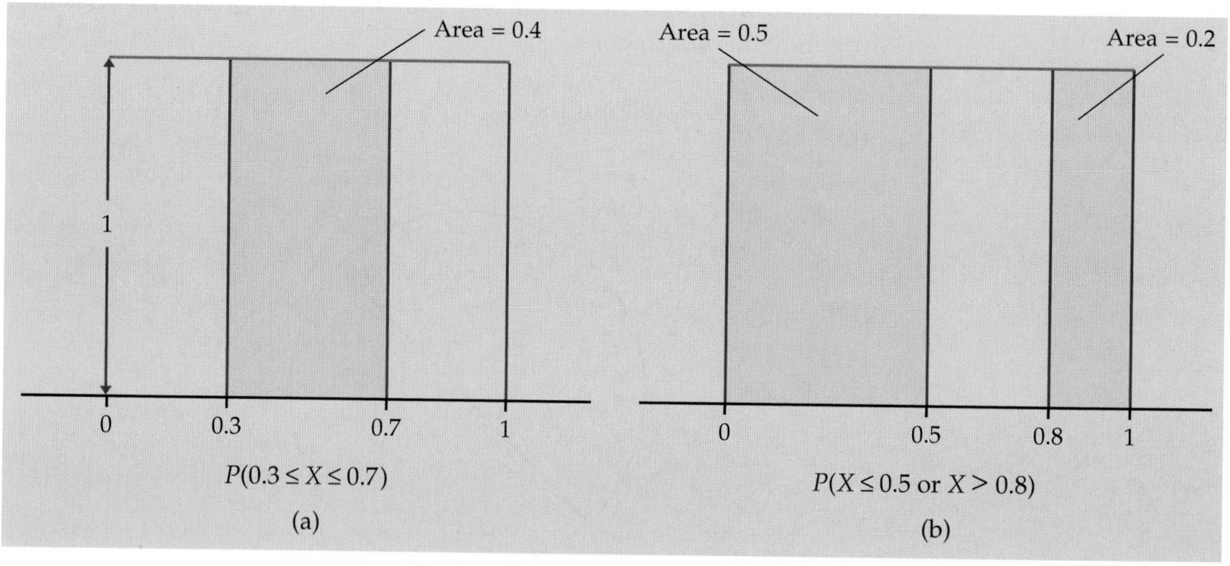

FIGURE 4.9 Assigning probabilities for generating a random number between 0 and 1. The probability of any interval of numbers is the area above the interval and under the curve.

Notice that the last event consists of two nonoverlapping intervals, so the total area above the event is found by adding two areas, as illustrated by Figure 4.9(b). This assignment of probabilities obeys all of our rules for probability.

Probability as area under a density curve is a second important way of assigning probabilities to events. Figure 4.10 illustrates this idea in general form. We call X in Example 4.17 a *continuous random variable* because its values are not isolated numbers but an entire interval of numbers.

Continuous Random Variable

A **continuous random variable** X takes all values in an interval of numbers. The **probability distribution** of X is described by a density curve. The probability of any event is the area under the density curve and above the values of X that make up the event.

The probability model for a continuous random variable assigns probabilities to intervals of outcomes rather than to individual outcomes. In fact, **all continuous probability distributions assign probability 0 to every individual outcome.** Only intervals of values have positive probability. To see that this is true, consider a specific outcome such as $P(X = 0.8)$ in Example 4.17. The probability of any interval is the same as its length. The point 0.8 has no length so its probability is 0.

Although this fact may seem odd, it makes intuitive as well as mathematical sense. The random number generator produces a number between 0.79 and 0.81 with probability 0.02. An outcome between 0.799 and 0.801 has probability 0.002. A result between 0.799999 and 0.800001 has probability 0.000002. You see that as we home in on 0.8 the probability gets closer to 0.

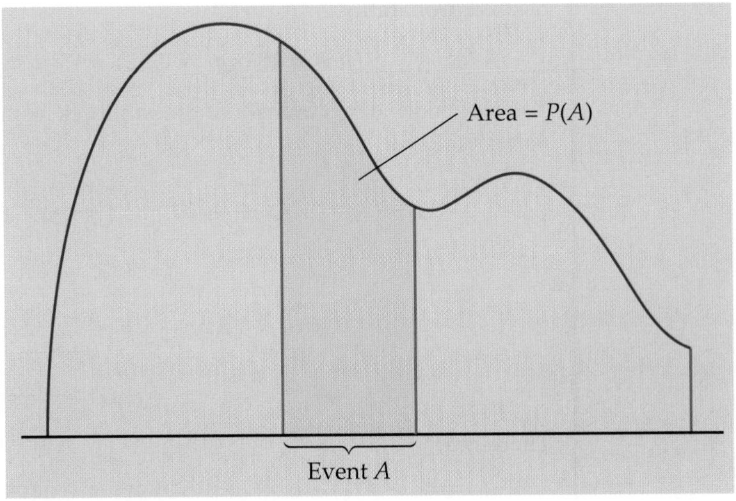

FIGURE 4.10 The probability distribution of a continuous random variable assigns probabilities as areas under a density curve.

To be consistent, the probability of outcome *exactly* equal to 0.8 must be 0. Because there is no probability exactly at $X = 0.8$, the two events $\{X > 0.8\}$ and $\{X \geq 0.8\}$ have the same probability. We can ignore the distinction between $>$ and $\geq$ when finding probabilities for continuous (but not discrete) random variables.

Normal distributions as probability distributions

The density curves that are most familiar to us are the normal curves. (We discussed normal curves in Section 1.3.) Because any density curve describes an assignment of probabilities, *normal distributions are probability distributions*. Recall that $N(\mu, \sigma)$ is our shorthand notation for the normal distribution having mean μ and standard deviation σ. In the language of random variables, if X has the $N(\mu, \sigma)$ distribution, then the standardized variable

$$Z = \frac{X - \mu}{\sigma}$$

is a standard normal random variable having the distribution $N(0, 1)$.

EXAMPLE 4.18

An opinion poll asks an SRS of 1500 American adults what they consider to be the most serious problem facing our schools. Suppose that if we could ask all adults this question, 30% would say "drugs." The proportion $p = 0.3$ is a population parameter. The proportion $\hat{p}$ of the sample who answer "drugs" is a statistic used to estimate p. We will see in the next chapter that $\hat{p}$ is a random variable that has approximately the $N(0.3, 0.0118)$ distribution. The mean 0.3 of this distribution is the same as the population parameter because $\hat{p}$ is an unbiased estimate of p. The standard deviation is controlled mainly by the sample size, which is 1500 in this case.

What is the probability that the poll result differs from the truth about the population by more than two percentage points? Figure 4.11 shows this probability as an area under a normal density curve. By the addition rule for disjoint events, the desired probability is

$$P(\hat{p} < 0.28 \text{ or } \hat{p} > 0.32) = P(\hat{p} < 0.28) + P(\hat{p} > 0.32)$$

You can find the two individual probabilities from software or by standardizing and using Table A.

$$P(\hat{p} < 0.28) = P\left(\frac{\hat{p} - 0.3}{0.0118} < \frac{0.28 - 0.3}{0.0118}\right)$$

$$= P(Z < -1.69) = 0.0455$$

$$P(\hat{p} > 0.32) = P\left(\frac{\hat{p} - 0.3}{0.0118} > \frac{0.32 - 0.3}{0.0118}\right)$$

$$= P(Z > 1.69) = 0.0455$$

Therefore,

$$P(\hat{p} < 0.28 \text{ or } \hat{p} > 0.32) = 0.0455 + 0.0455 = 0.0910$$

The probability that the sample result will miss the truth by more than two percentage points is 0.091. The arrangement of this calculation is familiar from our earlier work with normal distributions. Only the language of probability is new.

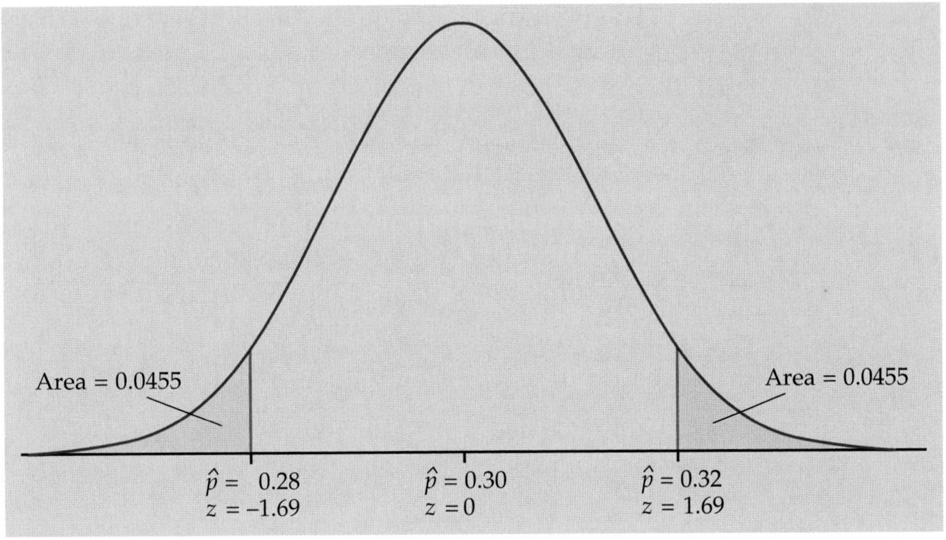

FIGURE 4.11 Probability in Example 4.18 as area under a normal density curve.

We could also do the calculation by first finding the probability of the complement:

$$P(0.28 \le \hat{p} \le 0.32) = P\left(\frac{0.28 - 0.3}{0.0118} \le Z \le \frac{0.32 - 0.3}{0.0118}\right)$$
$$= P(-1.69 \le Z \le 1.69)$$
$$= 0.9545 - 0.0455 = 0.9090$$

Then by the complement rule,

$$P(\hat{p} < 0.28 \text{ or } \hat{p} > 0.32) = 1 - P(0.28 \le \hat{p} \le 0.32)$$
$$= 1 - 0.9090 = 0.0910$$

There is often more than one correct way to use the rules of probability to answer a question.

We began this chapter with a general discussion of the idea of probability and the properties of probability models. Two very useful specific types of probability models are distributions of discrete and continuous random variables. In our study of statistics we will employ only these two types of probability models.

SUMMARY

A **random variable** is a variable taking numerical values determined by the outcome of a random phenomenon. The **probability distribution** of a random variable X tells us what the possible values of X are and how probabilities are assigned to those values.

A random variable X and its distribution can be discrete or continuous.

A **discrete random variable** has finitely many possible values. The probability distribution assigns each of these values a probability between 0 and 1 such that the sum of all the probabilities is exactly 1. The probability of any event is the sum of the probabilities of all the values that make up the event.

A **continuous random variable** takes all values in some interval of numbers. A **density curve** describes the probability distribution of a continuous random variable. The probability of any event is the area under the curve above the values that make up the event.

Normal distributions are one type of continuous probability distribution.

You can picture a probability distribution by drawing a **probability histogram** in the discrete case or by graphing the density curve in the continuous case.

SECTION 4.3 EXERCISES

4.40 Exercise 4.10 (page 298) asks you to give sample spaces for several kinds of measurement on a randomly chosen person. Some of these measurements are random variables, although we did not use the term. Is the random variable in each of parts (b), (c), and (d) of Exercise 4.10 discrete or continuous? Explain your choice in each case.

4.41 If a carefully made die is rolled once, it is reasonable to assign probability 1/6 to each of the six faces. What is the probability of rolling a number less than 3?

4.42 A couple plans to have three children. There are 8 possible arrangements of girls and boys. For example, GGB means the first two children are girls and the third child is a boy. All 8 arrangements are (approximately) equally likely.

(a) Write down all 8 arrangements of the sexes of three children. What is the probability of any one of these arrangements?

(b) Let X be the number of girls the couple has. What is the probability that $X = 2$?

(c) Starting from your work in (a), find the distribution of X. That is, what values can X take, and what are the probabilities for each value?

4.43 How do rented housing units differ from units occupied by their owner? Here are the distributions of the number of rooms for owner-occupied units and renter-occupied units in San Jose, California:[8]

Rooms	1	2	3	4	5	6	7	8	9	10
Owned	0.003	0.002	0.023	0.104	0.210	0.224	0.197	0.149	0.053	0.035
Rented	0.008	0.027	0.287	0.363	0.164	0.093	0.039	0.013	0.003	0.003

Make probability histograms of these two distributions, using the same scales. What are the most important differences between the distributions for owner-occupied and rented housing units?

4.44 In government data, a household consists of all occupants of a dwelling unit, while a family consists of two or more persons who live together and are related by blood or marriage. So all families form households, but some households are not families. Here are the distributions of household size and of family size in the United States:

Number of persons	1	2	3	4	5	6	7
Household probability	0.25	0.32	0.17	0.15	0.07	0.03	0.01
Family probability	0	0.42	0.23	0.21	0.09	0.03	0.02

Make probability histograms for these two discrete distributions, using the same scales. What are the most important differences between the sizes of households and families?

4.45 Let the random variable X be the number of rooms in a randomly chosen owner-occupied housing unit in San Jose, California. Exercise 4.43 gives the distribution of X.

(a) Express "the unit has 5 or more rooms" in terms of X. What is the probability of this event?

(b) Express the event $\{X > 5\}$ in words. What is its probability?

(c) What important fact about discrete random variables does comparing your answers to (a) and (b) illustrate?

4.46 Choose an American household at random and let the random variable Y be the number of persons living in the household. Exercise 4.44 gives the distribution of Y.

(a) Express "more than one person lives in this household" in terms of Y. What is the probability of this event?

(b) What is $P(2 < Y \leq 4)$?

(c) What is $P(Y \neq 2)$?

4.47 Choose an American household at random and let the random variable X be the number of cars (including SUVs and light trucks) they own. Here is the probability model if we ignore the few households that own more than 5 cars:

Number of cars X	0	1	2	3	4	5
Probability	0.09	0.36	0.35	0.13	0.05	0.02

(a) Verify that this is a legitimate discrete distribution. Display the distribution in a probability histogram.

(b) Say in words what the event $\{X \geq 1\}$ is. Find $P(X \geq 1)$.

(c) A housing company builds houses with two-car garages. What percent of households have more cars than the garage can hold?

4.48 Some games of chance rely on tossing two dice. Each die has six faces, marked with 1, 2, . . . , 6 spots called pips. The dice used in casinos are carefully balanced so that each face is equally likely to come up. When two dice are tossed, each of the 36 possible pairs of faces is equally likely to come up. The outcome of interest to a gambler is the sum of the pips on the two up-faces. Call this random variable X.

(a) Write down all 36 possible pairs of faces.

(b) If all pairs have the same probability, what must be the probability of each pair?

(c) Write the value of X next to each pair of faces and use this information with the result of (b) to give the probability distribution of X. Draw a probability histogram to display the distribution.

(d) One bet available in craps wins if a 7 or an 11 comes up on the next roll of two dice. What is the probability of rolling a 7 or an 11 on the next roll?

(e) Several bets in craps lose if a 7 is rolled. If any outcome other than 7 occurs, these bets either win or continue to the next roll. What is the probability that anything other than a 7 is rolled?

4.49 Nonstandard dice can produce interesting distributions of outcomes. You have two balanced, six-sided dice. One is a standard die, with faces having 1, 2, 3, 4, 5, and 6 spots. The other die has three faces with 0 spots and three faces with 6 spots. Find the probability distribution for the total number of spots Y on the up-faces when you roll these two dice.

4.50 Weary of the low turnout in student elections, a college administration decides to choose an SRS of three students to form an advisory board that represents student opinion. Suppose that 40% of all students oppose the use of student fees to fund student interest groups, and that the opinions of the three students on the board are independent. Then the probability is 0.4 that each opposes the funding of interest groups.

(a) Call the three students A, B, and C. What is the probability that A and B support funding and C opposes it?

(b) List all possible combinations of opinions that can be held by students A, B, and C. (*Hint:* There are eight possibilities.) Then give the probability of each of these outcomes. Note that they are not equally likely.

(c) Let the random variable X be the number of student representatives who oppose the funding of interest groups. Give the probability distribution of X.

(d) Express the event "a majority of the advisory board opposes funding" in terms of X and find its probability.

4.51 Let X be a random number between 0 and 1 produced by the idealized uniform random number generator described in Example 4.17 and Figure 4.9. Find the following probabilities:

(a) $P(X < 0.5)$

(b) $P(X \leq 0.5)$

(c) What important fact about continuous random variables does comparing your answers to (a) and (b) illustrate?

4.52 Let the random variable X be a random number with the uniform density curve in Figure 4.9. Find the following probabilities:

(a) $P(X \geq 0.27)$

(b) $P(X = 0.27)$

(c) $P(0.27 < X < 1.27)$

(d) $P(0.1 \leq X \leq 0.2 \text{ or } 0.8 \leq X \leq 0.9)$

(e) The probability that X is not in the interval 0.3 to 0.8.

4.53 Many random number generators allow users to specify the range of the random numbers to be produced. Suppose that you specify that the range is to be all numbers between 0 and 2. Call the random number generated Y. Then the density curve of the random variable Y has constant height between 0 and 2, and height 0 elsewhere.

(a) What is the height of the density curve between 0 and 2? Draw a graph of the density curve.

(b) Use your graph from (a) and the fact that probability is area under the curve to find $P(Y \leq 1)$.

(c) Find $P(0.5 < Y < 1.3)$.

(d) Find $P(Y \geq 0.8)$.

4.54 Generate *two* random numbers between 0 and 1 and take Y to be their sum. Then Y is a continuous random variable that can take any value between 0 and 2. The density curve of Y is the triangle shown in Figure 4.12.

(a) Verify by geometry that the area under this curve is 1.

(b) What is the probability that Y is less than 1? (Sketch the density curve, shade the area that represents the probability, then find that area. Do this for (c) also.)

(c) What is the probability that Y is less than 0.5?

4.55 An SRS of 400 American adults is asked, "What do you think is the most serious problem facing our schools?" Suppose that in fact 30% of all adults would answer "drugs" if asked this question. The proportion $\hat{p}$ of the sample who answer "drugs" will vary in repeated sampling. In fact, we can assign probabilities to values of $\hat{p}$ using the normal density curve with mean 0.3 and standard deviation 0.023. Use this density curve to find the probabilities of the following events:

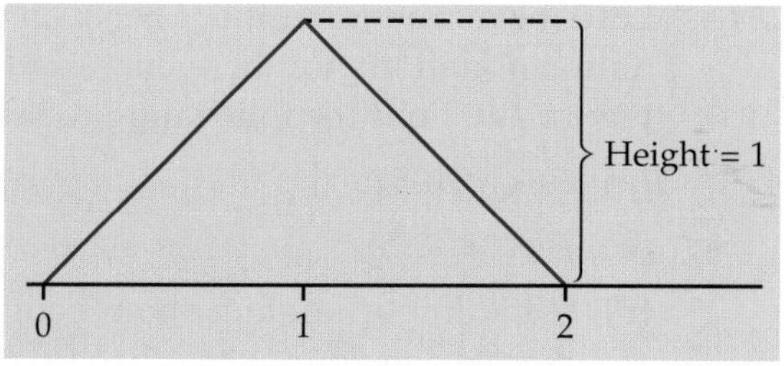

FIGURE 4.12 The density curve for the sum Y of two random numbers, for Exercise 4.54.

(a) At least half of the sample believes that drugs are the schools' most serious problem.

(b) Less than 25% of the sample believes that drugs are the most serious problem.

(c) The sample proportion is between 0.25 and 0.35.

4.56 An opinion poll asks an SRS of 1500 adults, "Do you happen to jog?" Suppose that the population proportion who jog (a parameter) is $p = 0.15$. To estimate p, we use the proportion $\hat{p}$ in the sample who answer "Yes." The statistic $\hat{p}$ is a random variable that is approximately normally distributed with mean $\mu = 0.15$ and standard deviation $\sigma = 0.0092$. Find the following probabilities:

(a) $P(\hat{p} \geq 0.16)$

(b) $P(0.14 \leq \hat{p} \leq 0.16)$

4.4 Means and Variances of Random Variables

Probability is the mathematical language that describes the long-run regular behavior of random phenomena. The probability distribution of a random variable is an idealized relative frequency distribution. The probability histograms and density curves that picture probability distributions resemble our earlier pictures of distributions of data. In describing data, we moved from graphs to numerical measures such as means and standard deviations. Now we will make the same move to expand our descriptions of the distributions of random variables. We can speak of the mean winnings in a game of chance or the standard deviation of the randomly varying number of calls a travel agency receives in an hour. In this section we will learn more about how to compute these descriptive measures and about the laws they obey.

The mean of a random variable

The mean $\bar{x}$ of a set of observations is their ordinary average. The mean of a random variable X is also an average of the possible values of X, but with an essential change to take into account the fact that not all outcomes need be equally likely. An example will show what we must do.

EXAMPLE 4.19

Most states and Canadian provinces have government-sponsored lotteries. Here is a simple lottery wager, from the Tri-State Pick 3 game that New Hampshire shares with Maine and Vermont. You choose a three-digit number; the state chooses a three-digit winning number at random and pays you \$500 if your number is chosen. Because there are 1000 three-digit numbers, you have probability 1/1000 of winning. Taking X to be the amount your ticket pays you, the probability distribution of X is

Payoff X	\$0	\$500
Probability	0.999	0.001

What is your average payoff from many tickets? The ordinary average of the two possible outcomes \$0 and \$500 is \$250, but that makes no sense as the average because \$500 is much less likely than \$0. In the long run you receive \$500 once in every 1000 tickets and \$0 on the remaining 999 of 1000 tickets. The long-run average payoff is

$$\$500\frac{1}{1000} + \$0\frac{999}{1000} = \$0.50$$

or fifty cents. That number is the mean of the random variable X. (Tickets cost \$1, so in the long run the state keeps half the money you wager.)

If you play Tri-State Pick 3 several times, we would as usual call the mean of the actual amounts you win $\overline{x}$. The mean in Example 4.19 is a different quantity—it is the long-run average winnings you expect if you play a very large number of times. Just as probabilities are an idealized description of long-run proportions, the mean of a probability distribution describes the long-run average outcome. We can't call this mean $\overline{x}$, so we need a different symbol. The common symbol for the **mean of a probability distribution** is μ, the Greek letter mu. We used μ in Chapter 1 for the mean of a normal distribution, so this is not a new notation. We will often be interested in several random variables, each having a different probability distribution with a different mean. To remind ourselves that we are talking about the mean of X, we often write μ_X rather than simply μ. In Example 4.19, $\mu_X = \$0.50$. Notice that, as often happens, the mean is not a possible value of X. You will often find the mean of a random variable X called the **expected value** of X. This term may be misleading, for we don't necessarily expect one observation on X to be close to its expected value.

mean μ

expected value

The mean of any discrete random variable is found just as in Example 4.19. It is an average of the possible outcomes, but a weighted average in which each outcome is weighted by its probability. Because the probabilities add to 1, we have total weight 1 to distribute among the outcomes. An outcome that occurs half the time has probability one-half and gets one-half the weight in calculating the mean. Here is the general definition.

Mean of a Discrete Random Variable

Suppose that X is a discrete random variable whose distribution is

Value of X	x_1	x_2	x_3	$\cdots$	x_k
Probability	p_1	p_2	p_3	$\cdots$	p_k

To find the **mean** of X, multiply each possible value by its probability, then add all the products:

$$\begin{aligned}\mu_X &= x_1p_1 + x_2p_2 + \cdots + x_kp_k \\ &= \sum x_i p_i\end{aligned}$$

EXAMPLE 4.20

If first digits in a set of data appear "at random," the nine possible digits 1 to 9 all have the same probability. The probability distribution of the first digit X is then

First digit X	1	2	3	4	5	6	7	8	9
Probability	1/9	1/9	1/9	1/9	1/9	1/9	1/9	1/9	1/9

The mean of this distribution is

$$\begin{aligned}\mu_X &= 1\frac{1}{9} + 2\frac{1}{9} + 3\frac{1}{9} + 4\frac{1}{9} + 5\frac{1}{9} + 6\frac{1}{9} + 7\frac{1}{9} + 8\frac{1}{9} + 9\frac{1}{9} \\ &= 45 \times \frac{1}{9} = 5\end{aligned}$$

If, on the other hand, the data obey Benford's Law, the distribution of the first digit V is

First digit V	1	2	3	4	5	6	7	8	9
Probability	0.301	0.176	0.125	0.097	0.079	0.067	0.058	0.051	0.046

The mean of V is

$$\begin{aligned}\mu_V &= (1)(0.301) + (2)(0.176) + (3)(0.125) + (4)(0.097) + (5)(0.079) \\ &\quad + (6)(0.067) + (7)(0.058) + (8)(0.051) + (9)(0.046) \\ &= 3.441\end{aligned}$$

The means reflect the greater probability of smaller first digits under Benford's Law.

Figure 4.13 locates the means of X and V on the two probability histograms. Because the discrete uniform distribution of Figure 4.13(a) is symmetric, the mean lies at the center of symmetry. We can't locate the mean of the right-skewed distribution of Figure 4.13(b) by eye—calculation is needed.

What about continuous random variables? The probability distribution of a continuous random variable X is described by a density curve. Chapter 1

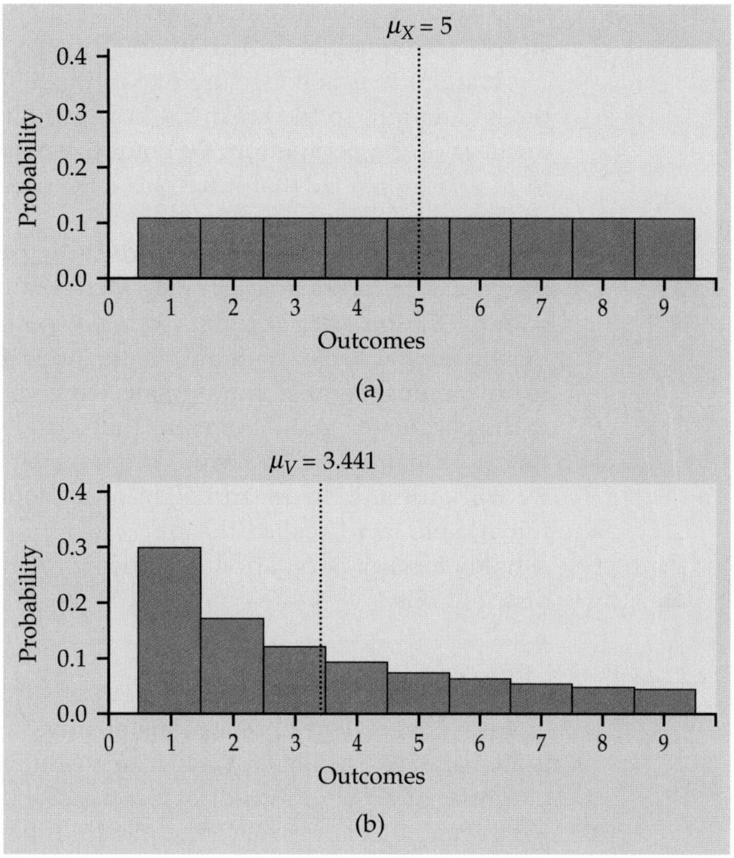

FIGURE 4.13 Locating the mean of a discrete random variable on the probability histogram for (a) digits between 1 and 9 chosen at random; (b) digits between 1 and 9 chosen from records that obey Benford's Law.

showed how to find the mean of the distribution: it is the point at which the area under the density curve would balance if it were made out of solid material. The mean lies at the center of symmetric density curves such as the normal curves. Exact calculation of the mean of a distribution with a skewed density curve requires advanced mathematics.[9] The idea that the mean is the balance point of the distribution applies to discrete random variables as well, but in the discrete case we have a formula that gives us this point.

Statistical estimation and the law of large numbers

We would like to estimate the mean height μ of the population of all American women between the ages of 18 and 24 years. This μ is the mean μ_X of the random variable X obtained by choosing a young woman at random and measuring her height. To estimate μ, we choose an SRS of young women and use the sample mean $\bar{x}$ to estimate the unknown population mean μ. In the language of Section 3.4 (page 261), μ is a *parameter* and $\bar{x}$ is a *statistic*. Statistics obtained from probability samples are random variables, because their values would vary in repeated sampling. The sampling distributions of statistics are just the probability distributions of these random variables.

It seems reasonable to use $\overline{x}$ to estimate μ. An SRS should fairly represent the population, so the mean $\overline{x}$ of the sample should be somewhere near the mean μ of the population. Of course, we don't expect $\overline{x}$ to be exactly equal to μ, and we realize that if we choose another SRS, the luck of the draw will probably produce a different $\overline{x}$.

If $\overline{x}$ is rarely exactly right and varies from sample to sample, why is it nonetheless a reasonable estimate of the population mean μ? We gave one answer in Section 3.4: $\overline{x}$ is unbiased and we can control its variability by choosing the sample size. Here is another answer: if we keep on adding observations to our random sample, the statistic $\overline{x}$ is *guaranteed* to get as close as we wish to the parameter μ and then stay that close. We have the comfort of knowing that if we can afford to keep on measuring more young women, eventually we will estimate the mean height of all young women very accurately. This remarkable fact is called the *law of large numbers*. It is remarkable because it holds for *any* population, not just for some special class such as normal distributions.

Law of Large Numbers

Draw independent observations at random from any population with finite mean μ. Decide how accurately you would like to estimate μ. As the number of observations drawn increases, the mean $\overline{x}$ of the observed values eventually approaches the mean μ of the population as closely as you specified and then stays that close.

The behavior of $\overline{x}$ is similar to the idea of probability. In the long run, the proportion of outcomes taking any value gets close to the probability of that value, and the average outcome gets close to the distribution mean. Figure 4.1 (page 282) shows how proportions approach probability in one example. Here is an example of how sample means approach the distribution mean.

EXAMPLE 4.21

The distribution of the heights of all young women is close to the normal distribution with mean 64.5 inches and standard deviation 2.5 inches. Suppose that $\mu = 64.5$ were exactly true. Figure 4.14 shows the behavior of the mean height $\overline{x}$ of n women chosen at random from a population whose heights follow the $N(64.5, 2.5)$ distribution. The graph plots the values of $\overline{x}$ as we add women to our sample. The first woman drawn had height 64.21 inches, so the line starts there. The second had height 64.35 inches, so for $n = 2$ the mean is

$$\overline{x} = \frac{64.21 + 64.35}{2} = 64.28$$

This is the second point on the line in the graph.

At first, the graph shows that the mean of the sample changes as we take more observations. Eventually, however, the mean of the observations gets close to the population mean $\mu = 64.5$ and settles down at that value. The law of large numbers says that this *always* happens.

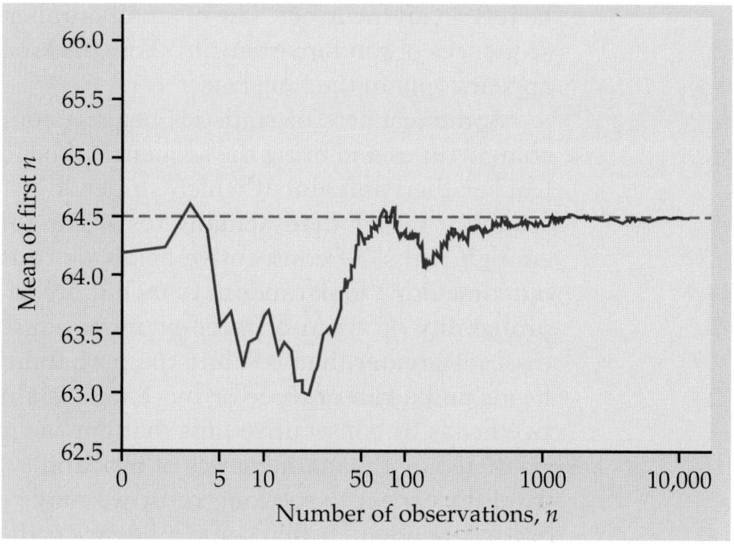

FIGURE 4.14 The law of large numbers in action. As we take more observations, the sample mean $\overline{x}$ always approaches the mean μ of the population.

The mean μ of a random variable is the average value of the variable in two senses. By its definition, μ is the average of the possible values, weighted by their probability of occurring. The law of large numbers says that μ is also the long-run average of many independent observations on the variable. The law of large numbers can be proved mathematically starting from the basic laws of probability.*

Thinking about the law of large numbers

The law of large numbers says broadly that the average results of many independent observations are stable and predictable. Casinos are not the only businesses that base forecasts on this fact. An insurance company deciding how much to charge for life insurance and a fast-food restaurant deciding how many beef patties to prepare rely on the fact that averaging over many individuals produces a stable result. It is worth the effort to think a bit more closely about so important a fact.

The "law of small numbers" Both the rules of probability and the law of large numbers describe the regular behavior of chance phenomena *in the long run*. Psychologists have discovered that the popular understanding of randomness is quite different from the true laws of chance.[10] Most people

*The earliest version of the law of large numbers was proved by the Swiss mathematician Jacob Bernoulli (1654–1705). The Bernoullis were a remarkable mathematical family; five of them contributed to the early study of probability.

believe in an incorrect "law of small numbers." That is, we expect even short sequences of random events to show the kind of average behavior that in fact appears only in the long run.

Some teachers of statistics begin a course by asking students to toss a coin 50 times and bring the sequence of heads and tails to the next class. The teacher then announces which students just wrote down a random-looking sequence rather than actually tossing a coin. The faked tosses don't have enough "runs" of consecutive heads or consecutive tails. Runs of the same outcome don't look random to us but are in fact common. For example, the probability of a run of three or more consecutive heads or tails in just 10 tosses is greater than 0.8, and the probability of *both* a run of three or more heads and a run of three or more tails is almost 0.2.[11] The runs of consecutive heads or consecutive tails that appear in real coin tossing (and that are predicted by the mathematics of probability) seem surprising to us. Because we don't expect to see long runs, we may conclude that the coin tosses are not independent or that some influence is disturbing the random behavior of the coin.

EXAMPLE 4.22

Belief in the law of small numbers influences behavior. If a basketball player makes several consecutive shots, both the fans and her teammates believe that she has a "hot hand" and is more likely to make the next shot. This is doubtful. Careful study suggests that runs of baskets made or missed are no more frequent in basketball than would be expected if each shot were independent of the player's previous shots. Baskets made or missed are just like heads and tails in tossing a coin. (Of course, some players make 30% of their shots in the long run and others make 50%, so a coin-toss model for basketball must allow coins with different probabilities of a head.) Our perception of hot or cold streaks simply shows that we don't perceive random behavior very well.[12]

Our inability to accurately distinguish random behavior from systematic influences points out once more the need for statistical inference to supplement exploratory analysis of data. Probability calculations can help verify that what we see in the data is more than a random pattern.

How large is a large number? The law of large numbers says that the actual mean outcome of many trials gets close to the distribution mean μ as more trials are made. It doesn't say how many trials are needed to guarantee a mean outcome close to μ. That depends on the *variability* of the random outcomes. The more variable the outcomes, the more trials are needed to ensure that the mean outcome $\bar{x}$ is close to the distribution mean μ.

The law of large numbers is the foundation of such business enterprises as gambling casinos and insurance companies. Games of chance must be quite variable if they are to hold the interest of gamblers. Even a long evening in a casino has an unpredictable outcome. Gambles with extremely variable outcomes, like state lottos with their very large but very improbable jackpots, require impossibly large numbers of trials to ensure that the average outcome is close to the expected value. Though most forms of gambling are less variable than lotto, the layman's answer to the applicability of the law of large numbers is usually that the house plays often enough to rely on it, but you don't. Much of the psychological allure of gambling is its unpredictability for the player. The business of gambling rests on the fact that the result is not unpredictable for the house. The average winnings of the house on tens of thousands of bets will be very close to the mean of the distribution of winnings. Needless to say, this mean guarantees the house a profit.

Exploring the Web

Your first stop on the Web should be the *What Is Probability* and *Expected Value* applets at `www.whfreeman.com/ips`. These animated explorations make the idea of probability and the law of large numbers clearer than any amount of writing. For entertainment, look at the *Wizard of Odds* site, `www.thewizardofodds.com`, for probability-based information about most casino games. The material under "Miscellaneous" has lots of actual probabilities. It also contains the Wizard's first commandment of gambling: expect to lose.

Beyond the basics ▸ more laws of large numbers

The law of large numbers is one of the central facts about probability. It helps us understand the mean μ of a random variable. It explains why gambling casinos and insurance companies make money. It assures us that statistical estimation will be accurate if we can afford enough observations. The basic law of large numbers applies to independent observations that all have the same distribution. Mathematicians have extended the law to many more general settings. Here are two of these.

Is there a winning system for gambling? Serious gamblers often follow a system of betting in which the amount bet on each play depends on the outcome of previous plays. You might, for example, double your bet on each spin of the roulette wheel until you win—or, of course, until your fortune is exhausted. Such a system tries to take advantage of the fact that you have a memory even though the roulette wheel does not. Can you beat the odds with a system based on the outcomes of past plays? No. Mathematicians have established a stronger version of the law of large numbers that says that if you do not have an infinite fortune to gamble with, your long-run average winnings μ remain the same as long as successive trials of the game (such as spins of the roulette wheel) are independent.

What if observations are not independent? You are in charge of a process that manufactures video screens for computer monitors. Your equipment measures the tension on the metal mesh that lies behind each screen and is critical to its image quality. You want to estimate the mean tension μ for the process by the average $\overline{x}$ of the measurements. Alas, the tension measurements are not independent. If the tension on one screen is a bit too high, the tension on the next is more likely to also be high. Many real-world processes are like this—the process stays stable in the long run, but observations made close together are likely to be both above or both below the long-run mean. Again the mathematicians come to the rescue: as long as the dependence dies out fast enough as we take measurements farther and farther apart in time, the law of large numbers still holds.

Rules for means

You are studying flaws in the painted finish of refrigerators made by your firm. Dimples and paint sags are two kinds of surface flaw. Not all refrigerators have the same number of dimples: many have none, some have one, some two, and so on. You ask for the average number of imperfections on a refrigerator. The inspectors report finding an average of 0.7 dimples and 1.4 sags per refrigerator. How many total imperfections of both kinds (on the average) are there on a refrigerator? That's easy: if the average number of dimples is 0.7 and the average number of sags is 1.4, then counting both gives an average of $0.7 + 1.4 = 2.1$ flaws.

In more formal language, the number of dimples on a refrigerator is a random variable X that varies as we inspect one refrigerator after another. We know only that the mean number of dimples is $\mu_X = 0.7$. The number of paint sags is a second random variable Y having mean $\mu_Y = 1.4$. (As usual, the subscripts keep straight which variable we are talking about.) The total number of both dimples and sags is another random variable, the sum $X + Y$. Its mean μ_{X+Y} is the average number of dimples and sags together. It is just the sum of the individual means μ_X and μ_Y. That's an important rule for how means of random variables behave.

Here's another rule. The crickets living in a field have mean length 1.2 inches. What is the mean in centimeters? There are 2.54 centimeters in an inch, so the length of a cricket in centimeters is 2.54 times its length in inches. If we multiply every observation by 2.54, we also multiply their average by 2.54. The mean in centimeters must be 2.54×1.2, or about 3.05 centimeters. More formally, the length in inches of a cricket chosen at random from the field is a random variable X with mean μ_X. The diameter in centimeters is $2.54X$, and this new random variable has mean $2.54\mu_X$.

The point of these examples is that means behave like averages. Here are the rules we need.

Rules for Means

Rule 1. If X is a random variable and a and b are fixed numbers, then

$$\mu_{a+bX} = a + b\mu_X$$

Rule 2. If X and Y are random variables, then

$$\mu_{X+Y} = \mu_X + \mu_Y$$

EXAMPLE 4.23

Gain Communications sells aircraft communications units to both the military and the civilian markets. Next year's sales depend on market conditions that cannot be predicted exactly. Gain follows the modern practice of using probability estimates of sales. The military division estimates its sales as follows:

Units sold	1000	3000	5000	10,000
Probability	0.1	0.3	0.4	0.2

These are personal probabilities that express the informed opinion of Gain's executives. The corresponding sales estimates for the civilian division are

Units sold	300	500	750
Probability	0.4	0.5	0.1

Take X to be the number of military units sold and Y the number of civilian units. From the probability distributions we compute that

$$\begin{aligned}\mu_X &= (1000)(0.1) + (3000)(0.3) + (5000)(0.4) + (10{,}000)(0.2)\\ &= 100 + 900 + 2000 + 2000\\ &= 5000 \text{ units}\\ \mu_Y &= (300)(0.4) + (500)(0.5) + (750)(0.1)\\ &= 120 + 250 + 75\\ &= 445 \text{ units}\end{aligned}$$

Gain makes a profit of \$2000 on each military unit sold and \$3500 on each civilian unit. Next year's profit from military sales will be $2000X$, \$2000 times the number X of units sold. By Rule 1 for means, the mean military profit is

$$\mu_{2000X} = 2000\mu_X = (2000)(5000) = \$10{,}000{,}000$$

Similarly, the civilian profit is $3500Y$, and the mean profit from civilian sales is

$$\mu_{3500Y} = 3500\mu_Y = (3500)(445) = \$1{,}557{,}500$$

The total profit is the sum of the military and civilian profit:

$$Z = 2000X + 3500Y$$

Rule 2 for means says that the mean of this sum of two variables is the sum of the two individual means:

$$\begin{aligned}\mu_Z &= \mu_{2000X} + \mu_{3500Y}\\ &= \$10{,}000{,}000 + \$1{,}557{,}500\\ &= \$11{,}557{,}500\end{aligned}$$

This mean is the company's best estimate of next year's profit, combining the probability estimates of the two divisions. We can do this calculation more quickly by combining Rules 1 and 2:

$$\begin{aligned}\mu_Z &= \mu_{2000X + 3500Y} \\ &= 2000\mu_X + 3500\mu_Y \\ &= (2000)(5000) + (3500)(445) = \$11{,}557{,}500\end{aligned}$$

The variance of a random variable

The mean is a measure of the center of a distribution. Even the most basic numerical description requires in addition a measure of the spread or variability of the distribution. The variance and the standard deviation are the measures of spread that accompany the choice of the mean to measure center. Just as for the mean, we need a distinct symbol to distinguish the variance of a random variable from the variance s^2 of a data set. We write the variance of a random variable X as σ_X^2. Once again the subscript reminds us which variable we have in mind. The definition of the variance σ_X^2 of a random variable is similar to the definition of the sample variance s^2 given in Chapter 1. That is, the variance is an average of the squared deviation $(X - \mu_X)^2$ of the variable X from its mean μ_X. As for the mean, the average we use is a weighted average in which each outcome is weighted by its probability in order to take account of outcomes that are not equally likely. Calculating this weighted average is straightforward for discrete random variables but requires advanced mathematics in the continuous case. Here is the definition.

Variance of a Discrete Random Variable

Suppose that X is a discrete random variable whose distribution is

Value of X	x_1	x_2	x_3	$\cdots$	x_k
Probability	p_1	p_2	p_3	$\cdots$	p_k

and that μ is the mean of X. The **variance** of X is

$$\begin{aligned}\sigma_X^2 &= (x_1 - \mu_X)^2 p_1 + (x_2 - \mu_X)^2 p_2 + \cdots + (x_k - \mu_X)^2 p_k \\ &= \sum (x_i - \mu_X)^2 p_i\end{aligned}$$

The **standard deviation** σ_X of X is the square root of the variance.

EXAMPLE 4.24 In Example 4.23 we saw that the number X of communications units that the Gain Communications military division hopes to sell has distribution

Units sold	1000	3000	5000	10,000
Probability	0.1	0.3	0.4	0.2

We can find the mean and variance of X by arranging the calculation in the form of a table. Both μ_X and σ_X^2 are sums of columns in this table.

x_i	p_i	$x_i p_i$	$(x_i - \mu_X)^2 p_i$
1,000	0.1	100	$(1{,}000 - 5{,}000)^2(0.1) = 1{,}600{,}000$
3,000	0.3	900	$(3{,}000 - 5{,}000)^2(0.3) = 1{,}200{,}000$
5,000	0.4	2,000	$(5{,}000 - 5{,}000)^2(0.4) = 0$
10,000	0.2	2,000	$(10{,}000 - 5{,}000)^2(0.2) = 5{,}000{,}000$
		$\mu_X = 5{,}000$	$\sigma_X^2 = 7{,}800{,}000$

We see that $\sigma_X^2 = 7{,}800{,}000$. The standard deviation of X is $\sigma_X = \sqrt{7{,}800{,}000} = 2792.8$. The standard deviation is a measure of how variable the number of units sold is. As in the case of distributions for data, the standard deviation of a probability distribution is easiest to understand for normal distributions.

Rules for variances

What are the facts for variances that parallel Rules 1 and 2 for means? The mean of a sum of random variables is always the sum of their means, but this addition rule is not always true for variances. To understand why, take X to be the percent of a family's after-tax income that is spent and Y the percent that is saved. When X increases, Y decreases by the same amount. Though X and Y may vary widely from year to year, their sum $X + Y$ is always 100% and does not vary at all. It is the association between the variables X and Y that prevents their variances from adding. If random variables are independent, this kind of association between their values is ruled out and their variances do add. Two random variables X and Y are **independent** if knowing that any event involving X alone did or did not occur tells us nothing about the occurrence of any event involving Y alone. Probability models often assume independence when the random variables describe outcomes that appear unrelated to each other. You should ask in each instance whether the assumption of independence seems reasonable.

independence

When random variables are not independent, the variance of their sum depends on the **correlation** between them as well as on their individual variances. In Chapter 2, we met the correlation r between two observed variables measured on the same individuals. We defined (page 127) the correlation r as an average of the products of the standardized x and y observations. The correlation between two random variables is defined in the same way, once again using a weighted average with probabilities as weights. We won't give the details—it is enough to know that the correlation between two random variables has the same basic properties as the correlation r calculated from data. We use ρ, the Greek letter rho, for the correlation between two random variables. The correlation ρ is a number between -1 and 1 that measures the direction and strength of the linear relationship between two variables. **The correlation between two independent random variables is zero.**

correlation

Returning to family finances, if X is the percent of a family's after-tax income that is spent and Y the percent that is saved, then $Y = 100 - X$. This is a perfect linear relationship with a negative slope, so the correlation between X and Y is $\rho = -1$. With the correlation at hand, we can state the rules for manipulating variances.

Rules for Variances

Rule 1. If X is a random variable and a and b are fixed numbers, then

$$\sigma^2_{a+bX} = b^2\sigma^2_X$$

Rule 2. If X and Y are independent random variables, then

$$\sigma^2_{X+Y} = \sigma^2_X + \sigma^2_Y$$
$$\sigma^2_{X-Y} = \sigma^2_X + \sigma^2_Y$$

This is the **addition rule for variances of independent random variables.**

Rule 3. If X and Y have correlation ρ, then

$$\sigma^2_{X+Y} = \sigma^2_X + \sigma^2_Y + 2\rho\sigma_X\sigma_Y$$
$$\sigma^2_{X-Y} = \sigma^2_X + \sigma^2_Y - 2\rho\sigma_X\sigma_Y$$

This is the **general addition rule for variances of random variables.**

Notice that because a variance is the average of *squared* deviations from the mean, multiplying X by a constant b multiplies σ^2_X by the *square* of the constant. Adding a constant a to a random variable changes its mean but does not change its variability. The variance of $X + a$ is therefore the same as the variance of X. Because the square of -1 is 1, the addition rule says that the variance of a difference of independent random variables is the *sum* of the variances. For independent random variables, the difference $X - Y$ is more variable than either X or Y alone because variations in both X and Y contribute to variation in their difference.

As with data, we prefer the standard deviation to the variance as a measure of the variability of a random variable. Rule 2 for variances implies that standard deviations of independent random variables do *not* add. To combine standard deviations, use the rules for variances rather than trying to remember separate rules for standard deviations. For example, the standard deviations of $2X$ and $-2X$ are both equal to $2\sigma_X$ because this is the square root of the variance $4\sigma^2_X$.

EXAMPLE 4.25

The payoff X of a \$1 ticket in the Tri-State Pick 3 game is \$500 with probability 1/1000 and 0 the rest of the time. Here is the combined calculation of mean and variance:

x_i	p_i	$x_i p_i$	$(x_i - \mu_X)^2 p_i$
0	0.999	0	$(0 - 0.5)^2(0.999) = 0.24975$
500	0.001	0.5	$(500 - 0.5)^2(0.001) = 249.50025$
		$\mu_X = 0.5$	$\sigma^2_X = 249.75$

The standard deviation is $\sigma_X = \sqrt{249.75} = \$15.80$. It is usual for games of chance to have large standard deviations because large variability makes gambling exciting.

If you buy a Pick 3 ticket, your winnings are $W = X - 1$ because the dollar you paid for the ticket must be subtracted from the payoff. By the rules for means, the mean amount you win is

$$\mu_W = \mu_X - 1 = -\$0.50$$

That is, you lose an average of 50 cents on a ticket. The rules for variances remind us that the variance and standard deviation of the winnings $W = X - 1$ are the same as those of X. Subtracting a fixed number changes the mean but not the variance.

Suppose now that you buy a \$1 ticket on each of two different days. The payoffs X and Y on the two tickets are independent because separate drawings are held each day. Your total payoff $X + Y$ has mean

$$\mu_{X+Y} = \mu_X + \mu_Y = \$0.50 + \$0.50 = \$1.00$$

Because X and Y are independent, the variance of $X + Y$ is

$$\sigma^2_{X+Y} = \sigma^2_X + \sigma^2_Y = 249.75 + 249.75 = 499.5$$

The standard deviation of the total payoff is

$$\sigma_{X+Y} = \sqrt{499.5} = \$22.35$$

This is not the same as the sum of the individual standard deviations, which is \$15.80 + \$15.80 = \$31.60. Variances of independent random variables add; standard deviations do not.

If you buy a ticket every day (365 tickets in a year), your mean payoff is the sum of 365 daily payoffs. That's 365 times 50 cents, or \$182.50. Of course, it cost \$365 to play, so the state's mean take from a daily Pick 3 player is \$182.50. Results for individual players will vary, but the law of large numbers assures the state its profit.

EXAMPLE 4.26

A college uses SAT scores as one criterion for admission. Experience has shown that the distribution of SAT scores among the college's entire population of applicants is such that

SAT math score X	$\mu_X = 625$	$\sigma_X = 90$
SAT verbal score Y	$\mu_Y = 590$	$\sigma_Y = 100$

What are the mean and standard deviation of the total score $X + Y$ among students applying to this college?

The mean overall SAT score is

$$\mu_{X+Y} = \mu_X + \mu_Y = 625 + 590 = 1215$$

The variance and standard deviation of the total *cannot be computed* from the information given. SAT verbal and math scores are not independent, because students who score high on one exam tend to score high on the other also. Therefore, Rule 2 does not apply and we need to know ρ, the correlation between X and Y, to apply Rule 3.

Nationally, the correlation between SAT math and verbal scores is about $\rho = 0.7$. If this is true for these students,

$$\begin{aligned}\sigma^2_{X+Y} &= \sigma^2_X + \sigma^2_Y + 2\rho\sigma_X\sigma_Y \\ &= (90)^2 + (100)^2 + (2)(0.7)(90)(100) \\ &= 30{,}700\end{aligned}$$

The variance of the sum $X + Y$ is greater than the sum of the variances $\sigma_X^2 + \sigma_Y^2$ because of the positive correlation between SAT math scores and SAT verbal scores. That is, X and Y tend to move up together and down together, which increases the variability of their sum. We find the standard deviation from the variance,

$$\sigma_{X+Y} = \sqrt{30{,}700} = 175$$

EXAMPLE 4.27 Zadie has invested 20% of her funds in Treasury bills and 80% in an "index fund" that represents all U.S. common stocks. The rate of return of an investment over a time period is the percent change in the price during the time period, plus any income received. If X is the annual return on T-bills and Y the annual return on stocks, the portfolio rate of return is

$$R = 0.2X + 0.8Y$$

The returns X and Y are random variables because they vary from year to year. Based on annual returns between 1950 and 2000, we have[13]

X = annual return on T-bills	$\mu_X = 5.2\%$	$\sigma_X = 2.9\%$
Y = annual return on stocks	$\mu_Y = 13.3\%$	$\sigma_Y = 17.0\%$
Correlation between X and Y	$\rho = -0.1$	

Stocks had higher returns than Treasury bills on the average, but the standard deviations show that returns on stocks varied much more from year to year. That is, the risk of investing in stocks is greater than the risk for Treasury bills because their returns are less predictable.

For the return R on Zadie's portfolio of 20% T-bills and 80% stocks,

$$\begin{aligned} R &= 0.2X + 0.8Y \\ \mu_R &= 0.2\mu_X + 0.8\mu_Y \\ &= (0.2 \times 5.2) + (0.8 \times 13.3) = 11.68\% \end{aligned}$$

To find the variance of the portfolio return, combine Rules 1 and 3:

$$\begin{aligned} \sigma_R^2 &= \sigma_{0.2X}^2 + \sigma_{0.8Y}^2 + 2\rho\sigma_{0.2X}\sigma_{0.8Y} \\ &= (0.2)^2\sigma_X^2 + (0.8)^2\sigma_Y^2 + 2\rho(0.2 \times \sigma_X)(0.8 \times \sigma_Y) \\ &= (0.2)^2(2.9)^2 + (0.8)^2(17.0)^2 + (2)(-0.1)(0.2 \times 2.9)(0.8 \times 17.0) \\ &= 183.719 \\ \sigma_R &= \sqrt{183.719} = 13.55\% \end{aligned}$$

The portfolio has a smaller mean return that an all-stock portfolio, but it is also less risky. As a proportion of the all-stock values, the reduction in standard deviation is greater than the reduction in mean return. That's why Zadie put some funds into Treasury bills.

SUMMARY

The probability distribution of a random variable X, like a distribution of data, has a **mean** $\boldsymbol{\mu_X}$ and a **standard deviation** $\boldsymbol{\sigma_X}$.

The **law of large numbers** says that the average of the values of X observed in many trials must approach μ.

The **mean μ** is the balance point of the probability histogram or density curve. If X is discrete with possible values x_i having probabilities p_i, the mean is the average of the values of X, each weighted by its probability:

$$\mu_X = x_1p_1 + x_2p_2 + \cdots + x_kp_k$$

The **variance σ_X^2** is the average squared deviation of the values of the variable from their mean. For a discrete random variable,

$$\sigma_X^2 = (x_1 - \mu)^2p_1 + (x_2 - \mu)^2p_2 + \cdots + (x_k - \mu)^2p_k$$

The **standard deviation σ_X** is the square root of the variance. The standard deviation measures the variability of the distribution about the mean. It is easiest to interpret for normal distributions.

The mean and variance of a continuous random variable can be computed from the density curve, but to do so requires more advanced mathematics.

The means and variances of random variables obey the following rules. If a and b are fixed numbers, then

$$\begin{aligned} \mu_{a+bX} &= a + b\mu_X \\ \sigma_{a+bX}^2 &= b^2\sigma_X^2 \end{aligned}$$

If X and Y are any two random variables having correlation ρ, then

$$\begin{aligned} \mu_{X+Y} &= \mu_X + \mu_Y \\ \sigma_{X+Y}^2 &= \sigma_X^2 + \sigma_Y^2 + 2\rho\sigma_X\sigma_Y \\ \sigma_{X-Y}^2 &= \sigma_X^2 + \sigma_Y^2 - 2\rho\sigma_X\sigma_Y \end{aligned}$$

If X and Y are **independent,** then $\rho = 0$. In this case,

$$\begin{aligned} \sigma_{X+Y}^2 &= \sigma_X^2 + \sigma_Y^2 \\ \sigma_{X-Y}^2 &= \sigma_X^2 + \sigma_Y^2 \end{aligned}$$

SECTION 4.4 EXERCISES

4.57 How do rented housing units differ from units occupied by their owner? Exercise 4.43 (page 314) gives the distributions of the number of rooms for owner-occupied units and renter-occupied units in San Jose, California. Find the mean number of rooms for both types of housing unit. How do the means reflect the differences between the distributions that you found in Exercise 4.43?

4.58 Exercise 4.44 (page 315) gives the distributions of the number of people in households and in families in the United States. An important difference is that many households consist of one person living alone, whereas a family must have at least two members. Some households may contain families along with other people, and so will be larger than the family. These differences make it hard to compare the distributions without calculations. Find the mean and standard deviation of both household size and family size. Combine these with your descriptions from Exercise 4.44 to give a comparison of the two distributions.

4.59 Which of the two distributions for room counts in Exercises 4.43 and 4.57 appears more spread out in the probability histograms? Why? Find the standard deviation for both distributions. The standard deviation provides a numerical measure of spread.

4.60 Example 4.15 gives the distribution of grades (A = 4, B = 3, and so on) in Accounting 210 at North Carolina State University as

Grade	0	1	2	3	4
Probability	0.07	0.09	0.34	0.32	0.18

Find the average (that is, the mean) grade in this course.

4.61 You purchase a hot stock for $1000. The stock either gains 30% or loses 25% each day, each with probability 0.5. Its returns on consecutive days are independent of each other. You plan to sell the stock after two days.

(a) What are the possible values of the stock after two days, and what is the probability for each value? What is the probability that the stock is worth more after two days than the $1000 you paid for it?

(b) What is the mean value of the stock after two days? You see that these two criteria give different answers to the question, "Should I invest?"

4.62 It is easier to use Benford's Law (Example 4.9, page 292) to spot suspicious patterns when you have very many items (for example, many invoices from the same vendor) than when you have only a few. Explain why this is true.

4.63 You have two balanced, six-sided dice. The first has 1, 3, 4, 5, 6, and 8 spots on its six faces. The second die has 1, 2, 2, 3, 3, and 4 spots on its faces.

(a) What is the mean number of spots on the up-face when you roll each of these dice?

(b) Write the probability model for the outcomes when you roll both dice independently. From this, find the probability distribution of the sum of the spots on the up-faces of the two dice.

(c) Find the mean number of spots on the two up-faces in two ways: from the distribution you found in (b) and by applying the addition rule to your results in (a). You should of course get the same answer.

4.64 For each of the following situations, would you expect the random variables X and Y to be independent? Explain your answers.

(a) X is the rainfall (in inches) on November 6 of this year and Y is the rainfall at the same location on November 6 of next year.

(b) X is the amount of rainfall today and Y is the rainfall at the same location tomorrow.

(c) X is today's rainfall at the Orlando, Florida, airport, and Y is today's rainfall at Disney World just outside Orlando.

4.65 In which of the following games of chance would you be willing to assume independence of X and Y in making a probability model? Explain your answer in each case.

(a) In blackjack, you are dealt two cards and examine the total points X on the cards (face cards count 10 points). You can choose to be dealt another card and compete based on the total points Y on all three cards.

(b) In craps, the betting is based on successive rolls of two dice. X is the sum of the faces on the first roll, and Y the sum of the faces on the next roll.

4.66 **(a)** A gambler knows that red and black are equally likely to occur on each spin of a roulette wheel. He observes five consecutive reds and bets heavily on red at the next spin. Asked why, he says that "red is hot" and that the run of reds is likely to continue. Explain to the gambler what is wrong with this reasoning.

(b) After hearing you explain why red and black remain equally probable after five reds on the roulette wheel, the gambler moves to a poker game. He is dealt five straight red cards. He remembers what you said and assumes that the next card dealt in the same hand is equally likely to be red or black. Is the gambler right or wrong? Why?

4.67 A time and motion study measures the time required for an assembly line worker to perform a repetitive task. The data show that the time required to bring a part from a bin to its position on an automobile chassis varies from car to car with mean 11 seconds and standard deviation 2 seconds. The time required to attach the part to the chassis varies with mean 20 seconds and standard deviation 4 seconds.

(a) What is the mean time required for the entire operation of positioning and attaching the part?

(b) If the variation in the worker's performance is reduced by better training, the standard deviations will decrease. Will this decrease change the mean you found in (a) if the mean times for the two steps remain as before?

(c) The study finds that the times required for the two steps are independent. A part that takes a long time to position, for example, does not take more or less time to attach than other parts. How would your answer in (a) change if the two variables were dependent with correlation 0.8? With correlation 0.3?

4.68 Laboratory data show that the time required to complete two chemical reactions in a production process varies. The first reaction has a mean time of 40 minutes and a standard deviation of 2 minutes; the second has a mean time of 25 minutes and a standard deviation of 1 minute. The two reactions are run in sequence during production. There is a fixed period of 5 minutes between them as the product of the first reaction is pumped into the vessel where the second reaction will take place. What is the mean time required for the entire process?

4.69 Find the standard deviation of the time required for the two-step assembly operation studied in Exercise 4.67, assuming that the study shows the two times to be independent. Redo the calculation assuming that the two times are dependent, with correlation 0.3. Can you explain in nontechnical language why positive correlation increases the variability of the total time?

4.70 The times for the two reactions in the chemical production process described in Exercise 4.68 are independent. Find the standard deviation of the time required to complete the process.

4.71 Here is a simple way to create a random variable X that has mean μ and standard deviation σ: X takes only the two values $\mu - \sigma$ and $\mu + \sigma$, each with probability 0.5. Use the definition of the mean and variance for discrete random variables to show that X does have mean μ and standard deviation σ.

4.72 We know that variances add if the random variables involved are uncorrelated ($\rho = 0$), but not otherwise. The opposite extreme is perfect positive correlation ($\rho = 1$). Show by using the general addition rule for variances that in this case the standard deviations add. That is, $\sigma_{X+Y} = \sigma_X + \sigma_Y$ if $\rho_{XY} = 1$.

4.73 Examples 4.23 (page 327) and 4.24 (page 328) concern a probabilistic projection of sales and profits by an electronics firm, Gain Communications.

(a) Find the variance and standard deviation of the estimated sales X of Gain's civilian division, using the distribution and mean from Example 4.23.

(b) Because the military budget and the civilian economy are not closely linked, Gain is willing to assume that its military and civilian sales vary independently. Combine your result from (a) with the results for the military division from Example 4.24 to obtain the standard deviation of the total sales $X + Y$.

(c) Find the standard deviation of the estimated profit, $Z = 2000X + 3500Y$.

4.74 In a process for manufacturing glassware, glass stems are sealed by heating them in a flame. The temperature of the flame varies a bit. Here is the distribution of the temperature X measured in degrees Celsius:

Temperature	540°	545°	550°	555°	560°
Probability	0.1	0.25	0.3	0.25	0.1

(a) Find the mean temperature μ_X and the standard deviation σ_X.

(b) The target temperature is 550° C. What are the mean and standard deviation of the number of degrees off target, $X - 550$?

(c) A manager asks for results in degrees Fahrenheit. The conversion of X into degrees Fahrenheit is given by

$$Y = \frac{9}{5}X + 32$$

What are the mean μ_Y and standard deviation σ_Y of the temperature of the flame in the Fahrenheit scale?

4.75 You have two scales for measuring weights in a chemistry lab. Both scales give answers that vary a bit in repeated weighings of the same item. If the true weight of a compound is 2 grams (g), the first scale produces readings X that have mean 2.000 g and standard deviation 0.002 g. The second scale's readings Y have mean 2.001 g and standard deviation 0.001 g.

(a) What are the mean and standard deviation of the difference $Y - X$ between the readings? (The readings X and Y are independent.)

(b) You measure once with each scale and average the readings. Your result is $Z = (X + Y)/2$. What are μ_Z and σ_Z? Is the average Z more or less variable than the reading Y of the less variable scale?

4.76 One consequence of the law of large numbers is that once we have a probability distribution for a random variable, we can find its mean by simulating many outcomes and averaging them. The law of large numbers says that if we take enough outcomes, their average value is sure to approach the mean of the distribution.

I have a little bet to offer you. Toss a coin ten times. If there is no run of three or more straight heads or tails in the ten outcomes, I'll pay you \$2. If there is a run of three or more, you pay me just \$1. Surely you will want to take advantage of me and play this game?

Simulate enough plays of this game (the outcomes are +\$2 if you win and −\$1 if you lose) to estimate the mean outcome. Is it to your advantage to play?

Insurance. *The business of selling insurance is based on probability and the law of large numbers. Consumers (including businesses) buy insurance because we all face risks that are unlikely but carry high cost. Think of a fire destroying your home. So we form a group to share the risk: we all pay a small amount, and the insurance policy pays a large amount to those few of us whose homes burn down. The insurance company sells many policies, so it can rely on the law of large numbers. Exercises 4.77 to 4.82 explore aspects of insurance.*

4.77 An insurance company looks at the records for millions of homeowners and sees that the mean loss from fire in a year is $\mu = \$250$ per person. (Most of us have no loss, but a few lose their homes. The \$250 is the average loss.) The company plans to sell fire insurance for \$250 plus enough to cover its costs and profit. Explain clearly why it would be stupid to sell only

12 policies. Then explain why selling thousands of such policies is a safe business.

4.78 In fact, the insurance company sees that in the entire population of homeowners, the mean loss from fire is $\mu = \$250$ and the standard deviation of the loss is $\sigma = \$1000$. What are the mean and standard deviation of the total loss for 12 policies? (Losses on separate policies are independent.) What are the mean and standard deviation of the average loss for 12 policies?

4.79 A life insurance company sells a term insurance policy to a 21-year-old male that pays \$100,000 if the insured dies within the next 5 years. The probability that a randomly chosen male will die each year can be found in mortality tables. The company collects a premium of \$250 each year as payment for the insurance. The amount X that the company earns on this policy is \$250 per year, less the \$100,000 that it must pay if the insured dies. Here is the distribution of X. Fill in the missing probability in the table and calculate the mean earnings μ_X.

Age at death	21	22	23	24	25	≥ 26
Earnings X	−\$99,750	−\$99,500	−\$99,250	−\$99,000	−\$98,750	\$1250
Probability	0.00183	0.00186	0.00189	0.00191	0.00193	?

4.80 It would be quite risky for you to insure the life of a 21-year-old friend under the terms of Exercise 4.79. There is a high probability that your friend would live and you would gain \$1250 in premiums. But if he were to die, you would lose almost \$100,000. Explain carefully why selling insurance is not risky for an insurance company that insures many thousands of 21-year-old men.

4.81 We have seen that the risk of an investment is often measured by the standard deviation of the return on the investment. The more variable the return is (the larger σ is), the riskier the investment. We can measure the great risk of insuring one person's life in Exercise 4.79 by computing the standard deviation of the income X that the insurer will receive. Find σ_X, using the distribution and mean you found in Exercise 4.79.

4.82 The risk of insuring one person's life is reduced if we insure many people. Use the result of the previous exercise and the rules for means and variances to answer the following questions.

(a) Suppose that we insure two 21-year-old males, and that their ages at death are independent. If X and Y are the insurer's income from the two insurance policies, the insurer's average income on the two policies is

$$Z = \frac{X + Y}{2} = 0.5X + 0.5Y$$

Find the mean and standard deviation of Z. You see that the mean income is the same as for a single policy but the standard deviation is less.

(b) If four 21-year-old men are insured, the insurer's average income is

$$Z = \frac{1}{4}(X_1 + X_2 + X_3 + X_4)$$

where X_i is the income from insuring one man. The X_i are independent and each has the same distribution as before. Find the mean and standard deviation of Z. Compare your results with the results of (a). We see that averaging over many insured individuals reduces risk.

Portfolio analysis. *Here are the means, standard deviations, and correlations for the monthly returns from three Fidelity mutual funds for the 36 months ending in December 2000.*[14] *Because there are now three random variables, there are three correlations. We use subscripts to show which pair of random variables a correlation refers to.*

W = monthly return on Magellan Fund	$\mu_W = 1.14\%$	$\sigma_W = 4.64\%$
X = monthly return on Real Estate Fund	$\mu_X = 0.16\%$	$\sigma_X = 3.61\%$
Y = monthly return on Japan Fund	$\mu_Y = 1.59\%$	$\sigma_Y = 6.75\%$

Correlations

$$\rho_{WX} = 0.19 \qquad \rho_{WY} = 0.54 \qquad \rho_{XY} = -0.17$$

Exercises 4.83 to 4.85 make use of these historical data.

4.83 Many advisors recommend using roughly 20% foreign stocks to diversify portfolios of U.S. stocks. Michael owns Fidelity Magellan Fund, which concentrates on stocks of large American companies. He decides to move to a portfolio of 80% Magellan and 20% Fidelity Japan Fund. Show that (based on historical data) this portfolio has both a *higher* mean return and *less* volatility than Magellan alone. This illustrates the beneficial effects of diversifying among investments.

4.84 Diversification works better when the investments in a portfolio have small correlations. To demonstrate this, suppose that returns on Magellan Fund and Japan Fund had the means and standard deviations we have given but were uncorrelated ($\rho_{WY} = 0$). Show that the standard deviation of a portfolio that combines 80% Magellan with 20% Japan is then smaller than your result from the previous exercise. What happens to the mean return if the correlation is 0?

4.85 Portfolios often contain more than two investments. The rules for means and variances continue to apply, though the arithmetic gets messier. A portfolio containing proportions a of Magellan Fund, b of Real Estate

Fund, and c of Japan Fund has return $R = aW + bX + cY$. Because a, b, and c are the proportions invested in the three funds, $a + b + c = 1$. The mean and variance of the portfolio return R are

$$\mu_R = a\mu_W + b\mu_X + c\mu_Y$$
$$\sigma_R^2 = a^2\sigma_W^2 + b^2\sigma_X^2 + c^2\sigma_Y^2 + 2ab\rho_{WX}\sigma_W\sigma_X + 2ac\rho_{WY}\sigma_W\sigma_Y + 2bc\rho_{XY}\sigma_X\sigma_Y$$

Having seen the advantages of diversification, Michael decides to invest his funds 60% in Magellan, 20% in Real Estate, and 20% in Japan. What are the (historical) mean and standard deviation of the monthly returns for this portfolio?

4.5 General Probability Rules*

Our study of probability has concentrated on random variables and their distributions. Now we return to the laws that govern any assignment of probabilities. The purpose of learning more laws of probability is to be able to give probability models for more complex random phenomena. We have already met and used five rules.

Rules of Probability

Rule 1. $0 \le P(A) \le 1$ for any event A

Rule 2. $P(S) = 1$

Rule 3. Complement rule: For any event A,

$$P(A^c) = 1 - P(A)$$

Rule 4. Addition rule: If A and B are **disjoint** events, then

$$P(A \text{ or } B) = P(A) + P(B)$$

Rule 5. Multiplication rule: If A and B are **independent** events, then

$$P(A \text{ and } B) = P(A)P(B)$$

General addition rules

Probability has the property that if A and B are disjoint events, then $P(A \text{ or } B) = P(A) + P(B)$. What if there are more than two events, or if the events are not disjoint? These circumstances are covered by more general addition rules for probability.

*This section extends the rules of probability discussed in Section 4.2. This material is not needed for understanding the statistical methods in later chapters. It can therefore be omitted if desired.

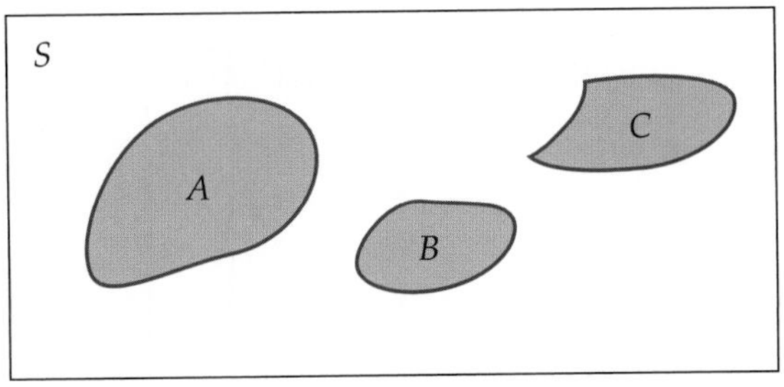

FIGURE 4.15 The addition rule for disjoint events: $P(A \text{ or } B \text{ or } C) = P(A) + P(B) + P(C)$ when events A, B, and C are disjoint.

Union

The **union** of any collection of events is the event that at least one of the collection occurs.

For two events A and B, the union is the event $\{A \text{ or } B\}$ that A or B or both occur. From the addition rule for two disjoint events we can obtain rules for more general unions. Suppose first that we have several events—say A, B, and C—that are disjoint in pairs. That is, no two can occur simultaneously. The Venn diagram in Figure 4.15 illustrates three disjoint events. The addition rule for two disjoint events extends to the following law:

Addition Rule for Disjoint Events

If events A, B, and C are disjoint in the sense that no two have any outcomes in common, then

$$P(\text{one or more of } A, B, C) = P(A) + P(B) + P(C)$$

This rule extends to any number of disjoint events.

EXAMPLE 4.28 Generate a random number X between 0 and 1. What is the probability that the first digit will be odd? The random number X is a continuous random variable whose density curve has constant height 1 between 0 and 1 and is 0 elsewhere. The event that the first digit of X is odd is the union of five disjoint events. These events are

$$0.10 \le X < 0.20$$
$$0.30 \le X < 0.40$$
$$0.50 \le X < 0.60$$
$$0.70 \le X < 0.80$$
$$0.90 \le X < 1.00$$

Figure 4.16 illustrates the probabilities of these events as areas under the density curve. Each has probability 0.1 equal to its length. The union of the five therefore has probability equal to the sum, or 0.5. As we should expect, a random number is equally likely to begin with an odd or an even digit.

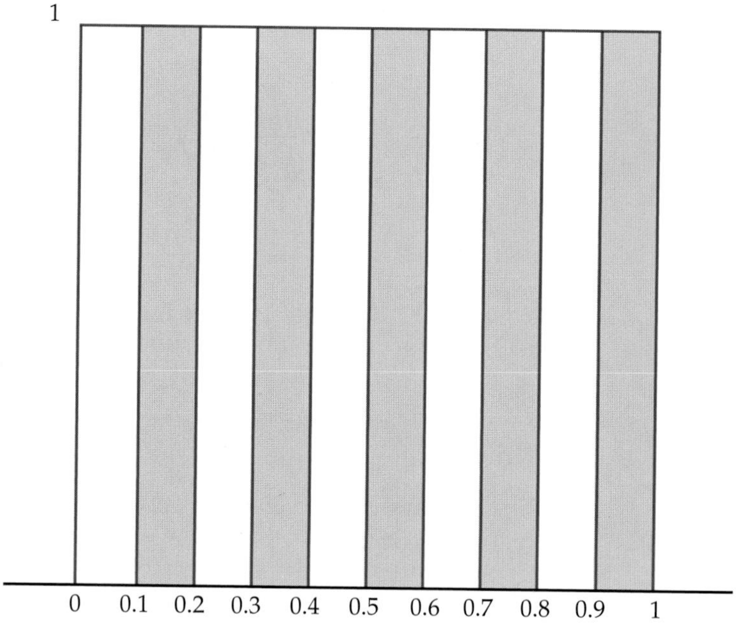

FIGURE 4.16 The probability that the first digit of a random number is odd is the sum of the probabilities of the 5 disjoint events shown.

If events A and B are not disjoint, they can occur simultaneously. The probability of their union is then *less* than the sum of their probabilities. As Figure 4.17 suggests, the outcomes common to both are counted twice when we add probabilities, so we must subtract this probability once. Here is the addition rule for the union of any two events, disjoint or not.

General Addition Rule for Unions of Two Events

For any two events A and B,

$$P(A \text{ or } B) = P(A) + P(B) - P(A \text{ and } B)$$

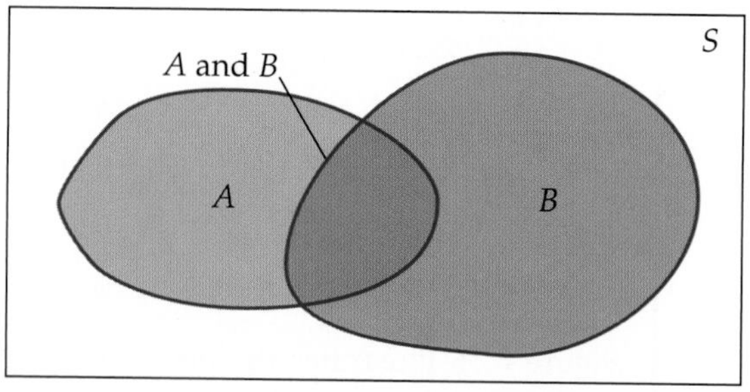

FIGURE 4.17 The general addition rule: $P(A \text{ or } B) = P(A) + P(B) - P(A \text{ and } B)$ for any events A and B.

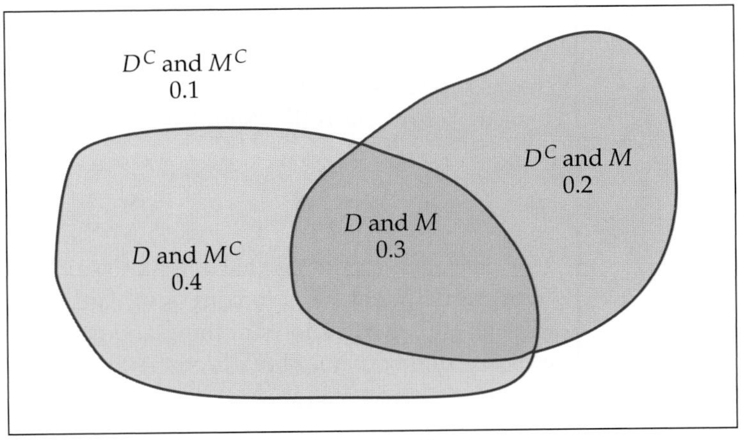

D = Deborah is made partner
M = Matthew is made partner

FIGURE 4.18 Venn diagram and probabilities for Example 4.29.

If A and B are disjoint, the event $\{A \text{ and } B\}$ that both occur has no outcomes in it. This *empty event* is the complement of the sample space S and must have probability 0. So the general addition rule includes Rule 4, the addition rule for disjoint events.

EXAMPLE 4.29

Deborah and Matthew are anxiously awaiting word on whether they have been made partners of their law firm. Deborah guesses that her probability of making partner is 0.7 and that Matthew's is 0.5. (These are personal probabilities reflecting Deborah's assessment of chance.) This assignment of probabilities does not give us enough information to compute the probability that at least one of the two is promoted. In particular, adding the individual probabilities of promotion gives the impossible result 1.2. If Deborah also guesses that the probability that *both* she and Matthew are made partners is 0.3, then by the addition rule for unions

$$P(\text{at least one is promoted}) = 0.7 + 0.5 - 0.3 = 0.9$$

The probability that *neither* is promoted is then 0.1 by the complement rule.

Venn diagrams are a great help in finding probabilities for unions, because you can just think of adding and subtracting areas. Figure 4.18 shows some events and their probabilities for Example 4.29. What is the probability that Deborah is promoted and Matthew is not? The Venn diagram shows that this is the probability that Deborah is promoted minus the probability that both are promoted, $0.7 - 0.3 = 0.4$. Similarly, the probability that Matthew is promoted and Deborah is not is $0.5 - 0.3 = 0.2$. The four probabilities that appear in the figure add to 1 because they refer to four disjoint events whose union is the entire sample space.

Conditional probability

The probability we assign to an event can change if we know that some other event has occurred. This idea is the key to many applications of probability.

EXAMPLE 4.30

Slim is a professional poker player. He stares at the dealer, who prepares to deal. What is the probability that the card dealt to Slim is an ace? There are 52 cards in the deck. Because the deck was carefully shuffled, the next card dealt is equally likely to be any of the cards that Slim has not seen. Four of the 52 cards are aces. So

$$P(\text{ace}) = \frac{4}{52} = \frac{1}{13}$$

This calculation assumes that Slim knows nothing about any cards already dealt. Suppose now that he is looking at 4 cards already in his hand, and that one of them is an ace. He knows nothing about the other 48 cards except that exactly 3 aces are among them. Slim's probability of being dealt an ace *given what he knows* is now

$$P(\text{ace} \mid 1 \text{ ace in 4 visible cards}) = \frac{3}{48} = \frac{1}{16}$$

Knowing that there is 1 ace among the 4 cards Slim can see changes the probability that the next card dealt is an ace.

conditional probability

The new notation $P(A \mid B)$ is a **conditional probability.** That is, it gives the probability of one event (the next card dealt is an ace) under the condition that we know another event (exactly 1 of the 4 visible cards is an ace). You can read the bar | as "given the information that."

In Example 4.30 we could find probabilities because we are willing to use an equally likely probability model for a shuffled deck of cards. Here is an example based on data.

EXAMPLE 4.31

Table 4.1 shows the marital status of adult women broken down by age group. We are interested in the probability that a randomly chosen woman is married. It is common sense that knowing her age group will change the probability: many young women have not married, most middle-aged women are married, and older women are more likely to be widows. To help us think carefully, let's define two events:

A = the woman chosen is young, ages 18 to 29

B = the woman chosen is married

TABLE 4.1 Age and marital status of women (thousands of women)

	Age			
	18 to 29	30 to 64	65 and over	Total
Married	7,842	43,808	8,270	59,920
Never married	13,930	7,184	751	21,865
Widowed	36	2,523	8,385	10,944
Divorced	704	9,174	1,263	11,141
Total	22,512	62,689	18,669	103,870

There are (in thousands) 103,870 adult women in the United States. Of these women, 22,512 are aged 18 to 29. Choosing at random gives each woman an equal chance, so the probability of choosing a young woman is

$$P(A) = \frac{22{,}512}{103{,}870} = 0.217$$

The table shows that there are 7,842 thousand young married women. So the probability that we choose a woman who is both young and married is

$$P(A \text{ and } B) = \frac{7{,}842}{103{,}870} = 0.075$$

To find the *conditional* probability that a woman is married *given the information* that she is young, look only at the "18 to 29" column. The young women are all in this column, so the information given says that only this column is relevant. The conditional probability is

$$P(B \mid A) = \frac{7{,}842}{22{,}512} = 0.348$$

You can verify that the conditional probability that a woman is married when we know she is under age 30 is much lower than the probability for a randomly chosen woman.

It is easy to confuse the three probabilities in Example 4.31. Look carefully at Table 4.1 and be sure you understand the example. There is a relationship among these three probabilities. The probability that a woman is both young *and* married is the product of the probabilities that she is young and that she is married *given* that she is young. That is,

$$\begin{aligned} P(A \text{ and } B) &= P(A) \times P(B \mid A) \\ &= \frac{22{,}512}{103{,}870} \times \frac{7{,}842}{22{,}512} \\ &= \frac{7{,}842}{103{,}870} = 0.075 \quad \text{(as before)} \end{aligned}$$

Try to think your way through this in words: First, the woman is young; then, given that she is young, she is married. We have just discovered the fundamental multiplication rule of probability.

Multiplication Rule

The probability that both of two events A and B happen together can be found by

$$P(A \text{ and } B) = P(A)P(B \mid A)$$

Here $P(B \mid A)$ is the conditional probability that B occurs given the information that A occurs.

EXAMPLE 4.32

Slim is still at the poker table. At the moment, he wants very much to draw two diamonds in a row. As he sits at the table looking at his hand and at the upturned cards on the table, Slim sees 11 cards. Of these, 4 are diamonds. The full deck contains

13 diamonds among its 52 cards, so 9 of the 41 unseen cards are diamonds. To find Slim's probability of drawing two diamonds, first calculate

$$P(\text{first card diamond}) = \frac{9}{41}$$

$$P(\text{second card diamond}|\text{first card diamond}) = \frac{8}{40}$$

Slim finds both probabilities by counting cards. The probability that the first card drawn is a diamond is 9/41 because 9 of the 41 unseen cards are diamonds. If the first card is a diamond, that leaves 8 diamonds among the 40 remaining cards. So the *conditional* probability of another diamond is 8/40. The multiplication rule now says that

$$P(\text{both cards diamonds}) = \frac{9}{41} \times \frac{8}{40} = 0.044$$

Slim will need luck to draw his diamonds.

If $P(A)$ and $P(A \text{ and } B)$ are given, we can rearrange the multiplication rule to produce a *definition* of the conditional probability $P(B|A)$ in terms of unconditional probabilities.

Definition of Conditional Probability

When $P(A) > 0$, the **conditional probability** of B given A is

$$P(B|A) = \frac{P(A \text{ and } B)}{P(A)}$$

Be sure to keep in mind the distinct roles in $P(B|A)$ of the event B whose probability we are computing and the event A that represents the information we are given. The conditional probability $P(B|A)$ makes no sense if the event A can never occur, so we require that $P(A) > 0$ whenever we talk about $P(B|A)$.

EXAMPLE 4.33

What is the conditional probability that a woman is a widow, given that she is at least 65 years old? We see from Table 4.1 that

$$P(\text{at least 65}) = \frac{18{,}669}{103{,}870} = 0.180$$

$$P(\text{widowed } \textit{and} \text{ at least 65}) = \frac{8{,}385}{103{,}870} = 0.081$$

The conditional probability is therefore

$$P(\text{widowed} \mid \text{at least 65}) = \frac{P(\text{widowed } \textit{and} \text{ at least 65})}{P(\text{at least 65})} = \frac{0.081}{0.180} = 0.450$$

Check that this agrees (up to roundoff error) with the result obtained from the "65 and over" column of Table 4.1:

$$P(\text{widowed} \mid \text{at least } 65) = \frac{8{,}385}{18{,}669} = 0.449$$

General multiplication rules

The definition of conditional probability reminds us that in principle all probabilities, including conditional probabilities, can be found from the assignment of probabilities to events that describes a random phenomenon. More often, however, conditional probabilities are part of the information given to us in a probability model, and the multiplication rule is used to compute $P(A \text{ and } B)$. This rule extends to more than two events.

The union of a collection of events is the event that *any* of them occur. Here is the corresponding term for the event that *all* of them occur.

Intersection

The **intersection** of any collection of events is the event that *all* of the events occur.

To extend the multiplication rule to the probability that all of several events occur, the key is to condition each event on the occurrence of *all* of the preceding events. For example, the intersection of three events A, B, and C has probability

$$P(A \text{ and } B \text{ and } C) = P(A)P(B \mid A)P(C \mid A \text{ and } B)$$

EXAMPLE 4.34

Only 5% of male high school basketball, baseball, and football players go on to play at the college level. Of these, only 1.7% enter major league professional sports. About 40% of the athletes who compete in college and then reach the pros have a career of more than 3 years.[15] Define these events:

$$\begin{aligned} A &= \{\text{competes in college}\} \\ B &= \{\text{competes professionally}\} \\ C &= \{\text{pro career longer than 3 years}\} \end{aligned}$$

What is the probability that a high school athlete competes in college and then goes on to have a pro career of more than 3 years? We know that

$$\begin{aligned} P(A) &= 0.05 \\ P(B \mid A) &= 0.017 \\ P(C \mid A \text{ and } B) &= 0.4 \end{aligned}$$

The probability we want is therefore

$$\begin{aligned} P(A \text{ and } B \text{ and } C) &= P(A)P(B \mid A)P(C \mid A \text{ and } B) \\ &= 0.05 \times 0.017 \times 0.4 = 0.00034 \end{aligned}$$

Only about 3 of every 10,000 high school athletes can expect to compete in college and have a professional career of more than 3 years. High school students would be wise to concentrate on studies rather than on unrealistic hopes of fortune from pro sports.

Tree diagrams

Probability problems often require us to combine several of the basic rules into a more elaborate calculation. Here is an example that illustrates how to solve problems that have several stages.

EXAMPLE 4.35

tree diagram

What is the probability that a male high school athlete will go on to professional sports? In the notation of Example 4.34, this is $P(B)$. To find $P(B)$ from the information in Example 4.34, use the **tree diagram** in Figure 4.19 to organize your thinking.

Each segment in the tree is one stage of the problem. Each complete branch shows a path that an athlete can take. The probability written on each segment is the conditional probability that an athlete follows that segment given that he has reached the point from which it branches. Starting at the left, high school athletes either do or do not compete in college. We know that the probability of competing in college is $P(A) = 0.05$, so the probability of not competing is $P(A^c) = 0.95$. These probabilities mark the leftmost branches in the tree.

Conditional on competing in college, the probability of playing professionally is $P(B \mid A) = 0.017$. So the conditional probability of *not* playing professionally is

$$P(B^c \mid A) = 1 - P(B \mid A) = 1 - 0.017 = 0.983$$

These conditional probabilities mark the paths branching out from A in Figure 4.19.

The lower half of the tree diagram describes athletes who do not compete in college (A^c). It is unusual for these athletes to play professionally, but a few go straight

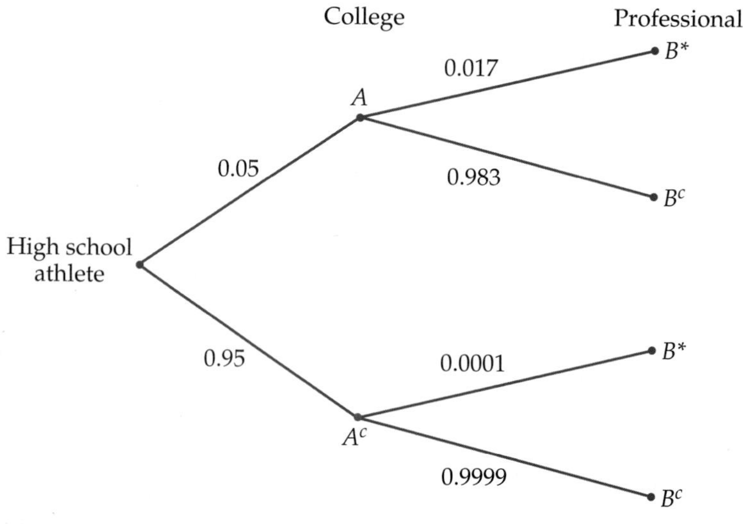

FIGURE 4.19 Tree diagram for Example 4.35. The probability $P(B)$ is the sum of the probabilities of the two branches marked with asterisks (*).

from high school to professional leagues. Suppose that the conditional probability that a high school athlete reaches professional play given that he does not compete in college is $P(B \mid A^c) = 0.0001$. We can now mark the two paths branching from A^c in Figure 4.19.

There are two disjoint paths to B (professional play). By the addition rule, $P(B)$ is the sum of their probabilities. The probability of reaching B through college (top half of the tree) is

$$\begin{aligned} P(B \text{ and } A) &= P(A)P(B \mid A) \\ &= 0.05 \times 0.017 = 0.00085 \end{aligned}$$

The probability of reaching B without college is

$$\begin{aligned} P(B \text{ and } A^c) &= P(A^c)P(B \mid A^c) \\ &= 0.95 \times 0.0001 = 0.000095 \end{aligned}$$

The final result is

$$P(B) = 0.00085 + 0.000095 = 0.000945$$

About 9 high school athletes out of 10,000 will play professional sports.

Tree diagrams combine the addition and multiplication rules. The multiplication rule says that the probability of reaching the end of any complete branch is the product of the probabilities written on its segments. The probability of any outcome, such as the event B that an athlete reaches professional sports, is then found by adding the probabilities of all branches that are part of that event.

Bayes's rule

There is another kind of probability question that we might ask in the context of studies of athletes. Our earlier calculations look forward toward professional sports as the final stage of an athlete's career. Now let's concentrate on professional athletes and look back at their earlier careers.

EXAMPLE 4.36

What proportion of professional athletes competed in college? In the notation of Examples 4.34 and 4.35 this is the conditional probability $P(A \mid B)$. We start from the definition of conditional probability and then apply the results of Example 4.35:

$$\begin{aligned} P(A \mid B) &= \frac{P(A \text{ and } B)}{P(B)} \\ &= \frac{0.00085}{0.000945} = 0.8995 \end{aligned}$$

Almost 90% of professional athletes competed in college.

We know the probabilities $P(A)$ and $P(A^c)$ that a high school athlete does and does not compete in college. We also know the conditional probabilities $P(B \mid A)$ and $P(B \mid A^c)$ that an athlete from each group reaches professional sports. Example 4.35 shows how to use this information to calculate $P(B)$.

The method can be summarized in a single expression that adds the probabilities of the two paths to B in the tree diagram:

$$P(B) = P(A)P(B \mid A) + P(A^c)P(B \mid A^c)$$

In Example 4.36 we calculated the "reverse" conditional probability $P(A \mid B)$. The denominator 0.000945 in that example came from the expression just above. Put in this general notation, we have another probability law.

Bayes's Rule

If A and B are any events whose probabilities are not 0 or 1,

$$P(A \mid B) = \frac{P(B \mid A)P(A)}{P(B \mid A)P(A) + P(B \mid A^c)P(A^c)}$$

Bayes's rule is named after Thomas Bayes, who wrestled with arguing from outcomes like B back to antecedents like A in a book published in 1763. It is far better to think your way through problems like Examples 4.35 and 4.36 rather than memorize these formal expressions.

Independence again

The conditional probability $P(B \mid A)$ is generally not equal to the unconditional probability $P(B)$. That is because the occurrence of event A generally gives us some additional information about whether or not event B occurs. If knowing that A occurs gives no additional information about B, then A and B are independent events. The formal definition of independence is expressed in terms of conditional probability.

Independent Events

Two events A and B that both have positive probability are **independent** if

$$P(B \mid A) = P(B)$$

This definition makes precise the informal description of independence given in Section 4.1. We now see that the multiplication rule for independent events, $P(A \text{ and } B) = P(A)P(B)$, is a special case of the general multiplication rule, $P(A \text{ and } B) = P(A)P(B \mid A)$, just as the addition rule for disjoint events is a special case of the general addition rule.

Decision analysis

One kind of decision making in the presence of uncertainty seeks to make the probability of a favorable outcome as large as possible. Here is an example that illustrates how the multiplication and addition rules, organized with the help of a tree diagram, apply to a decision problem.

EXAMPLE 4.37

Lynn has end-stage kidney disease: her kidneys have failed, so that she cannot survive unaided. Only about 52% of patients survive for three years with kidney dialysis. Fortunately, a kidney is available for transplant. Lynn's doctor gives her the following information for patients in her condition.

Transplant operations usually succeed. After one month, 96% of the transplanted kidneys are functioning. Three percent fail to function, and the patient must return to dialysis. The remaining 1% of the patients die within a month. Patients who return to dialysis have the same chance (52%) of surviving three years as if they had not attempted a transplant.

Of the successful transplants, however, only 82% continue to function for three years. Another 8% of these patients must return to dialysis, and 70% of these survive to the three-year mark. The remaining 10% of "successful" patients die without returning to dialysis.[16]

There is too much information here to sort through without a tree diagram. The key is to realize that most of the percents that Lynn's doctor gives her are conditional probabilities given that a patient has some specific prior history. Figure 4.20 is a tree diagram that organizes the information.

Each path through the tree represents a possible outcome of Lynn's case. The probability written beside each branch after the first stage is the conditional probability of the next step given that Lynn has reached this point. For example, 0.82 is the conditional probability that a patient whose transplant succeeded survives three years with the transplant still functioning. The conditional probabilities of the other three-year outcomes for a successful transplant are 0.08 and 0.10. They appear on the other branches from the "Success" node. These three conditional probabilities add to 1 because these are all the possible outcomes following a successful transplant. Study the tree to convince yourself that it organizes all the information available.

The multiplication rule says that the probability of reaching the end of any path is the product of all the probabilities along that path. For example,

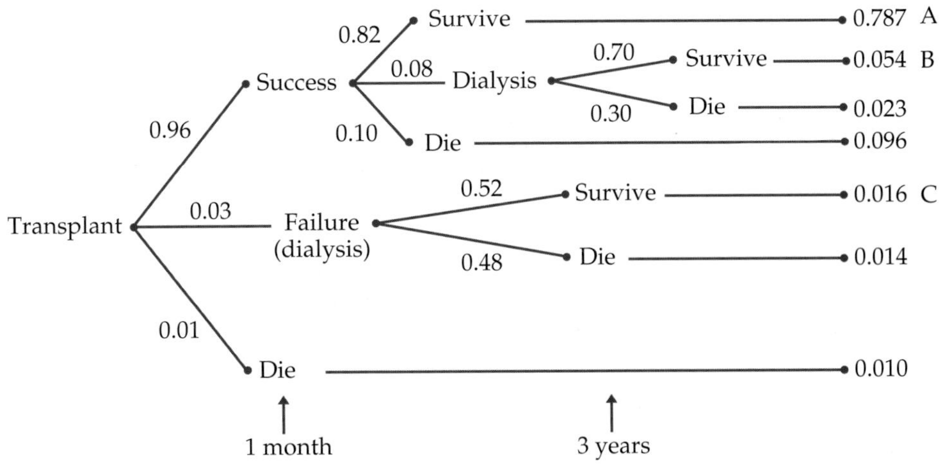

FIGURE 4.20 Tree diagram for the outcomes of a kidney transplant in Example 4.37.

look at the path marked A at the right of Figure 4.20. The probability that a transplant succeeds and endures for three years is

$$\begin{aligned} P(\text{succeeds and lasts 3 years}) &= P(\text{succeeds})P(\text{lasts 3 years} \mid \text{succeeds}) \\ &= (0.96)(0.82) = 0.787 \end{aligned}$$

Similarly, the path marked B is the event that a patient's transplant succeeds at the one-month stage, fails before three years, and the patient nonetheless survives to three years after returning to dialysis. The probability of this is

$$P(B) = (0.96)(0.08)(0.70) = 0.054$$

The probabilities at the end of all the paths in Figure 4.20 add to 1 because these are all the possible three-year outcomes.

What is the probability that Lynn will survive for three years if she has a transplant? This is the union of the three disjoint events marked A, B, and C in Figure 4.20. By the addition rule,

$$\begin{aligned} P(\text{survive}) &= P(A) + P(B) + P(C) \\ &= 0.787 + 0.054 + 0.016 = 0.857 \end{aligned}$$

Lynn's decision is easy: 0.857 is much higher than the probability 0.52 of surviving three years on dialysis. She will elect the transplant.

Where do the conditional probabilities in Example 4.37 come from? They are based in part on data—that is, on studies of many patients with kidney disease. But an individual's chances of survival depend on her age, general health, and other factors. Lynn's doctor considered her individual situation before giving her these particular probabilities. It is characteristic of most decision analysis problems that *personal probabilities* are used to describe the uncertainty of an informed decision maker.

SUMMARY

The **complement** A^c of an event A contains all outcomes that are not in A. The **union** $\{A \text{ or } B\}$ of events A and B contains all outcomes in A, in B, or in both A and B. The **intersection** $\{A \text{ and } B\}$ contains all outcomes that are in both A and B, but not outcomes in A alone or B alone.

The **conditional probability** $P(B \mid A)$ of an event B given an event A is defined by

$$P(B \mid A) = \frac{P(A \text{ and } B)}{P(A)}$$

when $P(A) > 0$. In practice, conditional probabilities are most often found from directly available information.

The essential general rules of elementary probability are

Legitimate values: $0 \leq P(A) \leq 1$ for any event A

Total probability 1: $P(S) = 1$

Complement rule: $P(A^c) = 1 - P(A)$

Addition rule: $P(A \text{ or } B) = P(A) + P(B) - P(A \text{ and } B)$

Multiplication rule: $P(A \text{ and } B) = P(A)P(B \mid A)$

If A and B are **disjoint,** then $P(A \text{ and } B) = 0$. The general addition rule for unions then becomes the special addition rule, $P(A \text{ or } B) = P(A) + P(B)$.

A and B are **independent** when $P(B \mid A) = P(B)$. The multiplication rule for intersections then becomes $P(A \text{ and } B) = P(A)P(B)$.

In problems with several stages, draw a **tree diagram** to organize use of the multiplication and addition rules.

SECTION 4.5 EXERCISES

4.86 Call a household prosperous if its income exceeds \$100,000. Call the household educated if the householder completed college. Select an American household at random, and let A be the event that the selected household is prosperous and B the event that it is educated. According to the Current Population Survey, $P(A) = 0.134$, $P(B) = 0.254$, and the probability that a household is both prosperous and educated is $P(A \text{ and } B) = 0.080$. What is the probability $P(A \text{ or } B)$ that the household selected is either prosperous or educated?

4.87 In the setting of the previous exercise, what is the conditional probability that a household is prosperous, given that it is educated? Explain why your result shows that events A and B are not independent.

4.88 Draw a Venn diagram that shows the relation between the events A and B in Exercise 4.86. Indicate each of the following events on your diagram and use the information in Exercise 4.86 to calculate the probability of each event. Finally, describe in words what each event is.

(a) $\{A \text{ and } B\}$

(b) $\{A \text{ and } B^c\}$

(c) $\{A^c \text{ and } B\}$

(d) $\{A^c \text{ and } B^c\}$

4.89 Consolidated Builders has bid on two large construction projects. The company president believes that the probability of winning the first contract (event A) is 0.6, that the probability of winning the second (event B) is 0.5, and that the probability of winning both jobs (event $\{A \text{ and } B\}$) is 0.3. What is the probability of the event $\{A \text{ or } B\}$ that Consolidated will win at least one of the jobs?

4.90 In the setting of the previous exercise, are events A and B independent? Do a calculation that proves your answer.

4.91 Draw a Venn diagram that illustrates the relation between events A and B in Exercise 4.89. Write each of the following events in terms of A, B, A^c, and B^c. Indicate the events on your diagram and use the information in Exercise 4.89 to calculate the probability of each.

(a) Consolidated wins both jobs.

(b) Consolidated wins the first job but not the second.

(c) Consolidated does not win the first job but does win the second.

(d) Consolidated does not win either job.

4.92 Choose an adult American woman at random. Table 4.1 (page 344) describes the population from which we draw. Use the information in that table to answer the following questions.

(a) What is the probability that the woman chosen is 65 years old or older?

(b) What is the conditional probability that the woman chosen is married, given that she is 65 or over?

(c) How many women are *both* married and in the 65 and over age group? What is the probability that the woman we choose is a married woman at least 65 years old?

(d) Verify that the three probabilities you found in (a), (b), and (c) satisfy the multiplication rule.

4.93 Choose an adult American woman at random. Table 4.1 describes the population from which we draw.

(a) What is the conditional probability that the woman chosen is 18 to 29 years old, given that she is married?

(b) In Example 4.31 we found that $P(\text{married} \mid \text{age 18 to 29}) = 0.348$. Complete this sentence: 0.348 is the proportion of women who are ________ among those women who are ________.

(c) In (a), you found $P(\text{age 18 to 29} \mid \text{married})$. Write a sentence of the form given in (b) that describes the meaning of this result. The two conditional probabilities give us very different information.

4.94 Here are the counts (in thousands) of earned degrees in the United States in the 2001–2002 academic year, classified by level and by the sex of the degree recipient:[17]

	Bachelor's	Master's	Professional	Doctorate	Total
Female	645	227	32	18	922
Male	505	161	40	26	732
Total	1150	388	72	44	1654

(a) If you choose a degree recipient at random, what is the probability that the person you choose is a woman?

(b) What is the conditional probability that you choose a woman, given that the person chosen received a professional degree?

(c) Are the events "choose a woman" and "choose a professional degree recipient" independent? How do you know?

4.95 The previous exercise gives the counts (in thousands) of earned degrees in the United States in the 2001–2002 academic year. Use these data to answer the following questions.

(a) What is the probability that a randomly chosen degree recipient is a man?

(b) What is the conditional probability that the person chosen received a bachelor's degree, given that he is a man?

(c) Use the multiplication rule to find the probability of choosing a male bachelor's degree recipient. Check your result by finding this probability directly from the table of counts.

4.96 Common sources of caffeine in the diet are coffee, tea, and cola drinks. Suppose that

55% of adults drink coffee

25% of adults drink tea

45% of adults drink cola

and also that

15% drink both coffee and tea

5% drink all three beverages

25% drink both coffee and cola

5% drink only tea

Draw a Venn diagram marked with this information. Use it along with the addition rules to answer the following questions.

(a) What percent of adults drink only cola?

(b) What percent drink none of these beverages?

Working. *In the language of government statistics, you are "in the labor force" if you are available for work and either working or actively seeking work. The unemployment rate is the proportion of the labor force (not of the entire population) who are unemployed. Here are data from the Current Population Survey for the civilian population aged 25 years and over. The table entries are counts in thousands of people. Exercises 4.97 to 4.99 concern these data.*

Highest education	Total population	In labor force	Employed
Did not finish high school	27,325	12,073	11,139
High school but no college	57,221	36,855	35,137
Less than bachelor's degree	45,471	33,331	31,975
College graduate	47,371	37,281	36,259

4.97 Find the unemployment rate for people with each level of education. How does the unemployment rate change with education? Explain carefully why your results show that level of education and being employed are not independent.

4.98 **(a)** What is the probability that a randomly chosen person 25 years of age or older is in the labor force?

(b) If you know that the person chosen is a college graduate, what is the conditional probability that he or she is in the labor force?

(c) Are the events "in the labor force" and "college graduate" independent? How do you know?

4.99 You know that a person is employed. What is the conditional probability that he or she is a college graduate? You know that a second person is a college graduate. What is the conditional probability that he or she is employed?

4.100 Functional Robotics Corporation buys electrical controllers from a Japanese supplier. The company's treasurer thinks that there is probability 0.4 that the dollar will fall in value against the Japanese yen in the next month. The probability that the supplier will demand that the contract be renegotiated is 0.8 if the dollar falls, and 0.2 if the dollar does not fall. What is the probability that the supplier will demand renegotiation? (Use a tree diagram to organize the information given.)

4.101 A telemarketing company calls telephone numbers chosen at random. It finds that 70% of calls are not completed (the party does not answer or refuses to talk), that 20% result in talking to a woman, and that 10% result in talking to a man. After that point, 30% of the women and 20% of the men actually buy something. What percent of calls result in a sale?

4.102 The voters in a large city are 40% white, 40% black, and 20% Hispanic. (Hispanics may be of any race in official statistics, but in this case we are speaking of political blocs.) A black mayoral candidate anticipates attracting 30% of the white vote, 90% of the black vote, and 50% of the Hispanic vote. Draw a tree diagram with probabilities for the race (white, black, or Hispanic) and vote (for or against the candidate) of a randomly chosen voter. What percent of the overall vote does the candidate expect to get?

4.103 In the setting of Exercise 4.101, what percent of sales are made to women? (Write this as a conditional probability.)

4.104 In the election described in Exercise 4.102, what percent of the candidate's votes come from black voters? (Write this as a conditional probability.)

Genetic counseling. *Conditional probabilities and Bayes's rule are a basis for counseling people who may have genetic defects that can be passed to their children. Exercises 4.105 to 4.109 concern genetic counseling settings.*

4.105 **Albinism.** People with albinism have little pigment in their skin, hair, and eyes. The gene that governs albinism has two forms (called alleles), which we denote by a and A. Each person has a pair of these genes, one inherited from each parent. A child inherits one of each parent's two alleles, independently with probability 0.5. Albinism is a recessive trait, so a person is albino only if the inherited pair is aa.

(a) Beth's parents are not albino but she has an albino brother. This implies that both of Beth's parents have type Aa. Why?

(b) Which of the types aa, Aa, AA could a child of Beth's parents have? What is the probability of each type?

(c) Beth is not albino. What are the conditional probabilities for Beth's possible genetic types, given this fact? (Use the definition of conditional probability.)

4.106 **Albinism, continued.** Beth knows the probabilities for her genetic types from part (c) of the previous exercise. She marries Bob, who is albino. Bob's genetic type must be aa.

(a) What is the conditional probability that a child of Beth and Bob is non-albino if Beth has type Aa? What is the conditional probability of a non-albino child if Beth has type AA?

(b) Beth and Bob's first child is non-albino. What is the conditional probability that Beth is a carrier, type Aa?

4.107 **Cystic fibrosis.** Cystic fibrosis is a lung disorder that often results in death. It is inherited but can be inherited only if both parents are carriers of an abnormal gene. In 1989, the CF gene that is abnormal in carriers of cystic fibrosis was identified. The probability that a randomly chosen person of European ancestry carries an abnormal CF gene is 1/25. (The probability is less in other ethnic groups.) The CF20m test detects most but not all harmful mutations of the CF gene. The test is positive for 90% of people who are carriers. It is (ignoring human error) never positive for people who are not carriers. Jason tests positive. What is the probability that he is a carrier?

4.108 **Cystic fibrosis, continued.** Jason knows that he is a carrier of cystic fibrosis. His wife, Julianne, has a brother with cystic fibrosis, which means the probability is 2/3 that she is a carrier. If Julianne is a carrier, each child she has with Jason has probability 1/4 of having cystic fibrosis. If she is not a carrier, her children cannot have the disease. Jason and Julianne have one child, who does not have cystic fibrosis. This information reduces the probability that Julianne is a carrier. Use Bayes's rule to find the conditional probability that Julianne is a carrier, given that she and Jason have one child who does not have cystic fibrosis.

4.109 **Muscular dystrophy.** Muscular dystrophy is an incurable muscle-wasting disease. The most common and serious type, called DMD, is caused by a sex-linked recessive mutation. Specifically: women can be carriers but do not get the disease; a son of a carrier has probability 0.5 of having DMD; a daughter has probability 0.5 of being a carrier. As many as 1/3 of DMD cases, however, are due to spontaneous mutations in sons of mothers who are not carriers. Toni has one son, who has DMD.

In the absence of other information, the probability is 1/3 that the son is the victim of a spontaneous mutation and 2/3 that Toni is a carrier. There is a screening test called the CK test that is positive with probability 0.7 if a woman is a carrier and with probability 0.1 if she is not. Toni's CK test is positive. What is the probability that she is a carrier?

4.110 John has coronary artery disease. He and his doctor must decide between medical management of the disease and coronary bypass surgery. Because John has been quite active, he is concerned about his quality of life as well as the length of life. He wants to make the decision that will maximize the probability of the event A that he survives for 5 years and is able to carry on moderate activity during that time. The doctor makes the following probability estimates for patients of John's age and condition:[18]

- Under medical management, $P(A) = 0.7$.
- There is probability 0.05 that John will not survive bypass surgery, probability 0.10 that he will survive with serious complications, and probability 0.85 that he will survive the surgery without complications.
- If he survives with complications, the conditional probability of the desired outcome A is 0.73. If there are no serious complications, the conditional probability of A is 0.76.

Draw a tree diagram that summarizes this information. Then calculate $P(A)$ assuming that John chooses the surgery. Does surgery or medical management offer him the better chance of achieving his goal?

4.111 Zipdrive, Inc., has developed a new disk drive for small computers. The demand for the new product is uncertain but can be described as "high" or "low" in any one year. After 4 years, the product is expected to be obsolete. Management must decide whether to build a plant or to contract with a factory in Hong Kong to manufacture the new drive. Building a plant will be profitable if demand remains high but could lead to a loss if demand drops in future years.

After careful study of the market and of all relevant costs, Zipdrive's planning office provides the following information. Let A be the event that the first year's demand is high, and B be the event that the following

3 years' demand is high. The marketing division's best estimate of the probabilities is

$$P(A) = 0.9$$
$$P(B \mid A) = 0.36$$
$$P(B \mid A^c) = 0$$

The probability that building a plant is more profitable than contracting the production to Hong Kong is 0.95 if demand is high all 4 years, 0.3 if demand is high only in the first year, and 0.1 if demand is low all 4 years.

Draw a tree diagram that organizes this information. The tree will have three stages: first year's demand, next 3 years' demand, and whether building or contracting is more profitable. Which decision has the higher probability of being more profitable? (When decision analysis is used for investment decisions like this, firms in fact compare the mean profits rather than the probability of a profit. We ignore this complication.)

CHAPTER 4 EXERCISES

4.112 An ancient Korean drinking game involves a 14-sided die. The players roll the die in turn and must submit to whatever humiliation is written on the up-face: something like, "Keep still when tickled on face." Six of the 14 faces are squares. Let's call them A, B, C, D, E, and F for short. The other eight faces are triangles, which we will call 1, 2, 3, 4, 5, 6, 7, and 8. Each of the squares is equally likely. Each of the triangles is also equally likely, but the triangle probability differs from the square probability. The probability of getting a square is 0.72. Give the probability model for the 14 possible outcomes.

4.113 Deal a five-card poker hand from a shuffled deck. The probabilities of several types of hand are approximately as follows:

Hand	Worthless	One pair	Two pairs	Better hands
Probability	0.50	0.42	0.05	?

What must be the probability of getting a hand better than two pairs? What is the probability of getting a hand that is not worthless?

4.114 Rotter Partners is planning a major investment. The amount of profit X is uncertain but a probabilistic estimate gives the following distribution (in millions of dollars):

Profit	1	1.5	2	4	10
Probability	0.1	0.2	0.4	0.2	0.1

(a) Find the mean profit μ_X and the standard deviation σ_X of the profit.

(b) Rotter Partners owes its source of capital a fee of \$200,000 plus 10% of the profits X. So the firm actually retains

$$Y = 0.9X - 0.2$$

from the investment. Find the mean and standard deviation of Y.

4.115 A grocery store gives its customers cards that may win them a prize when matched with other cards. The back of the card announces the following probabilities of winning various amounts if a customer visits the store 10 times:

Amount	\$1000	\$200	\$50	\$10
Probability	1/10,000	1/1000	1/100	1/20

(a) What is the probability of winning nothing?

(b) What is the mean amount won?

(c) What is the standard deviation of the amount won?

4.116 The College Board finds that the distribution of students' SAT scores depends on the level of education their parents have. Children of parents who did not finish high school have SAT math scores X with mean 442 and standard deviation 108. Scores Y of children of parents with graduate degrees have mean 566 and standard deviation 109. Perhaps we should standardize to a common scale for equity. Find positive numbers a, b, c, and d such that $a + bX$ and $c + dY$ both have mean 500 and standard deviation 100.

4.117 You are playing a board game in which the severity of a penalty is determined by rolling three dice and adding the spots on the up-faces. The dice are all balanced so that each face is equally likely, and the three dice fall independently. If X_1, X_2, and X_3 are the number of spots on the up-faces of the three dice, then $X = X_1 + X_2 + X_3$. Use this fact to find the mean μ_X and the standard deviation σ_X without finding the distribution of X. (Start with the distribution of each of the X_i.)

4.118 Slot machines are now video games, with winning determined by electronic random number generators. In the old days, slot machines were like this: you pull the lever to spin three wheels; each wheel has 20 symbols, all equally likely to show when the wheel stops spinning; the three wheels are independent of each other. Suppose that the middle wheel has 9 bells among its 20 symbols, and the left and right wheels have 1 bell each.

(a) You win the jackpot if all three wheels show bells. What is the probability of winning the jackpot?

(b) What is the probability that the wheels stop with exactly 2 bells showing?

4.119 Ann and Bob are playing the game Two-Finger Morra. Each player shows either one or two fingers and at the same time calls out a guess for the number of fingers the other player will show. If a player guesses correctly and the other player does not, the player wins a number of dollars equal to the total number of fingers shown by both players. If both or neither guesses correctly, no money changes hands. On each play both Ann and Bob choose one of the following options:

Choice	Show	Guess
A	1	1
B	1	2
C	2	1
D	2	2

(a) Give the sample space S by writing all possible choices for both players on a single play of this game.

(b) Let X be Ann's winnings on a play. (If Ann loses \$2, then $X = -2$; when no money changes hands, $X = 0$.) Write the value of the random variable X next to each of the outcomes you listed in (a). This is another choice of sample space.

(c) Now assume that Ann and Bob choose independently of each other. Moreover, they both play so that all four choices listed above are equally likely. Find the probability distribution of X.

(d) If the game is fair, X should have mean zero. Does it? What is the standard deviation of X?

The following exercises require familiarity with the material presented in the optional Section 4.5.

4.120 Independence of events is not always obvious. Toss two balanced coins independently. The four possible combinations of heads and tails in order each have probability 0.25. The events

$$A = \text{head on the first toss}$$

$$B = \text{both tosses have the same outcome}$$

may seem intuitively related. Show that $P(B \mid A) = P(B)$ so that A and B are in fact independent.

4.121 Choose a point at random in the square with sides $0 \le x \le 1$ and $0 \le y \le 1$. This means that the probability that the point falls in any region within the square is the area of that region. Let X be the x coordinate and Y the y coordinate of the point chosen. Find the conditional probability $P(Y < 1/2 \mid Y > X)$. (*Hint:* Sketch the square and the events $Y < 1/2$ and $Y > X$.)

4.122 In 1999, the Internal Revenue Service received 127,075,145 individual tax returns. Of these, 9,534,653 reported an adjusted gross income of at least \$100,000 and 205,124 reported at least \$1 million.

(a) What is the probability that a randomly chosen individual tax return reports an income of at least \$100,000? At least \$1 million?

(b) If you know that the return chosen shows an income of \$100,000 or more, what is the conditional probability that the income is at least \$1 million?

4.123 You have torn a tendon and are facing surgery to repair it. The orthopedic surgeon explains the risks to you. Infection occurs in 3% of such operations, the repair fails in 14%, and both infection and failure occur together in 1%. What percent of these operations succeed and are free from infection?

4.124 It is difficult to conduct sample surveys on sensitive issues because many people will not answer questions if the answers might embarrass them. "Randomized response" is an effective way to guarantee anonymity while collecting information on topics such as student cheating or sexual behavior. Here is the idea. To ask a sample of students whether they have plagiarized a term paper while in college, have each student toss a coin in private. If the coin lands heads *and* they have not plagiarized, they are to answer "No." Otherwise they are to give "Yes" as their answer. Only the student knows whether the answer reflects the truth or just the coin toss, but the researchers can use a proper random sample with follow-up for nonresponse and other good sampling practices.

Suppose that in fact the probability is 0.3 that a randomly chosen student has plagiarized a paper. Draw a tree diagram in which the first stage is tossing the coin and the second is the truth about plagiarism. The outcome at the end of each branch is the answer given to the randomized-response question. What is the probability of a "No" answer in the randomized-response poll? If the probability of plagiarism were 0.2, what would be the probability of a "No" response on the poll? Now suppose that you get 39% "No" answers in a randomized-response poll of a large sample of students at your college. What do you estimate to be the percent of the population who have plagiarized a paper?

4.125 Enzyme immunoassay (EIA) tests are used to screen blood specimens for the presence of antibodies to HIV, the virus that causes AIDS. Antibodies indicate the presence of the virus. The test is quite accurate but is not always correct. Here are approximate probabilities of positive and negative EIA outcomes when the blood tested does and does not actually contain antibodies to HIV:[19]

	Test result	
	+	−
Antibodies present	0.9985	0.0015
Antibodies absent	0.0060	0.9940

Suppose that 1% of a large population carries antibodies to HIV in their blood.

(a) Draw a tree diagram for selecting a person from this population (outcomes: antibodies present or absent) and for testing his or her blood (outcomes: EIA positive or negative).

(b) What is the probability that the EIA is positive for a randomly chosen person from this population?

(c) What is the probability that a person has the antibody, given that the EIA test is positive?

(This exercise illustrates a fact that is important when considering proposals for widespread testing for HIV, illegal drugs, or agents of biological warfare: if the condition being tested is uncommon in the population, many positives will be false positives.)

4.126 The previous exercise gives data on the results of EIA tests for the presence of antibodies to HIV. Repeat part (c) of that exercise for two different populations:

(a) Blood donors are prescreened for HIV risk factors, so perhaps only 0.1% (0.001) of this population carries HIV antibodies.

(b) Clients of a drug rehab clinic are a high-risk group, so perhaps 10% of this population carries HIV antibodies.

(c) What general lesson do your calculations illustrate?

Prelude: the house edge, and the real house edge

It's no secret that the expected values of games of chance are set so that the house wins on the average in the long run. In roulette, for example, the expected value of a \$1 bet is \$0.947. The house keeps the other 5.3 cents per dollar bet. Because the house plays hundreds of thousands of times each month, it will keep almost exactly 5.3 cents for every dollar bet. That's a consequence of the law of large numbers. In this chapter, we will say much more. Another big fact of probability theory is the *central limit theorem*, which says that the house's monthly winnings follow a normal distribution quite closely. Tell me how many bets will be placed, and I can use normal probability calculations to put good bounds on the house's take.

What *is* a secret, at least to naive gamblers, is that in the real world a casino does much better than our calculations suggest. In fact, casinos keep not 5% but about 20% of the money gamblers spend on roulette chips. That's because players who win keep on playing. Think of a player who gets back exactly 95% of each dollar bet. After one bet, he has 95 cents; after two bets, he has 95% of that, or 90.25 cents; after three bets, he has 95% of that, or 85.7 cents. The longer he keeps recycling his original dollar, the more of it the casino keeps. Real gamblers don't get a fixed percent back on each bet, but even the luckiest will lose his stake if he plays long enough. The casino keeps 5.3 cents of every dollar bet but 20 cents of every dollar that walks in the door.

That brings us to baccarat, the preferred game of the idle rich. Baccarat is a card game that requires no decisions and no skill, but it looks elegant. Some players bet \$100,000 per hand. There are only a few high-stakes players, and they don't play many hands by the standards of more plebeian games like roulette. So "the long run" that the casino profits from is not as long in baccarat. Runs of luck average out over a month or so in roulette but not in baccarat. The quarterly profit statements of some casino companies have suffered from bad luck at the baccarat tables.

CHAPTER 5

Robert Indiana, *The Small Diamond Demuth Five*, 1963.

Sampling Distributions

Introduction

Statistical inference draws conclusions about a population or process on the basis of data. The data are summarized by *statistics* such as means, proportions, and the slopes of least-squares regression lines. When the data are produced by random sampling or randomized experimentation, a statistic is a random variable that obeys the laws of probability theory. The link between probability and data is formed by the *sampling distributions* of statistics. A sampling distribution shows how a statistic would vary in repeated data production. That is, a sampling distribution is a probability distribution that answers the question "What would happen if we did this many times?" In Section 3.4 we looked at several sampling distributions empirically by simulating 1000 random samples. Considered as a probability distribution, a sampling distribution is an idealized mathematical description of the results of an indefinitely large number of samples rather than the results of a particular 1000 samples.

The Distribution of a Statistic

A statistic from a random sample or randomized experiment is a random variable. The probability distribution of the statistic is its **sampling distribution.**

Probability distributions also play a second role in statistical inference. Any quantity that can be measured for each member of a population is described by the distribution of its values for all members of the population. This is the context in which we first met distributions, as models for the overall pattern of data. Imagine choosing one individual at random from the population. The results of repeated choices have a probability distribution that is the same as the distribution for the population.

EXAMPLE 5.1

The distribution of heights of women between the ages of 18 and 24 is approximately normal with mean 64.5 inches and standard deviation 2.5 inches. Select a woman at random and measure her height. The result is a random variable X. We don't know the height of a randomly chosen woman, but we do know that in repeated sampling X will have the same $N(64.5, 2.5)$ distribution that describes the pattern of heights in the entire population. We call $N(64.5, 2.5)$ the *population distribution*.

Population Distribution

The **population distribution** of a variable is the distribution of its values for all members of the population. The population distribution is also the probability distribution of the variable when we choose one individual from the population at random.

The population of all women between the ages of 18 and 24 actually exists, so that we can in principle draw an SRS from it. Sometimes the population that we want to draw conclusions about does not actually exist in the physical world. For example, we might observe an industrial process producing rods with lengths that vary according to a normal distribution. The population contains all rods that would be produced if the process continued forever in its present state. That population does not actually exist, so we cannot draw an SRS from it. Yet we are often justified in treating the rods as if they were independent observations randomly chosen from this hypothetical population. Once again we describe the population (or the process, if you prefer) by the normal distribution that governs individual observations.

To progress from discussing probability as a topic in itself to probability as a foundation for inference, we must study the sampling distributions of some common statistics. We will meet other sampling distributions later in our study of inference. In each case, the nature of the sampling distribution depends on both the nature of the population distribution and the way we collect the data from the population.

5.1 Sampling Distributions for Counts and Proportions

A sample survey asks 2500 adults whether they agree that "I like buying new clothes, but shopping is often frustrating and time-consuming." The number who say "Yes" is a random variable X. A quality control engineer selects a sample of 10 switches from a supplier's shipment for detailed inspection. A number X fail to meet the specifications. In both of these examples, the random variable X is a **count** of the occurrences of some outcome in a fixed number of observations. There are 2500 observations in the first example and 10 observations in the second. If the number of observations is n, then the **sample proportion** is $\hat{p} = X/n$. For example, if 1650 of the 2500 shoppers say "Yes," the sample proportion is

count

sample proportion

$$\hat{p} = \frac{1650}{2500} = 0.66$$

Sample counts and sample proportions are common statistics. Finding their sampling distributions is the goal of this section.

The binomial distributions for sample counts

The distribution of a count X depends on how the data are produced. Here is a simple but common situation.

The Binomial Setting

1. There are a fixed number n of observations.
2. The n observations are all independent.
3. Each observation falls into one of just two categories, which for convenience we call "success" and "failure."
4. The probability of a success, call it p, is the same for each observation.

Think of tossing a coin n times as an example of the binomial setting. Each toss gives either heads or tails. The outcomes of successive tosses are independent. If we call heads a success, then p is the probability of a head and remains the same as long as we toss the same coin. The number of heads we count is a random variable X. The distribution of X, and more generally of the count of successes in any binomial setting, is completely determined by the number of observations n and the success probability p.

Binomial Distributions

The distribution of the count X of successes in the binomial setting is called the **binomial distribution** with parameters n and p. The parameter n is the number of observations, and p is the probability of a success on any one observation. The possible values of X are the whole numbers from 0 to n. As an abbreviation, we say that X is $B(n, p)$.

The binomial distributions are an important class of discrete probability distributions. Later in this section we will learn how to assign probabilities to outcomes and how to find the mean and standard deviation of binomial distributions. But the most important aspect of using binomial distributions is the ability to recognize situations to which they apply.

EXAMPLE 5.2

(a) Toss a balanced coin 10 times and count the number X of heads. There are $n = 10$ observations. Successive tosses are independent. If the coin is balanced, the probability of a head is $p = 0.5$ on each toss. The number of heads we observe has the binomial distribution $B(10, 0.5)$.

(b) Deal 10 cards from a shuffled deck and count the number X of red cards. There are 10 observations, and each gives either a red or a black card. But the observations are *not* independent. If the first card is black, the second is more likely to be red because there are now more red cards than black cards remaining in the deck. The count X does *not* have a binomial distribution.

(c) Genetics says that children receive genes from their parents independently. Each child of a particular pair of parents has probability 0.25 of having type O blood. If these parents have 5 children, the number who have type O blood is the count X of successes in 5 independent trials with probability 0.25 of a success on each trial. So X has the $B(5, 0.25)$ distribution.

(d) Engineers define reliability as the probability that an item will perform its function under specific conditions for a specific period of time. If an aircraft engine turbine has probability 0.999 of performing properly for an hour of flight, the number of turbines in a fleet of 350 engines that fly for an hour without failure has the $B(350, 0.999)$ distribution. This binomial distribution is obtained by assuming, as seems reasonable, that the turbines fail independently of each other. A common cause of failure, such as sabotage, would destroy the independence and make the binomial model inappropriate.

Binomial distributions in statistical sampling

The binomial distributions are important in statistics when we wish to make inferences about the proportion p of "successes" in a population. Here is a typical example.

EXAMPLE 5.3

A quality engineer chooses an SRS of 10 switches from a large shipment. Unknown to the engineer, 10% of the switches in the shipment do not conform to the specifications. The engineer counts the number X of nonconforming switches in the sample. *This is not quite a binomial setting.* Removing one switch changes the proportion of bad switches in the remaining population, so the state of the second switch chosen is not independent of the first.

If the shipment is large, however, removing a few items has a very small effect on the composition of the remaining population. Successive inspection results are very nearly independent. Suppose that the shipment contains 10,000 switches, of which 1000 are nonconforming. The proportion nonconforming is

$$p = \frac{1{,}000}{10{,}000} = 0.1$$

If the first switch chosen is bad, the proportion of bad switches remaining is 999/9999 = 0.0999. If the first switch is good, the proportion of bad switches left in the shipment is 1000/9999 = 0.10001. These proportions are so close to 0.1 that for practical purposes we can act as if removing one switch has no effect on the proportion of bad switches remaining. We act as if the count X of switches in the sample that fail inspection has the binomial distribution $B(10, 0.1)$.

In Example 5.3 and similar settings, the population distribution is just the distribution of successes and failures in the population, described by the single parameter p. The sampling distribution of the count of successes in an SRS from the population is nearly binomial if the population is large enough to allow us to ignore the small dependence between observations. The accuracy of this approximation improves as the size of the population increases relative to the size of the sample.

Sampling Distribution of a Count

When the population is much larger than the sample, the count X of successes in an SRS of size n has approximately the $B(n, p)$ distribution if the population proportion of successes is p.

As a rule of thumb, we will use the binomial sampling distribution for counts when the population is at least 10 times as large as the sample.

Finding binomial probabilities: tables

We will later give a formula for the probability that a binomial random variable takes any of its values. In practice, you will rarely have to use this formula for calculations. Some calculators and most statistical software packages calculate binomial probabilities. If you do not have suitable computing facilities, you can still shorten the work of calculating binomial probabilities for some values of n and p by looking up probabilities in Table C in the back of this book. The entries in the table are the probabilities $P(X = k)$ of individual outcomes for a binomial random variable X.

EXAMPLE 5.4

A quality engineer selects an SRS of 10 switches from a large shipment for detailed inspection. Unknown to the engineer, 10% of the switches in the shipment fail to meet the specifications. What is the probability that no more than 1 of the 10 switches in the sample fails inspection?

Example 5.3 reminds us that the count X of bad switches in the sample has approximately the $B(10, 0.1)$ distribution. Figure 5.1 is a probability histogram for this distribution. The distribution is strongly skewed. Although X can take any whole-number value from 0 to 10, the probabilities of values larger than 5 are so small that they do not appear in the histogram.

		p
n	k	.10
10	0	.3487
	1	.3874
	2	.1937
	3	.0574
	4	.0112
	5	.0015
	6	.0001
	7	.0000
	8	.0000
	9	.0000
	10	.0000

We want to calculate

$$P(X \leq 1) = P(X = 1) + P(X = 0)$$

when X is $B(10, 0.1)$. Your software may do this—look for the key word "binomial." To use Table C for this calculation, look opposite $n = 10$ and under $p = 0.10$. This part of the table appears at the left. The entry opposite each k is $P(X = k)$. We find

$$\begin{aligned} P(X \leq 1) &= P(X = 1) + P(X = 0) \\ &= 0.3874 + 0.3487 = 0.7361 \end{aligned}$$

About 74% of all samples will contain no more than 1 bad switch. In fact, 35% of the samples will contain no bad switches. A sample of size 10 cannot be trusted to alert the engineer to the presence of unacceptable items in the shipment.

The excerpt from Table C contains the full $B(10, 0.1)$ distribution. The probabilities are rounded to four decimal places. Outcomes larger than 6 do not have probability exactly 0, but their probabilities are so small that the rounded values are 0.0000. Check that the sum of the probabilities given is 1, as it should be.

The values of p that appear in Table C are all 0.5 or smaller. When the probability of a success is greater than 0.5, restate the problem in terms of the number of failures. The probability of a failure is less than 0.5 when the probability of a success exceeds 0.5. When using the table, always stop to ask whether you must count successes or failures.

EXAMPLE 5.5

Corinne is a basketball player who makes 75% of her free throws over the course of a season. In a key game, Corinne shoots 12 free throws and misses 5 of them. The fans think that she failed because she was nervous. Is it unusual for Corinne to perform this poorly?

To answer this question, assume that free throws are independent with probability 0.75 of a success on each shot. (Studies of long sequences of free throws have

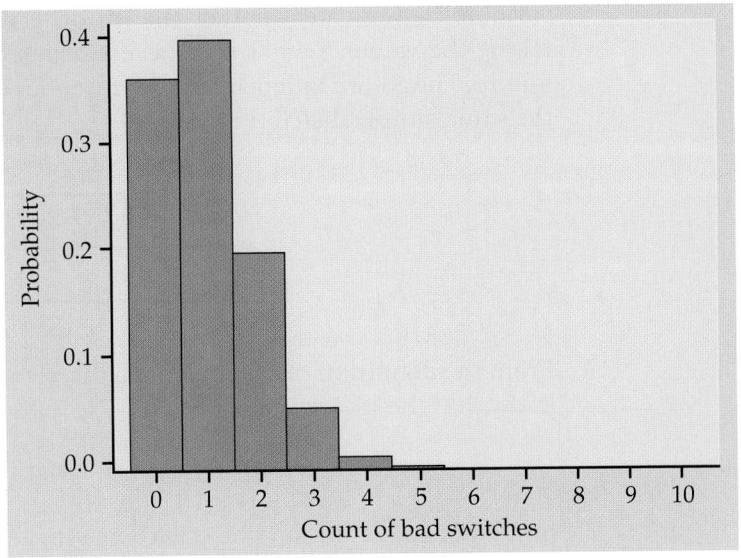

FIGURE 5.1 Probability histogram for the binomial distribution with $n = 10$ and $p = 0.1$, for Example 5.4.

found no evidence that they are dependent, so this is a reasonable assumption.) Because the probability of making a free throw is greater than 0.5, we count misses in order to use Table C. The probability of a miss is $1 - 0.75$, or 0.25. The number X of misses in 12 attempts has the $B(12, 0.25)$ distribution.

We want the probability of missing 5 or more. This is

$$P(X \geq 5) = P(X = 5) + P(X = 6) + \cdots + P(X = 12)$$
$$= 0.1032 + 0.0401 + \cdots + 0.0000 = 0.1576$$

Corinne will miss 5 or more out of 12 free throws about 16% of the time, or roughly one of every six games. While below her average level, this performance is well within the range of the usual chance variation in her shooting.

Binomial mean and standard deviation

If a count X is $B(n, p)$, what are the mean μ_X and the standard deviation σ_X? We can guess the mean. If Corinne makes 75% of her free throws, the mean number made in 12 tries should be 75% of 12, or 9. That's μ_X when X is $B(12, 0.75)$. Intuition suggests more generally that the mean of the $B(n, p)$ distribution should be np. Can we show that this is correct and also obtain a short formula for the standard deviation? Because binomial distributions are discrete probability distributions, we could find the mean and variance by using the definitions in Section 4.4. For example, we could obtain the mean by multiplying each outcome by its probability and adding over all outcomes. Here is an easier way.

A binomial random variable X is the count of successes in n independent observations that each have the same probability p of success. Let the random variable S_i indicate whether the ith observation is a success or failure by taking the values $S_i = 1$ if a success occurs and $S_i = 0$ if the outcome is a failure. The S_i are independent because the observations are, and each S_i has the same simple distribution:

Outcome	1	0
Probability	p	$1-p$

From the definition of the mean of a discrete random variable (page 320), we know that the mean of each S_i is

$$\mu_S = (1)(p) + (0)(1-p) = p$$

Similarly, the definition of the variance (page 328) shows that $\sigma_S^2 = p(1-p)$. Because each S_i is 1 for a hit and 0 for a miss, the total number of hits X is just the sum of the S_i's:

$$X = S_1 + S_2 + \cdots + S_n$$

Apply the addition rules for means and variances to this sum. The mean of X is the sum of the means of the S_i's:

$$\begin{aligned}\mu_X &= \mu_{S_1} + \mu_{S_2} + \cdots + \mu_{S_n} \\ &= n\mu_S = np\end{aligned}$$

Similarly, the variance is n times the variance of a single S, so that $\sigma_X^2 = np(1-p)$. The standard deviation σ_X is the square root of the variance. Here is the result.

Binomial Mean and Standard Deviation

If a count X has the binomial distribution $B(n, p)$, then

$$\begin{aligned}\mu_X &= np \\ \sigma_X &= \sqrt{np(1-p)}\end{aligned}$$

EXAMPLE 5.6

The Helsinki Heart Study asked whether the anticholesterol drug gemfibrozil reduces heart attacks. In planning such an experiment, the researchers must be confident that the sample sizes are large enough to enable them to observe enough heart attacks. The Helsinki study planned to give gemfibrozil to about 2000 men aged 40 to 55 and a placebo to another 2000. The probability of a heart attack during the five-year period of the study for men this age is about 0.04. What are the mean and standard deviation of the number of heart attacks that will be observed in one group if the treatment does not change this probability?

There are 2000 independent observations, each having probability $p = 0.04$ of a heart attack. The count X of heart attacks is $B(2000, 0.04)$, so that

$$\mu_X = np = (2000)(0.04) = 80$$
$$\sigma_X = \sqrt{np(1-p)} = \sqrt{(2000)(0.04)(0.96)} = 8.76$$

The expected number of heart attacks is large enough to permit conclusions about the effectiveness of the drug. In fact, there were 84 heart attacks among the 2035 men actually assigned to the placebo, quite close to the mean. The gemfibrozil group of 2046 men suffered only 56 heart attacks. This is evidence that the drug does reduce the chance of a heart attack.

Sample proportions

What proportion of a large lot of switches fail to meet specifications? What percent of adults favor stronger laws restricting firearms? In statistical sampling we often want to estimate the proportion p of "successes" in a population. Our estimator is the sample proportion of successes:

$$\hat{p} = \frac{\text{count of successes in sample}}{\text{size of sample}} = \frac{X}{n}$$

Be sure to distinguish between the *proportion* $\hat{p}$ and the *count* X. The count takes whole-number values between 0 and n, but a proportion is always a number between 0 and 1. In the binomial setting, the count X has a binomial distribution. The proportion $\hat{p}$ does *not* have a binomial distribution. We can, however, do probability calculations about $\hat{p}$ by restating them in terms of the count X and using binomial methods. As an example, we return to the sample survey that we studied by simulation in Example 3.22 (page 262).

Exploring the Web

The Web abounds in applets that simulate various random phenomena. One amusing probability problem is named *Buffon's needle*. Draw lines 1 inch apart on a piece of paper, then drop a 1-inch-long needle on the paper. What is the probability that the needle crosses a line? You can find both a solution by mathematics and a simulation at George Reese's site, `www.mste.uiuc.edu/reese/buffon/buffon.html`. Personal sites sometimes vanish; a search on "Buffon's needle" will turn up alternative sites. The probability turns out to be $2/\pi$, where π is the famous number that is the ratio of any circle's circumference to its diameter. So the simulation is also a way to calculate π.

EXAMPLE 5.7

A recent survey asked a nationwide random sample of 2500 adults if they agreed or disagreed that "I like buying new clothes, but shopping is often frustrating and time-consuming." Suppose that 60% of all adults would agree if asked this question. What is the probability that the sample proportion who agree is at least 58%?

The count X who agree has the binomial distribution $B(2500, 0.6)$. The sample proportion $\hat{p} = X/2500$ does *not* have a binomial distribution, because it is not a count. We therefore translate any question about the sample proportion $\hat{p}$ into a question about the count X. Because 58% of 2500 is 1450,

$$\begin{aligned} P(\hat{p} \geq 0.58) &= P(X \geq 1450) \\ &= P(X = 1450) + P(X = 1451) + \cdots + P(X = 2500) \end{aligned}$$

This is a rather elaborate calculation. We must add more than 1000 binomial probabilities. Software tells us that $P(\hat{p} \geq 0.58) = 0.9802$. Because many software packages cannot handle an n as large as 2500, we need another way to do this calculation.

As a first step, we can obtain the mean and standard deviation of the sample proportion from the mean and standard deviation of the sample count using the rules from Section 4.4 for the mean and variance of a constant times a random variable. Here is the result.

Mean and Standard Deviation of a Sample Proportion

Let $\hat{p}$ be the sample proportion of successes in an SRS of size n drawn from a large population having population proportion p of successes. The mean and standard deviation of $\hat{p}$ are

$$\mu_{\hat{p}} = p$$

$$\sigma_{\hat{p}} = \sqrt{\frac{p(1-p)}{n}}$$

The formula for $\sigma_{\hat{p}}$ is correct in the binomial setting. It is approximately correct for an SRS from a large population. We will use it when the population is at least 10 times as large as the sample.

EXAMPLE 5.8

The mean and standard deviation of the proportion of the survey respondents in Example 5.7 who find shopping frustrating are

$$\mu_{\hat{p}} = p = 0.6$$

$$\sigma_{\hat{p}} = \sqrt{\frac{p(1-p)}{n}} = \sqrt{\frac{(0.6)(0.4)}{2500}} = 0.0098$$

The fact that the mean of $\hat{p}$ is p states in statistical language that the sample proportion $\hat{p}$ in an SRS is an *unbiased estimator* of the population proportion p. When a sample is drawn from a new population having a different value of the population proportion p, the sampling distribution of the unbiased estimator $\hat{p}$ changes so that its mean moves to the new value of p. We observed this fact empirically in Section 3.4 and have now verified it from the laws of probability.

The variability of $\hat{p}$ about its mean, as described by the variance or standard deviation, decreases as the sample size increases. So a sample proportion from a large sample will usually lie quite close to the population proportion p. We observed this in the simulation experiment on page 262 in Section 3.4. Now we have discovered exactly how the variability decreases: the standard deviation is $\sqrt{p(1-p)/n}$, or $\sqrt{p(1-p)}/\sqrt{n}$. The $\sqrt{n}$ in the denominator means that the sample size must be multiplied by 4 if we wish to divide the standard deviation in half.

Normal approximation for counts and proportions

We discovered empirically in Section 3.4 that the sampling distribution of a sample proportion $\hat{p}$ is close to normal. Now we have deduced that the distribution of $\hat{p}$ is that of a binomial count divided by the sample size n. This seems at first to be a contradiction. To clear up the matter, look at Figure 5.2. This is a probability histogram of the exact distribution of the proportion of frustrated shoppers $\hat{p}$, based on the binomial distribution $B(2500, 0.6)$. The

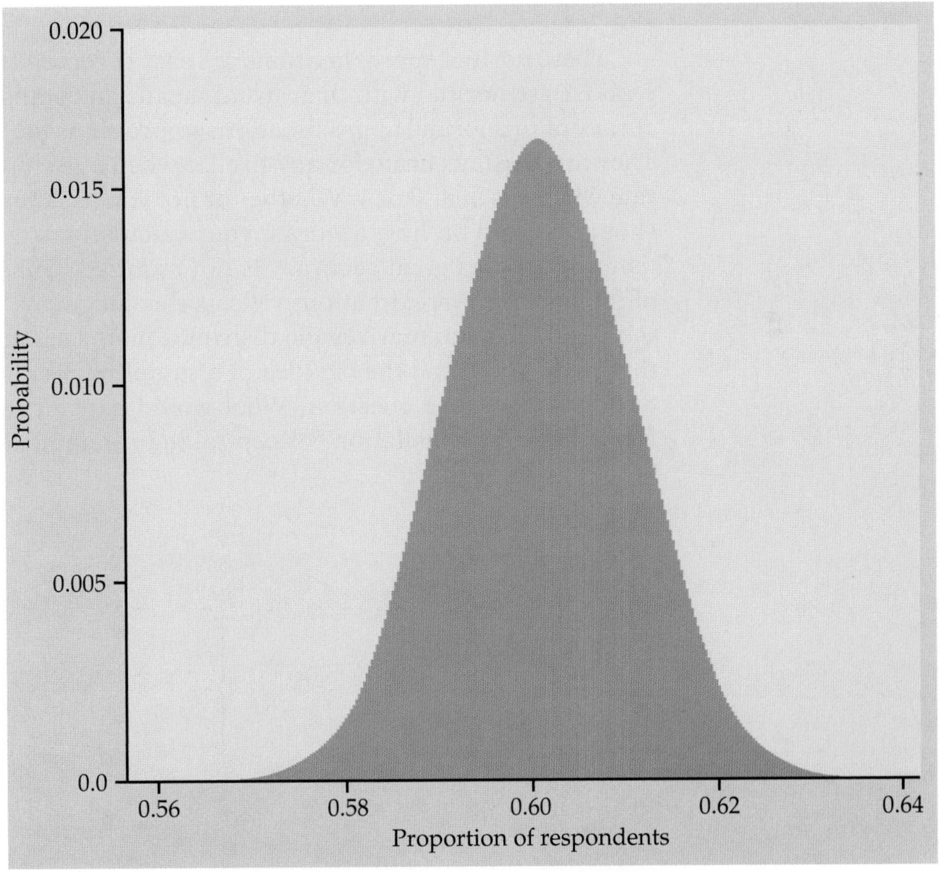

FIGURE 5.2 Probability histogram of the sample proportion $\hat{p}$ based on a binomial count with $n = 2500$ and $p = 0.6$. The distribution is very close to normal.

histogram shows 201 bars, for 201 of the 2501 possible values of $\hat{p}$. The other values have probabilities too small to show in a graph. *The probability histogram looks very normal!* In fact, both the count X and the sample proportion $\hat{p}$ are approximately normal in large samples.

Normal Approximation for Counts and Proportions

Draw an SRS of size n from a large population having population proportion p of successes. Let X be the count of successes in the sample and $\hat{p} = X/n$ the sample proportion of successes. When n is large, the sampling distributions of these statistics are approximately normal:

$$X \text{ is approximately } N\left(np, \sqrt{np(1-p)}\right)$$

$$\hat{p} \text{ is approximately } N\left(p, \sqrt{\frac{p(1-p)}{n}}\right)$$

As a rule of thumb, we will use this approximation for values of n and p that satisfy $np \geq 10$ and $n(1-p) \geq 10$.

These normal approximations are easy to remember because they say that $\hat{p}$ and X are normal with their usual means and standard deviations. The accuracy of the normal approximations improves as the sample size n increases. They are most accurate for any fixed n when p is close to 1/2, and least accurate when p is near 0 or 1. Whether or not you use the normal approximations should depend on how accurate your calculations need to be. For most statistical purposes great accuracy is not required. Our "rule of thumb" for use of the normal approximations reflects this judgment.

Figure 5.3 summarizes the distribution of a sample proportion in a form that helps you recall the big idea of a sampling distribution. Sampling distributions answer the question "What would happen if we took many samples from the same population?" Keep taking random samples of size n from a

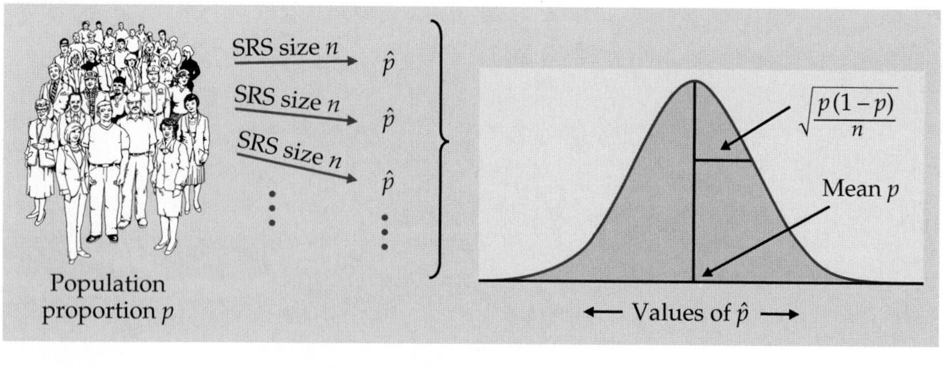

FIGURE 5.3 The sampling distribution of a sample proportion $\hat{p}$ is approximately normal with mean p and standard deviation $\sqrt{p(1-p)/n}$.

population that contains proportion p of successes. Find the sample proportion $\hat{p}$ for each sample. Collect all the $\hat{p}$'s and display their distribution. That's the sampling distribution of $\hat{p}$.

EXAMPLE 5.9

Let's compare the normal approximation for the calculation of Example 5.7 with the exact calculation from software. We want to calculate $P(\hat{p} \geq 0.58)$ when the sample size is $n = 2500$ and the population proportion is $p = 0.6$. Example 5.8 shows that

$$\mu_{\hat{p}} = p = 0.6$$

$$\sigma_{\hat{p}} = \sqrt{\frac{p(1-p)}{n}} = 0.0098$$

Act as if $\hat{p}$ were normal with mean 0.6 and standard deviation 0.0098. The approximate probability, as illustrated in Figure 5.4, is

$$P(\hat{p} \geq 0.58) = P\left(\frac{\hat{p} - 0.6}{0.0098} \geq \frac{0.58 - 0.6}{0.0098}\right)$$

$$\doteq P(Z \geq -2.04) = 0.9793$$

That is, about 98% of all samples have a sample proportion that is at least 0.58. Because the sample was large, this normal approximation is quite accurate. It misses the software value 0.9802 by only 0.0009.

EXAMPLE 5.10

The quality engineer of Example 5.4 realizes that a sample of 10 switches is inadequate and decides to inspect 100 switches from a large shipment. If in fact 10%

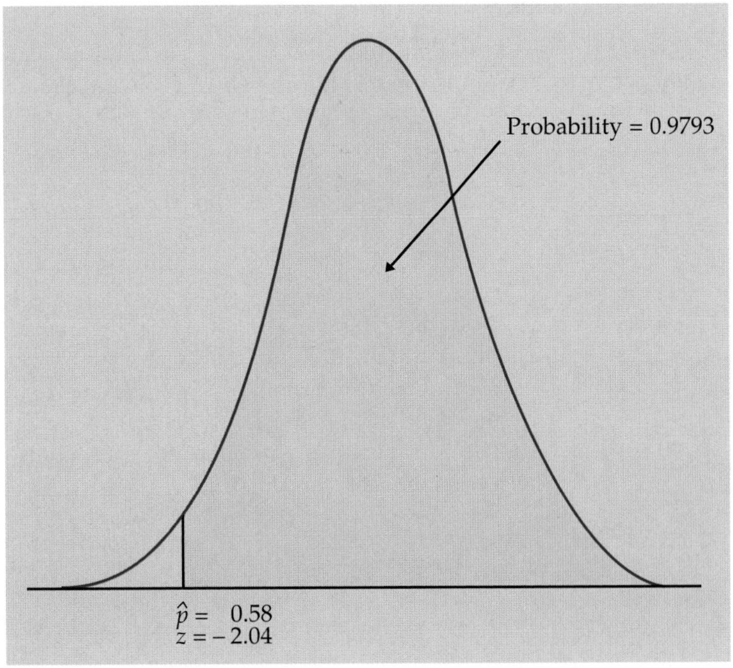

FIGURE 5.4 The normal probability calculation for Example 5.9.

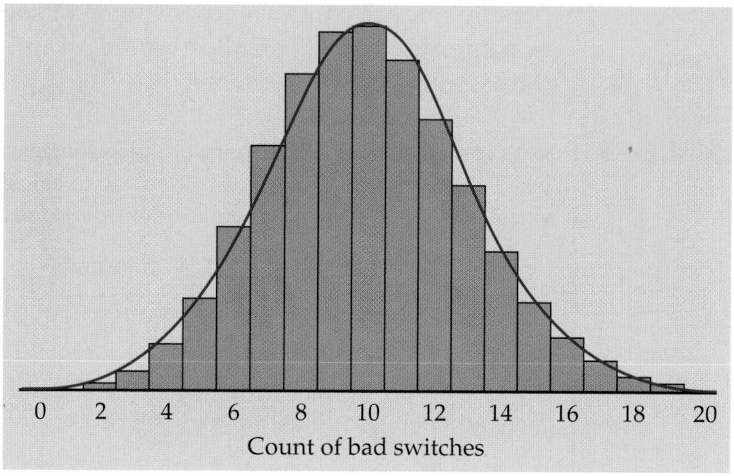

FIGURE 5.5 Probability histogram and normal approximation for the binomial distribution with $n = 100$ and $p = 0.1$, for Example 5.10.

of the switches in the shipment fail to meet the specifications, then the count X of bad switches in the sample has approximately the $B(100, 0.1)$ distribution. The distribution of the count of bad switches in a sample of 10 is distinctly nonnormal, as Figure 5.1 showed. When the sample size is increased to 100, however, the shape of the binomial distribution becomes roughly normal.

Software tells us that the actual binomial probability that no more than 9 of the switches in the sample fail inspection is $P(X \leq 9) = 0.4513$. How accurate is the normal approximation?

According to the normal approximation to the binomial distributions, the count X is approximately normal with mean and standard deviation

$$\mu_X = np = (100)(0.1) = 10$$

$$\sigma_X = \sqrt{np(1-p)} = \sqrt{(100)(0.1)(0.9)} = 3$$

Figure 5.5 displays the probability histogram of the binomial distribution with the density curve of the approximating normal distribution superimposed. Both distributions have the same mean and standard deviation, and both the area under the histogram and the area under the curve are 1. (Although the binomial random variable X can take values as large as 100, the probabilities of outcomes larger than 20 are so small that the bars would have no visible height in the figure.)

The normal approximation to the probability of no more than 9 bad switches is the area to the left of $X = 9$ under the normal curve (Figure 5.6). Using Table A,

$$P(X \leq 9) = P\left(\frac{X - 10}{3} \leq \frac{9 - 10}{3}\right)$$

$$\doteq P(Z \leq -0.33) = 0.3707$$

The approximation 0.37 to the binomial probability 0.45 is not very accurate. Notice that $np = (100)(0.1) = 10$, so that this combination of n and p is on the border of the values for which we are willing to use the approximation.

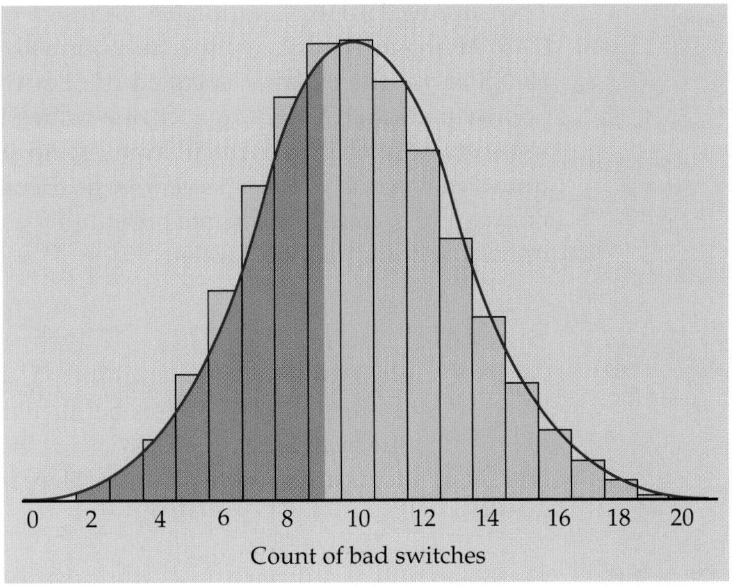

FIGURE 5.6 Area under the normal approximation curve for the probability in Example 5.10.

The continuity correction*

Figure 5.6 suggests an idea that greatly improves the accuracy of the normal approximation to binomial probabilities. The discrete binomial distribution in Example 5.10 puts probability exactly on $X = 9$ and $X = 10$ and no probability between these whole numbers. The normal distribution spreads its probability continuously. The bar for $X = 9$ in the probability histogram of Figure 5.6 extends from 8.5 to 9.5, but the normal calculation of $P(X \leq 9)$ includes only the area to the left of the center of this bar. The normal approximation is more accurate if we consider $X = 9$ to extend from 8.5 to 9.5, $X = 10$ to extend from 9.5 to 10.5, and so on.

When we want to include the outcome $X = 9$, we include the entire interval from 8.5 to 9.5 that is the base of the $X = 9$ bar in the histogram. So $P(X \leq 9)$ is calculated as $P(X \leq 9.5)$. On the other hand, $P(X < 9)$ excludes the outcome $X = 9$, so we exclude the entire interval from 8.5 to 9.5 and calculate $P(X \leq 8.5)$ from the normal table. Here is the result of the normal calculation in Example 5.10 improved in this way:

$$
\begin{aligned}
P(X \leq 9) &= P(X \leq 9.5) \\
&= P\left(\frac{X - 10}{3} \leq \frac{9.5 - 10}{3}\right) \\
&\doteq P(Z \leq -0.17) = 0.4325
\end{aligned}
$$

*This material can be omitted if desired.

continuity correction

The improved approximation 0.43 is much closer to the binomial probability 0.45. Acting as though a whole number occupies the interval from 0.5 below to 0.5 above the number is called the **continuity correction** to the normal approximation. If you need accurate values for binomial probabilities, try to use software to do exact calculations. If no software is available, use the continuity correction unless n is very large. Because most statistical purposes do not require extremely accurate probability calculations, we do not emphasize use of the continuity correction.

Binomial formulas*

We can find a formula for the probability that a binomial random variable takes any value by adding probabilities for the different ways of getting exactly that many successes in n observations. Here is the example we will use to show the idea.

EXAMPLE 5.11

Each child born to a particular set of parents has probability 0.25 of having blood type O. If these parents have 5 children, what is the probability that exactly 2 of them have type O blood?

The count of children with type O blood is a binomial random variable X with $n = 5$ tries and probability $p = 0.25$ of a success on each try. We want $P(X = 2)$.

Because the method doesn't depend on the specific example, we will use "S" for success and "F" for failure for short. In Example 5.11, "S" would stand for type O blood. Do the work in two steps.

Step 1: Find the probability that a specific 2 of the 5 tries give successes, say the first and the third. This is the outcome SFSFF. The multiplication rule for independent events tells us that

$$\begin{aligned} P(\text{SFSFF}) &= P(\text{S})P(\text{F})P(\text{S})P(\text{F})P(\text{F}) \\ &= (0.25)(0.75)(0.25)(0.75)(0.75) \\ &= (0.25)^2(0.75)^3 \end{aligned}$$

Step 2: Observe that the probability of *any one* arrangement of 2 S's and 3 F's has this same probability. That's true because we multiply together 0.25 twice and 0.75 three times whenever we have 2 S's and 3 F's. The probability that $X = 2$ is the probability of getting 2 S's and 3 F's in any arrangement whatsoever. Here are all the possible arrangements:

SSFFF	SFSFF	SFFSF	SFFFS	FSSFF
FSFSF	FSFFS	FFSSF	FFSFS	FFFSS

*The formula for binomial probabilities is useful in many settings, but we will not need it in our study of statistical inference. This section can therefore be omitted if desired.

There are 10 of them, all with the same probability. The overall probability of 2 successes is therefore

$$P(X = 2) = 10(0.25)^2(0.75)^3 = 0.2637$$

The pattern of this calculation works for any binomial probability. To use it, we need to be able to count the number of arrangements of k successes in n observations without actually listing them. We use the following fact to do the counting.

Binomial Coefficient

The number of ways of arranging k successes among n observations is given by the **binomial coefficient**

$$\binom{n}{k} = \frac{n!}{k!\,(n-k)!}$$

for $k = 0, 1, 2, \ldots, n$.

factorial

The formula for binomial coefficients uses the **factorial** notation. The factorial $n!$ for any positive whole number n is

$$n! = n \times (n-1) \times (n-2) \times \cdots \times 3 \times 2 \times 1$$

Also, $0! = 1$. Notice that the larger of the two factorials in the denominator of a binomial coefficient will cancel much of the $n!$ in the numerator. For example, the binomial coefficient we need for Example 5.11 is

$$\begin{aligned}\binom{5}{2} &= \frac{5!}{2!\,3!} \\ &= \frac{(5)(4)(3)(2)(1)}{(2)(1) \times (3)(2)(1)} \\ &= \frac{(5)(4)}{(2)(1)} = \frac{20}{2} = 10\end{aligned}$$

This agrees with our earlier enumeration.

The notation $\binom{n}{k}$ is *not* related to the fraction $\frac{n}{k}$. A helpful way to remember its meaning is to read it as "binomial coefficient n choose k." Binomial coefficients have many uses in mathematics, but we are interested in them only as an aid to finding binomial probabilities. The binomial coefficient $\binom{n}{k}$ counts the number of ways in which k successes can be distributed among n observations. The binomial probability $P(X = k)$ is this count multiplied by the probability of any specific arrangement of the k successes. Here is the formula we seek.

Binomial Probability

If X has the binomial distribution $B(n, p)$ with n observations and probability p of success on each observation, the possible values of X are $0, 1, 2, \ldots, n$. If k is any one of these values, the **binomial probability** is

$$P(X = k) = \binom{n}{k} p^k (1-p)^{n-k}$$

Here is an example of the use of the binomial probability formula.

EXAMPLE 5.12

The number X of switches that fail inspection in Example 5.4 has the $B(10, 0.1)$ distribution. The probability that no more than 1 switch fails is

$$\begin{aligned} P(X \leq 1) &= P(X = 1) + P(X = 0) \\ &= \binom{10}{1}(0.1)^1(0.9)^9 + \binom{10}{0}(0.1)^0(0.9)^{10} \\ &= \frac{10!}{1!\,9!}(0.1)(0.3874) + \frac{10!}{0!\,10!}(1)(0.3487) \\ &= (10)(0.1)(0.3874) + (1)(1)(0.3487) \\ &= 0.3874 + 0.3487 = 0.7361 \end{aligned}$$

The calculation used the facts that $0! = 1$ and that $a^0 = 1$ for any number $a \neq 0$. The result agrees with that obtained from Table C in Example 5.4.

SUMMARY

A **count** X of successes has the **binomial distribution** $B(n, p)$ in the **binomial setting:** there are n trials, all independent, each resulting in a success or a failure, and each having the same probability p of a success.

Binomial probabilities are most easily found by software. There is an exact formula that is practical for calculations when n is small. Table C contains binomial probabilities for some values of n and p. For large n, you can use the normal approximation.

The binomial distribution $B(n, p)$ is a good approximation to the **sampling distribution of the count of successes** in an SRS of size n from a large population containing proportion p of successes. We will use this approximation when the population is at least 10 times larger than the sample.

The mean and standard deviation of a **binomial count** X and a **sample proportion** of successes $\hat{p} = X/n$ are

$$\mu_X = np \qquad \mu_{\hat{p}} = p$$

$$\sigma_X = \sqrt{np(1-p)} \qquad \sigma_{\hat{p}} = \sqrt{\frac{p(1-p)}{n}}$$

The sample proportion $\hat{p}$ is therefore an unbiased estimator of the population proportion p.

The **normal approximation** to the binomial distribution says that if X is a count having the $B(n, p)$ distribution, then when n is large,

$$X \text{ is approximately } N\left(np, \sqrt{np(1-p)}\right)$$

$$\hat{p} \text{ is approximately } N\left(p, \sqrt{\frac{p(1-p)}{n}}\right)$$

We will use these approximations when $np \geq 10$ and $n(1-p) \geq 10$. If the **continuity correction** is employed to improve the accuracy, the normal approximations can be used more freely.

The exact **binomial probability formula** is

$$P(X = k) = \binom{n}{k} p^k (1-p)^{n-k}$$

where the possible values of X are $k = 0, 1, \ldots, n$. The binomial probability formula uses the **binomial coefficient**

$$\binom{n}{k} = \frac{n!}{k!\,(n-k)!}$$

Here the **factorial** $n!$ is

$$n! = n \times (n-1) \times (n-2) \times \cdots \times 3 \times 2 \times 1$$

for positive whole numbers n and $0! = 1$. The binomial coefficient counts the number of ways of distributing k successes among n trials.

SECTION 5.1 EXERCISES

All of the binomial probability calculations required in these exercises can be done by using Table C or the normal approximation. Your instructor may request that you use the binomial probability formula or software. In exercises requiring the normal approximation, you should use the continuity correction only if you studied that topic.

5.1 For each of the following situations, indicate whether a binomial distribution is a reasonable probability model for the random variable X. Give your reasons in each case.

(a) You observe the sex of the next 50 children born at a local hospital; X is the number of girls among them.

(b) A couple decides to continue to have children until their first girl is born; X is the total number of children the couple has.

(c) You want to know what percent of married people believe that mothers of young children should not be employed outside the home. You plan to interview 50 people, and for the sake of convenience you decide to interview both the husband and the wife in 25 married couples. The random variable X is the number among the 50 persons interviewed who think mothers should not be employed.

5.2 In each situation below, is it reasonable to use a binomial distribution for the random variable X? Give reasons for your answer in each case.

(a) An auto manufacturer chooses one car from each hour's production for a detailed quality inspection. One variable recorded is the count X of finish defects (dimples, ripples, etc.) in the car's paint.

(b) The pool of potential jurors for a murder case contains 100 persons chosen at random from the adult residents of a large city. Each person in the pool is asked whether he or she opposes the death penalty; X is the number who say "Yes."

(c) Joe buys a ticket in his state's "Pick 3" lottery game every week; X is the number of times in a year that he wins a prize.

5.3 In each of the following cases, decide whether or not a binomial distribution is an appropriate model, and give your reasons.

(a) Fifty students are taught about binomial distributions by a television program. After completing their study, all students take the same examination. The number of students who pass is counted.

(b) A student studies binomial distributions using computer-assisted instruction. After the initial instruction is completed, the computer presents 10 problems. The student solves each problem and enters the answer; the computer gives additional instruction between problems if the student's answer is wrong. The number of problems that the student solves correctly is counted.

(c) A chemist repeats a solubility test 10 times on the same substance. Each test is conducted at a temperature 10° higher than the previous test. She counts the number of times that the substance dissolves completely.

5.4 Some of the methods in this section are approximations rather than exact probability results. We have given rules of thumb for safe use of these approximations.

(a) You are interested in attitudes toward drinking among the 75 members of a fraternity. You choose 25 members at random to interview. One question is "Have you had five or more drinks at one time during the last week?" Suppose that in fact 20% of the 75 members would say

"Yes." Explain why you *cannot* safely use the $B(25, 0.2)$ distribution for the count X in your sample who say "Yes."

(b) The National AIDS Behavioral Surveys found that 0.2% (that's 0.002 as a decimal fraction) of adult heterosexuals had both received a blood transfusion and had a sexual partner from a group at high risk of AIDS. Suppose that this national proportion holds for your region. Explain why you *cannot* safely use the normal approximation for the sample proportion who fall in this group when you interview an SRS of 500 adults.

5.5 A factory employs several thousand workers, of whom 30% are Hispanic. If the 15 members of the union executive committee were chosen from the workers at random, the number of Hispanics on the committee would have the binomial distribution with $n = 15$ and $p = 0.3$.

(a) What is the probability that exactly 3 members of the committee are Hispanic?

(b) What is the probability that 3 or fewer members of the committee are Hispanic?

5.6 A university that is better known for its basketball program than for its academic strength claims that 80% of its basketball players get degrees. An investigation examines the fate of all 20 players who entered the program over a period of several years that ended 6 years ago. Of these players, 11 graduated and the remaining 9 are no longer in school. If the university's claim is true, the number of players who graduate among the 20 studied should have the $B(20, 0.8)$ distribution.

(a) Find the probability that exactly 11 of the 20 players graduate.

(b) Find the probability that 11 or fewer players graduate. This probability is so small that it casts doubt on the university's claim.

5.7 **(a)** What is the mean number of Hispanics on randomly chosen committees of 15 workers in Exercise 5.5?

(b) What is the standard deviation σ of the count X of Hispanic members?

(c) Suppose that 10% of the factory workers were Hispanic. Then $p = 0.1$. What is σ in this case? What is σ if $p = 0.01$? What does your work show about the behavior of the standard deviation of a binomial distribution as the probability of a success gets closer to 0?

5.8 **(a)** Find the mean number of graduates out of 20 players in the setting of Exercise 5.6 if the university's claim is true.

(b) Find the standard deviation σ of the count X.

(c) Suppose that the 20 players came from a population of which $p = 0.9$ graduated. What is the standard deviation σ of the count of graduates? If $p = 0.99$, what is σ? What does your work show about the behavior of the standard deviation of a binomial distribution as the probability p of success gets closer to 1?

5.9 Each entry in a table of random digits like Table B has probability 0.1 of being a 0, and digits are independent of each other.

(a) What is the probability that a group of five digits from the table will contain at least one 0?

(b) What is the mean number of 0s in lines 40 digits long?

applet

5.10 The *What Is Probability?* applet available at `www.whfreeman.com/ips` simulates tosses of a coin. You can choose the number of tosses n and the probability p of a head. That is, you can use the applet to simulate binomial random variables.

The count of nonconforming switches in Example 5.12 (page 382) has the binomial distribution with $n = 10$ and $p = 0.1$. Set these values for the number of tosses and probability of heads in the applet. The example calculates that the probability of getting a sample with exactly 1 bad switch is 0.3874. Of course, when we inspect only a few lots, the proportion of samples with exactly 1 bad switch will differ from this probability. Click "Toss" and "Reset" repeatedly to simulate inspecting 20 lots. Record the number of bad switches (the count of heads) in each of the 20 samples. What proportion of the 20 lots had exactly 1 bad switch? Remember that probability tells us only what happens in the long run.

5.11 In 1998, Mark McGwire of the St. Louis Cardinals hit 70 home runs, a new major league record. Was this feat as surprising as most of us thought? In the three seasons before 1998, McGwire hit a home run in 11.6% of his times at bat. He went to bat 509 times in 1998. If he continues his past performance, McGwire's home run count in 509 times at bat has approximately the binomial distribution with $n = 509$ and $p = 0.116$.

(a) What is the mean number of home runs McGwire will hit in 509 times at bat?

(b) What is the probability that he hits 70 or more home runs?

(c) In 2001, Barry Bonds of the San Francisco Giants hit 73 home runs, breaking McGwire's record. This was surprising. In the three previous seasons, Bonds hit a home run in 8.65% of his times at bat. He batted 476 times in 2001. Considering his home run count as a binomial random variable with $n = 476$ and $p = 0.0865$, what is the probability of 73 or more home runs?

5.12 You operate a restaurant. You read that a sample survey by the National Restaurant Association shows that 40% of adults are committed to eating nutritious food when eating away from home. To help plan your menu, you decide to conduct a sample survey in your own area. You will use random digit dialing to contact an SRS of 200 households by telephone.

(a) If the national result holds in your area, it is reasonable to use the $B(200, 0.4)$ distribution to describe the count X of respondents who seek nutritious food when eating out. Explain why.

(b) What is the mean number of nutrition-conscious people in your sample if $p = 0.4$ is true? What is the probability that X lies between 75 and 85?

(c) You find 100 of your 200 respondents concerned about nutrition. Is this reason to believe that the percent in your area is higher than the national 40%? To answer this question, find the probability that X is 100 or larger if $p = 0.4$ is true. If this probability is very small, that is reason to think that p is actually greater than 0.4.

5.13 The Harvard College Alcohol Study finds that 67% of college students support efforts to "crack down on underage drinking." The study took a sample of almost 15,000 students, so the population proportion who support a crackdown is very close to $p = 0.67$.[1] The administration of your college surveys an SRS of 100 students and finds that 62 support a crackdown on underage drinking.

(a) What is the sample proportion who support a crackdown on underage drinking?

(b) If in fact the proportion of all students on your campus who support a crackdown is the same as the national 67%, what is the probability that the proportion in an SRS of 100 students is as small or smaller than the result of the administration's sample?

(c) A writer in the student paper says that support for a crackdown is lower on your campus than nationally. Write a short letter to the editor explaining why the survey does not support this conclusion.

5.14 "How would you describe your own physical health at this time? Would you say your physical health is—excellent, good, only fair, or poor?" The Gallup Poll asked this question of 1005 randomly selected adults, of whom 29% said "excellent."[2] Suppose that in fact the proportion of the adult population who say their health is excellent is $p = 0.29$.

(a) What is the probability that the sample proportion $\hat{p}$ of an SRS of size $n = 1000$ who say their health is excellent lies between 26% and 32%? (That is, within ±3% of the truth about the population.)

(b) Repeat the probability calculation of (a) for SRSs of sizes $n = 250$ and $n = 4000$. What general conclusion can you draw from your calculations?

5.15 "How would you describe your own personal weight situation right now—very overweight, somewhat overweight, about right, somewhat underweight, or very underweight?" When the Gallup Poll asked an SRS of 1005 adults this question, 51% answered "about right."[3] Suppose that in fact 51% of the entire adult population think their weight is about right.

(a) Many opinion polls announce a "margin of error" of about ±3%. What is the probability that an SRS of size 1005 has a sample proportion $\hat{p}$ that is within ±3% (±0.03) of the population proportion $p = 0.51$?

(b) Answer the same question if the population proportion is $p = 0.06$. (This is the proportion who told Gallup that they were "very overweight.") How does the probability change as p moves from near 0.5 to near zero?

5.16 A student organization is planning to ask a sample of 50 students if they have noticed alcohol abuse brochures on campus. The sample percentage who say "Yes" will be reported. The organization's statistical advisor says that the standard deviation of this percentage will be about 7%.

(a) What would the standard deviation be if the sample contained 100 students rather than 50?

(b) How large a sample is required to reduce the standard deviation of the percentage who say "Yes" from 7% to 3.5%? Explain to someone who knows no statistics the advantage of taking a larger sample in a survey of opinion.

5.17 Children inherit their blood type from their parents, with probabilities that reflect the parents' genetic makeup. Children of Juan and Maria each have probability 1/4 of having blood type A and inherit independently of each other. Juan and Maria plan to have 4 children; let X be the number who have blood type A.

(a) What are n and p in the binomial distribution of X?

(b) Find the probability of each possible value of X, and draw a probability histogram for this distribution.

(c) Find the mean number of children with type A blood, and mark the location of the mean on your probability histogram.

5.18 A believer in the "random walk" theory of the behavior of stock prices thinks that an index of stock prices has probability 0.65 of increasing in any year. Moreover, the change in the index in any given year is not influenced by whether it rose or fell in earlier years. Let X be the number of years among the next 6 years in which the index rises.

(a) What are n and p in the binomial distribution of X?

(b) Give the possible values that X can take and the probability of each value. Draw a probability histogram for the distribution of X.

(c) Find the mean of the number X of years in which the stock price index rises and mark the mean on your probability histogram.

(d) Find the standard deviation of X. What is the probability that X takes a value within one standard deviation of its mean?

5.19 In a test for ESP (extrasensory perception), the experimenter looks at cards that are hidden from the subject. Each card contains either a star, a circle, a wave, or a square. As the experimenter looks at each of 20 cards in turn, the subject names the shape on the card.

(a) If a subject simply guesses the shape on each card, what is the probability of a successful guess on a single card? Because the cards are independent, the count of successes in 20 cards has a binomial distribution.

(b) What is the probability that a subject correctly guesses at least 10 of the 20 shapes?

(c) In many repetitions of this experiment with a subject who is guessing, how many cards will the subject guess correctly on the average? What is the standard deviation of the number of correct guesses?

(d) A standard ESP deck actually contains 25 cards. There are five different shapes, each of which appears on 5 cards. The subject knows that the deck has this makeup. Is a binomial model still appropriate for the count of correct guesses in one pass through this deck? If so, what are n and p? If not, why not?

5.20 A sociology professor asks her class to observe cars having a man and a woman in the front seat and record which of the two is the driver.

(a) Explain why it is reasonable to use the binomial distribution for the number of male drivers in n cars if all observations are made in the same location at the same time of day.

(b) Explain why the binomial model may not apply if half the observations are made outside a church on Sunday morning and half are made on campus after a dance.

(c) The professor requires students to observe 10 cars during business hours in a retail district close to campus. Past observations have shown that the man is driving in about 85% of cars in this location. What is the probability that the man is driving in 8 or fewer of the 10 cars?

(d) The class has 10 students, who will observe 100 cars in all. What is the probability that the man is driving in 80 or fewer of these?

5.21 When the ESP study of Exercise 5.19 discovers a subject whose performance appears to be better than guessing, the study continues at greater length. The experimenter looks at many cards bearing one of five shapes (star, square, circle, wave, and cross) in an order determined by random numbers. The subject cannot see the experimenter as he looks at each card in turn, in order to avoid any possible nonverbal clues. The answers of a subject who does not have ESP should be independent observations, each with probability 1/5 of success. We record 1000 attempts.

(a) What are the mean and the standard deviation of the count of successes?

(b) What are the mean and standard deviation of the proportion of successes among the 1000 attempts?

(c) What is the probability that a subject without ESP will be successful in at least 24% of 1000 attempts?

(d) The researcher considers evidence of ESP to be a proportion of successes so large that there is only probability 0.01 that a subject could do this well or better by guessing. What proportion of successes must a subject have to meet this standard? (*Hint:* Example 1.28, page 77, shows how to do a normal calculation of the type required here.)

5.22 A study by a federal agency concludes that polygraph (lie detector) tests given to truthful persons have probability about 0.2 of suggesting that the person is deceptive.[4]

(a) A firm asks 12 job applicants about thefts from previous employers, using a polygraph to assess their truthfulness. Suppose that all 12 answer truthfully. What is the probability that the polygraph says at least 1 is deceptive?

(b) What is the mean number among 12 truthful persons who will be classified as deceptive? What is the standard deviation of this number?

(c) What is the probability that the number classified as deceptive is less than the mean?

5.23 One way of checking the effect of undercoverage, nonresponse, and other sources of error in a sample survey is to compare the sample with known demographic facts about the population. The 2000 census found that 23,772,494 of the 209,128,094 adults (age 18 and over) in the United States called themselves "Black or African American."

(a) What is the population proportion p of blacks among American adults?

(b) An opinion poll chooses 1500 adults at random. What is the mean number of blacks in such samples? (Explain the reasoning behind your calculation.)

(c) Use a normal approximation to find the probability that such a sample will contain 170 or fewer blacks. Be sure to check that you can safely use the approximation.

5.24 A selective college would like to have an entering class of 1200 students. Because not all students who are offered admission accept, the college admits more than 1200 students. Past experience shows that about 70% of the students admitted will accept. The college decides to admit 1500 students. Assuming that students make their decisions independently, the number who accept has the $B(1500, 0.7)$ distribution. If this number is less than 1200, the college will admit students from its waiting list.

(a) What are the mean and the standard deviation of the number X of students who accept?

(b) Use the normal approximation to find the probability that at least 1000 students accept.

(c) The college does not want more than 1200 students. What is the probability that more than 1200 will accept?

(d) If the college decides to increase the number of admission offers to 1700, what is the probability that more than 1200 will accept?

5.25 Here is a simple probability model for multiple-choice tests. Suppose that each student has probability p of correctly answering a question chosen at random from a universe of possible questions. (A strong student has a higher p than a weak student.) The correctness of answers to different questions are independent. Jodi is a good student for whom $p = 0.75$.

(a) Use the normal approximation to find the probability that Jodi scores 70% or lower on a 100-question test.

(b) If the test contains 250 questions, what is the probability that Jodi will score 70% or lower?

(c) How many questions must the test contain in order to reduce the standard deviation of Jodi's proportion of correct answers to half its value for a 100-item test?

(d) Laura is a weaker student for whom $p = 0.6$. Does the answer you gave in (c) for the standard deviation of Jodi's score apply to Laura's standard deviation also?

5.26 Use the definition of binomial coefficients to show that each of the following facts is true. Then restate each fact in words in terms of the number of ways that k successes can be distributed among n observations.

(a) $\binom{n}{n} = 1$ for any whole number $n \geq 1$.

(b) $\binom{n}{n-1} = n$ for any whole number $n \geq 1$.

(c) $\binom{n}{k} = \binom{n}{n-k}$ for any n and k with $k \leq n$.

5.27 You are tossing a balanced die that has probability 1/6 of coming up 1 on each toss. Tosses are independent. We are interested in how long we must wait to get the first 1.

(a) The probability of a 1 on the first toss is 1/6. What is the probability that the first toss is not a 1 and the second toss is a 1?

(b) What is the probability that the first two tosses are not 1s and the third toss is a 1? This is the probability that the first 1 occurs on the third toss.

(c) Now you see the pattern. What is the probability that the first 1 occurs on the fourth toss? On the fifth toss? Give the general result: what is the probability that the first 1 occurs on the kth toss?

geometric distribution

Comment: The distribution of the number of trials to the first success is called a **geometric distribution.** In this problem you have found geometric distribution probabilities when the probability of a success on each trial is $p = 1/6$. The same idea works for any p.

5.2 The Sampling Distribution of a Sample Mean

Counts and proportions are discrete random variables that describe categorical data. The statistics most often used to describe quantitative data, on the other hand, are continuous random variables. The sample mean, percentiles, and standard deviation are examples of statistics based on quantitative data. Statistical theory describes the sampling distributions of these statistics. In this section we will concentrate on the sample mean. Because sample means are just averages of observations, they are among the most common statistics.

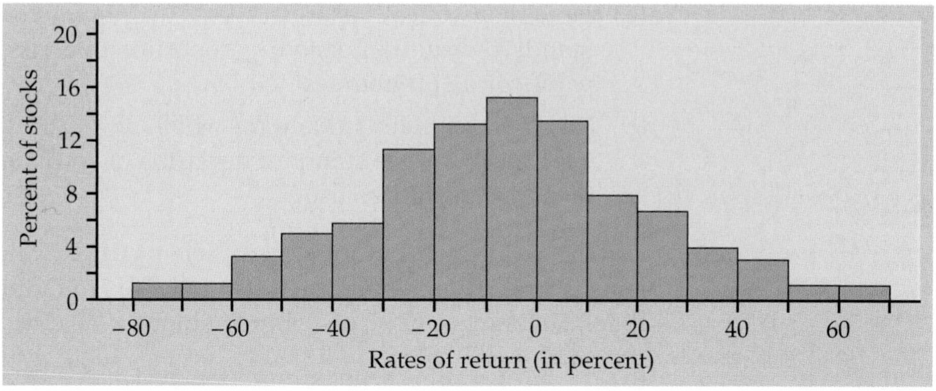

FIGURE 5.7(a) The distribution of returns for New York Stock Exchange common stocks in 1987.

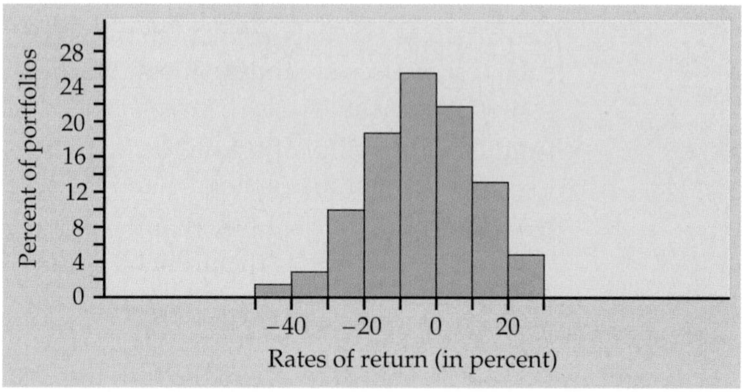

FIGURE 5.7(b) The distribution of returns for portfolios of 5 stocks in 1987. (Figure 5.7 is taken with permission from John K. Ford, "A method for grading 1987 stock recommendations," *American Association of Individual Investors Journal*, March 1988, pp. 16–17.)

EXAMPLE 5.13

A basic principle of investment is that diversification reduces risk. That is, buying several securities rather than just one reduces the variability of the return on an investment. Figure 5.7 illustrates this principle in the case of common stocks listed on the New York Stock Exchange. Figure 5.7(a) shows the distribution of returns for all 1815 stocks listed on the Exchange for the entire year 1987. This was a year of extreme swings in stock prices, including a record loss of more than 22% in a single day, October 19, 1987. The mean return for all 1815 stocks was −3.5%, and the distribution shows a very wide spread.

Figure 5.7(b) shows the distribution of returns for all possible portfolios that invested equal amounts in each of 5 stocks in 1987. A portfolio is just a sample of 5 stocks, and its return is the average return for the 5 stocks chosen. The mean return for all portfolios is still −3.5%, but the variation among portfolios is much less than the variation among individual stocks. For example, 11% of all individual stocks had a loss of more than 40%, but only 1% of the portfolios had a loss that large.

The histograms in Figure 5.7 illustrate a fact that we will make precise in this section: **averages are less variable than individual observations.** More detailed examination of the distributions, by normal quantile plots, would point to a second fact: **averages are more normal than individual observations.** These two facts contribute to the popularity of sample means in statistical inference.

The mean and standard deviation of $\bar{x}$

The sample mean $\bar{x}$ from a sample or an experiment is an estimate of the mean μ of the underlying population, just as a sample proportion $\hat{p}$ is an estimate of a population proportion p. The sampling distribution of $\bar{x}$ is determined by the design used to produce the data, the sample size n, and the population distribution.

Select an SRS of size n from a population, and measure a variable X on each individual in the sample. The data consist of observations on n random variables $X_1, X_2, \ldots, X_n$. A single X_i is a measurement on one individual selected at random from the population and therefore has the distribution of the population. If the population is large relative to the sample, we can consider $X_1, X_2, \ldots, X_n$ to be independent random variables each having the same distribution. This is our probability model for measurements on each individual in an SRS.

The sample mean of an SRS of size n is

$$\bar{x} = \frac{1}{n}(X_1 + X_2 + \cdots + X_n)$$

If the population has mean μ, then μ is the mean of each observation X_i. Therefore, by the addition rule for means of random variables,

$$\begin{aligned} \mu_{\bar{x}} &= \frac{1}{n}(\mu_{X_1} + \mu_{X_2} + \cdots + \mu_{X_n}) \\ &= \frac{1}{n}(\mu + \mu + \cdots + \mu) = \mu \end{aligned}$$

That is, *the mean of $\bar{x}$ is the same as the mean of the population*. The sample mean $\bar{x}$ is therefore an unbiased estimator of the unknown population mean μ.

The observations are independent, so the addition rule for variances also applies:

$$\begin{aligned} \sigma^2_{\bar{x}} &= \left(\frac{1}{n}\right)^2 (\sigma^2_{X_1} + \sigma^2_{X_2} + \cdots + \sigma^2_{X_n}) \\ &= \left(\frac{1}{n}\right)^2 (\sigma^2 + \sigma^2 + \cdots + \sigma^2) \\ &= \frac{\sigma^2}{n} \end{aligned}$$

Just as in the case of a sample proportion $\hat{p}$, the variability of the sampling distribution of a sample mean decreases as the sample size grows. Because the standard deviation of $\bar{x}$ is $\sigma/\sqrt{n}$, it is again true that the standard deviation of the statistic decreases in proportion to the square root of the sample size. Here is a summary of these facts.

Mean and Standard Deviation of a Sample Mean

Let $\bar{x}$ be the mean of an SRS of size n from a population having mean μ and standard deviation σ. The mean and standard deviation of $\bar{x}$ are

$$\mu_{\bar{x}} = \mu$$

$$\sigma_{\bar{x}} = \frac{\sigma}{\sqrt{n}}$$

EXAMPLE 5.14

The height X of a single randomly chosen young woman varies according to the $N(64.5, 2.5)$ distribution. If a medical study asked the height of an SRS of 100 young women, the sampling distribution of the sample mean height $\bar{x}$ would have mean and standard deviation

$$\mu_{\bar{x}} = \mu = 64.5 \text{ inches}$$

$$\sigma_{\bar{x}} = \frac{\sigma}{\sqrt{n}} = \frac{2.5}{\sqrt{100}} = 0.25 \text{ inch}$$

The heights of individual women vary widely about the population mean ($\sigma = 2.5$ inches), but the average height of a sample of 100 women has a standard deviation only one-tenth as large ($\sigma_{\bar{x}} = 0.25$ inch).

EXAMPLE 5.15

The fact that a mean of several measurements is less variable than a single measurement is important in science. When Simon Newcomb set out to measure the speed of light, he took repeated measurements of the time required for a beam of light to travel a known distance. Newcomb's 66 measurements appear in Table 1.1 (page 8). We saw that the data contain two outliers that may not be observations from the same population. Disregarding these leaves 64 observations. These are the values of 64 independent random variables, each with a probability distribution that describes the population of all measurements made with Newcomb's apparatus. If Newcomb's measurements have no bias, the population mean μ is the true passage time of light. The population variability reflects the random variation in the measurements due to small changes in the environment, the equipment, and the procedure. Suppose that the standard deviation for this population is $\sigma = 5$ nanoseconds.

If Newcomb had taken a single measurement, the standard deviation of the result would be 5, so that a second observation might have a quite different value. Taking 10 measurements and reporting their mean $\bar{x}$ reduces the standard deviation to

$$\sigma_{\bar{x}} = \frac{5}{\sqrt{10}} = 1.58 \text{ nanoseconds}$$

The sample mean of any number of measurements is an unbiased estimate of the population mean. Averaging over more measurements reduces the variability and makes it more likely that $\bar{x}$ is close to μ. The mean $\bar{x}$ of all 64 measurements has standard deviation

$$\sigma_{\overline{x}} = \frac{5}{\sqrt{64}} = 0.625 \text{ nanosecond}$$

Newcomb's observed value of this random variable, $\overline{x} = 27.75$, is a much more reliable estimate of the true passage time than is a single measurement. Increasing the number of measurements from 1 to 64 divides the standard deviation by 8, the square root of 64.

The sampling distribution of $\overline{x}$

We have described the center and spread of the probability distribution of a sample mean $\overline{x}$, but not its shape. The shape of the distribution of $\overline{x}$ depends on the shape of the population distribution. Here is one important case: if the population distribution is normal, then so is the distribution of the sample mean.

Sampling Distribution of a Sample Mean

If a population has the $N(\mu, \sigma)$ distribution, then the sample mean $\overline{x}$ of n independent observations has the $N(\mu, \sigma/\sqrt{n})$ distribution.

EXAMPLE 5.16

Newcomb's population consists of all measurements he might make. A normal quantile plot of his 64 observations (Figure 1.32 on page 81) shows that individual measurements are close to normal. A single measurement has the $N(\mu, 5)$ distribution, where μ is the unknown population mean. By the 95 part of the 68–95–99.7 rule, 95% of these measurements will lie within 2×5, or 10 nanoseconds, of the unknown μ.

The mean $\overline{x}$ of 10 measurements also has a normal distribution, but its standard deviation (Example 5.15) is only 1.58 nanoseconds. So 95% of the time the observed $\overline{x}$ falls within $2 \times 1.58 = 3.16$ nanoseconds of μ. Figure 5.8 compares the distributions of a single measurement and the sample mean of 10 measurements. The picture makes dramatically clear the advantages of averaging over many observations.

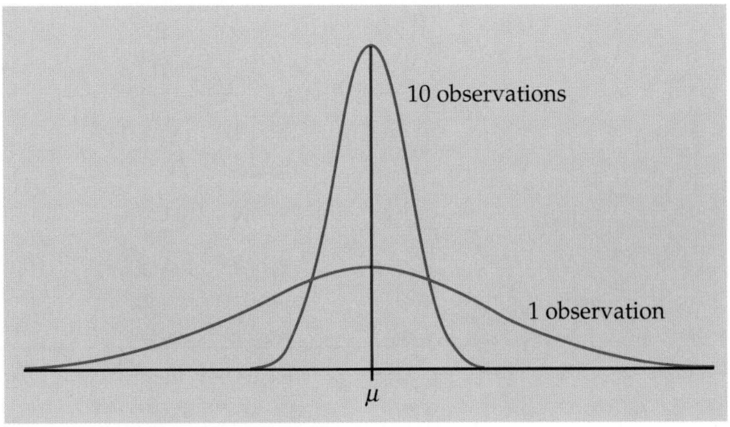

FIGURE 5.8 The sampling distribution of $\overline{x}$ for samples of size 10 compared with the distribution of a single observation. See Example 5.16.

The fact that the sample mean of an SRS from a normal population has a normal distribution is a special case of a more general fact: **any linear combination of independent normal random variables is also normally distributed.** That is, if X and Y are independent normal random variables and a and b are any fixed numbers, $aX + bY$ is also normally distributed, and so it is for any number of normal variables. In particular, the sum or difference of independent normal random variables has a normal distribution. The mean and standard deviation of $aX + bY$ are found as usual from the addition rules for means and variances. These facts are often used in statistical calculations.

EXAMPLE 5.17

Tom and George are playing in the club golf tournament. Their scores vary as they play the course repeatedly. Tom's score X has the $N(110, 10)$ distribution, and George's score Y varies from round to round according to the $N(100, 8)$ distribution. If they play independently, what is the probability that Tom will score lower than George and thus do better in the tournament? The difference $X - Y$ between their scores is normally distributed, with mean and variance

$$\mu_{X-Y} = \mu_X - \mu_Y = 110 - 100 = 10$$

$$\sigma^2_{X-Y} = \sigma^2_X + \sigma^2_Y = 10^2 + 8^2 = 164$$

Because $\sqrt{164} = 12.8$, $X - Y$ has the $N(10, 12.8)$ distribution. Figure 5.9 illustrates the probability computation:

$$\begin{aligned} P(X < Y) &= P(X - Y < 0) \\ &= P\left(\frac{(X - Y) - 10}{12.8} < \frac{0 - 10}{12.8}\right) \\ &= P(Z < -0.78) = 0.2177 \end{aligned}$$

Although George's score is 10 strokes lower on the average, Tom will have the lower score in about one of every five matches.

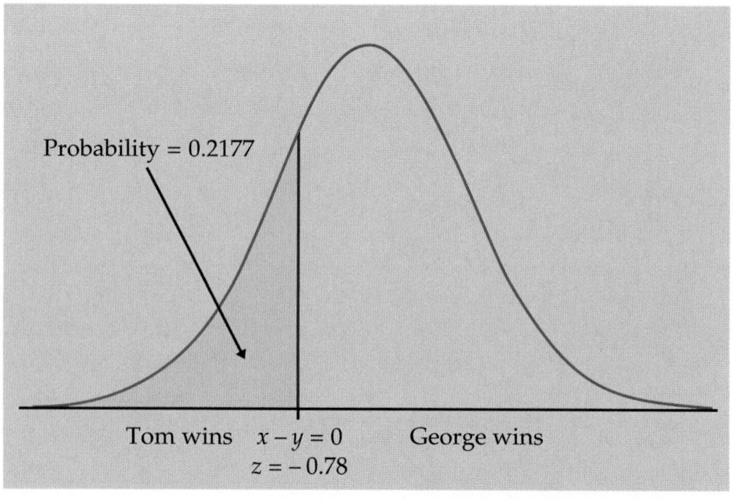

FIGURE 5.9 The normal probability calculation for Example 5.17.

The central limit theorem

The sampling distribution of $\bar{x}$ is normal if the underlying population itself has a normal distribution. What happens when the population distribution is not normal? It turns out that as the sample size increases, the distribution of $\bar{x}$ gets closer to a normal distribution. This is true no matter what shape the population distribution has, as long as the population has a finite standard deviation σ. This famous fact of probability theory is called the *central limit theorem*. It is much more useful than the fact that the distribution of $\bar{x}$ is exactly normal if the population is exactly normal.* For large sample size n, we can regard $\bar{x}$ as having the $N(\mu, \sigma/\sqrt{n})$ distribution.

Central Limit Theorem

Draw an SRS of size n from any population with mean μ and finite standard deviation σ. When n is large, the sampling distribution of the sample mean $\bar{x}$ is approximately normal:

$$\bar{x} \text{ is approximately } N\left(\mu, \frac{\sigma}{\sqrt{n}}\right)$$

More general versions of the central limit theorem say that the distribution of a sum or average of many small random quantities is close to normal. This is true even if the quantities are not independent (as long as they are not too highly correlated) and even if they have different distributions (as long as no one random quantity is so large that it dominates the others). The central limit theorem suggests why the normal distributions are common models for observed data. Any variable that is a sum of many small influences will have approximately a normal distribution.

How large a sample size n is needed for $\bar{x}$ to be close to normal depends on the population distribution. More observations are required if the shape of the population distribution is far from normal.

EXAMPLE 5.18

exponential distribution

Figure 5.10 shows the central limit theorem in action for a very nonnormal population. Figure 5.10(a) displays the density curve of a single observation, that is, of the population. The distribution is strongly right-skewed, and the most probable outcomes are near 0. The mean μ of this distribution is 1, and its standard deviation σ is also 1. This particular continuous distribution is called an **exponential distribution.** Exponential distributions are used as models for the lifetime in service of electronic components and for the time required to serve a customer or repair a machine.

Figures 5.10(b), (c), and (d) are the density curves of the sample means of 2, 10, and 25 observations from this population. As n increases, the shape becomes more normal. The mean remains at $\mu = 1$, but the standard deviation decreases, taking the value $1/\sqrt{n}$. The density curve for 10 observations is still somewhat skewed to

*The first general version of the central limit theorem was established in 1810 by the French mathematician Pierre Simon Laplace (1749–1827).

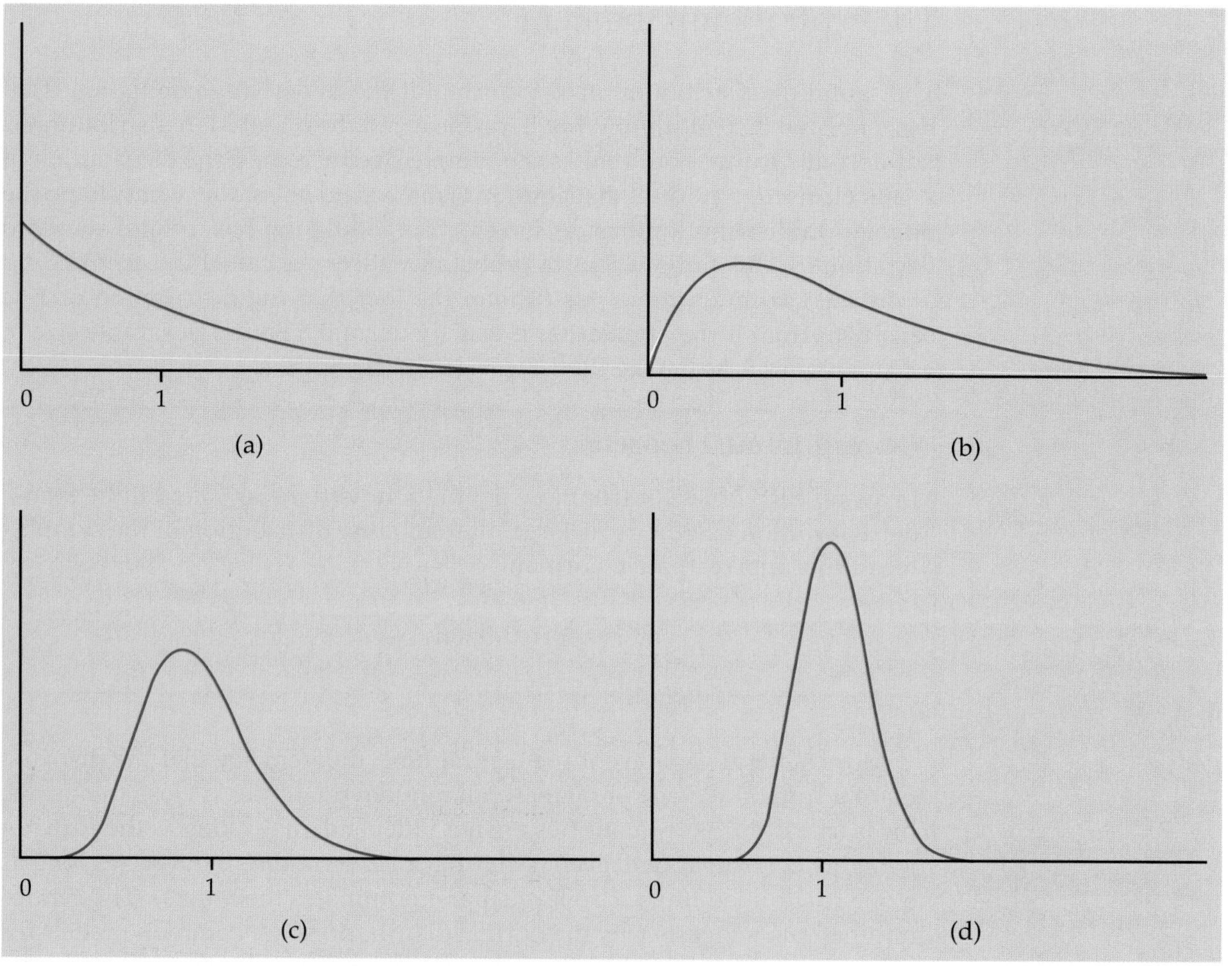

FIGURE 5.10 The central limit theorem in action: the distribution of sample means from a strongly nonnormal population becomes more normal as the sample size increases. (a) The distribution of 1 observation. (b) The distribution of $\bar{x}$ for 2 observations. (c) The distribution of $\bar{x}$ for 10 observations. (d) The distribution of $\bar{x}$ for 25 observations.

the right but already resembles a normal curve having $\mu = 1$ and $\sigma = 1/\sqrt{10} = 0.32$. The density curve for $n = 25$ is yet more normal. The contrast between the shapes of the population distribution and of the distribution of the mean of 10 or 25 observations is striking.

The central limit theorem allows us to use normal probability calculations to answer questions about sample means from many observations even when the population distribution is not normal.

EXAMPLE 5.19

The time X that a technician requires to perform preventive maintenance on an air-conditioning unit is governed by the exponential distribution whose density curve appears in Figure 5.10(a). The mean time is $\mu = 1$ hour and the standard deviation is $\sigma = 1$ hour. Your company operates 70 of these units. What is the probability that their average maintenance time exceeds 50 minutes?

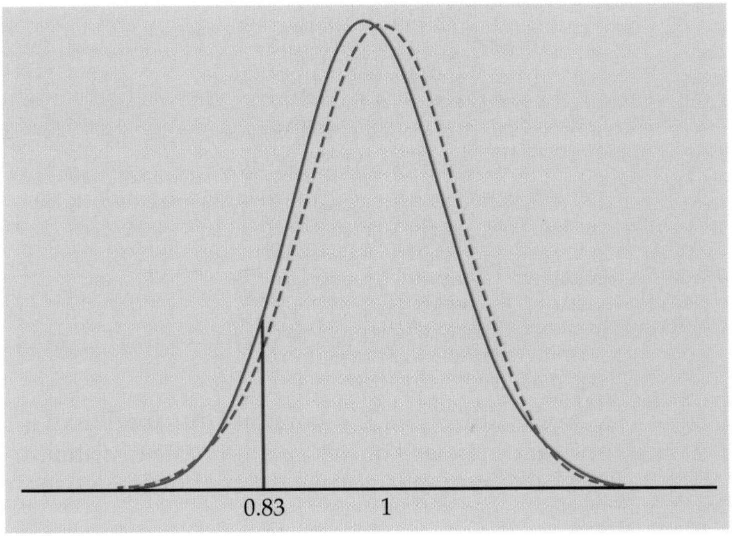

FIGURE 5.11 The exact distribution (*dashed*) and the normal approximation from the central limit theorem (*solid*) for the average time needed to maintain an air conditioner, for Example 5.19.

The central limit theorem says that the sample mean time $\bar{x}$ (in hours) spent working on 70 units has approximately the normal distribution with mean equal to the population mean $\mu = 1$ hour and standard deviation

$$\frac{\sigma}{\sqrt{70}} = \frac{1}{\sqrt{70}} = 0.12 \text{ hour}$$

The distribution of $\bar{x}$ is therefore approximately $N(1, 0.12)$. Figure 5.11 shows this normal curve (*solid*) and also the actual density curve of $\bar{x}$ (*dashed*).

Because 50 minutes is 50/60 of an hour, or 0.83 hour, the probability we want is $P(\bar{x} > 0.83)$. A normal distribution calculation gives this probability as 0.9222. This is the area to the right of 0.83 under the solid normal curve in Figure 5.11. The exactly correct probability is the area under the dashed density curve in the figure. It is 0.9294. The central limit theorem normal approximation is off by only about 0.007.

The normal approximation for sample proportions and counts is an important example of the central limit theorem. This is true because a sample proportion can be thought of as a sample mean. Recall the idea that we used to find the mean and variance of a $B(n, p)$ random variable X. We wrote the count X as a sum

$$X = S_1 + S_2 + \cdots + S_n$$

of random variables S_i that take the value 1 if a success occurs on the ith trial and the value 0 otherwise. The variables S_i take only the values 0 and 1 and are far from normal. The proportion $\hat{p} = X/n$ is the sample mean of the S_i and, like all sample means, is approximately normal when n is large.

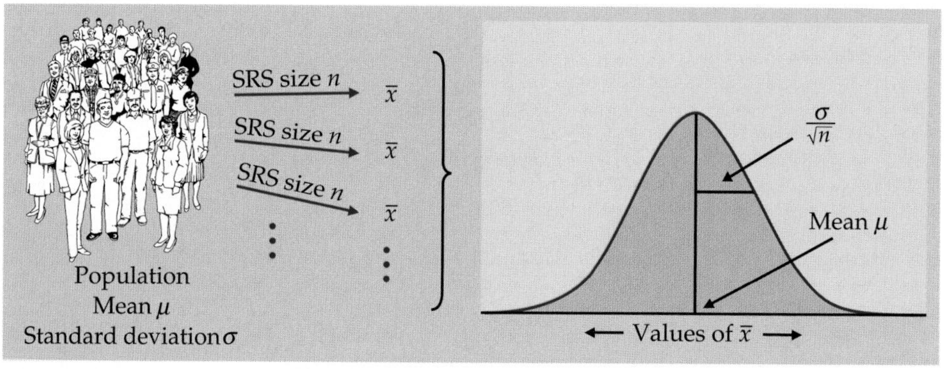

FIGURE 5.12 The sampling distribution of a sample mean $\bar{x}$ has mean μ and standard deviation $\sigma/\sqrt{n}$. The distribution is normal if the population distribution is normal; it is approximately normal for large samples in any case.

Figure 5.12 summarizes the facts about the sampling distribution of $\bar{x}$ in a way that reminds us of the big idea of a sampling distribution. Keep taking random samples of size n from a population with mean μ. Find the sample mean $\bar{x}$ for each sample. Collect all the $\bar{x}$'s and display their distribution. That's the sampling distribution of $\bar{x}$. Sampling distributions are the key to understanding statistical inference. Keep this figure in mind as you go forward.

Beyond the basics ▸ Weibull distributions

Our discussion of sampling distributions has concentrated on the binomial model for count data and the normal model for quantitative variables. These models are important in statistical practice, but simplicity also contributes to their popularity. The parameters p in the binomial model and μ in the normal model are easy to understand. To estimate them from data we use statistics $\hat{p}$ and $\bar{x}$ that are also easy to understand and that have simple sampling distributions.

There are many other probability distributions that are used to model data in various circumstances. Estimation and sampling distributions are often more complex than for the binomial and normal distributions. Here is an example.

EXAMPLE 5.20

Weibull distributions

The normal distributions are not suitable models for all quantitative variables. The time that a product, such as a computer disk drive, lasts before failing rarely has a normal distribution, for example. Another class of continuous distributions, the **Weibull distributions,** is used to model time to failure. For engineers studying the reliability of products, Weibull distributions are more common than normal distributions.

Figure 5.13 shows the density curves of three members of the Weibull family. Each describes a different type of distribution for the time to failure of a product. Figure 5.13(a) is a model for *infant mortality*. Many of these products fail immediately. If they do not fail at once, then most last a long time. The manufacturer tests these products and only ships the ones that do not fail immediately.

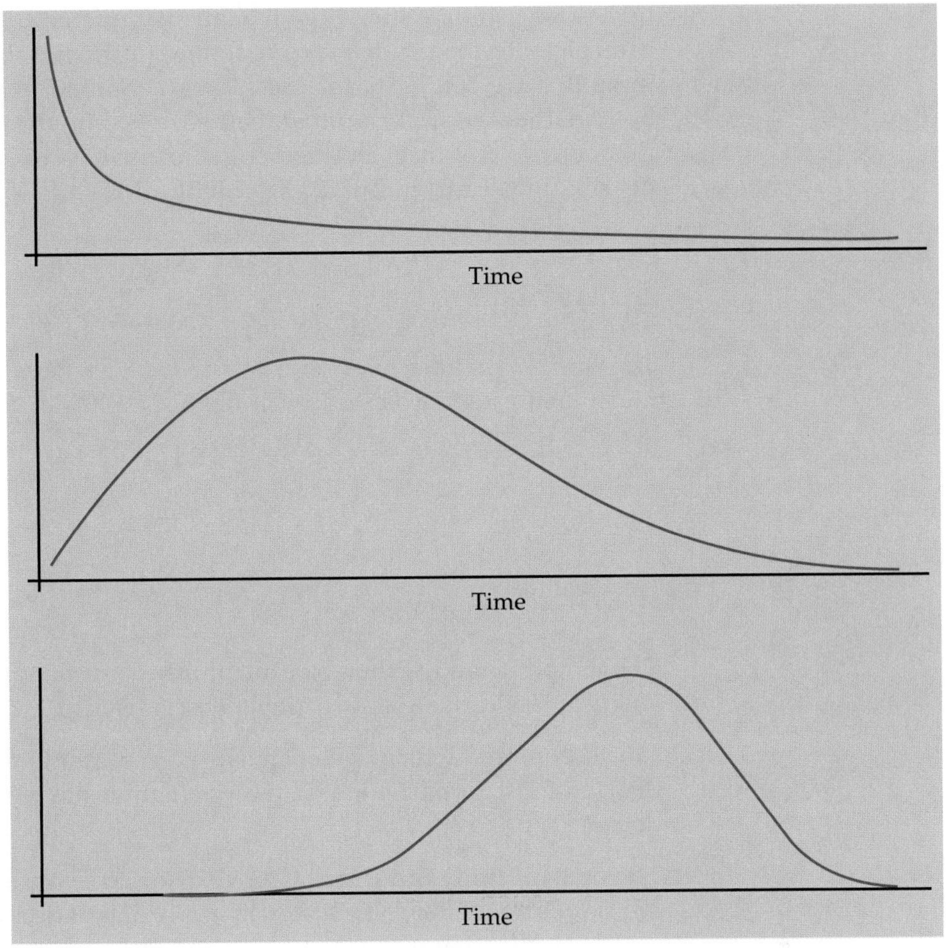

FIGURE 5.13 Density curves for three members of the Weibull family of distributions, for Example 5.20.

Figure 5.13(b) is a model for *early failure*. These products do not fail immediately, but many fail early in their lives after they are in the hands of customers. This is disastrous—the product or the process that makes it must be changed at once.

Figure 5.13(c) is a model for *old-age wearout*. Very few of these products fail until they begin to wear out, and then many fail at about the same age.

A manufacturer certainly wants to know to which of these classes a new product belongs. To find out, engineers operate a random sample of products until they fail. From the failure time data we can estimate the parameter (called the "shape parameter") that distinguishes among the three Weibull distributions in Figure 5.13. The shape parameter has no simple definition like that of a population proportion or mean, and it cannot be estimated by a simple statistic such as $\hat{p}$ or $\bar{x}$.

Two things save the situation. First, statistical theory provides general approaches for finding good estimates of any parameter. These general methods tell us to use $\hat{p}$ and $\bar{x}$ in the binomial and normal settings, and they also tell us how to estimate the Weibull shape parameter. Second, modern software can calculate the estimate from data even though there is no algebraic formula that we can write for the estimate. Statistical practice often relies on both mathematical theory and methods of computation more elaborate than the ones we will meet in this book. Fortunately, big ideas such as sampling distributions carry over to more complicated situations.[5]

SUMMARY

The **sample mean** $\bar{x}$ of an SRS of size n drawn from a large population with mean μ and standard deviation σ has a sampling distribution with mean and standard deviation

$$\mu_{\bar{x}} = \mu$$

$$\sigma_{\bar{x}} = \frac{\sigma}{\sqrt{n}}$$

The sample mean $\bar{x}$ is therefore an unbiased estimator of the population mean μ and is less variable than a single observation.

Linear combinations of independent normal random variables have normal distributions. In particular, if the population has a normal distribution, so does $\bar{x}$.

The **central limit theorem** states that for large n the sampling distribution of $\bar{x}$ is approximately $N(\mu, \sigma/\sqrt{n})$ for any population with mean μ and finite standard deviation σ.

SECTION 5.2 EXERCISES

5.28 Antonio makes a measurement in a chemistry laboratory and records the result in his lab report. The standard deviation of students' lab measurements is $\sigma = 10$ milligrams. Antonio repeats the measurement 3 times and records the mean $\bar{x}$ of his 3 measurements.

(a) What is the standard deviation $\sigma_{\bar{x}}$ of Antonio's mean result?

(b) How many times must Antonio repeat the measurement to reduce the standard deviation of $\bar{x}$ to 5? Explain to someone who knows no statistics the advantage of reporting the average of several measurements rather than the result of a single measurement.

5.29 A study of the health of teenagers plans to measure the blood cholesterol level of an SRS of youth of ages 13 to 16 years. The researchers will report the mean $\bar{x}$ from their sample as an estimate of the mean cholesterol level μ in this population.

(a) Explain to someone who knows no statistics what it means to say that $\bar{x}$ is an "unbiased" estimator of μ.

(b) The sample result $\overline{x}$ is an unbiased estimator of the population truth μ no matter what size SRS the study chooses. Explain to someone who knows no statistics why a large sample gives more trustworthy results than a small sample.

5.30 Newly manufactured automobile radiators may have small leaks. Most have no leaks, but some have 1, 2, or more. The number of leaks in radiators made by one supplier has mean 0.15 and standard deviation 0.4. The distribution of number of leaks cannot be normal because only whole-number counts are possible. The supplier ships 400 radiators per day to an auto assembly plant. Take $\overline{x}$ to be the mean number of leaks in these 400 radiators. Over several years of daily shipments, what range of values will contain the middle 95% of the many $\overline{x}$'s?

5.31 The scores of 12th-grade students on the National Assessment of Educational Progress year 2000 mathematics test have a distribution that is approximately normal with mean $\mu = 300$ and standard deviation $\sigma = 35$.

(a) Choose one 12th-grader at random. What is the probability that his or her score is higher than 300? Higher than 335?

(b) Now choose an SRS of four 12th-graders. What is the probability that their mean score is higher than 300? Higher than 335?

5.32 The scores of students on the ACT college entrance examination in 2001 had mean $\mu = 21.0$ and standard deviation $\sigma = 4.7$. The distribution of scores is only roughly normal.

(a) What is the approximate probability that a single student randomly chosen from all those taking the test scores 23 or higher?

(b) Now take an SRS of 50 students who took the test. What are the mean and standard deviation of the sample mean score $\overline{x}$ of these 50 students?

(c) What is the approximate probability that the mean score $\overline{x}$ of these students is 23 or higher?

(d) Which of your two normal probability calculations in (a) and (c) is more accurate? Why?

5.33 North Carolina State University posts the grade distributions for its courses online.[6] You can find that the distribution of grades in Accounting 210 in the spring 2001 semester was

Grade	A	B	C	D	F
Probability	0.18	0.32	0.34	0.09	0.07

(a) Using the common scale A = 4, B = 3, C = 2, D = 1, F = 0, take X to be the grade of a randomly chosen Accounting 201 student. Use the definitions of the mean (page 320) and standard deviation (page 328) for discrete random variables to find the mean μ and the standard deviation σ of grades in this course.

(b) Accounting 201 is a large course. We can take the grades of an SRS of 50 students to be independent of each other. If $\overline{x}$ is the average of these 50 grades, what are the mean and standard deviation of $\overline{x}$?

(c) What is the probability $P(X \geq 3)$ that a randomly chosen Accounting 201 student gets a B or better? What is the approximate probability $P(\overline{x} \geq 3)$ that the grade point average for 50 randomly chosen Accounting 201 students is B or better?

5.34 An automatic grinding machine in an auto parts plant prepares axles with a target diameter $\mu = 40.125$ millimeters (mm). The machine has some variability, so the standard deviation of the diameters is $\sigma = 0.002$ mm. A sample of 4 axles is inspected each hour for process control purposes, and records are kept of the sample mean diameter. What will be the mean and standard deviation of the numbers recorded?

5.35 Sheila's doctor is concerned that she may suffer from gestational diabetes (high blood glucose levels during pregnancy). There is variation both in the actual glucose level and in the blood test that measures the level. A patient is classified as having gestational diabetes if the glucose level is above 140 milligrams per deciliter (mg/dl) one hour after a sugary drink is ingested. Sheila's measured glucose level one hour after ingesting the sugary drink varies according to the normal distribution with $\mu = 125$ mg/dl and $\sigma = 10$ mg/dl.

(a) If a single glucose measurement is made, what is the probability that Sheila is diagnosed as having gestational diabetes?

(b) If measurements are made instead on 4 separate days and the mean result is compared with the criterion 140 mg/dl, what is the probability that Sheila is diagnosed as having gestational diabetes?

5.36 A roulette wheel has 38 slots, of which 18 are black, 18 are red, and 2 are green. When the wheel is spun, the ball is equally likely to come to rest in any of the slots. Gamblers can place a number of different bets in roulette. One of the simplest wagers chooses red or black. A bet of \$1 on red will pay off an additional dollar if the ball lands in a red slot. Otherwise, the player loses his dollar. When gamblers bet on red or black, the two green slots belong to the house.

(a) A gambler's winnings on a \$1 bet are either \$1 or $-\$1$. Give the probabilities of these outcomes. Find the mean and standard deviation of the gambler's winnings.

(b) Explain briefly what the law of large numbers tells us about what will happen if the gambler makes a large number of bets on red.

(c) The central limit theorem tells us the approximate distribution of the gambler's mean winnings in 50 bets. What is this distribution? Use the 68–95–99.7 rule to give the range in which the mean winnings will fall 95% of the time. Multiply by 50 to get the middle 95% of the distribution of the gambler's winnings on nights when he places 50 bets.

(d) What is the probability that the gambler will lose money if he makes 50 bets? (This is the probability that the mean is less than 0.)

(e) The casino takes the other side of these bets. If 100,000 bets on red are placed in a week at the casino, what is the distribution of the mean winnings of gamblers on these bets? What range covers the middle 95% of mean winnings in 100,000 bets? Multiply by 100,000 to get the range of gamblers' losses. (Gamblers' losses are the casino's winnings. Part (c) shows that a gambler gets excitement. Now we see that the casino has a business.)

5.37 In Exercise 5.35, Sheila's measured glucose level one hour after ingesting the sugary drink varies according to the normal distribution with $\mu = 125$ mg/dl and $\sigma = 10$ mg/dl. What is the level L such that there is probability only 0.05 that the mean glucose level of 4 test results falls above L for Sheila's glucose level distribution?

5.38 A company that owns and services a fleet of cars for its sales force has found that the service lifetime of disc brake pads varies from car to car according to a normal distribution with mean $\mu = 55,000$ miles and standard deviation $\sigma = 4500$ miles. The company installs a new brand of brake pads on 8 cars.

(a) If the new brand has the same lifetime distribution as the previous type, what is the distribution of the sample mean lifetime for the 8 cars?

(b) The average life of the pads on these 8 cars turns out to be $\bar{x} = 51{,}800$ miles. What is the probability that the sample mean lifetime is 51,800 miles or less if the lifetime distribution is unchanged? The company takes this probability as evidence that the average lifetime of the new brand of pads is less than 55,000 miles.

5.39 The number of flaws per square yard in a type of carpet material varies with mean 1.6 flaws per square yard and standard deviation 1.2 flaws per square yard. This population distribution cannot be normal, because a count takes only whole-number values. An inspector studies 200 square yards of the material, records the number of flaws found in each square yard, and calculates $\bar{x}$, the mean number of flaws per square yard inspected. Use the central limit theorem to find the approximate probability that the mean number of flaws exceeds 2 per square yard.

5.40 The number of accidents per week at a hazardous intersection varies with mean 2.2 and standard deviation 1.4. This distribution takes only whole-number values, so it is certainly not normal.

(a) Let $\bar{x}$ be the mean number of accidents per week at the intersection during a year (52 weeks). What is the approximate distribution of $\bar{x}$ according to the central limit theorem?

(b) What is the approximate probability that $\bar{x}$ is less than 2?

(c) What is the approximate probability that there are fewer than 100 accidents at the intersection in a year? (*Hint:* Restate this event in terms of $\bar{x}$.)

5.41 The idea of insurance is that we all face risks that are unlikely but carry high cost. Think of a fire destroying your home. So we form a group to share the risk: we all pay a small amount, and the insurance policy pays a large amount to those few of us whose homes burn down. An insurance company looks at the records for millions of homeowners and sees that the mean loss from fire in a year is $\mu = \$250$ per house and that the standard deviation of the loss is $\sigma = \$1000$. (The distribution of losses is extremely right-skewed: most people have \$0 loss, but a few have large losses.) The company plans to sell fire insurance for \$250 plus enough to cover its costs and profit.

(a) Explain clearly why it would be unwise to sell only 12 policies. Then explain why selling many thousands of such policies is a safe business.

(b) If the company sells 10,000 policies, what is the approximate probability that the average loss in a year will be greater than \$275?

5.42 The distribution of annual returns on common stocks is roughly symmetric, but extreme observations are more frequent than in a normal distribution. Because the distribution is not strongly nonnormal, the mean return over even a moderate number of years is close to normal. In the long run, annual real returns on common stocks have varied with mean about 9% and standard deviation about 28%. Andrew plans to retire in 45 years and is considering investing in stocks. What is the probability (assuming that the past pattern of variation continues) that the mean annual return on common stocks over the next 45 years will exceed 15%? What is the probability that the mean return will be less than 5%?

5.43 Children in kindergarten are sometimes given the Ravin Progressive Matrices Test (RPMT) to assess their readiness for learning. Experience at Southwark Elementary School suggests that the RPMT scores for its kindergarten pupils have mean 13.6 and standard deviation 3.1. The distribution is close to normal. Mr. Lavin has 22 children in his kindergarten class this year. He suspects that their RPMT scores will be unusually low because the test was interrupted by a fire drill. To check this suspicion, he wants to find the level L such that there is probability only 0.05 that the mean score of 22 children falls below L when the usual Southwark distribution remains true. What is the value of L?

5.44 The design of an electronic circuit calls for a 100-ohm resistor and a 250-ohm resistor connected in series so that their resistances add. The components used are not perfectly uniform, so that the actual resistances vary independently according to normal distributions. The resistance of 100-ohm resistors has mean 100 ohms and standard deviation 2.5 ohms, while that of 250-ohm resistors has mean 250 ohms and standard deviation 2.8 ohms.

(a) What is the distribution of the total resistance of the two components in series?

(b) What is the probability that the total resistance lies between 345 and 355 ohms?

5.45 An experiment to compare the nutritive value of normal corn and high-lysine corn divides 40 chicks at random into two groups of 20. One group is fed a diet based on normal corn while the other receives high-lysine corn. At the end of the experiment, inference about which diet is superior is based on the difference $\bar{y} - \bar{x}$ between the mean weight gain $\bar{y}$ of the 20 chicks in the high-lysine group and the mean weight gain $\bar{x}$ of the 20 in the normal-corn group. Because of the randomization, the two sample means are independent.

(a) Suppose that $\mu_X = 360$ grams (g) and $\sigma_X = 55$ g in the population of all chicks fed normal corn, and that $\mu_Y = 385$ g and $\sigma_Y = 50$ g in the high-lysine population. What are the mean and standard deviation of $\bar{y} - \bar{x}$?

(b) The weight gains are normally distributed in both populations. What is the distribution of $\bar{x}$? Of $\bar{y}$? What is the distribution of $\bar{y} - \bar{x}$?

(c) What is the probability that the mean weight gain in the high-lysine group exceeds the mean weight gain in the normal-corn group by 25 g or more?

5.46 An experiment on the teaching of reading compares two methods, A and B. The response variable is the Degree of Reading Power (DRP) score. The experimenter uses Method A in a class of 26 students and Method B in a comparable class of 24 students. The classes are assigned to the teaching methods at random. Suppose that in the population of all children of this age the DRP score has the $N(34, 12)$ distribution if Method A is used and the $N(37, 11)$ distribution if Method B is used.

(a) What is the distribution of the mean DRP score $\bar{x}$ for the 26 students in the A group? (Assume that this group can be regarded as an SRS from the population of all children of this age.)

(b) What is the distribution of the mean score $\bar{y}$ for the 24 students in the B group?

(c) Use the results of (a) and (b), keeping in mind that $\bar{x}$ and $\bar{y}$ are independent, to find the distribution of the difference $\bar{y} - \bar{x}$ between the mean scores in the two groups.

(d) What is the probability that the mean score for the B group will be at least 4 points higher than the mean score for the A group?

5.47 The two previous exercises illustrate a common setting for statistical inference. This exercise gives the general form of the sampling distribution needed in this setting. We have a sample of n observations from a treatment group and an independent sample of m observations from a control group. Suppose that the response to the treatment has the $N(\mu_X, \sigma_X)$ distribution and that the response of control subjects has the $N(\mu_Y, \sigma_Y)$ distribution. Inference about the difference $\mu_Y - \mu_X$ between the population means is based on the difference $\bar{y} - \bar{x}$ between the sample means in the two groups.

(a) Under the assumptions given, what is the distribution of $\bar{y}$? Of $\bar{x}$?

(b) What is the distribution of $\bar{y} - \bar{x}$?

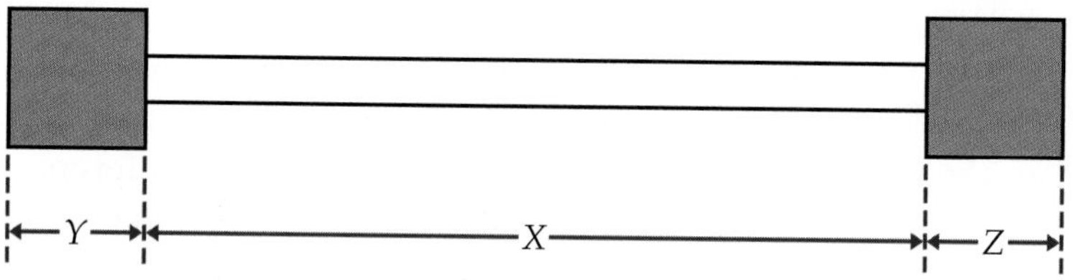

FIGURE 5.14 The dimensions of a mechanical assembly, for Exercise 5.48.

5.48 A mechanical assembly (Figure 5.14) consists of a shaft with a bearing at each end. The total length of the assembly is the sum $X + Y + Z$ of the shaft length X and the lengths Y and Z of the bearings. These lengths vary from part to part in production, independently of each other and with normal distributions. The shaft length X has mean 11.2 inches and standard deviation 0.002 inch, while each bearing length Y and Z has mean 0.4 inch and standard deviation 0.001 inch.

(a) According to the 68–95–99.7 rule, about 95% of all shafts have lengths in the range $11.2 \pm d_1$ inches. What is the value of d_1? Similarly, about 95% of the bearing lengths fall in the range $0.4 \pm d_2$. What is the value of d_2?

(b) It is common practice in industry to state the "natural tolerance" of parts in the form used in (a). An engineer who knows no statistics thinks that tolerances add, so that the natural tolerance for the total length of the assembly (shaft and two bearings) is $12 \pm d$ inches, where $d = d_1 + 2d_2$. Find the standard deviation of the total length $X + Y + Z$. Then find the value d such that about 95% of all assemblies have lengths in the range $12 \pm d$. Was the engineer correct?

5.49 Leona and Fred are friendly competitors in high school. Both are about to take the ACT college entrance examination. They agree that if one of them scores 5 or more points better than the other, the loser will buy the winner a pizza. Suppose that in fact Fred and Leona have equal ability, so that each score varies normally with mean 24 and standard deviation 2. (The variation is due to luck in guessing and the accident of the specific questions being familiar to the student.) The two scores are independent. What is the probability that the scores differ by 5 or more points in either direction?

5.50 ACT, Inc., the producer of the ACT test of readiness for college work, also produces tests for 8th- and 9th-grade students designed to help them plan for the future. Two of these tests measure reading and mathematics achievement. Each has scores that range from 1 to 25. For students tested in the fall of their 8th-grade year, the reading test has mean 13.9 and standard deviation 3.63. The mean score on the math test is 14.4 and the standard deviation is 3.46. Scores roughly follow a normal distribution.[7]

(a) If a student's reading score X and mathematics score Y were independent, what would be the distribution of total $X + Y$?

(b) Using the distribution you found in (a), what percent of 8th-graders have a total score of 30 or higher?

(c) In fact, the X and Y scores are strongly correlated. In this case, does the mean of $X + Y$ still have the value you found in (a)? Does the standard deviation still have the value you found in (a)?

5.51 A study of working couples measures the income X of the husband and the income Y of the wife in a large number of couples in which both partners are employed. Suppose that you knew the means μ_X and μ_Y and the variances σ_X^2 and σ_Y^2 of both variables in the population.

(a) Is it reasonable to take the mean of the total income $X + Y$ to be $\mu_X + \mu_Y$? Explain your answer.

(b) Is it reasonable to take the variance of the total income to be $\sigma_X^2 + \sigma_Y^2$? Explain your answer.

5.52 Table 1.5 (page 32) gives the survival times of 72 guinea pigs in a medical experiment. The distribution of survival times is strongly skewed to the right. Sampling from this small population can demonstrate how averaging reduces variability and creates a more normal distribution.

(a) Use software to choose 100 SRSs of size 12 from this population. Find the mean survival time $\bar{x}$ for each of the 100 samples.

(b) Find the mean μ of the population. Make a histogram of the 100 $\bar{x}$'s and find their mean. Is the sampling distribution of $\bar{x}$ centered near μ?

(c) Find the standard deviation σ of the population. Then find the standard deviation of the 100 $\bar{x}$'s. In the long run, what do you expect to be the standard deviation of the mean from samples of size $n = 12$? How close to this value did your 100 samples come?

(d) Make normal quantile plots of both the population and the 100 $\bar{x}$'s. Is the sampling distribution of the $\bar{x}$'s markedly closer to normal than the distribution of the population?

CHAPTER 5 EXERCISES

5.53 The probability that a randomly chosen driver will be involved in an accident in the next year is about 0.2. This is based on the proportion of millions of drivers who have accidents. "Accident" includes things like crumpling a fender in your own driveway, not just highway accidents. David, Carlos, Jermaine, Ramon, Sean, and Scott are college students who live together in an off-campus apartment. Last year, 3 of the 6 had accidents. What is the probability that 3 or more of 6 randomly chosen drivers have an accident in the same year? Why does your calculation not apply to drivers like the 6 students?

5.54 The distribution of scores for persons over 16 years of age on the Wechsler Adult Intelligence Scale (WAIS) is approximately normal with mean 100 and standard deviation 15. The WAIS is one of the most common "IQ tests" for adults.

(a) What is the probability that a randomly chosen individual has a WAIS score of 105 or higher?

(b) What are the mean and standard deviation of the average WAIS score $\overline{x}$ for an SRS of 60 people?

(c) What is the probability that the average WAIS score of an SRS of 60 people is 105 or higher?

(d) Would your answers to any of (a), (b), or (c) be affected if the distribution of WAIS scores in the adult population were distinctly nonnormal?

5.55 High school dropouts make up 12.1% of all Americans aged 18 to 24. A vocational school that wants to attract dropouts mails an advertising flyer to 25,000 persons between the ages of 18 and 24.

(a) If the mailing list can be considered a random sample of the population, what is the mean number of high school dropouts who will receive the flyer? What is the standard deviation of this number?

(b) What is the probability that at least 3500 dropouts will receive the flyer?

5.56 The Census Bureau says that the 10 most common names in the United States are (in order) Smith, Johnson, Williams, Jones, Brown, Davis, Miller, Wilson, Moore, and Taylor. These names account for 5.6% of all U.S. residents. Out of curiosity, you look at the authors of the textbooks for your current courses. There are 9 authors in all. Would you be surprised if none of the names of these authors were among the 10 most common? Give a probability to support your answer and explain the reasoning behind your calculation.

5.57 A political activist is gathering signatures on a petition by going door-to-door asking citizens to sign. She wants 100 signatures. Suppose that the probability of getting a signature at each household is 1/10 and let the random variable X be the number of households visited to collect exactly 100 signatures. Is it reasonable to use a binomial distribution for X? If so, give n and p. If not, explain why not.

5.58 According to genetic theory, the blossom color in the second generation of a certain cross of sweet peas should be red or white in a 3:1 ratio. That is, each plant has probability 3/4 of having red blossoms, and the blossom colors of separate plants are independent.

(a) What is the probability that exactly 6 out of 8 of these plants have red blossoms?

(b) What is the mean number of red-blossomed plants when 80 plants of this type are grown from seeds?

(c) What is the probability of obtaining at least 50 red-blossomed plants when 80 plants are grown from seeds?

5.59 The weight of the eggs produced by a certain breed of hen is normally distributed with mean 65 grams (g) and standard deviation 5 g. If cartons of such eggs can be considered to be SRSs of size 12 from the population of all eggs, what is the probability that the weight of a carton falls between 750 g and 825 g?

5.60 According to a market research firm, 52% of all residential telephone numbers in Los Angeles are unlisted. A telemarketing company uses random digit dialing equipment that dials residential numbers at random, regardless of whether they are listed in the telephone directory. The firm calls 500 numbers in Los Angeles.

(a) What is the exact distribution of the number X of unlisted numbers that are called?

(b) Use a suitable approximation to calculate the probability that at least half of the numbers called are unlisted.

5.61 A study of rush hour traffic in San Francisco records the number of people in each car entering a freeway at a suburban interchange. Suppose that this number X has mean 1.5 and standard deviation 0.75 in the population of all cars that enter at this interchange during rush hours.

(a) Does the count X have a binomial distribution? Why or why not?

(b) Could the exact distribution of X be normal? Why or why not?

(c) Traffic engineers estimate that the capacity of the interchange is 700 cars per hour. According to the central limit theorem, what is the approximate distribution of the mean number of persons $\bar{x}$ in 700 randomly selected cars at this interchange?

(d) The count of people in 700 cars is $700\bar{x}$. Use your result from (c) to give an approximate distribution for the count. What is the probability that 700 cars will carry more than 1075 people?

5.62 Serving in a bomber crew in World War II was dangerous. The British estimated that the probability of an aircraft loss due to enemy action was 1/20 for each mission. A tour of duty for British airmen in Bomber Command was 30 missions. What is the probability that an airman would complete a tour of duty without being on an aircraft lost from enemy action?

5.63 An opinion poll asks a sample of 500 adults whether they favor giving parents of school-age children vouchers that can be exchanged for education at any public or private school of their choice. Each school would be paid by the government on the basis of how many vouchers it collected. Suppose that in fact 45% of the population favor this idea. What is the probability that at least half of the sample are in favor? (Assume that the sample is an SRS.)

5.64 A machine fastens plastic screw-on caps onto containers of motor oil. If the machine applies more torque than the cap can withstand, the cap will break. Both the torque applied and the strength of the caps vary. The capping-machine torque has the normal distribution with mean 7 inch-pounds and standard deviation 0.9 inch-pounds. The cap strength (the torque that would break the cap) has the normal distribution with mean 10 inch-pounds and standard deviation 1.2 inch-pounds.

(a) Explain why it is reasonable to assume that the cap strength and the torque applied by the machine are independent.

(b) What is the probability that a cap will break while being fastened by the capping machine?

5.65 An important step in the manufacture of integrated circuit chips is etching the lines that will conduct current between components on the chip. The chips contain a line width test pattern that is used for process control measurements. Past data show that the line width varies in production according to a normal distribution with mean 2.829 micrometers (μm) and standard deviation 0.1516 μm. The company uses statistical process control to check the stability of the process. Samples of 5 chips are selected at regular intervals during production. The mean line width $\overline{x}$ on these chips is compared with "control limits" given by the 99.7 part of the 68–95–99.7 rule for normal distributions, namely $\mu_{\overline{x}} \pm 3\sigma_{\overline{x}}$. What are the numerical values of these control limits for $\overline{x}$?

5.66 In an experiment on learning foreign languages, researchers studied the effect of delaying oral practice when beginning language study. The researchers randomly assigned 23 beginning students of Russian to an experimental group and another 23 to a control group. The control group began speaking practice immediately while the experimental group delayed speaking for 4 weeks. At the end of the semester both groups took a standard test of comprehension of spoken Russian. Suppose that in the population of all beginning students, the test scores under the control method vary according to the $N(32, 6)$ distribution. The population distribution when oral practice is delayed is $N(29, 5)$.

(a) What is the sampling distribution of the mean score $\overline{x}$ in the control group in many repetitions of the experiment? What is the sampling distribution of the mean score $\overline{y}$ in the experimental group?

(b) If the experiment were repeated many times, what would be the sampling distribution of the difference $\overline{y} - \overline{x}$ between the mean scores in the two groups?

(c) What is the probability that the experiment will find (misleadingly) that the experimental group has a mean at least as large as that of the control group?

5.67 The World Health Organization MONICA Project collected health information from random samples of adults in many nations. One question asked was, "Have you had your blood pressure measured in the past year?" Of 186 American males aged 25 to 34, 153 said "Yes." For females in the same age group, 235 of the sample of size 248 said "Yes."[8] Let us suppose that for the entire population in this age group, 82% of men and 95% of women have had their blood pressure measured in the past year.

(a) What is the approximate distribution of the proportion $\hat{p}_1$ of "Yes" responses in an SRS of 186 men? Of the corresponding proportion $\hat{p}_2$ for an SRS of 248 women?

(b) The samples of women and men are of course independent. What is the approximate distribution of the difference $\hat{p}_2 - \hat{p}_1$?

(c) What is the approximate probability that the female proportion exceeds the male proportion by at least 10 percentage points?

Prelude: reacting to numbers

Ivan is the product manager in charge of blenders for a manufacturer of household appliances. Last month's sales report showed a 4% increase in blender sales to retailers, so Ivan treated the sales staff to lunch. This month, blender sales decline by 3%. Ivan is mystified—he can find no reason why sales should fall. Perhaps the sales force is getting lazy. Ivan expresses his disappointment, the sales staff are defensive, morale declines. . . .

Then Caroline, a student intern who has taken a statistics class, asks about the source of the sales data. They are, she is told, estimates based on a sample of customer orders. Complete sales data aren't available until orders are filled and paid for. Caroline looks at the month-to-month variation in past estimates and at the relationship of estimates to actual orders in the same month from all customers. Her report notes many reasons why the estimates will vary, including both normal month-to-month variation in blender orders and the additional variation due to using data from just a sample. Her report shows that changes of 3% or 4% up or down often occur simply because the data are from only a sample of orders. She even describes how large a change in the estimates provides good evidence that actual orders are really different from last month's total.

Ivan took every number as fixed and solid. He failed to grasp a statistical truism: we expect data to vary "just by chance." Only variation larger than typical chance variation is good evidence of a real change.

CHAPTER 6

Hans Hoffman, *Golden Splendor*.

Introduction to Inference

6.1 Estimating with Confidence
6.2 Tests of Significance
6.3 Use and Abuse of Tests
6.4 Power and Inference as a Decision

Introduction

The purpose of statistical inference is to draw conclusions from data. We have examined data and arrived at conclusions many times previously. Formal inference adds an emphasis on substantiating our conclusions by probability calculations.

Probability allows us to take chance variation into account and so to correct our judgment by calculation. Was the first Vietnam era draft lottery biased? The correlation between birth date and draft number was $r = -0.226$. Because any two variables will have some chance association in practice, this rather small correlation does not seem convincing. Calculation that a correlation this far from zero has probability less than 0.001 in a truly random lottery gives strong evidence that the lottery was in fact unfair. Our unaided judgment can also err in the opposite direction, seeing a systematic effect when only chance variation is at work. Give a new drug and a placebo to 20 patients each; 12 of those taking the drug show improvement, but only 8 of the placebo patients improve. Is the drug more effective than the placebo? Perhaps, but a difference this large or larger between the results in the two groups would occur about one time in five simply because of chance variation. An effect that could so easily be just chance is not convincing.

In this chapter we introduce the two most prominent types of formal statistical inference. Section 6.1 concerns *confidence intervals* for estimating the value of a population parameter. Section 6.2 presents *tests of significance*, which assess the evidence for a claim. Both types of inference are based on the sampling distributions of statistics. That is, both report probabilities that state *what would happen if we used the inference method many times*. This kind of probability statement is characteristic of standard statistical inference. Users of statistics must understand the nature of the reasoning employed and the meaning of the probability statements that appear, for example, on computer output for statistical procedures.

Because the methods of formal inference are based on sampling distributions, they require a probability model for the data. Trustworthy probability models can arise in many ways, but the model is most secure and inference is most reliable when the data are produced by a properly randomized design. *When you use statistical inference you are acting as if the data come from a random sample or a randomized experiment.* If this is not true, your conclusions may be open to challenge. Do not be overly impressed by the complex details of formal inference. This elaborate machinery cannot remedy basic flaws in producing the data such as voluntary response samples and confounded experiments. Use the common sense developed in your study of the first three chapters of this book, and proceed to detailed formal inference only when you are satisfied that the data deserve such analysis.

The primary purpose of this chapter is to describe the reasoning used in statistical inference. We will discuss only a few specific inference techniques, and these require an unrealistic assumption: that we know the standard deviation σ. Later chapters will present inference methods for use in most of the settings we met in learning to explore data. There are libraries—both of books and of computer software—full of more elaborate statistical techniques. Informed use of any of these methods requires an understanding of the underlying reasoning. A computer will do the arithmetic, but you must still exercise judgment based on understanding.

6.1 Estimating with Confidence

The SAT* tests are widely used measures of readiness for college study. There are two parts, one for verbal reasoning ability (SATV) and one for mathematical reasoning ability (SATM). In April 1995, the scores were *recentered* so that the mean is approximately 500 in a large "standardization group." This scale is maintained from year to year so that scores have a constant interpretation. In 1996, 1,084,725 college-bound seniors took the SAT. Their mean SATV score was 505 with a standard deviation of 110. For the SATM the mean was 508 with a standard deviation of 112.

EXAMPLE 6.1

You want to estimate the mean SATM score for the more than 250,000 high school seniors in California. You know better than to trust data from the students who choose to take the SAT. Only about 45% of California students take the SAT. These self-selected students are planning to attend college and are not representative of all California seniors. At considerable effort and expense, you give the test to a simple random sample (SRS) of 500 California high school seniors. The mean score for your sample is $\bar{x} = 461$. What can you say about the mean score μ in the population of all 250,000 seniors?

The sample mean $\bar{x}$ is the natural estimator of the unknown population mean μ. We know that $\bar{x}$ is an unbiased estimator of μ. More important, the law of large numbers says that the sample mean must approach the population mean as the size of the sample grows. The value $\bar{x} = 461$ therefore appears to be a reasonable estimate of the mean score μ that all 250,000 students would achieve if they took the test. But how reliable is this estimate? A second sample would surely not give 461 again. Unbiasedness says only that there is no systematic tendency to underestimate or overestimate the truth. Could we plausibly get a sample mean of 530 or 410 on repeated samples? An estimate without an indication of its variability is of little value.

Statistical confidence

Just as unbiasedness of an estimator concerns the center of its sampling distribution, questions about variation are answered by looking at the spread. We know that if the entire population of SAT scores has mean μ and standard deviation σ, then in repeated samples of size 500 the sample mean $\bar{x}$ has a $N(\mu, \sigma/\sqrt{500})$ distribution. Let us suppose that we know that the standard deviation σ of SATM scores in our California population is $\sigma = 100$. (This is not realistic. We will see in the next chapter how to proceed when σ is not known. For now, we are more interested in statistical reasoning than in

*These tests have been revised and are now called SAT I—Reasoning tests.

details of realistic methods.) In repeated sampling the sample mean $\bar{x}$ follows the normal distribution centered at the unknown population mean μ and having standard deviation

$$\sigma_{\bar{x}} = \frac{100}{\sqrt{500}} \doteq 4.5$$

Now we are in business. Consider this line of thought, which is illustrated by Figure 6.1:

- The 68–95–99.7 rule says that the probability is about 0.95 that $\bar{x}$ will be within 9 points (two standard deviations of $\bar{x}$) of the population mean score μ.
- To say that $\bar{x}$ lies within 9 points of μ is the same as saying that μ is within 9 points of $\bar{x}$.
- So 95% of all samples will capture the true μ in the interval from $\bar{x} - 9$ to $\bar{x} + 9$.

We have simply restated a fact about the sampling distribution of $\bar{x}$. *The language of statistical inference uses this fact about what would happen in the long run to express our confidence in the results of any one sample.* Our sample gave $\bar{x} = 461$. We say that we are *95% confident* that the unknown mean score for all California seniors lies between

$$\bar{x} - 9 = 461 - 9 = 452$$

and

$$\bar{x} + 9 = 461 + 9 = 470$$

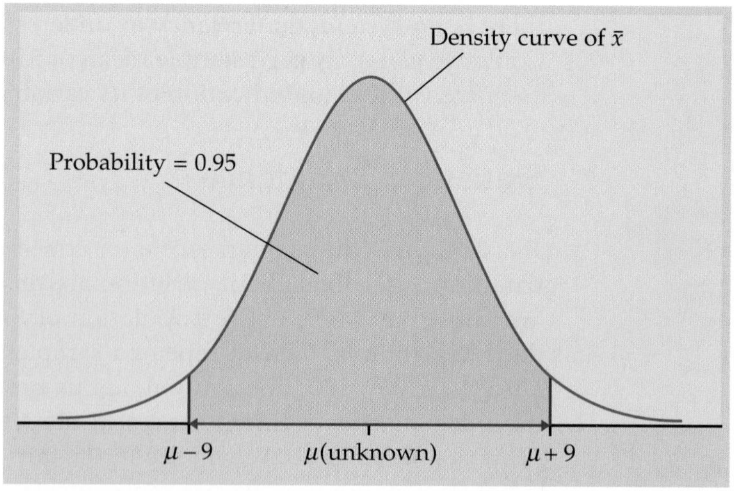

FIGURE 6.1 $\bar{x}$ lies within ± 9 of μ in 95% of all samples, so μ also lies within ± 9 of $\bar{x}$ in those samples.

Be sure you understand the grounds for our confidence. There are only two possibilities:

1. The interval between 452 and 470 contains the true μ.
2. Our SRS was one of the few samples for which $\overline{x}$ is not within 9 points of the true μ. Only 5% of all samples give such inaccurate results.

We cannot know whether our sample is one of the 95% for which the interval $\overline{x} \pm 9$ catches μ or one of the unlucky 5%. The statement that we are 95% confident that the unknown μ lies between 452 and 470 is shorthand for saying, "We arrived at these numbers by a method that gives correct results 95% of the time."

Confidence intervals

The interval of numbers between the values $\overline{x} \pm 9$ is called a *95% confidence interval* for μ. Like most confidence intervals we will meet, this one has the form

$$\text{estimate} \pm \text{margin of error}$$

margin of error

The estimate ($\overline{x}$ in this case) is our guess for the value of the unknown parameter. The **margin of error** ± 9 shows how accurate we believe our guess is, based on the variability of the estimate. The confidence level shows how confident we are that the procedure will catch the true population mean μ.

Figure 6.2 illustrates the behavior of 95% confidence intervals in repeated sampling. The center of each interval is at $\overline{x}$ and therefore varies from sample to sample. The sampling distribution of $\overline{x}$ appears at the top of the figure to show the long-term pattern of this variation. The 95% confidence intervals $\overline{x} \pm 9$ from 25 SRSs appear below. The center $\overline{x}$ of each interval is marked by a dot. The arrows on either side of the dot span the confidence interval. All except one of the 25 intervals cover the true value of μ. In a very large number of samples, 95% of the confidence intervals would contain μ.*

Statisticians have constructed confidence intervals for many different parameters based on a variety of designs for data collection. We will meet a number of these in later chapters. You need to know two important things about a confidence interval:

1. It is an interval of the form (a, b), where a and b are numbers computed from the data.
2. It has a property called a confidence level that gives the probability that the interval covers the parameter.

*Confidence intervals as a systematic method were invented in 1937 by Jerzy Neyman (1894–1981). Neyman was a Pole who moved to England in 1934 and spent the last half of his long life at the University of California at Berkeley. He remained active until his death, almost doubling the list of his publications after his official retirement in 1961.

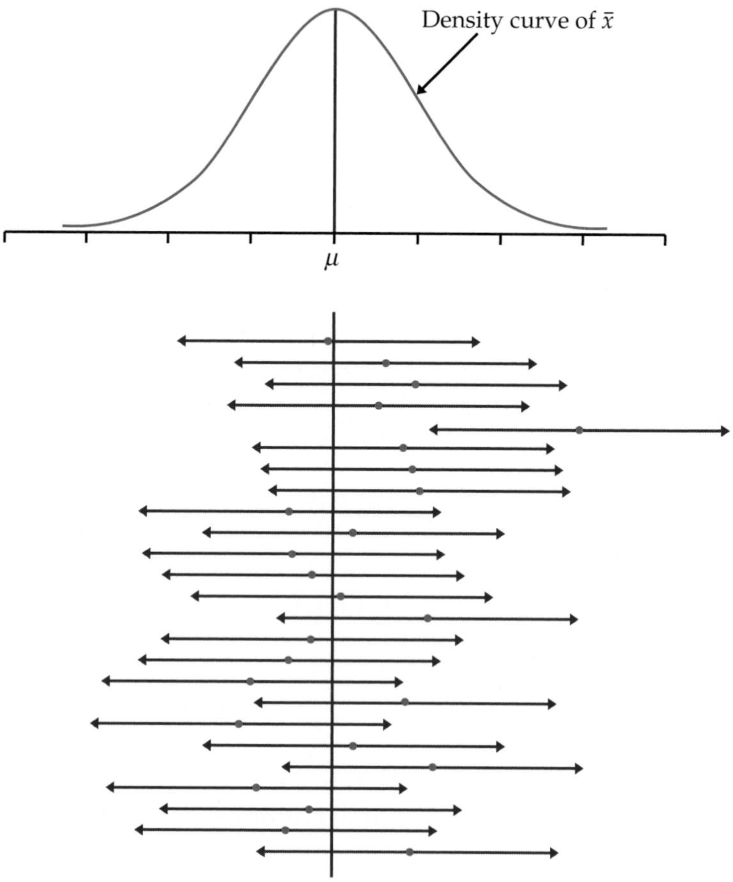

FIGURE 6.2 Twenty-five samples from the same population gave these 95% confidence intervals. In the long run, 95% of all samples give an interval that covers μ.

Users can choose the confidence level, most often 90% or higher because we most often want to be quite sure of our conclusions. We will use C to stand for the confidence level in decimal form. For example, a 95% confidence level corresponds to $C = 0.95$.

> **Confidence Interval**
>
> A level C confidence interval for a parameter is an interval computed from sample data by a method that has probability C of producing an interval containing the true value of the parameter.

Confidence interval for a population mean

We will now construct a level C confidence interval for the mean μ of a population when the data are an SRS of size n. The construction is based on the sampling distribution of the sample mean $\bar{x}$. This distribution is exactly

$N(\mu, \sigma/\sqrt{n})$ when the population has the $N(\mu, \sigma)$ distribution. The central limit theorem says that this same sampling distribution is approximately correct for large samples whenever the population mean and standard deviation are μ and σ.

Our construction of a 95% confidence interval for the mean SAT score began by noting that any normal distribution has probability about 0.95 within ± 2 standard deviations of its mean. To construct a level C confidence interval we first catch the central C area under a normal curve. That is, we must find the number z^* such that any normal distribution has probability C within $\pm z^*$ standard deviations of its mean. Because all normal distributions have the same standardized form, we can obtain everything we need from the standard normal curve. Figure 6.3 shows how C and z^* are related. Values of z^* for many choices of C appear in the row labeled z^* in Table D at the back of the book. Here are the most important entries from that part of the table:

z^*	1.645	1.960	2.576
C	90%	95%	99%

Any normal curve has probability C between the point z^* standard deviations below the mean and the point z^* above the mean, as Figure 6.3 reminds us. The sample mean $\bar{x}$ has the normal distribution with mean μ and standard deviation $\sigma/\sqrt{n}$. So there is probability C that $\bar{x}$ lies between

$$\mu - z^* \frac{\sigma}{\sqrt{n}} \quad \text{and} \quad \mu + z^* \frac{\sigma}{\sqrt{n}}$$

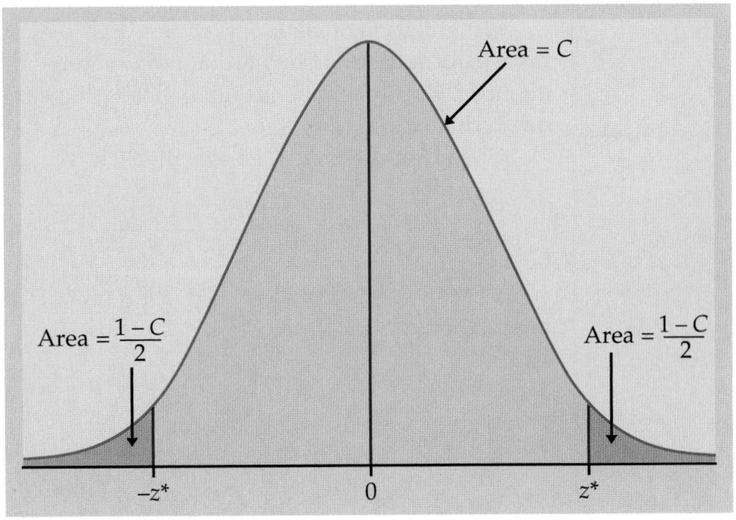

FIGURE 6.3 The area between $-z^*$ and z^* under the standard normal curve is C.

This is exactly the same as saying that the unknown population mean μ lies between

$$\bar{x} - z^* \frac{\sigma}{\sqrt{n}} \quad \text{and} \quad \bar{x} + z^* \frac{\sigma}{\sqrt{n}}$$

That is, there is probability C that the interval $\bar{x} \pm z^*\sigma/\sqrt{n}$ contains μ. That is our confidence interval. The estimate of the unknown μ is $\bar{x}$, and the margin of error is $z^*\sigma/\sqrt{n}$.

Confidence Interval for a Population Mean

Choose an SRS of size n from a population having unknown mean μ and known standard deviation σ. A level C confidence interval for μ is

$$\bar{x} \pm z^* \frac{\sigma}{\sqrt{n}}$$

Here z^* is the value on the standard normal curve with area C between $-z^*$ and z^*. This interval is exact when the population distribution is normal and is approximately correct for large n in other cases.

EXAMPLE 6.2

Tim Kelley has been weighing himself once a week for several years. Last month his four measurements (in pounds) were

190.5 189.0 195.5 187.0

Give a 90% confidence interval for his mean weight for last month.

We treat the four measurements as an SRS of all possible measurements that Tim could have taken last month. These are estimates of μ, his true mean weight for last month.

Examination of Tim's past data reveals that over relatively short periods of time, his weight measurements are approximately normal with a standard deviation of about 3. For our confidence interval we will use this value as the true standard deviation, that is, $\sigma = 3$.

The mean of Tim's weight readings is

$$\bar{x} = \frac{190.5 + 189.0 + 195.5 + 187.0}{4} = 190.5$$

For 90% confidence, we see from Table D that $z^* = 1.645$. A 90% confidence interval for μ is therefore

$$\begin{aligned} \bar{x} \pm z^* \frac{\sigma}{\sqrt{n}} &= 190.5 \pm 1.645 \frac{3}{\sqrt{4}} \\ &= 190.5 \pm 2.5 \\ &= (188.0, 193.0) \end{aligned}$$

We are 90% confident that Tim's mean weight last month was between 188 and 193 pounds.

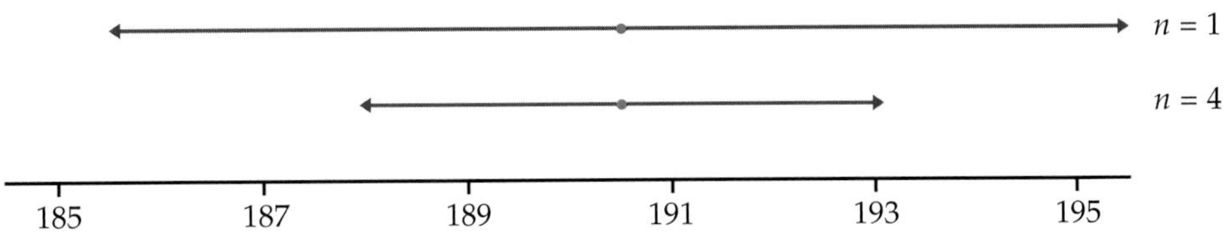

FIGURE 6.4 Confidence intervals for $n = 4$ and $n = 1$, for Examples 6.2 and 6.3.

In this example we used a value for σ based on past experience. This is a useful choice, particularly when the sample size is very small. Under these circumstances, the sample standard deviation can be less accurate than a value based on past data. When $n = 1$, we have an extreme case where it is not possible to calculate the sample standard deviation. Why not?

EXAMPLE 6.3

Suppose Tim Kelley had weighed himself only once last month, and that his one observation was $x = 190.5$, the same as the mean $\bar{x}$ in Example 6.2. Repeating the calculation with $n = 1$ shows that the 90% confidence interval based on a single measurement is

$$\begin{aligned} \bar{x} \pm z^* \frac{\sigma}{\sqrt{1}} &= 190.5 \pm (1.645)(3) \\ &= 190.5 \pm 4.9 \\ &= (185.6, 195.4) \end{aligned}$$

The mean of four measurements gives a margin of error that is half the size of the margin of error for a sample of size one. Figure 6.4 illustrates the gain from using four observations.

The argument leading to the form of confidence intervals for the population mean μ rested on the fact that the statistic $\bar{x}$ used to estimate μ has a normal distribution. Because many sample estimates have normal distributions (at least approximately), it is useful to notice that the confidence interval has the form

$$\text{estimate} \pm z^* \sigma_{\text{estimate}}$$

The estimate based on the sample is the center of the confidence interval. The margin of error is $z^* \sigma_{\text{estimate}}$. The desired confidence level determines z^* from Table D. The standard deviation of the estimate is found from a knowledge of the sampling distribution in a particular case. When the estimate is $\bar{x}$ from an SRS, the standard deviation of the estimate is $\sigma/\sqrt{n}$.

How confidence intervals behave

The margin of error $z^* \sigma/\sqrt{n}$ for the mean of a normal population illustrates several important properties that are shared by all confidence intervals in common use. The user chooses the confidence level, and the margin of error follows from this choice. High confidence is desirable and so is a small margin of error. High confidence says that our method almost always gives correct answers. A small margin of error says that we have pinned down the parameter quite precisely.

Suppose that you calculate a margin of error and decide that it is too large. Here are your choices to reduce it:

- Use a lower level of confidence (smaller C).
- Increase the sample size (larger n).
- Reduce σ.

For most problems you would choose a confidence level of 90%, 95%, or 99%. So z^* can be 1.645, 1.960, or 2.576. A look at Figure 6.3 will convince you that z^* will be smaller for lower confidence (smaller C). Table D shows that this is indeed the case. If n and σ are unchanged, a smaller z^* leads to a smaller margin of error. Similarly, increasing the sample size n reduces the margin of error for any fixed confidence level. The square root in the formula implies that we must multiply the number of observations by 4 in order to cut the margin of error in half. The standard deviation σ measures the variation in the population. You can think of the variation among individuals in the population as noise that obscures the average value μ. It is harder to pin down the mean μ of a highly variable population; that is why the margin of error of a confidence interval increases with σ. In practice, we can sometimes reduce σ by carefully controlling the measurement process or by restricting our attention to only part of a large population.

EXAMPLE 6.4

Suppose that Tim Kelley in Example 6.2 wants 99% confidence rather than 90%. Table D tells us that for 99% confidence, $z^* = 2.576$. The margin of error for 99% confidence based on four repeated measurements is

$$\begin{aligned} z^*\frac{\sigma}{\sqrt{n}} &= 2.576\frac{3}{\sqrt{4}} \\ &= 3.9 \end{aligned}$$

The 99% confidence interval is

$$\begin{aligned} \bar{x} \pm \text{margin of error} &= 190.5 \pm 3.9 \\ &= (186.6, 194.4) \end{aligned}$$

Requiring 99%, rather than 90%, confidence has increased the margin of error from 2.5 to 3.9. Figure 6.5 compares the two intervals.

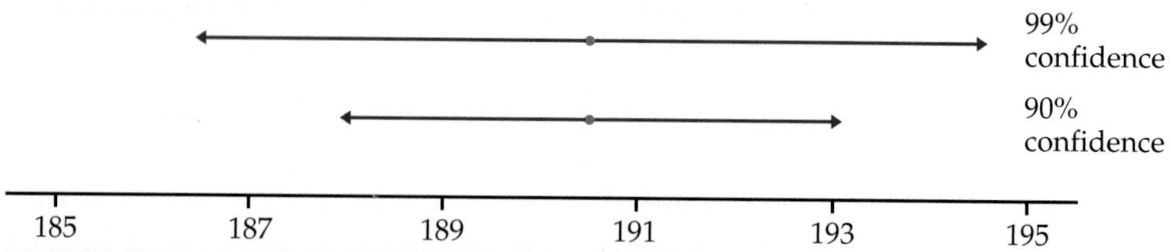

FIGURE 6.5 Confidence intervals for Examples 6.2 and 6.4.

Choosing the sample size

A wise user of statistics never plans data collection without at the same time planning the inference. You can arrange to have both high confidence and a small margin of error. The margin of error of the confidence interval $\bar{x} \pm z^*\sigma/\sqrt{n}$ for a normal mean is

$$z^* \frac{\sigma}{\sqrt{n}}$$

To obtain a desired margin of error m, just set this expression equal to m, substitute the value of z^* for your desired confidence level, and solve for the sample size n. Here is the result.

> **Sample Size for Desired Margin of Error**
>
> The confidence interval for a population mean will have a specified margin of error m when the sample size is
>
> $$n = \left(\frac{z^*\sigma}{m}\right)^2$$

This formula is not the proverbial free lunch. In practice, taking observations costs time and money. The required sample size may be impossibly expensive. Do notice once again that it is the size of the *sample* that determines the margin of error. The size of the *population* (as long as the population is much larger than the sample) does not influence the sample size we need.

EXAMPLE 6.5

Tim Kelley of Example 6.2 has decided that he wants his estimate of his monthly weight accurate to within 2 or 3 pounds with 95% confidence. How many measurements must he take to achieve these margins of error?

For 95% confidence, Table D gives $z^* = 1.960$. For a margin of error of 2 pounds we have

$$n = \left(\frac{z^*\sigma}{m}\right)^2 = \left(\frac{1.96 \times 3}{2}\right)^2 = 8.6$$

Because 8 measurements will give a slightly wider interval than desired and 9 measurements a slightly narrower interval, Tim must take 9 measurements for his estimate to be within 2 pounds of the true value with 95% confidence. (Always round *up* to the next higher whole number when finding n.) For a margin of error of 3 pounds, the formula gives

$$n = \left(\frac{z^*\sigma}{m}\right)^2 = \left(\frac{1.96 \times 3}{3}\right)^2 = 3.8$$

Four observations per month would be sufficient. Tim chooses this option.

Some cautions

We have already seen that small margins of error and high confidence can require large numbers of observations. You should also be keenly aware that *any formula for inference is correct only in specific circumstances*. If the government required statistical procedures to carry warning labels like those on drugs, most inference methods would have long labels indeed. Our handy formula $\bar{x} \pm z^*\sigma/\sqrt{n}$ for estimating a normal mean comes with the following list of warnings for the user:

- The data must be an SRS from the population. We are completely safe if we actually did a randomization and drew an SRS. We are not in great danger if the data can plausibly be thought of as independent observations from a population. That is the case in Examples 6.2 to 6.5, where we have in mind the population that would result if Tim Kelley weighed himself very many times.
- The formula is not correct for probability sampling designs more complex than an SRS. Correct methods for other designs are available. We will not discuss confidence intervals based on multistage or stratified samples. If you plan such samples, be sure that you (or your statistical consultant) know how to carry out the inference you desire.
- There is no correct method for inference from data haphazardly collected with bias of unknown size. Fancy formulas cannot rescue badly produced data.
- Because $\bar{x}$ is not resistant, outliers can have a large effect on the confidence interval. You should search for outliers and try to correct them or justify their removal before computing the interval. If the outliers cannot be removed, ask your statistical consultant about procedures that are not sensitive to outliers.
- If the sample size is small and the population is not normal, the true confidence level will be different from the value C used in computing the interval. Examine your data carefully for skewness and other signs of nonnormality. The interval relies only on the distribution of $\bar{x}$, which even for quite small sample sizes is much closer to normal than that of the individual observations. When $n \geq 15$, the confidence level is not greatly disturbed by nonnormal populations unless extreme outliers or quite strong skewness are present. We will discuss this issue in more detail in the next chapter.
- You must know the standard deviation σ of the population. This unrealistic requirement renders the interval $\bar{x} \pm z^*\sigma/\sqrt{n}$ of little use in statistical practice. We will learn in the next chapter what to do when σ is unknown. If, however, the sample is large, the sample standard deviation s will be close to the unknown σ. The recipe $\bar{x} \pm z^*s/\sqrt{n}$ is then an approximate confidence interval for μ.

The most important caution concerning confidence intervals is a consequence of the first of these warnings. *The margin of error in a confidence inter-*

val covers only random sampling errors. The margin of error is obtained from the sampling distribution and indicates how much error can be expected because of chance variation in randomized data production. Practical difficulties such as undercoverage and nonresponse in a sample survey cause additional errors. These errors can be larger than the random sampling error, particularly when the sample size is large (so that $\sigma/\sqrt{n}$ is small). Remember this unpleasant fact when reading the results of an opinion poll or other sample survey. The practical conduct of the survey influences the trustworthiness of its results in ways that are not included in the announced margin of error.

Every inference procedure that we will meet has its own list of warnings. Because many of the warnings are similar to those above, we will not print the full warning label each time. It is easy to state (from the mathematics of probability) conditions under which a method of inference is exactly correct. These conditions are *never* fully met in practice. For example, no population is exactly normal. Deciding when a statistical procedure should be used in practice often requires judgment assisted by exploratory analysis of the data. Mathematical facts are therefore only a part of statistics. The difference between statistics and mathematics can be stated thus: mathematical theorems are true; statistical methods are often effective when used with skill.

Finally, you should understand what statistical confidence does not say. We are 95% confident that the mean SATM score for the California students in Example 6.1 lies between 452 and 470. This says that these numbers were calculated by a method that gives correct results in 95% of all possible samples. It does *not* say that the probability is 95% that the true mean falls between 452 and 470. No randomness remains after we draw a particular sample and get from it a particular interval. The true mean either is or is not between 452 and 470. Probability in its interpretation as long-term relative frequency makes no sense in this situation. The probability calculations of standard statistical inference describe how often the *method* gives correct answers.

Beyond the basics ▸ the bootstrap

Confidence intervals are based on sampling distributions. In this section we have used the fact that the sampling distribution of $\overline{x}$ is $N(\mu, \sigma/\sqrt{n})$ when the data are an SRS from a $N(\mu, \sigma)$ population. If the data are not normal, the central limit theorem tells us that this sampling distribution is still a reasonable approximation as long as the distribution of the data is not strongly skewed and there are no outliers. We used the sampling distribution of $\overline{x}$ to conclude that, for example, 95% of the time, $\overline{x}$ and μ will not differ by more than $1.96\sigma/\sqrt{n}$.

What if the population does not appear to be normal and we have only a small sample? Then we do not know what the sampling distribution of $\overline{x}$ looks like. The **bootstrap** is a new procedure for approximating sampling distributions when theory cannot tell us their shape.[1]

bootstrap

The basic idea is to act as if our sample were the population. We take many samples from it. Each of these is called a **resample.** We calculate the mean $\overline{x}$ for each resample. We get different results from different resamples because

resample

we sample *with replacement*. An individual observation in the original sample can appear more than once in the resample.

For example, if we start with Tim Kelley's weight observations from Example 6.2,

190.5 189.0 195.5 187.0

one resample with replacement might be

189.0 195.5 187.0 189.0

with $\overline{x} = 190.125$. Collect the $\overline{x}$'s from 1000 such resamples. Their distribution will be close to what we would get if we took 1000 samples from the entire population. We treat the distribution of $\overline{x}$'s from our 1000 resamples as if it were the sampling distribution. If we want a 95% confidence interval, for example, we use the middle 95% of this distribution.

The bootstrap is practical only when you can use a computer to take 1000 or more samples quickly. It is an example of how the use of fast and easy computing is changing the way we do statistics.

SUMMARY

The purpose of a **confidence interval** is to estimate an unknown parameter with an indication of how accurate the estimate is and of how confident we are that the result is correct.

Any confidence interval has two parts: an interval computed from the data and a confidence level. The interval often has the form

estimate ± margin of error

The **confidence level** states the probability that the method will give a correct answer. That is, if you use 95% confidence intervals often, in the long run 95% of your intervals will contain the true parameter value. When you apply the method once, you do not know if your interval gave a correct value (this happens 95% of the time) or not (this happens 5% of the time).

A level C confidence interval for the mean μ of a normal population with known standard deviation σ, based on an SRS of size n, is given by

$$\overline{x} \pm z^* \frac{\sigma}{\sqrt{n}}$$

Here z^* is obtained from the bottom row in Table D. The probability is C that a standard normal random variable takes a value between $-z^*$ and z^*.

Other things being equal, the margin of error of a confidence interval decreases as

- the confidence level C decreases,
- the sample size n increases, and
- the population standard deviation σ decreases.

The sample size required to obtain a confidence interval of specified margin of error m for a normal mean is

$$n = \left(\frac{z^*\sigma}{m}\right)^2$$

where z^* is the critical point for the desired level of confidence.

A specific confidence interval recipe is correct only under specific conditions. The most important conditions concern the method used to produce the data. Other factors such as the form of the population distribution may also be important.

SECTION 6.1 EXERCISES

6.1 A survey of users of the Internet found that males outnumbered females by nearly 2 to 1. This was a surprise, because earlier surveys had put the ratio of men to women closer to 9 to 1. Later in the article we find this information:

> *Detailed surveys were sent to more than 13,000 organizations on the Internet; 1,468 usable responses were received. According to Mr. Quarterman, the margin of error is 2.8 percent, with a confidence level of 95 percent.*[2]

(a) What was the *response rate* for this survey? (The response rate is the percent of the planned sample that responded.)

(b) Do you think that the small margin of error is a good measure of the accuracy of the survey's results? Explain your answer.

6.2 You want to rent an unfurnished one-bedroom apartment for next semester. The mean monthly rent for a random sample of 10 apartments advertised in the local newspaper is \$540. Assume that the standard deviation is \$80. Find a 95% confidence interval for the mean monthly rent for unfurnished one-bedroom apartments available for rent in this community.

6.3 A questionnaire about study habits was given to a random sample of students taking a large introductory statistics class. The sample of 25 students reported that they spent an average of 110 minutes per week studying statistics. Assume that the standard deviation is 40 minutes.

(a) Give a 95% confidence interval for the mean time spent studying statistics by students in this class.

(b) Is it true that 95% of the students in the class have weekly study times that lie in the interval you found in part (a)? Explain your answer.

6.4 Refer to the previous exercise.

(a) Give the mean and standard deviation in hours.

(b) Calculate the 95% confidence interval in hours from your answer to part (a).

(c) Explain how you could have directly calculated this interval from the 95% interval that you calculated in the previous exercise.

6.5 The mean amount μ for all of the invoices for your company last month is not known. Based on your past experience, you are willing to assume that the standard deviation of invoice amounts is about \$200.

(a) If you take a random sample of 100 invoices, what is the value of the standard deviation for $\bar{x}$?

(b) The 68–95–99.7 rule says that the probability is about 0.95 that $\bar{x}$ is within ________ of the population mean μ. Fill in the blank.

(c) About 95% of all samples will capture the true mean of all of the invoices in the interval $\bar{x}$ plus or minus ________. Fill in the blank.

6.6 You measure the weights of 24 male runners. You do not actually choose an SRS, but you are willing to assume that these runners are a random sample from the population of male runners in your town or city. Here are their weights in kilograms:

67.8	61.9	63.0	53.1	62.3	59.7	55.4	58.9
60.9	69.2	63.7	68.3	64.7	65.6	56.0	57.8
66.0	62.9	53.6	65.0	55.8	60.4	69.3	61.7

Suppose that the standard deviation of the population is known to be $\sigma = 4.5$ kg.

(a) What is $\sigma_{\bar{x}}$, the standard deviation of $\bar{x}$?

(b) Give a 95% confidence interval for μ, the mean of the population from which the sample is drawn. Are you quite sure that the average weight of the population of runners is less than 65 kg?

6.7 Crop researchers plant 15 plots with a new variety of corn. The yields in bushels per acre are

138.0	139.1	113.0	132.5	140.7	109.7	118.9	134.8
109.6	127.3	115.6	130.4	130.2	111.7	105.5	

Assume that $\sigma = 10$ bushels per acre.

(a) Find the 90% confidence interval for the mean yield μ for this variety of corn.

(b) Find the 95% confidence interval.

(c) Find the 99% confidence interval.

(d) How do the margins of error in (a), (b), and (c) change as the confidence level increases?

6.8 Suppose that you had measured the weights of the runners in Exercise 6.6 in pounds rather than kilograms. Use your answers to Exercise 6.6 and the fact that 1 kilogram equals 2.2 pounds to answer these questions.

(a) What is the mean weight of these runners?

(b) What is the standard deviation of the mean weight?

(c) Give a 95% confidence interval for the mean weight of the population of runners that these runners represent.

6.9 Find a 99% confidence interval for the rent μ of the one-bedroom apartments in Exercise 6.2. Is the 99% confidence interval wider or narrower than the 95% interval found in Exercise 6.2? Explain in plain language why this is true.

6.10 Suppose that the crop researchers in Exercise 6.7 obtained the same value of $\bar{x}$ from a sample of 50 plots rather than 15.

(a) Compute the 95% confidence interval for the mean yield μ.

(b) Is the margin of error larger or smaller than the margin of error found for the sample of 15 plots in Exercise 6.7? Explain in plain language why the change occurs.

(c) Will the 90% and 99% intervals for a sample of size 50 be wider or narrower than those for $n = 15$? (You need not actually calculate these intervals.)

6.11 A test for the level of potassium in the blood is not perfectly precise. Moreover, the actual level of potassium in a person's blood varies slightly from day to day. Suppose that repeated measurements for the same person on different days vary normally with $\sigma = 0.2$.

(a) Julie's potassium level is measured once. The result is $x = 3.4$. Give a 90% confidence interval for her mean potassium level.

(b) If three measurements were taken on different days and the mean result is $\bar{x} = 3.4$, what is a 90% confidence interval for Julie's mean blood potassium level?

6.12 A study of the career paths of hotel general managers sent questionnaires to an SRS of 160 hotels belonging to major U.S. hotel chains. There were 114 responses. The average time these 114 general managers had spent with their current company was 11.78 years. Give a 99% confidence interval for the mean number of years general managers of major-chain hotels have spent with their current company. (Take it as known that the standard deviation of time with the company for all general managers is 3.2 years.)

6.13 Here are the Degree of Reading Power (DRP) scores for a sample of 44 third-grade students:

40	26	39	14	42	18	25	43	46	27	19
47	19	26	35	34	15	44	40	38	31	46
52	25	35	35	33	29	34	41	49	28	52
47	35	48	22	33	41	51	27	14	54	45

Suppose that the standard deviation of the population of DRP scores is known to be $\sigma = 11$. Give a 95% confidence interval for the population mean score.

6.14 The *Confidence Intervals* applet lets you simulate large numbers of confidence intervals quickly. Select 95% confidence and then sample 50 intervals. Record the number of intervals that cover the true value (this appears in the "hit" box in the applet). Press the reset button and repeat 30 times. Make a stemplot of the results and find the mean. Describe the results. If you repeated this experiment very many times, what would you expect the average number of hits to be?

6.15 Repeat the previous exercise for 80% confidence.

6.16 You are planning a survey of starting salaries for recent liberal arts major graduates from your college. From a pilot study you estimate that the standard deviation is about \$8000. What sample size do you need to have a margin of error equal to \$500 with 95% confidence?

6.17 Suppose that in the setting of the previous exercise you are willing to settle for a margin of error of \$1000. Will the required sample size be larger or smaller? Verify your answer by performing the calculations.

6.18 How large a sample of one-bedroom apartments in Exercise 6.2 would be needed to estimate the mean μ within $\pm$\$20 with 90% confidence?

6.19 Researchers planning a study of the reading ability of third-grade children want to obtain a 95% confidence interval for the population mean score on a reading test, with margin of error no greater than 3 points. They carry out a small pilot study to estimate the variability of test scores. The sample standard deviation is $s = 12$ points in the pilot study, so in preliminary calculations the researchers take the population standard deviation to be $\sigma = 12$.

(a) The study budget will allow as many as 100 students. Calculate the margin of error of the 95% confidence interval for the population mean based on $n = 100$.

(b) There are many other demands on the research budget. If all of these demands were met, there would be funds to measure only 10 children. What is the margin of error of the confidence interval based on $n = 10$ measurements?

(c) Find the smallest value of n that would satisfy the goal of a 95% confidence interval with margin of error 3 or less. Is this sample size within the limits of the budget?

6.20 How large a sample of Julie's potassium levels in Exercise 6.11 would be needed to estimate her mean μ within ± 0.06 with 95% confidence?

6.21 In Exercises 6.7 and 6.10, we compared confidence intervals based on corn yields from 15 and 50 small plots of ground. How large a sample is required to estimate the mean yield within ± 6 bushels per acre with 90% confidence?

6.22 To assess the accuracy of a laboratory scale, a standard weight known to weigh 10 grams is weighed repeatedly. The scale readings are normally distributed with unknown mean (this mean is 10 grams if the scale has no bias). The standard deviation of the scale readings is known to be 0.0002 gram.

(a) The weight is weighed five times. The mean result is 10.0023 grams. Give a 98% confidence interval for the mean of repeated measurements of the weight.

(b) How many measurements must be averaged to get a margin of error of ± 0.0001 with 98% confidence?

6.23 The Gallup Poll asked 1571 adults what they considered to be the most serious problem facing the nation's public schools; 30% said drugs. This sample percent is an estimate of the percent of all adults who think that drugs are the schools' most serious problem. The news article reporting the poll result adds, "The poll has a margin of error—the measure of its statistical accuracy—of three percentage points in either direction; aside from this imprecision inherent in using a sample to represent the whole, such practical factors as the wording of questions can affect how closely a poll reflects the opinion of the public in general."

The Gallup Poll uses a complex multistage sample design, but the sample percent has approximately a normal distribution. Moreover, it is standard practice to announce the margin of error for a 95% confidence interval unless a different confidence level is stated.

(a) The announced poll result was $30\% \pm 3\%$. Can we be certain that the true population percent falls in this interval?

(b) Explain to someone who knows no statistics what the announced result $30\% \pm 3\%$ means.

(c) This confidence interval has the same form we have met earlier:

$$\text{estimate} \pm z^* \sigma_{\text{estimate}}$$

(Actually σ is estimated from the data, but we ignore this for now.) What is the standard deviation σ_{estimate} of the estimated percent?

(d) Does the announced margin of error include errors due to practical problems such as undercoverage and nonresponse?

6.24 When the statistic that estimates an unknown parameter has a normal distribution, a confidence interval for the parameter has the form

$$\text{estimate} \pm z^* \sigma_{\text{estimate}}$$

In a complex sample survey design, the appropriate unbiased estimate of the population mean and the standard deviation of this estimate may require elaborate computations. But when the estimate is known to have a normal distribution and its standard deviation is given, we can calculate a confidence interval for μ from complex sample designs without knowing the formulas that led to the numbers given.

A report based on the Current Population Survey estimates the 1999 median annual earnings of households as \$40,816 and also estimates that the standard deviation of this estimate is \$191. The Current Population Survey uses an elaborate multistage sampling design to select a sample of about 50,000 households. The sampling distribution of the estimated median income is approximately normal. Give a 95% confidence interval for the 1999 median annual earnings of households.

6.25 The previous problem reports data on the median household income for the entire United States. In a detailed report based on the same sample survey, you find that the estimated median income for four-person families in Michigan is \$65,467. Is the margin of error for this estimate with 95% confidence greater or less than the margin of error for the national median. Why?

6.26 As we prepare to take a sample and compute a 95% confidence interval, we know that the probability that the interval we compute will cover the parameter is 0.95. That's the meaning of 95% confidence. If we use several such intervals, however, our confidence that *all* give correct results is less than 95%.

Suppose we are interested in confidence intervals for the median household incomes for three states. We compute a 95% interval for each of the three, based on independent samples in the three states.

(a) What is the probability that all three intervals cover the true median incomes? This probability (expressed as a percent) is our overall confidence level for the three simultaneous statements.

(b) What is the probability that at least two of the three intervals cover the true median incomes?

6.27 The Bowl Championship Series (BCS) was designed to select the top two teams in college football for a final championship game. The teams are selected by a complicated formula. In 2001, the University of Miami Hurricanes and the University of Nebraska Cornhuskers played for the championship. However, many football fans thought that Nebraska should not have played in the game because it was rated only fourth in both major opinion polls. Third-ranked University of Colorado fans were particularly upset because Colorado soundly beat the Cornhuskers late in the season. A new CNN/*USA Today*/Gallup poll reports that a majority of fans would prefer a national championship play-off as an alternative to the BCS. The news media polled a random sample of 1019 adults 18 years of age or older. A summary of the results states that 54% prefer the play-off, and the margin of error is 3% for 95% confidence.

(a) Give the 95% confidence interval.

(b) Do you think that a newspaper headline stating that a majority of fans prefer a play-off is justified by the results of this study? Explain your answer.

6.28 An advertisement in the student newspaper asks you to consider working for a telemarketing company. The ad states, "Earn between \$500 and \$1000 per week." Do you think that the ad is describing a confidence interval? Explain your answer.

6.29 A *New York Times* poll on women's issues interviewed 1025 women and 472 men randomly selected from the United States, excluding Alaska and Hawaii. The poll found that 47% of the women said they do not get enough time for themselves.

(a) The poll announced a margin of error of ± 3 percentage points for 95% confidence in conclusions about women. Explain to someone who knows no statistics why we can't just say that 47% of all adult women do not get enough time for themselves.

(b) Then explain clearly what "95% confidence" means.

(c) The margin of error for results concerning men was ± 4 percentage points. Why is this larger than the margin of error for women?

6.30 A radio talk show invites listeners to enter a dispute about a proposed pay increase for city council members. "What yearly pay do you think council members should get? Call us with your number." In all, 958 people call. The mean pay they suggest is $\bar{x} = \$9740$ per year, and the standard deviation of the responses is $s = \$1125$. For a large sample such as this, s is very close to the unknown population σ. The station calculates the 95% confidence interval for the mean pay μ that all citizens would propose for council members to be \$9669 to \$9811. Is this result trustworthy? Explain your answer.

6.2 Tests of Significance

Confidence intervals are one of the two most common types of formal statistical inference. They are appropriate when our goal is to estimate a population parameter. The second common type of inference is directed at a quite different goal: to assess the evidence provided by the data in favor of some claim about the population.

The reasoning of significance tests

A significance test is a formal procedure for comparing observed data with a hypothesis whose truth we want to assess. The hypothesis is a statement about the parameters in a population or model. The results of a test are expressed in terms of a probability that measures how well the data and the hypothesis agree. We use the following example to illustrate these ideas. We start by focusing on the concepts. Don't worry about the details; we will deal with them later.

EXAMPLE 6.6

Tim Kelley of Example 6.2 has a driver's license that gives his weight as 187 pounds. He did not change this information the last time he renewed his license although he suspects that he may have gained a little weight over the past few years. We will examine the truth of the weight on Tim's license using a test of significance. Recall that the four weights recorded this month were

190.5 189.0 195.5 187.0

with sample mean $\overline{x} = 190.5$. As before, we assume that $\sigma = 3$ and that the data are normal.

Are these observations compatible with a true mean weight of $\mu = 187$ pounds? To answer this question, we ask another: what is the probability of observing a sample mean of 190.5 or larger from a population with a true mean of 187? The answer is 0.01. Because this probability is small, we conclude that Tim has gained weight.

What are the key steps in this example? We first started with a question. Is the mean of Tim's four weight measurements last month compatible with a true mean of 187 pounds, or is his suspicion that he has gained weight correct?

Next, we compare the data he collected ($\overline{x} = 190.5$ pounds) with the mean specified in his question ($\mu = 187$ pounds). The answer is given as a probability: if $\mu = 187$, then a sample mean $\overline{x} = 190.5$ or more based on $n = 4$ observations has probability 0.01.

How does the calculated probability of 0.01 lead to the conclusion that Tim's weight has increased? If the true mean is $\mu = 187$ pounds, we would rarely see a sample mean $\overline{x}$ as large as 190.5 pounds. In fact, we would get a result this large only 1 time in 100. So, either

1. we have observed something that is unusual, or
2. our assumption about Tim's true weight ($\mu = 187$ pounds) is not correct—Tim has gained weight.

Each of these possibilities leads to a conclusion. The first accepts the data as compatible with $\mu = 187$ pounds. The second says that Tim's suspicions are true; he has gained weight. We use the calculated probability, 0.01 in this case, to guide our choice. Because this probability is small, we prefer the second conclusion: Tim has gained weight. The chance of observing a weight of

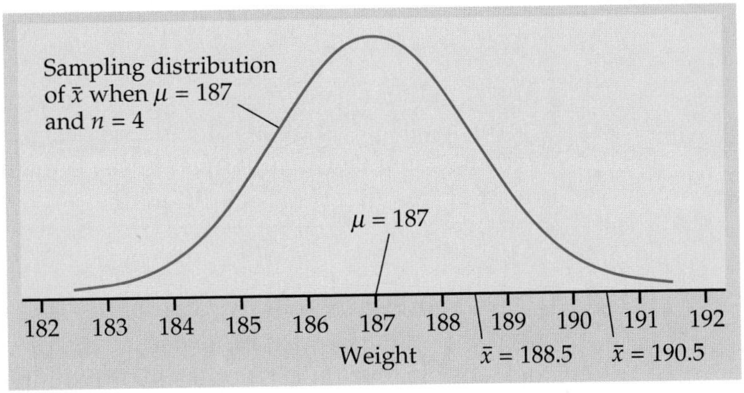

FIGURE 6.6 Comparison of the means in Examples 6.6 and 6.7 relative to the null hypothesized value $\mu = 187$.

$\bar{x} = 190.5$ or higher when the true mean weight is $\mu = 187$ is less than 1 in 100. The following example illustrates a scenario where the data lead to a different conclusion.

EXAMPLE 6.7

Suppose the mean of Tim Kelley's four observations last month was $\bar{x} = 188.5$ pounds. This sample mean is closer to the value $\mu = 187$ pounds given on his driver's license. For a test of significance in this case, we need to calculate the probability that the mean of a sample of size $n = 4$ from a normal population with mean $\mu = 187$ and standard deviation $\sigma = 3$ is greater than or equal to the observed value of $\bar{x} = 188.5$ pounds. The answer is 0.16. Because this probability is not particularly small, we do not conclude that Kelley has gained weight.

The probabilities in Examples 6.6 and 6.7 are measures of the compatibility of the data ($\bar{x} = 188.5$ and $\bar{x} = 190.5$) with the *null hypothesis* ($\mu = 187$). Figure 6.6 compares the two results graphically. The normal curve centered at 187 is the sampling distribution of $\bar{x}$ when $\mu = 187$. You can see that we are not particularly surprised to observe $\bar{x} = 188.5$, but $\bar{x} = 190.5$ is an unusual observation. We will now consider some of the formal aspects of significance testing and the technical details.

Stating hypotheses

In Example 6.6, we asked whether the data Tim Kelley collected are reasonable if, in fact, his true weight (μ) is 187 pounds. To answer this, we begin by supposing that the statement following the "if" in the previous sentence is true. In other words, we suppose $\mu = 187$. We then ask whether the data provide evidence against the supposition we have made. If so, we have evidence in favor of the effect (Tim has gained weight) we are seeking. The first step in a test of significance is to state a claim that we will try to find evidence *against*.

Null Hypothesis

The statement being tested in a test of significance is called the **null hypothesis.** The test of significance is designed to assess the strength of the evidence against the null hypothesis. Usually the null hypothesis is a statement of "no effect" or "no difference."

We abbreviate "null hypothesis" as H_0. A null hypothesis is a statement about a population, expressed in terms of some parameter or parameters. For example, our null hypothesis for Example 6.6 is

$$H_0: \mu = 187$$

alternative hypothesis

It is convenient also to give a name to the statement we hope or suspect is true instead of H_0. This is called the **alternative hypothesis** and is abbreviated as H_a. In Example 6.6, the alternative hypothesis states that Tim's weight has increased. We write this as

$$H_a: \mu > 187$$

Hypotheses always refer to some population or model, not to a particular outcome. For this reason, we must state H_0 and H_a in terms of population parameters.

one-sided and two-sided alternatives

Because H_a expresses the effect that we hope to find evidence *for*, we often begin with H_a and then set up H_0 as the statement that the hoped-for effect is not present. Stating H_a is often the more difficult task. It is not always clear, in particular, whether H_a should be **one-sided** or **two-sided.**

The alternative H_a: $\mu > 187$ in the weight example is one-sided. Because people tend to gain weight as they age, we are only interested in detecting an increase in μ. On the other hand, it is easy to imagine scenarios where the two-sided alternative is appropriate. Suppose, for example, that μ is the mean time required to process orders for the customers of your company. Here you would be interested in knowing whether this mean has increased or decreased. If the historical value is 3.8 days, we would test H_0: $\mu = 3.8$ days versus the alternative H_a: $\mu \neq 3.8$ days.

The alternative hypothesis should express the hopes or suspicions we bring to the data. It is cheating to first look at the data and then frame H_a to fit what the data show. If you do not have a specific direction firmly in mind in advance, use a two-sided alternative. Moreover, some users of statistics argue that we should *always* work with the two-sided alternative.

The choice of the hypotheses in Example 6.6 as

$$H_0: \mu = 187$$

$$H_a: \mu > 187$$

deserves a final comment. We are not concerned with the possibility that Tim may have lost weight. However, we can allow for the possibility that μ is less than 187 pounds by including this case in the null hypothesis. Then we would write

$$H_0: \mu \leq 187$$

$$H_a: \mu > 187$$

This statement is logically satisfying because the hypotheses account for all possible values of μ. However, only the parameter value in H_0 that is closest to H_a influences the form of the test in all common significance-testing situations. We will therefore take H_0 to be the simpler statement that the parameter has a specific value, in this case H_0: $\mu = 187$.

Test statistics

We will learn the form of significance tests in a number of common situations. Here are some principles that apply to most tests and that help in understanding the form of tests:

- The test is based on a statistic that estimates the parameter that appears in the hypotheses. Usually this is the same estimate we would use in a confidence interval for the parameter. When H_0 is true, we expect the estimate to take a value near the parameter value specified by H_0.
- Values of the estimate far from the parameter value specified by H_0 give evidence against H_0. The alternative hypothesis determines which directions count against H_0.

EXAMPLE 6.8

In Example 6.6, the hypotheses are stated in terms of Kelley's weight as given on his driver's license:

$$H_0: \mu = 187$$

$$H_a: \mu > 187$$

The estimate of μ is the sample mean $\overline{x}$. Because H_a is one-sided on the high side, only large values of $\overline{x}$ count as evidence against the null hypothesis.

test statistic

A **test statistic** measures compatibility between the null hypothesis and the data. We use it for the probability calculation that we need for our test of significance. It is a random variable with a distribution that we know.

EXAMPLE 6.9

In Example 6.6, the null hypothesis is H_0: $\mu = 187$, and $\overline{x} = 190.5$. Recall (page 395) that $\overline{x}$ has the $N(\mu, \sigma/\sqrt{n})$ distribution when the data are normal. The test statistic for this problem is the standardized version of $\overline{x}$:

$$z = \frac{\overline{x} - \mu}{\sigma/\sqrt{n}}$$

This random variable has the standard normal distribution $N(0, 1)$ when we use $\mu = 187$ and the null hypothesis is true. For the data we have, its value is

$$z = \frac{190.5 - 187}{3/\sqrt{4}} = 2.33$$

In this example we were willing to assume that the data are normal based on past experience. However, the central limit theorem (page 397) tells us that z will still be approximately normal when the sample size is large. We will use facts about the normal distribution in what follows. Do you recall the 68–95–99.7 rule and the calculations that we did with the standard normal distribution (Table A)? If not, this would be a good time to refresh your memory by reviewing Section 1.3.

P-values

A test of significance assesses the evidence against the null hypothesis in terms of probability. If the observed outcome is unlikely under the supposition that the null hypothesis is true but is more probable if the alternative hypothesis is true, that outcome is evidence against H_0 in favor of H_a. The less probable the outcome is, the stronger the evidence that H_0 is false. Usually, any *specific* outcome has low probability. We do not expect the mean of Kelley's four weight measurements to be *exactly* 187 pounds, even when H_0: $\mu = 187$ is true. On the other hand, if we observe $\bar{x} = 187$ or something less, we clearly do not have evidence to reject H_0 in favor of H_a: $\mu > 187$.

What about $\bar{x} = 188.5$ or 190 or 191.5 or 202? Clearly, as $\bar{x}$ moves farther away from μ we have stronger evidence that H_0 is not true. Recall that the standard deviation of $\bar{x}$ is $\sigma/\sqrt{n} = 3/\sqrt{4} = 1.5$ pounds. Therefore, if $\bar{x} = 187 + 1.5 = 188.5$, the difference is one standard deviation ($z = 1$); if $\bar{x} = 187 + 2(1.5) = 190.0$, the difference is two standard deviations ($z = 2$); if $\bar{x} = 187 + 3(1.5) = 191.5$, the difference is three standard deviations ($z = 3$); and if $\bar{x} = 187 + 10(1.5) = 202.0$, the difference is ten standard deviations ($z = 10$). Our conclusions are easy in the first and last cases. From our work with the standard normal distribution, we know that $z = 1$ is not unusual and that $z = 10$ essentially never happens. So if $z = 1$, we do not have evidence to reject H_0; and if $z = 10$, we are on firm ground if we reject H_0. The Supreme Court of the United States has said that "two or three standard deviations" ($z = 2$ or 3) is its criterion for rejecting H_0 (see Exercise 8.81 on page 607), and this is the criterion used in most applications involving the law.

Because not all test statistics are z's (random variables with the $N(0, 1)$ distribution), we translate the value of test statistics into a common language, the language of probability.

A test of significance finds the probability of getting an outcome *as extreme or more extreme than the actually observed outcome*. "Extreme" means "far from what we would expect if H_0 were true." The direction or directions that count as "far from what we would expect" are determined by H_a as well as H_0.

In Example 6.6 we want to know if Kelley's weight has increased, a one-sided H_a. So the evidence against H_0 is measured by the probability

$$P(\bar{x} \geq 190.5)$$

that his sample mean weight would be *as high or higher than* the observed value. This probability is calculated under the assumption that the population mean is $\mu = 187$ pounds.

> ***P*-Value**
>
> The probability, computed assuming that H_0 is true, that the test statistic would take a value as extreme or more extreme than that actually observed is called the ***P*-value** of the test. The smaller the P-value, the stronger the evidence against H_0 provided by the data.

The key to calculating the *P*-value is the sampling distribution of the test statistic. For the problems we consider in this chapter, we need only the standard normal distribution for the test statistic z.

EXAMPLE 6.10

In Example 6.6 the observations are an SRS of size $n = 4$ from a normal population with $\sigma = 3$. The observed average weight is $\bar{x} = 190.5$. In Example 6.9, we found that the test statistic for testing

$$H_0\colon \mu = 187 \quad \text{versus} \quad H_a\colon \mu > 187$$

is

$$z = \frac{190.5 - 187}{3/\sqrt{4}} = 2.33$$

If H_0 is true, then z is a single observation from the standard normal, $N(0, 1)$, distribution. Figure 6.7 illustrates this calculation. The *P*-value is the probability of observing a value of z at least as large as the one that we observed, $z = 2.33$. From Table A, our table of standard normal probabilities, we find

$$P(Z \geq 2.33) = 1 - 0.9901 = 0.01$$

This is the value that was reported on page 436.

Statistical significance

We have finished our discussion of the details of tests of significance. These details are needed to translate $\bar{x}$ and the μ that we specified in our null hypothesis, H_0, into a test statistic z and a *P*-value. The details are easy and can be programmed into a computer. One important final step is needed. What do we conclude?

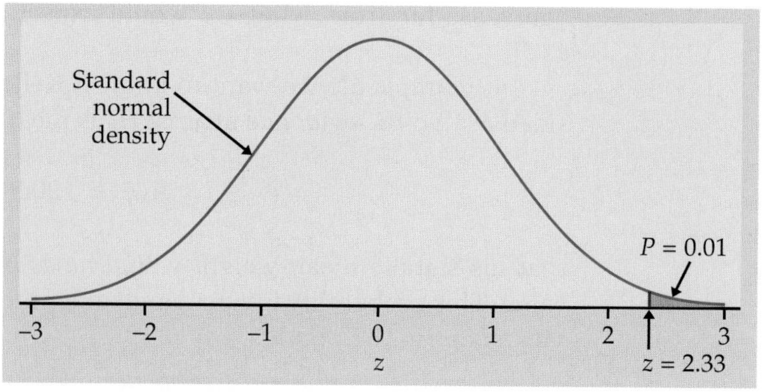

FIGURE 6.7 The *P*-value for Example 6.10. The *P*-value here is the probability (when H_0 is true) that $\bar{x}$ takes a value as large as or larger than the actually observed value.

significance level

We can compare the *P*-value we calculated with a fixed value that we regard as decisive. This amounts to announcing in advance how much evidence against H_0 we will require to reject H_0. The decisive value of *P* is called the **significance level.** It is denoted by α, the Greek letter alpha. If we choose $\alpha = 0.05$, we are requiring that the data give evidence against H_0 so strong that it would happen no more than 5% of the time (1 time in 20) when H_0 is true. If we choose $\alpha = 0.01$, we are insisting on stronger evidence against H_0, evidence so strong that it would appear only 1% of the time (1 time in 100) if H_0 is in fact true. The Supreme Court's criterion ($z = 2$ or 3) translates into $\alpha = 0.0445$ or $\alpha = 0.0027$ for a two-sided test.

Statistical Significance

If the *P*-value is as small or smaller than α, we say that the data are **statistically significant at level α.**

"Significant" in the statistical sense does not mean "important." The original meaning of the word is "signifying something." In statistics* the term is used to indicate only that the evidence against the null hypothesis reached the standard set by α. Significance at level 0.01 is often expressed by the statement "The results were significant ($P < 0.01$)." Here *P* stands for the *P*-value. The *P*-value is more informative than a statement of significance because we

*The reasoning of significance tests has been used sporadically at least since Laplace in the 1820s. A clear exposition by the English statistician Francis Y. Edgeworth (1845–1926) appeared in 1885 in the context of a study of social statistics. Edgeworth introduced the term "significant" as meaning "corresponds to a real difference in fact."

can then assess significance at any level we choose. For example, a result with $P = 0.03$ is significant at the $\alpha = 0.05$ level but is not significant at the $\alpha = 0.01$ level.

A test of significance is a recipe for assessing the significance of the evidence provided by data against a null hypothesis. The four steps common to all tests of significance are as follows:

1. State the *null hypothesis* H_0 and the *alternative hypothesis* H_a. The test is designed to assess the strength of the evidence against H_0; H_a is the statement that we will accept if the evidence enables us to reject H_0.
2. Calculate the value of the *test statistic* on which the test will be based. This statistic usually measures how far the data are from H_0.
3. Find the *P-value* for the observed data. This is the probability, calculated assuming that H_0 is true, that the test statistic will weigh against H_0 at least as strongly as it does for these data.
4. State a conclusion. One way to do this is to choose a *significance level* α, how much evidence against H_0 you regard as decisive. If the P-value is less than or equal to α, you conclude that the alternative hypothesis is true; if it is greater than α, you conclude that the data do not provide sufficient evidence to reject the null hypothesis. Your conclusion is a sentence that summarizes what you have found by using a test of significance.

We will learn the details of many tests of significance in the following chapters. In most cases, Steps 2 and 3 of our outline are almost automatic once you have practiced a bit. The proper test statistic is determined by the hypotheses and the data collection design. We use computer software or a calculator to find its numerical value and the P-value. The computer will not formulate your hypotheses for you, however. Nor will it decide if significance testing is appropriate or help you to interpret the P-value that it presents to you. The most difficult and important step is the last one: stating a conclusion. For Tim Kelley's question about his weight we could say, "Tim has gained weight ($z = 2.33$, $P < 0.01$)." A confidence interval such as the one we gave in Example 6.3 would also be useful.

Exploring the Web

Confidence intervals and statistical significance appear frequently in reports of studies in many fields. Most scientific journals now have Web sites that display at least summaries of articles appearing in the journals. Choose a major journal in your field of study. Use a web search engine to find its web site by searching on the name of the journal. Look at the summaries of recent papers. If your field does studies that produce data, you will surely find phrases like "95% confidence" and "significant ($P = 0.01$)."

Tests for a population mean

Our discussion has focused on the main ideas of statistical tests, and we have developed the details for one specific procedure. Here is a summary. We have an SRS of size n drawn from a normal population with unknown mean μ. We want to test the hypothesis that μ has a specified value. Call the specified value μ_0. The null hypothesis is

$$H_0\colon \mu = \mu_0$$

The test is based on the sample mean $\bar{x}$. Because the standard scale is easy to interpret, we use as our test statistic the *standardized* sample mean z:

$$z = \frac{\bar{x} - \mu_0}{\sigma/\sqrt{n}}$$

The statistic z has the standard normal distribution when H_0 is true. If the alternative is one-sided on the high side

$$H_a\colon \mu > \mu_0$$

then the P-value is the probability that a standard normal random variable Z takes a value at least as large as the observed z. That is,

$$P = P(Z \geq z)$$

Example 6.10 applied this test. There, $\mu_0 = 187$, the standardized sample mean was $z = 2.33$, and the P-value was $P(Z \geq 2.33) = 0.01$.

Similar reasoning applies when the alternative hypothesis states that the true μ lies below the hypothesized μ_0 (one-sided). When H_a states that μ is simply unequal to μ_0 (two-sided), values of z away from zero in either direction count against the null hypothesis. The P-value is the probability that a standard normal Z is at least as far from zero as the observed z. For example, if the test statistic is $z = 1.7$, the two-sided P-value is the probability that $Z \leq -1.7$ or $Z \geq 1.7$. Because the standard normal distribution is symmetric, we calculate this probability by finding $P(Z \geq 1.7)$ and *doubling* it:

$$\begin{aligned} P(Z \leq -1.7 \text{ or } Z \geq 1.7) &= 2P(Z \geq 1.7) \\ &= 2(1 - 0.9554) = 0.0892 \end{aligned}$$

We would make exactly the same calculation if we observed $z = -1.7$. It is the absolute value $|z|$ that matters, not whether z is positive or negative. Here is a statement of the test in general terms.

z Test for a Population Mean

To test the hypothesis H_0: $\mu = \mu_0$ based on an SRS of size n from a population with unknown mean μ and known standard deviation σ, compute the test statistic

$$z = \frac{\overline{x} - \mu_0}{\sigma/\sqrt{n}}$$

In terms of a standard normal random variable Z, the P-value for a test of H_0 against

H_a: $\mu > \mu_0$ is $P(Z \geq z)$

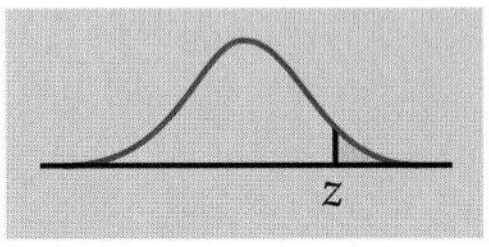

H_a: $\mu < \mu_0$ is $P(Z \leq z)$

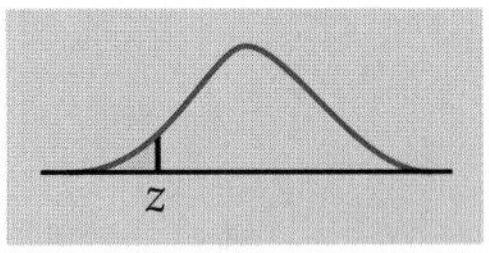

H_a: $\mu \neq \mu_0$ is $2P(Z \geq |z|)$

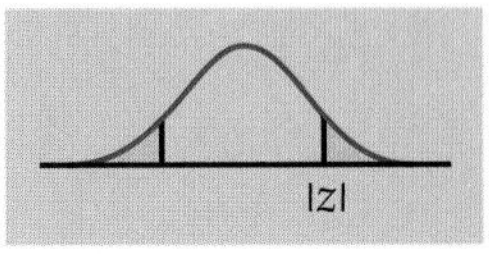

These P-values are exact if the population distribution is normal and are approximately correct for large n in other cases.

EXAMPLE 6.11

Do middle-aged male executives have different average blood pressure than the general population? The National Center for Health Statistics reports that the mean systolic blood pressure for males 35 to 44 years of age is 128 and the standard deviation in this population is 15. The medical director of a company looks at the medical records of 72 company executives in this age group and finds that the mean systolic blood pressure in this sample is $\overline{x} = 126.07$. Is this evidence that executive blood pressures differ from the national average?

The null hypothesis is "no difference" from the national mean $\mu_0 = 128$. The alternative is two-sided, because the medical director did not have a particular direction in mind before examining the data. So the hypotheses about the unknown mean μ of the executive population are

$$H_0: \mu = 128$$

$$H_a: \mu \neq 128$$

As usual in this chapter, we make the unrealistic assumption that the population standard deviation is known, in this case that executives have the same $\sigma = 15$ as the general population of middle-aged males. The z test requires that the 72 executives in the sample are an SRS from the population of all middle-aged male executives in the company. We check this assumption by asking how the data were produced. If medical records are available only for executives with recent medical problems, for example, the data are of little value for our purpose. It turns out that all executives are given a free annual medical exam and that the medical director selected 72 exam results at random.

We can now proceed to compute the test statistic:

$$z = \frac{\bar{x} - \mu_0}{\sigma/\sqrt{n}} = \frac{126.07 - 128}{15/\sqrt{72}}$$
$$= -1.09$$

Figure 6.8 illustrates the P-value, which is the probability that a standard normal variable Z takes a value at least 1.09 away from zero. From Table A we find that this probability is

$$P = 2P(Z \geq |-1.09|) = 2P(Z \geq 1.09)$$
$$= 2(1 - 0.8621) = 0.2758$$

That is, more than 27% of the time an SRS of size 72 from the general male population would have a mean blood pressure at least as far from 128 as that of the executive sample. The observed $\bar{x} = 126.07$ is therefore not good evidence that executives differ from other men.

The data in Example 6.11 do *not* establish that the mean blood pressure μ for this company's middle-aged male executives is 128. We sought evidence that μ differed from 128 and failed to find convincing evidence. That is all we can say. No doubt the mean blood pressure of the entire executive population is not exactly equal to 128. A large enough sample would give evidence of the difference, even if it is very small. Tests of significance assess the evidence *against* H_0. If the evidence is strong, we can confidently reject H_0 in favor of the alternative. Failing to find evidence against H_0 means only that the data are consistent with H_0, not that we have clear evidence that H_0 is true.

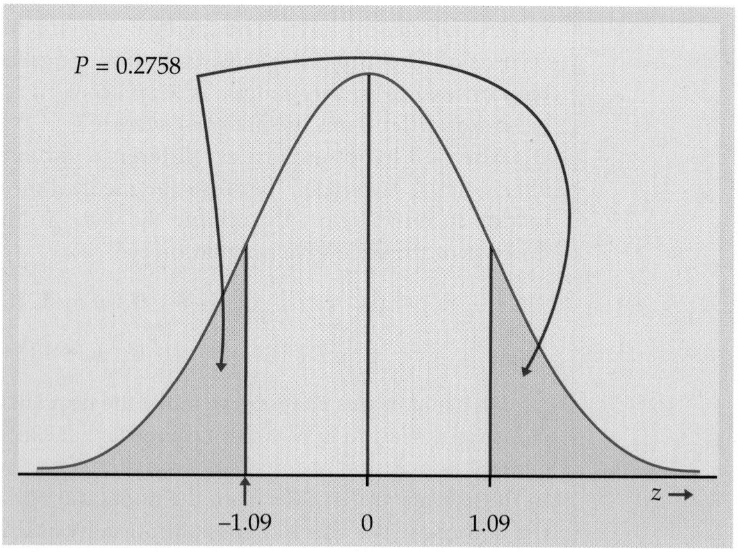

FIGURE 6.8 The P-value for the two-sided test in Example 6.11.

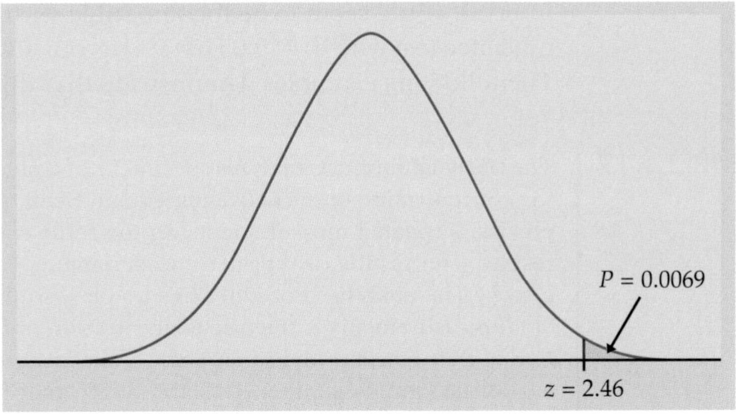

FIGURE 6.9 The P-value for the one-sided test in Example 6.12.

EXAMPLE 6.12

In a discussion of SAT scores, someone comments: "Because only a minority of high school students take the test, the scores overestimate the ability of typical high school seniors. The mean SAT mathematics score is about 508, but I think that if all seniors took the test, the mean score would be no more than 450." You gave the test to an SRS of 500 seniors from California (Example 6.1). These students had a mean score of $\bar{x} = 461$. Is this good evidence against the claim that the mean for all California seniors is no more than 450? The hypotheses are

$$H_0: \mu = 450$$

$$H_a: \mu > 450$$

As in Example 6.1, we assume that $\sigma = 100$. The z statistic is

$$z = \frac{\bar{x} - \mu_0}{\sigma/\sqrt{n}} = \frac{461 - 450}{100/\sqrt{500}}$$

$$= 2.46$$

Because H_a is one-sided on the high side, large values of z count against H_0. From Table A, we find that the P-value is

$$P = P(Z \geq 2.46) = 1 - 0.9931 = 0.0069$$

Figure 6.9 illustrates this P-value. A mean score as large as that observed would occur fewer than seven times in 1000 samples if the population mean were 450. This is convincing evidence that the mean SATM score for all California high school seniors is higher than 450.

Two-sided significance tests and confidence intervals

Recall the confidence intervals that we discussed in the first section of this chapter. The basic idea was that we constructed an interval that would include the true value of μ with a specified probability C. Suppose we use a 95% confidence interval ($C = 0.95$). Then the values of μ that are not in our interval would seem to be incompatible with the data. This sounds like a significance test with $\alpha = 0.05$ (or 5%) as our standard for drawing a conclusion. The following examples demonstrate that this is correct.

EXAMPLE 6.13

The Deely Laboratory analyzes specimens of a pharmaceutical product to determine the concentration of the active ingredient. Such chemical analyses are not perfectly precise. Repeated measurements on the same specimen will give slightly different results. The results of repeated measurements follow a normal distribution quite closely. The analysis procedure has no bias, so that the mean μ of the population of all measurements is the true concentration in the specimen. The standard deviation of this distribution is a property of the analytical procedure and is known to be $\sigma = 0.0068$ grams per liter. The laboratory analyzes each specimen three times and reports the mean result.

The Deely Laboratory has been asked to evaluate the claim that the concentration of the active ingredient in a specimen is 0.86%. The true concentration is the mean μ of the population of repeated analyses. The hypotheses are

$$H_0: \mu = 0.86$$

$$H_a: \mu \neq 0.86$$

The lab chooses the 1% level of significance, $\alpha = 0.01$.

Three analyses of one specimen give concentrations

$$0.8403 \quad 0.8363 \quad 0.8447$$

The sample mean of these readings is

$$\bar{x} = \frac{0.8403 + 0.8363 + 0.8447}{3} = 0.8404$$

The test statistic is

$$z = \frac{\bar{x} - \mu_0}{\sigma/\sqrt{n}} = \frac{0.8404 - 0.86}{0.0068/\sqrt{3}} = -4.99 \text{ standard deviations}$$

Because the alternative is two-sided, the P-value is

$$P = 2P(Z \geq |-4.99|) = 2P(Z \geq 4.99)$$

We cannot find this probability in Table A. The largest value of z in that table is 3.49. All that we can say from Table A is that P is less than $2P(Z \geq 3.49) = 2(1 - 0.9998) = 0.0004$. If we use the bottom row of Table D, we find that the largest value of z^* is 3.291, corresponding to a P-value of $1 - 0.999 = 0.001$. Software could be used to give an accurate value of the P-value. However, because the P-value is clearly less than the company's standard of 1%, we reject H_0.

Suppose we compute a 99% confidence interval for the same data.

EXAMPLE 6.14

The 99% confidence interval for μ in Example 6.13 is

$$\bar{x} \pm z^* \frac{\sigma}{\sqrt{n}} = 0.8404 \pm 0.0101$$
$$= (0.8303, 0.8505)$$

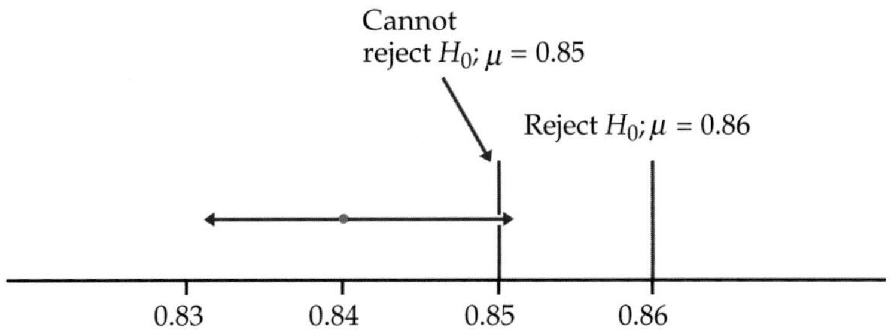

FIGURE 6.10 Values of μ falling outside a 99% confidence interval can be rejected at the 1% significance level; values falling inside the interval cannot be rejected.

The hypothesized value $\mu_0 = 0.86$ in Example 6.13 falls outside the confidence interval we computed in Example 6.14. We are therefore 99% confident that μ is *not* equal to 0.86, so we can reject

$$H_0\colon \mu = 0.86$$

at the 1% significance level. On the other hand, we cannot reject

$$H_0\colon \mu = 0.85$$

at the 1% level in favor of the two-sided alternative $H_a\colon \mu \neq 0.85$, because 0.85 lies inside the 99% confidence interval for μ. Figure 6.10 illustrates both cases.

The calculation in Example 6.13 for a 1% significance test is very similar to the calculation for a 99% confidence interval. In fact, a two-sided test at significance level α can be carried out directly from a confidence interval with confidence level $C = 1 - \alpha$.

> **Confidence Intervals and Two-Sided Tests**
>
> A level α two-sided significance test rejects a hypothesis $H_0\colon \mu = \mu_0$ exactly when the value μ_0 falls outside a level $1 - \alpha$ confidence interval for μ.

P-values versus fixed α

The observed result in Example 6.13 was $z = -4.99$. The conclusion that this result is significant at the 1% level does not tell the whole story. The observed z is far beyond the z corresponding to 1%, and the evidence against H_0 is far stronger than 1% significance suggests. The P-value

$$2P(Z \geq 4.99) = 0.0000006$$

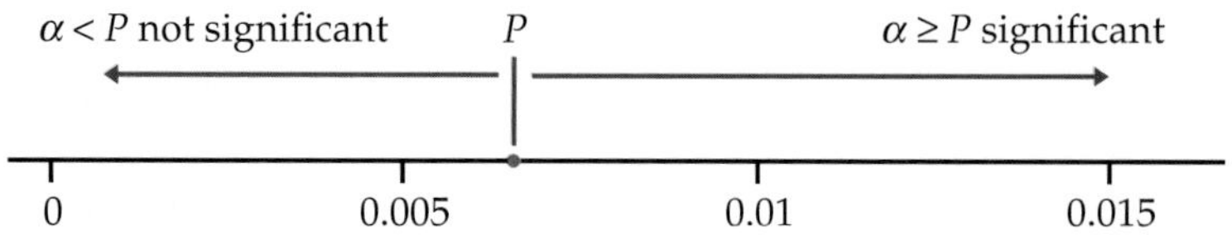

FIGURE 6.11 An outcome with P-value P is significant at all levels α at or above P and is not significant at smaller levels α.

gives a better sense of how strong the evidence is. *The P-value is the smallest level α at which the data are significant.* Knowing the P-value allows us to assess significance at any level.

EXAMPLE 6.15

In Example 6.12, we tested the hypotheses

$$H_0\colon \mu = 450$$
$$H_a\colon \mu > 450$$

concerning the mean SAT mathematics score μ of California high school seniors. The test had the P-value $P = 0.0069$. This result is significant at the $\alpha = 0.01$ level because $0.0069 \le 0.01$. It is not significant at the $\alpha = 0.005$ level, because the P-value is larger than 0.005. See Figure 6.11.

critical value

A P-value is more informative than a reject-or-not finding at a fixed significance level. But assessing significance at a fixed level α is easier, because no probability calculation is required. You need only look up a number in a table. A value z^* with a specified area to its right under the standard normal curve is called a **critical value** of the standard normal distribution. Because the practice of statistics almost always employs computer software that calculates P-values automatically, the use of tables of critical values is becoming outdated. We include the usual tables of critical values (such as Table D) at the end of the book for learning purposes and to rescue students without good computing facilities. The tables can be used directly to carry out fixed α tests. They also allow us to approximate P-values quickly without a probability calculation. The following example illustrates the use of Table D to find an approximate P-value.

EXAMPLE 6.16

Bottles of a popular cola drink are supposed to contain 300 milliliters (ml) of cola. There is some variation from bottle to bottle because the filling machinery is not perfectly precise. The distribution of the contents is normal with standard deviation $\sigma = 3$ ml. A student who suspects that the bottler is underfilling measures the contents of six bottles. The results are

299.4 297.7 301.0 298.9 300.2 297.0

Is this convincing evidence that the mean contents of cola bottles is less than the advertised 300 ml? The hypotheses are

$$H_0\colon \mu = 300$$
$$H_a\colon \mu < 300$$

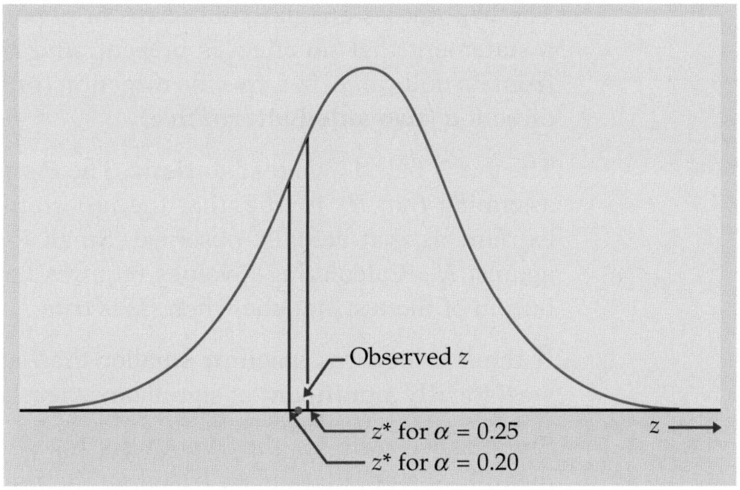

FIGURE 6.12 The P-value of a z statistic can be approximated by noting which levels from Table D it falls between. Here, P lies between 0.20 and 0.25.

The sample mean contents of the six bottles measured is $\bar{x} = 299.03$ ml. The z test statistic is therefore

$$z = \frac{\bar{x} - \mu_0}{\sigma/\sqrt{n}} = \frac{299.03 - 300}{3/\sqrt{6}}$$
$$= -0.792$$

Small values of z count against H_0 because H_a is one-sided on the low side. The P-value is

$$P(Z \leq -0.792)$$

Rather than compute this probability, compare $z = -0.792$ with the critical values in Table D. According to this table, we would

$$\text{reject } H_0 \text{ at level } \alpha = 0.25 \text{ if } z \leq -0.674$$

$$\text{reject } H_0 \text{ at level } \alpha = 0.20 \text{ if } z \leq -0.841$$

As Figure 6.12 illustrates, the observed $z = -0.792$ lies between these two critical points. We can reject at $\alpha = 0.25$ but not at $\alpha = 0.20$. That is the same as saying that the P-value lies between 0.25 and 0.20. A sample mean at least as small as that observed will occur in between 20% and 25% of all samples if the population mean is $\mu = 300$ ml. There is no convincing evidence that the mean is below 300, and there is no need to calculate the P-value more exactly.

SUMMARY

A **test of significance** is intended to assess the evidence provided by data against a **null hypothesis** $\boldsymbol{H_0}$ and in favor of an **alternative hypothesis** $\boldsymbol{H_a}$.

The hypotheses are stated in terms of population parameters. Usually H_0 is a statement that no effect is present, and H_a says that a parameter differs from its null value, in a specific direction (**one-sided alternative**) or in either direction (**two-sided alternative**).

The test is based on a **test statistic.** The ***P*-value** is the probability, computed assuming that H_0 is true, that the test statistic will take a value at least as extreme as that actually observed. Small *P*-values indicate strong evidence against H_0. Calculating *P*-values requires knowledge of the sampling distribution of the test statistic when H_0 is true.

If the *P*-value is as small or smaller than a specified value α, the data are **statistically significant** at significance level α.

Significance tests for the hypothesis H_0: $\mu = \mu_0$ concerning the unknown mean μ of a population are based on the z **statistic:**

$$z = \frac{\bar{x} - \mu_0}{\sigma/\sqrt{n}}$$

The z test assumes an SRS of size n, known population standard deviation σ, and either a normal population or a large sample. *P*-values are computed from the normal distribution (Table A). Fixed α tests use the table of **standard normal critical values** (Table D).

SECTION 6.2 EXERCISES

6.31 Each of the following situations requires a significance test about a population mean μ. State the appropriate null hypothesis H_0 and alternative hypothesis H_a in each case.

(a) A dual X-ray absorptiometry (DXA) scanner is used to measure bone mineral density for people who may be at risk for osteoporosis. To be sure that the measurements are accurate, an object called a "phantom" that has known mineral density $\mu = 1.4$ grams per square centimeter is measured. This phantom is scanned 10 times.

(b) Feedback from your customers shows that many think it takes too long to fill out the online order form for your products. You redesign the form and survey a random sample of customers to determine whether or not they think that the new form is actually an improvement. The response uses a five-point scale: -2 if the new form takes much less time than the old form; -1 if the new form takes a little less time; 0 if the new form takes about the same time; +1 if the new form takes a little more time; and +2 if the new form takes much more time.

(c) You purchase a shipment of 60-watt light bulbs to be used in a variety of your products. If the wattage is too low or two high, your product will not look good. You measure the wattage of a random sample of 20 bulbs.

6.32 In each of the following situations, a significance test for a population mean μ is called for. State the null hypothesis H_0 and the alternative hypothesis H_a in each case.

(a) Experiments on learning in animals sometimes measure how long it takes a mouse to find its way through a maze. The mean time is 18 seconds for one particular maze. A researcher thinks that a loud noise will cause the mice to complete the maze faster. She measures how long each of 10 mice takes with a noise as stimulus.

(b) The examinations in a large history class are scaled after grading so that the mean score is 50. A self-confident teaching assistant thinks that his students have a higher mean score than the class as a whole. His students this semester can be considered a sample from the population of all students he might teach, so he compares their mean score with 50.

(c) The Census Bureau reports that households spend an average of 31% of their total spending on housing. A homebuilders association in Cleveland wonders if the national finding applies in their area. They interview a sample of 40 households in the Cleveland metropolitan area to learn what percent of their spending goes toward housing.

6.33 In each of the following situations, state an appropriate null hypothesis H_0 and alternative hypothesis H_a. Be sure to identify the parameters that you use to state the hypotheses. (We have not yet learned how to test these hypotheses.)

(a) A sociologist asks a large sample of high school students which academic subject they like best. She suspects that a higher percent of males than of females will name mathematics as their favorite subject.

(b) An education researcher randomly divides sixth-grade students into two groups for physical education class. He teaches both groups basketball skills, using the same methods of instruction in both classes. He encourages Group A with compliments and other positive behavior but acts cool and neutral toward Group B. He hopes to show that positive teacher attitudes result in a higher mean score on a test of basketball skills than do neutral attitudes.

(c) An economist believes that among employed young adults there is a positive correlation between income and the percent of disposable income that is saved. To test this, she gathers income and savings data from a sample of employed persons in her city aged 25 to 34.

6.34 Translate each of the following research questions into appropriate H_0 and H_a.

(a) Census Bureau data show that the mean household income in the area served by a shopping mall is \$62,500 per year. A market research firm questions shoppers at the mall to find out whether the mean household income of mall shoppers is higher than that of the general population.

(b) Last year, your company's service technicians took an average of 2.6 hours to respond to trouble calls from business customers who had purchased service contracts. Do this year's data show a different average response time?

6.35 A test of the null hypothesis H_0: $\mu = \mu_0$ gives test statistic $z = 1.8$.

(a) What is the *P*-value if the alternative is H_a: $\mu > \mu_0$?

(b) What is the *P*-value if the alternative is H_a: $\mu < \mu_0$?

(c) What is the *P*-value if the alternative is H_a: $\mu \neq \mu_0$?

6.36 The *P*-value for a two-sided test of the null hypothesis H_0: $\mu = 10$ is 0.06.

(a) Does the 95% confidence interval include the value 10? Why?

(b) Does the 90% confidence interval include the value 10? Why?

6.37 A 95% confidence interval for a population mean is (28, 35).

(a) Can you reject the null hypothesis that $\mu = 34$ at the 5% significance level? Why?

(b) Can you reject the null hypothesis that $\mu = 36$ at the 5% significance level? Why?

6.38 A new supplier offers a good price on a catalyst used in your production process. You compare the purity of this catalyst with that from your current supplier. The *P*-value for a test of "no difference" is 0.15. Can you be confident that the purity of the new product is the same as the purity of the product that you have been using? Discuss.

6.39 We often see televised reports of brushfires threatening homes in California. Some people argue that the modern practice of quickly putting out small fires allows fuel to accumulate and so increases the damage done by large fires. A detailed study of historical data suggests that this is wrong—the damage has risen simply because there are more houses in risky areas.[3] As usual, the study report gives statistical information tersely. Here is the summary of a regression of number of fires on decade (9 data points, for the 1910s to the 1990s):

> *Collectively, since 1910, there has been a highly significant increase ($r^2 = 0.61$, $P < 0.01$) in the number of fires per decade.*

How would you explain this statement to someone who knows no statistics? Include an explanation of both the description given by r^2 and the statistical significance.

6.40 A randomized comparative experiment examined whether a calcium supplement in the diet reduces the blood pressure of healthy men. The subjects received either a calcium supplement or a placebo for 12 weeks. The statistical analysis was quite complex, but one conclusion was that "the calcium group had lower seated systolic blood pressure ($P = 0.008$) compared with the placebo group."[4] Explain this conclusion, especially the P-value, as if you were speaking to a doctor who knows no statistics.

6.41 A social psychologist reports that "ethnocentrism was significantly higher ($P < 0.05$) among church attenders than among nonattenders." Explain what this means in language understandable to someone who knows no statistics. Do not use the word "significance" in your answer.

6.42 A study examined the effect of exercise on how students perform on their final exam in statistics. The P-value was given as 0.87.

(a) State null and alternative hypotheses that could have been used for this study. (Note that there is more than one correct answer.)

(b) Do you reject the null hypothesis?

(c) What is your conclusion?

(d) What other facts about the study would you like to know for a proper interpretation of the results?

6.43 The financial aid office of a university asks a sample of students about their employment and earnings. The report says that "for academic year earnings, a significant difference ($P = 0.038$) was found between the sexes, with men earning more on the average. No difference ($P = 0.476$) was found between the earnings of black and white students."[5] Explain both of these conclusions, for the effects of sex and of race on mean earnings, in language understandable to someone who knows no statistics.

6.44 Statistics can help decide the authorship of literary works. Sonnets by a certain Elizabethan poet are known to contain an average of $\mu = 6.9$ new words (words not used in the poet's other works). The standard deviation of the number of new words is $\sigma = 2.7$. Now a manuscript with 5 new sonnets has come to light, and scholars are debating whether it is the poet's work. The new sonnets contain an average of $\overline{x} = 11.2$ words not used in the poet's known works. We expect poems by another author to contain more new words, so to see if we have evidence that the new sonnets are not by our poet we test

$$H_0: \mu = 6.9$$

$$H_a: \mu > 6.9$$

Give the z test statistic and its P-value. What do you conclude about the authorship of the new poems?

6.45 The Survey of Study Habits and Attitudes (SSHA) is a psychological test that measures the motivation, attitude toward school, and study habits of students. Scores range from 0 to 200. The mean score for U.S. college students is about 115, and the standard deviation is about 30. A teacher who suspects that older students have better attitudes toward school gives the SSHA to 20 students who are at least 30 years of age. Their mean score is $\overline{x} = 135.2$.

(a) Assuming that $\sigma = 30$ for the population of older students, carry out a test of

$$H_0\colon \mu = 115$$
$$H_a\colon \mu > 115$$

Report the P-value of your test, and state your conclusion clearly.

(b) Your test in (a) required two important assumptions in addition to the assumption that the value of σ is known. What are they? Which of these assumptions is most important to the validity of your conclusion in (a)?

6.46 The mean yield of corn in the United States is about 120 bushels per acre. A survey of 40 farmers this year gives a sample mean yield of $\bar{x} = 123.8$ bushels per acre. We want to know whether this is good evidence that the national mean this year is not 120 bushels per acre. Assume that the farmers surveyed are an SRS from the population of all commercial corn growers and that the standard deviation of the yield in this population is $\sigma = 10$ bushels per acre. Give the P-value for the test of

$$H_0\colon \mu = 120$$
$$H_a\colon \mu \neq 120$$

Are you convinced that the population mean is not 120 bushels per acre? Is your conclusion correct if the distribution of corn yields is somewhat nonnormal? Why?

6.47 In the past, the mean score of the seniors at South High on the American College Testing (ACT) college entrance examination has been 20. This year a special preparation course is offered, and all 53 seniors planning to take the ACT test enroll in the course. The mean of their 53 ACT scores is 22.1. The principal believes that the new course has improved the students' ACT scores.

(a) Assume that ACT scores vary normally with standard deviation 6. Is the outcome $\bar{x} = 22.1$ good evidence that the population mean score is greater than 20? State H_0 and H_a, compute the test statistic and the P-value, and answer the question by interpreting your result.

(b) The results are in any case inconclusive because of the design of the study. The effects of the new course are confounded with any change from past years, such as other new courses or higher standards. Briefly outline the design of a better study of the effect of the new course on ACT scores.

6.48 The level of calcium in the blood in healthy young adults varies with mean about 9.5 milligrams per deciliter and standard deviation about $\sigma = 0.4$. A clinic in rural Guatemala measures the blood calcium level of 180 healthy pregnant women at their first visit for prenatal care. The mean is $\bar{x} = 9.58$. Is this an indication that the mean calcium level in the population from which these women come differs from 9.5?

(a) State H_0 and H_a.

(b) Carry out the test and give the *P*-value, assuming that $\sigma = 0.4$ in this population. Report your conclusion.

(c) Give a 95% confidence interval for the mean calcium level μ in this population. We are confident that μ lies quite close to 9.5. This illustrates the fact that a test based on a large sample ($n = 180$ here) will often declare even a small deviation from H_0 to be statistically significant.

6.49 Here are the Degree of Reading Power (DRP) scores for a sample of 44 third-grade students:

40	26	39	14	42	18	25	43	46	27	19
47	19	26	35	34	15	44	40	38	31	46
52	25	35	35	33	29	34	41	49	28	52
47	35	48	22	33	41	51	27	14	54	45

These students can be considered to be an SRS of the third-graders in a suburban school district. DRP scores are approximately normal. Suppose that the standard deviation of scores in this school district is known to be $\sigma = 11$. The researcher believes that the mean score μ of all third-graders in this district is higher than the national mean, which is 32.

(a) State the appropriate H_0 and H_a to test this suspicion.

(b) Carry out the test. Give the *P*-value, and then interpret the result in plain language.

applet

6.50 The *Test of Significance* applet lets you compare a null hypothesis with simulated data. The null hypothesis is that a basketball shooter makes 80% of his free throws. You can vary the number of shots and you can change players. Different players have different true probabilities of making free throws.

(a) Click the shoot button and record the sample proportion of free throws made. Do you think that the null hypothesis is true? Record your response. Then click the box that says "show true %." Record this percent and whether or not the null hypothesis was true.

(b) Repeat part (a) for 10 different shooters.

(c) Summarize what you have learned from this exercise.

applet

6.51 Refer to the previous exercise. Change the number of shots to 100 and repeat the exercise.

6.52 A computer has a random number generator designed to produce random numbers that are uniformly distributed on the interval from 0 to 1. If this is true, the numbers generated come from a population with $\mu = 0.5$ and $\sigma = 0.2887$. A command to generate 100 random numbers gives outcomes with mean $\bar{x} = 0.4365$. Assume that the population σ remains fixed. We want to test

$$H_0\colon \mu = 0.5$$
$$H_a\colon \mu \neq 0.5$$

(a) Calculate the value z of the z test statistic.

(b) Is the result significant at the 5% level ($\alpha = 0.05$)?

(c) Is the result significant at the 1% level ($\alpha = 0.01$)?

6.53 To determine whether the mean nicotine content of a brand of cigarettes is greater than the advertised value of 1.4 milligrams, a health advocacy group tests

$$H_0\colon \mu = 1.4$$
$$H_a\colon \mu > 1.4$$

The calculated value of the test statistic is $z = 2.42$.

(a) Is the result significant at the 5% level?

(b) Is the result significant at the 1% level?

6.54 There are other z statistics that we have not yet studied. We can use Table D to assess the significance of any z statistic. A study compares the habits of students who are on academic probation with students whose grades are satisfactory. One variable measured is the hours spent watching television last week. The null hypothesis is "no difference" between the means for the two populations. The alternative hypothesis is two-sided. The value of the test statistic is $z = -1.37$.

(a) Is the result significant at the 5% level?

(b) Is the result significant at the 1% level?

6.55 Explain in plain language why a significance test that is significant at the 1% level must always be significant at the 5% level.

6.56 You are told that a significance test is significant at the 5% level. From this information can you determine whether or not it is significant at the 1% level? Explain your answer.

6.57 Consider a significance test for a null hypothesis versus a two-sided alternative with a z test statistic. Give a value of z that will give a result significant at the 1% level but not at the 0.5% level.

6.58 You have performed a two-sided test of significance and obtained a value of $z = -3.3$. Use Table D to find the approximate P-value for this test.

6.59 You have performed a one-sided test of significance and obtained a value of $z = 0.315$. Use Table D to find the approximate P-value for this test.

6.60 You will perform a significance test of

$$H_0\colon \mu = 0$$

versus

$$H_a\colon \mu > 0$$

(a) What values of z would lead you to reject H_0 at the 5% level?

(b) If the alternative hypothesis was

$$H_a\colon \mu \neq 0$$

what values of z would lead you to reject H_0 at the 5% level?

(c) Explain why your answers to parts (a) and (b) are different.

6.61 Consider a significance test for a null hypothesis versus a two-sided alternative. Between what values from Table D does the P-value for an outcome $z = 1.12$ lie? Calculate the P-value using Table A, and verify that it lies between the values you found from Table D.

6.62 Repeat the previous exercise for $z = -1.12$.

6.63 Radon is a colorless, odorless gas that is naturally released by rocks and soils and may concentrate in tightly closed houses. Because radon is slightly radioactive, there is some concern that it may be a health hazard. Radon detectors are sold to homeowners worried about this risk, but the detectors may be inaccurate. University researchers placed 12 detectors in a chamber where they were exposed to 105 picocuries per liter (pCi/l) of radon over 3 days. Here are the readings given by the detectors:[6]

91.9	97.8	111.4	122.3	105.4	95.0
103.8	99.6	96.6	119.3	104.8	101.7

Assume (unrealistically) that you know that the standard deviation of readings for all detectors of this type is $\sigma = 9$.

(a) Give a 95% confidence interval for the mean reading μ for this type of detector.

(b) Is there significant evidence at the 5% level that the mean reading differs from the true value 105? State hypotheses and base a test on your confidence interval from (a).

6.64 You measure the weights of 24 male runners. These runners are not a random sample from a population, but you are willing to assume that their weights represent the weights of similar runners. Here are their weights in kilograms:

67.8	61.9	63.0	53.1	62.3	59.7	55.4	58.9
60.9	69.2	63.7	68.3	64.7	65.6	56.0	57.8
66.0	62.9	53.6	65.0	55.8	60.4	69.3	61.7

Exercise 6.6 (page 430) asks you to find a 95% confidence interval for the mean weight of the population of all such runners, assuming that the population standard deviation is $\sigma = 4.5$ kg.

(a) Give the confidence interval from that exercise, or calculate the interval if you did not do the exercise.

(b) Based on this confidence interval, does a test of

$$H_0\colon \mu = 61.3 \text{ kg}$$

$$H_a\colon \mu \neq 61.3 \text{ kg}$$

reject H_0 at the 5% significance level?

(c) Would $H_0\colon \mu = 63$ be rejected at the 5% level if tested against a two-sided alternative?

6.65 An understanding of cockroach biology may lead to an effective control strategy for these annoying insects. Researchers studying the absorption of sugar by insects feed cockroaches a diet containing measured amounts of a particular sugar. After 10 hours, the cockroaches are killed and the concentration of the sugar in various body parts is determined by a chemical analysis. The paper that reports the research states that a 95% confidence interval for the mean amount (in milligrams) of the sugar in the hindguts of the cockroaches is 4.2 ± 2.3.[7]

(a) Does this paper give evidence that the mean amount of sugar in the hindguts under these conditions is not equal to 7 mg? State H_0 and H_a and base a test on the confidence interval.

(b) Would the hypothesis that $\mu = 5$ mg be rejected at the 5% level in favor of a two-sided alternative?

6.66 Market pioneers, companies that are among the first to develop a new product or service, tend to have higher market shares than latecomers to the market. What accounts for this advantage? Here is an excerpt from the conclusions of a study of a sample of 1209 manufacturers of industrial goods:

> Can patent protection explain pioneer share advantages? Only 21% of the pioneers claim a significant benefit from either a product patent or a trade secret. Though their average share is two points higher than that of pioneers without this benefit, the increase is not statistically significant ($z = 1.13$). Thus, at least in mature industrial markets, product patents and trade secrets have little connection to pioneer share advantages.[8]

Find the P-value for the given z. Then explain to someone who knows no statistics what "not statistically significant" in the study's conclusion means. Why does the author conclude that patents and trade secrets don't help, even though they contributed 2 percentage points to average market share?

6.67 An old farmer claims to be able to detect the presence of water with a forked stick. In a test of this claim, he is presented with 5 identical barrels, some containing water and some not. He is right in 4 of the 5 cases.

(a) Suppose the farmer has probability p of being correct. If he is just guessing, $p = 0.5$. State an appropriate H_0 and H_a in terms of p for a test of whether he does better than guessing.

(b) If the farmer is simply guessing, what is the distribution of the number X of correct answers in 5 tries?

(c) The observed outcome is $X = 4$. What is the P-value of the test that takes large values of X to be evidence against H_0?

6.3 Use and Abuse of Tests

Carrying out a test of significance is often quite simple, especially if a fixed significance level α is used or if the P-value is given effortlessly by a computer. Using tests wisely is not so simple. Each test is valid only in certain circumstances, with properly produced data being particularly important. The z test, for example, should bear the same warning label that was attached in Section 6.1 to the corresponding confidence interval. Similar warnings accompany the other tests that we will learn. There are additional caveats that concern tests more than confidence intervals, enough to warrant this separate section. Some hesitation about the unthinking use of significance tests is a sign of statistical maturity.

The reasoning of significance tests has appealed to researchers in many fields, so that tests are widely used to report research results. In this setting H_a is a "research hypothesis" asserting that some effect or difference is present. The null hypothesis H_0 says that there is no effect or no difference. A low P-value represents good evidence that the research hypothesis is true. Here are some comments on the use of significance tests, with emphasis on their use in reporting scientific research.

Choosing a level of significance

The spirit of a test of significance is to give a clear statement of the degree of evidence provided by the sample against the null hypothesis. The P-value does this. But sometimes you will make some decision or take some action if your evidence reaches a certain standard. You can set such a standard by giving a level of significance α. Perhaps you will announce a new scientific finding if your data are significant at the $\alpha = 0.05$ level. Or perhaps you will recommend using a new method of teaching reading if the evidence of its superiority is significant at the $\alpha = 0.01$ level.

Making a decision is different in spirit from testing significance, though the two are often mixed in practice. Choosing a level α in advance makes sense if you must make a decision, but not if you wish only to describe the strength of your evidence. Using tests with fixed α for decision making is discussed at greater length at the end of this chapter.

If you do use a fixed α significance test to make a decision, choose α by asking how much evidence is required to reject H_0. This depends first on how plausible H_0 is. If H_0 represents an assumption that everyone in your field has believed for years, strong evidence (small α) will be needed to reject it. Second, the level of evidence required to reject H_0 depends on the consequences of such a decision. If rejecting H_0 in favor of H_a means making an expensive changeover from one medical therapy or instructional method to another, strong evidence is needed. Both the plausibility of H_0 and H_a and the consequences of any action that rejection may lead to are somewhat subjective. Different persons may feel that different levels of significance are appropriate. It is better to report the P-value, which allows each of us to decide individually if the evidence is sufficiently strong.

Users of statistics have often emphasized certain standard levels of significance, such as 10%, 5%, and 1%. This emphasis reflects the time when tables of critical points rather than computer programs dominated statistical practice. The 5% level ($\alpha = 0.05$) is particularly common. Significance at that level is still a widely accepted criterion for meaningful evidence in research work. *There is no sharp border between "significant" and "insignificant," only increasingly strong evidence as the P-value decreases.* There is no practical distinction between the P-values 0.049 and 0.051. It makes no sense to treat $\alpha = 0.05$ as a universal rule for what is significant.

There is a reason for the common use of $\alpha = 0.05$—the great influence of Sir R. A. Fisher.* Fisher did not originate tests of significance. But because his writings organized statistics, especially as a tool of scientific research, his views on tests were enormously influential. Here is his opinion on choosing a level of significance:

> . . . it is convenient to draw the line at about the level at which we can say: "Either there is something in the treatment, or a coincidence has occurred such as does not occur more than once in twenty trials. . . ."
>
> If one in twenty does not seem high enough odds, we may, if we prefer it, draw the line at one in fifty (the 2 percent point), or one in a hundred (the 1 percent point). Personally, the writer prefers to set a low standard of significance at the 5 percent point, and ignore entirely all results which fail to reach that level. A scientific fact should be regarded as experimentally established only if a properly designed experiment *rarely fails* to give this level of significance.[9]

There you have it. Fisher thought 5% was about right, and who was to disagree with the master? Fisher was of course not an advocate of blind use of significance at the 5% level as a yes-or-no criterion. The last sentence quoted above shows an experienced scientist's feeling for the repeated studies and variable results that mark the advance of knowledge.[10]

---

*We have met Fisher as the inventor of randomized experimental designs. He also originated many other statistical techniques, derived the distributions of many common statistics, and introduced such basic terms as "parameter" and "statistic" into statistical writing.

What statistical significance doesn't mean

When a null hypothesis ("no effect" or "no difference") can be rejected at the usual levels, $\alpha = 0.05$ or $\alpha = 0.01$, there is good evidence that an effect is present. But that effect may be extremely small. When large samples are available, even tiny deviations from the null hypothesis will be significant. For example, suppose that we are testing the hypothesis of no correlation between two variables. With 1000 observations, an observed correlation of only $r = 0.08$ is significant evidence at the $\alpha = 0.01$ level that the correlation in the population is not zero but positive. The low significance level does not mean there is a strong association, only that there is strong evidence of some association. The true population correlation is probably quite close to the observed sample value, $r = 0.08$. We might well conclude that for practical purposes we can ignore the association between these variables, even though we are confident (at the 1% level) that the correlation is positive. Remember the wise saying: *statistical significance is not the same as practical significance.* On the other hand, if we fail to reject the null hypothesis, it may be because H_0 is true or because our sample size is insufficient to detect the alternative. Exercise 6.74 demonstrates in detail the effect on P of increasing the sample size.

The remedy for attaching too much importance to statistical significance is to pay attention to the actual experimental results as well as to the P-value. Plot your data and examine them carefully. Are there outliers or other deviations from a consistent pattern? A few outlying observations can produce highly significant results if you blindly apply common tests of significance. Outliers can also destroy the significance of otherwise convincing data. The foolish user of statistics who feeds the data to a computer without exploratory analysis will often be embarrassed. Is the effect you are seeking visible in your plots? If not, ask yourself if the effect is large enough to be practically important. It is usually wise to give a confidence interval for the parameter in which you are interested. A confidence interval actually estimates the size of an effect rather than simply asking if it is too large to reasonably occur by chance alone. Confidence intervals are not used as often as they should be, while tests of significance are perhaps overused.

Don't ignore lack of significance

Researchers typically have in mind the research hypothesis that some effect exists. Following the peculiar logic of tests of significance, they set up as H_0 the null hypothesis that no such effect exists and try their best to get evidence against H_0. A perverse legacy of Fisher's opinion on $\alpha = 0.05$ is that research in some fields has rarely been published unless significance at that level is attained. For example, a survey of four journals published by the American Psychological Association showed that of 294 articles using statistical tests, only 8 reported results that did not attain the 5% significance level.[11]

Such a publication policy impedes the spread of knowledge, and not only by declaring that a *P*-value of 0.051 is "not significant." If a researcher has good reason to suspect that an effect is present and then fails to find significant evidence of it, that may be interesting news—perhaps more interesting than if evidence in favor of the effect at the 5% level had been found. Keeping silent about negative results may condemn other researchers to repeat the attempt to find an effect that isn't there.

Of course, an experiment that fails only causes a stir if it is clear that the experiment would have detected the effect if it were really there. An important aspect of planning a study is to verify that the test you plan to use does have high probability of detecting an effect of the size you hope to find. This probability is the *power* of the test. Power calculations are discussed later in this chapter.

Tests of statistical significance are routinely used to assess the results of research in agriculture, education, engineering, medicine, psychology, and sociology and increasingly in other fields as well. Any tool used routinely is often used unthinkingly. We therefore offer some comments for the thinking researcher on possible abuses of this tool. Thinking consumers of research findings (such as students) should also ponder these comments.

Statistical inference is not valid for all sets of data

We learned long ago that badly designed surveys or experiments often produce invalid results. Formal statistical inference cannot correct basic flaws in the design. There is no doubt a significant difference in English vocabulary scores between high school seniors who have studied a foreign language and those who have not. But because the effect of actually studying a language is confounded with the differences between students who choose language study and those who do not, this statistical significance is hard to interpret. It does indicate that the difference in English scores is greater than would often arise by chance alone. That leaves unsettled the issue of *what* other than chance caused the difference. The most plausible explanation is that students who were already good at English chose to study another language. A randomized comparative experiment would isolate the actual effect of language study and so make significance meaningful.

Tests of significance and confidence intervals are based on the laws of probability. Randomization in sampling or experimentation ensures that these laws apply. But we must often analyze data that do not arise from randomized samples or experiments. To apply statistical inference to such data, we must have confidence in a probability model for the data. The diameters of successive holes bored in auto engine blocks during production, for example, may behave like independent observations on a normal distribution. We can check this probability model by examining the data. If the model appears correct, we can apply the recipes of this chapter to do inference about the process mean diameter μ. Do ask how the data were produced, and don't be too impressed by *P*-values on a printout until you are confident that the data deserve a formal analysis.

Beware of searching for significance

Statistical significance is a commodity much sought after by researchers. It means (or ought to mean) that you have found an effect that you were looking for. *The reasoning behind statistical significance works well if you decide what effect you are seeking, design an experiment or sample to search for it, and use a test of significance to weigh the evidence you get*. But because a successful search for a new scientific phenomenon often ends with statistical significance, it is all too tempting to make significance itself the object of the search. There are several ways to do this, none of them acceptable in polite scientific society.

One tactic is to make many tests on the same data. Once upon a time three psychiatrists studied a sample of schizophrenic persons and a sample of nonschizophrenic persons. They measured 77 variables for each subject—religion, family background, childhood experiences, and so on. Their goal was to discover what distinguishes persons who later become schizophrenic. Having measured 77 variables, they made 77 separate tests of the significance of the differences between the two groups of subjects. Pause for a moment of reflection. If you made 77 tests at the 5% level, you would expect a few of them to be significant by chance alone. After all, results significant at the 5% level do occur 5 times in 100 in the long run even when H_0 is true. The psychiatrists found 2 of their 77 tests significant at the 5% level and immediately published this exciting news.[12] Running one test and reaching the $\alpha = 0.05$ level is reasonably good evidence that you have found something; running 77 tests and reaching that level only twice is not.

The case of the 77 tests happened long ago. Such crimes are rarer now—or at least better concealed. The computer has freed us from the labor of doing arithmetic. This is surely a blessing in statistics, where the arithmetic can be long and complicated indeed. Comprehensive statistical software systems are everywhere available, so a few simple commands will set the machine to work performing all manner of complicated tests and operations on your data. The result can be much like the 77 tests of old. We will state it as a law that any large set of data—even several pages of a table of random digits—contains some unusual pattern. Sufficient computer time will discover that pattern, and when you test specifically for the pattern that turned up, the result will be significant. It also will mean exactly nothing.

One lesson here is not to be overawed by the computer. The computer has greatly extended the range of statistical inference, allowing us to handle larger sets of data and to carry out more complex analyses. But it has changed the logic of inference not one bit. Doing 77 tests and finding 2 significant at the $\alpha = 0.05$ level is not evidence of a real discovery. Neither is doing factor analysis followed by discriminant analysis followed by multiple regression and at last discovering a significant pattern in the data. Fancy computer programs are no remedy for bad scientific logic. It is convincing to hypothesize that an effect or pattern will be present, design a study to look for it, and find it at a low significance level. It is not convincing to search for any effect or pattern whatever and find one.

We do not mean that searching data for suggestive patterns is not proper scientific work. It certainly is. Many important discoveries have been made by accident rather than by design. Exploratory analysis of data is an essential part of statistics. We do mean that the usual reasoning of statistical inference does not apply when the search for a pattern is successful. You cannot legitimately test a hypothesis on the same data that first suggested that hypothesis. The remedy is clear. Once you have a hypothesis, design a study to search specifically for the effect you now think is there. If the result of this study is statistically significant, you have real evidence at last.

Exploring the Web

The report of the American Psychological Association's Task Force on Statistical Inference is an excellent brief introduction to wise use of inference. The report appeared in the journal *American Psychologist* in 1999. You can find it on the Web in the list of "Selected Articles" from this journal at `www.apa.org/journals/amp.html`.

SUMMARY

P-values are more informative than the reject-or-not result of a fixed level α test. Beware of placing too much weight on traditional values of α, such as $\alpha = 0.05$.

Very small effects can be highly significant (small *P*), especially when a test is based on a large sample. A statistically significant effect need not be practically important. Plot the data to display the effect you are seeking, and use confidence intervals to estimate the actual value of parameters.

On the other hand, lack of significance does not imply that H_0 is true, especially when the test has low power.

Significance tests are not always valid. Faulty data collection, outliers in the data, and testing a hypothesis on the same data that suggested the hypothesis can invalidate a test. Many tests run at once will probably produce some significant results by chance alone, even if all the null hypotheses are true.

SECTION 6.3 EXERCISES

6.68 Give an example of a set of data for which statistical inference is not valid.

6.69 More than 200,000 people worldwide take the GMAT examination each year as they apply for MBA programs. Their scores vary normally with mean about $\mu = 525$ and standard deviation about $\sigma = 100$. One hundred students go through a rigorous training program designed to raise their

GMAT scores. Test the hypotheses

$$H_0: \mu = 525$$
$$H_a: \mu > 525$$

in each of the following situations:

(a) The students' average score is $\bar{x} = 541.4$. Is this result significant at the 5% level?

(b) The average score is $\bar{x} = 541.5$. Is this result significant at the 5% level?

The difference between the two outcomes in (a) and (b) is of no importance. Beware attempts to treat $\alpha = 0.05$ as sacred.

6.70 How much education children get is strongly associated with the wealth and social status of their parents. In social science jargon, this is "socioeconomic status," or SES. But the SES of parents has little influence on whether children who have graduated from college go on to yet more education. One study looked at whether college graduates took the graduate admissions tests for business, law, and other graduate programs. The effects of the parents' SES on taking the LSAT test for law school were "both statistically insignificant and small."

(a) What does "statistically insignificant" mean?

(b) Why is it important that the effects were small in size as well as insignificant?

6.71 What distinguishes managerial trainees who eventually become executives from those who, after expensive training, don't succeed and leave the company? We have abundant data on past trainees—data on their personalities and goals, their college preparation and performance, even their family backgrounds and their hobbies. Statistical software makes it easy to perform dozens of significance tests on these dozens of variables to see which ones best predict later success. We find that future executives are significantly more likely than washouts to have an urban or suburban upbringing and an undergraduate degree in a technical field.

Explain clearly why using these "significant" variables to select future trainees is not wise. Then suggest a follow-up study using this year's trainees as subjects that should clarify the importance of the variables identified by the first study.

6.72 Which of the following questions does a test of significance answer?

(a) Is the sample or experiment properly designed?

(b) Is the observed effect due to chance?

(c) Is the observed effect important?

6.73 In a study of the suggestion that taking vitamin C will prevent colds, 400 subjects are assigned at random to one of two groups. The experimental group takes a vitamin C tablet daily, while the control group takes a placebo. At the end of the experiment, the researchers calculate the difference between the percents of subjects in the two groups who were free of colds. This difference is statistically significant ($P = 0.03$) in favor of the vitamin C group. Can we conclude that vitamin C has a strong effect in preventing colds? Explain your answer.

6.74 Every user of statistics should understand the distinction between statistical significance and practical importance. A sufficiently large sample will declare very small effects statistically significant. Let us suppose that SAT mathematics (SATM) scores in the absence of coaching vary normally with mean $\mu = 475$ and $\sigma = 100$. Suppose further that coaching may change μ but does not change σ. An increase in the SATM score from 475 to 478 is of no importance in seeking admission to college, but this unimportant change can be statistically very significant. To see this, calculate the P-value for the test of

$$H_0: \mu = 475$$
$$H_a: \mu > 475$$

in each of the following situations:

(a) A coaching service coaches 100 students; their SATM scores average $\bar{x} = 478$.

(b) By the next year, the service has coached 1000 students; their SATM scores average $\bar{x} = 478$.

(c) An advertising campaign brings the number of students coached to 10,000; their average score is still $\bar{x} = 478$.

6.75 Give a 99% confidence interval for the mean SATM score μ after coaching in each part of the previous exercise. For large samples, the confidence interval says, "Yes, the mean score is higher after coaching, but only by a small amount."

6.76 As in the previous exercises, suppose that SATM scores vary normally with $\sigma = 100$. One hundred students go through a rigorous training program designed to raise their SATM scores by improving their mathematics skills. Carry out a test of

$$H_0: \mu = 475$$
$$H_a: \mu > 475$$

in each of the following situations:

(a) The students' average score is $\bar{x} = 491.4$. Is this result significant at the 5% level?

(b) The average score is $\bar{x} = 491.5$. Is this result significant at the 5% level?

The difference between the two outcomes in (a) and (b) is of no importance. Beware attempts to treat $\alpha = 0.05$ as sacred.

6.77 A local television station announces a question for a call-in opinion poll on the six o'clock news and then gives the response on the eleven o'clock news. Today's question concerns a proposed increase in funds for student loans. Of the 2372 calls received, 1921 oppose the increase. The station, following standard statistical practice, makes a confidence statement: "81% of the Channel 13 Pulse Poll sample oppose the increase. We can be 95% confident that the proportion of all viewers who oppose the increase is within 1.6% of the sample result." Is the station's conclusion justified? Explain your answer.

6.78 A researcher looking for evidence of extrasensory perception (ESP) tests 500 subjects. Four of these subjects do significantly better ($P < 0.01$) than random guessing.

(a) Is it proper to conclude that these four people have ESP? Explain your answer.

(b) What should the researcher now do to test whether any of these four subjects have ESP?

Bonferroni procedure

6.79 One way to deal with the problem of misleading P-values when performing more than one significance test is to adjust the criterion you use for statistical significance. The **Bonferroni procedure** does this in a simple way. If you perform 2 tests and want to use the $\alpha = 5\%$ significance level, you would require a P-value of $0.05/2 = 0.025$ to declare either one of the null hypotheses significant. In general, if you perform k tests and want protection at level α, use α/k as your cutoff for statistical significance. You perform 6 tests and calculate P-values of 0.476, 0.032, 0.241, 0.008, 0.010, 0.001. Which of these are statistically significant using the Bonferroni procedure with $\alpha = 0.05$?

6.80 Refer to the previous problem. A researcher has performed 12 tests of significance and wants to apply the Bonferroni procedure with $\alpha = 0.05$. The calculated P-values are 0.041, 0.569, 0.050, 0.416, 0.001, 0.004, 0.256, 0.041, 0.888, 0.010, 0.002, 0.433. Which of the null hypotheses are rejected with this procedure?

6.81 The text cites an example in which researchers carried out 77 separate significance tests, of which 2 were significant at the 5% level. Suppose that these tests are independent of each other. (In fact they were not independent, because all involved the same subjects.) If all of the null hypotheses are true, each test has probability 0.05 of being significant at the 5% level.

(a) What is the distribution of the number X of tests that are significant?

(b) Find the probability that 2 or more of the tests are significant.

6.4 Power and Inference as a Decision*

Although we prefer to use P-values rather than the reject-or-not view of the fixed α significance test, the latter view is very important for planning studies and for understanding statistical decision theory. We will discuss these two topics in this section.

*Although the topics in this section are important in planning and interpreting significance tests, they can be omitted without loss of continuity.

Power

In examining the usefulness of a confidence interval, we are concerned with both the level of confidence and the margin of error. The confidence level tells us how reliable the method is in repeated use. The margin of error tells us how sensitive the method is, that is, how closely the interval pins down the parameter being estimated. Fixed level α significance tests are closely related to confidence intervals—in fact, we saw that a two-sided test can be carried out directly from a confidence interval. The significance level, like the confidence level, says how reliable the method is in repeated use. If we use 5% significance tests repeatedly when H_0 is in fact true, we will be wrong (the test will reject H_0) 5% of the time and right (the test will fail to reject H_0) 95% of the time.

High confidence is of little value if the interval is so wide that few values of the parameter are excluded. Similarly, a test with a small level of α is of little value if it almost never rejects H_0 even when the true parameter value is far from the hypothesized value. We must be concerned with the ability of a test to detect that H_0 is false, just as we are concerned with the margin of error of a confidence interval. This ability is measured by the probability that the test will reject H_0 when an alternative is true. The higher this probability is, the more sensitive the test is.

Power

The probability that a fixed level α significance test will reject H_0 when a particular alternative value of the parameter is true is called the **power** of the test to detect that alternative.

EXAMPLE 6.17

Can a 6-month exercise program increase the total body bone mineral content (TBBMC) of young women? A team of researchers is planning a study to examine this question. Based on the results of a previous study, they are willing to assume that $\sigma = 2$ for the percent change in TBBMC over the 6-month period. A change in TBBMC of 1% would be considered important, and the researchers would like to have a reasonable chance of detecting a change this large or larger. Is 25 subjects a large enough sample for this project?

We will answer this question by calculating the power of the significance test that will be used to evaluate the data to be collected. The calculation consists of three steps:

1. State H_0, H_a, the particular alternative we want to detect, and the significance level α.
2. Find the values of $\bar{x}$ that will lead us to reject H_0.
3. Calculate the probability of observing these values of $\bar{x}$ when the alternative is true.

Step 1

The null hypothesis is that the exercise program has no effect on TBBMC. In other words, the mean percent change is zero. The alternative is that exercise is beneficial; that is, the mean change is positive. Formally, we have

$$H_0: \mu = 0$$
$$H_a: \mu > 0$$

The alternative of interest is $\mu = 1\%$ increase in TBBMC. A 5% test of significance will be used.

Step 2

The z test rejects H_0 at the $\alpha = 0.05$ level whenever

$$z = \frac{\bar{x} - \mu_0}{\sigma/\sqrt{n}} = \frac{\bar{x} - 0}{2/\sqrt{25}} \geq 1.645$$

Be sure you understand why we use 1.645. Rewrite this in terms of $\bar{x}$:

$$\bar{x} \geq 1.645\frac{2}{\sqrt{25}}$$
$$\bar{x} \geq 0.658$$

Because the significance level is $\alpha = 0.05$, this event has probability 0.05 of occurring *when the population mean* μ *is 0*.

Step 3

The power against the alternative $\mu = 1\%$ is the probability that H_0 will be rejected *when in fact* $\mu = 1\%$. We calculate this probability by standardizing $\bar{x}$, using the value $\mu = 1$, the population standard deviation $\sigma = 2$, and the sample size $n = 25$. The power is

$$P(\bar{x} \geq 0.658 \text{ when } \mu = 1) = P\left(\frac{\bar{x} - \mu}{\sigma/\sqrt{n}} \geq \frac{0.658 - 1}{2/\sqrt{25}}\right)$$
$$= P(Z \geq -0.855) = 0.80$$

Figure 6.13 illustrates the power with the sampling distribution of $\bar{x}$ when $\mu = 1$. This significance test rejects the null hypothesis that exercise has no effect on TBBMC 80% of the time if the true effect of exercise is a 1% increase in TBBMC. If the true effect of exercise is a greater percent increase, the test will have greater power; it will reject with a higher probability.

High power is desirable. Along with 95% confidence intervals and 5% significance tests, 80% power is becoming a standard. Many U.S. government agencies that provide research funds require that the sample size for

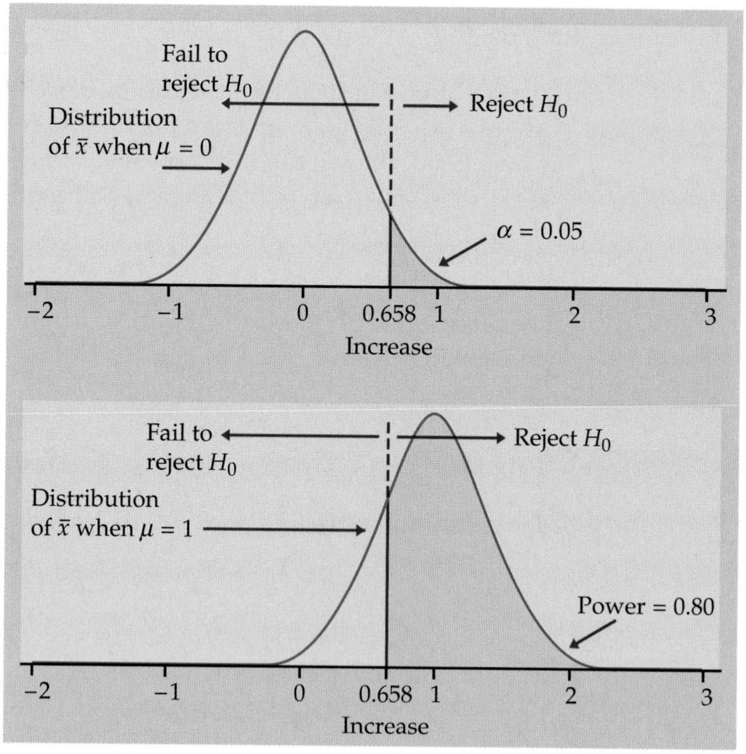

FIGURE 6.13 The sampling distributions of $\bar{x}$ when $\mu = 0$ and when $\mu = 1$ with the α and the power. Power is the probability that the test rejects H_0 when the alternative is true.

the funded studies be sufficient to detect important results 80% of the time using a 5% test of significance.

Increasing the power

Suppose you have performed a power calculation and found that the power is too small. What can you do to increase it? Here are four ways:

- Increase α. A 5% test of significance will have a greater chance of rejecting the alternative than a 1% test because the strength of evidence required for rejection is less.
- Consider a particular alternative that is farther away from μ_0. Values of μ that are in H_a but lie close to the hypothesized value μ_0 are harder to detect (lower power) than values of μ that are far from μ_0.
- Increase the sample size. More data will provide more information about $\bar{x}$ so we have a better chance of distinguishing values of μ.
- Decrease σ. This has the same effect as increasing the sample size: more information about μ. Improving the measurement process and restricting attention to a subpopulation are two common ways to decrease σ.

Power calculations are important in planning studies. Using a significance test with low power makes it unlikely that you will find a significant effect even if the truth is far from the null hypothesis. A null hypothesis that is in fact false can become widely believed if repeated attempts to find evidence against it fail because of low power. The following example illustrates this point.

EXAMPLE 6.18

The "efficient market hypothesis" for the time series of stock prices says that future stock prices (when adjusted for inflation) show only random variation. No information available now will help us predict stock prices in the future, because the efficient working of the market has already incorporated all available information in the present price. Many studies have tested the claim that one or another kind of information is helpful. In these studies, the efficient market hypothesis is H_0, and the claim that prediction is possible is H_a. Almost all the studies have failed to find good evidence against H_0. As a result, the efficient market theory is quite popular. But an examination of the significance tests employed finds that the power is generally low. Failure to reject H_0 when using tests of low power is not evidence that H_0 is true. As one expert says, "The widespread impression that there is strong evidence for market efficiency may be due just to a lack of appreciation of the low power of many statistical tests."[13]

Here is another example of a power calculation, this time for a two-sided z test.

EXAMPLE 6.19

Example 6.13 (page 448) presented a test of

$$H_0\colon \mu = 0.86$$
$$H_a\colon \mu \neq 0.86$$

at the 1% level of significance. What is the power of this test against the specific alternative $\mu = 0.845$?

The test rejects H_0 when $|z| \geq 2.576$. The test statistic is

$$z = \frac{\bar{x} - 0.86}{0.0068/\sqrt{3}}$$

Some arithmetic shows that the test rejects when either of the following is true:

$$z \geq 2.576 \qquad \text{(in other words, } \bar{x} \geq 0.870\text{)}$$
$$z \leq -2.576 \qquad \text{(in other words, } \bar{x} \leq 0.850\text{)}$$

These are disjoint events, so the power is the sum of their probabilities, *computed assuming that the alternative $\mu = 0.845$ is true*. We find that

$$P(\bar{x} \geq 0.87) = P\left(\frac{\bar{x} - \mu}{\sigma/\sqrt{n}} \geq \frac{0.87 - 0.845}{0.0068/\sqrt{3}}\right)$$
$$= P(Z \geq 6.37) \doteq 0$$

$$P(\bar{x} \leq 0.85) = P\left(\frac{\bar{x} - \mu}{\sigma/\sqrt{n}} \leq \frac{0.85 - 0.845}{0.0068/\sqrt{3}}\right)$$
$$= P(Z \leq 1.27) = 0.8980$$

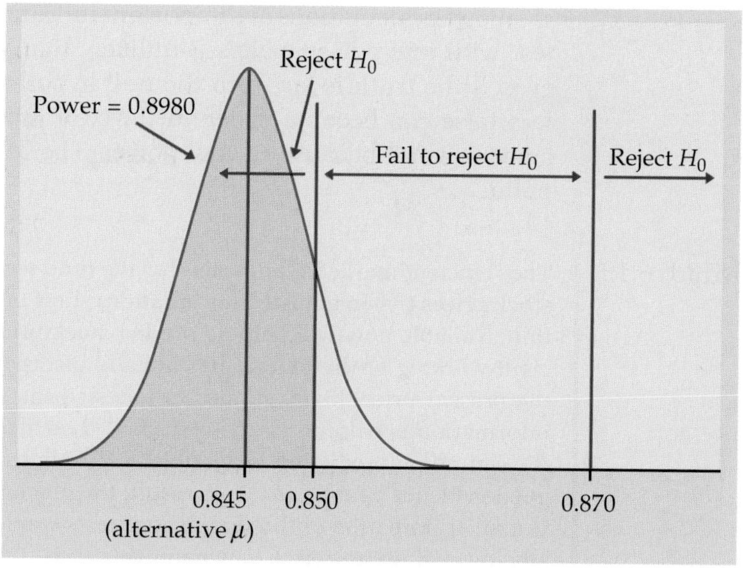

FIGURE 6.14 The power for Example 6.19.

Figure 6.14 illustrates this calculation. Because the power is about 0.9, we are quite confident that the test will reject H_0 when this alternative is true.

Inference as decision*

We have presented tests of significance as methods for assessing the strength of evidence against the null hypothesis. This assessment is made by the P-value, which is a probability computed under the assumption that H_0 is true. The alternative hypothesis (the statement we seek evidence for) enters the test only to help us see what outcomes count against the null hypothesis. Such is the reasoning of tests of significance as advocated by Fisher and as practiced by many users of statistics.

But signs of another way of thinking were present in the discussion of significance tests with fixed level α. A level of significance α chosen in advance points to the outcome of the test as a *decision*. If the P-value is less than α, we reject H_0 in favor of H_a. Otherwise we fail to reject H_0. The transition from measuring the strength of evidence to making a decision is not a small step. Many statisticians agree with Fisher that making decisions is too grand a goal, especially in scientific inference. A decision is reached only after the evidence of many studies is weighed. Indeed, the goal of research is not "decision" but a gradually evolving understanding. Statistical inference should content itself with confidence intervals and tests of significance. Many users of statistics are content with such methods. It is rare to set up a level α in advance as a rule

*The purpose of this section is to clarify the reasoning of significance tests by contrast with a related type of reasoning. It can be omitted without loss of continuity.

for making a decision in a scientific problem. More commonly, users think of significance at level 0.05 as a description of good evidence. This is made clearer by giving the P-value.

acceptance sampling

Yet there are circumstances that call for a decision or action as the end result of inference. **Acceptance sampling** is one such circumstance. A producer of bearings and the consumer of the bearings agree that each carload lot shall meet certain quality standards. When a carload arrives, the consumer chooses a sample of bearings to be inspected. On the basis of the sample outcome, the consumer will either accept or reject the carload. Fisher agreed that this is a genuine decision problem. But he insisted that acceptance sampling is completely different from scientific inference. Other eminent statisticians have argued that if "decision" is given a broad meaning, almost all problems of statistical inference can be posed as problems of making decisions in the presence of uncertainty. We will not venture further into the arguments over how we ought to think about inference. We do want to show how a different concept—inference as decision—changes the reasoning used in tests of significance.

Two types of error

Tests of significance concentrate on H_0, the null hypothesis. If a decision is called for, however, there is no reason to single out H_0. There are simply two hypotheses, and we must accept one and reject the other. It is convenient to call the two hypotheses H_0 and H_a, but H_0 no longer has the special status (the statement we try to find evidence against) that it had in tests of significance. In the acceptance sampling problem, we must decide between

H_0: the lot of bearings meets standards

H_a: the lot does not meet standards

on the basis of a sample of bearings.

We hope that our decision will be correct, but sometimes it will be wrong. There are two types of incorrect decisions. We can accept a bad lot of bearings, or we can reject a good lot. Accepting a bad lot injures the consumer, while rejecting a good lot hurts the producer. To help distinguish these two types of error, we give them specific names.

Type I and Type II Errors

If we reject H_0 (accept H_a) when in fact H_0 is true, this is a **Type I error.** If we accept H_0 (reject H_a) when in fact H_a is true, this is a **Type II error.**

The possibilities are summed up in Figure 6.15. If H_0 is true, our decision either is correct (if we accept H_0) or is a Type I error. If H_a is true, our decision either is correct or is a Type II error. Only one error is possible at one time. Figure 6.16 applies these ideas to the acceptance sampling example.

Decision based on sample	Truth about the population: H_0 true	Truth about the population: H_a true
Reject H_0	Type I error	Correct decision
Accept H_0	Correct decision	Type II error

FIGURE 6.15 The two types of error in testing hypotheses.

Decision based on sample	Truth about the lot: Does meet standards	Truth about the lot: Does not meet standards
Reject the lot	Type I error	Correct decision
Accept the lot	Correct decision	Type II error

FIGURE 6.16 The two types of error in the acceptance sampling setting.

Error probabilities

Any rule for making decisions is assessed in terms of the probabilities of the two types of error. This is in keeping with the idea that statistical inference is based on probability. We cannot (short of inspecting the whole lot) guarantee that good lots of bearings will never be rejected and bad lots never be accepted. But by random sampling and the laws of probability, we can say what the probabilities of both kinds of error are.

Significance tests with fixed level α give a rule for making decisions, because the test either rejects H_0 or fails to reject it. If we adopt the decision-making way of thought, failing to reject H_0 means deciding that H_0 is true. We can then describe the performance of a test by the probabilities of Type I and Type II errors.

EXAMPLE 6.20

The mean diameter of a type of bearing is supposed to be 2.000 centimeters (cm). The bearing diameters vary normally with standard deviation $\sigma = 0.010$ cm. When a lot of the bearings arrives, the consumer takes an SRS of 5 bearings from the lot and measures their diameters. The consumer rejects the bearings if the sample mean diameter is significantly different from 2 at the 5% significance level.

This is a test of the hypotheses

$$H_0: \mu = 2$$

$$H_a: \mu \neq 2$$

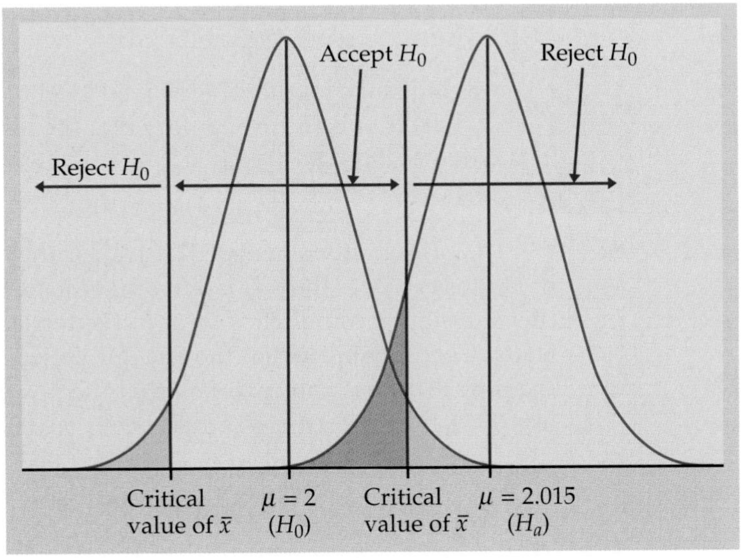

FIGURE 6.17 The two error probabilities for Example 6.20. The probability of a Type I error (*light tan area*) is the probability of rejecting H_0: $\mu = 2$ when in fact $\mu = 2$. The probability of a Type II error (*dark tan area*) is the probability of accepting H_0 when in fact $\mu = 2.015$.

To carry out the test, the consumer computes the z statistic:

$$z = \frac{\bar{x} - 2}{0.01/\sqrt{5}}$$

and rejects H_0 if

$$z < -1.96 \text{ or } z > 1.96$$

A Type I error is to reject H_0 when in fact $\mu = 2$.

What about Type II errors? Because there are many values of μ in H_a, we will concentrate on one value. The producer and the consumer agree that a lot of bearings with mean 0.015 cm away from the desired mean 2.000 should be rejected. So a particular Type II error is to accept H_0 when in fact $\mu = 2.015$.

Figure 6.17 shows how the two probabilities of error are obtained from the two sampling distributions of $\bar{x}$, for $\mu = 2$ and for $\mu = 2.015$. When $\mu = 2$, H_0 is true and to reject H_0 is a Type I error. When $\mu = 2.015$, accepting H_0 is a Type II error. We will now calculate these error probabilities.

The probability of a Type I error is the probability of rejecting H_0 when it is really true. In Example 6.20, this is the probability that $|z| \geq 1.96$ when $\mu = 2$. But this is exactly the significance level of the test. The critical value 1.96 was chosen to make this probability 0.05, so we do not have to compute it again. The definition of "significant at level 0.05" is that sample outcomes this extreme will occur with probability 0.05 when H_0 is true.

Significance and Type I Error

The significance level α of any fixed level test is the probability of a Type I error. That is, α is the probability that the test will reject the null hypothesis H_0 when H_0 is in fact true.

The probability of a Type II error for the particular alternative $\mu = 2.015$ in Example 6.20 is the probability that the test will fail to reject H_0 when μ has this alternative value. The *power* of the test against the alternative $\mu = 2.015$ is just the probability that the test *does* reject H_0. By following the method of Example 6.19, we can calculate that the power is about 0.92. The probability of a Type II error is therefore $1 - 0.92$, or 0.08.

Power and Type II Error

The power of a fixed level test against a particular alternative is 1 minus the probability of a Type II error for that alternative.

The two types of error and their probabilities give another interpretation of the significance level and power of a test. The distinction between tests of significance and tests as rules for deciding between two hypotheses does not lie in the calculations but in the reasoning that motivates the calculations. In a test of significance we focus on a single hypothesis (H_0) and a single probability (the P-value). The goal is to measure the strength of the sample evidence against H_0. Calculations of power are done to check the sensitivity of the test. If we cannot reject H_0, we conclude only that there is not sufficient evidence against H_0, not that H_0 is actually true. If the same inference problem is thought of as a decision problem, we focus on two hypotheses and give a rule for deciding between them based on the sample evidence. We therefore must focus equally on two probabilities, the probabilities of the two types of error. We must choose one or the other hypothesis and cannot abstain on grounds of insufficient evidence.

The common practice of testing hypotheses

Such a clear distinction between the two ways of thinking is helpful for understanding. In practice, the two approaches often merge. We continued to call one of the hypotheses in a decision problem H_0. The common practice of *testing hypotheses* mixes the reasoning of significance tests and decision rules as follows:

1. State H_0 and H_a just as in a test of significance.
2. Think of the problem as a decision problem, so that the probabilities of Type I and Type II errors are relevant.
3. Because of Step 1, Type I errors are more serious. So choose an α (significance level) and consider only tests with probability of Type I error no greater than α.

4. Among these tests, select one that makes the probability of a Type II error as small as possible (that is, power as large as possible). If this probability is too large, you will have to take a larger sample to reduce the chance of an error.

Testing hypotheses may seem to be a hybrid approach. It was, historically, the effective beginning of decision-oriented ideas in statistics. An impressive mathematical theory of hypothesis testing was developed between 1928 and 1938 by Jerzy Neyman and the English statistician Egon Pearson. The decision-making approach came later (1940s). Because decision theory in its pure form leaves you with two error probabilities and no simple rule on how to balance them, it has been used less often than either tests of significance or tests of hypotheses. Decision ideas have been applied in testing problems mainly by way of the Neyman-Pearson hypothesis-testing theory. That theory asks you first to choose α, and the influence of Fisher often has led users of hypothesis testing comfortably back to $\alpha = 0.05$ or $\alpha = 0.01$. Fisher, who was exceedingly argumentative, violently attacked the Neyman-Pearson decision-oriented ideas, and the argument still continues.

SUMMARY

The **power** of a significance test measures its ability to detect an alternative hypothesis. Power against a specific alternative is calculated as the probability that the test will reject H_0 when that alternative is true. This calculation requires knowledge of the sampling distribution of the test statistic under the alternative hypothesis. Increasing the size of the sample increases the power when the significance level remains fixed.

An alternative to significance testing regards H_0 and H_a as two statements of equal status that we must decide between. This **decision theory** point of view regards statistical inference in general as giving rules for making decisions in the presence of uncertainty.

In the case of testing H_0 versus H_a, decision analysis chooses a decision rule on the basis of the probabilities of two types of error. A **Type I error** occurs if H_0 is rejected when it is in fact true. A **Type II error** occurs if H_0 is accepted when in fact H_a is true.

In a fixed level α significance test, the significance level α is the probability of a Type I error, and the power against a specific alternative is 1 minus the probability of a Type II error for that alternative.

SECTION 6.4 EXERCISES

6.82 You want to see if a redesign of the cover of a mail-order catalog will increase sales. A very large number of customers will receive the original catalog, and a random sample of customers will receive the one with the new cover. For planning purposes, you are willing to assume that the sales from the new catalog will be approximately normal with $\sigma = 50$ dollars and that the mean for the original catalog will be $\mu = 25$ dollars. You decide to use a sample size of $n = 900$. You wish to test

$$H_0: \mu = 25$$
$$H_a: \mu > 25$$

You decide to reject H_0 if $\overline{x} > 26$ and to accept H_0 otherwise.

(a) Find the probability of a Type I error, that is, the probability that your test rejects H_0 when in fact $\mu = 25$ dollars.

(b) Find the probability of a Type II error when $\mu = 28$ dollars. This is the probability that your test accepts H_0 when in fact $\mu = 28$.

(c) Find the probability of a Type II error when $\mu = 30$.

(d) The distribution of sales is not normal because many customers buy nothing. Why is it nonetheless reasonable in this circumstance to assume that the mean will be approximately normal?

6.83 Example 6.12 gives a test of a hypothesis about the SAT scores of California high school students based on an SRS of 500 students. The hypotheses are

$$H_0: \mu = 450$$
$$H_a: \mu > 450$$

Assume that the population standard deviation is $\sigma = 100$. The test rejects H_0 at the 1% level of significance when $z \geq 2.326$, where

$$z = \frac{\overline{x} - 450}{100/\sqrt{500}}$$

Is this test sufficiently sensitive to usually detect an increase of 12 points in the population mean SAT score? Answer this question by calculating the power of the test against the alternative $\mu = 462$.

6.84 Example 6.16 discusses a test about the mean contents of cola bottles. The hypotheses are

$$H_0: \mu = 300$$
$$H_a: \mu < 300$$

The sample size is $n = 6$, and the population is assumed to have a normal distribution with $\sigma = 3$. A 5% significance test rejects H_0 if $z \leq -1.645$, where the test statistic z is

$$z = \frac{\overline{x} - 300}{3/\sqrt{6}}$$

Power calculations help us see how large a shortfall in the bottle contents the test can be expected to detect.

(a) Find the power of this test against the alternative $\mu = 298$.

(b) Find the power against the alternative $\mu = 294$.

(c) Is the power against $\mu = 296$ higher or lower than the value you found in (b)? Explain why this result makes sense.

6.85 Increasing the sample size increases the power of a test when the level α is unchanged. Suppose that in the previous exercise a sample of n bottles had been measured. In that exercise, $n = 6$. The 5% significance test still rejects H_0 when $z \leq -1.645$, but the z statistic is now

$$z = \frac{\bar{x} - 300}{3/\sqrt{n}}$$

where we substitute the sample size for n.

(a) Find the power of this test against the alternative $\mu = 298$ when $n = 30$.

(b) Find the power against $\mu = 298$ when $n = 120$.

6.86 In Example 6.11, a company medical director failed to find significant evidence that the mean blood pressure of a population of executives differed from the national mean $\mu = 128$. The medical director now wonders if the test used would detect an important difference if one were present. For the SRS of size 72 from a population with standard deviation $\sigma = 15$, the z statistic is

$$z = \frac{\bar{x} - 128}{15/\sqrt{72}}$$

The two-sided test rejects

$$H_0\colon \mu = 128$$

at the 5% level of significance when $|z| \geq 1.96$.

(a) Find the power of the test against the alternative $\mu = 135$.

(b) Find the power of the test against $\mu = 121$. Can the test be relied on to detect a mean that differs from 128 by 7?

(c) If the alternative were farther from H_0, say $\mu = 138$, would the power be higher or lower than the values calculated in (a) and (b)?

6.87 You have an SRS of size $n = 16$ from a normal distribution with $\sigma = 1$. You wish to test

$$H_0\colon \mu = 0$$

$$H_a\colon \mu > 0$$

You decide to reject H_0 if $\bar{x} > 0$ and to accept H_0 otherwise.

(a) Find the probability of a Type I error, that is, the probability that your test rejects H_0 when in fact $\mu = 0$.

(b) Find the probability of a Type II error when $\mu = 0.2$. This is the probability that your test accepts H_0 when in fact $\mu = 0.2$.

(c) Find the probability of a Type II error when $\mu = 0.6$.

6.88 Use the result of Exercise 6.84 to give the probabilities of Type I and Type II errors for the test discussed there. Take the alternative hypothesis to be $\mu = 294$.

6.89 Use the result of Exercise 6.83 to give the probability of a Type I error and the probability of a Type II error for the test in that exercise when the alternative is $\mu = 462$.

6.90 You must decide which of two discrete distributions a random variable X has. We will call the distributions p_0 and p_1. Here are the probabilities they assign to the values x of X:

x	0	1	2	3	4	5	6
p_0	0.1	0.1	0.1	0.1	0.2	0.1	0.3
p_1	0.2	0.1	0.1	0.2	0.2	0.1	0.1

You have a single observation on X and wish to test

$$H_0: p_0 \text{ is correct}$$

$$H_a: p_1 \text{ is correct}$$

One possible decision procedure is to accept H_0 if $X = 4$ or $X = 6$ and reject H_0 otherwise.

(a) Find the probability of a Type I error, that is, the probability that you reject H_0 when p_0 is the correct distribution.

(b) Find the probability of a Type II error.

6.91 You are designing a computerized medical diagnostic program. The program will scan the results of routine medical tests (pulse rate, blood pressure, urinalysis, etc.) and either clear the patient or refer the case to a doctor. The program will be used as part of a preventive-medicine system to screen many thousands of persons who do not have specific medical complaints. The program makes a decision about each patient.

(a) What are the two hypotheses and the two types of error that the program can make? Describe the two types of error in terms of "false-positive" and "false-negative" test results.

(b) The program can be adjusted to decrease one error probability, at the cost of an increase in the other error probability. Which error probability would you choose to make smaller, and why? (This is a matter of judgment. There is no single correct answer.)

6.92 **(Optional)** The acceptance sampling test in Example 6.20 has probability 0.05 of rejecting a good lot of bearings and probability 0.08 of accepting a bad lot. The consumer of the bearings may imagine that acceptance sampling guarantees that most accepted lots are good. Alas, it is not so. Suppose that 90% of all lots shipped by the producer are bad.

(a) Draw a tree diagram for shipping a lot (the branches are "bad" and "good") and then inspecting it (the branches at this stage are "accept" and "reject").

(b) Write the appropriate probabilities on the branches, and find the probability that a lot shipped is accepted.

(c) Use the definition of conditional probability or Bayes's formula (page 350) to find the probability that a lot is bad, given that the lot is accepted. This is the proportion of bad lots among the lots that the sampling plan accepts.

CHAPTER 6 EXERCISES

6.93 Patients with chronic kidney failure may be treated by dialysis, using a machine that removes toxic wastes from the blood, a function normally performed by the kidneys. Kidney failure and dialysis can cause other changes, such as retention of phosphorus, that must be corrected by changes in diet. A study of the nutrition of dialysis patients measured the level of phosphorus in the blood of several patients on six occasions. Here are the data for one patient (in milligrams of phosphorus per deciliter of blood):[14]

5.6 5.1 4.6 4.8 5.7 6.4

The measurements are separated in time and can be considered an SRS of the patient's blood phosphorus level. Assuming that this level varies normally with $\sigma = 0.9$ mg/dl, give a 90% confidence interval for the mean blood phosphorus level.

6.94 The normal range of phosphorus in the blood is considered to be 2.6 to 4.8 mg/dl. Is there strong evidence that the patient in the previous exercise has a mean phosphorus level that exceeds 4.8?

6.95 Because sulfur compounds cause "off-odors" in wine, oenologists (wine experts) have determined the odor threshold, the lowest concentration of a compound that the human nose can detect. For example, the odor threshold for dimethyl sulfide (DMS) is given in the oenology literature as 25 micrograms per liter of wine (μg/l). Untrained noses may be less sensitive, however. Here are the DMS odor thresholds for 10 beginning students of oenology:

31 31 43 36 23 34 32 30 20 24

Assume (this is not realistic) that the standard deviation of the odor threshold for untrained noses is known to be $\sigma = 7$ μg/l.

(a) Make a stemplot to verify that the distribution is roughly symmetric with no outliers. (A normal quantile plot confirms that there are no systematic departures from normality.)

(b) Give a 95% confidence interval for the mean DMS odor threshold among all beginning oenology students.

(c) Are you convinced that the mean odor threshold for beginning students is higher than the published threshold, 25 μg/l? Carry out a significance test to justify your answer.

6.96 A government report gives a 90% confidence interval for the 1999 median annual household income as \$40,816 $\pm$ \$314. This result was calculated by advanced methods from the Current Population Survey, a multistage random sample of about 50,000 households.

(a) Would a 95% confidence interval be wider or narrower? Explain your answer.

(b) Would the null hypothesis that the 1999 median household income was \$40,000 be rejected at the 10% significance level in favor of the two-sided alternative?

6.97 Refer to the previous problem. Give a 90% confidence interval for the 1999 median *weekly* household income. Use 52.14 as the number of weeks in a year.

6.98 An agronomist examines the cellulose content of a variety of alfalfa hay. Suppose that the cellulose content in the population has standard deviation $\sigma = 8$ milligrams per gram (mg/g). A sample of 15 cuttings has mean cellulose content $\overline{x} = 145$ mg/g.

(a) Give a 90% confidence interval for the mean cellulose content in the population.

(b) A previous study claimed that the mean cellulose content was $\mu =$ 140 mg/g, but the agronomist believes that the mean is higher than that figure. State H_0 and H_a and carry out a significance test to see if the new data support this belief.

(c) The statistical procedures used in (a) and (b) are valid when several assumptions are met. What are these assumptions?

6.99 In a study of possible iron deficiency in infants, researchers compared several groups of infants who were following different feeding patterns. One group of 26 infants was being breast-fed. At 6 months of age, these children had a mean hemoglobin level of $\overline{x} = 12.9$ grams per 100 milliliters of blood and a standard deviation of 1.6. Taking the standard deviation to be the population value σ, give a 95% confidence interval for the mean hemoglobin level of breast-fed infants. What assumptions are required for the validity of the method you used to get the confidence interval?

6.100 Consumers can purchase nonprescription medications at food stores, mass merchandise stores such as Kmart and Wal-Mart, or pharmacies. About

45% of consumers make such purchases at pharmacies. What accounts for the popularity of pharmacies, which often charge higher prices?

A study examined consumers' perceptions of overall performance of the three types of stores, using a long questionnaire that asked about such things as "neat and attractive store," "knowledgeable staff," and "assistance in choosing among various types of nonprescription medication." A performance score was based on 27 such questions. The subjects were 201 people chosen at random from the Indianapolis telephone directory. Here are the means and standard deviations of the performance scores for the sample:[15]

Store type	$\bar{x}$	s
Food stores	18.67	24.95
Mass merchandisers	32.38	33.37
Pharmacies	48.60	35.62

We do not know the population standard deviations, but a sample standard deviation s from so large a sample is usually close to σ. Use s in place of the unknown σ in this exercise.

(a) What population do you think the authors of the study want to draw conclusions about? What population are you certain they can draw conclusions about?

(b) Give 95% confidence intervals for the mean performance for each type of store.

(c) Based on these confidence intervals, are you convinced that consumers think that pharmacies offer higher performance than the other types of stores? Note that in Chapter 12, we will study a statistical method for comparing means of several groups.

6.101 A study of the pay of corporate chief executive officers (CEOs) examined the increase in cash compensation of the CEOs of 104 companies, adjusted for inflation, in a recent year. The mean increase in real compensation was $\bar{x} = 6.9\%$, and the standard deviation of the increases was $s = 55\%$. Is this good evidence that the mean real compensation μ of all CEOs increased that year? The hypotheses are

$$H_0\colon \mu = 0 \quad \text{(no increase)}$$

$$H_a\colon \mu > 0 \quad \text{(an increase)}$$

Because the sample size is large, the sample s is close to the population σ, so take $\sigma = 55\%$.

(a) Sketch the normal curve for the sampling distribution of $\bar{x}$ when H_0 is true. Shade the area that represents the P-value for the observed outcome $\bar{x} = 6.9\%$.

(b) Calculate the P-value.

(c) Is the result significant at the $\alpha = 0.05$ level? Do you think the study gives strong evidence that the mean compensation of all CEOs went up?

6.102 Statisticians prefer large samples. Describe briefly the effect of increasing the size of a sample (or the number of subjects in an experiment) on each of the following:

(a) The width of a level C confidence interval.

(b) The P-value of a test, when H_0 is false and all facts about the population remain unchanged as n increases.

(c) The power of a fixed level α test, when α, the alternative hypothesis, and all facts about the population remain unchanged.

6.103 A roulette wheel has 18 red slots among its 38 slots. You observe many spins and record the number of times that red occurs. Now you want to use these data to test whether the probability of a red has the value that is correct for a fair roulette wheel. State the hypotheses H_0 and H_a that you will test. (We will describe the test for this situation in Chapter 8.)

6.104 When asked to explain the meaning of "statistically significant at the $\alpha = 0.05$ level," a student says, "This means there is only probability 0.05 that the null hypothesis is true." Is this an essentially correct explanation of statistical significance? Explain your answer.

6.105 Another student, when asked why statistical significance appears so often in research reports, says, "Because saying that results are significant tells us that they cannot easily be explained by chance variation alone." Do you think that this statement is essentially correct? Explain your answer.

6.106 Use a computer to generate $n = 9$ observations from a normal distribution with mean 15 and standard deviation 6: $N(15, 6)$. Find the 95% confidence interval for μ. Repeat this process 100 times and then count the number of times that the confidence interval includes the value $\mu = 15$. Explain your results.

6.107 Use a computer to generate $n = 9$ observations from a normal distribution with mean 15 and standard deviation 6: $N(15, 6)$. Test the null hypothesis that $\mu = 15$ using a two-sided significance test. Repeat this process 100 times and then count the number of times that you reject H_0. Explain your results.

6.108 Use the same procedure for generating data as in the previous exercise. Now test the null hypothesis that $\mu = 18$. Explain your results.

6.109 Figure 6.2 (page 420) demonstrates the behavior of a confidence interval in repeated sampling by showing the results of 25 samples from the same population. Now you will do a similar demonstration. Suppose that (unknown to the researcher) the mean SATM score of all California high school seniors is $\mu = 475$, and that the standard deviation is known to be $\sigma = 100$. The scores vary normally.

(a) Simulate the drawing of 50 SRSs of size $n = 100$ from this population.

(b) The 95% confidence interval for the population mean μ has the form $\bar{x} \pm m$. What is the margin of error m? (Remember that we know $\sigma = 100$.)

(c) Use your software to calculate the 95% confidence interval for μ when $\sigma = 100$ for each of your 50 samples. Verify the computer's calculations by checking the interval given for the first sample against your result in (b). Use the $\bar{x}$ reported by the software.

(d) How many of the 50 confidence intervals contain the true mean $\mu = 475$? If you repeated the simulation, would you expect exactly the same number of intervals to contain μ? In a very large number of samples, what percent of the confidence intervals would contain μ?

6.110 In the previous exercise you simulated the SATM scores of 50 SRSs of 100 California seniors. Now use these samples to demonstrate the behavior of a significance test. We know that the population of all SATM scores is normal with standard deviation $\sigma = 100$.

(a) Use your software to carry out a test of

$$H_0: \mu = 475$$

$$H_a: \mu \neq 475$$

for each of the 50 samples.

(b) Verify the computer's calculations by using Table A to find the P-value of the test for the first of your samples. Use the $\bar{x}$ reported by your software.

(c) How many of your 50 tests reject the null hypothesis at the $\alpha = 0.05$ significance level? (That is, how many have P-value 0.05 or smaller?) Because the simulation was done with $\mu = 475$, samples that lead to rejecting H_0 produce the wrong conclusion. In a very large number of samples, what percent would falsely reject the hypothesis?

6.111 Suppose that in fact the mean SATM score of California high school seniors is $\mu = 500$. Would the test in the previous exercise usually detect a mean this far from the hypothesized value? This is a question about the power of the test.

(a) Simulate the drawing of 50 SRSs from a normal population with mean $\mu = 500$ and $\sigma = 100$. These represent the results of sampling when in fact the alternative $\mu = 500$ is true.

(b) Repeat on these new data the test of

$$H_0: \mu = 475$$

$$H_a: \mu \neq 475$$

that you did in the previous exercise. How many of the 50 tests have P-values 0.05 or smaller? These tests reject the null hypothesis at the $\alpha = 0.05$ significance level, which is the correct conclusion.

(c) The power of the test against the alternative $\mu = 500$ is the probability that the test will reject H_0: $\mu = 475$ when in fact $\mu = 500$. Calculate this power. In a very large number of samples from a population with mean 500, what percent would reject H_0?

6.112 Persons aged 55 and over represented 21.3% of the U.S. population in the year 2000. This group is expected to increase to 30.5% by 2025. In terms of actual numbers of people, the increase is from 58.6 million to 101.4 million. Restauranteurs have found this market to be important and would like to make their businesses attractive to older customers. One study used a questionnaire to collect data from people aged 50 and over.[16] For one part of the analysis, individuals were classified into two age groups: 50–64 and 65–79. There were 267 people in the first group and 263 in the second. One set of items concerned ambience, menu design, and service. A series of questions were rated on a 1 to 5 scale with 1 representing "strongly disagree" and 5 representing "strongly agree." In some cases the wording of questions has been shortened in the table below. Here are the means:

Question	50–64	65–79
Ambience		
Most restaurants are too dark	2.75	2.93
Most restaurants are too noisy	3.33	3.43
Background music is often too loud	3.27	3.55
Restaurants are too smoky	3.17	3.12
Tables are too small	3.00	3.19
Tables are too close together	3.79	3.81
Menu design		
Print size is not large enough	3.68	3.77
Glare makes menus difficult to read	2.81	3.01
Colors of menus make them difficult to read	2.53	2.72
Service		
It is difficult to hear the service staff	2.65	3.00
I would rather be served than serve myself	4.23	4.14
I would rather pay the server than a cashier	3.88	3.48
Service is too slow	3.13	3.10

First examine the means of the people who are 50 to 64. Order the statements according to the means and describe the results. Then do the same for the older group. For each question compute the z statistic and the associated P-value for the comparison between the two groups. For these calculations you can assume that the standard deviation of the difference is 0.08, so z is simply the difference in the means divided by 0.08. Note that you are performing 13 significance tests in this exercise. Keep this in mind when you interpret your results. Write a report summarizing your work.

Prelude: diversify your investments

A basic principle of investing is "don't put all your eggs in one basket." In previous chapters we learned that $\bar{x}$ has variance σ^2/n when the data are an SRS from a population. The variance goes down as the number of observations goes up. More general versions of the same theory tell us that averages of many random quantities are less variable than averages of only a few. This applies to investing: diversification reduces the variability of returns, and so reduces the investor's risk. If you put all of your retirement money in the stock of one company, the company may be very successful and you will be rich. On the other hand, the company may go bankrupt and you will be very poor. If you invest in several stocks, you are unlikely to be very rich or very poor.

Reputable investment advisors accept the idea that diversification should guide the investment strategy of individuals who are investing for retirement. An investor with a portfolio of several hundred thousand dollars sued his broker and brokerage firm because they did not diversify his investments. The conflict was settled by an arbitration panel that gave "substantial damages" to the investor.[1] Statistical methods were used to show that this investor's return was substantially worse than the Standard & Poor's (S&P) average return.

CHAPTER 7

Frank Stella, *Flin Flon III*, 1969.

Inference for Distributions

Introduction

We began our study of data analysis in Chapter 1 by learning graphical and numerical tools for describing the distribution of a single variable and for comparing several distributions. Our study of the practice of statistical inference begins in the same way, with inference about a single distribution and comparison of two distributions. Comparing more than two distributions requires more elaborate methods, which are presented in Chapters 12 and 13.

Two important aspects of any distribution are its center and spread. If the distribution is normal, we describe its center by the mean μ and its spread by the standard deviation σ. In this chapter, we will meet confidence intervals and significance tests for inference about a population mean μ and for comparing the means of two populations. The previous chapter emphasized the reasoning of tests and confidence intervals; now we emphasize statistical practice, so we no longer assume that population standard deviations are known. The t procedures for inference about means are among the most common statistical methods. Inference about the spread of a population, as we will see, poses some difficult practical problems.

7.1 Inference for the Mean of a Population

Both confidence intervals and tests of significance for the mean μ of a normal population are based on the sample mean $\bar{x}$, which estimates the unknown μ. The sampling distribution of $\bar{x}$ depends on σ. This fact causes no difficulty when σ is known. When σ is unknown, however, we must estimate σ even though we are primarily interested in μ. The sample standard deviation s is used to estimate the population standard deviation σ.

The *t* distributions

Suppose that we have a simple random sample (SRS) of size n from a normally distributed population with mean μ and standard deviation σ. The sample mean $\bar{x}$ then has the normal distribution with mean μ and standard deviation $\sigma/\sqrt{n}$. When σ is not known, we estimate it with the sample standard deviation s, and then we estimate the standard deviation of $\bar{x}$ by $s/\sqrt{n}$. This quantity is called the *standard error* of the sample mean $\bar{x}$ and we denote it by $\text{SE}_{\bar{x}}$.

Standard Error

When the standard deviation of a statistic is estimated from the data, the result is called the **standard error** of the statistic. The standard error of the sample mean is

$$\text{SE}_{\bar{x}} = \frac{s}{\sqrt{n}}$$

The term "standard error" is sometimes used for the actual standard deviation of a statistic, $\sigma/\sqrt{n}$ in the case of $\bar{x}$. The estimated value $s/\sqrt{n}$ is then called the "estimated standard error." In this book we will use the term "standard error" only when the standard deviation of a statistic is estimated from the data. The term has this meaning in the output of many statistical computer packages and in reports of research in many fields that apply statistical methods.

The standardized sample mean, or one-sample z statistic,

$$z = \frac{\bar{x} - \mu}{\sigma/\sqrt{n}}$$

is the basis of the z procedures for inference about μ when σ is known. This statistic has the standard normal distribution $N(0, 1)$. When we substitute the standard error $s/\sqrt{n}$ for the standard deviation $\sigma/\sqrt{n}$ of $\bar{x}$, the statistic does *not* have a normal distribution. It has a distribution that is new to us, called a *t distribution*.

The *t* Distributions

Suppose that an SRS of size n is drawn from an $N(\mu, \sigma)$ population. Then the **one-sample *t* statistic**

$$t = \frac{\bar{x} - \mu}{s/\sqrt{n}}$$

has the ***t* distribution** with $n - 1$ **degrees of freedom.**

degrees of freedom

There is a different t distribution for each sample size. A particular t distribution is specified by giving the **degrees of freedom.** The degrees of freedom for this t statistic come from the sample standard deviation s in the denominator of t. We saw in Chapter 1 (page 50) that s has $n - 1$ degrees of freedom. There are other t statistics with different degrees of freedom, some of which we will meet later in this chapter.

We use $t(k)$ to stand for the t distribution with k degrees of freedom.* The density curves of the $t(k)$ distributions are similar in shape to the standard normal curve. That is, they are symmetric about 0 and are bell-shaped. The spread of the t distributions is a bit greater than that of the standard normal distribution. This is due to the extra variability caused by substituting the random variable s for the fixed parameter σ. As the degrees of freedom k increase, the $t(k)$ density curve approaches the $N(0, 1)$ curve ever more closely. This reflects the fact that s approaches σ as the sample size increases. Figure 7.1 compares the density curves of the standard normal distribution and the t distribution with 5 degrees of freedom. The similarity in shape is apparent, as is the fact that the t distribution has more probability in the tails and less in the center than does the standard normal distribution.

---

*The t distributions were discovered in 1908 by William S. Gosset. Gosset was a statistician employed by the Guinness brewing company, which required that he not publish his discoveries under his own name. He therefore wrote under the pen name "Student." The t distribution is often called "Student's t" in his honor.

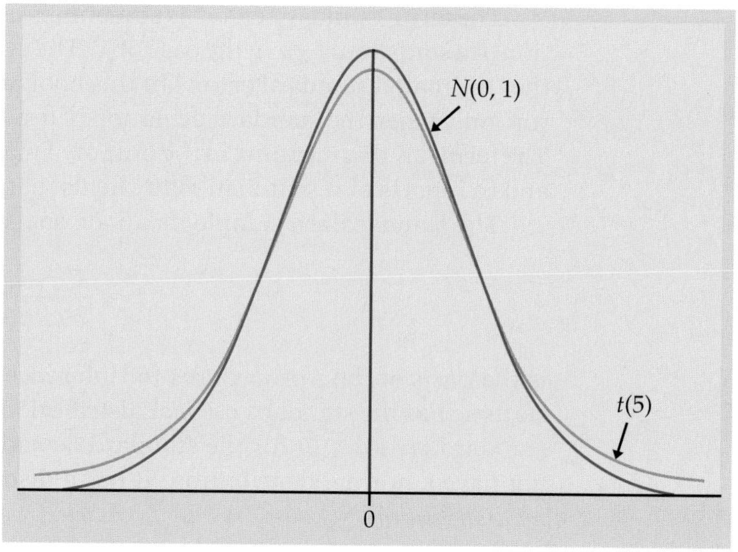

FIGURE 7.1 Density curves for the standard normal and $t(5)$ distributions. Both are symmetric with center 0. The t distributions have more probability in the tails than does the standard normal distribution.

Table D in the back of the book gives critical values for the t distributions. For convenience, we have labeled the table entries by both the critical value p needed for significance tests, and by the confidence level C (in percent) required for confidence intervals. The standard normal critical values in the bottom row of entries are labeled z^*. As in the case of the normal table (Table A), computer software often makes Table D unnecessary.

The one-sample *t* confidence interval

With the t distributions to help us, we can analyze samples from normal populations with unknown σ by replacing the standard deviation $\sigma/\sqrt{n}$ of $\bar{x}$ by its standard error $s/\sqrt{n}$ in the z procedures of Chapter 6. The z statistic then becomes the one-sample t statistic. We must now employ P-values or critical values from t in place of the corresponding normal values (z). Here are the details for the confidence interval.

The One-Sample *t* Confidence Interval

Suppose that an SRS of size n is drawn from a population having unknown mean μ. A level C **confidence interval** for μ is

$$\bar{x} \pm t^* \frac{s}{\sqrt{n}}$$

where t^* is the value for the $t(n-1)$ density curve with area C between $-t^*$ and t^*. This interval is exact when the population distribution is normal and is approximately correct for large n in other cases.

The one-sample t confidence interval is similar in both reasoning and computational detail to the z confidence interval of Chapter 6. There, the margin of error for the population mean was $z^*\sigma/\sqrt{n}$. Here, we replace σ by its estimate s and z^* by t^*. So the **margin of error** for the population mean when we use the data to estimate σ is $t^*s/\sqrt{n}$.

margin of error

EXAMPLE 7.1

In fiscal year 1996, the U.S. Agency for International Development provided 238,300 metric tons of corn soy blend (CSB) for development programs and emergency relief in countries throughout the world. CSB is a highly nutritious, low-cost fortified food that is partially precooked and can be incorporated into different food preparations by the recipients. As part of a study to evaluate appropriate vitamin C levels in this commodity, measurements were taken on samples of CSB produced in a factory.[2]

The following data are the amounts of vitamin C, measured in milligrams per 100 grams (mg/100 g) of blend (dry basis), for a random sample of size 8 from a production run:

26 31 23 22 11 22 14 31

We want to find a 95% confidence interval for μ, the mean vitamin C content of the CSB produced during this run.

The sample mean is $\overline{x} = 22.50$ and the standard deviation is $s = 7.19$ with degrees of freedom $n - 1 = 7$. The standard error is

$$\mathrm{SE}_{\overline{x}} = s/\sqrt{n} = 7.19/\sqrt{8} = 2.54$$

From Table D we find $t^* = 2.365$. The 95% confidence interval is

$$\begin{aligned}\overline{x} \pm t^*\frac{s}{\sqrt{n}} &= 22.50 \pm 2.365\frac{7.19}{\sqrt{8}}\\ &= 22.5 \pm (2.365)(2.54)\\ &= 22.5 \pm 6.0\\ &= (16.5,\ 28.5)\end{aligned}$$

We are 95% confident that the mean vitamin C content of the CSB for this run is between 16.5 and 28.5 mg/100 g.

In this example we have given the actual interval (16.5, 28.5) as our answer. Sometimes, we prefer to report the mean and margin of error: the mean vitamin C content is 22.5 mg/100 g with a margin of error of 6.0 mg/100 g.

The use of the t confidence interval in Example 7.1 rests on assumptions that cannot easily be checked but are reasonable in this case. The samples from the production run were taken at regular time intervals, and experience has shown that this procedure gives data that can be treated as random samples from a normal population. The assumption that the population distribution is normal cannot be effectively checked with only 8 observations. On the other hand, when we make a stemplot, we can see clearly that there are no outliers:

```
3 | 1 1
2 | 2 2 3 6
1 | 1 4
```

The one-sample t test

Significance tests using the estimated standard error are very similar to the z test that we studied in the last chapter.

The One-Sample t Test

Suppose that an SRS of size n is drawn from a population having unknown mean μ. To test the hypothesis H_0: $\mu = \mu_0$ based on an SRS of size n, compute the one-sample t statistic

$$t = \frac{\overline{x} - \mu_0}{s/\sqrt{n}}$$

In terms of a random variable T having the $t(n-1)$ distribution, the P-value for a test of H_0 against

H_a: $\mu > \mu_0$ is $P(T \geq t)$

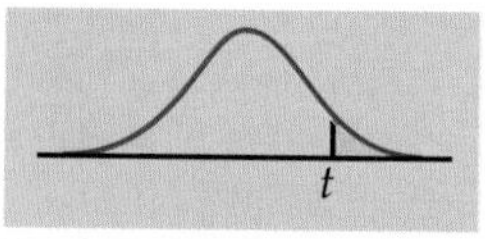

H_a: $\mu < \mu_0$ is $P(T \leq t)$

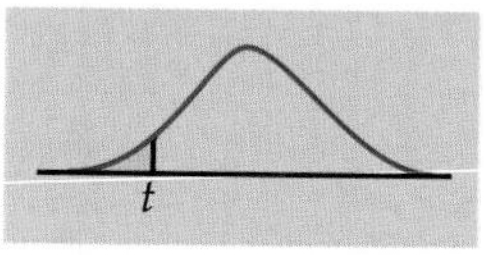

H_a: $\mu \neq \mu_0$ is $2P(T \geq |t|)$

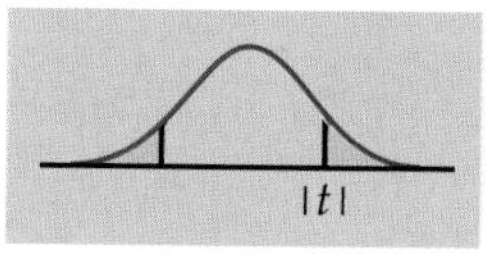

These P-values are exact if the population distribution is normal and are approximately correct for large n in other cases.

EXAMPLE 7.2

The specifications for the CSB described in Example 7.1 state that the mixture should contain 2 pounds of vitamin premix for every 2000 pounds of product. These specifications are designed to produce a mean (μ) vitamin C content in the final product of 40 mg/100 g. We can test a null hypothesis that the mean vitamin C content of the production run in Example 7.1 conforms to these specifications. Specifically, we test

$$H_0: \mu = 40$$

$$H_a: \mu \neq 40$$

Recall that $n = 8$, $\overline{x} = 22.50$, and $s = 7.19$. The t test statistic is

$$t = \frac{\overline{x} - \mu_0}{s/\sqrt{n}} = \frac{22.5 - 40}{7.2/\sqrt{8}} = -6.88$$

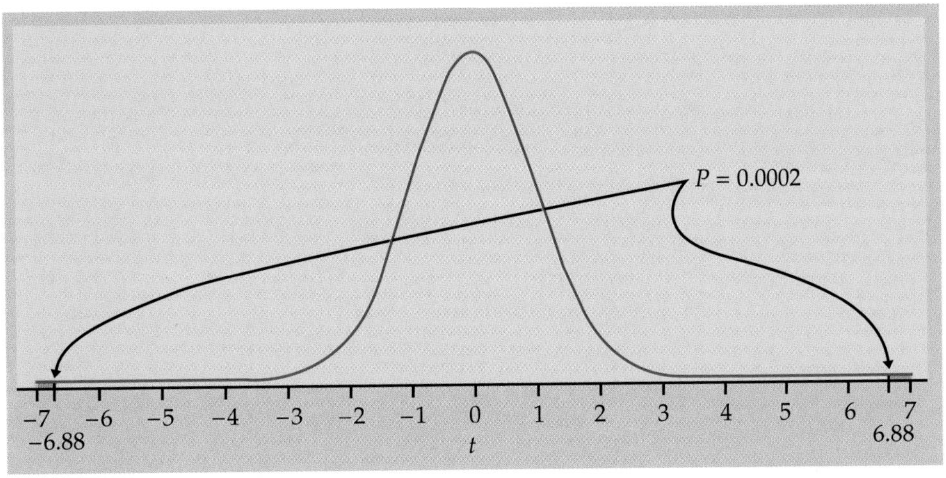

FIGURE 7.2 The P-value for Example 7.2.

df = 7		
p	0.001	0.0005
t^*	4.785	5.408

Because the degrees of freedom are $n - 1 = 7$, this t statistic has the $t(7)$ distribution. Figure 7.2 shows that the P-value is $2P(T \geq 6.88)$, where T has the $t(7)$ distribution. From Table D we see that $P(T \geq 5.408) = 0.0005$. Therefore, we conclude that the P-value is less than 2×0.0005. Software gives the exact value as $P = 0.0002$. Clearly, these data are incompatible with a process mean of $\mu = 40$ mg/100 g. We reject H_0 and conclude that the vitamin C content for this run is below the specifications.

In this example we tested the specifications $\mu = 40$ mg/100 g against the two-sided alternative $\mu \neq 40$ mg/100 g because we had no prior suspicion that the production run would produce CSB with too much or too little vitamin C. If we knew that the proper amount of vitamin C was added to the mixture but suspected that some of the vitamin might be lost or destroyed by the production process, we should use a one-sided test. It is *wrong*, however, to examine the data first and then decide to do a one-sided test in the direction indicated by the data. If in doubt, use a two-sided test.

EXAMPLE 7.3

For the vitamin C problem described in the previous example, we want to test whether or not vitamin C is lost or destroyed in the production process. Here we test

$$H_0\colon \mu = 40$$

versus

$$H_a\colon \mu < 40$$

The t test statistic does not change: $t = -6.88$. As Figure 7.3 illustrates, however, the P-value is now $P(T \leq -6.88)$, half of the value in the previous exercise. From Table D we can determine that $P < 0.0005$; software gives the exact value as $P = 0.0001$. We conclude that the production process has lost or destroyed some of the vitamin C.

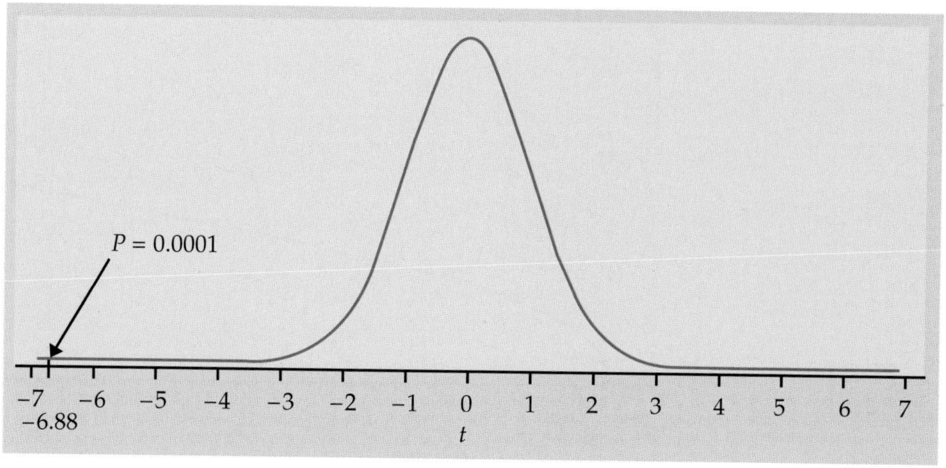

FIGURE 7.3 The *P*-value for Example 7.3.

Note that the conclusion drawn in this example relies heavily on the assumption that the proper amount of vitamin C was added to the blend. If this assumption is not correct, it is still true that the CSB produced has a vitamin C content below the specification, but the explanation that this is caused by loss or destruction of the vitamin C in the production process is not necessarily true.

For small data sets, such as the one in Example 7.1, it is easy to perform the computations for confidence intervals and significance tests with an ordinary calculator. For larger data sets, however, we prefer to use software or a statistical calculator.

EXAMPLE 7.4

The first data that we looked at were Simon Newcomb's measurements of the passage time of light. Table 1.1 (page 8) records his 66 measurements. A normal quantile plot (Figure 7.4) reminds us that when the 2 outliers to the left are omitted, the remaining 64 observations follow a normal distribution quite closely. What result should Newcomb report from these 64 observations?

We want to estimate the mean μ of the distribution of measurements from which Newcomb's 64 are a sample. The sample mean $\bar{x} = 27.750$ estimates μ, but to indicate the precision of this estimate we will give a 99% confidence interval. The sample standard deviation is $s = 5.083$ with $n - 1 = 63$ degrees of freedom.

Figure 7.5 displays the outputs from Minitab and the TI-83. They give the results in slightly different forms but it is easy to find what we need. You should have no trouble reading the output from other software systems as well.

The 99% confidence interval for Newcomb's estimated passage time of light is (26.06, 29.44).

We can also use software or a statistical calculator to perform significance tests easily.

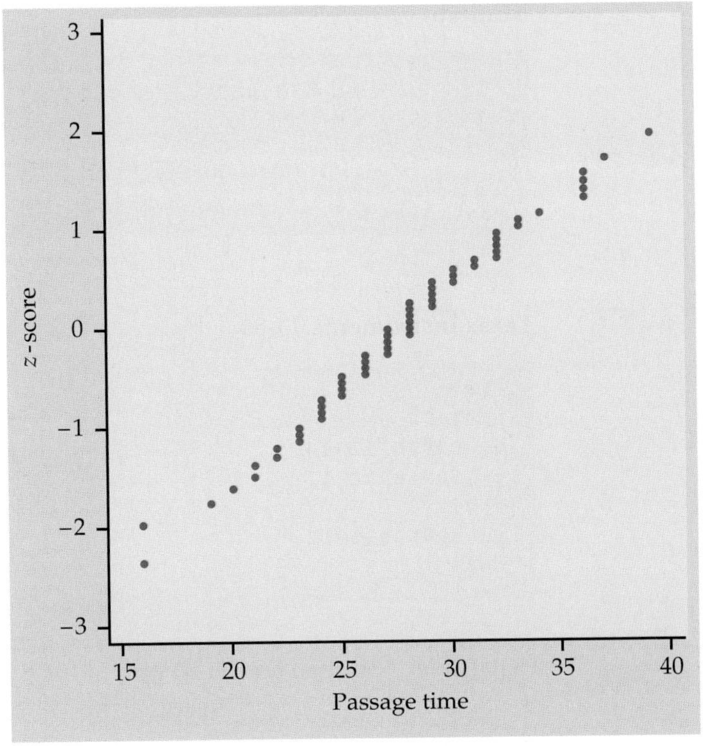

FIGURE 7.4 Normal quantile plot for Newcomb's passage time data with the outliers omitted, for Example 7.4.

Minitab

	N	MEAN	STDEV	SE MEAN	99.0 PERCENT C.I.
C1	64	27.750	5.083	0.635	(26.062, 29.438)

Texas Instruments TI-83

```
TInterval
 (26.062,29.438
 x=27.75
 Sx=5.083430912
 n=64
```

FIGURE 7.5 Minitab and TI-83 output for Example 7.4.

EXAMPLE 7.5

The best modern measurements of the speed of light correspond to a passage time of 33.02 in Newcomb's experiment. Is his result significantly different from this modern value? To answer this question, we test

$$H_0\colon \mu = 33.02$$
$$H_a\colon \mu \neq 33.02$$

Minitab

```
TTEST OF MU = 33.020 VS MU N.E. 33.020

         N      MEAN      STDEV     SE MEAN       T     P VALUE
C1      64    27.750      5.083       0.635   -8.29      0.0000
```

Texas Instruments TI-83

```
T-Test
 μ≠33.02
 t=-82936112889
 p=1.094873E-11
 x̄=27.75
 Sx=5.083430912
 n=64
```

FIGURE 7.6 Minitab and TI-83 output for Example 7.5.

The outputs from Minitab and the TI-83 are given in Figure 7.6. These are also easy to read. Here is one way to report the conclusion: the mean of Newcomb's passage times, $\bar{x} = 27.75$, is lower than the value calculated by modern methods, $\mu = 33.02$ ($t = -8.29$, df $= 63$, $P < 0.0001$).

We have concluded that Newcomb's mean is lower than the hypothesized value even though we have tested the two-sided alternative. At the end of our conclusion statement we give the test statistic, degrees of freedom, and the P-value in parentheses. The software outputs do not provide the degrees of freedom; we need to know that it is $n - 1 = 64 - 1 = 63$ in this case. We reported $P < 0.0001$. Minitab gave the value as 0.0000, a rounded value that represents anything less than 0.00005. TI-83 gave $P = 1.094873\text{E}{-11}$. The E$-$11 means that we move the decimal point 11 places to the left in the number given—that is, $P = 0.00000000001094873$, or one in one hundred billion. We have chosen to report $P < 0.0001$; most people would accept 1 in ten thousand as an unlikely event. The very small P-value is not the most important result here. The message in this example is that Newcomb's measurements averaged 27.75 when the modern value is 33.02 and that we have sufficient data to conclude that the difference is statistically significant.

Newcomb's measurements can be regarded as independent observations drawn from the population of all measurements he might make. We verified the normality of the measurements with the normal quantile plot in Figure 7.4. Moreover, the sample mean of 64 observations will have a distribution that is nearly normal even if the population is not normal. Only normality of $\bar{x}$ is required by the t procedures. Use of these procedures in this example is very well justified.

The following example reminds us of the relationship between confidence intervals and significance tests.

EXAMPLE 7.6

In Example 7.4 we found a 99% confidence interval for μ, the mean passage time using modern statistical methods. It was (26.06, 29.44). Example 7.5 gave us the results of the hypothesis test that Newcomb was trying to estimate $\mu = 33.02$. Because the confidence interval does not include the hypothesized value, $\mu = 33.02$, we can reject the null hypothesis with $\alpha = 0.01$.

Matched pairs *t* procedures

Newcomb wanted to estimate a constant of nature, so only a single population was involved. We saw in discussing data production that comparative studies are usually preferred to single-sample investigations because of the protection they offer against confounding. For that reason, inference about a parameter of a single distribution is less common than comparative inference. One common comparative design, however, makes use of single-sample procedures. In a *matched pairs* study, subjects are matched in pairs and the outcomes are compared within each matched pair. The experimenter can toss a coin to assign two treatments to the two subjects in each pair. Matched pairs are also common when randomization is not possible. One situation calling for matched pairs is before-and-after observations on the same subjects, as illustrated in the next example.

EXAMPLE 7.7

The National Endowment for the Humanities sponsors summer institutes to improve the skills of high school teachers of foreign languages. One such institute hosted 20 French teachers for 4 weeks. At the beginning of the period, the teachers were given the Modern Language Association's listening test of understanding of spoken French. After 4 weeks of immersion in French in and out of class, the listening test was given again. (The actual French spoken in the two tests was different, so that simply taking the first test should not improve the score on the second test.) Table 7.1 gives the pretest and posttest scores. The maximum possible score on the test is 36.[3]

TABLE 7.1 Modern Language Association listening scores for French teachers

Teacher	Pretest	Posttest	Gain	Teacher	Pretest	Posttest	Gain
1	32	34	2	11	30	36	6
2	31	31	0	12	20	26	6
3	29	35	6	13	24	27	3
4	10	16	6	14	24	24	0
5	30	33	3	15	31	32	1
6	33	36	3	16	30	31	1
7	22	24	2	17	15	15	0
8	25	28	3	18	32	34	2
9	32	26	−6	19	23	26	3
10	20	26	6	20	23	26	3

To analyze these data, we first subtract the pretest score from the posttest score to obtain the improvement for each student. These 20 differences form a single sample. They appear in the "Gain" columns in Table 7.1. The first teacher, for example, improved from 32 to 34, so the gain is $34 - 32 = 2$.

To assess whether the institute significantly improved the teachers' comprehension of spoken French, we test

$$H_0: \mu = 0$$

$$H_a: \mu > 0$$

Here μ is the mean improvement that would be achieved if the entire population of French teachers attended a summer institute. The null hypothesis says that no improvement occurs, and H_a says that posttest scores are higher on the average.

The 20 differences have

$$\bar{x} = 2.5 \text{ and } s = 2.893$$

The one-sample t statistic is therefore

$$t = \frac{\bar{x} - 0}{s/\sqrt{n}} = \frac{2.5}{2.893/\sqrt{20}}$$
$$= 3.86$$

	df = 19	
p	0.001	0.0005
t^*	3.579	3.883

The P-value is found from the $t(19)$ distribution (remember that the degrees of freedom are 1 less than the sample size). Table D shows that 3.86 lies between the upper 0.001 and 0.0005 critical values of the $t(19)$ distribution. The P-value therefore lies between these values. Software gives the value $P = 0.00053$. The improvement in listening scores is very unlikely to be due to chance alone. We have strong evidence that the institute was effective in raising scores. In scholarly publications, the details of routine statistical procedures are omitted; our test would be reported in the form: "The improvement in scores was significant ($t = 3.86$, df = 19, $P = 0.00053$)."

The results of the significance test allow us to conclude that the mean scores have improved. A statistically significant but very small improvement would not be sufficient grounds to say that the summer institute substantially improved the skills of the teachers who participated. A confidence interval tells us how much they improved with a margin of error.

EXAMPLE 7.8

A 90% confidence interval for the mean improvement in the entire population requires the critical value $t^* = 1.729$ from Table D. The confidence interval is

$$\bar{x} \pm t^* \frac{s}{\sqrt{n}} = 2.5 \pm 1.729 \frac{2.893}{\sqrt{20}}$$
$$= 2.5 \pm 1.12$$
$$= (1.38,\ 3.62)$$

The estimated average improvement is 2.5 points, with margin of error 1.12 for 90% confidence. Though statistically significant, the effect of the institute was rather small.

The following are key points to remember concerning matched pairs:

1. A matched pairs analysis is needed when there are two measurements or observations on each individual and we want to examine the change from the first to the second. Typically, the observations are "before" and "after" measures in some sense.
2. For each individual, subtract the "before" measure from the "after" measure.
3. Analyze the difference using the one-sample confidence interval and significance-testing procedures that we learned in this section.

Use of the t procedures in Examples 7.7 and 7.8 faces several difficulties. First, the teachers are not an SRS from the population of high school French teachers. There is some selection bias in favor of energetic, committed teachers who are willing to give up 4 weeks of their summer vacation. It is therefore not clear to what population the results apply. Second, a look at the data shows that several of the teachers had pretest scores close to the maximum of 36. They could not improve their scores very much even if their mastery of French increased substantially. This is a weakness in the listening test that is the measuring instrument in this study. The differences in scores may not adequately indicate the effectiveness of the institute. This is one reason why the average increase was small.

A final difficulty is that the data show departures from normality. In a matched pairs analysis, we assume that the population of differences has a normal distribution because the t procedures are applied to the differences. In these examples, one teacher actually lost 6 points between the pretest and the posttest. This one subject lowered the sample mean from 2.95 for the other 19 subjects to 2.5 for all 20. A normal quantile plot (Figure 7.7) displays this

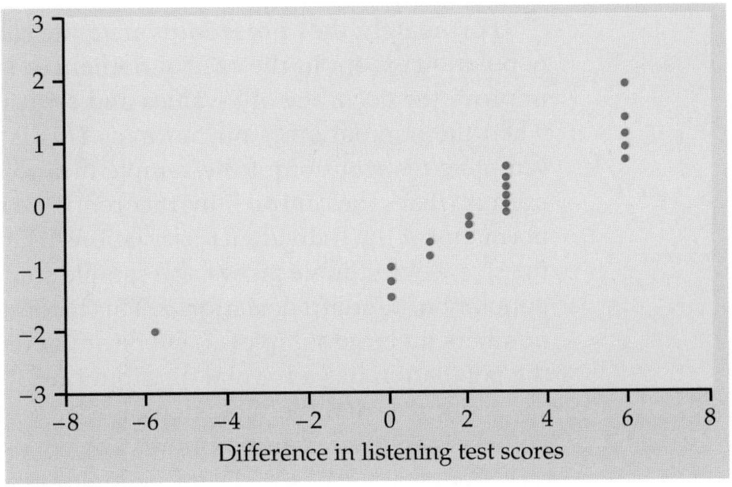

FIGURE 7.7 Normal quantile plot for the change in French listening scores, for Example 7.7.

outlier as well as granularity due to the fact that only whole-number scores are possible. The overall pattern of the plot is otherwise roughly straight. Does this nonnormality forbid use of the t test? The behavior of the t procedures when the population does not have a normal distribution is one of their most important properties.

Robustness of the *t* procedures

The results of one-sample t procedures are exactly correct only when the population is normal. Real populations are never exactly normal. The usefulness of the t procedures in practice therefore depends on how strongly they are affected by nonnormality. Procedures that are not strongly affected are called *robust*.

Robust Procedures

A statistical inference procedure is called **robust** if the probability calculations required are insensitive to violations of the assumptions made.

The assumption that the population is normal rules out outliers, so the presence of outliers shows that this assumption is not valid. The t procedures are not robust against outliers, because $\bar{x}$ and s are not resistant to outliers. If we dropped the single outlier in Example 7.7, the test statistic would change from $t = 3.86$ to $t = 5.98$, and the P-value would be much smaller. In this case, the outlier makes the test result *less* significant and the margin of error of a confidence interval *larger* than they would otherwise be. The results of the t procedures in Example 7.7 are conservative in the sense that the conclusions show a smaller effect than would be the case if the outlier were not present.

Fortunately, the t procedures are quite robust against nonnormality of the population except in the case of outliers or strong skewness. Larger samples improve the accuracy of P-values and critical values from the t distributions when the population is not normal. This is true for two reasons. First, the sampling distribution of the sample mean $\bar{x}$ from a large sample is close to normal (that's the central limit theorem). We need be less concerned about the normality of the individual observations when the sample is large. Second, as the size n of a sample grows, the sample standard deviation s approaches the population standard deviation σ. This fact is closely related to the law of large numbers. In large samples, s will be an accurate estimate of σ whether or not the population has a normal distribution.

A normal quantile plot, stemplot, or boxplot to check for skewness and outliers is an important preliminary to the use of t procedures for small samples. For most purposes, the one-sample t procedures can be safely used when $n \geq 15$ unless an outlier or clearly marked skewness is present. Except in the case of small samples, the assumption that the data are an SRS from the population of interest is more crucial than the assumption that the population distribution is normal. Here are practical guidelines for inference on a single mean:[4]

- *Sample size less than 15:* Use t procedures if the data are close to normal. If the data are clearly nonnormal or if outliers are present, do not use t.
- *Sample size at least 15:* The t procedures can be used except in the presence of outliers or strong skewness.
- *Large samples:* The t procedures can be used even for clearly skewed distributions when the sample is large, roughly $n \geq 40$.

Consider, for example, some of the data we studied in Chapter 1. Newcomb's data on the passage time of light in Figure 1.31 (page 80) contain two outliers, which make the use of t procedures risky even though the sample size is $n = 66$. When the outliers are removed (Figure 1.32 on page 81), the data are quite normal, and the t procedures are well justified. We would not employ t for inference about the city and highway gas mileage data on page 385, a small sample ($n = 17$) because of the outlier. On the other hand, the grocery spending data in Figure 1.33 (page 81), although clearly skewed, have no outliers and a moderate sample size ($n = 50$). We would apply the t procedures in this case.

The power of the *t* test*

The power of a statistical test measures its ability to detect deviations from the null hypothesis. In practice, we carry out the test in the hope of showing that the null hypothesis is false, so high power is important. The power of the one-sample t test against a specific alternative value of the population mean μ is the probability that the test will reject the null hypothesis when the alternative value of the mean is true. To calculate the power, we assume a fixed level of significance, often $\alpha = 0.05$.

Calculation of the exact power of the t test takes into account the estimation of σ by s and is a bit complex. But an approximate calculation that acts as if σ were known is almost always adequate for planning a study. This calculation is very much like that for the z test, presented in Section 6.4: write the event that the test rejects H_0 in terms of $\bar{x}$, and then find the probability of this event when the population mean has the alternative value.

EXAMPLE 7.9

It is the winter before the summer language institute of Example 7.7. The director of the institute, thinking ahead to the report he must write, hopes that the planned 20 students will enable him to be quite certain of detecting an average improvement of 2 points in the mean listening score. Is this realistic?

We wish to compute the power of the t test for

$$H_0: \mu = 0$$

$$H_a: \mu > 0$$

*This section can be omitted without loss of continuity.

against the alternative $\mu = 2$ when $n = 20$. We must have a rough guess of the size of σ in order to compute the power. In planning a large study, a pilot study is often run for this and other purposes. In this case, listening-score improvements in past summer language institutes have had sample standard deviations of about 3. We therefore take both $\sigma = 3$ and $s = 3$ in our approximate calculation.

The t test with 20 observations rejects H_0 at the 5% significance level if the t statistic

$$t = \frac{\overline{x} - 0}{s/\sqrt{20}}$$

exceeds the upper 5% point of $t(19)$, which is 1.729. Taking $s = 3$, the event that the test rejects H_0 is therefore

$$t = \frac{\overline{x}}{3/\sqrt{20}} \geq 1.729$$

$$\overline{x} \geq 1.729\frac{3}{\sqrt{20}}$$

$$\overline{x} \geq 1.160$$

The power is the probability that $\overline{x} \geq 1.160$ when $\mu = 2$. Taking $\sigma = 3$, this probability is found by standardizing $\overline{x}$:

$$P(\overline{x} \geq 1.160 \text{ when } \mu = 2) = P\left(\frac{\overline{x} - 2}{3/\sqrt{20}} \geq \frac{1.160 - 2}{3/\sqrt{20}}\right)$$

$$= P(Z \geq -1.252)$$

$$= 1 - 0.1056 = 0.8944$$

A true difference of 2 points in the population mean scores will produce significance at the 5% level in 89% of all possible samples. The director can be reasonably confident of detecting a difference this large.

Inference for nonnormal populations*

We have not discussed how to do inference about the mean of a clearly nonnormal distribution based on a small sample. If you face this problem, you should consult an expert. Three general strategies are available:

- In some cases a distribution other than a normal distribution will describe the data well. There are many nonnormal models for data, and inference procedures for these models are available.
- Because skewness is the chief barrier to the use of t procedures on data without outliers, you can attempt to transform skewed data so that the distribution is symmetric and as close to normal as possible. Confidence levels and P-values from the t procedures applied to the transformed data will be quite accurate for even moderate sample sizes.

*This section can be omitted without loss of continuity.

distribution-free procedures

nonparametric procedures.

- The third strategy is to use a **distribution-free** inference procedure. Such procedures do not assume that the population distribution has any specific form, such as normal. Distribution-free procedures are often called **nonparametric procedures.**

Each of these strategies can be effective, but each quickly carries us beyond the basic practice of statistics. We emphasize procedures based on normal distributions because they are the most common in practice, because their robustness makes them widely useful, and (most important) because we are first of all concerned with understanding the principles of inference. We will therefore not discuss procedures for nonnormal continuous distributions and will present only one of the many distribution-free procedures that do not require the population to have any specific type of distribution. We will be content with illustrating by example the use of a transformation and of a simple distribution-free procedure.

log transformation

Transforming data When the distribution of a variable is skewed, it often happens that a simple transformation results in a variable whose distribution is symmetric and even close to normal. The most common transformation is the **logarithm,** or **log.** The logarithm tends to pull in the right tail of a distribution. For example, the data 2, 3, 4, 20 show an outlier in the right tail. Their logarithms 0.30, 0.48, 0.60, 1.30 are much less skewed. Taking logarithms is a possible remedy for right-skewness. Instead of analyzing values of the original variable X, we first compute their logarithms and analyze the values of $\log X$. Here is an example of this approach.

EXAMPLE 7.10

Table 7.2 presents data on the amounts (in grams per mile) of three pollutants in the exhaust of 46 vehicles of the same type, measured under standard conditions prescribed by the Environmental Protection Agency. We will concentrate on emissions of carbon monoxide (CO). We would like to give a confidence interval for the mean emissions μ for this vehicle type.[5]

A normal quantile plot of the CO data (Figure 7.8) from Table 7.2 shows that the distribution is skewed to the right. Because there are no extreme outliers, the sample

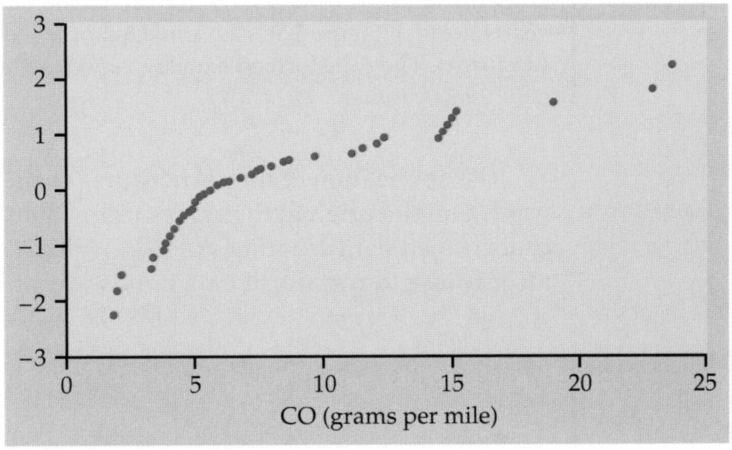

FIGURE 7.8 Normal quantile plot for CO emissions, for Example 7.10. The distribution is skewed to the right.

TABLE 7.2 Amounts of three pollutants emitted by light-duty engines (grams per mile)

EN	HC	CO	NOX	EN	HC	CO	NOX
1	0.50	5.01	1.28	2	0.65	14.67	0.72
3	0.46	8.60	1.17	4	0.41	4.42	1.31
5	0.41	4.95	1.16	6	0.39	7.24	1.45
7	0.44	7.51	1.08	8	0.55	12.30	1.22
9	0.72	14.59	0.60	10	0.64	7.98	1.32
11	0.83	11.53	1.32	12	0.38	4.10	1.47
13	0.38	5.21	1.24	14	0.50	12.10	1.44
15	0.60	9.62	0.71	16	0.73	14.97	0.51
17	0.83	15.13	0.49	18	0.57	5.04	1.49
19	0.34	3.95	1.38	20	0.41	3.38	1.33
21	0.37	4.12	1.20	22	1.02	23.53	0.86
23	0.87	19.00	0.78	24	1.10	22.92	0.57
25	0.65	11.20	0.95	26	0.43	3.81	1.79
27	0.48	3.45	2.20	28	0.41	1.85	2.27
29	0.51	4.10	1.78	30	0.41	2.26	1.87
31	0.47	4.74	1.83	32	0.52	4.29	2.94
33	0.56	5.36	1.26	34	0.70	14.83	1.16
35	0.51	5.69	1.73	36	0.52	6.35	1.45
37	0.57	6.02	1.31	38	0.51	5.79	1.51
39	0.36	2.03	1.80	40	0.48	4.62	1.47
41	0.52	6.78	1.15	42	0.61	8.43	1.06
43	0.58	6.02	0.97	44	0.46	3.99	2.01
45	0.47	5.22	1.12	46	0.55	7.47	1.39

mean of 46 observations will nonetheless have an approximately normal sampling distribution. The t procedures could be used for approximate inference. For more exact inference, we will seek to transform the data so that the distribution is more nearly normal. Figure 7.9 is a normal quantile plot of the logarithms of the CO measurements. The transformed data are very close to normal, so t procedures will give quite exact results.

The application of the t procedures to the transformed data is straightforward. Call the original CO values from Table 7.2 values of the variable X. The transformed data are values of $X_{new} = \log X$. In most software packages, it is an easy task to transform data in this way and then analyze the new variable.

EXAMPLE 7.11

Analysis of the logs of the CO values in Minitab produces the following output:

```
 N        MEAN         STDEV        SE MEAN      95.0 PERCENT C.I.
46      0.8198        0.2654         0.0391      (0.7409, 0.8986)
```

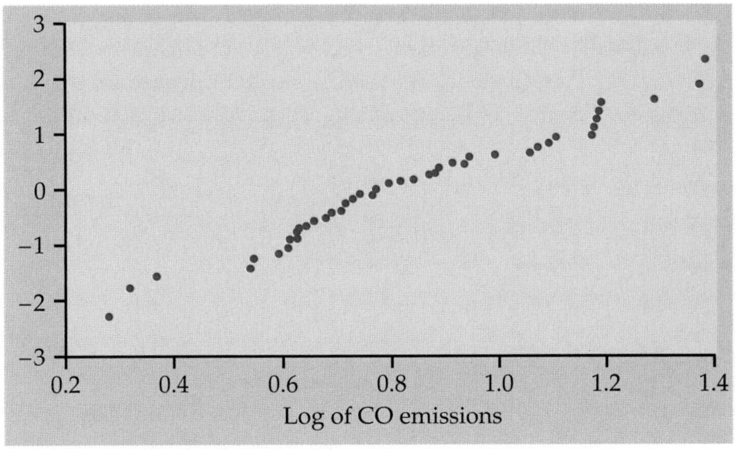

FIGURE 7.9 Normal quantile plot for the logarithms of the CO emissions, for Example 7.10. This distribution is close to normal.

For comparison, the 95% t confidence interval for the original mean μ is found from the original data as follows:

N	MEAN	STDEV	SE MEAN	95.0 PERCENT C.I.
46	7.960	5.261	0.776	(6.398, 9.523)

The advantage of analyzing transformed data is that use of procedures based on the normal distributions is better justified and the results are more exact. The disadvantage is that a confidence interval for the mean μ in the original scale (in our example, emissions measured in grams per mile) cannot be recovered from the confidence interval for the mean of the logs.

The sign test Perhaps the most straightforward way to cope with nonnormal data is to use a *distribution-free* procedure. As the name indicates, these procedures do not require the population distribution to have any specific form, such as normal. Distribution-free significance tests are quite simple and are available in most statistical software systems. Distribution-free tests have two drawbacks. First, they are generally less powerful than tests designed for use with a specific distribution, such as the t test. Second, we must often modify the statement of the hypotheses in order to use a distribution-free test. A distribution-free test concerning the center of a distribution, for example, is usually stated in terms of the median rather than the mean. This is sensible when the distribution may be skewed. But the distribution-free test does not ask the same question (Has the mean changed?) that the t test does. The simplest distribution-free test, and one of the most useful, is the **sign test.** The following example illustrates this test.

sign test

EXAMPLE 7.12

Return to the data of Example 7.7 (page 501) showing the improvement in French listening scores after attending a summer institute. In that example we used the one-sample t test on these data, despite granularity and an outlier that make the P-value only roughly correct. The sign test is based on the following simple observation: of the 17 teachers whose scores changed, 16 improved and only 1 did more poorly. This is evidence that the institute improved French listening skills.

To perform a significance test based on the count of teachers whose scores improved, let p be the probability that a randomly chosen teacher would improve if she attended the institute. The null hypothesis of "no effect" says that the posttest is just a repeat of the pretest with no change in ability, so a teacher is equally likely to do better on either test. We therefore want to test

$$H_0\colon p = 1/2$$
$$H_a\colon p > 1/2$$

The 17 teachers whose scores changed are 17 independent trials, so the number who improve has the binomial distribution $B(17, 1/2)$ if H_0 is true. The P-value for the observed count 16 is therefore $P(X \geq 16)$, where X has the $B(17, 1/2)$ distribution. You can compute this probability with software or from the binomial probability formula (page 381):

$$\begin{aligned} P(X \geq 16) &= P(X = 16) + P(X = 17) \\ &= \binom{17}{16}\left(\frac{1}{2}\right)^{16}\left(\frac{1}{2}\right)^{1} + \binom{17}{17}\left(\frac{1}{2}\right)^{17}\left(\frac{1}{2}\right)^{0} \\ &= (17)\left(\frac{1}{2}\right)^{17} + \left(\frac{1}{2}\right)^{17} \\ &= 0.00014 \end{aligned}$$

As in Example 7.7, there is very strong evidence that participation in the institute has improved performance on the listening test.

There are several varieties of sign test, all based on counts and the binomial distribution. The sign test for matched pairs (Example 7.12) is the most useful. The null hypothesis of "no effect" is then always $H_0\colon p = 1/2$. The alternative can be one-sided in either direction or two-sided, depending on the type of change we are looking for. The test gets its name from the fact that we look only at the signs of the differences, not their actual values.

The Sign Test for Matched Pairs

Ignore pairs with difference 0; the number of trials n is the count of the remaining pairs. The test statistic is the count X of pairs with a positive difference. P-values for X are based on the binomial $B(n, 1/2)$ distribution.

The matched pairs t test in Example 7.7 tested the hypothesis that the mean of the distribution of differences (score after the institute minus score before) is 0. The sign test in Example 7.12 is in fact testing the hypothesis that the *median* of the differences is 0. If p is the probability that a difference is positive, then $p = 1/2$ when the median is 0. This is true because the median of the distribution is the point with probability $1/2$ lying to its right. As Figure 7.10 illustrates, $p > 1/2$ when the median is greater than 0, again

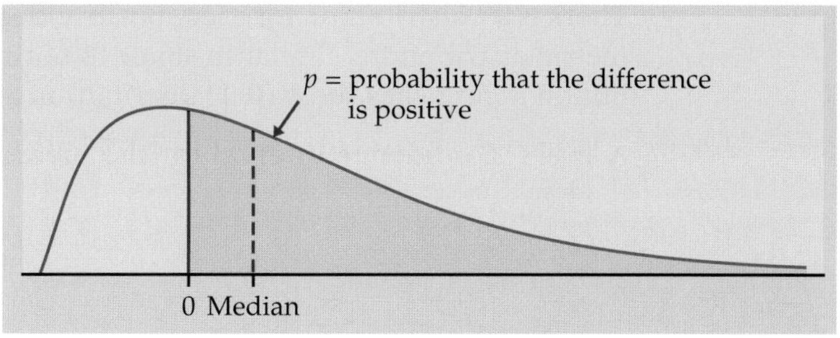

FIGURE 7.10 Why the sign test tests the median difference: when the median is greater than 0, the probability p of a positive difference is greater than $1/2$, and vice versa.

because the probability to the right of the median is always $1/2$. The sign test of H_0: $p = 1/2$ against H_a: $p > 1/2$ is a test of

$$H_0\text{: population median} = 0$$

$$H_a\text{: population median} > 0$$

The sign test in Example 7.12 makes no use of the actual scores—it just counts how many teachers improved. The teachers whose scores did not change were ignored altogether. Because the sign test uses so little of the available information, it is much less powerful than the t test when the population is close to normal. There are other distribution-free tests that are more powerful than the sign test.[6]

SUMMARY

Significance tests and confidence intervals for the mean μ of a normal population are based on the sample mean $\bar{x}$ of an SRS. Because of the central limit theorem, the resulting procedures are approximately correct for other population distributions when the sample is large.

The standardized sample mean, or **one-sample z statistic,**

$$z = \frac{\bar{x} - \mu}{\sigma/\sqrt{n}}$$

has the $N(0, 1)$ distribution. If the standard deviation $\sigma/\sqrt{n}$ of $\bar{x}$ is replaced by the **standard error** $s/\sqrt{n}$, the **one-sample t statistic**

$$t = \frac{\bar{x} - \mu}{s/\sqrt{n}}$$

has the ***t* distribution** with $n - 1$ degrees of freedom.

There is a t distribution for every positive **degrees of freedom k.** All are symmetric distributions similar in shape to normal distributions. The $t(k)$ distribution approaches the $N(0, 1)$ distribution as k increases.

A level C **confidence interval for the mean μ** of a normal population is

$$\bar{x} \pm t^* \frac{s}{\sqrt{n}}$$

where t^* is the value for the $t(n-1)$ density curve with area C between $-t^*$ and t^*. The quantity

$$t^* \frac{s}{\sqrt{n}}$$

is the **margin of error.**

Significance tests for H_0: $\mu = \mu_0$ are based on the t statistic. P-values or fixed significance levels are computed from the $t(n-1)$ distribution.

These one-sample procedures are used to analyze **matched pairs** data by first taking the differences within the matched pairs to produce a single sample.

The t procedures are relatively **robust** against nonnormal populations, especially for larger sample sizes. The t procedures are useful for nonnormal data when $n \geq 15$ unless the data show outliers or strong skewness.

The power of the t test is calculated like that of the z test, using an approximate value for both σ and s.

Small samples from skewed populations can sometimes be analyzed by first applying a **transformation** (such as the logarithm) to obtain an approximately normally distributed variable. The t procedures then apply to the transformed data.

The **sign test** is a **distribution-free test** because it uses probability calculations that are correct for a wide range of population distributions.

The sign test for "no treatment effect" in matched pairs counts the number of positive differences. The P-value is computed from the $B(n, 1/2)$ distribution, where n is the number of non-0 differences. The sign test is less powerful than the t test in cases where use of the t test is justified.

SECTION 7.1 EXERCISES

7.1 Your local newspaper contains a large number of advertisements for unfurnished one-bedroom apartments. You choose 10 at random and calculate that their mean monthly rent is \$540 and that the standard deviation of their rents is \$80.

(a) What is the standard error of the mean?

(b) What are the degrees of freedom for a one-sample t statistic?

7.2 You want to rent an unfurnished one-bedroom apartment for next semester. You take a random sample of 10 apartments advertised in the local newspaper and record the rental rates. Here are the rents (in dollars per month):

500, 650, 600, 505, 450, 550, 515, 495, 650, 395

Find a 95% confidence interval for the mean monthly rent for unfurnished one-bedroom apartments available for rent in this community.

7.3 If you chose 99% rather than 95% confidence, would your margin of error in the previous exercise be larger or smaller? Explain your answer and verify it by doing the calculations.

7.4 A random sample of 10 one-bedroom apartments from your local newspaper has these monthly rents (dollars):

500, 650, 600, 505, 450, 550, 515, 495, 650, 395

Do these data give good reason to believe that the mean rent of all advertised apartments is greater than \$500 per month? State hypotheses, find the t statistic and its P-value, and state your conclusion.

7.5 A test of a null hypothesis versus a two-sided alternative gives $t = 2.55$.

(a) The sample size is 25. Is the test result significant at the 5% level? Explain how you obtained your answer.

(b) The sample size is 5. Is the test result significant at the 5% level? What general fact about statistical tests do your two answers illustrate?

7.6 You will have complete sales information for last month in a week but right now you have data from a random sample of 50 stores. The mean change in sales in the sample is +4.8% and the standard deviation of the changes is 15%. Are average sales for all stores different from last month?

(a) State appropriate null and alternative hypotheses. Explain how you decided between the one- and two-sided alternatives.

(b) Find the t statistic and its P-value. State your conclusion.

(c) If the test gives strong evidence against the null hypothesis, would you conclude that sales are up in every one of your stores? Explain your answer.

7.7 You are using software to calculate t procedures for a large sample, n greater than 5000.

(a) For a two-sided test, how large must the t statistic be to reach significance at the 1% level? (Note that either a large positive number or a large negative number will be significant.)

(b) In which row of Table D did you find your result in part (a)? What general fact about the t distributions does this illustrate?

7.8 National Fuelsaver Corporation manufactures the Platinum Gasaver, a device they claim "may increase gas mileage by 22%." Here are the percent changes in gas mileage for 15 identical vehicles, as presented in one of the company's advertisements:

48.3	46.9	46.8	44.6	40.2	38.5	34.6	33.7
28.7	28.7	24.8	10.8	10.4	6.9	−12.4	

Would you recommend use of a t confidence interval to estimate the mean fuel savings in the population of all such vehicles? Explain your answer.

7.9 A manufacturer of small appliances employs a market research firm to estimate retail sales of its products. Here are last month's sales of electric can openers from an SRS of 50 stores in the Midwest sales region:

19	19	16	19	25	26	24	63	22	16
13	26	34	10	48	16	20	14	13	24
34	14	25	16	26	25	25	26	11	79
17	25	18	15	13	35	17	15	21	12
19	20	32	19	24	19	17	41	24	27

(a) Make a stemplot of the data. The distribution is skewed to the right and has several high outliers. The *bootstrap* (page 427) is a modern computer-intensive tool for getting accurate confidence intervals without the normality condition. Three bootstrap simulations, each with 10,000 repetitions, give these 95% confidence intervals for mean sales in the entire region: (20.42, 27.26), (20.40, 27.18), and (20.48, 27.28).

(b) Find the 95% t confidence interval for the mean. It is essentially the same as the bootstrap intervals. The lesson is that for sample sizes as large as $n = 50$, t procedures are very robust.

7.10 What critical value t^* from Table D should be used for a confidence interval for the mean of the population in each of the following situations?

(a) A 95% confidence interval based on $n = 15$ observations.

(b) A 90% confidence interval from an SRS of 25 observations.

(c) An 80% confidence interval from a sample of size 30.

7.11 Use software to find the critical values t^* that you would use for 95% confidence intervals for sample sizes of 10, 20, 30, 40, 50, 100, 200, and 500. Plot the values of t^* versus the sample size and describe the relationship.

7.12 The one-sample t statistic for testing

$$H_0: \mu = 0$$

$$H_a: \mu > 0$$

from a sample of $n = 20$ observations has the value $t = 1.87$.

(a) What are the degrees of freedom for this statistic?

(b) Give the two critical values t^* from Table D that bracket t.

(c) What are the right-tail probabilities p for these two entries?

(d) Between what two values does the P-value of the test fall?

(e) Is the value $t = 1.87$ significant at the 5% level? Is it significant at the 1% level?

(f) If you have software available, find the exact P-value.

7.13 The one-sample t statistic from a sample of $n = 50$ observations for the two-sided test of

$$H_0: \mu = 50$$

$$H_a: \mu \neq 50$$

has the value $t = 1.65$.

(a) What are the degrees of freedom for t?

(b) Locate the two critical values t^* from Table D that bracket t. What are the right-tail probabilities p for these two values?

(c) How would you report the P-value for this test?

(d) Is the value $t = 1.65$ statistically significant at the 10% level? At the 5% level?

(e) If you have software available, find the exact P-value.

7.14 The one-sample t statistic for a test of

$$H_0: \mu = 30$$

$$H_a: \mu < 30$$

based on $n = 10$ observations has the value $t = -4.45$.

(a) What are the degrees of freedom for this statistic?

(b) How would you report the P-value based on Table D?

(c) If you have software available, find the exact P-value.

7.15 The scores of four roommates on the Law School Admission Test (LSAT) are

628, 593, 455, 503

Find the mean, the standard deviation, and the standard error of the mean. Is it appropriate to calculate a confidence interval for these data? Explain why or why not.

7.16 Here are estimates of the daily intakes of calcium (in milligrams) for 38 women between the ages of 18 and 24 years who participated in a study of women's bone health:

808	882	1062	970	909	802	374	416	784	997
651	716	438	1420	1425	948	1050	976	572	403
626	774	1253	549	1325	446	465	1269	671	696
1156	684	1933	748	1203	2433	1255	1100		

(a) Display the data using a stemplot and make a normal quantile plot. Describe the distribution of calcium intakes for these women.

(b) Calculate the mean, the standard deviation, and the standard error.

(c) Find a 95% confidence interval for the mean.

7.17 Refer to the previous exercise. Eliminate the two largest values and answer parts (a), (b), and (c).

7.18 Refer to Exercise 7.16. Suppose that the recommended daily allowance (RDA) of calcium for women in this age range is 1300 milligrams (this value is changed from time to time on the basis of the statistical analysis of new data). We want to express the results in terms of percent of the RDA.

(a) Divide each intake by the RDA, multiply by 100, and compute the 95% confidence interval from the transformed data.

(b) Verify that you can obtain the same result by similarly transforming the interval you calculated in Exercise 7.16.

7.19 Refer to Exercises 7.16 and 7.18. You want to compare the average calcium intake of these women with the RDA using a significance test.

(a) State appropriate null and alternative hypotheses.

(b) Give the test statistic, the degrees of freedom, and the *P*-value.

(c) State your conclusion.

(d) Repeat the calculations without the two largest values. Does your conclusion depend on whether or not these observations are included in the analysis?

7.20 How much do users pay for Internet service? Here are the monthly fees (in dollars) paid by a random sample of 50 users of commercial Internet service providers in August 2000:[7]

20	40	22	22	21	21	20	10	20	20
20	13	18	50	20	18	15	8	22	25
22	10	20	22	22	21	15	23	30	12
9	20	40	22	29	19	15	20	20	20
20	15	19	21	14	22	21	35	20	22

(a) Make a stemplot of the data. Also make a normal quantile plot if your software permits. The data are not normal: there are stacks of observations taking the same values, and the distribution is more spread out in both directions and somewhat skewed to the right. The

t procedures are nonetheless approximately correct because $n = 50$ and there are no extreme outliers.

(b) Give a 95% confidence interval for the mean monthly cost of Internet access in August 2000.

7.21 The data in the previous exercise show that many people paid \$20 per month for Internet access, presumably because major providers such as AOL charged this amount. Do the data give good reason to think that the mean cost for all Internet users differs from \$20 per month?

7.22 Refer to the two previous exercises concerning fees paid for Internet access by a national random sample of clients of Internet service providers in 2000. The Census Bureau estimates that 44 million households had Internet access in 2000. Use the confidence interval that you found to give a 95% confidence interval for the total amount these households paid in Internet access fees. This is one aspect of the national economic impact of the Internet.

7.23 Refer to Exercise 7.20. Suppose you are interested in the cost per year rather than the cost per month. Find a 95% confidence interval for the mean yearly cost of Internet access. How does this interval relate to the one that you found in Exercise 7.20?

7.24 Refer to Exercise 7.9. Each electric can opener sold generates a profit of \$2.50 for the manufacturer.

(a) What is the mean profit per store in the Midwest sales region?

(b) Transform the confidence interval you found in the previous exercise into an interval for the mean profit for stores in the Midwest region.

7.25 Refer to Exercise 7.9 and the previous exercise. There are 4325 stores that sell can openers manufactured by this company.

(a) Estimate the total profit for sales last month in the Midwest region.

(b) Give a 95% confidence interval for the total profit for sales last month in the Midwest region.

7.26 The U.S. Department of Agriculture (USDA) uses many types of surveys to obtain important economic estimates. In one pilot study they used 22 observations to estimate the price received by farmers for corn sold in October. They reported the mean price as \$2.08 per bushel with a standard error of \$0.176 per bushel. Give a 95% confidence interval for the mean price received by farmers for corn sold in October.[8]

7.27 The cost of health care is the subject of many studies that use statistical methods. One such study estimated that the average length of service for home health care among people over the age of 65 who use this type of service is 96.0 days with a standard error of 5.1 days. Assuming that the degrees of freedom are large, calculate a 90% confidence interval for the true mean length of service.[9]

7.28 Many organizations are doing surveys to determine the satisfaction of their customers. Attitudes toward various aspects of campus life were the subject of one such study conducted at Purdue University. Each question was rated on a 1 to 5 scale, with 5 being the highest rating. The average response of 1406 first-year students to "Feeling welcomed at Purdue" was 3.9 with a standard deviation of 0.98. Assuming that the respondents are an SRS, give a 99% confidence interval for the mean of all first-year students.

7.29 Do piano lessons improve the spatial-temporal reasoning of preschool children? Neurobiological arguments suggest that this may be true. A study designed to test this hypothesis measured the spatial-temporal reasoning of 34 preschool children before and after 6 months of piano lessons.[10] (The study also included children who took computer lessons and a control group; but we are not concerned with those here.) The changes in the reasoning scores are

2	5	7	–2	2	7	4	1	0	7	3	4	3	4	9	4	5
2	9	6	0	3	6	–1	3	4	6	7	–2	7	–3	3	4	4

(a) Display the data and summarize the distribution.

(b) Find the mean, the standard deviation, and the standard error of the mean.

(c) Give a 95% confidence interval for the mean improvement in reasoning scores.

7.30 Refer to the previous exercise. Test the null hypothesis that there is no improvement versus the alternative suggested by the neurobiological arguments. State the hypotheses, and give the test statistic with degrees of freedom and the *P*-value. What do you conclude? From your answer to part (c) of the previous exercise what can be concluded from this significance test?

7.31 The researchers studying vitamin C in CSB in Example 7.1 were also interested in a similar commodity called wheat soy blend (WSB). Both of these commodities are mixed with other ingredients and cooked. Loss of vitamin C as a result of this process was another concern of the researchers. One preparation used in Haiti called gruel (or "bouillie" in Creole) can be made from WSB, salt, sugar, milk, banana, and other optional items to improve the taste. Samples of gruel prepared in Haitian households were collected. The vitamin C content (in milligrams per 100 grams of blend, dry basis) was measured before and after cooking. Here are the results:

Sample	1	2	3	4	5
Before	73	79	86	88	78
After	20	27	29	36	17

Set up appropriate hypotheses and carry out a significance test for these data. (It is not possible for cooking to increase the amount of vitamin C.)

7.32 Refer to the previous exercise. The fact that vitamin C is destroyed by cooking is neither new nor surprising; so the significance test you performed in the previous exercise simply confirms that this fact is evident in the small sample of data collected. The real question here concerns how much of the vitamin C is lost.

(a) Give a 95% confidence interval for the amount of vitamin C lost by preparing and cooking gruel in Haiti.

(b) The units for your interval are mg/100 g of blend. To meaningfully interpret the amount lost, it is necessary to know something about the amount that we started with. For example, a loss of 50 mg/100 g would probably not be of much concern if we started with 5000 mg/100 g. The specifications call for the blend to contain 98 mg/100 g (dry basis). The difference between this specification and the "before" values above is due to sample variation in the manufacturing process and the handling of the product from the time it was manufactured until it was used to prepare gruel in these Haitian homes. Express the "after" data as percent of specification and give a 95% confidence interval for the mean percent.

7.33 A bank wonders whether omitting the annual credit card fee for customers who charge at least $3000 in a year would increase the amount charged on its credit card. The bank makes this offer to an SRS of 500 of its existing credit card customers. It then compares how much these customers charge this year with the amount that they charged last year. The mean increase is $565, and the standard deviation is $267.

(a) Is there significant evidence at the 1% level that the mean amount charged increases under the no-fee offer? State H_0 and H_a and carry out a t test.

(b) Give a 95% confidence interval for the mean amount of the increase.

(c) The distribution of the amount charged is skewed to the right, but outliers are prevented by the credit limit that the bank enforces on each card. Use of the t procedures is justified in this case even though the population distribution is not normal. Explain why.

(d) A critic points out that the customers would probably have charged more this year than last even without the new offer because the economy is more prosperous and interest rates are lower. Briefly describe the design of an experiment to study the effect of the no-fee offer that would avoid this criticism.

7.34 The level of various substances in the blood of kidney dialysis patients is of concern because kidney failure and dialysis can lead to nutritional problems. A researcher performed blood tests on several dialysis patients on 6 consecutive clinic visits. One variable measured was the level of phosphate in the blood. Phosphate levels for an individual tend to vary normally over time. The data on one patient, in milligrams of phosphate per deciliter (mg/dl) of blood, are given below:[11]

5.6 5.1 4.6 4.8 5.7 6.4

(a) Calculate the sample mean $\overline{x}$ and its standard error.

(b) Use the t procedures to give a 90% confidence interval for this patient's mean phosphate level.

7.35 The normal range of values for blood phosphate levels is 2.6 to 4.8 mg/dl. The sample mean for the patient in the previous exercise falls above this range. Is this good evidence that the patient's mean level in fact falls above 4.8? State H_0 and H_a and use the data in the previous exercise to carry out a t test. Between which levels from Table D does the P-value lie? Are you convinced that the patient's phosphate level is higher than normal?

7.36 In a randomized comparative experiment on the effect of dietary calcium on blood pressure, 54 healthy white males were divided at random into two groups. One group received calcium; the other, a placebo. At the beginning of the study, the researchers measured many variables on the subjects. The paper reporting the study gives $\bar{x} = 114.9$ and $s = 9.3$ for the seated systolic blood pressure of the 27 members of the placebo group.

(a) Give a 95% confidence interval for the mean blood pressure of the population from which the subjects were recruited.

(b) What assumptions about the population and the study design are required by the procedure you used in (a)? Which of these assumptions are important for the validity of the procedure in this case?

7.37 How accurate are radon detectors of a type sold to homeowners? To answer this question, university researchers placed 12 detectors in a chamber that exposed them to 105 picocuries per liter (pCi/l) of radon.[12] The detector readings were as follows:

91.9	97.8	111.4	122.3	105.4	95.0
103.8	99.6	96.6	119.3	104.8	101.7

(a) Make a stemplot of the data. The distribution is somewhat skewed to the right, but not strongly enough to forbid use of the t procedures.

(b) Is there convincing evidence that the mean reading of all detectors of this type differs from the true value of 105? Carry out a test in detail and write a brief conclusion.

7.38 The embryos of brine shrimp can enter a dormant phase in which metabolic activity drops to a low level. Researchers studying this dormant phase measured the level of several compounds important to normal metabolism. The results were reported in a table, with the note, "Values are means ± SEM for three independent samples." The table entry for the compound ATP was 0.84 ± 0.01. Biologists reading the article are presumed to be able to decipher this.[13]

(a) What does the abbreviation "SEM" stand for?

(b) The researchers made three measurements of ATP, which had $\bar{x} = 0.84$. What was the sample standard deviation s for these measurements?

(c) Give a 90% confidence interval for the mean ATP level in dormant brine shrimp embryos.

7.39 The researchers studying vitamin C in CSB in Example 7.1 were also interested in a similar commodity called wheat soy blend (WSB). A major concern was the possibility that some of the vitamin C content would be destroyed as a result of storage and shipment of the commodity to its final destination. The researchers specially marked a collection of bags at the factory and took a sample from each of these to determine the vitamin C content. Five months later in Haiti they found the specially marked bags and took samples. The data consist of two vitamin C measures for each bag, one at the time of production in the factory and the other five months later in Haiti. The units are mg/100 g as in Example 7.1. Here are the data:

Factory	Haiti	Factory	Haiti	Factory	Haiti
44	40	45	38	39	43
50	37	32	40	52	38
48	39	47	35	45	38
44	35	40	38	37	38
42	35	38	34	38	41
47	41	41	35	44	40
49	37	43	37	43	35
50	37	40	34	39	38
39	34	37	40	44	36

(a) Set up hypotheses to examine the question of interest to these researchers.

(b) Perform the significance test and summarize your results.

(c) Find 95% confidence intervals for the mean at the factory, the mean five months later in Haiti, and for the change.

7.40 The design of controls and instruments has a large effect on how easily people can use them. A student project investigated this effect by asking 25 right-handed students to turn a knob (with their right hands) that moved an indicator by screw action. There were two identical instruments, one with a right-hand thread (the knob turns clockwise) and the other with a left-hand thread (the knob turns counterclockwise). The table below gives the times required (in seconds) to move the indicator a fixed distance:[14]

Subject	Right thread	Left thread	Subject	Right thread	Left thread
1	113	137	14	107	87
2	105	105	15	118	166
3	130	133	16	103	146
4	101	108	17	111	123
5	138	115	18	104	135
6	118	170	19	111	112
7	87	103	20	89	93
8	116	145	21	78	76
9	75	78	22	100	116
10	96	107	23	89	78
11	122	84	24	85	101
12	103	148	25	88	123
13	116	147			

(a) Each of the 25 students used both instruments. Discuss briefly how the experiment should be arranged and how randomization should be used.

(b) The project hoped to show that right-handed people find right-hand threads easier to use. State the appropriate H_0 and H_a about the mean time required to complete the task.

(c) Carry out a test of your hypotheses. Give the *P*-value and report your conclusions.

7.41 Give a 90% confidence interval for the mean time advantage of right-hand over left-hand threads in the setting of the previous exercise. Do you think that the time saved would be of practical importance if the task were performed many times—for example, by an assembly-line worker? To help answer this question, find the mean time for right-hand threads as a percent of the mean time for left-hand threads.

7.42 The table below gives the pretest and posttest scores on the MLA listening test in Spanish for 20 high school Spanish teachers who attended an intensive summer course in Spanish. The setting is identical to the one described in Example 7.7.[15]

Subject	Pretest	Posttest	Subject	Pretest	Posttest
1	30	29	11	30	32
2	28	30	12	29	28
3	31	32	13	31	34
4	26	30	14	29	32
5	20	16	15	34	32
6	30	25	16	20	27
7	34	31	17	26	28
8	15	18	18	25	29
9	28	33	19	31	32
10	20	25	20	29	32

(a) We hope to show that attending the institute improves listening skills. State an appropriate H_0 and H_a. Be sure to identify the parameters appearing in the hypotheses.

(b) Make a graphical check for outliers or strong skewness in the data that you will use in your statistical test, and report your conclusions on the validity of the test.

(c) Carry out a test. Can you reject H_0 at the 5% significance level? At the 1% significance level?

(d) Give a 90% confidence interval for the mean increase in listening score due to attending the summer institute.

7.43 An agricultural field trial compares the yield of two varieties of tomatoes for commercial use. The researchers divide in half each of 10 small plots of land in different locations and plant each tomato variety on one half of each plot. After harvest, they compare the yields in pounds per plant at each location. The 10 differences (Variety A − Variety B) give the following statistics: $\bar{x} = 0.46$ and $s = 0.92$. Is there convincing evidence that Variety A has the higher mean yield? State H_0 and H_a, and give a *P*-value to answer this question.

7.44 The following situations all require inference about a mean or means. Identify each as (1) a single sample, (2) matched pairs, or (3) two independent samples. The procedures of this section apply to cases (1) and (2). We will learn procedures for (3) in the next section.

(a) An education researcher wants to learn whether inserting questions before or after introducing a new concept in an elementary school mathematics text is more effective. He prepares two text segments that teach the concept, one with motivating questions before and the other with review questions after. Each text segment is used to teach a different group of children, and their scores on a test over the material are compared.

(b) Another researcher approaches the same problem differently. She prepares text segments on two unrelated topics. Each segment comes in two versions, one with questions before and the other with questions after. Each of a group of children is taught both topics, one topic (chosen at random) with questions before and the other with questions after. Each child's test scores on the two topics are compared to see which topic he or she learned better.

(c) To evaluate a new analytical method, a chemist obtains a reference specimen of known concentration from the National Institute of Standards and Technology. She then makes 20 measurements of the concentration of this specimen with the new method and checks for bias by comparing the mean result with the known concentration.

(d) Another chemist is evaluating the same new method. He has no reference specimen, but a familiar analytic method is available. He wants to know if the new and old methods agree. He takes a specimen of unknown concentration and measures the concentration 10 times with the new method and 10 times with the old method.

7.45 Table 1.3 (page 31) gives the number of medical doctors per 100,000 people for each of the 50 states. It does not make sense to use the *t* procedures (or any other statistical procedures) to give a 95% confidence interval for the mean number of medical doctors per 100,000 people in the population of the American states. Explain why not.

The following exercises concern the optional material in the sections on the power of the t test and on nonnormal populations.

7.46 Exercise 7.31 gives data on the amount of vitamin C in gruel made from wheat soy blend in 5 Haitian households before and after cooking. Is there evidence that the median amount of vitamin C is less after cooking? State hypotheses, carry out a sign test, and report your conclusion.

7.47 The bank in Exercise 7.33 tested a new idea on a sample of 500 customers. Suppose that the bank wanted to be quite certain of detecting a mean increase of $\mu = \$150$ in the amount charged, at the $\alpha = 0.01$ significance level. Perhaps a sample of only $n = 100$ customers would accomplish this. Find the approximate power of the test with $n = 100$ against the alternative $\mu = \$150$ as follows:

(a) What is the t critical value for the one-sided test with $\alpha = 0.01$ and $n = 100$?

(b) Write the criterion for rejecting H_0: $\mu = 0$ in terms of the t statistic. Then take $s = 267$ (an estimate based on the data in Exercise 7.33) and state the rejection criterion in terms of $\bar{x}$.

(c) Assume that $\mu = 150$ (the given alternative) and that $\sigma = 267$. The approximate power is the probability of the event you found in (b), calculated under these assumptions. Find the power. Would you recommend that the bank do a test on 100 customers, or should more customers be included?

7.48 The tomato experts who carried out the field trial described in Exercise 7.43 suspect that the relative lack of significance there is due to low power. They would like to be able to detect a mean difference in yields of 0.6 pound per plant at the 0.05 significance level. Based on the previous study, use 0.92 as an estimate of both the population σ and the value of s in future samples.

(a) What is the power of the test from Exercise 7.43 with $n = 12$ against the alternative $\mu = 0.6$?

(b) If the sample size is increased to $n = 30$ plots of land, what will be the power against the same alternative?

7.49 Apply the sign test to the data in Exercise 7.29 to assess the effects of piano lessons on spatial-temporal reasoning.

(a) State the hypotheses two ways, in terms of a population median and in terms of the probability of an improvement in the test score.

(b) Carry out the sign test. Find the approximate P-value using the normal approximation to the binomial distributions, and report your conclusion.

7.50 Use the sign test to assess whether the summer institute of Exercise 7.42 improves Spanish listening skills. State the hypotheses, give the P-value using the binomial table (Table C), and report your conclusion.

7.51 A manufacturer of electric motors tests insulation at a high temperature (250° C) and records the number of hours until the insulation fails.[16] The data for 5 specimens are

300 324 372 372 444

The small sample size makes judgment from the data difficult, but engineering experience suggests that the logarithm of the failure time will have a normal distribution. Take the logarithms of the 5 observations, and use t procedures to give a 90% confidence interval for the mean of the log failure time for insulation of this type.

7.52 The calcium intake data in Exercise 7.16 contain two large observations and we have some concern about the use of the t procedures because of this. In Exercise 7.19 we compared the mean of the data with 1300 milligrams, the RDA. We can use a version of the sign test to compare the median intake with the RDA. First subtract 1300 from each intake. If the population median is 1300, we expect approximately half of the observations to be above the median and half to be below it. The number of observations that will be above the median is binomial with $n = 38$ and $p = 0.5$. Carry out the sign test and summarize your results.

7.2 Comparing Two Means

A medical researcher is interested in the effect on blood pressure of added calcium in our diet. She conducts a randomized comparative experiment in which one group of subjects receives a calcium supplement and a control group gets a placebo. A psychologist develops a test that measures social insight. He compares the social insight of male college students with that of female college students by giving the test to a large group of students of each sex. A bank wants to know which of two incentive plans will most increase the use of its credit cards. It offers each incentive to a random sample of credit card customers and compares the amount charged during the following six months. Two-sample problems such as these are among the most common situations encountered in statistical practice.

Two-Sample Problems

- The goal of inference is to compare the responses in two groups.
- Each group is considered to be a sample from a distinct population.
- The responses in each group are independent of those in the other group.

A two-sample problem can arise from a randomized comparative experiment that randomly divides the subjects into two groups and exposes each group to a different treatment. Comparing random samples separately selected from two populations is also a two-sample problem. Unlike the matched pairs designs studied earlier, there is no matching of the units in the two samples, and the two samples may be of different sizes. Inference procedures for two-sample data differ from those for matched pairs.

We can present two-sample data graphically by a back-to-back stemplot (for small samples) or by side-by-side boxplots (for larger samples). Now we will apply the ideas of formal inference in this setting. When both population distributions are symmetric, and especially when they are at least approximately normal, a comparison of the mean responses in the two populations is most often the goal of inference.

We have two independent samples, from two distinct populations (such as subjects given a treatment and those given a placebo). The same variable is measured for both samples. We will call the variable x_1 in the first population and x_2 in the second because the variable may have different distributions in the two populations. Here is the notation that we will use to describe the two populations:

Population	Variable	Mean	Standard deviation
1	x_1	μ_1	σ_1
2	x_2	μ_2	σ_2

We want to compare the two population means, either by giving a confidence interval for $\mu_1 - \mu_2$ or by testing the hypothesis of no difference, H_0: $\mu_1 = \mu_2$.

Inference is based on two independent SRSs, one from each population. Here is the notation that describes the samples:

Population	Sample size	Sample mean	Sample standard deviation
1	n_1	$\overline{x}_1$	s_1
2	n_2	$\overline{x}_2$	s_2

Throughout this section, the subscripts 1 and 2 show the population to which a parameter or a sample statistic refers.

The two-sample z statistic

The natural estimator of the difference $\mu_1 - \mu_2$ is the difference between the sample means, $\overline{x}_1 - \overline{x}_2$. If we are to base inference on this statistic, we must know its sampling distribution. Our knowledge of probability is equal to the task. First, the mean of the difference $\overline{x}_1 - \overline{x}_2$ is the difference of the means $\mu_1 - \mu_2$. This follows from the addition rule for means (page 326) and the fact that the mean of any $\overline{x}$ is the same as the mean of the population. Because the samples are independent, their sample means $\overline{x}_1$ and $\overline{x}_2$ are independent random variables. The addition rule for variances (page 330) says that the variance of the difference $\overline{x}_1 - \overline{x}_2$ is the sum of their variances, which is

$$\frac{\sigma_1^2}{n_1} + \frac{\sigma_2^2}{n_2}$$

We now know the mean and variance of the distribution of $\overline{x}_1 - \overline{x}_2$ in terms of the parameters of the two populations. If the two population distributions are both normal, then the distribution of $\overline{x}_1 - \overline{x}_2$ is also normal. This is true because each sample mean alone is normally distributed and because a difference of independent normal random variables is also normal.

EXAMPLE 7.13

A fourth-grade class has 10 girls and 7 boys. The children's heights are recorded on their tenth birthdays. What is the chance that the girls are taller than the boys? Of course, it is very unlikely that all of the girls are taller than all of the boys. We translate the question into the following: what is the probability that the mean height of the girls is greater than the mean height of the boys?

Based on information from the National Center for Health Statistics, we assume that the heights (in inches) of 10-year-old girls are $N(54.5, 2.7)$ and the heights of 10-year-old boys are $N(54.1, 2.4)$. The heights of the students in our class are assumed to be random samples from these populations. The two distributions are shown in Figure 7.11(a).

The difference $\bar{x}_1 - \bar{x}_2$ between the female and male mean heights varies in different random samples. The sampling distribution has mean

$$\mu_1 - \mu_2 = 54.5 - 54.1 = 0.4 \text{ inch}$$

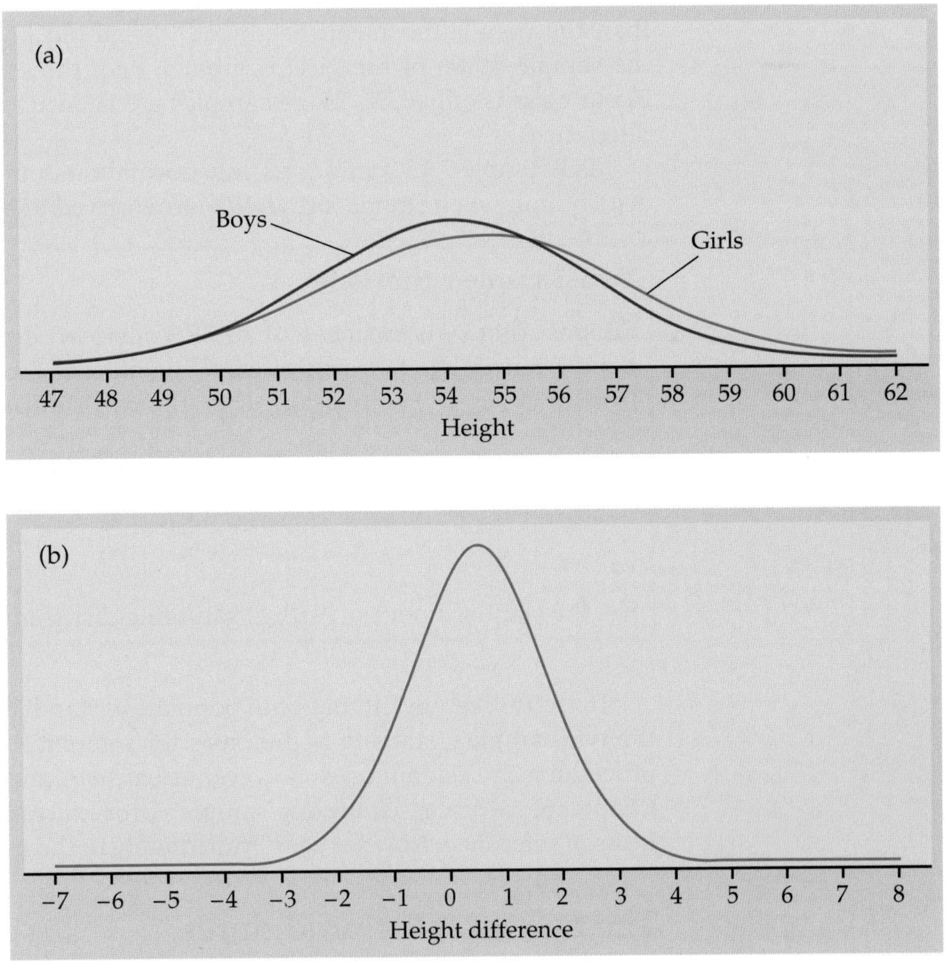

FIGURE 7.11 Distributions for Example 7.13. (a) Distributions of heights of 10-year-old boys and girls. (b) Distribution of the difference between means of 10 girls and 7 boys.

and variance

$$\frac{\sigma_1^2}{n_1} + \frac{\sigma_2^2}{n_2} = \frac{2.7^2}{10} + \frac{2.4^2}{7}$$
$$= 1.55$$

The standard deviation of the difference in sample means is therefore $\sqrt{1.55} = 1.25$ inches.

If the heights vary normally, the difference in sample means is also normally distributed. The distribution of the difference in heights is shown in Figure 7.11(b). We standardize $\bar{x}_1 - \bar{x}_2$ by subtracting its mean (0.4) and dividing by its standard deviation (1.25). Therefore, the probability that the girls are taller than the boys is

$$P(\bar{x}_1 - \bar{x}_2 > 0) = P\left(\frac{(\bar{x}_1 - \bar{x}_2) - 0.4}{1.25} > \frac{0 - 0.4}{1.25}\right)$$
$$= P(Z > -0.32) = 0.63$$

Even though the population mean height of 10-year-old girls is greater than the population mean height of 10-year-old boys, the probability that the sample mean of the girls is greater than the sample mean of the boys in our class is only 63%. Large samples are needed to see the effects of small differences.

As Example 7.13 reminds us, any normal random variable has the $N(0, 1)$ distribution when standardized. We have arrived at a new z statistic.

Two-Sample z Statistic

Suppose that $\bar{x}_1$ is the mean of an SRS of size n_1 drawn from an $N(\mu_1, \sigma_1)$ population and that $\bar{x}_2$ is the mean of an independent SRS of size n_2 drawn from an $N(\mu_2, \sigma_2)$ population. Then the **two-sample z statistic**

$$z = \frac{(\bar{x}_1 - \bar{x}_2) - (\mu_1 - \mu_2)}{\sqrt{\frac{\sigma_1^2}{n_1} + \frac{\sigma_2^2}{n_2}}}$$

has the standard normal $N(0, 1)$ sampling distribution.

In the unlikely event that both population standard deviations are known, the two-sample z statistic is the basis for inference about $\mu_1 - \mu_2$. Exact z procedures are seldom used, however, because σ_1 and σ_2 are rarely known. In Chapter 6, we discussed the one-sample z procedures in order to introduce the ideas of inference. Now we pass immediately to the more useful t procedures.

The two-sample t procedures

Suppose now that the population standard deviations σ_1 and σ_2 are not known. We estimate them by the sample standard deviations s_1 and s_2 from our two samples. Following the pattern of the one-sample case, we substitute the standard errors $s_i/\sqrt{n_i}$ for the standard deviations $\sigma_i/\sqrt{n_i}$ used in the two-sample z statistic. The result is the *two-sample t statistic:*

$$t = \frac{(\bar{x}_1 - \bar{x}_2) - (\mu_1 - \mu_2)}{\sqrt{\dfrac{s_1^2}{n_1} + \dfrac{s_2^2}{n_2}}}$$

Unfortunately, this statistic does *not* have a t distribution. A t distribution replaces an $N(0, 1)$ distribution only when a single standard deviation (σ) in a z statistic is replaced by a standard error (s). In this case, we replaced two standard deviations (σ_1 and σ_2) by their estimates (s_1 and s_2), which does not produce a statistic having a t distribution.

approximations for the degrees of freedom

Nonetheless, we can approximate the distribution of the two-sample t statistic by using the $t(k)$ distribution with an **approximation for the degrees of freedom k.** We use these approximations to find approximate values of t^* for confidence intervals and to find approximate P-values for significance tests. Here are two approximations:

1. Use a value of k that is calculated from the data. In general it will not be a whole number.
2. Use k equal to the smaller of $n_1 - 1$ and $n_2 - 1$.

Most statistical software uses the first option to approximate the $t(k)$ distribution for two-sample problems unless the user requests another method. Use of this approximation without software is a bit complicated; we will give the details later in this section. If you are not using software, the second approximation is preferred. This approximation is appealing because it is conservative.[17] Margins of error for confidence intervals are a bit larger than they need to be, so the true confidence level is larger than C; P-values are a bit smaller, so we are a little less likely to reject H_0 when it is true. In practice, the choice of approximation almost never makes a difference in our conclusion.

The two-sample *t* significance test

The Two-Sample *t* Significance Test

Suppose that an SRS of size n_1 is drawn from a normal population with unknown mean μ_1 and that an independent SRS of size n_2 is drawn from another normal population with unknown mean μ_2. To test the hypothesis H_0: $\mu_1 = \mu_2$, compute the **two-sample *t* statistic**

$$t = \frac{\bar{x}_1 - \bar{x}_2}{\sqrt{\dfrac{s_1^2}{n_1} + \dfrac{s_2^2}{n_2}}}$$

and use P-values or critical values for the $t(k)$ distribution, where the degrees of freedom k are either approximated by software or are the smaller of $n_1 - 1$ and $n_2 - 1$.

TABLE 7.3 DRP scores for third-graders

Treatment group				Control group			
24	61	59	46	42	33	46	37
43	44	52	43	43	41	10	42
58	67	62	57	55	19	17	55
71	49	54		26	54	60	28
43	53	57		62	20	53	48
49	56	33		37	85	42	

EXAMPLE 7.14

An educator believes that new directed reading activities in the classroom will help elementary school pupils improve some aspects of their reading ability. She arranges for a third-grade class of 21 students to take part in these activities for an eight-week period. A control classroom of 23 third-graders follows the same curriculum without the activities. At the end of the eight weeks, all students are given a Degree of Reading Power (DRP) test, which measures the aspects of reading ability that the treatment is designed to improve. The data appear in Table 7.3.[18]

First examine the data:

Control		Treatment
9 7 0	1	
8 6 0	2	4
7 7 3	3	3
8 6 3 2 2 2 1	4	3 3 3 4 6 9 9
5 5 4 3	5	2 3 4 6 7 7 8 9
2 0	6	1 2 7
	7	1
5	8	

A back-to-back stemplot suggests that there is a mild outlier in the control group but no deviation from normality serious enough to forbid use of t procedures. Separate normal quantile plots for both groups (Figure 7.12) confirm that both are approximately normal. The scores of the treatment group appear to be somewhat higher than those of the control group. The summary statistics are

Group	n	$\overline{x}$	s
Treatment	21	51.48	11.01
Control	23	41.52	17.15

Because we hope to show that the treatment (Group 1) is better than the control (Group 2), the hypotheses are

$$H_0: \mu_1 = \mu_2$$

$$H_a: \mu_1 > \mu_2$$

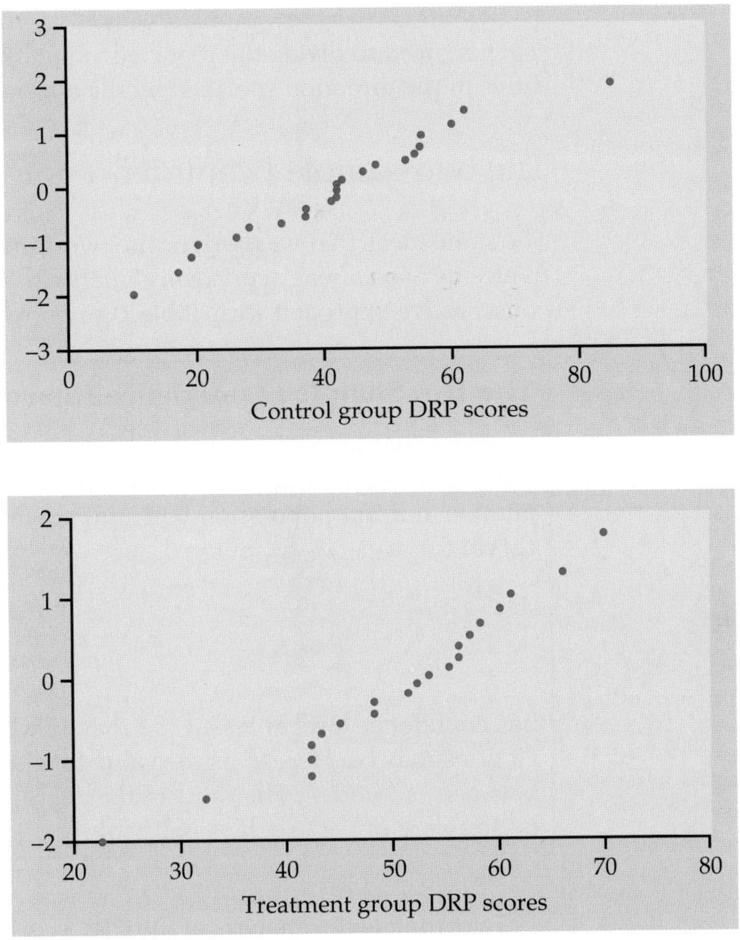

FIGURE 7.12 Normal quantile plots for the DRP scores in Table 7.3.

The two-sample t test statistic is

$$t = \frac{\bar{x}_1 - \bar{x}_2}{\sqrt{\frac{s_1^2}{n_1} + \frac{s_2^2}{n_2}}}$$

$$= \frac{51.48 - 41.52}{\sqrt{\frac{11.01^2}{21} + \frac{17.15^2}{23}}} = 2.31$$

The P-value for the one-sided test is $P(T \geq 2.31)$. Software gives the approximate P-value as 0.0132 and uses 37.9 as the degrees of freedom. For the second approximation, the degrees of freedom k are equal to the smaller of

$$n_1 - 1 = 21 - 1 = 20 \text{ and } n_2 - 1 = 23 - 1 = 22$$

df = 20		
p	0.02	0.01
t^*	2.197	2.528

Comparing 2.31 with the entries in Table D for 20 degrees of freedom, we see that P lies between 0.02 and 0.01. The data strongly suggest that directed reading activity improves the DRP score ($t = 2.31$, df $= 20$, $P < 0.02$).

If your software gives only P-values for the two-sided alternative, $2P(T \geq |t|)$, you need to divide the reported value by 2 after checking that the means differ in the direction specified by the alternative hypothesis.

The two-sample *t* confidence interval

The same ideas that we used for the two-sample t significance tests also apply to give us *two-sample t confidence intervals*. We can use either software or the conservative approach with Table D to approximate the value of t^*.

The Two-Sample *t* Confidence Interval

Suppose that an SRS of size n_1 is drawn from a normal population with unknown mean μ_1 and that an independent SRS of size n_2 is drawn from another normal population with unknown mean μ_2. The **confidence interval for $\mu_1 - \mu_2$** given by

$$(\bar{x}_1 - \bar{x}_2) \pm t^* \sqrt{\frac{s_1^2}{n_1} + \frac{s_2^2}{n_2}}$$

has confidence level at least C no matter what the population standard deviations may be. Here, t^* is the value for the $t(k)$ density curve with area C between $-t^*$ and t^*. The value of the degrees of freedom k is approximated by software or we use the smaller of $n_1 - 1$ and $n_2 - 1$.

To complete the analysis of the DRP scores we examined in Example 7.14, we need to describe the size of the treatment effect. We do this with a confidence interval for the difference between the treatment group and the control group means.

EXAMPLE 7.15

We will find a 95% confidence interval for the mean improvement in the entire population of third-graders. The interval is

$$\begin{aligned}(\bar{x}_1 - \bar{x}_2) \pm t^* \sqrt{\frac{s_1^2}{n_1} + \frac{s_2^2}{n_2}} &= (51.48 - 41.52) \pm t^* \sqrt{\frac{11.01^2}{21} + \frac{17.15^2}{23}} \\ &= 9.96 \pm 4.31t^*\end{aligned}$$

Using software, the degrees of freedom are 37.9 and $t^* = 2.025$. This approximation gives

$$9.96 \pm (4.31 \times 2.025) = 9.96 \pm 8.72 = (1.2,\ 18.7)$$

The conservative approach uses the $t(20)$ distribution. Table D gives $t^* = 2.086$. With this approximation we have

$$9.96 \pm (4.31 \times 2.086) = 9.96 \pm 8.99 = (1.0,\ 18.9)$$

We can see that the conservative approach does, in fact, give a larger interval than the more accurate approximation used by software. However, the difference is rather small.

We estimate the mean improvement to be about 10 points, but with a margin of error of almost 9 points. Although we have good evidence of some improvement, the data do not allow a very precise estimate of the size of the average improvement.

The design of the study in Example 7.14 is not ideal. Random assignment of students was not possible in a school environment, so existing third-grade classes were used. The effect of the reading programs is therefore confounded with any other differences between the two classes. The classes were chosen to be as similar as possible—for example, in the social and economic status of the students. Extensive pretesting showed that the two classes were on the average quite similar in reading ability at the beginning of the experiment. To avoid the effect of two different teachers, the researcher herself taught reading in both classes during the eight-week period of the experiment. We can therefore be somewhat confident that the two-sample test is detecting the effect of the treatment and not some other difference between the classes. This example is typical of many situations in which an experiment is carried out but randomization is not possible.

Robustness of the two-sample procedures

The two-sample t procedures are more robust than the one-sample t methods. When the sizes of the two samples are equal and the distributions of the two populations being compared have similar shapes, probability values from the t table are quite accurate for a broad range of distributions when the sample sizes are as small as $n_1 = n_2 = 5$.[19] When the two population distributions have different shapes, larger samples are needed. The guidelines given on page 505 for the use of one-sample t procedures can be adapted to two-sample procedures by replacing "sample size" with the "sum of the sample sizes" $n_1 + n_2$. These guidelines are rather conservative, especially when the two samples are of equal size. In planning a two-sample study, you should usually choose equal sample sizes. The two-sample t procedures are most robust against nonnormality in this case, and the conservative probability values are most accurate.

Here is an example with large sample sizes. Even if the distributions are not normal, we are confident that the sample means will be approximately normal. The two-sample t test is very robust in this case.

EXAMPLE 7.16

The Chapin Social Insight Test is a psychological test designed to measure how accurately the subject appraises other people. The possible scores on the test range from 0 to 41. During the development of the Chapin test, it was given to several different groups of people. Here are the results for male and female college students majoring in the liberal arts:[20]

Group	Sex	n	$\bar{x}$	s
1	Male	133	25.34	5.05
2	Female	162	24.94	5.44

Do these data support the contention that female and male students differ in average social insight? Because no specific direction for the male/female difference was hypothesized before looking at the data, we choose a two-sided alternative. The hypotheses are

$$H_0: \mu_1 = \mu_2$$

$$H_a: \mu_1 \neq \mu_2$$

The two-sample t statistic is

$$t = \frac{\bar{x}_1 - \bar{x}_2}{\sqrt{\frac{s_1^2}{n_1} + \frac{s_2^2}{n_2}}}$$

$$= \frac{25.34 - 24.94}{\sqrt{\frac{(5.05)^2}{133} + \frac{(5.44)^2}{162}}}$$

$$= 0.654$$

	df = 100	
p	0.25	0.20
t^*	0.677	0.845

Using the conservative approach we find the P-value by comparing 0.654 to critical values for the $t(132)$ distribution and then doubling p because the alternative is two-sided. Table D (for 100 degrees of freedom) shows that 0.654 does not reach the 0.25 critical value, which is the largest entry for p in this table. The P-value is therefore greater than 0.50. The data give no evidence of a gender difference in mean social insight score ($t = 0.654$, df $= 132$, $P > 0.50$). Software uses the $t(289)$ distribution and gives $P = 0.051$.

The researcher in Example 7.16 did not do an experiment but compared samples from two populations. The large samples imply that the assumption that the populations have normal distributions is of little importance. The sample means will be nearly normal in any case. The major question concerns the population to which the conclusions apply. The student subjects are certainly not an SRS of all liberal arts majors. If they are volunteers from a single college, the sample results may not extend to a wider population.

Inference for small samples

Small samples require special care. We do not have enough observations for boxplots or normal quantile plots, and only the most extreme outliers stand out. Because we have so few observations, we are very reluctant to discard any of them. Furthermore, the power of our significance test tends to be low, and the margin of error of our confidence interval tends to be large. Despite all of these difficulties, we can often draw important conclusions from studies with small sample sizes. If the size of an effect is large, it will be evident even if the n's are small.

EXAMPLE 7.17

The pesticide DDT causes tremors and convulsions if it is ingested by humans or other mammals. Researchers seek to understand how the convulsions are caused. In a randomized comparative experiment, 6 white rats poisoned with DDT were compared with a control group of 6 unpoisoned rats. Electrical measurements of nerve activity are the main clue to the nature of DDT poisoning. When a nerve is stimulated, its electrical response shows a sharp spike followed by a much smaller second spike.

Researchers found that the second spike is larger in rats fed DDT than in normal rats. This observation helps biologists understand how DDT causes tremors.[21]

The researchers measured the amplitude of the second spike as a percentage of the first spike when a nerve in the rat's leg was stimulated. For the poisoned rats the results were

12.207 16.869 25.050 22.429 8.456 20.589

The control group data were

11.074 9.686 12.064 9.351 8.182 6.642

Normal quantile plots (Figure 7.13) show no evidence of outliers or strong skewness. Both populations are reasonably normal, as far as can be judged from 6 observations. The difference in means is quite large, but in such small samples the sample mean is highly variable. A significance test can help confirm that we are seeing a real effect. Because the researchers did not conjecture in advance that the size of the second spike would increase in rats fed DDT, we test

$$H_0: \mu_1 = \mu_2$$

$$H_a: \mu_1 \neq \mu_2$$

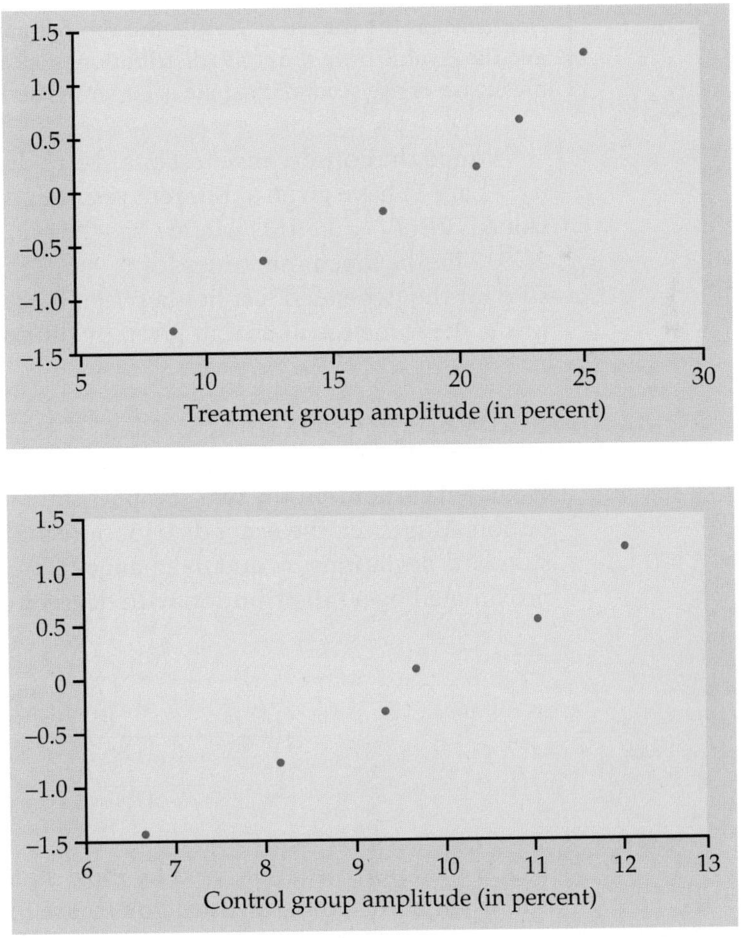

FIGURE 7.13 Normal quantile plots for the amplitude of the second spike as a percent of the first spike in Example 7.17.

Here is the output from the SAS statistical software system for these data:

```
                      TTEST PROCEDURE

Variable:  SPIKE

GROUP          N              Mean          Std Dev       Std Error
-------------------------------------------------------------------
DDT            6       17.60000000       6.34014839      2.58835474
CONTROL        6        9.49983333       1.95005932      0.79610839

Variances          T          DF       Prob>|T|
-----------------------------------------------
Unequal       2.9912         5.9         0.0247
Equal         2.9912        10.0         0.0135
```

SAS reports the results of two t procedures: the general two-sample procedure ("Unequal" variances) and a special procedure that assumes the two population variances to be equal. For now, we are interested in the first of these procedures. We will discuss the second in the section on pooled two-sample t procedures. The two-sample t statistic has the value $t = 2.9912$, the approximate degrees of freedom are df $= 5.9$, and the P-value from the $t(5.9)$ distribution is 0.0247. There is good evidence that the mean size of the secondary spike is larger in rats fed DDT.

Would the conservative test based on 5 degrees of freedom (both $n_1 - 1$ and $n_2 - 1$ are 5) have given a different result in this example? The conservative P-value is $2P(T \geq 2.9912)$ where t has the $t(5)$ distribution. Table D shows that 2.9912 lies between the values for $p = 0.02$ and 0.01 for the $t(5)$ distribution, so P for the two-sided test lies between 0.02 and 0.04. For practical purposes this is the same result as that given by the software.

Software approximation for the degrees of freedom*

We noted earlier that the two-sample t statistic does not have an exact t distribution. Moreover, the exact distribution changes as the unknown population standard deviations σ_1 and σ_2 change. However, the distribution can be approximated by a t distribution with degrees of freedom given by

$$\text{df} = \frac{\left(\dfrac{s_1^2}{n_1} + \dfrac{s_2^2}{n_2}\right)^2}{\dfrac{1}{n_1 - 1}\left(\dfrac{s_1^2}{n_1}\right)^2 + \dfrac{1}{n_2 - 1}\left(\dfrac{s_2^2}{n_2}\right)^2}$$

This is the approximation used by most statistical software. It is quite accurate when both sample sizes n_1 and n_2 are 5 or larger.

*This material can be omitted unless you are using statistical software and wish to understand what the software does.

EXAMPLE 7.18

For the DRP study of Example 7.14, the following table summarizes the data:

Group	n	$\bar{x}$	s
1	21	51.48	11.01
2	23	41.52	17.15

For greatest accuracy, we will use critical points from the t distribution with degrees of freedom given by the equation above:

$$\text{df} = \frac{\left(\frac{11.01^2}{21} + \frac{17.15^2}{23}\right)^2}{\frac{1}{20}\left(\frac{11.01^2}{21}\right)^2 + \frac{1}{22}\left(\frac{17.15^2}{23}\right)^2}$$

$$= \frac{344.486}{9.099} = 37.86$$

This is the value that we reported in Examples 7.14 and 7.15 where we gave the results produced by software.

The number df given by the above approximation is always at least as large as the smaller of $n_1 - 1$ and $n_2 - 1$. On the other hand, df is never larger than the sum $n_1 + n_2 - 2$ of the two individual degrees of freedom. The number of degrees of freedom is generally not a whole number. There is a t distribution with any positive degrees of freedom, even though Table D contains entries only for whole-number degrees of freedom. When df is small and is not a whole number, interpolation between entries in Table D may be needed to obtain an accurate critical value or P-value. Because of this and the need to calculate df, we do not recommend regular use of this approximation if a computer is not doing the arithmetic. With a computer, however, the more accurate procedures are painless.

The pooled two-sample *t* procedures*

There is one situation in which a t statistic for comparing two means has exactly a t distribution. Suppose that the two normal population distributions have the *same* standard deviation. In this case we need substitute only a single standard error in a z statistic, and the resulting t statistic has a t distribution. We will develop the z statistic first, as usual, and from it the t statistic.

Call the common—and still unknown—standard deviation of both populations σ. Both sample variances s_1^2 and s_2^2 estimate σ^2. The best way to combine these two estimates is to average them with weights equal to their degrees of freedom. This gives more weight to the information from the larger sample, which is reasonable. The resulting estimator of σ^2 is

$$s_p^2 = \frac{(n_1 - 1)s_1^2 + (n_2 - 1)s_2^2}{n_1 + n_2 - 2}$$

*This section can be omitted if desired, but it should be read if you plan to read Chapters 12 and 13.

pooled estimator of σ^2

This is called the **pooled estimator of σ^2** because it combines the information in both samples.

When both populations have variance σ^2, the addition rule for variances says that $\bar{x}_1 - \bar{x}_2$ has variance equal to the *sum* of the individual variances, which is

$$\frac{\sigma^2}{n_1} + \frac{\sigma^2}{n_2} = \sigma^2\left(\frac{1}{n_1} + \frac{1}{n_2}\right)$$

The standardized difference of means in this equal-variance case is therefore

$$z = \frac{(\bar{x}_1 - \bar{x}_2) - (\mu_1 - \mu_2)}{\sigma\sqrt{\dfrac{1}{n_1} + \dfrac{1}{n_2}}}$$

This is a special two-sample z statistic for the case in which the populations have the same σ. Replacing the unknown σ by the estimate s_p gives a t statistic. The degrees of freedom are $n_1 + n_2 - 2$, the sum of the degrees of freedom of the two sample variances. This statistic is the basis of the pooled two-sample t inference procedures.

The Pooled Two-Sample *t* Procedures

Suppose that an SRS of size n_1 is drawn from a normal population with unknown mean μ_1 and that an independent SRS of size n_2 is drawn from another normal population with unknown mean μ_2. Suppose also that the two populations have the same standard deviation. A level C confidence interval for $\mu_1 - \mu_2$ is

$$(\bar{x}_1 - \bar{x}_2) \pm t^* s_p \sqrt{\frac{1}{n_1} + \frac{1}{n_2}}$$

Here t^* is the value for the $t(n_1 + n_2 - 2)$ density curve with area C between $-t^*$ and t^*.

To test the hypothesis H_0: $\mu_1 = \mu_2$, compute the pooled two-sample t statistic

$$t = \frac{\bar{x}_1 - \bar{x}_2}{s_p\sqrt{\dfrac{1}{n_1} + \dfrac{1}{n_2}}}$$

In terms of a random variable T having the $t(n_1 + n_2 - 2)$ distribution, the P-value for a test of H_0 against

H_a: $\mu_1 > \mu_2$ is $P(T \geq t)$

H_a: $\mu_1 < \mu_2$ is $P(T \leq t)$

H_a: $\mu_1 \neq \mu_2$ is $2P(T \geq |t|)$

TABLE 7.4 Seated systolic blood pressure

Calcium group			Placebo group		
Begin	End	Decrease	Begin	End	Decrease
107	100	7	123	124	−1
110	114	−4	109	97	12
123	105	18	112	113	−1
129	112	17	102	105	−3
112	115	−3	98	95	3
111	116	−5	114	119	−5
107	106	1	119	114	5
112	102	10	114	112	2
136	125	11	110	121	−11
102	104	−2	117	118	−1
			130	133	−3

EXAMPLE 7.19

Does increasing the amount of calcium in our diet reduce blood pressure? Examination of a large sample of people revealed a relationship between calcium intake and blood pressure, but such observational studies do not establish causation. Animal experiments then showed that calcium supplements do reduce blood pressure in rats, justifying an experiment with human subjects. A randomized comparative experiment gave one group of 10 black men a calcium supplement for 12 weeks. The control group of 11 black men received a placebo that appeared identical. (In fact, a block design with black and white men as the blocks was used. We will look only at the results for blacks, because the earlier survey suggested that calcium is more effective for blacks.) The experiment was double-blind. Table 7.4 gives the seated systolic (heart contracted) blood pressure for all subjects at the beginning and end of the 12-week period, in millimeters (mm) of mercury. Because the researchers were interested in decreasing blood pressure, Table 7.4 also shows the decrease for each subject. An increase appears as a negative entry.[22]

As usual, we first examine the data. To compare the effects of the two treatments, take the response variable to be the amount of the decrease in blood pressure. Inspection of the data reveals that there are no outliers. Normal quantile plots (Figure 7.14) give a more detailed picture. The calcium group has a somewhat short left tail, but there are no departures from normality that will prevent use of t procedures. To examine the question of the researchers who collected these data, we perform a significance test.

EXAMPLE 7.20

Take Group 1 to be the calcium group and Group 2 to be the placebo group. The evidence that calcium lowers blood pressure more than a placebo is assessed by testing

$$H_0: \mu_1 = \mu_2$$

$$H_a: \mu_1 > \mu_2$$

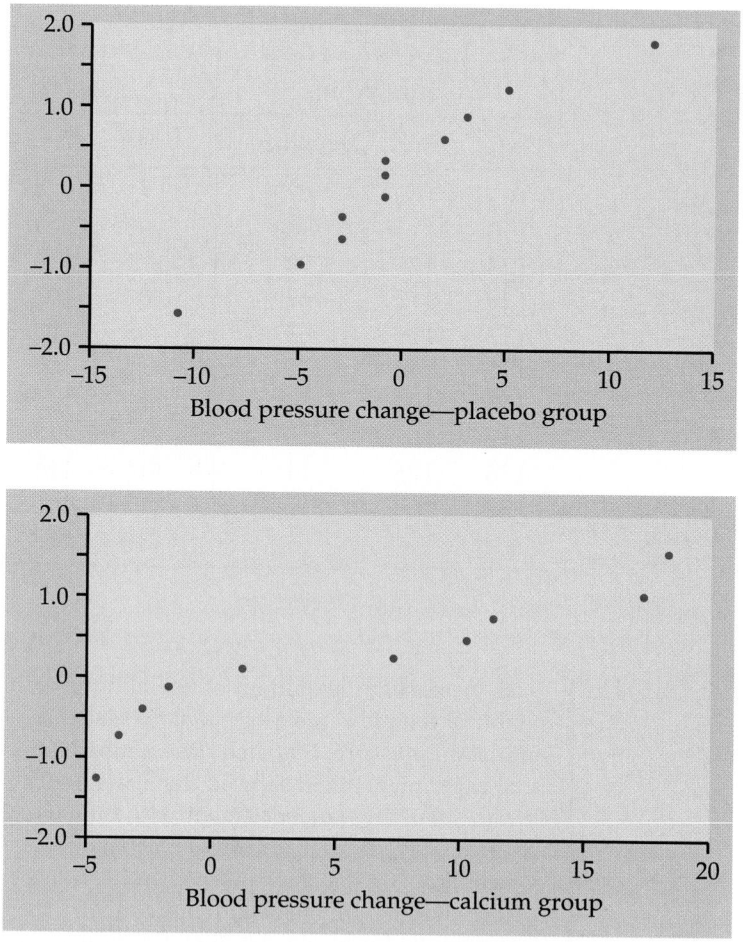

FIGURE 7.14 Normal quantile plots for the decrease in blood pressure from Table 7.4.

Here are the summary statistics for the decrease in blood pressure:

Group	Treatment	n	$\bar{x}$	s
1	Calcium	10	5.000	8.743
2	Placebo	11	−0.273	5.901

The calcium group shows a drop in blood pressure, and the placebo group has a small increase. The sample standard deviations do not rule out equal population standard deviations. A difference this large will often arise by chance in samples this small. We are willing to assume equal population standard deviations. The pooled sample variance is

$$s_p^2 = \frac{(n_1 - 1)s_1^2 + (n_2 - 1)s_2^2}{n_1 + n_2 - 2}$$

$$= \frac{(9)(8.743)^2 + (10)(5.901)^2}{10 + 11 - 2} = 54.536$$

so that

$$s_p = \sqrt{54.536} = 7.385$$

The pooled two-sample t statistic is

$$\begin{aligned} t &= \frac{\overline{x}_1 - \overline{x}_2}{s_p\sqrt{\frac{1}{n_1} + \frac{1}{n_2}}} \\ &= \frac{5.000 - (-0.273)}{7.385\sqrt{\frac{1}{10} + \frac{1}{11}}} \\ &= \frac{5.273}{3.227} = 1.634 \end{aligned}$$

df = 19		
p	0.10	0.05
t^*	1.328	1.729

The P-value is $P(T \geq 1.634)$, where T has the $t(19)$ distribution. From Table D we can see that P falls between the $\alpha = 0.10$ and $\alpha = 0.05$ levels. Statistical software gives the exact value $P = 0.059$. The experiment found evidence that calcium reduces blood pressure, but the evidence falls a bit short of the traditional 5% and 1% levels.

Sample size strongly influences the P-value of a test. An effect that fails to be significant at a specified level α in a small sample can be significant in a larger sample. In the light of the rather small samples in Example 7.20, the evidence for some effect of calcium on blood pressure is rather good. The published account of the study combined these results for blacks with the results for whites and adjusted for pretest differences among the subjects. Using this more detailed analysis, the researchers were able to report the P-value $P = 0.008$.

Of course, a P-value is almost never the last part of a statistical analysis. To make a judgment regarding the size of the effect of calcium on blood pressure, we need a confidence interval.

EXAMPLE 7.21

We estimate that the effect of calcium supplementation is the difference between the sample means of the calcium and the placebo groups, $\overline{x}_1 - \overline{x}_2 = 5.273$ mm. A 90% confidence interval for $\mu_1 - \mu_2$ uses the critical value $t^* = 1.729$ from the $t(19)$ distribution. The interval is

$$\begin{aligned} (\overline{x}_1 - \overline{x}_2) \pm t^* s_p\sqrt{\frac{1}{n_1} + \frac{1}{n_2}} &= [5.000 - (-0.273)] \pm (1.729)(7.385)\sqrt{\frac{1}{10} + \frac{1}{11}} \\ &= 5.273 \pm 5.579 \end{aligned}$$

We are 90% confident that the difference in means is in the interval $(-0.306, 10.852)$. The calcium treatment reduced blood pressure by about 5.3 mm more than a placebo on the average, but the margin of error for this estimate is 5.6 mm.

The pooled two-sample t procedures are anchored in statistical theory and so have long been the standard version of the two-sample t in textbooks. But they require the assumption that the two unknown population standard deviations are equal. As we shall see in Section 7.3, this assumption is hard to verify. The pooled t procedures are therefore a bit risky. They are reasonably robust against both nonnormality and unequal standard deviations when the sample sizes are nearly the same. When the samples are quite different in size, the pooled t procedures become sensitive to unequal standard deviations and should be used with caution unless the samples are large. Unequal standard deviations are quite common. In particular, it is common for the spread of data to increase when the center moves up, as happened in Example 7.17. Statistical software usually calculates both the pooled and the unpooled t statistics, as in Example 7.17. The very unequal sample standard deviations in that example suggest that we should not use the pooled procedures.

SUMMARY

Significance tests and confidence intervals for the difference of the means μ_1 and μ_2 of two normal populations are based on the difference $\bar{x}_1 - \bar{x}_2$ of the sample means from two independent SRSs. Because of the central limit theorem, the resulting procedures are approximately correct for other population distributions when the sample sizes are large.

When independent SRSs of sizes n_1 and n_2 are drawn from two normal populations with parameters μ_1, σ_1 and μ_2, σ_2 the **two-sample z statistic**

$$z = \frac{(\bar{x}_1 - \bar{x}_2) - (\mu_1 - \mu_2)}{\sqrt{\dfrac{\sigma_1^2}{n_1} + \dfrac{\sigma_2^2}{n_2}}}$$

has the $N(0, 1)$ distribution.

The **two-sample t statistic**

$$t = \frac{(\bar{x}_1 - \bar{x}_2) - (\mu_1 - \mu_2)}{\sqrt{\dfrac{s_1^2}{n_1} + \dfrac{s_2^2}{n_2}}}$$

does *not* have a t distribution. However, good approximations are available.

Conservative inference procedures for comparing μ_1 and μ_2 are obtained from the two-sample t statistic by using the $t(k)$ distribution with degrees of freedom k equal to the smaller of $n_1 - 1$ and $n_2 - 1$.

More accurate probability values can be obtained by estimating the degrees of freedom from the data. This is the usual procedure for statistical software.

An approximate level C **confidence interval** for $\mu_1 - \mu_2$ is given by

$$(\bar{x}_1 - \bar{x}_2) \pm t^* \sqrt{\frac{s_1^2}{n_1} + \frac{s_2^2}{n_2}}$$

Here, t^* is the value for the $t(k)$ density curve with area C between $-t^*$ and t^*, where k is computed from the data by software or is the smaller of $n_1 - 1$ and $n_2 - 1$. The quantity

$$t^* \sqrt{\frac{s_1^2}{n_1} + \frac{s_2^2}{n_2}}$$

is the **margin of error.**

Significance tests for $H_0\colon \mu_1 = \mu_2$ use the **two-sample *t* statistic**

$$t = \frac{\bar{x}_1 - \bar{x}_2}{\sqrt{\frac{s_1^2}{n_1} + \frac{s_2^2}{n_2}}}$$

The *P*-value is approximated using the $t(k)$ where k is estimated from the data using software or is the smaller of $n_1 - 1$ and $n_2 - 1$.

The guidelines for practical use of two-sample t procedures are similar to those for one-sample t procedures. Equal sample sizes are recommended.

If we can assume that the two populations have equal variances, **pooled two-sample *t* procedures** can be used. These are based on the **pooled estimator**

$$s_p^2 = \frac{(n_1 - 1)s_1^2 + (n_2 - 1)s_2^2}{n_1 + n_2 - 2}$$

of the unknown common variance and the $t(n_1 + n_2 - 2)$ distribution.

SECTION 7.2 EXERCISES

In exercises that call for two-sample t procedures, you may use either of the two approximations for the degrees of freedom that we have discussed: the value given by your software or the smaller of $n_1 - 1$ and $n_2 - 1$. Be sure to state clearly which approximation you have used.

7.53 You want to compare the daily sales for two different designs of Web pages for your Internet business. You assign the next 60 days to either design A or design B, 30 days to each.

(a) Would you use a one-sided or two-sided significance test for this problem? Explain your choice.

(b) If you use Table D to find the critical value, what are the degrees of freedom?

(c) The t statistic for comparing the mean sales is 3.15. If you use Table D, what *P*-value would you report? What would you conclude?

7.54 If you perform the significance test in the previous exercise using level $\alpha = 0.05$, how large (positive or negative) must the t statistic be to reject the null hypothesis that the two designs give the same average sales?

7.55 Refer to the previous two exercises. You realize that day of the week may affect online sales. To compare two Web page designs, you choose two successive weeks in the middle of a month. You flip a coin to assign one Monday to the first design and the other Monday to the second. You repeat this for each of the seven days of the week. You now have 7 sales amounts for each design. It is *incorrect* to use the two-sample t test to see if the mean sales differ for the two designs. Carefully explain why.

7.56 The purchasing department has suggested that all new computer monitors for your company should be flat screens. You want data to assure you that employees will like the new screens. The next 20 employees needing a new computer are the subjects for an experiment.

(a) Label the employees 01 to 20. Randomly choose 10 to receive flat screens. The remaining 10 get standard monitors.

(b) After a month of use, employees express their satisfaction with their new monitors by responding to the statement "I like my new monitor" on a scale from 1 to 5, where 1 represents "strongly disagree," 2 is "disagree," 3 is "neutral," 4 is "agree," and 5 stands for "strongly agree." The employees with the flat screens have average satisfaction 4.6 with standard deviation 0.7. The employees with the standard monitors have average 3.2 with standard deviation 1.8. Give a 95% confidence interval for the difference in the mean satisfaction scores for all employees.

(c) Would you reject the null hypothesis that the mean satisfaction for the two types of monitors is the same versus the two-sided alternative at significance level 0.05? Use your confidence interval to answer this question. Explain why you do not need to calculate the test statistic.

7.57 Refer to the previous exercise. A coworker suggested that you give the flat screens to the next 10 employees who need new screens and the standard monitor to the following 10. Explain why your randomized design is better.

7.58 Pat wants to compare the cost of one- and two-bedroom apartments in the area of your campus. She collects data for a random sample of 10 advertisements of each type. Here are the rents for the two-bedroom apartments (in dollars per month):

595, 500, 580, 650, 675, 675, 750, 500, 495, 670

Here are the rents for the one-bedroom apartments:

500, 650, 600, 505, 450, 550, 515, 495, 650, 395

Find a 95% confidence interval for the additional cost of a second bedroom.

7.59 Pat wonders if two-bedroom apartments rent for significantly more than one-bedroom apartments. Use the data in the previous exercise to find out.

(a) State appropriate null and alternative hypotheses.

(b) Report the test statistic, its degrees of freedom, and the *P*-value. What do you conclude?

(c) Can you conclude that every one-bedroom apartment costs less than every two-bedroom apartment?

(d) In the previous exercise you found a confidence interval. In this exercise you performed a significance test. Which do you think is more useful to someone planning to rent an apartment? Why?

7.60 Does bread lose its vitamins when stored? Small loaves of bread were prepared with flour that was fortified with a fixed amount of vitamins. After baking, the vitamin C content of two loaves was measured. Another two loaves were baked at the same time, stored for three days, and then the vitamin C content was measured. The units are milligrams per hundred grams of flour (mg/100 g).[23] Here are the data:

Immediately after baking:	47.62,	49.79
Three days after baking:	21.25,	22.34

(a) When bread is stored, does it lose vitamin C? To answer this question, perform a two-sample t test for these data. Be sure to state your hypotheses, the test statistic with degrees of freedom, and the *P*-value.

(b) Give a 90% confidence interval for the amount of vitamin C lost.

7.61 Suppose that the researchers in the previous exercise could have measured the same two loaves of bread immediately after baking and again after three days. Assume that the data given had come from this study design. (Assume that the values given in Exercise 7.60 are for first loaf and second loaf from left to right.)

(a) Explain carefully why your analysis in the previous exercise is *not correct* now, even though the data are the same.

(b) Redo the analysis for the design based on measuring the same loaves twice.

7.62 Refer to Exercise 7.60. The amount of vitamin E (in mg/100 g of flour) in the same loaves was also measured. Here are the data:

Immediately after baking:	94.6,	96.0
Three days after baking:	97.4,	94.3

(a) When bread is stored, does it lose vitamin E? To answer this question, perform a two-sample t test for these data. Be sure to state your hypotheses, the test statistic with degrees of freedom, and the *P*-value.

(b) Give a 90% confidence interval for the amount of vitamin E lost.

7.63 Refer to Exercises 7.60 and 7.62. Some people claim that significance tests with very small samples never lead to rejection of the null hypothesis. Discuss this claim using the results of these two exercises.

7.64 You use statistical software to perform a significance test of the null hypothesis that two means are equal. The software reports *P*-values for the two-sided alternative. Your alternative is that the first mean is greater than the second mean.

(a) The software reports $t = 2.07$ with a *P*-value of 0.06. Would you reject H_0 with $\alpha = 0.05$? Explain your answer.

(b) The software reports $t = -2.07$ with a *P*-value of 0.06. Would you reject H_0 with $\alpha = 0.05$? Explain your answer.

7.65 Do piano lessons improve the spatial-temporal reasoning of preschool children? We examined this question in Exercises 7.29 and 7.30 (page 518) by analyzing the change in spatial-temporal reasoning of 34 preschool children after six months of piano lessons. Here we examine the same question by comparing the changes of those students with the changes of 44 children in a control group.[24] Here are the data for the children who took piano lessons:

2	5	7	−2	2	7	4	1	0	7	3	4	3	4	9	4	5
2	9	6	0	3	6	−1	3	4	6	7	−2	7	−3	3	4	4

The control group scores are

1	−1	0	1	−4	0	0	1	0	−1	0	1	1	−3	−2	4
−1	2	4	2	2	2	−3	−3	0	2	0	−1	3	−1	5	−1
7	0	4	0	2	1	−6	0	2	−1	0	−2				

(a) Display the data and summarize the distributions.

(b) Make a table with the sample size, the mean, the standard deviation, and the standard error of the mean for each of the two groups.

(c) Translate the question of interest into hypotheses, test them, and summarize your conclusions.

7.66 Refer to the previous exercise. Give a 95% confidence interval that describes the comparison between the children who took piano lessons and the controls.

7.67 Refer to Exercises 7.29 and 7.30 (page 518) and the previous two exercises. We have used four ways to address the question of interest. Discuss the relative merits of each approach.

7.68 In what ways are companies that fail different from those that continue to do business? A study compared various characteristics of 68 healthy and 33 failed firms. One of the variables was the ratio of current assets to

TABLE 7.5 Ratio of current assets to current liabilities

Healthy firms						Failed firms		
1.50	0.10	1.76	1.14	1.84	2.21	0.82	0.89	1.31
2.08	1.43	0.68	3.15	1.24	2.03	0.05	0.83	0.90
2.23	2.50	2.02	1.44	1.39	1.64	1.68	0.99	0.62
0.89	0.23	1.20	2.16	1.80	1.87	0.91	0.52	1.45
1.91	1.67	1.87	1.21	2.05	1.06	1.16	1.32	1.17
0.93	2.17	2.61	3.05	1.52	1.93	0.42	0.48	0.93
1.95	2.61	1.11	0.95	0.96	2.25	0.88	1.10	0.23
2.73	1.56	2.73	0.90	2.12	1.42	1.11	0.19	0.13
1.62	1.76	2.22	2.80	1.85	0.96	2.03	0.51	1.12
1.71	1.02	2.50	1.55	1.69	1.64	0.92	0.26	1.15
1.03	1.80	0.67	2.44	2.30	2.21	0.13	0.88	0.09
1.96	1.81							

current liabilities. Roughly speaking, this is the amount that the firm is worth divided by what it owes.[25] The data are given in Table 7.5.

(a) Display the data so that the two distributions can be compared. Describe the shapes of the distributions and any important characteristics.

(b) We expect that failed firms will have a lower ratio. Describe and test appropriate hypotheses for these data. What do you conclude?

(c) It is not possible to do a randomized experiment for this kind of question. Explain why.

7.69 Physical fitness is related to personality characteristics. In one study of this relationship, middle-aged college faculty who had volunteered for a fitness program were divided into low-fitness and high-fitness groups based on a physical examination. The subjects then took the Cattell Sixteen Personality Factor Questionnaire.[26] Here are the data for the "ego strength" personality factor:

Low fitness			High fitness		
4.99	5.53	3.12	6.68	5.93	5.71
4.24	4.12	3.77	6.42	7.08	6.20
4.74	5.10	5.09	7.32	6.37	6.04
4.93	4.47	5.40	6.38	6.53	6.51
4.16	5.30		6.16	6.68	

(a) Is the difference in mean ego strength significant at the 5% level? At the 1% level? Be sure to state H_0 and H_a.

(b) You should be hesitant to generalize these results to the population of all middle-aged men. Explain why.

7.70 Does cocaine use by pregnant women cause their babies to have low birth weight? To study this question, birth weights of babies of women who tested positive for cocaine/crack during a drug-screening test were compared with the birth weights for women who either tested negative or were not tested, a group we call "other." Here are the summary statistics. The birth weights are measured in grams.[27]

Group	n	$\bar{x}$	s
Positive test	134	2733	599
Other	5974	3118	672

(a) Formulate appropriate hypotheses and carry out the test of significance for these data.

(b) Give a 95% confidence interval for the mean difference in birth weights.

(c) Discuss the limitations of the study design. What do you believe can be concluded from this study?

7.71 The U.S. Department of Agriculture (USDA) uses many types of surveys to obtain important economic estimates. In one pilot study they estimated wheat prices in July and in September using independent samples. Here is a brief summary from the report:[28]

Month	n	$\bar{x}$	$s/\sqrt{n}$
July	90	\$2.95	\$0.023
September	45	\$3.61	\$0.029

(a) Note that the report gave standard errors. Find the standard deviation for each of the samples.

(b) Use a significance test to examine whether or not the price of wheat was the same in July and September. Be sure to give details and carefully state your conclusion.

7.72 Refer to the previous exercise. Give a 95% confidence interval for the increase in price between July and September.

7.73 A market research firm supplies manufacturers with estimates of the retail sales of their products from samples of retail stores. Marketing managers are prone to look at the estimate and ignore sampling error. Suppose that an SRS of 75 stores this month shows mean sales of 52 units of a small appliance, with standard deviation 13 units. During the same month last year, an SRS of 53 stores gave mean sales of 49 units, with standard deviation 11 units. An increase from 49 to 52 is a rise of 6%. The marketing manager is happy, because sales are up 6%.

(a) Use the two-sample t procedure to give a 95% confidence interval for the difference in mean number of units sold at all retail stores.

(b) Explain in language that the manager can understand why he cannot be certain that sales rose by 6%, and that in fact sales may even have dropped.

7.74 A study of iron deficiency among infants compared samples of infants following different feeding regimens. One group contained breast-fed infants, while the children in another group were fed a standard baby formula without any iron supplements. Here are summary results on blood hemoglobin levels at 12 months of age:[29]

Group	n	$\bar{x}$	s
Breast-fed	23	13.3	1.7
Formula	19	12.4	1.8

(a) Is there significant evidence that the mean hemoglobin level is higher among breast-fed babies? State H_0 and H_a and carry out a t test. Give the P-value. What is your conclusion?

(b) Give a 95% confidence interval for the mean difference in hemoglobin level between the two populations of infants.

(c) State the assumptions that your procedures in (a) and (b) require in order to be valid.

7.75 In a study of heart surgery, one issue was the effect of drugs called beta-blockers on the pulse rate of patients during surgery. The available subjects were divided at random into two groups of 30 patients each. One group received a beta-blocker; the other, a placebo. The pulse rate of each patient at a critical point during the operation was recorded. The treatment group had mean 65.2 and standard deviation 7.8. For the control group, the mean was 70.3 and the standard deviation was 8.3.

(a) Do beta-blockers reduce the pulse rate? State the hypotheses and do a t test. Is the result significant at the 5% level? At the 1% level?

(b) Give a 99% confidence interval for the difference in mean pulse rates.

7.76 The Survey of Study Habits and Attitudes (SSHA) is a psychological test designed to measure the motivation, study habits, and attitudes toward learning of college students. These factors, along with ability, are important in explaining success in school. Scores on the SSHA range from 0 to 200. A selective private college gives the SSHA to an SRS of both male and female first-year students. The data for the women are as follows:

154	109	137	115	152	140	154	178	101
103	126	126	137	165	165	129	200	148

Here are the scores of the men:

108	140	114	91	180	115	126	92	169	146
109	132	75	88	113	151	70	115	187	104

(a) Examine each sample graphically, with special attention to outliers and skewness. Is use of a t procedure acceptable for these data?

(b) Most studies have found that the mean SSHA score for men is lower than the mean score in a comparable group of women. Test this supposition here. That is, state hypotheses, carry out the test and obtain a P-value, and give your conclusions.

(c) Give a 90% confidence interval for the mean difference between the SSHA scores of male and female first-year students at this college.

7.77 Table 7.4 (page 539) gives data on the blood pressure before and after treatment for two groups of black males. One group took a calcium supplement, and the other group received a placebo. Example 7.20 compares the decrease in blood pressure in the two groups using pooled two-sample t procedures.

(a) Repeat the significance test using a two-sample t test that does not require equal population standard deviations. Compare your P-value with the result $P = 0.059$ for the pooled t test.

(b) Give a 90% confidence interval for the difference in means, again using a procedure that does not require equal standard deviations. How does the margin of error of your interval compare with that in Example 7.21?

7.78 Table 1.9 (page 59) gives data on the calories in a sample of brands of each of three kinds of hot dogs. Poultry hot dogs have fewer calories than either beef or meat hot dogs.

(a) Give a 95% confidence interval for the difference in mean calorie content between beef and poultry hot dogs.

(b) Based on your confidence interval, can the hypothesis that the population means are equal be rejected at the 5% significance level? Explain your answer.

(c) What assumptions does your statistical procedure in (a) require? Which of these assumptions are justified or not important in this case? Are any of the assumptions doubtful in this case?

7.79 Researchers studying the learning of speech often compare measurements made on the recorded speech of adults and children. One variable of interest is called the voice onset time (VOT). Here are the results for 6-year-old children and adults asked to pronounce the word "bees." The VOT is measured in milliseconds and can be either positive or negative.[30]

Group	n	$\bar{x}$	s
Children	10	−3.67	33.89
Adults	20	−23.17	50.74

(a) What is the standard error of the sample mean VOT for the 20 adult subjects? What is the standard error of the difference $\bar{x}_1 - \bar{x}_2$ between the mean VOT for children and adults?

(b) The researchers were investigating whether VOT distinguishes adults from children. State H_0 and H_a and carry out a two-sample t test. Give a P-value and report your conclusions.

(c) Give a 95% confidence interval for the difference in mean VOTs when pronouncing the word "bees." Explain why you knew from your result in (b) that this interval would contain 0 (no difference).

7.80 The researchers in the study discussed in the previous exercise looked at VOTs for adults and children pronouncing several different words. Explain why they should not perform a separate two-sample t test for each word and conclude that the words with a significant difference (say, $P < 0.05$) distinguish children from adults. (The researchers did not make this mistake.)

7.81 College financial aid offices expect students to use summer earnings to help pay for college. But how large are these earnings? One college studied this question by asking a sample of students how much they earned. Omitting students who were not employed, 1296 responses were received.[31] Here are the data in summary form:

Group	n	$\bar{x}$	s
Males	675	\$3297.91	\$2394.65
Females	621	\$2380.68	\$1815.55

(a) Use the two-sample t procedures to give a 90% confidence interval for the difference between the mean summer earnings of male and female students.

(b) The distribution of earnings is strongly skewed to the right. Nevertheless, use of t procedures is justified. Why?

(c) Once the sample size was decided, the sample was chosen by taking every kth name from an alphabetical list of undergraduates. Is it reasonable to consider the sample as two SRSs chosen from the male and female undergraduate populations?

(d) What other information about the study would you request before accepting the results as describing all undergraduates?

The following exercises concern optional material on the pooled two-sample t procedures and on the power of tests.

7.82 Exercise 7.56 (page 544) compares flat-screen monitors with standard screens. Reanalyze the data using the pooled procedure. Does the conclusion depend on the choice of the method? The standard deviations are quite different for these data, so we do not recommend use of the pooled procedures in this case.

7.83 The analysis of the loss of vitamin C when bread is stored (Exercise 7.60, page 545) is a rather unusual case involving very small sample sizes. There are only two observations per condition (immediately after baking and three days later). When the samples are so small, we have very little information to make a judgment about whether the population standard deviations are equal. The potential gain from pooling is large when the sample sizes are very small. Assume that we will perform a two-sided test using the 5% significance level.

(a) Find the critical value for the unpooled t test statistic that does not assume equal variances. Use the minimum of $n_1 - 1$ and $n_2 - 1$ for the degrees of freedom.

(b) Find the critical value for the pooled t test statistic.

(c) How does comparing these critical values show an advantage of the pooled test?

7.84 Suppose that in the setting of the previous exercise you are interested in 95% confidence intervals for the difference rather than significance testing. Find the widths of the intervals for the two procedures (assuming or not assuming equal standard deviations). How do they compare?

7.85 Example 7.17 (page 534) reports the analysis of some biological data. Starting from the computer's results for $\overline{x}_i$ and s_i, verify the values given for the test statistic $t = 2.99$ and the degrees of freedom df $= 5.9$.

7.86 Example 7.16 (page 533) analyzed the Chapin Social Insight Test scores of male and female college students. The sample standard deviations s_i for the two samples are almost identical. This suggests that the two populations have similar standard deviations σ_i. The use of the pooled two-sample t procedures is justified in this case. Repeat the analysis in Example 7.16 using the pooled procedures. Compare your results with those obtained in the example.

7.87 The pooled two-sample t procedures can be used to analyze the hemoglobin data in Exercise 7.74 because the very similar sample standard deviations suggest that the assumption of equal population standard deviations is justified. Repeat parts (a) and (b) of that exercise using the pooled t procedures. Compare your results with those you obtained with the two-sample t procedures in Exercise 7.74.

7.88 Repeat the comparison of mean VOTs for children and adults in Exercise 7.79 using a pooled t procedure. (In practice, we would not pool in this case, because the data suggest some difference in the population standard deviations.)

(a) Carry out the significance test, and give a P-value.

(b) Give a 95% confidence interval for the difference in population means.

(c) How similar are your results to those you obtained in Exercise 7.79 from the two-sample t procedures?

7.3 Optional Topics in Comparing Distributions*

In this section we discuss three topics that are related to the material that we have already covered in this chapter. If we can do inference for means, it is natural to ask if we can do something similar for spread. The answer is yes, but there are many cautions. We also discuss robustness and show how to find the power for the two-sample t test. If you plan to design studies, you should become familiar with this last topic.

Inference for population spread

The two most basic descriptive features of a distribution are its center and spread. In a normal population, these aspects are measured by the mean and the standard deviation. We have described procedures for inference about population means for normal populations and found that these procedures are often useful for nonnormal populations as well. It is natural to turn next to inference about the standard deviations of normal populations. Our recommendation here is short and clear: Don't do it without expert advice.

There are indeed inference procedures appropriate for the standard deviations of normal populations. We will describe the most common such procedure, the F test for comparing the spread of two normal populations. Unlike the t procedures for means, the F test and other procedures for standard deviations are extremely sensitive to nonnormal distributions. This lack of robustness does not improve in large samples. It is difficult in practice to tell whether a significant F-value is evidence of unequal population spreads or simply evidence that the populations are not normal.

The deeper difficulty that underlies the very poor robustness of normal population procedures for inference about spread already appeared in our work on describing data. The standard deviation is a natural measure of spread for normal distributions but not for distributions in general. In fact, because skewed distributions have unequally spread tails, no single numerical measure is adequate to describe the spread of a skewed distribution. Thus, the standard deviation is not always a useful parameter, and even when it is, the results of inference about it are not trustworthy. Consequently, we do not recommend use of inference about population standard deviations in basic statistical practice.[32]

It was once common to test equality of standard deviations as a preliminary to performing the pooled two-sample t test for equality of two population means. It is better practice to check the distributions graphically, with special attention to skewness and outliers, and to use the software-based two-sample t that does not require equal standard deviations.

Chapters 12 and 13 discuss analysis of variance procedures for comparing several means. These procedures are extensions of the pooled t test and require that the populations have a common standard deviation. Fortunately, like the t test, analysis of variance comparisons of means are quite robust.

*This section can be omitted without loss of continuity.

(Analysis of variance uses F statistics, but these are not the same as the F statistic for comparing two population standard deviations.) Formal tests for the assumption of equal standard deviations are not very robust and will often give misleading results. In the words of one distinguished statistician, "To make a preliminary test on variances is rather like putting to sea in a rowing boat to find out whether conditions are sufficiently calm for an ocean liner to leave port!"[33]

The *F* test for equality of spread

Because of the limited usefulness of procedures for inference about the standard deviations of normal distributions, we will present only one such procedure. Suppose that we have independent SRSs from two normal populations, a sample of size n_1 from $N(\mu_1, \sigma_1)$ and a sample of size n_2 from $N(\mu_2, \sigma_2)$. The population means and standard deviations are all unknown. The hypothesis of equal spread,

$$H_0: \sigma_1 = \sigma_2$$

$$H_a: \sigma_1 \neq \sigma_2$$

is tested by a simple statistic, the ratio of sample variances.

The *F* Statistic and *F* Distributions

When s_1^2 and s_2^2 are sample variances from independent SRSs of sizes n_1 and n_2 drawn from normal populations, the F statistic

$$F = \frac{s_1^2}{s_2^2}$$

has the F distribution with $n_1 - 1$ and $n_2 - 1$ degrees of freedom when $H_0: \sigma_1 = \sigma_2$ is true.

F distributions

The ***F* distributions** are a family of distributions with two parameters, the degrees of freedom of the sample variances in the numerator and denominator of the F statistic.* The numerator degrees of freedom are always mentioned first. Interchanging the degrees of freedom changes the distribution, so the order is important. Our brief notation will be $F(j, k)$ for the F distribution with j degrees of freedom in the numerator and k in the denominator. The F distributions are not symmetric but are right-skewed. The density curve in Figure 7.15 illustrates the shape. Because sample variances cannot be negative, the F statistic takes only positive values and the F distribution has no

*The F distributions are another of R. A. Fisher's contributions to statistics and are called F in his honor. Fisher introduced F statistics for comparing several means. We will meet these useful statistics in later chapters.

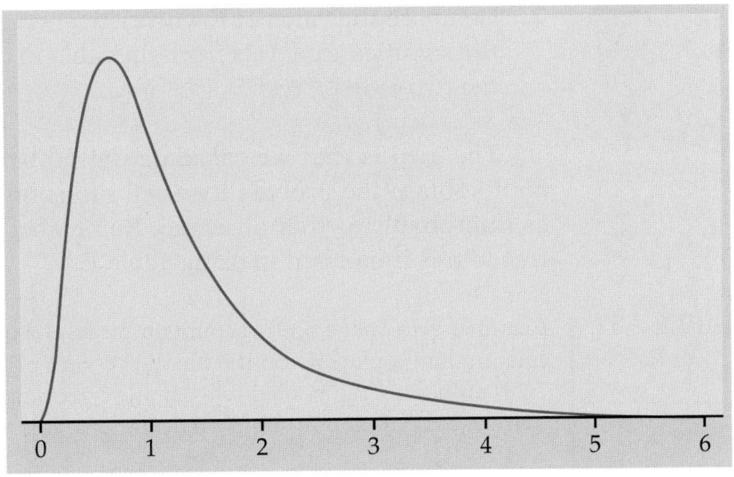

FIGURE 7.15 The density curve for the $F(9, 10)$ distribution. The F distributions are skewed to the right.

probability below 0. The peak of the F density curve is near 1; values far from 1 in either direction provide evidence against the hypothesis of equal standard deviations.

Tables of F critical values are awkward, because a separate table is needed for every pair of degrees of freedom j and k. Table E in the back of the book gives upper p critical values of the F distributions for $p = 0.10, 0.05, 0.025, 0.01$, and 0.001. For example, these critical values for the $F(9, 10)$ distribution shown in Figure 7.15 are

p	0.10	0.05	0.025	0.01	0.001
F^*	2.35	3.02	3.78	4.94	8.96

The skewness of the F distributions causes additional complications. In the symmetric normal and t distributions, the point with probability 0.05 below it is just the negative of the point with probability 0.05 above it. This is not true for F distributions. We therefore require either tables of both the upper and lower tails or means of eliminating the need for lower-tail critical values. Statistical software that eliminates the need for tables is plainly very convenient. If you do not use statistical software, arrange the F test as follows:

1. Take the test statistic to be

$$F = \frac{\text{larger } s^2}{\text{smaller } s^2}$$

This amounts to naming the populations so that s_1^2 is the larger of the observed sample variances. The resulting F is always 1 or greater.

2. Compare the value of F with critical values from Table E. Then *double* the significance levels from the table to obtain the significance level for the two-sided F test.

The idea is that we calculate the probability in the upper tail and double to obtain the probability of all ratios on either side of 1 that are at least as improbable as that observed. Remember that the order of the degrees of freedom is important in using Table E.

EXAMPLE 7.22

Example 7.19 (page 539) recounts a medical experiment comparing the effects of calcium and a placebo on the blood pressure of black men. The analysis (Example 7.20) employed the pooled two-sample t procedures. Because these procedures require equal population standard deviations, it is tempting to first test

$$H_0: \sigma_1 = \sigma_2$$

$$H_a: \sigma_1 \neq \sigma_2$$

The larger of the two sample standard deviations is $s = 8.743$ from 10 observations. The other is $s = 5.901$ from 11 observations. The two-sided test statistic is therefore

$$F = \frac{\text{larger } s^2}{\text{smaller } s^2} = \frac{8.743^2}{5.901^2} = 2.195$$

We compare the calculated value $F = 2.20$ with critical points for the $F(9, 10)$ distribution. Table E shows that 2.20 is *less* than the 0.10 critical value of the $F(9, 10)$ distribution, which is $F^* = 2.35$. Doubling 0.10, we know that the observed F falls short of the 0.20 critical value. The results are not significant at the 20% level (or any lower level). Statistical software shows that the exact upper-tail probability is 0.118, and hence $P = 0.236$. *If* the populations were normal, the observed standard deviations would give little reason to suspect unequal population standard deviations. Because one of the populations shows some nonnormality, however, we cannot be fully confident of this conclusion.

Robustness of normal inference procedures

We have claimed that

- The t procedures for inference about means are quite robust against nonnormal population distributions. These procedures are particularly robust when the population distributions are symmetric and (for the two-sample case) when the two sample sizes are equal.
- The F test and other procedures for inference about variances are so lacking in robustness as to be of little use in practice.

Simulations with a large variety of nonnormal distributions support these claims. Note 4 at the end of this chapter gives references. One set of simulations was carried out with samples of size 25 and used significance tests with fixed level $\alpha = 0.05$. The three types of tests studied were the one-sample and pooled two-sample t tests and the F test for comparing two variances.

The robustness of the one-sample and two-sample t procedures is remarkable. The true significance level remains between about 4% and 6% for a large range of populations. The t test and the corresponding confidence intervals are among the most reliable tools that statisticians use. Remember, however, that outliers can greatly disturb the t procedures. Also, two-sample procedures are less robust when the sample sizes are not similar.

The lack of robustness of the tests for variances is equally remarkable. The true significance levels depart rapidly from the target 5% as the population distribution departs from normality. The two-sided F test carried out with 5% critical values can have a true level of less than 1% or greater than 11% even in symmetric populations with no outliers. Results such as these are the basis for our recommendation that these procedures not be used.

The power of the two-sample *t* test

The two-sample t test is one of the most used statistical procedures. Unfortunately, because of inadequate planning, users frequently fail to find evidence for the effects that they believe to be true. Power calculations should be part of the planning of any statistical study. Information from a pilot study or previous research is needed.

In Section 7.1 (optional material), we learned how to find an approximation for the power of the one-sample t test. The basic concepts for the two-sample case are the same. Here, we give the exact method, which involves a new distribution, the **noncentral *t* distribution.** To perform the calculations, we simply need software to calculate probabilities for this distribution.

noncentral t distribution

We first present the method for the pooled two-sample t test, where the parameters are μ_1, μ_2, and the common standard deviation σ. Modifications to get approximate results when we do not pool are then described.

To find the power for the pooled two-sample t test, use the following steps. We consider only the case where the null hypothesis is $\mu_1 - \mu_2 = 0$.

1. Specify
   (a) an alternative that you consider important to detect; that is, a value for $\mu_1 - \mu_2$;
   (b) the sample sizes, n_1 and n_2;
   (c) the Type I error for a fixed significance level, α;
   (d) a guess at the standard deviation, σ.
2. Find the degrees of freedom df $= n_1 + n_2 - 2$ and the value of t^* that will lead to rejection of H_0.
3. Calculate the **noncentrality parameter**

noncentrality parameter

$$\delta = \frac{|\mu_1 - \mu_2|}{\sigma\sqrt{\dfrac{1}{n_1} + \dfrac{1}{n_2}}}$$

4. Find the power as the probability that a noncentral t random variable with degrees of freedom df and noncentrality parameter δ will be less than t^*. In SAS this is 1-PROBT(tstar,df,delta). If you do not have software that can perform this calculation, you can approximate the power as the probability that a standard normal random variable is greater than $t^* - \delta$, that is, $P(z > t^* - \delta)$, and use Table A.

Note that the denominator in the noncentrality parameter,

$$\sigma\sqrt{\frac{1}{n_1} + \frac{1}{n_2}}$$

is our guess at the standard error for the difference in the sample means. Therefore, if we wanted to assess a possible study in terms of the margin of error for the estimated difference, we would examine t^* times this quantity.

If we do not assume that the standard deviations are equal, we need to guess both standard deviations and then combine these for our guess at the standard error:

$$\sqrt{\frac{\sigma_1^2}{n_1} + \frac{\sigma_2^2}{n_2}}$$

This guess is then used in the denominator of the noncentrality parameter. For the degrees of freedom, the conservative approximation is appropriate.

EXAMPLE 7.23

In Example 7.20 we examined the effect of calcium on blood pressure by comparing the means of a treatment group and a placebo group using a two-sample t test. The P-value was 0.059, failing to achieve the usual standard of 0.05 for statistical significance. Suppose that we wanted to plan a new study that would provide convincing evidence, say at the 0.01 level, with high probability. Let's examine a study design with 45 subjects in each group ($n_1 = n_2 = 45$). Based on our previous results we choose $\mu_1 - \mu_2 = 5$ as an alternative that we would like to be able to detect with $\alpha = 0.01$. For σ we use 7.4, our pooled estimate from Example 7.20. The degrees of freedom are $n_1 + n_2 - 2 = 88$ and $t^* = 2.37$ for the significance test. The noncentrality parameter is

$$\delta = \frac{5}{7.4\sqrt{\frac{1}{45} + \frac{1}{45}}} = \frac{5}{1.56} = 3.21$$

Software gives the power as 0.7965, or 80%. The normal approximation gives 0.7983, a very accurate result. With this choice of sample sizes we would expect the margin of error for a 95% confidence interval ($t^* = 1.99$) for the difference in means to be

$$t^* \times 7.4\sqrt{\frac{1}{45} + \frac{1}{45}} = 1.99 \times 1.56 = 3.1$$

With software it is very easy to examine the effects of variations on a study design. In the above example, we might want to examine the power for $\alpha = 0.05$ and the effects of reducing the sample sizes.

SUMMARY

Inference procedures for comparing the standard deviations of two normal populations are based on the ***F* statistic,** which is the ratio of sample variances:

$$F = \frac{s_1^2}{s_2^2}$$

If an SRS of size n_1 is drawn from the x_1 population and an independent SRS of size n_2 is drawn from the x_2 population, the F statistic has the ***F* distribution** $F(n_1 - 1, n_2 - 1)$ if the two population standard deviations σ_1 and σ_2 are in fact equal.

The ***F* test for equality of standard deviations** tests H_0: $\sigma_1 = \sigma_2$ versus H_a: $\sigma_1 \neq \sigma_2$ using the statistic

$$F = \frac{\text{larger } s^2}{\text{smaller } s^2}$$

and doubles the upper-tail probability to obtain the P-value.

The t procedures are quite **robust** when the distributions are not normal. The F tests and other procedures for inference about the spread of one or more normal distributions are so strongly affected by nonnormality that we do not recommend them for regular use.

The **power** of the pooled two-sample t test is found by first computing the critical value for the significance test, the degrees of freedom, and the **noncentrality parameter** for the alternative of interest. These are used to find the power from the **noncentral *t* distribution.** A normal approximation works quite well. Calculating margins of error for various study designs and assumptions is an alternative procedure for evaluating designs.

SECTION 7.3 EXERCISES

In all exercises calling for use of the F test, assume that both population distributions are very close to normal. The actual data are not always sufficiently normal to justify use of the F test.

7.89 An F statistic will be used to compare two variances. The sample sizes are both 20. How large does the ratio of the largest to the smallest variance need to be for the significance test to reject the null hypothesis that the population variances are the same?

7.90 The F statistic $F = s_1^2/s_2^2$ is calculated from samples of size $n_1 = 16$ and $n_2 = 20$. (Remember that n_1 is the numerator sample size.)

(a) What is the upper 5% critical value for this F?

(b) In a test of equality of standard deviations against the two-sided alternative, this statistic has the value $F = 2.71$. Is this value significant at the 5% level? Is it significant at the 1% level?

7.91 The F statistic for equality of standard deviations based on samples of sizes $n_1 = 31$ and $n_2 = 28$ takes the value $F = 1.72$.

(a) Is this significant evidence of unequal population standard deviations at the 5% level?

(b) Use Table E to give an upper and a lower bound for the P-value.

7.92 Exercise 7.58 (page 544) compares the rents of one-bedroom and two-bedroom apartments. Is there any evidence in the data that would lead us to conclude that the standard deviations are different? State the appropriate hypotheses, calculate the test statistic, and write a short summary of the results.

7.93 In Exercise 7.56 (page 544) you compared employee satisfaction with two types of computer monitor. Is the variability of the data the same in the two groups? Write null and alternative hypotheses for this question. Perform the significance test and state your conclusion.

7.94 Refer to Exercise 7.65 (page 546), where we examined the effect of piano lessons on spatial-temporal reasoning. Do the data provide evidence that would cause us to suspect that the standard deviation of the children who took piano lessons is different from that of the controls? Set up the hypotheses, perform the significance test, and summarize the results.

7.95 A USDA survey used to estimate wheat prices in July and September is described in Exercise 7.71 (page 548). Using the standard deviations you calculated there, perform the test for equality of standard deviations and summarize your conclusion.

7.96 We studied the loss of vitamin C when bread is stored in Exercise 7.60 (page 545). Recall that two loaves were measured immediately after baking and another two loaves were measured after three days of storage. These are very small sample sizes.

(a) Use Table E to find the value that the ratio of variances would have to exceed for us to reject the null hypothesis (at the 5% level) that the standard deviations are equal. What does this suggest about the power of the test?

(b) Perform the test and state your conclusion.

7.97 Return to the SSHA data in Exercise 7.76 (page 549). SSHA scores are generally less variable among women than among men. We want to know whether this is true for this college.

(a) State H_0 and H_a. Note that H_a is one-sided in this case.

(b) Because Table E contains only upper critical values for F, a one-sided test requires that in calculating F the numerator s^2 belongs to the group that H_a claims to have the larger σ. Calculate this F.

(c) Compare F to the entries in Table E (no doubling of p) to obtain the P-value. Be sure the degrees of freedom are in the proper order. What do you conclude about the variation in SSHA scores?

7.98 The data for VOTs of children and adults in Exercise 7.79 (page 550) show quite different sample standard deviations. How statistically significant is the observed inequality?

7.99 In Exercise 7.70 (page 548) data on cocaine use and birth weight are summarized. The study has been criticized because of several design problems. Suppose that you are designing a new study. Based on the results in Exercise 7.70, you think that the true difference in mean birth weights may be about 300 grams (g); a difference this large is clinically important. For planning purposes assume that you will have 80 women in each group and that the common standard deviation is 650 g, a guess that is between the two standard deviations in Exercise 7.70. If you use a pooled two-sample t test with a Type I error of 0.05, what is the power of the test for this design?

7.100 Refer to the previous exercise. Repeat the power calculation, for 20, 40, 60, 80, 100, and 120 women in each group. Plot the power versus the sample size and write a short summary of the results.

7.101 Refer to the previous two exercises. For each of the sample sizes considered, what is your guess at the margin of error for the 95% confidence interval for the difference in mean weights? Display these results with a graph or a sketch.

7.102 Suppose that you wanted to compare intramural basketball players and intramural soccer players on the "ego strength" personality factor described in Exercise 7.69 (page 547). With the data from that exercise, you will use $\sigma = 0.7$ for planning purposes. The pooled two-sample t test with $\alpha = 0.05$ will be used to make the comparison. Based on Exercise 7.69, you judge a difference of 0.5 points to be of interest. Pick several values of n and find the power. Plot the power versus n and use the plot to find a value of n that will give approximately 80% power. Calculate the power for the value of n that you found.

CHAPTER 7 EXERCISES

7.103 A retailer entered into an exclusive agreement with a supplier who guaranteed to provide all products at competitive prices. The retailer eventually began to purchase supplies from other vendors who offered better prices. The original supplier filed a legal action claiming violation of the agreement. In defense, the retailer had an audit performed on a random sample of invoices. For each audited invoice, all purchases made from other suppliers were examined and the prices were compared with those offered by the original supplier. For each invoice, the percent of purchases for which the alternate supplier offered a lower price than the original supplier was recorded.[34] Here are the data:

100	0	0	100	33	45	100	34	78	100	77	33	100
69	100	89	100	100	100	100	100	100	100	100	100	

Report the average of the percents with a 95% margin of error. Do the sample invoices suggest that the original supplier's prices are not competitive on the average?

7.104 A manufacturer of electric motors tests insulation at a high temperature (250° C) and records the number of hours until the insulation fails. The data for 5 specimens are[35]

300 324 372 372 444

The small sample size makes judgment from the data difficult, but engineering experience suggests that the logarithm of the failure time will have a normal distribution. Take the logarithms of the 5 observations, and use t procedures to give a 90% confidence interval for the mean of the log failure time for insulation of this type.

7.105 Some research suggests that women perform better than men in school but men score higher on standardized tests. Table 1.6 (page 33) presents data on a measure of school performance, grade point average (GPA), and a standardized test, IQ, for 78 seventh-grade students. Do these data lend further support to the previously found gender differences? Give graphical displays of the data and describe the distributions. Use significance tests and confidence intervals to examine this question, and prepare a short report summarizing your findings.

7.106 Table 1.10 (page 97) gives positions and weights (in pounds) for players on a major college football team. Assume that these players can be viewed as a random sample of players from similar teams. Use a significance test and a confidence interval to compare the mean weights of offensive line (OL) and defensive line (DL) players. Give a summary of your results with a graphical display of the data.

7.107 On the morning of March 5, 1996, a train with 14 tankers of propane derailed near the center of the small Wisconsin town of Weyauwega. Six of the tankers were ruptured and burning when the 1700 residents were ordered to evacuate the town. Researchers study disasters like this so that effective relief efforts can be designed for future disasters. About half of the households with pets did not evacuate all of their pets. A study conducted after the derailment focused on problems associated with retrieval of the pets after the evacuation and characteristics of the pet owners. One of the scales measured "commitment to adult animals," and the people who evacuated all or some of their pets were compared with those who did not evacuate any of their pets. Higher scores indicate that the pet owner is more likely to take actions that benefit the pet.[36] Here are the data summaries:

Group	n	$\bar{x}$	s
Evacuated all or some pets	116	7.95	3.62
Did not evacuate any pets	125	6.26	3.56

Analyze the data and prepare a short report describing the results.

7.108 In a study of the effectiveness of weight-loss programs, 47 subjects who were at least 20% overweight took part in a group support program for 10 weeks. Private weighings determined each subject's weight at the beginning of the program and 6 months after the program's end. The matched pairs t test was used to assess the significance of the average weight loss. The paper reporting the study said, "The subjects lost a significant amount of weight over time, $t(46) = 4.68, p < 0.01$." It is common to report the results of statistical tests in this abbreviated style.[37]

(a) Why was the matched pairs statistic appropriate?

(b) Explain to someone who knows no statistics but is interested in weight-loss programs what the practical conclusion is.

(c) The paper follows the tradition of reporting significance only at fixed levels such as $\alpha = 0.01$. In fact, the results are more significant than "$p < 0.01$" suggests. What can you say about the P-value of the t test?

7.109 Nitrites are often added to meat products as preservatives. In a study of the effect of these chemicals on bacteria, the rate of uptake of a radiolabeled amino acid was measured for a number of cultures of bacteria, some growing in a medium to which nitrites had been added. Here are the summary statistics from this study:

Group	n	$\bar{x}$	s
Nitrite	30	7880	1115
Control	30	8112	1250

Carry out a test of the research hypothesis that nitrites decrease amino acid uptake, and report your results.

7.110 The one-hole test is used to test the manipulative skill of job applicants. This test requires subjects to grasp a pin, move it to a hole, insert it, and return for another pin. The score on the test is the number of pins inserted in a fixed time interval. In one study, male college students were compared with experienced female industrial workers. Here are the data for the first minute of the test:[38]

Group	n	$\overline{x}$	s
Students	750	35.12	4.31
Workers	412	37.32	3.83

(a) It was expected that the experienced workers would outperform the students, at least during the first minute, before learning occurs. State the hypotheses for a statistical test of this expectation and perform the test. Give a P-value and state your conclusions.

(b) The distribution of scores is slightly skewed to the left. Explain why the procedure you used in (a) is nonetheless acceptable.

(c) One purpose of the study was to develop performance norms for job applicants. Based on the data above, what is the range that covers the middle 95% of experienced workers? (Be careful! This is not the same as a 95% confidence interval for the mean score of experienced workers.)

(d) The five-number summary of the distribution of scores among the workers is

23 33.5 37 40.5 46

for the first minute and

32 39 44 49 59

for the fifteenth minute of the test. Display these facts graphically, and describe briefly the differences between the distributions of scores in the first and fifteenth minute.

7.111 The composition of the earth's atmosphere may have changed over time. One attempt to discover the nature of the atmosphere long ago studies the gas trapped in bubbles inside ancient amber. Amber is tree resin that has hardened and been trapped in rocks. The gas in bubbles within amber should be a sample of the atmosphere at the time the amber was formed. Measurements on specimens of amber from the late Cretaceous era (75 to 95 million years ago) give these percents of nitrogen:

63.4 65.0 64.4 63.3 54.8 64.5 60.8 49.1 51.0

These values are quite different from the present 78.1% of nitrogen in the atmosphere. Assume (this is not yet agreed on by experts) that these observations are an SRS from the late Cretaceous atmosphere.[39]

(a) Graph the data, and comment on skewness and outliers.

(b) The t procedures will be only approximate in this case. Give a 90% t confidence interval for the mean percent of nitrogen in ancient air.

7.112 Do various occupational groups differ in their diets? A British study of this question compared 98 drivers and 83 conductors of London double-decker buses.[40] The conductors' jobs require more physical activity. The article reporting the study gives the data as "Mean daily consumption ($\pm$ se)." Some of the study results appear below:

	Drivers	Conductors
Total calories	2821 ± 44	2844 ± 48
Alcohol (grams)	0.24 ± 0.06	0.39 ± 0.11

(a) What does "se" stand for? Give $\bar{x}$ and s for each of the four sets of measurements.

(b) Is there significant evidence at the 5% level that conductors consume more calories per day than do drivers? Use the two-sample t method to give a P-value, and then assess significance.

(c) How significant is the observed difference in mean alcohol consumption? Use two-sample t methods to obtain the P-value.

(d) Give a 95% confidence interval for the mean daily alcohol consumption of London double-decker bus conductors.

(e) Give a 99% confidence interval for the difference in mean daily alcohol consumption between drivers and conductors.

7.113 Use of the pooled two-sample t test is justified in part (b) of the previous exercise. Explain why. Find the P-value for the pooled t statistic, and compare with your result in the previous exercise.

7.114 The report cited in Exercise 7.112 says that the distribution of alcohol consumption among the individuals studied is "grossly skew."

(a) Do you think that this skewness prevents the use of the two-sample t test for equality of means? Explain your answer.

(b) Do you think that the skewness of the distributions prevents the use of the F test for equality of standard deviations? Explain your answer.

7.115 Table 1.3 (page 31) gives the number of medical doctors per 100,000 population by state. Is it proper to apply the one-sample t method to these data to give a 95% confidence interval for the mean number of medical doctors per 100,000 population per state? Explain your answer.

7.116 The amount of lead in a certain type of soil, when released by a standard extraction method, averages 86 parts per million (ppm). A new extraction method is tried on 40 specimens of the soil, yielding a mean of 83 ppm lead and a standard deviation of 10 ppm.

(a) Is there significant evidence at the 5% level that the new method frees less lead from the soil? What about the 1% level?

(b) A critic argues that because of variations in the soil, the effectiveness of the new method is confounded with characteristics of the particular soil specimens used. Briefly describe a better data production design that avoids this criticism.

7.117 High levels of cholesterol in the blood are not healthy in either humans or dogs. Because a diet rich in saturated fats raises the cholesterol level, it is plausible that dogs owned as pets have higher cholesterol levels than dogs owned by a veterinary research clinic. "Normal" levels of cholesterol based on the clinic's dogs would then be misleading. A clinic compared healthy dogs it owned with healthy pets brought to the clinic to be neutered.[41] The summary statistics for blood cholesterol levels (milligrams per deciliter of blood) appear below:

Group	n	$\bar{x}$	s
Pets	26	193	68
Clinic	23	174	44

(a) Is there strong evidence that pets have a higher mean cholesterol level than clinic dogs? State the H_0 and H_a and carry out an appropriate test. Give the P-value and state your conclusion.

(b) Give a 95% confidence interval for the difference in mean cholesterol levels between pets and clinic dogs.

(c) Give a 95% confidence interval for the mean cholesterol level in clinic dogs.

(d) What assumptions must be satisfied to justify the procedures you used in (a), (b), and (c)? Assuming that the cholesterol measurements have no outliers and are not strongly skewed, what is the chief threat to the validity of the results of this study?

7.118 Exercise 2.73 (page 175) and Table 2.12 (page 176) give data on the concentration of airborne particulate matter in a rural area upwind from a small city and in the center of the city. In that exercise, the focus was on using the rural readings to predict the city readings for the same day. Now we want to compare the mean level of particulates in the city and in the rural area. We suspect that pollution is higher in the city and hope to find evidence for this suspicion.

(a) State H_0 and H_a.

(b) Which type of t procedure is appropriate: one-sample, matched pairs, or two-sample?

(c) Make a graph to check for outliers or strong skewness that might prevent the use of t procedures. Your graph should reflect the type of procedure that you will use.

(d) Carry out the appropriate t test. Give the P-value and report your conclusion.

(e) Give a 95% confidence interval for the mean amount by which the city particulate level exceeds the rural level.

7.119 The sign test allows us to assess whether city particulate levels are higher than nearby rural levels on the same day without the use of normal distributions. Carry out a sign test for the data used in the previous exercise. State H_0 and H_a and give the P-value and your conclusion.

7.120 Exercise 1.27 (page 30) gives 29 measurements of the density of the earth made in 1798 by Henry Cavendish. Display the data graphically to check for skewness and outliers. Then give an estimate for the density of the earth from Cavendish's data and a margin of error for your estimate.

7.121 Elite distance runners are thinner than the rest of us. Here are data on skinfold thickness, which indirectly measures body fat, for 20 elite runners and 95 ordinary men in the same age group.[42] The data are in millimeters and are given in the form "mean (standard deviation)."

	Runners	Others
Abdomen	7.1 (1.0)	20.6 (9.0)
Thigh	6.1 (1.8)	17.4 (6.6)

Use confidence intervals to describe the difference between runners and typical young men.

7.122 Table 7.2 (page 508) gives the levels of three pollutants in the exhaust of 46 randomly selected vehicles of the same type. You will investigate emissions of nitrogen oxides (NOX).

(a) Make a stemplot and, if your software allows, a normal quantile plot of the NOX levels. Do the plots suggest that the distribution of NOX emissions is approximately normal? Can you safely employ t procedures to analyze these data?

(b) Give a 95% confidence interval for the mean NOX level in vehicles of this type.

(c) Your supervisor hopes the average NOX level is less than 1 gram per mile. You will have to tell him that it's not so. Carry out a significance test to assess the strength of the evidence that the mean NOX level is greater than 1 and then write a short report to your supervisor based on your work in (b) and (c). (Your supervisor has never heard of P-values, so you must use plain language.)

7.123 Is there a difference between the average SAT scores of males and females? The CSDATA data set gives the math (SATM) and verbal (SATV) scores for a group of 224 computer science majors. The variable SEX indicates whether each individual is male or female.

(a) Compare the two distributions graphically, and then use the two-sample t test to compare the average SATM scores of males and females. Is it appropriate to use the pooled t test for this comparison? Write a brief summary of your results and conclusions that refers to both versions of the t test and to the F test for equality of standard deviations. Also give a 95% confidence interval for the difference in the means.

(b) Answer part (a) for the SATV score.

(c) The students in the CSDATA data set were all computer science majors who began college during a particular year. To what extent do you think that your results would generalize to (*i*) computer science students entering in different years, (*ii*) computer science majors at other colleges and universities, and (*iii*) college students in general?

7.124 In the READING data set the response variable Post3 is to be compared for three methods of teaching reading. The Basal method is the standard, or control, method, and the two new methods are DRTA and Strat. We can use the methods of this chapter to compare Basal with DRTA and Basal with Strat. Note that to make comparisons among three treatments it is more appropriate to use the procedures that we will learn in Chapter 12.

(a) Is the mean reading score with the DRTA method higher than that for the Basal method? Perform an analysis to answer this question and summarize your results.

(b) Answer part (a) for the Strat method in place of DRTA.

7.125 Example 7.13 (page 527) tells us that the mean height of 10-year-old girls is $N(54.5, 2.7)$ and for boys it is $N(54.1, 2.4)$. The null hypothesis that the mean heights of 10-year-old boys and girls are equal is clearly false. The difference in mean heights is $54.5 - 54.1 = 0.4$ inch. Small differences such as this can require large sample sizes to detect. To simplify our calculations, let's assume that the standard deviations are the same, say $\sigma = 2.5$, and that we will measure the heights of an equal number of girls and boys. How many would we need to measure to have a 90% chance of detecting the (true) alternative hypothesis?

7.126 In Exercise 7.102 (page 561) you found the power for a study designed to compare the "ego strengths" of intramural basketball players and soccer players. Here you will design a study to compare intramural basketball players and volleyball players. Assume the same value of $\sigma = 0.7$ and use $\alpha = 0.05$. You are planning to have 25 players in each group. Calculate the power of the pooled two-sample t test that compares the mean ego strengths of players who participate in these two sports for different values of the true difference. Include values that have a very small chance of being detected and some that are virtually certain to be seen in your sample. Plot the power versus the true difference and write a short summary of what you have found.

7.127 As the degrees of freedom increase, the t distributions get closer and closer to the z ($N(0, 1)$) distribution. One way to see this is to look at how the value of t^* for a 95% confidence interval changes with the degrees of freedom. Make a plot with degrees of freedom from 2 to 100 on the x axis and t^* on the y axis. Draw a horizontal line on the plot corresponding to the value of $z^* = 1.96$. Summarize the main features of the plot.

7.128 Refer to the previous exercise. Make a similar plot and summarize its features for the value of t^* for a 90% confidence interval.

7.129 The margin of error for a confidence interval depends on the confidence level, the standard deviation, and the sample size. Fix the confidence level at 95% and the standard deviation at 1 to examine the effect of the sample size. Find the margin of error for sample sizes of 5 to 100 by 5's—that is, let

$n = 5, 10, 15, \ldots, 100$. Plot the margins of error versus the sample size and summarize the relationship.

7.130 Refer to the previous exercise. Make a similar plot and summarize its features for a 99% confidence interval.

7.131 How much more would you expect to pay for a home that has four bedrooms than for a home that has three? Here are some data for West Lafayette, Indiana.[43] These are the selling prices (in dollars) that the owners of the homes are asking.

Four-bedroom homes:

121,900	139,900	157,000	159,900	176,900	224,900	235,000
245,000	294,000					

Three-bedroom homes:

65,500	79,900	79,900	79,900	82,900	87,900	94,000
97,500	105,000	111,900	116,900	117,900	119,900	122,900
124,000	125,000	126,900	127,900	127,900	127,900	132,900
145,000	145,500	157,500	194,000	205,900	259,900	265,000

(a) Plot the selling prices for the two sets of homes and describe the two distributions.

(b) Test the null hypothesis that the mean asking prices for the two sets of homes are equal versus the two-sided alternative. Give the test statistic with degrees of freedom, the *P*-value, and your conclusion.

(c) Would you consider using a one-sided alternative for this analysis? Explain why or why not.

(d) Give a 95% confidence interval for the difference in mean selling prices.

(e) These data are not SRSs from a population. Give a justification for use of the two-sample t procedures in this case.

7.132 Go to the Website `www.realtor.com` and select two geographical areas of interest to you. You will compare the prices of similar types of homes in these two areas. State clearly how you define the areas and the type of homes. For example, you can use city names or zip codes to define the area and you can select single-family homes or condominiums. We view these homes as representative of the asking prices of homes for these areas at the time of your search. If the search gives a large number of homes, select a random sample. Be sure to explain exactly how you do this. Use the methods you have learned in this chapter to compare the selling prices. Be sure to include a graphical summary.

Prelude: jury selection

In a small border county in Texas, a so-called key man system was used to select citizens to serve on juries. Under this procedure, lists of persons in the county were compiled by jury commissioners and these were interviewed under oath by the county judge, who determined whether or not they were qualified. Qualifications included literacy and "good moral character." This system was challenged in a case called *Castaneda v. Partida* on the basis that it reduced the proportion of Hispanic citizens serving on juries. A statistical comparison was at the heart of the case: 79% of the county's adult population had Spanish surnames, but only 39% of the grand jurors did. This comparison gave a z statistic of 29 with an extremely small P-value. The United States Supreme Court concluded that this disparity established a prima facie case.[1] Procedures such as the key man system are no longer used to select juries in the United States. The Court interpreted the z statistic as the number of standard deviations by which the two percents differed: 79% differed from 39% by 29 standard deviations. The decision stated that a difference *greater than 2 or 3 standard deviations* is sufficient. This criterion is sometimes called the *Castaneda rule*. These values of z correspond to P-values of 0.0456 and 0.0026.

CHAPTER 8

Johannes Itten, *Kind Light.*

Inference for Proportions

Introduction

Many statistical studies produce counts rather than measurements. An opinion poll asks a sample of adults whether they approve of the president's conduct of his office; the data are the counts of "Yes," "No," and "Don't know." An experiment compares the effectiveness of four treatments to prevent the common cold; the data are the number of subjects given each treatment who catch a cold during the next month. A researcher classifies each of a sample of students according to his or her major field of study and political preference; the data are the counts of students in each cell of a two-way table. This chapter presents procedures for statistical inference in these settings.

The parameters we want to do inference about are population proportions. Just as in the case of inference about population means, we may be concerned with a single population or with comparing two populations. Inference about proportions in these one-sample and two-sample settings is very similar to inference about means, which we discussed in Chapter 7.

We begin in Section 8.1 with inference about a single population proportion. The statistical model for a count is then the binomial distribution, which we studied in Section 5.1. Section 8.2 concerns methods for comparing two proportions. Binomial distributions again play an important role.

8.1 Inference for a Single Proportion

We want to estimate the proportion p of some characteristic, such as approving the president's conduct of his office, among the members of a large population. We select a simple random sample (SRS) of size n from the population and record the count X of "successes" (such as "Yes" answers to a question about the president). We will use "success" as shorthand for whatever characteristic we are interested in. The sample proportion of successes $\hat{p} = X/n$ estimates the unknown population proportion p. If the population is much larger than the sample (say, at least 10 times as large), the individual responses are nearly independent and the count X has approximately the binomial distribution $B(n, p)$. In statistical terms, we are concerned with inference about the probability p of a success in the binomial setting.

If the sample size n is small, we must base tests and confidence intervals for p on the binomial distributions. These are awkward to work with because of the discreteness of the binomial distributions.[2] But we know that when the sample is large, both the count X and the sample proportion $\hat{p}$ are approximately normal. We will consider only inference procedures based on the normal approximation. These procedures are similar to those for inference about the mean of a normal distribution.

Confidence interval for a single proportion

The unknown population proportion p is estimated by the sample proportion $\hat{p} = X/n$. We know (Chapter 5, page 376) that if the sample size n is sufficiently large, $\hat{p}$ has approximately the normal distribution, with mean $\mu_{\hat{p}} = p$ and standard deviation $\sigma_{\hat{p}} = \sqrt{p(1-p)/n}$.

The traditional confidence interval for p is based on the normal approximation to the distribution of $\hat{p}$ with p replaced by $\hat{p}$ in the expression for $\sigma_{\hat{p}}$. Unfortunately, modern computer studies reveal that confidence intervals based on this approach can be quite inaccurate, even for large samples. We can get a hint of the difficulties by asking what we would do if we tossed a coin $n = 3$ times and obtained $X = 3$ heads. The estimate of the coin's probability of heads is $\hat{p} = 3/3 = 1$, and using this value to estimate the standard error gives $\hat{p}(1-\hat{p})/n = (1)(0)/3 = 0$. In other words, we would estimate that our coin has a probability of 1 of coming up heads with a standard error of 0. This isn't plausible. Both computer studies and careful mathematics show that we can do better by moving the sample proportion $\hat{p}$ slightly away from 0 and 1.[3] There are several ways to do this. We will use a simple adjustment that works very well in practice.

The adjustment is based on the following idea: act as if we have 4 additional observations, 2 of which are successes and 2 of which are failures. The new sample size is $n + 4$ and the count of successes is $X + 2$. The estimator of the population proportion based on this *add 2 successes and 2 failures* rule is

$$\tilde{p} = \frac{X+2}{n+4}$$

Wilson estimate

Because this estimate was first suggested by Edwin Bidwell Wilson in 1927 (though rarely used in practice until recently), we call it the **Wilson estimate.** We base a confidence interval on the z statistic obtained by standardizing the Wilson estimate $\tilde{p}$. Because $\tilde{p}$ is the sample proportion for our modified sample of size $n + 4$, it isn't surprising that the distribution of $\tilde{p}$ is close to the normal distribution with mean p and standard deviation $\sqrt{p(1-p)/(n+4)}$. To get a confidence interval, we estimate p by $\tilde{p}$ in this standard deviation to get the standard error of $\tilde{p}$. Here is the final result.

Confidence Interval for a Population Proportion

Choose an SRS of size n from a large population with unknown proportion p of successes. The **Wilson estimate of the population proportion** is

$$\tilde{p} = \frac{X+2}{n+4}$$

The **standard error of $\tilde{p}$** is

$$SE_{\tilde{p}} = \sqrt{\frac{\tilde{p}(1-\tilde{p})}{n+4}}$$

An **approximate level C confidence interval** for p is

$$\tilde{p} \pm z^* SE_{\tilde{p}}$$

where z^* is the value for the standard normal density curve with area C between $-z^*$ and z^*. The **margin of error** is

$$m = z^* SE_{\tilde{p}}$$

Use this interval when the sample size is at least $n = 5$ and the confidence level is 90%, 95%, or 99%.

EXAMPLE 8.1

Alcohol abuse has been described by college presidents as the number one problem on campus, and it is an important cause of death in young adults. How common is it? A survey of 17,096 students in U.S. four-year colleges collected information on drinking behavior and alcohol-related problems.[4] The researchers defined "frequent binge drinking" as having five or more drinks in a row three or more times in the past two weeks. According to this definition, 3314 students were classified as frequent binge drinkers. The Wilson estimate of the proportion of drinkers is

$$\tilde{p} = \frac{3314 + 2}{17{,}096 + 4} = 0.194$$

To find a 95% confidence interval, first compute the standard error:

$$\begin{aligned} SE_{\tilde{p}} &= \sqrt{\frac{\tilde{p}(1-\tilde{p})}{n+4}} \\ &= \sqrt{\frac{(0.194)(1-0.194)}{17{,}096+4}} \\ &= 0.0302 \end{aligned}$$

From Table D we find the value of z^* to be 1.960. So the confidence interval is

$$\begin{aligned} \tilde{p} \pm z^* SE_{\tilde{p}} &= 0.194 \pm (1.960)(0.0302) \\ &= 0.194 \pm 0.006 \\ &= (0.188,\ 0.200) \end{aligned}$$

We estimate with 95% confidence that between 18.8% and 20.0% of college students are frequent binge drinkers. Here is an alternative summary. We estimate that 19.4% of college students are frequent binge drinkers. For 95% confidence, the margin of error is 0.6%.

Note that adding 2 successes and 2 failures has had virtually no effect on our estimates in this example. This will happen whenever the sample sizes are large.

Because the calculations for statistical inference for a single proportion are relatively straightforward, many software packages do not include them. Moreover, newer methods such as the interval based on the Wilson estimate have only slowly begun to appear in software. Figure 8.1 gives output from Minitab for Example 8.1. We added 2 successes and 2 failures to the actual data; this forces the software to use the Wilson estimate. That's why the output gives the sample size as 17,100. As usual, the output reports more digits than are useful. When you use software, be sure to think about how many digits are meaningful for your purposes.

Remember that the margin of error in this confidence interval includes only random sampling error. There are other sources of error that are not taken into account. This survey used a design where the number of students sampled was proportional to the size of the college they attended; in this way

```
Sample      X       N     Sample p           95.0 % CI
1        3316   17100     0.193918      (0.187992, 0.199844)
```

FIGURE 8.1 Minitab output for Example 8.1.

we can treat the data as if we had an SRS. However, as is the case with many such surveys, we are forced to assume that the respondents provided accurate information. If the students did not answer the questions honestly, the results may be biased. Furthermore, we also have the typical problem of nonresponse. The response rate for this survey was 69%, an excellent rate for surveys of this type. Do the students who did not respond have different drinking habits than those who did? If so, this is another source of bias.

Significance test for a single proportion

Recall that the sample proportion $\hat{p} = X/n$ is approximately normal, with mean $\mu_{\hat{p}} = p$ and standard deviation $\sigma_{\hat{p}} = \sqrt{p(1-p)/n}$. For confidence intervals, we substitute $\hat{p}$ for p in the last expression to obtain the standard error. When performing a significance test, however, the null hypothesis specifies a value for p, and we assume that this is the true value when calculating the P-value. Therefore, when we test $H_0\colon p = p_0$, we substitute this p_0 into the expression for $\sigma_{\hat{p}}$ and then standardize $\hat{p}$. Here are the details.

Large-Sample Significance Test for a Population Proportion

Draw an SRS of size n from a large population with unknown proportion p of successes. To test the hypothesis $H_0\colon p = p_0$, compute the z **statistic**

$$z = \frac{\hat{p} - p_0}{\sqrt{\dfrac{p_0(1-p_0)}{n}}}$$

In terms of a standard normal random variable Z, the approximate P-value for a test of H_0 against

$H_a\colon p > p_0$ is $P(Z \geq z)$

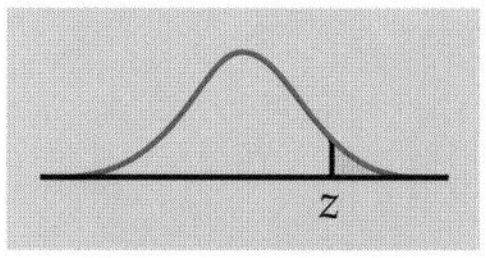

$H_a\colon p < p_0$ is $P(Z \leq z)$

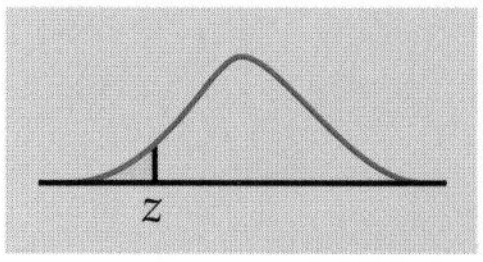

$H_a\colon p \neq p_0$ is $2P(Z \geq |z|)$

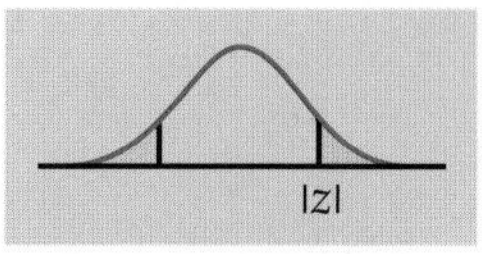

We recommend the large-sample z significance test as long as the expected number of successes (np_0) and the expected number of failures ($n(1-p_0)$) are both greater than 10. If this rule of thumb is not met, or if the population is less than 10 times as large as the sample, other procedures should be used. Here is an example.

EXAMPLE 8.2 The French naturalist Count Buffon once tossed a coin 4040 times and obtained 2048 heads. This is a binomial experiment with $n = 4040$. The sample proportion is

$$\hat{p} = \frac{2048}{4040} = 0.5069$$

If Buffon's coin was balanced, then the probability of obtaining heads on any toss is 0.5. To assess whether the data provide evidence that the coin was not balanced, we test

$$H_0: p = 0.5$$
$$H_a: p \neq 0.5$$

The test statistic is

$$z = \frac{\hat{p} - 0.5}{\sqrt{\frac{(0.5)(0.5)}{4040}}} = \frac{0.5069 - 0.5}{\sqrt{\frac{(0.5)(0.5)}{4040}}} = 0.88$$

Figure 8.2 illustrates the calculation of the P-value. From Table A we find $P(Z \leq 0.88) = 0.8106$. Therefore, the probability in each tail is $1 - 0.8106 = 0.1894$, and the P-value is $P = 2 \times 0.1894 = 0.38$. Figure 8.3 gives output from Minitab for this problem. Because these calculations are so simple, some software packages do not provide them. The data are compatible with the balanced-coin hypothesis ($\hat{p} = 0.5069$, $z = 0.88$, $P = 0.38$).

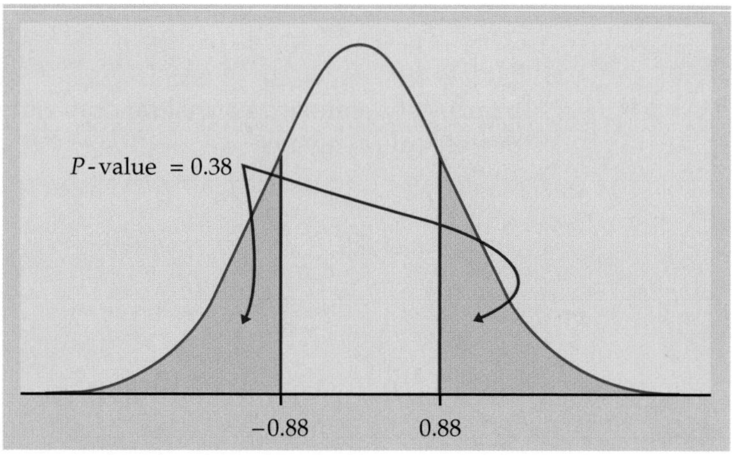

FIGURE 8.2 The P-value for Example 8.2.

```
Test of p = 0.5 vs p not = 0.5

Sample      X       N     Sample p      Z-Value     P-Value
1        2048    4040     0.506931         0.88       0.378
```

FIGURE 8.3 Minitab output for Example 8.2.

In this example we have arbitrarily chosen to associate the outcome "heads" with success and "tails" with failure. Suppose we reversed the choice. If we observed 2048 heads, then the other $4040 - 2048 = 1992$ tosses were tails. Let's repeat the significance test with "tails" as the success outcome.

EXAMPLE 8.3

A coin was tossed $n = 4040$ times and we observed $X = 1992$ tails. We want to test the null hypothesis that the coin is fair—that is, that the probability of tails is 0.5. So p is the probability that the coin comes up tails and we test

$$H_0: p = 0.5$$

$$H_a: p \neq 0.5$$

The test statistic is

$$z = \frac{\hat{p} - 0.5}{\sqrt{\frac{(0.5)(0.5)}{4040}}} = \frac{0.4931 - 0.5}{\sqrt{\frac{(0.5)(0.5)}{4040}}} = -0.88$$

Using Table A, we find that $P = 0.38$.

When we interchanged heads and tails (or success and failure), we simply changed the sign of the test statistic z. The P-value remained the same. These facts are true in general. Our conclusion does not depend on an arbitrary choice of success and failure.

Unless a coin is perfectly balanced, its true probability of heads will not be *exactly* 0.5. The significance test in Example 8.2 demonstrates only that Buffon's coin is statistically indistinguishable from a fair coin on the basis of 4040 tosses.

Confidence intervals provide additional information

To see what other values of p are compatible with the sample results, we will calculate a confidence interval.

EXAMPLE 8.4

For Buffon's coin-tossing experiment, the Wilson standard error for the proportion of heads is

$$\begin{aligned} SE_{\tilde{p}} &= \sqrt{\frac{\tilde{p}(1-\tilde{p})}{n+4}} \\ &= \sqrt{\frac{(0.5069)(1-0.5069)}{4044}} \\ &= 0.00786 \end{aligned}$$

We will calculate a 99% confidence interval. From Table D we find $z^* = 2.576$, so the margin of error is

$$z^* SE_{\tilde{p}} = (2.576)(0.00786) = 0.0202$$

So the 99% confidence interval for the probability of heads is

$$\begin{aligned} \tilde{p} \pm z^*\text{SE}_{\tilde{p}} &= 0.507 \pm 0.0202 \\ &= (0.4867,\ 0.5271) \end{aligned}$$

We estimate that Buffon's coin comes up heads 50.7% of the time. The margin of error for 99% confidence is 2%. The 99% confidence interval is 48.7% to 52.7%.

The confidence interval of Example 8.4 is much more informative than the significance test of Example 8.2. We have determined the values of p that are consistent with the observed results. This is the range of p_0's that would not be rejected by a test at the $\alpha = 0.01$ level of significance. We would not be surprised if the true probability of heads for Buffon's coin were something like 0.51.

We do not often use significance tests for a single proportion because it is uncommon to have a situation where there is a precise p_0 that we want to test. For physical experiments such as coin tossing or drawing cards from a well-shuffled deck, probability arguments lead to an ideal p_0, as in Example 8.2. Even here, however, it can be argued that no real coin has a probability of heads *exactly* equal to 0.5. Data from past large samples can sometimes provide a p_0 for the null hypothesis of a significance test. In some types of epidemiology research, for example, "historical controls" from past studies serve as the benchmark for evaluating new treatments. Medical researchers argue about the validity of these approaches, because the past never quite resembles the present. In general, we prefer comparative studies whenever possible.

Choosing a sample size

In Chapter 6, we showed how to choose the sample size n to obtain a confidence interval with specified margin of error m for a normal mean. Because we are using a normal approximation for inference about a population proportion, sample size selection proceeds in much the same way.

Recall that the margin of error for the large-sample confidence interval for a population proportion is

$$m = z^*\text{SE}_{\tilde{p}} = z^*\sqrt{\frac{\tilde{p}(1-\tilde{p})}{n+4}}$$

Choosing a confidence level C fixes the critical value z^*. The margin of error also depends on the value of $\tilde{p}$ and the sample size n. Because we don't know the value of $\tilde{p}$ until we gather the data, we must guess a value to use in the calculations. We will call the guessed value p^*. Here are two ways to get p^*:

- Use the sample estimate from a pilot study or from similar studies done earlier.
- Use $p^* = 0.5$. Because the margin of error is largest when $\tilde{p} = 0.5$, this choice gives a sample size that is somewhat larger than we really need for the confidence level we choose. It is a safe choice no matter what the data later show.

Once we have chosen p^* and the margin of error m that we want, we can find the n we need to achieve this margin of error. Here is the result.

Sample Size for Desired Margin of Error

The level C confidence interval for a proportion p will have a margin of error approximately equal to a specified value m when the sample size satisfies

$$n + 4 = \left(\frac{z^*}{m}\right)^2 p^*(1 - p^*)$$

Here z^* is the critical value for confidence C, and p^* is a guessed value for the proportion of successes in the future sample.

The margin of error will be less than or equal to m if p^* is chosen to be 0.5. The sample size required is then given by

$$n + 4 = \left(\frac{z^*}{2m}\right)^2$$

The value of n obtained by this method is not particularly sensitive to the choice of p^* when p^* is fairly close to 0.5. However, if the value of p is smaller than about 0.3 or larger than about 0.7, use of $p^* = 0.5$ may select a sample size that is much larger than needed.

EXAMPLE 8.5

You are doing a sample survey to estimate the proportion of voters who plan to vote for candidate Garcia in the coming election. Let p be the true proportion of voters who intend to vote for her. Because you judge that the value of p is not very different from 0.5, you use $p^* = 0.5$ for planning the sample size.

You want to estimate p with 95% confidence and a margin of error less than or equal to 3%, or 0.03. The sample size required is

$$n + 4 = \left(\frac{z^*}{2m}\right)^2 = \left[\frac{1.96}{(2)(0.03)}\right]^2 = 1067.1$$

which we round up to $n + 4 = 1068$, or $n = 1064$. (Rounding down would give a margin of error slightly greater than 0.03.)

Similarly, for a 2.5% margin of error we have (after rounding up)

$$n + 4 = \left[\frac{1.96}{(2)(0.025)}\right]^2 = 1537$$

or $n = 1533$. For a 2% margin of error a larger sample size is required:

$$n + 4 = \left[\frac{1.96}{(2)(0.02)}\right]^2 = 2401$$

or $n = 2397$.

News reports frequently describe the results of surveys with sample sizes between 1000 and 1500 and a margin of error of about 3%. These surveys generally use sampling procedures more complicated than simple random sampling, so the calculation of confidence intervals is more involved than what we have studied in this section. The calculations in Example 8.5 nonetheless show in principle how such surveys are planned.

In practice, many factors influence the choice of a sample size. The following example illustrates one set of factors.

EXAMPLE 8.6

An association of Christmas tree growers in Indiana sponsored a sample survey of Indiana households to help improve the marketing of Christmas trees.[5] The researchers decided to use a telephone survey and estimated that each telephone interview would take about 2 minutes. Nine trained students in agribusiness marketing were to make the phone calls between 1:00 P.M. and 8:00 P.M. on a Sunday. After discussing problems related to people not being at home or being unwilling to answer the questions, the survey team proposed a sample size of 500. Several of the questions asked demographic information about the household; the key questions of interest had responses of "Yes" or "No." For example, "Did you have a Christmas tree last year?"

The primary purpose of the survey was to estimate various sample proportions for Indiana households. Would the proposed sample size of $n = 500$ be adequate to provide the sponsors of the research with the information they required? To address this question, we calculate the margins of error of 95% confidence intervals for various values of $\hat{p}$.

EXAMPLE 8.7

In the Christmas tree market survey, the margin of error of a 95% confidence interval for any value of $\tilde{p}$ and $n = 500$ is

$$\begin{aligned} m &= z^*\text{SE}_{\tilde{p}} \\ &= 1.96\sqrt{\frac{\tilde{p}(1-\tilde{p})}{500+4}} \\ &= 0.088\sqrt{\tilde{p}(1-\tilde{p})} \end{aligned}$$

The results for various values of $\tilde{p}$ are

$\tilde{p}$	m	$\tilde{p}$	m
0.05	0.019	0.60	0.043
0.10	0.026	0.70	0.040
0.20	0.035	0.80	0.035
0.30	0.040	0.90	0.026
0.40	0.043	0.95	0.019
0.50	0.044		

The survey team judged these margins of error to be acceptable, and they used a sample size of 500 in their survey.

The table in Example 8.7 illustrates two points. First, the margins of error for $\tilde{p} = 0.05$ and $\tilde{p} = 0.95$ are the same. The margins of error will always be the same for $\tilde{p}$ and $1 - \tilde{p}$. This is a direct consequence of the form of the confidence interval. Second, the margin of error varies only between 0.040 and 0.044 as $\tilde{p}$ varies from 0.3 to 0.7, and the margin of error is greatest when

$\tilde{p} = 0.5$, as we claimed earlier. It is true in general that the margin of error will vary relatively little for values of $\tilde{p}$ between 0.3 and 0.7. Therefore, when planning a study, it is not necessary to have a very precise guess for p. If $p^* = 0.5$ is used and the observed $\tilde{p}$ is between 0.3 and 0.7, the actual interval will be a little shorter than needed but the difference will be small.

SUMMARY

Inference about a population proportion p from an SRS of size n is based on the **sample proportion** $\hat{p} = X/n$ and the **Wilson estimate** $\tilde{p} = (X + 2)/(n + 4)$, which is just the sample proportion when we add 2 successes and 2 failures to the data. When n is large, $\hat{p}$ has approximately the normal distribution with mean p and standard deviation $\sqrt{p(1-p)/n}$, and $\tilde{p}$ is approximately normal with mean p and standard deviation $\sqrt{p(1-p)/(n+4)}$.

An **approximate level *C* confidence interval** for p is

$$\tilde{p} \pm z^* \mathrm{SE}_{\tilde{p}}$$

where z^* is the value for the standard normal density curve with area C between $-z^*$ and z^*, and the **standard error of $\tilde{p}$** is

$$\mathrm{SE}_{\tilde{p}} = \sqrt{\frac{\tilde{p}(1-\tilde{p})}{n+4}}$$

The **sample size** required to obtain a confidence interval of approximate margin of error m for a proportion is found from

$$n + 4 = \left(\frac{z^*}{m}\right)^2 p^*(1-p^*)$$

where p^* is a guessed value for the proportion, and z^* is the standard normal critical value for the desired level of confidence. To ensure that the margin of error of the interval is less than or equal to m no matter what $\tilde{p}$ may be, use

$$n + 4 = \left(\frac{z^*}{2m}\right)^2$$

Tests of $H_0\colon p = p_0$ are based on the **z statistic**

$$z = \frac{\hat{p} - p_0}{\sqrt{\dfrac{p_0(1-p_0)}{n}}}$$

with P-values calculated from the $N(0, 1)$ distribution.

SECTION 8.1 EXERCISES

8.1 To profitably produce a planned upgrade of a software product you make, you must charge customers \$100. Are your customers willing to pay this much? You contact a random sample of 40 customers and find that 11 would pay \$100 for the upgrade. Find a 95% confidence interval for the proportion of all of your customers (the population) who would be willing to buy the upgrade for \$100.

8.2 A poll of 811 adults aged 18 or older asked about purchases that they intended to make for the upcoming holiday season.[6] One of the questions asked about what kind of gift they intended to buy for the person on whom they would spend the most. Clothing was the first choice of 487 people. Give a 99% confidence interval for the proportion of people in this population who intend to buy clothing as their first choice.

8.3 Yesterday, your top salesperson called on 5 customers and obtained orders for your new product from all 5. Suppose that it is reasonable to view these 5 customers as a random sample of all of her customers.

(a) Give the Wilson estimate of the proportion of her customers who would buy the new product. Notice that we don't estimate that all customers will buy, even though all 5 in the sample did.

(b) Give the margin of error for 95% confidence. (You may see that the upper endpoint of the confidence interval is greater than 1. In that case, take the upper endpoint to be 1.)

(c) Do the results apply to all of your sales force? Explain why or why not.

8.4 When trying to hire managers and executives, companies sometimes verify the academic credentials described by the applicants. One company that performs these checks summarized its findings for a six-month period. Of the 84 applicants whose credentials were checked, 15 lied about having a degree.[7]

(a) Find the proportion of applicants who lied about having a degree and the standard error.

(b) Consider these data to be a random sample of credentials from a large collection of similar applicants. Give a 95% confidence interval for the true proportion of applicants who lie about having a degree.

8.5 Refer to the previous exercise. Suppose that 10 applicants lied about their major. Can we conclude that a total of $25 = 15 + 10$ applicants lied about having a degree or about their major? Explain your answer.

8.6 In the United States approximately 900 people die in bicycle accidents each year. One study examined the records of 1711 bicyclists aged 15 or older who were fatally injured in bicycle accidents between 1987 and 1991 and were tested for alcohol. Of these, 542 tested positive for alcohol (blood alcohol concentration of 0.01% or higher).[8]

(a) Summarize the data with appropriate descriptive statistics.

(b) To do statistical inference for these data, we think in terms of a model where p is a parameter that represents the probability that a tested bicycle rider is positive for alcohol. Find a 99% confidence interval for p.

(c) Can you conclude from your statistical analysis of this study that alcohol causes fatal bicycle accidents?

8.7 Refer to the previous exercise. In this study 386 bicyclists had blood alcohol levels above 0.10%, a level defining legally drunk in many states. Give a 99% confidence interval for the proportion who were legally drunk according to this criterion.

8.8 One question in the Christmas tree market survey described in Examples 8.6 and 8.7 (page 580) was "Did you have a Christmas tree last year?" Of the 500 respondents, 421 answered "Yes."

(a) Find the sample proportion and its standard error.

(b) Give a 90% confidence interval for the proportion of Indiana households who had a Christmas tree last year.

8.9 As part of a quality improvement program, your mail-order company is studying the process of filling customer orders. According to company standards, an order is shipped on time if it is sent within 3 working days of the time it is received. You select an SRS of 200 of the 5000 orders received in the past month for an audit. The audit reveals that 185 of these orders were shipped on time. Find a 95% confidence interval for the true proportion of the month's orders that were shipped on time.

8.10 Large trees growing near power lines can cause power failures during storms when their branches fall on the lines. Power companies spend a great deal of time and money trimming and removing trees to prevent this problem. Researchers are developing hormone and chemical treatments that will stunt or slow tree growth. If the treatment is too severe, however, the tree will die. In one series of laboratory experiments on 216 sycamore trees, 41 trees died. Give a 95% confidence interval for the proportion of sycamore trees that would be expected to die from this particular treatment.

8.11 In recent years over 70% of first-year college students responding to a national survey have identified "being well-off financially" as an important personal goal. A state university finds that 103 of an SRS of 150 of its first-year students say that this goal is important. Give a 95% confidence interval for the proportion of all first-year students at the university who would identify being well-off as an important personal goal.

8.12 In Exercise 8.1 we found that 11 customers from a random sample of 40 would be willing to buy a software upgrade that costs \$100. If the upgrade is to be profitable, you will need to sell it to more than 20% of your customers. Do the sample data give good evidence that more than 20% are willing to buy?

(a) Formulate this problem as a hypothesis test. Give the null and alternative hypotheses. Will you use a one-sided or a two-sided alternative? Why?

(b) Carry out the significance test. Report the test statistic and the *P*-value.

(c) Should you proceed with plans to produce and market the upgrade?

8.13 In each of the following cases state whether or not the normal approximation to the binomial should be used for a significance test on the population proportion p.

(a) $n = 10$ and $H_0\colon p = 0.6$.

(b) $n = 100$ and $H_0\colon p = 0.4$.

(c) $n = 2000$ and $H_0\colon p = 0.996$.

(d) $n = 500$ and $H_0\colon p = 0.25$.

8.14 Refer to the Christmas tree market survey described in Examples 8.6 and 8.7 (page 580). Of the 500 responding households, 38% were from rural areas (including small towns), and the other 62% were from urban areas (including suburbs). According to the census, 36% of Indiana households are in rural areas, and the remaining 64% are in urban areas. Let p be the proportion of rural respondents. Set up hypotheses about p_0 and perform a test of significance to examine how well the sample represents the state in regard to rural versus urban residence. Summarize your results.

8.15 Refer to the previous exercise. There we arbitrarily chose to state the hypotheses in terms of the proportion of rural respondents. We could as easily have used the proportion of *urban* respondents.

(a) Write hypotheses in terms of the proportion of urban residents to examine how well the sample represents the state in regard to rural versus urban residence.

(b) Perform the test of significance and summarize the results.

(c) Compare your results with the results of the previous exercise. Summarize and generalize your conclusion.

8.16 A national opinion poll found that 44% of all American adults agree that parents should be given vouchers good for education at any public or private school of their choice. The result was based on a small sample. How large an SRS is required to obtain a margin of error of ± 0.03 (that is, $\pm 3\%$) in a 95% confidence interval? (Use the previous poll's result to obtain the guessed value p^*.)

8.17 An entomologist samples a field for egg masses of a harmful insect by placing a yard-square frame at random locations and carefully examining the ground within the frame. An SRS of 75 locations selected from a county's pastureland found egg masses in 13 locations. Give a 90% confidence interval for the proportion of all possible locations that are infested.

8.18 Of the 500 respondents in the Christmas tree market survey, 44% had no children at home and 56% had at least one child at home. The corresponding figures for the most recent census are 48% with no children and 52% with at least one child. Test the null hypothesis that the telephone survey technique has a probability of selecting a household with no children that is equal to the value obtained by the census. Give the z statistic and the P-value. What do you conclude?

8.19 The English mathematician John Kerrich tossed a coin 10,000 times and obtained 5067 heads.

(a) Is this significant evidence at the 5% level that the probability that Kerrich's coin comes up heads is not 0.5?

(b) Use a 95% confidence interval to find the range of probabilities of heads that would not be rejected at the 5% level.

8.20 A matched pairs experiment compares the taste of instant versus fresh-brewed coffee. Each subject tastes two unmarked cups of coffee, one of each type, in random order and states which he or she prefers. Of the 50 subjects who participate in the study, 19 prefer the instant coffee. Let p be the probability that a randomly chosen subject prefers freshly brewed coffee to instant coffee. (In practical terms, p is the proportion of the population who prefer fresh-brewed coffee.)

(a) Test the claim that a majority of people prefer the taste of fresh-brewed coffee. Report the z statistic and its P-value. Is your result significant at the 5% level? What is your practical conclusion?

(b) Find a 90% confidence interval for p.

8.21 Shereka, a starting player for a major college basketball team, made only 36.2% of her free throws last season. During the summer she worked on developing a softer shot in the hope of improving her free-throw accuracy. In the first eight games of this season Shereka made 22 free throws in 42 attempts. Let p be her probability of making each free throw she shoots this season.

(a) State the null hypothesis H_0 that Shereka's free-throw probability has remained the same as last year and the alternative H_a that her work in the summer resulted in a higher probability of success.

(b) Calculate the z statistic for testing H_0 versus H_a.

(c) Do you accept or reject H_0 for $\alpha = 0.05$? Find the P-value.

(d) Give a 90% confidence interval for Shereka's free-throw success probability for the new season. Are you convinced that she is now a better free-throw shooter than last season?

(e) What assumptions are needed for the validity of the test and confidence interval calculations that you performed?

8.22 One of your employees has suggested that your company develop a new product. You decide to take a random sample of your customers and ask whether or not there is interest in the new product. The response is on a

1 to 5 scale with 1 indicating "definitely would not purchase"; 2, "probably would not purchase"; 3, "not sure"; 4, "probably would purchase"; and 5, "definitely would purchase." For an initial analysis, you will record the responses 1, 2, and 3 as "No" and 4 and 5 as "Yes". What sample size would you use if you wanted the 95% margin of error to be 0.10 or less?

8.23 Refer to the previous exercise. Suppose that after reviewing the results of the previous survey, you proceeded with preliminary development of the product. Now you are at the stage where you need to decide whether or not to make a major investment to produce and market it. You will use another random sample of your customers but now you want the margin of error to be smaller. What sample size would you use if you wanted the 95% margin of error to be 0.05 or less?

8.24 You want to estimate the proportion of students at your college or university who are employed for 10 or more hours per week while classes are in session. You plan to present your results by a 95% confidence interval. Using the guessed value $p^* = 0.40$, find the sample size required if the interval is to have an approximate margin of error of $m = 0.06$.

8.25 Land's Beginning is a company that sells its merchandise through the mail. It is considering buying a list of addresses from a magazine. The magazine claims that at least 25% of its subscribers have high incomes (they define this to be household income in excess of $100,000). Land's Beginning would like to estimate the proportion of high-income people on the list. Checking income is very difficult and expensive but another company offers this service. Land's Beginning will pay to find incomes for an SRS of people on the magazine's list. They would like the margin of error of the 95% confidence interval for the proportion to be 0.05 or less. Use the guessed value $p^* = 0.25$ to find the required sample size.

8.26 Refer to the previous exercise. For each of the following variations on the design specifications, state whether the required sample size will be higher, lower, or the same as that found above.

(a) Use a 90% confidence interval.

(b) Change the allowable margin of error to 0.10.

(c) Use a planning value of $p^* = 0.30$.

(d) Use a different company to do the income checks.

8.27 A student organization wants to start a nightclub for students under the age of 21. To assess support for this proposal, they will select an SRS of students and ask each respondent if he or she would patronize this type of establishment. They expect that about 60% of the student body would respond favorably. What sample size is required to obtain a 95% confidence interval with an approximate margin of error of 0.08? Suppose that 50% of the sample responds favorably. Calculate the margin of error of the 95% confidence interval.

8.28 An automobile manufacturer would like to know what proportion of its customers are dissatisfied with the service received from their local dealer. The customer relations department will survey a random sample of customers and compute a 95% confidence interval for the proportion that are dissatisfied. From past studies, they believe that this proportion will be about 0.25. Find the sample size needed if the margin of error of the confidence interval is to be about 0.02. Suppose 15% of the sample say that they are dissatisfied. What is the margin of error of the 95% confidence interval?

8.29 You have been asked to survey students at a large college to determine the proportion who favor an increase in student fees to support an expansion of the student newspaper. Each student will be asked whether he or she is in favor of the proposed increase. Using records provided by the registrar you can select a random sample of students from the college. After careful consideration of your resources, you decide that it is reasonable to conduct a study with a sample of 10 students. For this sample size, construct a table of the margins of error for 95% confidence intervals when $\hat{p}$ takes the values 0.1, 0.2, 0.3, 0.4, 0.5, 0.6, 0.7, 0.8, and 0.9.

8.30 A former editor of the student newspaper agrees to underwrite the study in the previous exercise because she believes the results will demonstrate that most students support an increase in fees. She is willing to provide funds for a sample of size 400. Write a short summary for your benefactor of why the increased sample size will provide better results.

8.2 Comparing Two Proportions

Because comparative studies are so common, we often want to compare the proportions of two groups (such as men and women) that have some characteristic. We call the two groups being compared Population 1 and Population 2, and the two population proportions of "successes" p_1 and p_2. The data consist of two independent SRSs, of size n_1 from Population 1 and size n_2 from Population 2. The proportion of successes in each sample estimates the corresponding population proportion. Here is the notation we will use in this section:

Population	Population proportion	Sample size	Count of successes	Sample proportion
1	p_1	n_1	X_1	$\hat{p}_1 = X_1/n_1$
2	p_2	n_2	X_2	$\hat{p}_2 = X_2/n_2$

To compare the two populations, we use the difference between the two sample proportions:

$$D = \hat{p}_1 - \hat{p}_2$$

When both sample sizes are sufficiently large, the sampling distribution of the difference D is approximately normal.

Inference procedures for comparing proportions are z procedures based on the normal approximation and on standardizing the difference D. The first step is to obtain the mean and standard deviation of D. By the addition rule for means, the mean of D is the difference of the means:

$$\mu_D = \mu_{\hat{p}_1} - \mu_{\hat{p}_2} = p_1 - p_2$$

That is, the difference $D = \hat{p}_1 - \hat{p}_2$ between the sample proportions is an unbiased estimator of the population difference $p_1 - p_2$. Similarly, the addition rule for variances tells us that the variance of D is the *sum* of the variances:

$$\sigma_D^2 = \sigma_{\hat{p}_1}^2 + \sigma_{\hat{p}_2}^2 = \frac{p_1(1-p_1)}{n_1} + \frac{p_2(1-p_2)}{n_2}$$

Therefore, when n_1 and n_2 are large, D is approximately normal with mean $\mu_D = p_1 - p_2$ and standard deviation

$$\sigma_D = \sqrt{\frac{p_1(1-p_1)}{n_1} + \frac{p_2(1-p_2)}{n_2}}$$

Confidence intervals

Just as in the case of estimating a single proportion, a small modification of the sample proportions greatly improves the accuracy of confidence intervals.[9] As before, add 2 successes and 2 failures to the actual data, and now divide them equally between the two samples. That is, *add 1 success and 1 failure to each sample*. We will again call the estimates produced by adding hypothetical observations **Wilson estimates.** The Wilson estimates of the two population proportions are

Wilson estimates

$$\tilde{p}_1 = \frac{X_1 + 1}{n_1 + 2} \quad \text{and} \quad \tilde{p}_2 = \frac{X_2 + 1}{n_2 + 2}$$

The estimated difference between the populations is

$$\tilde{D} = \tilde{p}_1 - \tilde{p}_2$$

The standard deviation of $\tilde{D}$ is approximately

$$\sigma_{\tilde{D}} = \sqrt{\frac{p_1(1-p_1)}{n_1 + 2} + \frac{p_2(1-p_2)}{n_2 + 2}}$$

This is similar to the formula for σ_D, adjusted for the sizes of the modified samples.

To obtain a confidence interval for $p_1 - p_2$, we once again replace the unknown parameters in the standard deviation by estimates to obtain an estimated standard deviation, or standard error. Here is the confidence interval we want.

Confidence Intervals for Comparing Two Proportions

Choose an SRS of size n_1 from a large population having proportion p_1 of successes and an independent SRS of size n_2 from another population having proportion p_2 of successes. An **approximate level *C* confidence interval** for $p_1 - p_2$ is

$$(\tilde{p}_1 - \tilde{p}_2) \pm z^* \mathrm{SE}_{\tilde{D}}$$

where

$$\tilde{p}_1 = \frac{X_1 + 1}{n_1 + 2} \qquad \tilde{p}_2 = \frac{X_2 + 1}{n_2 + 2}$$

are the **Wilson estimates** of the population proportions and the **standard error of the difference** is

$$\mathrm{SE}_{\tilde{D}} = \sqrt{\frac{\tilde{p}_1(1 - \tilde{p}_1)}{n_1 + 2} + \frac{\tilde{p}_2(1 - \tilde{p}_2)}{n_2 + 2}}$$

and z^* is the value for the standard Normal density curve with area C between $-z^*$ and z^*. The **margin of error** is

$$m = \pm z^* \mathrm{SE}_{\tilde{D}}$$

Use this method when both sample sizes are at least 10 and the confidence level is 90%, 95%, or 99%.

In Example 8.1 (page 574) we estimated the proportion of college students who engage in frequent binge drinking. Are there characteristics of these students that relate to this behavior? For example, how similar is this behavior in men and women? This kind of follow-up question is natural in many studies such as this one. We will first use a confidence interval to examine it.

EXAMPLE 8.8

In the binge-drinking study, data were also summarized by gender.

Population	n	X	$\hat{p} = X/n$	$\tilde{p} = (X + 1)/(n + 2)$
1 (men)	7,180	1,630	0.227	0.227
2 (women)	9,916	1,684	0.170	0.170
Total	17,096	3,314	0.194	

```
Sample      X        N        Sample p
1        1631     7182        0.227096
2        1685     9918        0.169893

Estimate for p(1) - p(2): 0.0572024
95% CI for p(1) - p(2): (0.0450161, 0.0693887)
```

FIGURE 8.4 Minitab output for Example 8.8.

In this table the $\hat{p}$ column gives the sample proportions of frequent binge drinkers. The $\tilde{p}$ column gives the Wilson estimates that we will use to construct a confidence interval for the difference. The last line gives the totals that we studied in Example 8.1. Note that because of the large sample sizes, the adjustments made for the Wilson method make very little change in the estimates. We will find a 95% confidence interval for the difference between the proportions of men and of women who are frequent binge drinkers. Output from Minitab is given in Figure 8.4. To perform the computations using our formulas, we first calculate the standard error of the observed difference:

$$\begin{aligned} SE_{\tilde{D}} &= \sqrt{\frac{\tilde{p}_1(1-\tilde{p}_1)}{n_1+2} + \frac{\tilde{p}_2(1-\tilde{p}_2)}{n_2+2}} \\ &= \sqrt{\frac{(0.227)(0.773)}{7180+2} + \frac{(0.170)(0.830)}{9916+2}} \\ &= 0.00622 \end{aligned}$$

The 95% confidence interval is

$$\begin{aligned} (\tilde{p}_1 - \tilde{p}_2) \pm z^* SE_{\tilde{D}} &= (0.227 - 0.170) \pm (1.960)(0.00622) \\ &= 0.057 \pm 0.012 \\ &= (0.045,\ 0.069) \end{aligned}$$

With 95% confidence we can say that the difference in the proportions is between 0.045 and 0.069. Alternatively, we can report that the men are 5.7% more likely to be frequent binge drinkers, with a 95% margin of error of 1.2%.

In this example men and women were not sampled separately. The sample sizes are in fact random and reflect the gender distributions of the colleges that were randomly chosen. Two-sample significance tests and confidence intervals are still approximately correct in this situation. The authors of the report note that women are somewhat overrepresented partly because 6 of the 140 colleges in the study were all-women institutions.

In the example above we chose men to be the first population. Had we chosen women to be the first population, the estimate of the difference would be negative (−0.057). Because it is easier to discuss positive numbers, we generally choose the first population to be the one with the higher proportion.

Significance tests

Although we prefer to compare two proportions by giving a confidence interval for the difference between the two population proportions, it is sometimes useful to test the null hypothesis that the two population proportions are the same.

We standardize $D = \hat{p}_1 - \hat{p}_2$ by subtracting its mean $p_1 - p_2$ and then dividing by its standard deviation

$$\sigma_D = \sqrt{\frac{p_1(1-p_1)}{n_1} + \frac{p_2(1-p_2)}{n_2}}$$

If n_1 and n_2 are large, the standardized difference is approximately $N(0, 1)$. For the large-sample confidence interval we used sample estimates in place of the unknown population values in the expression for σ_D. Although this approach would lead to a valid significance test, we instead adopt the more common practice of replacing the unknown σ_D with an estimate that takes into account that our null hypothesis states that $p_1 = p_2$. If these two proportions are equal, then we can view all of the data as coming from a single population. Let p denote the common value of p_1 and p_2; then the standard deviation of $D = \hat{p}_1 - \hat{p}_2$ is

$$\begin{aligned}\sigma_D &= \sqrt{\frac{p(1-p)}{n_1} + \frac{p(1-p)}{n_2}} \\ &= \sqrt{p(1-p)\left(\frac{1}{n_1} + \frac{1}{n_2}\right)}\end{aligned}$$

We estimate the common value of p by the overall proportion of successes in the two samples:

$$\hat{p} = \frac{\text{number of successes in both samples}}{\text{number of observations in both samples}} = \frac{X_1 + X_2}{n_1 + n_2}$$

pooled estimate of p

This estimate of p is called the **pooled estimate** because it combines, or pools, the information from both samples.

To estimate σ_D under the null hypothesis, we substitute $\hat{p}$ for p in the expression for σ_D. The result is a standard error for D that assumes the truth of H_0: $p_1 = p_2$:

$$\text{SE}_{Dp} = \sqrt{\hat{p}(1-\hat{p})\left(\frac{1}{n_1} + \frac{1}{n_2}\right)}$$

The subscript on SE_{Dp} reminds us that we pooled data from the two samples to construct the estimate.

Significance Tests for Comparing Two Proportions

To test the hypothesis

$$H_0: p_1 = p_2$$

compute the z **statistic**

$$z = \frac{\hat{p}_1 - \hat{p}_2}{SE_{Dp}}$$

where the **pooled standard error** is

$$SE_{Dp} = \sqrt{\hat{p}(1-\hat{p})\left(\frac{1}{n_1} + \frac{1}{n_2}\right)}$$

and where

$$\hat{p} = \frac{X_1 + X_2}{n_1 + n_2}$$

In terms of a standard normal random variable Z, the P-value for a test of H_0 against

$H_a: p_1 > p_2$ is $P(Z \geq z)$

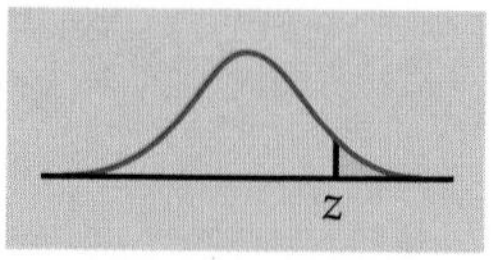

$H_a: p_1 < p_2$ is $P(Z \leq z)$

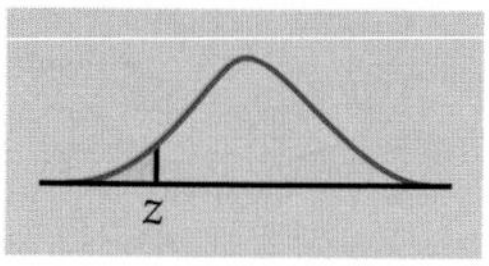

$H_a: p_1 \neq p_2$ is $2P(Z \geq |z|)$

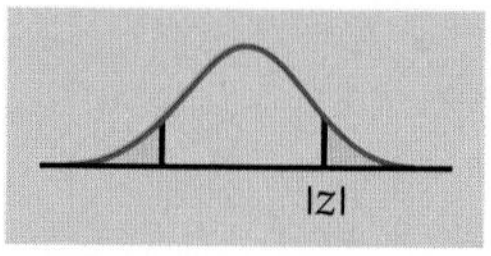

This z test is based on the normal approximation to the binomial distribution. As a general rule, we will use it when the number of successes and the number of failures in each of the samples is at least 5.

EXAMPLE 8.9

Are men and women college students equally likely to be frequent binge drinkers? We examine the survey data in Example 8.8 (page 589) to answer this question. Here is the data summary:

Population	n	X	$\hat{p} = X/n$
1 (men)	7,180	1,630	0.227
2 (women)	9,916	1,684	0.170
Total	17,096	3,314	0.194

```
Sample      X       N      Sample p
1        1630    7180      0.227019
2        1684    9916      0.169827

Test for p(1) - p(2) = 0 (vs not = 0): Z = 9.34  P-Value = 0.000
```

FIGURE 8.5 Minitab output for Example 8.9.

The sample proportions are certainly quite different, but we will perform a significance test to see if the difference is large enough to lead us to believe that the population proportions are not equal. Formally, we test the hypotheses

$$H_0: p_1 = p_2$$
$$H_a: p_1 \neq p_2$$

The pooled estimate of the common value of p is

$$\hat{p} = \frac{1630 + 1684}{7180 + 9916} = \frac{3314}{17,096} = 0.194$$

Note that this is the estimate on the bottom line of the data summary above.

The test statistic is calculated as follows:

$$\mathrm{SE}_{Dp} = \sqrt{(0.194)(0.806)\left(\frac{1}{7180} + \frac{1}{9916}\right)} = 0.006126$$

$$z = \frac{\hat{p}_1 - \hat{p}_2}{\mathrm{SE}_{Dp}} = \frac{0.227 - 0.170}{0.006126}$$
$$= 9.34$$

The P-value is $2P(Z \geq 9.34)$. The largest value of z in Table A is 3.49, so from this table we can conclude $P < 2 \times 0.0002 = 0.0004$. Most software reports this result as simply 0 or a very small number. Output from Minitab is given in Figure 8.5. It reports the P-value as 0.000. This means that the calculated value is something less than 0.0005; this is certainly a very small number. The exact value is not particularly important; it is clear that we should reject the null hypothesis. We report: among college students in the study, 22.7% of the men and 17.0% of the women were frequent binge drinkers; the difference is statistically significant ($z = 9.34$, $P < 0.0005$).

We could have argued that we expect the proportion to be higher for men than for women in this example. This would justify using the one-sided alternative $H_a: p_1 > p_2$. The P-value would be half of the value obtained for the two-sided test. Because the z statistic is so large, this distinction is of no practical importance.

Beyond the basics ▸ relative risk

We summarized the comparison of the frequent binge-drinking proportions for men and for women by reporting a confidence interval for the difference in Example 8.8. Another way to summarize the comparison is to use the ratio relative risk

of the two proportions, which is called the **relative risk (RR)** in many applications. Note that a relative risk of 1 means that the two proportions, $\hat{p}_1$ and $\hat{p}_2$, are equal. The procedure for calculating confidence intervals for relative risk is based on the same kind of principles that we have studied, but the details are somewhat more complicated. Fortunately, we can leave the details to software and concentrate on interpretation and communication of the results.

EXAMPLE 8.10

In Example 8.8 (page 589) we summarized the data on the proportions of men and women who are frequent binge drinkers with the following table:

Population	n	X	$\hat{p} = X/n$
1 (men)	7,180	1,630	0.227
2 (women)	9,916	1,684	0.170
Total	17,096	3,314	0.194

The relative risk is

$$\text{RR} = \frac{\hat{p}_1}{\hat{p}_2} = \frac{0.227}{0.170} = 1.34$$

Software (for example, PROC FREQ with the MEASURES option in SAS) gives the 95% confidence interval as 1.26 to 1.42. Men are 1.34 times as likely as women to be frequent binge drinkers; the 95% confidence interval is (1.26, 1.42).

In this example the confidence interval appears to be symmetric about the estimate: that is, 1.34 is the midpoint of 1.26 and 1.42. If we reported the results with more accuracy (RR = 1.337, 95% confidence interval = 1.258 to 1.421), we would see that the interval is *not* symmetric, and this is true in general.

Relative risk is particularly useful when the proportions that we are studying are small. In epidemiology and many areas of medical statistics it is routinely used. Here is another example.

EXAMPLE 8.11

There is much evidence that high blood pressure is associated with increased risk of death from cardiovascular disease. A major study of this association examined 3338 men with high blood pressure and 2676 men with low blood pressure. During the period of the study, 21 men in the low-blood-pressure and 55 in the high-blood-pressure group died from cardiovascular disease.[10] We choose the group with the higher proportion to be Population 1 so that the RR will be greater than 1. For men with low blood pressure, the proportion who died is 0.0078, while the proportion for those with high blood pressure is 0.0165. The difference is statistically significant ($z = 2.98$, $P = 0.0014$), and the 95% confidence interval for the difference in proportions is (0.0032, 0.0141). Let's see if relative risk helps us to communicate the results of this study more effectively. The relative risk is

$$\text{RR} = \frac{\hat{p}_1}{\hat{p}_2} = \frac{0.0165}{0.0078} = 2.1$$

Software gives the 95% confidence interval as (1.3, 3.5). The risk of dying from cardiovascular disease is doubled for men with high blood pressure compared with men who have low blood pressure (RR = 2.1, 95% confidence interval = 1.3 to 3.5).

SUMMARY

The approximate **level C confidence interval** for the difference in two proportions $p_1 - p_2$ is

$$(\tilde{p}_1 - \tilde{p}_2) \pm z^* \text{SE}_{\tilde{D}}$$

where the **Wilson estimates** of the population proportions are

$$\tilde{p}_1 = \frac{X_1 + 1}{n_1 + 2} \quad \text{and} \quad \tilde{p}_2 = \frac{X_2 + 1}{n_2 + 2}$$

and the **standard error of the difference** is

$$\text{SE}_{\tilde{D}} = \sqrt{\frac{\tilde{p}_1(1 - \tilde{p}_1)}{n_1 + 2} + \frac{\tilde{p}_2(1 - \tilde{p}_2)}{n_2 + 2}}$$

Significance tests of $H_0: p_1 = p_2$ use the **z statistic**

$$z = \frac{\hat{p}_1 - \hat{p}_2}{\text{SE}_{Dp}}$$

with P-values from the $N(0, 1)$ distribution. In this statistic,

$$\text{SE}_{Dp} = \sqrt{\hat{p}(1 - \hat{p})\left(\frac{1}{n_1} + \frac{1}{n_2}\right)}$$

and $\hat{p}$ is the **pooled estimate** of the common value of p_1 and p_2:

$$\hat{p} = \frac{X_1 + X_2}{n_1 + n_2}$$

Relative risk is the ratio of two sample proportions:

$$\text{RR} = \frac{\hat{p}_1}{\hat{p}_2}$$

Confidence intervals for relative risk are often used to summarize the comparison of two proportions.

SECTION 8.2 EXERCISES

8.31 Different kinds of companies compensate their key employees in different ways. Established companies may pay higher salaries, while new companies may offer stock options that will be valuable if the company succeeds. Do high-tech companies tend to offer stock options more often than other companies? One study looked at a random sample of 200 companies. Of these, 91 were listed in the *Directory of Public High Technology Corporations* and 109 were not listed. Treat these two groups as SRSs of high-tech and non-high-tech companies. Seventy-three of the high-tech companies and 75 of the non-high-tech companies offered incentive stock options to key employees.[11] Give a 95% confidence interval for the difference in the proportions of the two types of companies that offer stock options.

8.32 Refer to the previous exercise. Compare the two groups of companies with a significance test. Summarize your analysis and conclusions.

8.33 A study was designed to find reasons why patients leave a health maintenance organization (HMO).[12] Patients were classified as to whether or not they had filed a complaint with the HMO. We want to compare the proportion of complainers who leave the HMO with the proportion of those who do not file complaints. In the year of the study, 639 patients filed complaints, and 54 of these patients left the HMO voluntarily. For comparison, the HMO chose an SRS of 743 patients who had not filed complaints. Twenty-two of these patients left voluntarily. Give an estimate of the difference in the two proportions with a 95% confidence interval.

8.34 In the previous exercise you examined data from a study designed to find reasons why patients leave an HMO. There you compared the proportion of complainers who leave the HMO with the proportion of noncomplainers who leave. In the year of the study, 639 patients filed complaints and 54 of these patients left the HMO voluntarily. For comparison, the HMO chose an SRS of 743 patients who had not filed complaints. Twenty-two of those patients left voluntarily. We expect a higher proportion of complainers to leave. Do the data support this expectation? State hypotheses, find the test statistic and its *P*-value, and state your conclusion.

8.35 To what extent do syntax textbooks, which analyze the structure of sentences, illustrate gender bias? A study of this question sampled sentences from 10 texts.[13] One part of the study examined the use of the words "girl," "boy," "man," and "woman." We will call the first two words *juvenile* and the last two *adult*. Is the proportion of female references that are juvenile (girl) equal to the proportion of male references that are juvenile (boy)? Here are data from one of the texts:

Gender	n	X(juvenile)
Female	60	48
Male	132	52

(a) Find the proportion of juvenile references for females and its standard error. Do the same for the males.

(b) Give a 90% confidence interval for the difference and briefly summarize what the data show.

8.36 In Exercise 8.6 (page 582) we examined the percent of fatally injured bicyclists tested for alcohol who tested positive. Here we examine the same data with respect to gender.

Gender	n	X(tested positive)
Female	191	27
Male	1520	515

(a) Summarize the data by giving the Wilson estimates of the two population proportions and a 95% confidence interval for their difference.

(b) The standard error $\mathrm{SE}_{\tilde{D}}$ contains a contribution from each sample, $\tilde{p}_1(1-\tilde{p}_1)/(n_1+2)$ and $\tilde{p}_2(1-\tilde{p}_2)/(n_2+2)$. Which of these contributes the larger amount to the standard error of the difference? Explain why.

8.37 Is lying about credentials by job applicants changing? In Exercise 8.4 (page 582) we looked at the proportion of applicants who lied about having a degree in a six-month period. To see if there is a change over time, we can compare that period with the following six months. Here are the data:

Period	n	X(lied)
1	84	15
2	106	21

Use a 95% confidence interval to address the question of interest.

8.38 Exercise 8.35 addresses a question about gender bias with a confidence interval. Set up the problem as a significance test. Carry out the test and summarize the results.

8.39 The proportions of female and male fatally injured bicyclists were compared with a confidence interval in Exercise 8.36. Examine the same data with a test of significance.

8.40 Data on the proportion of applicants who lied about having a degree in two consecutive six-month periods are given in Exercise 8.37. Formulate appropriate null and alternative hypotheses that can be addressed with these data, carry out the significance test, and summarize the results.

8.41 In the Christmas tree survey introduced in Example 8.6 (page 580), respondents who had a tree during the holiday season were asked whether the tree was natural or artificial. Respondents were also asked if they lived in an urban area or in a rural area. Of the 421 households displaying a Christmas tree, 160 lived in rural areas and 261 were urban residents. The tree growers want to know if there is a difference in preference for natural trees versus artificial trees between urban and rural households. Here are the data:

Population	n	X(natural)
1 (rural)	160	64
2 (urban)	261	89

(a) Give the null and alternative hypotheses that are appropriate for this problem assuming that we have no prior information suggesting that one population would have a higher preference than the other.

(b) Test the null hypothesis. Give the test statistic and the P-value, and summarize the results.

(c) Give a 95% confidence interval for the difference in proportions.

8.42 In the 2000 regular baseball season, the World Series Champion New York Yankees played 80 games at home and 81 games away. They won 44 of their home games and 43 of the games played away. We can consider these games as samples from potentially large populations of games played at home and away. How much advantage does the Yankee home field provide?

(a) Find the Wilson estimate of proportion of wins for all home games. Do the same for away games.

(b) Find the standard error needed to compute a confidence interval for the difference in the proportions.

(c) Compute a 90% confidence interval for the difference between the probability that the Yankees win at home and the probability that they win when on the road. Are you convinced that the Yankees were more likely to win at home in 2000?

8.43 Refer to the New York Yankees baseball data in the previous exercise.

(a) Combining all of the games played, what proportion did the Yankees win?

(b) Find the standard error needed for testing that the probability of winning is the same at home and away.

(c) Most people think that it is easier to win at home than away. Formulate null and alternative hypotheses to examine this idea.

(d) Compute the z statistic and its P-value. What conclusion do you draw?

8.44 In the 2000 World Series the New York Yankees played the New York Mets. The previous two exercises examine the Yankees' home and away victories. During the regular season the Mets won 55 of the 84 home games that they played and 39 of the 81 games that they played away. Perform the same analyses for the Mets and write a short summary comparing these results with those you found for the Yankees.

8.45 The state agriculture department asked random samples of Indiana farmers in each county whether they favored a mandatory corn checkoff program to pay for corn product marketing and research. In Tippecanoe County, 263 farmers were in favor of the program and 252 were not. In neighboring Benton County, 260 were in favor and 377 were not.

(a) Find the proportions of farmers in favor of the program in each of the two counties.

(b) Find the standard error needed to compute a confidence interval for the difference in the proportions.

(c) Compute a 95% confidence interval for the difference between the proportions of farmers favoring the program in Tippecanoe County and in Benton County. Do you think opinions differed in the two counties?

8.46 Return to the survey of farmers described in the previous exercise.

(a) Combine the two samples and find the overall proportion of farmers who favor the corn checkoff program.

(b) Find the standard error needed for testing that the population proportions of farmers favoring the program are the same in the two counties.

(c) Formulate null and alternative hypotheses for comparing the two counties.

(d) Compute the z statistic and its P-value. What conclusion do you draw?

8.47 A major court case on the health effects of drinking contaminated water took place in the town of Woburn, Massachusetts. A town well in Woburn was contaminated by industrial chemicals. During the period that residents drank water from this well, there were 16 birth defects among 414 births. In years when the contaminated well was shut off and water was supplied from other wells, there were 3 birth defects among 228 births. The plaintiffs suing the firm responsible for the contamination claimed that these data show that the rate of birth defects was higher when the contaminated well was in use.[14] How statistically significant is the evidence? What assumptions does your analysis require? Do these assumptions seem reasonable in this case?

8.48 A study of chromosome abnormalities and criminality examined data on 4124 Danish males born in Copenhagen.[15] The study used the penal registers maintained in the offices of the local police chiefs and classified each man as having a criminal record or not. Each was also classified as having the normal male XY chromosome pair or one of the abnormalities XYY or XXY. Of the 4096 men with normal chromosomes, 381 had criminal records, while 8 of the 28 men with chromosome abnormalities had criminal records. Some experts believe that chromosome abnormalities are associated with increased criminality. Do these data lend support to this belief? Report your analysis and draw a conclusion.

8.49 A university financial aid office polled an SRS of undergraduate students to study their summer employment. Not all students were employed the previous summer. Here are the results for men and women:

	Men	Women
Employed	728	603
Not employed	89	149
Total	817	752

(a) Is there evidence that the proportion of male students employed during the summer differs from the proportion of female students who were employed? State H_0 and H_a, compute the test statistic, and give its P-value.

(b) Give a 95% confidence interval for the difference between the proportions of male and female students who were employed during the summer. Does the difference seem practically important to you?

8.50 A clinical trial examined the effectiveness of aspirin in the treatment of cerebral ischemia (stroke). Patients were randomized into treatment and control groups. The study was double-blind in the sense that neither the patients nor the physicians who evaluated the patients knew which patients received aspirin and which the placebo tablet.[16] After six months of treatment, the attending physicians evaluated each patient's progress as either favorable or unfavorable. Of the 78 patients in the aspirin group, 63 had favorable outcomes; 43 of the 77 control patients had favorable outcomes.

(a) Compute the sample proportions of patients having favorable outcomes in the two groups.

(b) Give a 90% confidence interval for the difference between the favorable proportions in the treatment and control groups.

(c) The physicians conducting the study had concluded from previous research that aspirin was likely to increase the chance of a favorable outcome. Carry out a significance test to confirm this conclusion. State hypotheses, find the P-value, and write a summary of your results.

8.51 The pesticide diazinon is in common use to treat infestations of the German cockroach, *Blattella germanica*. A study investigated the persistence of this pesticide on various types of surfaces.[17] Researchers applied a 0.5% emulsion of diazinon to glass and plasterboard. After 14 days, they placed 18 cockroaches on each surface and recorded the number that died within 48 hours. On glass, 9 cockroaches died, while on plasterboard, 13 died.

(a) Calculate the mortality rates (sample proportion that died) for the two surfaces.

(b) Find a 95% confidence interval for the difference in the two population proportions.

(c) Chemical analysis of the residues of diazinon suggests that it may persist longer on plasterboard than on glass because it binds to the paper covering on the plasterboard. The researchers therefore expected the mortality rate to be greater on plasterboard than on glass. Conduct a significance test to assess the evidence that this is true.

8.52 Refer to the study of undergraduate student summer employment described in Exercise 8.49. Similar results from a smaller number of students may not have the same statistical significance. Specifically, suppose that 73 of 82 men surveyed were employed and 60 of 75 women surveyed were employed. The sample proportions are essentially the same as in the earlier exercise.

(a) Compute the z statistic for these data and report the P-value. What do you conclude?

(b) Compare the results of this significance test with your results in Exercise 8.49. What do you observe about the effect of the sample size on the results of these significance tests?

8.53 Suppose that the experiment of Exercise 8.51 placed more cockroaches on each surface and observed similar mortality rates. Specifically, suppose that 36 cockroaches were placed on each surface and that 26 died on the plasterboard, while 18 died on the glass.

(a) Compute the z statistic for these data and report its P-value. What do you conclude?

(b) Compare the results of this significance test with those you gave in Exercise 8.51. What do you observe about the effect of the sample size on the results of these significance tests?

CHAPTER 8 EXERCISES

8.54 "The nature of work is changing at whirlwind speed. Perhaps now more than ever before, job stress poses a threat to the health of workers and, in turn, to the health of organizations."[18] So says the National Institute for Occupational Safety and Health. Employers are concerned about the effect of stress on their employees. Stress can lower morale and efficiency and increase medical costs. A large survey of restaurant employees found that 75% reported that work stress had a negative impact on their personal lives.[19] The human resources manager of a chain of restaurants is concerned that work stress may be affecting the chain's employees. She asks a random sample of 100 employees to respond Yes or No to the question "Does work stress have a negative impact on your personal life?" Of these, 68 say "Yes." Give a 95% confidence interval for the proportion of employees who work for this chain of restaurants who believe that work stress has a negative impact on their personal lives.

8.55 Refer to the previous exercise. Is there evidence to conclude that the proportion for this chain of restaurants differs from the value given for the national survey? For this exercise, assume that there is no error associated with the estimate for the national survey.

8.56 Following complaints about the working conditions in some apparel factories both in the United States and abroad, a joint government and industry commission recommended in 1998 that companies that monitor and enforce proper standards be allowed to display a "No Sweat" label on their products. Does the presence of these labels influence consumer behavior? A survey of U.S. residents aged 18 or older asked a series of questions about how likely they would be to purchase a garment under various conditions. For some conditions, it was stated that the garment had a "No Sweat" label; for others, there was no mention of such a label. On the basis of the responses, each person was classified as a "label user" or a "label nonuser."[20] There were 296 women surveyed. Of these, 63 were label users. On the other hand, 27 of 251 men were classified as users. Give a 95% confidence interval for the difference in the proportions.

8.57 Refer to the previous exercise. You would like to compare the women with the men. Set up appropriate hypotheses, and find the test statistic and the *P*-value. What do you conclude?

8.58 The Ping Company makes custom-built golf clubs and competes in the $4 billion golf equipment industry. To improve its business processes, Ping decided to seek ISO 9001 certification.[21] As part of this process, a study of the time it took to repair golf clubs sent to the company by mail determined that 16% of orders were sent back to the customers in 5 days or less. Ping examined the processing of repair orders and made changes. Following the changes, 90% of orders were completed within 5 days. Assume that each of the estimated percents is based on a random sample of 200 orders.

(a) How many orders were completed in 5 days or less before the changes? Give a 95% confidence interval for the proportion of orders completed in this time.

(b) Do the same for orders after the changes.

(c) Give a 95% confidence interval for the improvement. Express this both for a difference in proportions and for a difference in percents.

8.59 To devise effective marketing strategies it is helpful to know the characteristics of your customers. A study compared demographic characteristics of people who use the Internet for travel arrangements and of people who do not.[22] Of 1132 Internet users, 643 had completed college. Among the 852 nonusers, 349 had completed college.

(a) Do users and nonusers differ significantly in the proportion of college graduates?

(b) Give a 95% confidence interval for the difference in the proportions.

8.60 The study mentioned in the previous exercise also asked about income. Among Internet users, 493 reported income of less than $50,000 and 378 reported income of $50,000 or more. (Not everyone answered the income question.) The corresponding numbers for nonusers were 477 and 200.

Perform a significance test to compare the incomes of users with nonusers and also give an estimate of the difference in proportions with a 95% margin of error.

8.61 Refer to the previous two exercises. Give the total number of users and the total number of nonusers for the analysis of education. Do the same for the analysis of income. The difference is due to respondents who chose "Rather not say" for the income question. Give the proportions of "Rather not say" individuals for users and nonusers. Perform a significance test to compare these and give a 95% confidence interval for the difference. People are often reluctant to provide information about their income. Do you think that this type of nonresponse for the income is a serious limitation for this study?

8.62 According to literature on brand loyalty, consumers who are loyal to a brand are likely to consistently select the same product. This type of consistency could come from a positive childhood association. To examine brand loyalty among fans of the Chicago Cubs, 371 Cubs fans among patrons of a restaurant located in Wrigleyville were surveyed prior to a game at Wrigley Field, the Cubs home field.[23] The respondents were classified as "die-hard fans" or "less loyal fans." Of the 134 die-hard fans, 90.3% reported that they watched or listened to Cubs games when they were children. Among the 237 less loyal fans, 67.9% said that they watched or listened as children.

(a) Find the number of die-hard Cubs fans who watched or listened to games when they were children. Do the same for the less loyal fans.

(b) Use a significance test to compare the die-hard fans with the less loyal fans with respect to their childhood experiences relative to the team.

(c) Express the results with a 95% confidence interval for the difference in proportions.

8.63 The study mentioned in the previous exercise found that two-thirds of the die-hard fans attended Cubs games at least once a month, but only 20% of the less loyal fans attended this often. Analyze these data using a significance test and a confidence interval. Write a short summary of your findings.

8.64 A Gallup Poll used telephone interviews to survey a sample of 1025 U.S. residents over the age of 18 regarding their use of credit cards.[24] The poll reported that 76% of Americans said that they had at least one credit card. Give the 95% margin of error for this estimate.

8.65 The Gallup Poll in the previous exercise reported that 41% of those who have credit cards do not pay the full balance each month. Find the number of people in the survey who said that they had at least one credit card, using the information in the previous exercise. Combine this number with the reported 41% to give a margin of error for the proportion of credit card owners who do not pay their full balance.

8.66 Refer to the study of gender bias and stereotyping described in Exercise 8.35 (page 596). Here are the counts of "girl," "woman," "boy," and "man" for all of the grammar texts studied. The one we analyzed in Exercise 8.35 was number 6.

	Text number									
	1	2	3	4	5	6	7	8	9	10
Girl	2	5	25	11	2	48	38	5	48	13
Woman	3	2	31	65	1	12	2	13	24	5
Boy	7	18	14	19	12	52	70	6	128	32
Man	27	45	51	138	31	80	2	27	48	95

For each text perform the significance test to compare the proportions of juvenile references for females and males. Summarize the results of the significance tests for the 10 texts studied. The researchers who conducted the study note that the authors of the last three texts are women, while the other seven texts were written by men. Do you see any pattern that suggests that the gender of the author is associated with the results?

8.67 Refer to the previous exercise. Let us now combine the categories "girl" with "woman" and "boy" with "man." For each text calculate the proportion of male references and test the hypothesis that male and female references are equally likely (that is, the proportion of male references is equal to 0.5). Summarize the results of your 10 tests. Is there a pattern that suggests a relation with the gender of the author?

8.68 Many new products introduced into the market are targeted toward children. The choice behavior of children with regard to new products is of particular interest to companies that design marketing strategies for these products. As part of one study, children in different age groups were compared on their ability to sort new products into the correct product category (milk or juice).[25] Here are some of the data:

Age group	n	Number who sorted correctly
4- to 5-year-olds	50	10
6- to 7-year-olds	53	28

Test the null hypothesis that the two age groups are equally skilled at sorting. Justify your choice of an alternative hypothesis. Also, give a 90% confidence interval for the difference. Summarize your results with a short paragraph.

8.69 A television news program conducts a call-in poll about a proposed city ban on handgun ownership. Of the 2372 calls, 1921 oppose the ban. The station, following recommended practice, makes a confidence statement: "81% of the Channel 13 Pulse Poll sample opposed the ban. We can be 95%

confident that the true proportion of citizens opposing a handgun ban is within 1.6% of the sample result." Is this conclusion justified?

8.70 Eleven percent of the products produced by an industrial process over the past several months fail to conform to the specifications. The company modifies the process in an attempt to reduce the rate of nonconformities. In a trial run, the modified process produces 16 nonconforming items out of a total of 300 produced. Do these results demonstrate that the modification is effective? Support your conclusion with a clear statement of your assumptions and the results of your statistical calculations.

8.71 In the setting of the previous exercise, give a 95% confidence interval for the proportion of nonconforming items for the modified process. Then, taking $p_0 = 0.11$ to be the old proportion and p the proportion for the modified process, give a 95% confidence interval for $p - p_0$.

8.72 In a random sample of 950 students from a large public university, it was found that 444 of the students changed majors during their college years.

(a) Give a 95% confidence interval for the proportion of students at this university who change majors.

(b) Express your results from (a) in terms of the *percent* of students who change majors.

(c) University officials concerned with counseling students are interested in the number of students who change majors rather than the proportion. The university has 38,000 undergraduate students. Convert the confidence interval you found in (a) to a confidence interval for the *number* of students who change majors during their college years.

8.73 In a study on blood pressure and diet, a random sample of Seventh-Day Adventists were interviewed at a national meeting. Because many people who belong to this denomination are vegetarians, they are a very useful group for studying the effects of a meatless diet.[26] Blacks in the population as a whole have a higher average blood pressure than whites. A study of this type should therefore take race into account in the analysis. The 312 people in the sample were categorized by race and whether or not they were vegetarians. The data are given in the following table:

	Black	White
Vegetarian	42	135
Not vegetarian	47	88

Are the proportions of vegetarians the same among all black and white Seventh-Day Adventists who attended this meeting? Analyze the data, paying particular attention to this question. Summarize your analysis and conclusions. What can you infer about the proportions of vegetarians among black and white Seventh-Day Adventists in general? What about blacks and whites in general?

8.74 In Example 8.11 (page 594) we discussed a study that examined the association between high blood pressure and increased risk of death from cardiovascular disease. There were 2676 men with low blood pressure and 3338 men with high blood pressure. In the low-blood-pressure group, 21 men died from cardiovascular disease; in the high-blood-pressure group, 55 died.

(a) Verify the calculations in Example 8.11 by computing the 95% confidence interval for the difference in proportions.

(b) Do the study data confirm that death rates are higher among men with high blood pressure? State hypotheses, carry out a significance test, and give your conclusions.

8.75 An experiment designed to assess the effects of aspirin on cardiovascular disease studied 5139 male British medical doctors. The doctors were randomly assigned to two groups. One group of 3429 doctors took one aspirin daily, and the other group did not take aspirin. After 6 years, there were 148 deaths from heart attack or stroke in the first group and 79 in the second group. A similar experiment used male American medical doctors as subjects. These doctors were also randomly assigned to one of two groups. The 11,037 doctors in the first group took one aspirin every other day, and the 11,034 doctors in the second group took no aspirin. After nearly 5 years, there were 104 deaths from heart attacks in the first group and 189 in the second.[27] Analyze the data from these two studies and summarize the results. How do the conclusions of the two studies differ, and why?

8.76 Gastric freezing was once a recommended treatment for ulcers in the upper intestine. A randomized comparative experiment found that 28 of the 82 patients who were subjected to gastric freezing improved, while 30 of the 78 patients in the control group improved.

(a) State the appropriate hypothesis and a two-sided alternative and carry out a z test. What is the P-value?

(b) What do you conclude about the effectiveness of gastric freezing as a treatment for ulcers? (See Example 3.5 on page 231 for a discussion of gastric freezing.)

8.77 In this exercise we examine the effect of the sample size on the significance test for comparing two proportions. In each case suppose that $\hat{p}_1 = 0.7$ and $\hat{p}_2 = 0.5$, and take n to be the common value of n_1 and n_2. Use the z statistic to test H_0: $p_1 = p_2$ versus the alternative H_a: $p_1 \neq p_2$. Compute the statistic and the associated P-value for the following values of n: 10, 20, 40, 60, 80, 100, and 500. Summarize the results in a table. Explain what you observe about the effect of the sample size on statistical significance when the sample proportions $\hat{p}_1$ and $\hat{p}_2$ are unchanged.

8.78 In the first section of this chapter, we studied the effect of the sample size on the margin of error of the confidence interval for a single proportion. In this exercise we perform some calculations to observe this effect for the two-sample problem. Suppose that $\tilde{p}_1 = 0.7$ and $\tilde{p}_2 = 0.5$, and n represents the common value of n_1 and n_2. Compute the 95% margins of error for the difference in the two proportions for n = 16, 26, 56, 86, 106, and 506. Summarize and explain your results.

8.79 For a single proportion the margin of error of a confidence interval is largest for any given sample size n and confidence level C when $\hat{p} = 0.5$. This led us to use $p^* = 0.5$ for planning purposes. The same kind of result is true for the two-sample problem. The margin of error of the confidence interval for the difference between two proportions is largest when $\hat{p}_1 = \hat{p}_2 = 0.5$. Use these conservative values in the following calculations, and assume that the sample sizes n_1 and n_2 have the common value n. Calculate the margins of error of the 95% confidence intervals for the difference in two proportions for the following choices of n: 16, 26, 56, 86, 106, and 506. Present the results in a table or with a graph. Summarize your conclusions.

8.80 As the previous problem noted, using the guessed value 0.5 for both $\tilde{p}_1$ and $\tilde{p}_2$ gives a conservative margin of error in confidence intervals for the difference between two population proportions. You are planning a survey and will calculate a 95% confidence interval for the difference in two proportions when the data are collected. You would like the margin of error of the interval to be less than or equal to 0.05. You will use the same sample size n for both populations.

(a) How large a value of n is needed?

(b) Give a general formula for n in terms of the desired margin of error m and the critical value z^*.

8.81 *Castaneda v. Partida* is an important court case in which statistical methods were used as part of a legal argument.[28] When reviewing this case, the Supreme Court used the phrase "two or three standard deviations" as a criterion for statistical significance. This Supreme Court review has served as the basis for many subsequent applications of statistical methods in legal settings. (The two or three standard deviations referred to by the Court are values of the z statistic and correspond to P-values of approximately 0.05 and 0.0026.) In *Castaneda* the plaintiffs alleged that the method for selecting juries in a county in Texas was biased against Mexican Americans. For the period of time at issue, there were 181,535 persons eligible for jury duty, of whom 143,611 were Mexican Americans. Of the 870 people selected for jury duty, 339 were Mexican Americans.

(a) What proportion of eligible voters were Mexican Americans? Let this value be p_0.

(b) Let p be the probability that a randomly selected juror is a Mexican American. The null hypothesis to be tested is H_0: $p = p_0$. Find the value of $\hat{p}$ for this problem, compute the z statistic, and find the P-value. What do you conclude? (A finding of statistical significance in this circumstance does not constitute a proof of discrimination. It can be used, however, to establish a prima facie case. The burden of proof then shifts to the defense.)

(c) We can reformulate this exercise as a two-sample problem. Here we wish to compare the proportion of Mexican Americans among those selected as jurors with the proportion of Mexican Americans among those not selected as jurors. Let p_1 be the probability that a randomly selected juror is a Mexican American, and let p_2 be the probability that a randomly selected nonjuror is a Mexican American. Find the z statistic and its P-value. How do your answers compare with your results in (b)?

PART 3

Topics in Inference

Prelude: music and marketing

Music influences our moods and behavior. Compare, for instance, the music we hear during car chases with that played during love scenes in films and on television. Marketers try to choose background music that will influence consumers. Does the type of music we hear have an effect on the purchases we make? One study of this question varied the type of background music and observed the effect on wine sales. You see that this is a comparative experiment. The percent of French wine among all bottles sold was 52% when French music was playing and 36% when other types of music were playing.[1] The researchers used the methods of this chapter to determine that their results were statistically significant.

CHAPTER 9

Gene Davis, *Prince Albert*, 1978.

Analysis of Two-Way Tables

Introduction

We continue our study of methods for analyzing categorical data in this chapter. Inference about proportions in one-sample and two-sample settings was the focus of Chapter 8. We now study how to compare two or more populations when the response variable has two or more possible values and how to test whether two categorical variables are independent. A single statistical test handles both of these cases. The corresponding procedures for quantitative variables are more elaborate; they will appear in Chapters 12 and 13.

Of course, we need to do a careful job of describing the data before we apply our methods of statistical inference. Therefore, we begin our study with a discussion of data analysis methods for two-way tables. The second section of this chapter gives the basics of statistical inference that is appropriate in this setting. An optional third section provides some technical details.

9.1 Data Analysis for Two-Way Tables

When we studied inference for two proportions in Chapter 8, we started by summarizing the raw data by giving the number of observations in each group (n) and how many of these were classified as "successes" (X).

EXAMPLE 9.1

Here is a summary from Example 8.8 (page 589), where we compared frequent binge drinking of men and women by examining the proportions of each gender who engage in this activity.

Population	n	X
1 (men)	7,180	1,630
2 (women)	9,916	1,684
Total	17,096	3,314

The two-way table

two-way table

In this chapter we use a different summary of the data. Rather than recording just the count of binge drinkers, we record counts for all outcomes in a **two-way table.**

EXAMPLE 9.2

Here is the two-way table classifying students by gender and whether or not they are binge drinkers:

Two-way table for frequent binge drinking and gender

	Gender		
Frequent binge drinker	Men	Women	Total
Yes	1,630	1,684	3,314
No	5,550	8,232	13,782
Total	7,180	9,916	17,096

For this problem we could view gender as an explanatory variable and frequent binge drinking as a response variable. This is why we put gender in the columns (like the x axis in a regression) and frequent binge drinking in the rows (like the y axis in a regression). Be sure that you understand how this table is obtained from the one in Example 9.1. Most errors in the use of categorical-data methods come from a misunderstanding of how these tables are constructed.

In Chapter 2 we studied relationships in which at least the response variable was quantitative. In this section we have a similar goal; but here both of the variables are categorical. Some variables—such as gender, race, and occupation—are inherently categorical. Other categorical variables are created by grouping values of a quantitative variable into classes. Published data are often reported in grouped form to save space. To describe categorical data, we use the *counts* (frequencies) or *percents* (relative frequencies) of individuals that fall into various categories.

row and column variables

The two-way table in Example 9.2 describes two categorical variables, gender and binge drinking. We call binge drinking the **row variable** because each horizontal row in the table describes the drinking behavior. Gender is the **column variable** because each vertical column describes one gender group. Each combination of values for these two variables is called a **cell.** For example, the cell corresponding to men who are frequent binge drinkers contains the number 1630. This table is called a 2 × 2 table because there are 2 rows and 2 columns.

cell

Suppose we are interested in the men who are binge drinkers. The proportion of these is simply 1630 divided by 17,096, or 0.095. We would estimate that 9.5% of college students are male frequent binge drinkers. For each cell, a proportion is obtained by dividing the cell entry by the total sample size. The collection of these proportions is the **joint distribution** of the two categorical variables.

joint distribution

EXAMPLE 9.3

For the binge-drinking example, the joint distribution of binge drinking and gender is

Joint distribution of frequent binge drinking and gender

	Gender	
Frequent binge drinker	Men	Women
Yes	0.095	0.099
No	0.325	0.482

Because this is a distribution, the sum of the proportions should be 1. For this example the sum is 1.001. The difference is due to roundoff error.

From the joint distribution we see that the proportions of men and women frequent binge drinkers are similar in the population of college students. For the men we have 9.5%; the women are slightly higher at 9.9%. This suggests that the numbers of frequent binge drinkers in the population of college students is similar for men and women. Note, however, that the proportion for women who are not frequent binge drinkers is also higher than the proportion for men. One reason for this is that there are more women in the sample than men. To understand this set of data we will need to do some additional calculations. Let's look at the distribution of gender.

Marginal distributions

marginal distribution

When we examine the distribution of a single variable in a two-way table, we are looking at a **marginal distribution.** There are two marginal distributions, one for each categorical variable in the two-way table. They are very easy to compute.

EXAMPLE 9.4

Look at the table in Example 9.2. The total numbers of men and women are given in the bottom row, labeled "Total." Our sample has 7180 men and 9916 women. To find the marginal distribution of gender we simply divide these numbers by the total sample size, 17,096. The marginal distribution of gender is

Marginal distribution of gender		
	Men	Women
Proportion	0.420	0.580

Note that the proportions sum to 1; there is no roundoff error.

Often we prefer to use percents rather than proportions. Here is the marginal distribution of gender described with percents:

Marginal distribution of gender		
	Men	Women
Proportion	42.0%	58.0%

Which form do you prefer?

The other marginal distribution for this example is the distribution of binge drinking.

EXAMPLE 9.5

In the table for Example 9.2, the numbers of frequent binge drinkers and non-frequent binge drinkers are given in the last column. This is the other margin of the table. There are 3314 frequent binge drinkers. The other 13,782 students are not frequent binge drinkers. If we convert these counts to percents, we obtain the marginal distribution of the frequent binge drinking variable (in terms of percents):

Marginal distribution of frequent binge drinking		
	Yes	No
Percent	19.4%	80.6%

Each marginal distribution from a two-way table is a distribution for a single categorical variable. As we saw in Chapter 1 (pages 6 and 7), we can use a bar graph or a pie chart to display such a distribution. For our two-way table, we will be content with numerical summaries: 58% of these college students are women, and 19.4% of the students are frequent binge drinkers. When we have more rows or columns, the graphical displays are particularly useful. We will illustrate these for other examples in Figures 9.3 and 9.5 (pages 622 and 631).

Describing relations in two-way tables

The table in Exercise 9.2 contains much more information than the two marginal distributions of gender alone and frequent binge drinking alone. We need to do a little more work to examine the relationship. *Relationships among categorical variables are described by calculating appropriate percents from the counts given.* What percents do you think we should use to describe the relationship between gender and frequent binge drinking?

EXAMPLE 9.6

What percent of the women in our sample are frequent binge drinkers? This is the count of the women who are frequent binge drinkers as a percent of the number of women in the sample:

$$\frac{1684}{9916} = 0.170 = 17.0\%$$

A similar calculation gives the percent of men who are frequent binge drinkers:

$$\frac{1630}{7180} = 0.227 = 22.7\%$$

Recall that among all college students in the sample, 9.5% were male frequent binge drinkers and 9.9% were female frequent binge drinkers. These percents are based on the joint distribution. They are fairly similar because the counts for these two groups, 1630 and 1684, are close. The calculations that we just performed, however, give us a different view. When we look separately at women and men, we see that the proportions of frequent binge drinkers are somewhat different, 17.0% for women versus 22.7% for men.

Conditional distributions

conditional distribution

In Example 9.6 we first looked at the women alone and then examined the distribution of the other categorical variable, frequent binge drinking. Another way to say this is that we conditioned on the value of gender being female. Similarly we conditioned on the value of gender being male. When we condition on the value of one variable and calculate the distribution of the other variable, we obtain a **conditional distribution.** Note that in Example 9.6 we calculated the percents for only frequent binge drinking. The complete conditional distribution gives the proportions or percents for all possible values.

EXAMPLE 9.7

For women, the conditional distribution of the frequent binge-drinking variable in terms of percents is

Conditional distribution of frequent binge drinking for women

	Yes	No
Percent	17.0%	83.0%

Note that we have included the percents for both of the possible values, Yes and No, of the frequent binge-drinking variable. These percents sum to 100%. Similar calculations give the corresponding distribution for men:

Conditional distribution of frequent binge drinking for men

	Yes	No
Percent	22.7%	77.3%

Comparing the conditional distributions reveals the nature of the association between gender and frequent binge drinking. In this set of data the men are more likely to be frequent binge drinkers than the women.

Bar graphs can help us to see relationships between two categorical variables. We will use these in our example for the following section. No single graph (such as a scatterplot) portrays the form of the relationship between categorical variables, and no single numerical measure (such as the correlation) summarizes the strength of an association. Bar graphs are flexible enough to be helpful, but you must think about what comparisons you want to display. For numerical measures, we must rely on well-chosen percents or on more advanced statistical methods.[2]

A two-way table contains a great deal of information in compact form. Making that information clear almost always requires finding percents. You must decide which percents you need.

Of course, we prefer to use software to compute the joint, marginal, and conditional distributions.

EXAMPLE 9.8

Figure 9.1 gives SAS output for the data in Example 9.2. There are minor variations among software packages, but this is typical of what is usually produced. Each cell in the 2×2 table has four entries. The legend in the upper-left corner helps to explain the entries. "Frequency" is the count in the cell. The "Percent" entries give the joint distribution. The marginal distributions are labeled "Row Pct" and "Col Pct." Marginal totals and distributions are given in the rightmost column and the bottom row.

Note that we used Yes for the first row of the table, but the software used No. Most software packages order the row and column labels numerically or alphabetically. You should verify that the entries in Figure 9.1 correspond to the calculations that we performed in Examples 9.3 to 9.7. In addition, verify the calculations for the conditional distributions of gender for each value of the frequent binge-drinking variable.

```
Table of Binge by Gender

Binge        Gender

Frequency|
Percent  |
Row Pct  |
Col Pct  | Men       | Women    | Total
---------+-----------+----------+------
No       |   5550    |   8232   | 13782
         |  32.46    |  48.15   | 80.62
         |  40.27    |  59.73   |
         |  77.30    |  83.02   |
---------+-----------+----------+------
Yes      |   1630    |   1684   |  3314
         |   9.53    |   9.85   | 19.38
         |  49.19    |  50.81   |
         |  22.70    |  16.98   |
---------+-----------+----------+------
Total        7180       9916     17096
            42.00      58.00    100.00
```

FIGURE 9.1 Computer output for the binge-drinking study in Example 9.2.

Simpson's paradox

As is the case with quantitative variables, the effects of lurking variables can strongly influence relationships between two categorical variables. Here is an example that demonstrates the surprises that can await the unsuspecting consumer of data.

EXAMPLE 9.9 To help people make informed decisions about health care, the government releases data about patient outcomes in hospitals. You want to compare Hospital A and Hospital B, which serve your community. Here is a two-way table of data on the survival of patients after surgery in these two hospitals. All patients undergoing surgery in a recent time period are included. "Survived" means that the patient lived at least 6 weeks following surgery.

Patient survival by hospital

	Hospital	
Survival	A	B
Died	63	16
Survived	2037	784
Total	2100	800

The evidence seems clear: Hospital A loses 3% (63/2100) of its surgery patients, and Hospital B loses only 2% (16/800). It seems that you should choose Hospital B if you need surgery.

Let's take a more careful look at the problem. We know that not all surgery cases are equally serious. Later in the government report you find data on the outcome of surgery broken down by the condition of the patient before the operation.

EXAMPLE 9.10

Patients are classified as being in either "poor" or "good" condition. Here are the more detailed data. Check that the entries in the original two-way table are just the sums of the "poor" and "good" entries in this pair of tables.

Good condition		
	Hospital	
Survival	A	B
Died	6	8
Survived	594	592
Total	600	600

Poor condition		
	Hospital	
Survival	A	B
Died	57	8
Survived	1443	192
Total	1500	200

Hospital A beats Hospital B for patients in good condition: only 1% (6/600) died in Hospital A, compared with 1.3% (8/600) in Hospital B. Hospital A wins again for patients in poor condition, losing 3.8% (57/1500) to Hospital B's 4% (8/200). So Hospital A is safer for both patients in good condition and patients in poor condition. If you are facing surgery, you should choose Hospital A.

The patient's condition is a lurking variable when we compare the death rates at the two hospitals. When we ignore the lurking variable, Hospital B seems safer, even though Hospital A does better for both classes of patients. How can A do better in each group, yet do worse overall? Look at the data. Hospital A is a medical center that attracts seriously ill patients from a wide region. It had 1500 patients in poor condition. Hospital B had only 200 such cases. Because patients in poor condition are more likely to die, Hospital A has a higher death rate despite its superior performance for each class of patients. The original two-way table, which did not take account of the condition of the patients, was misleading. Examples 9.9 and 9.10 illustrate *Simpson's paradox*.

Simpson's Paradox

An association or comparison that holds for all of several groups can reverse direction when the data are combined to form a single group. This reversal is called **Simpson's paradox.**

The lurking variables in Simpson's paradox are categorical. That is, they break the individuals into groups, as when surgery patients are classified as "good condition" or "poor condition." Simpson's paradox is an extreme form of the fact that observed associations can be misleading when there are lurking variables. Although the counts in Examples 9.9 and 9.10 were made up, the setting is real. The hospital death rates compiled and released by the Federal Health Care Financing Agency do allow for each patient's condition and diagnosis. Hospitals continue to complain that there are differences among patients that can make the government's mortality data misleading.

The perils of aggregation

three-way table

The table of outcome by hospital by patient condition in Example 9.10 is a **three-way table** that reports the frequencies of each combination of levels of three categorical variables. We present a three-way table as two or more two-way tables, one for each level of the third variable. In Example 9.10, there is a separate table of patient outcome by hospital for each patient condition. We can obtain the original two-way table by adding the corresponding entries in these tables. This is called **aggregating** the data. Aggregation has the effect of ignoring patient condition, which then becomes a lurking variable. Conclusions that seem obvious when we look only at aggregated data often become debatable when we examine lurking variables. Here is a real-life example.

aggregation

EXAMPLE 9.11

A study by the National Science Foundation[3] found that the median salary of newly graduated female engineers and scientists was only 73% of the median salary for males. When the new graduates were broken down by field, however, the picture changed. Women's median salaries as a percent of the male median in the 16 fields studied were

94%	96%	98%	95%	85%	85%	84%	100%
103%	100%	107%	93%	104%	93%	106%	100%

How can women do nearly as well as men in every field, and better in some, yet fall far behind men when we look at all young engineers and scientists? Women are concentrated in the life and social sciences, where salaries are lower than in the physical sciences and engineering. When salaries from all fields are lumped together, the women place lower because the fields they favor pay less. Field of science is a lurking variable that influences the observed association between gender and salary.

SUMMARY

A **two-way table** of counts organizes data about two categorical variables. Values of the **row variable** label the rows that run across the table, and values of the **column variable** label the columns that run down the table. Two-way tables are often used to summarize large amounts of data by grouping outcomes into categories.

The **row totals** and **column totals** in a two-way table give the **marginal distributions** of the two variables separately. It is clearer to present these distributions as percents of the table total. Marginal distributions do not give any information about the relationship between the variables.

To find the **conditional distribution** of the row variable for one specific value of the column variable, look only at that one column in the table. Find each entry in the column as a percent of the column total.

There is a conditional distribution of the row variable for each column in the table. Comparing these conditional distributions is one way to describe the association between the row and the column variables. It is particularly useful when the column variable is the explanatory variable. When the row variable is explanatory, find the conditional distribution of the column variable for each row and compare these distributions.

Bar graphs are a flexible means of presenting categorical data. There is no single best way to describe an association between two categorical variables.

We present data on three categorical variables in a **three-way table,** printed as separate two-way tables for each level of the third variable. A comparison between two variables that holds for each level of a third variable can be changed or even reversed when the data are **aggregated** by summing over all levels of the third variable. **Simpson's paradox** refers to the reversal of a comparison by aggregation. It is an example of the potential effect of lurking variables on an observed association.

9.2 Inference for Two-Way Tables

In the previous section we used a 2×2 table to illustrate the basic ideas about joint, marginal, and conditional distributions. There we learned that the key to examining the relationship between two categorical variables is to look at conditional distributions. In this section we discuss the significance test that we use to examine the relationship.

EXAMPLE 9.12

Do men and women participate in sports for the same reasons? One goal for sports participants is social comparison—the desire to win or to do better than other people. Another is mastery—the desire to improve one's skills or to try one's best. A study on why students participate in sports collected data from 67 male and 67 female undergraduates at a large university.[4] Each student was classified into one of four categories based on his or her responses to a questionnaire about sports goals. The four categories were high social comparison–high mastery (HSC-HM), high social comparison–low mastery (HSC-LM), low social comparison–high mastery (LSC-HM), and low social comparison–low mastery (LSC-LM). One purpose of the study was to compare the goals of male and female students. Here are the data displayed in a two-way table:

Observed counts for sports goals

	Gender		
Goal	Female	Male	Total
HSC-HM	14	31	45
HSC-LM	7	18	25
LSC-HM	21	5	26
LSC-LM	25	13	38
Total	67	67	134

The entries in this 4×2 table are the numbers of students in each gender by sports goal category. For example, there are 14 women in the high social comparison–high mastery group. Note that we also give the marginal totals and the grand total.

The two categorical variables in this example are goal and gender. The objective in this example is to compare men and women. We view gender as an explanatory variable and therefore we make it the column variable. The row variable is a categorical response variable, type of sports goal.

EXAMPLE 9.13

Figure 9.2 shows the typical computer output for the data of Example 9.12. It is similar to Figure 9.1, which displayed the output for Example 9.2, but the new figure includes some additional information needed for statistical inference. Check

TABLE OF GOAL BY GENDER

GOAL	GENDER		
FREQUENCY EXPECTED PERCENT ROW PCT COL PCT	FEMALE	MALE	TOTAL
HSC-HM	14 22.5 10.45 31.11 20.90	31 22.5 23.13 68.89 46.27	45 33.58
HSC-LM	7 12.5 5.22 28.00 10.45	18 12.5 13.43 72.00 26.87	25 18.66
LSC-HM	21 13.0 15.67 80.77 31.34	5 13.0 3.73 19.23 7.46	26 19.40
LSC-LM	25 19.0 18.66 65.79 37.31	13 19.0 9.70 34.21 19.40	38 28.36
Total	67 50.00	67 50.00	134 100.00

STATISTICS FOR TABLE OF GOAL BY GENDER

STATISTIC	DF	VALUE	PROB
CHI-SQUARE	3	24.898	0.000

SAMPLE SIZE=134

FIGURE 9.2 Computer output for the sports goals study in Example 9.13.

Figure 9.2 carefully. Be sure that you can identify the joint distribution, the marginal distributions, and the conditional distributions.

In this example, we are interested in the effect of gender on the distribution of sports goals. To compare the men and women we examine conditional distribution of sports goals for each gender. These are the column percents.

Conditional distributions of sports goals		
	Gender	
Goal	Female	Male
HSC-HM	21	46
HSC-LM	10	27
LSC-HM	31	7
LSC-LM	37	19
Total	100	100

We rounded the percents from the computer output to get a clearer summary of the data. The Total row reminds us that these groups account for 100% of both men and women. (In fact, the sums differ slightly from 100% because of roundoff error.) The bar graphs in Figure 9.3 compare the male and female percents. The data reveal something interesting: it appears that females and males have different goals when they participate in recreational sports.

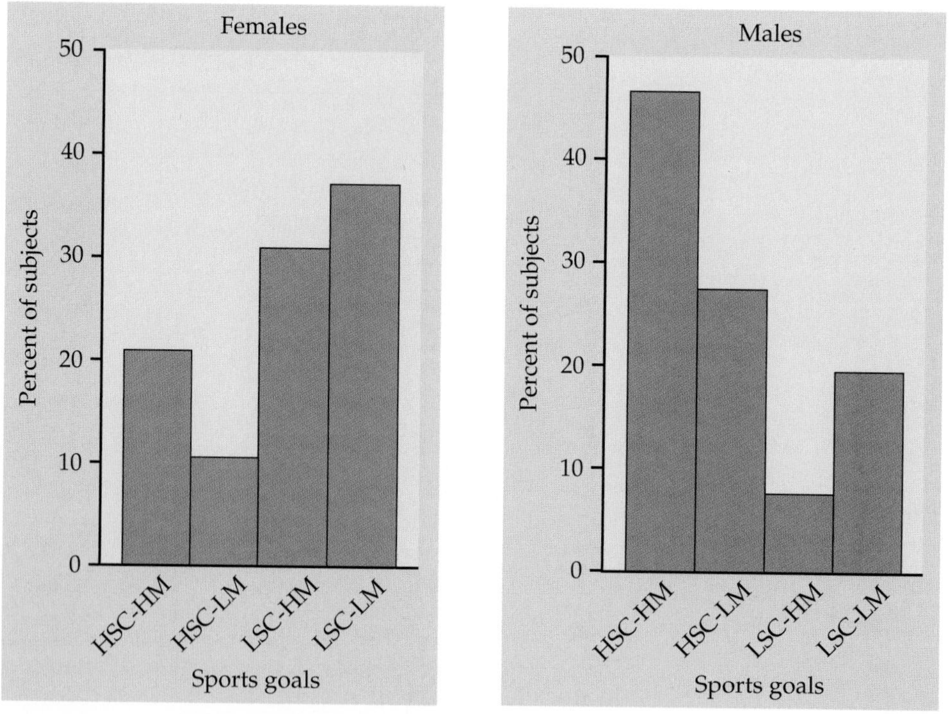

FIGURE 9.3 Comparison of sports goals distributions for females and males, for Example 9.13.

The differences between the distributions of male and female sports goals in the sample appear to be large. A statistical test will tell us whether or not these differences can be plausibly attributed to chance. Specifically, if the two population distributions were the same, how likely is it that a sample would show differences as large or larger than those displayed in Figure 9.3?

The hypothesis: no association

The null hypothesis H_0 of interest in a two-way table is that there is *no association* between the row variable and the column variable. In the sports goals example, this null hypothesis says that gender and sports goals are not related. The alternative hypothesis H_a is that there *is* an association between these two variables. The alternative H_a does not specify any particular direction for the association, but it includes possibilities such as "men rate social comparison higher as a goal than do women." We cannot describe H_a as either one-sided or two-sided, because it includes all of the many kinds of association that are possible.

In our example, the hypothesis H_0 that there is no association between gender and sports goals is equivalent to the statement that the distributions of sports goals in the male and female populations are the same. For $r \times c$ tables like that in Example 9.12, where the columns correspond to independent samples from distinct populations, there are c distributions for the row variable, one for each population. The null hypothesis then says that the c distributions of the row variable are identical. The alternative hypothesis is that the distributions are not all the same.

Expected cell counts

expected cell counts

To test the null hypothesis in $r \times c$ tables, we compare the observed cell counts with **expected cell counts** calculated under the assumption that the null hypothesis is true. A numerical summary of the comparison will be our test statistic.

EXAMPLE 9.14

The expected counts for the sports goals example appear in the computer output shown in Figure 9.2. They are labeled EXPECTED and are the second entry in each cell. For example, the expected number of high social comparison–high mastery females is 22.5.

How is this expected count obtained? Look at the percents in the right margin of the table in Figure 9.2. We see that 33.58% of all respondents (female and male together) are in the HSC-HM group. If the null hypothesis of no gender difference in sports goals is true, we expect this overall percentage to apply to both men and women. So we expect 33.58% of the 67 females in the study to be in this group. The expected count is therefore 33.58% of 67, which is 22.5. The other expected counts are calculated similarly. Because the number of males is the same as the number of females in this study, the expected counts for males are the same as the expected counts for females.

Let's examine the calculation of the expected cell counts a bit more carefully. To compute the expected count for HSC-HM females we multiplied the proportion of HSC-HM people in the data (45/134) by the number of females (67). From Figure 9.2 we see that the numbers 45 and 67 are the row and column totals for the cell of interest; 134 is n, the total number of observations for the table. The expected cell count is therefore the product of the row and column totals divided by the table total:

$$\text{expected cell count} = \frac{\text{row total} \times \text{column total}}{n}$$

The chi-square test

To test the H_0 that there is no association between the row and column classifications, we use a statistic that compares the entire set of observed counts with the set of expected counts. First, take the difference between each observed count and its corresponding expected count, and square these values so that they are all 0 or positive. A large difference means less if it comes from a cell we think will have a large count, so divide each squared difference by the expected count, a kind of standardization. Finally, sum over all cells. The result is called the *chi-square statistic* X^2.*

Chi-Square Statistic

The **chi-square statistic** is a measure of how much the observed cell counts in a two-way table diverge from the expected cell counts. The recipe for the statistic is

$$X^2 = \sum \frac{(\text{observed count} - \text{expected count})^2}{\text{expected count}}$$

where "observed" represents an observed sample count, "expected" represents the expected count for the same cell, and the sum is over all $r \times c$ cells in the table.

If the expected counts and the observed counts are very different, a large value of X^2 will result. Large values of X^2 provide evidence against the null hypothesis. To obtain a *P*-value for the test, we need the sampling distribution of X^2 under the assumption that H_0 (no association between the row and column variables) is true. We once again use an approximation, related to the normal approximation for binomial distributions. The result is a new distribution, the **chi-square distribution,** which we denote by χ^2 (χ is the lowercase Greek letter chi).

chi-square distribution χ^2

Like the t distributions, the χ^2 distributions form a family described by a single parameter, the degrees of freedom. We use χ^2(df) to indicate a particu-

*The chi-square statistic was invented by the English statistician Karl Pearson (1857–1936) in 1900, for purposes slightly different from ours. It is the oldest inference procedure still used in its original form. With the work of Pearson and his contemporaries at the beginning of this century, statistics first emerges as a separate discipline.

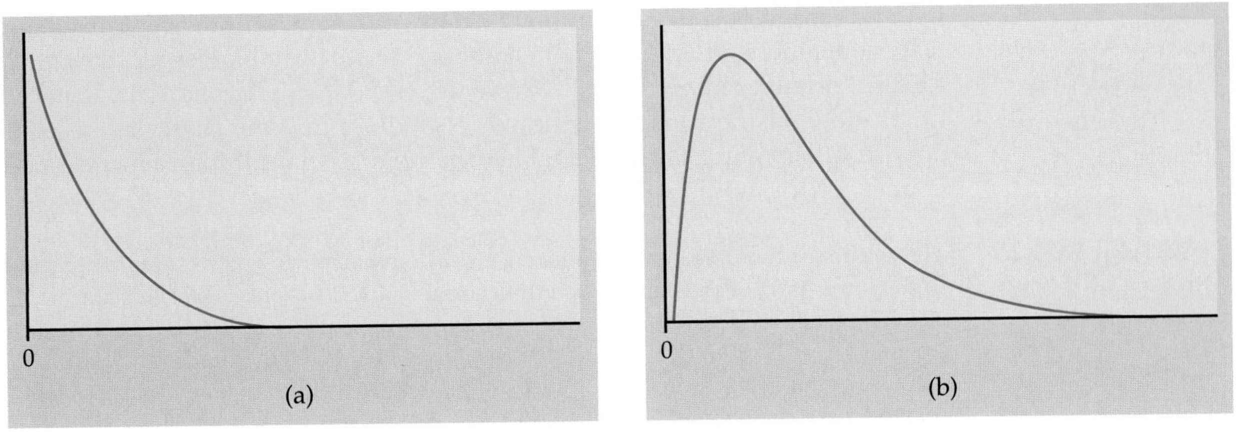

FIGURE 9.4 (a) The $\chi^2(2)$ density curve. (b) The $\chi^2(4)$ density curve.

lar member of this family. Figure 9.4 displays the density curves of the $\chi^2(2)$ and $\chi^2(4)$ distributions. As the figure suggests, χ^2 distributions take only positive values and are skewed to the right. Table F in the back of the book gives upper critical values for the χ^2 distributions.

Chi-Square Test for Two-Way Tables

The null hypothesis H_0 is that there is no association between the row and column variables in a two-way table. The alternative is that these variables are related.

If H_0 is true, the chi-square statistic X^2 has approximately a χ^2 distribution with $(r - 1)(c - 1)$ degrees of freedom.

The P-value for the chi-square test is

$$P(\chi^2 \geq X^2)$$

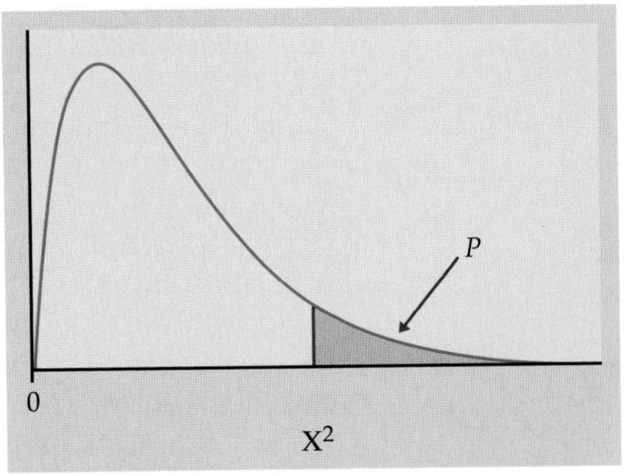

where χ^2 is a random variable having the $\chi^2(\text{df})$ distribution with df = $(r - 1)(c - 1)$.

The chi-square test always uses the upper tail of the χ^2 distribution, because any deviation from the null hypothesis makes the statistic larger. The approximation of the distribution of X^2 by χ^2 becomes more accurate as the cell counts increase. Moreover, it is more accurate for tables larger than 2×2 tables. For tables larger than 2×2, we will use this approximation whenever the average of the expected counts is 5 or more and the smallest expected count is 1 or more. For 2×2 tables, we require that all four expected cell counts be 5 or more.[5]

EXAMPLE 9.15

The results of the chi-square significance test for the sports goals example appear in the lower part of the computer output in Figure 9.2. Because all of the expected cell counts are moderately large, the χ^2 distribution provides accurate *P*-values. We see that $X^2 = 24.898$, df $= 3$, and the *P*-value is given as 0.000 under the heading PROB. Because the *P*-value is rounded to the nearest 0.001, we know that it is less than 0.0005. As a check we verify that the df are correct for a 4×2 table:

$$(r - 1)(c - 1) = (4 - 1)(2 - 1) = 3$$

The chi-square test confirms that the data contain clear evidence against the null hypothesis that female and male students have the same distributions of sports goals. Under H_0, the chance of obtaining a value of X^2 greater than or equal to the calculated value of 24.898 is very small—less than 0.0005.

The test indicates that the male and female distributions are not the same, but it does not say how they differ. Always combine the test with a description that shows what kind of relationship is present. From the percents in Figure 9.2 and the graphs in Figure 9.3, we see that the percent of males in each of the HSC goal classes is more than twice the percent of females. The HSC-HM group contains 46.27% of the males but only 20.90% of the females, and the HSC-LM group contains 26.87% of the males and 10.45% of the females. The pattern is reversed for the LSC goal classes. We conclude that males are more likely to be motivated by social comparison goals and females are more likely to be motivated by mastery goals.

The chi-square test and the z test

A comparison of the proportions of "successes" in two populations leads to a 2×2 table. We can compare two population proportions either by the chi-square test or by the two-sample z test from Section 8.2. In fact, *these tests always give exactly the same result*, because the χ^2 statistic is equal to the square of the z statistic, and $\chi^2(1)$ critical values are equal to the squares of the corresponding $N(0, 1)$ critical values. The advantage of the z test is that we can test either one-sided or two-sided alternatives. The chi-square test always tests the two-sided alternative. Of course, the chi-square test can compare more than two populations, whereas the z test compares only two.

Beyond the basics ▸ meta-analysis

Policymakers wanting to make decisions based on research are sometimes faced with the problem of summarizing the results of many studies. These studies may show effects of different magnitudes, some highly significant and meta-analysis

some not significant. What *overall conclusion* can we draw? **Meta-analysis** is a collection of statistical techniques designed to combine information from different but similar studies. Each individual study must be examined with care to ensure that its design and data quality are adequate. The basic idea is to compute a measure of the effect of interest for each study. These are then combined, usually by taking some sort of weighted average, to produce a summary measure for all of the studies. Of course, a confidence interval for the summary is included in the results. Here is an example.

EXAMPLE 9.16

Vitamin A is often given to young children in developing countries to prevent night blindness. It was observed that children receiving vitamin A appear to have reduced death rates. To investigate the possible relationship between vitamin A supplementation and death, a large field trial with over 25,000 children was undertaken in the Aceh province of Indonesia. About half of the children were given large doses of vitamin A, and the other half were controls. In 1986, the researchers reported a 34% reduction in mortality (deaths) for the treated children who were 1 to 6 years old compared with the controls. Several additional studies were then undertaken. Most of the results confirmed the association: treatment of young children in developing countries with vitamin A reduces the death rate; but the size of the effect varied quite a bit.

How can we use the results of these studies to guide policy decisions? To address this question, a meta-analysis was performed on data available from eight studies.[6] Although the designs varied, each study provided a two-way table of counts. Here is the table for Aceh. A total of $n = 25{,}200$ children were enrolled in the study. Approximately half received vitamin A supplements. One year after the start of the study, the number of children who died was determined.

	Vitamin A	Control
Dead	101	130
Alive	12,890	12,079
Total	12,991	12,209

relative risk

The summary measure chosen was the **relative risk:** the ratio formed by dividing the proportion of children who died in the vitamin A group by the proportion of children who died in the control group. For Aceh, the proportion who died in the vitamin A group was

$$\frac{101}{12{,}991} = 0.00777$$

or 7.7 per thousand; for the control group, the proportion who died was

$$\frac{130}{12{,}209} = 0.01065$$

or 10.6 per thousand. The relative risk is therefore

$$\frac{0.00777}{0.01065} = 0.73$$

Relative risk less than 1 means that the vitamin A group has the lower mortality rate.

The relative risks for the eight studies were

0.73	0.50	0.94	0.71	0.70	1.04	0.74	0.80

A meta-analysis combined these eight results to produce a relative risk estimate of 0.77 with a 95% confidence interval of (0.68, 0.88). That is, vitamin A supplementation reduced the mortality rate to 77% of its value in an untreated group. In other words, there is a 23% reduction in the mortality rate. The confidence interval does not include 1, so the null hypothesis of no effect (a relative risk of 1) can be clearly rejected. The researchers examined many variations of this meta-analysis, such as using different weights and leaving out one study at a time. These variations had little effect on the final estimate.

After these findings were published, large-scale programs to distribute high-potency vitamin A supplements were started. These programs have saved hundreds of thousands of lives since the meta-analysis was conducted and the arguments and uncertainties were resolved.

SUMMARY

The **null hypothesis** for $r \times c$ tables of count data is that there is no relationship between the row variable and the column variable.

Expected cell counts under the null hypothesis are computed using the formula

$$\text{expected count} = \frac{\text{row total} \times \text{column total}}{n}$$

The null hypothesis is tested by the **chi-square statistic,** which compares the observed counts with the expected counts:

$$X^2 = \sum \frac{(\text{observed} - \text{expected})^2}{\text{expected}}$$

Under the null hypothesis, X^2 has approximately the χ^2 distribution with $(r-1)(c-1)$ degrees of freedom. The P-value for the test is

$$P(\chi^2 \geq X^2)$$

where χ^2 is a random variable having the $\chi^2(\text{df})$ distribution with df = $(r-1)(c-1)$.

The chi-square approximation is adequate for practical use when the average expected cell count is 5 or greater and all individual expected counts are 1 or greater, except in the case of 2×2 tables. All four expected counts in a 2×2 table should be 5 or greater.

The section we just completed assumed that you have access to software or a statistical calculator. If you do not, you now need to study the material on computations in the following optional section. All exercises appear at the end of the chapter.

9.3 Formulas and Models for Two-Way Tables*

Computations

The calculations required to analyze a two-way table are straightforward but tedious. In practice, we recommend using software, but it is possible to do the work with a calculator, and some insight can be gained by examining the details. Here is an outline of the steps required.

Computations for Two-Way Tables

1. Calculate descriptive statistics that convey the important information in the table. Usually these will be column or row percents.
2. Find the expected counts and use these to compute the X^2 statistic.
3. Use chi-square critical values from Table F to find the approximate *P*-value.
4. Draw a conclusion about the association between the row and column variables.

The following example illustrates these steps. The two-way table in this example does not compare several populations. Instead, it arises by classifying observations on a single population in two ways.

EXAMPLE 9.17

In a study of heart disease in male federal employees, researchers classified 356 volunteer subjects according to their socioeconomic status (SES) and their smoking habits.[7] There were three categories of SES: high, middle, and low. Individuals were asked whether they were current smokers, former smokers, or had never smoked, producing three categories for smoking habits as well. Here is the two-way table that summarizes the data:

Observed counts for smoking and SES

	SES			
Smoking	High	Middle	Low	Total
Current	51	22	43	116
Former	92	21	28	141
Never	68	9	22	99
Total	211	52	93	356

*The analysis of two-way tables is based on computations that are a bit messy and on statistical models that require a fair amount of notation to describe. This section gives the details. By studying this material you will deepen your understanding of the methods described in this chapter, but this section is optional.

This is a 3 × 3 table, to which we have added the marginal totals obtained by summing across rows and columns. For example, the first-row total is 51 + 22 + 43 = 116. The grand total, the number of subjects in the study, can be computed by summing the row totals, 116 + 141 + 99 = 356, or the column totals, 211 + 52 + 93 = 356. It is good statistical practice to do both as a check on your arithmetic.

Computing conditional distributions

We start our analysis by computing descriptive statistics that summarize the observed relation between SES and smoking. The researchers suspected that SES helps explain smoking, so in this situation SES is the explanatory variable and smoking is the response variable. In general, the clearest description of the relationship is provided by comparing the conditional distributions of the response variable for each value of the explanatory variable. That is, compare the column percents that give the conditional distribution of smoking for each SES category.

EXAMPLE 9.18

We must calculate the column percents. For the high-SES group, there are 51 current smokers out of a total of 211 people. The column proportion for this cell is

$$\frac{51}{211} = 0.242$$

That is, 24.2% of the high-SES group are current smokers. Similarly, 92 of the 211 people in this group are former smokers. The column proportion is

$$\frac{92}{211} = 0.436$$

or 43.6%. In all, we must calculate nine percents. Here are the results:

Column percents for smoking and SES

	SES			
Smoking	High	Middle	Low	All
Current	24.2	42.3	46.2	32.6
Former	43.6	40.4	30.1	39.6
Never	32.2	17.3	23.7	27.8
Total	100.0	100.0	100.0	100.0

In addition to the distributions of smoking behavior in each of the three SES categories, the table gives the overall breakdown of smoking among all 356 subjects. These percents appear in the rightmost column, labeled "All." They make up the marginal distribution of the row variable "Smoking."

The sum of the percents in each column should be 100, except for possible small roundoff errors. It is good practice to calculate each percent separately and then sum each column as a check. In this way we can find arithmetic

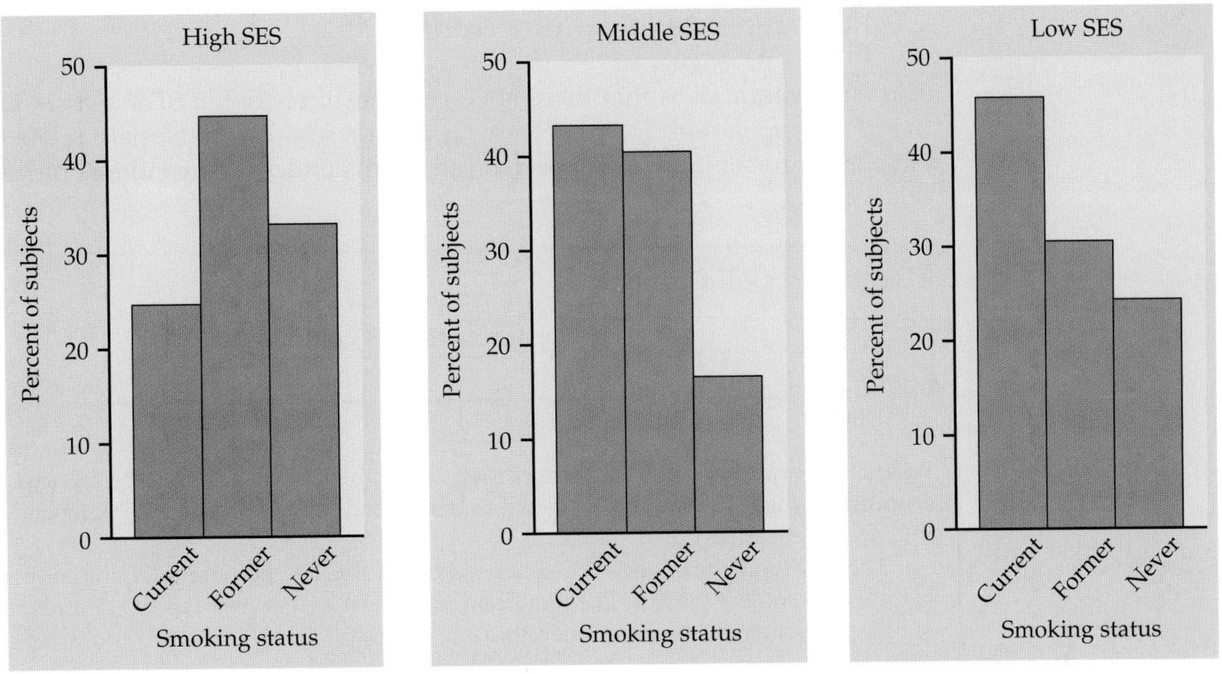

FIGURE 9.5 Comparison of the smoking-behavior distributions for high, middle, and low SES in Example 9.18.

errors that would not be uncovered if, for example, we calculated the column percent for the "Never" row by subtracting the sum of the percents for "Current" and "Former" from 100.

Figure 9.5 compares the distributions of smoking behavior in the three SES groups. The percent of current smokers decreases as SES increases from low to middle to high; in particular, relatively few high-SES subjects smoke. The percent of former smokers increases as SES increases, suggesting that higher-SES smokers were more likely to quit. The percent of people who never smoked is highest in the high-SES group, but the middle-SES group has a somewhat lower percent than the low-SES group. Overall, the column percents suggest that there is a negative association between smoking and SES: higher-SES people tend to smoke less.

The chi-square test assesses whether this observed association is statistically significant. That is, is the SES-smoking relationship in the sample sufficiently strong for us to conclude that it is due to a relationship between these two variables in the underlying population and not merely to chance? Note that the test only asks whether there is evidence of some relationship. To explore the direction or nature of the relationship we must examine the column or row percents. Note also that in using the chi-square test we are acting as if the subjects were a simple random sample from the population of interest. If the volunteers are a biased sample—for example, if smokers are reluctant to volunteer for a study of employee health—then conclusions about the entire population of employees are not justified.

Computing expected cell counts

The null hypothesis is that there is no relationship between SES and smoking in the population. The alternative is that these two variables are related. Here is the formula for the expected cell counts under the hypothesis of "no relationship."

Expected Cell Counts

$$\text{expected count} = \frac{\text{row total} \times \text{column total}}{n}$$

EXAMPLE 9.19

What is the expected count in the upper-left cell in the table of Example 9.17, corresponding to high-SES current smokers, under the null hypothesis that smoking and SES are independent?

The row total, the count of current smokers, is 116. The column total, the count of high-SES subjects, is 211. The total sample size is $n = 356$. The expected number of high-SES current smokers is therefore

$$\frac{(116)(211)}{356} = 68.75$$

Note that 116/356 is the proportion of subjects who are current smokers. The number of high-SES subjects is 211. Because we *expect* 116/356 (32.6%) of these to be current smokers, the expected count is $(116/356) \times 211$. This is what our formula specifies. You can find the expected counts for the other cells in the same way. We summarize these calculations in a table of expected counts:

Expected counts for smoking and SES

	SES			
Smoking	High	Middle	Low	Total
Current	68.75	16.94	30.30	115.99
Former	83.57	20.60	36.83	141.00
Never	58.68	14.46	25.86	99.00
Total	211.00	52.00	92.99	355.99

We can check our work by adding the expected counts to obtain the row and column totals, as in the table. These should be the same as those in the table of observed counts, except for small roundoff errors, such as 115.99 rather than 116 for the first-row total.

Computing the chi-square statistic

The expected counts are all large, so we proceed with the chi-square test. We compare the table of observed counts with the table of expected counts using the X^2 statistic (page 624).[8] We must calculate the term for each cell, then sum over all nine cells. For the high-SES current smokers, the observed count is 51 and the expected count is 68.75. The contribution to the X^2 statistic for this cell is therefore

$$\frac{(51 - 68.75)^2}{68.75} = 4.583$$

Similarly, the calculation for the middle-SES current smokers is

$$\frac{(22 - 16.94)^2}{16.94} = 1.511$$

The X^2 statistic is the sum of nine such terms:

$$\begin{aligned} X^2 &= \sum \frac{(\text{observed} - \text{expected})^2}{\text{expected}} \\ &= \frac{(51 - 68.75)^2}{68.75} + \frac{(22 - 16.94)^2}{16.94} + \frac{(43 - 30.30)^2}{30.30} \\ &\quad + \frac{(92 - 83.57)^2}{83.57} + \frac{(21 - 20.60)^2}{20.60} + \frac{(28 - 36.83)^2}{36.83} \\ &\quad + \frac{(68 - 58.68)^2}{58.68} + \frac{(9 - 14.46)^2}{14.46} + \frac{(22 - 25.86)^2}{25.86} \\ &= 4.583 + 1.511 + 5.323 + 0.850 + 0.008 + 2.117 + 1.480 + 2.062 + 0.576 \\ &= 18.51 \end{aligned}$$

Because there are $r = 3$ smoking categories and $c = 3$ SES groups, the degrees of freedom for this statistic are

$$(r - 1)(c - 1) = (3 - 1)(3 - 1) = 4$$

df = 4		
p	0.001	0.0005
χ^2	18.47	20.00

Under the null hypothesis that smoking and SES are independent, the test statistic X^2 has a $\chi^2(4)$ distribution. To obtain the P-value, refer to the row in Table F corresponding to 4 df. The calculated value $X^2 = 18.51$ lies between upper critical points corresponding to probabilities 0.001 and 0.0005. The P-value is therefore between 0.001 and 0.0005. Because the expected cell counts are all large, the P-value from Table F will be quite accurate. There is strong evidence ($X^2 = 18.51$, df $= 4$, $P < 0.001$) of an association between smoking and SES in the population of federal employees. The size and nature of this association are described by the table of percents examined in Example 9.18 and the display of these percents in Figure 9.5. The chi-square test assesses only the significance of the association, so the descriptive percents are essential to an understanding of the data.

Models for two-way tables

The chi-square test for the presence of a relationship between the two directions in a two-way table is valid for data produced from several different study designs. The precise statement of the null hypothesis "no relationship" in terms of population parameters is different for different designs. We now describe two of these settings in detail. The essential requirement is that *each experimental unit or subject is counted only once in the data table*. In the sports goals example, each student is either male or female and is classified as having only one sports goal. In the SES-smoking example, each federal employee fits into one SES-smoking cell. If, for example, the same individual is observed at several different times and so may appear in several cells of the table, more advanced statistical procedures are needed.

Comparing several populations: the first model

The first model for a two-way table is illustrated by the comparison of two proportions for the binge-drinking data discussed in Section 8.2 and Example 9.2 and also by the comparison of the sports goals of men and women in Example 9.12. We have *separate and independent random samples* from each of c populations. Each individual falls into one of r categories, and we wish to compare the distributions of these categories in the populations. In Section 8.2, we compared just two populations ($c = 2$) and we categorized each observation as a success or a failure ($r = 2$). The data could have been displayed there in a two-way 2×2 table. Because the counts of failures are known once we know the counts of successes, we simply compared the two proportions of successes and did not need the two-way table format. The $r \times c$ table allows us to compare more than two populations or to use more than two categories, or both.

EXAMPLE 9.20

In the sports goals example, we compare two populations (men and women) and assign each subject to one of four categories. We have a 4×2 table. The parameters of the model are the population proportions for each of the four categories, one set of proportions for each population. We have samples from two populations of subjects, men and women.

More generally, if we take an independent SRS from each of c populations and classify each outcome into one of r categories, we have an $r \times c$ table of population proportions. There are c different sets of proportions to be compared. There are c groups of subjects, and a single categorical variable with r possible values is measured for each individual.

EXAMPLE 9.21

The null hypothesis for the sports goals example is that there is no relationship between gender and goals. That is, the sports goals distributions for females and males are the same. Therefore, the probability that a randomly selected female is in the high social comparison–high mastery group is equal to the corresponding probability for a randomly selected male. The same is true for each of the other three goals. The null hypothesis H_0, expressed in terms of the population parameters, says that the two probability distributions are equal.

> **Model for Comparing Several Populations Using Two-Way Tables**
>
> Select independent SRSs from each of c populations, of sizes $n_1, n_2, \ldots, n_c$. Classify each individual in a sample according to a categorical response variable with r possible values. There are c different probability distributions, one for each population.
>
> The null hypothesis is that the distributions of the response variable are the same in all c populations. The alternative hypothesis says that these c distributions are not all the same.

Testing independence: the second model A second model for which our analysis of $r \times c$ tables is valid is illustrated by the SES-smoking study, Example 9.17. There, a *single* sample from a *single* population was classified according to two categorical variables.

EXAMPLE 9.22

In this study, researchers recorded the values of two categorical variables for each of the 356 federal employees that they studied. These variables were SES and smoking. Since each variable has three possible values, there are nine possible outcomes. We summarize the data in a 3×3 table of counts.

In the table of observed cell counts, only the table total is fixed before the data are observed. This total (356) is the number of subjects in the study. The row and column sums are random and would no doubt be different if we took another sample.

More generally, take an SRS from a single population and record the values of two categorical variables, one with r possible values and the other with c possible values. The data are summarized by recording the numbers of individuals for each possible combination of outcomes for the two random variables. This gives an $r \times c$ table of counts. Each of these $r \times c$ possible outcomes has its own probability. The probabilities give the **joint distribution** of the two categorical variables.

joint distribution

Each of the two categorical random variables has a distribution. These are the **marginal distributions** because they are the sums of the population proportions in the rows and columns.

marginal distributions

The null hypothesis "no relationship" now states that the row and column variables are independent. The multiplication rule for independent events (page 295) tells us that the joint probabilities are the products of the marginal probabilities.

EXAMPLE 9.23

The joint probability distribution gives a probability for each of the nine cells in our 3×3 table of smoking by SES. The marginal distribution for smoking gives probabilities for each of the three possible smoking categories; the marginal distribution for SES gives probabilities for each of the three possible SES categories.

Independence between smoking and SES implies that the joint distribution can be obtained by multiplying the appropriate terms from the two marginal distributions. For example, the probability that an individual is a high-SES current smoker is the probability that the individual is high-SES times the probability that the individual is a current smoker. The hypothesis that smoking and SES are independent says that the multiplication rule applies to *all* outcomes.

> **Model for Examining Independence in Two-Way Tables**
>
> Select an SRS of size n from a population. Measure two categorical variables for each individual.
>
> The null hypothesis is that the row and column variables are independent. The alternative hypothesis is that the row and column variables are dependent.

The formula that we gave for expected counts (page 632) can also be justified using the independence model. We illustrate the idea with the following example.

EXAMPLE 9.24

The expected number of high-SES current smokers is simply the sample size n times the probability that an individual is a high-SES current smoker:

$$n \times P(\text{high-SES current smoker})$$

In Example 9.23 we saw that the independence model implies that this probability is equal to the probability that the individual is high-SES times the probability that the individual is a current smoker. So,

$$P(\text{high-SES current smoker}) = P(\text{current smoker}) \times P(\text{high-SES})$$

and under the independence model, the expected number of high-SES current smokers is

$$n \times P(\text{current smoker}) \times P(\text{high-SES})$$

Using the estimated marginal distributions from the table of counts, we find that the estimated count is

$$356 \times \frac{116}{356} \times \frac{211}{356} = \frac{116 \times 211}{356} = 68.75$$

or

$$\text{expected count} = \frac{\text{row total} \times \text{column total}}{n}$$

Concluding remarks

You can distinguish between the two models by examining the design of the study. In the independence model, there is a single sample. The column totals and row totals are random variables. The total sample size n is set by the researcher; the column and row sums are known only after the data are collected. For the comparison-of-populations model, on the other hand, there is a sample from each of two or more populations. The column sums are the sample sizes selected at the design phase of the research. The null hypothesis in both models says that there is no relationship between the column variable and the row variable. The precise statement of the hypothesis differs, depending on the sampling design. Fortunately, *the test of the hypothesis of "no relationship" is the same for both models*; it is the chi-square test. There are yet other statistical models for two-way tables that justify the chi-square test of the null hypothesis "no relation," made precise in ways suitable for these models. Statistical methods related to the chi-square test also allow the analysis of three-way and higher-way tables of count data. You can find a discussion of these topics in advanced texts on categorical data.[9]

SUMMARY

For two-way tables we first compute percents or proportions that describe the relationship of interest. Then, we compute expected counts, the X^2 statistic, and the P-value.

Two different models for generating $r \times c$ tables lead to the chi-square test. In the first model, independent SRSs are drawn from each of c populations, and each observation is classified according to a categorical variable with r possible values. The null hypothesis is that the distributions of the row categorical variable are the same for all c populations. In the second model, a single SRS is drawn from a population, and observations are classified according to two categorical variables having r and c possible values. In this model, H_0 states that the row and column variables are independent.

CHAPTER 9 EXERCISES

9.1 The Census Bureau collects data on years of school completed by Americans of different ages. The following table gives the years of education for three different age groups. People under the age of 25 are not included because many have not yet completed their education. Note that the unit of measure for each entry in the table is thousands of persons.

Years of school completed, by age (thousands of persons)

	Age group			
Education	25 to 34	35 to 54	55 and over	Total
Did not complete high school	5,325	9,152	16,035	30,512
Completed high school	14,061	24,070	18,320	56,451
College, 1 to 3 years	11,659	19,926	9,662	41,247
College, 4 or more years	10,342	19,878	8,005	38,225
Total	41,388	73,028	52,022	166,438

(a) Give the joint distribution of education and age for this table.

(b) What is the marginal distribution of age?

(c) What is the marginal distribution of education?

9.2 Refer to the previous exercise. Find the conditional distribution of education for each of the three age categories. Make a bar graph for each distribution and summarize their differences and similarities.

9.3 Refer to the previous exercise. Compute the conditional distribution of age for each of the four education categories. Summarize the distributions graphically and write a short paragraph describing the distributions and how they differ.

9.4 The National Center for Education Statistics collects data on undergraduate students enrolled in U.S. colleges and universities. The following table gives counts of undergraduate enrollment in four different classifications:

Undergraduate college enrollment (thousands of students)				
Age	2-year full-time	2-year part-time	4-year full-time	4-year part-time
Under 18	36	98	75	37
18 to 21	1126	711	3270	223
22 to 34	634	1575	2267	1380
35 and over	221	1092	390	915
Total	2017	3476	6002	2555

(a) How many undergraduate students were enrolled in colleges and universities?

(b) What percent of all undergraduate students were under 18 years old?

(c) Find the percent of the undergraduates enrolled in each of the four types of program who were 22 to 34 years old. Make a bar graph to compare these percents.

(d) The 18 to 21 group is the traditional age group for college students. Briefly summarize what you have learned from the data about the extent to which this group predominates in different kinds of college programs.

9.5 Refer to the previous exercise. Find the marginal distribution of college type. Describe the distribution graphically and write a short summary.

9.6 Take the counts for the four college types that you used in the previous exercise and write these in a 2×2 table with year (2-year or 4-year) as the columns and time (full-time or part-time) as the rows. The marginal distribution that you calculated for the previous exercise is the joint distribution for this exercise. Compute the conditional distribution of time for each year category. Describe and contrast these distributions. Does there appear to be a relationship? If so, describe it.

9.7 For each type of college find the conditional distributions of age for the 4×3 table in Exercise 9.4. Display the distributions with bar graphs and describe how the age profile of the students varies with the type of college.

9.8 The sports goals example in Example 9.13 gives a 4 × 2 table of counts that produce the output in Figure 9.2 (page 621). Refer to this figure and answer the following questions.

(a) Give the joint distribution of gender and sports goals.

(b) What is the conditional distribution of gender for each sports goal category?

(c) What is the conditional distribution of sports goals for each gender?

(d) Of your answers in (a), (b), and (c), which is most useful for understanding this table of counts?

9.9 Here are the row and column totals for a two-way table with two rows and two columns:

a	b	100
c	d	100
100	100	200

Find *two different* sets of counts a, b, c, and d for the body of the table that give these same totals. This shows that the relationship between two variables cannot be obtained from the two individual distributions of the variables.

9.10 Construct a 2 × 3 table of counts where there is no apparent association between the row and column variables.

9.11 A marketing research firm conducted a survey of companies in its state. They mailed a questionnaire to 300 small companies, 300 medium-sized companies, and 300 large companies. The rate of nonresponse is important in deciding how reliable survey results are. Here are the data on response to this survey:

	Response		
Size of company	Yes	No	Total
Small	175	125	300
Medium	145	155	300
Large	120	180	300

(a) What was the overall percent of nonresponse?

(b) Describe how nonresponse is related to the size of the business. (Use percents to make your statements precise.)

(c) Draw a bar graph to compare the nonresponse percents for the three size categories.

(d) Using the total number of responses as a base, compute the percent of responses that come from each of small, medium, and large businesses.

(e) The sampling plan was designed to obtain equal numbers of responses from small, medium, and large companies. In preparing an analysis of the survey results, do you think it would be reasonable to proceed as if the responses represented companies of each size equally?

9.12 A study of the career plans of young women and men sent questionnaires to all 722 members of the senior class in the College of Business Administration at the University of Illinois. One question asked which major within the business program the student had chosen.[10] Here are the data from the students who responded:

	Gender	
Major	Female	Male
Accounting	68	56
Administration	91	40
Economics	5	6
Finance	61	59

(a) Describe the differences between the distributions of majors for women and men with percents, with a graph, and in words.

(b) What percent of the students did not respond to the questionnaire? The nonresponse weakens conclusions drawn from these data.

9.13 Cocaine addiction is hard to break. Addicts need cocaine to feel any pleasure, so perhaps giving them an antidepressant drug will help. A 3-year study with 72 chronic cocaine users compared an antidepressant drug called desipramine with lithium and a placebo. (Lithium is a standard drug to treat cocaine addiction. A placebo is a dummy drug, used so that the effect of being in the study but not taking any drug can be seen.) One-third of the subjects, chosen at random, received each treatment.[11] Here are the results:

	Cocaine relapse?	
Treatment	Yes	No
Desipramine	10	14
Lithium	18	6
Placebo	20	4

Compare the effectiveness of the three treatments in preventing relapse. Use percents and draw a bar graph. Write a brief summary of your conclusions.

9.14 Asia has become a major competitor of the United States and Western Europe in education as well as economics. Here are counts of first university degrees in science and engineering in the three regions:[12]

	Region		
Field	United States	Western Europe	Asia
Engineering	61,941	158,931	280,772
Natural science	111,158	140,126	242,879
Social science	182,166	116,353	236,018

Direct comparison of counts of degrees would require us to take into account Asia's much larger population. We can, however, compare the distribution of degrees by field of study in the three regions. Do this using calculations and graphs, and write a brief summary of your findings.

9.15 The following two-way table describes the age and marital status of American women. The table entries are in thousands of women.

	Marital status				
Age (years)	Never married	Married	Widowed	Divorced	Total
18–24	9,289	3,046	19	260	12,613
25–39	6,948	21,437	206	3,408	32,000
40–64	2,307	26,679	2,219	5,508	36,713
≥ 65	768	7,767	8,636	1,091	18,264
Total	19,312	58,931	11,080	10,266	99,588

(a) Find the sum of the entries in the "Married" column. Why does this sum differ from the "Total" entry for that column?

(b) Give the marginal distribution of marital status for all adult women (use percents). Draw a bar graph to display this distribution.

(c) Compare the conditional distributions of marital status for women aged 18 to 24 and women aged 40 to 64. Briefly describe the most important differences between the two groups of women, and back up your description with percents.

(d) You are planning a magazine aimed at women who have never been married. Find the conditional distribution of age among single women and display it in a bar graph. What age group or groups should your magazine aim to attract?

9.16 Here is a two-way table of suicides committed, categorized by the gender of the victim and the method used. ("Hanging" also includes suffocation.) Write a brief account of differences in suicide between males and females. Use calculations and a graph to justify your statements.

	Gender	
Method	Male	Female
Firearms	16,381	2,559
Poison	3,569	2,110
Hanging	3,824	803
Other	1,641	623

9.17 Mountain View University has professional schools in business and law. Here is a three-way table of applicants to these professional schools, categorized by gender, school, and admission decision.[13]

Business		
	Admit	
Gender	Yes	No
Male	400	200
Female	200	100

Law		
	Admit	
Gender	Yes	No
Male	90	110
Female	200	200

(a) Make a two-way table of gender by admission decision for the combined professional schools by summing entries in the three-way table.

(b) From your two-way table, compute separately the percents of male and female applicants admitted. Male applicants are admitted to Mountain View's professional schools at a higher rate than female applicants.

(c) Now compute separately the percents of male and female applicants admitted by the business school and by the law school.

(d) Explain carefully, as if speaking to a skeptical reporter, how it can happen that Mountain View appears to favor males when this is not true within each of the professional schools.

9.18 Refer to the previous exercise. Make up a similar table for a hypothetical university having three different schools that illustrates the same point. Carefully summarize your table with the appropriate percents.

9.19 Here are the numbers of flights on time and delayed for two airlines at five airports.[14] Overall on-time percents for each airline are often reported in the news. Lurking variables can make such reports misleading.

	Alaska Airlines		America West	
	On time	Delayed	On time	Delayed
Los Angeles	497	62	694	117
Phoenix	221	12	4840	415
San Diego	212	20	383	65
San Francisco	503	102	320	129
Seattle	1841	305	201	61

(a) What percent of all Alaska Airlines flights were delayed? What percent of all America West flights were delayed? These are the numbers usually reported.

(b) Now find the percent of delayed flights for Alaska Airlines at each of the five airports. Do the same for America West.

(c) America West does worse at *every one* of the five airports, yet does better overall. That sounds impossible. Explain carefully, referring to the data, how this can happen. (The weather in Phoenix and Seattle lies behind this example of Simpson's paradox.)

9.20 Market researchers know that background music can influence the mood and purchasing behavior of customers. One study in a supermarket in Northern Ireland compared three treatments: no music, French accordion music, and Italian string music. Under each condition, the researchers recorded the numbers of bottles of French, Italian, and other wine purchased.[15] Here is the two-way table that summarizes the data:

	Music			
Wine	None	French	Italian	Total
French	30	39	30	99
Italian	11	1	19	31
Other	43	35	35	113
Total	84	75	84	243

Is there a relationship between the type of wine purchased and the type of music that is playing? Give the details of the significance test that you perform to answer this question. Write a short summary of your conclusion and include appropriate graphical displays that show what you have found.

9.21 Following complaints about the working conditions in some apparel factories both in the United States and abroad, a joint government and industry commission recommended in 1998 that companies that monitor and enforce proper standards be allowed to display a "No Sweat" label on their products. Does the presence of these labels influence consumer behavior? A survey of U.S. residents aged 18 or older asked a series of questions about how likely they would be to purchase a garment under various conditions. For some conditions, it was stated that the garment had a "No Sweat" label; for others, there was no mention of such a label. On the basis of the responses, each person was classified as a "label user" or a "label nonuser."[16] There were 296 women surveyed. Of these, 63 were label users. On the other hand, 27 of 251 men were classified as users.

(a) Construct the 2×2 table of counts for this problem. Include the marginal totals for your table.

(b) Use a X^2 statistic to examine the question of whether or not there is a relationship between gender and use of No Sweat labels. Give the test statistic, degrees of freedom, the *P*-value, and your conclusion.

(c) You examined this question using the methods of the previous chapter in Exercise 8.57 (page 602). Verify that if you square the z statistic you calculated for that exercise, you obtain the X^2 statistic that you calculated for this exercise.

9.22 To be competitive in global markets, many U.S. corporations are undertaking major reorganizations. Often these involve "downsizing" or a "reduction in force" (RIF), where substantial numbers of employees are terminated. Federal and various state laws require that employees be treated equally regardless of their age. In particular, employees over the age of 40 years are in a "protected" class, and many allegations of discrimination focus on comparing employees over 40 with their younger coworkers. Here are the data for a recent RIF:

	Over 40	
Terminated	No	Yes
Yes	16	78
No	585	765

(a) Make a table that includes the following information for each group (over 40 or not): total number of employees, the proportion of employees who were terminated, and the standard error for the proportion.

(b) Perform the chi-square test for this two-way table. Give the test statistic, the degrees of freedom, the *P*-value, and your conclusion.

9.23 A major issue that arises in the kind of case described in the previous exercise concerns the extent to which the employees are similar. This often involves examining other variables. For example, suppose that the employees over 40 did not do as good a job as the younger workers. Let's examine the last performance appraisal to explore this idea. The possible values are as follows: "partially meets expectations," "fully meets expectations," "usually exceeds expectations," and "continually exceeds expectations." Because there were very few employees who partially exceeded expectations, we combine the first two categories. Here are the data:

	Over 40	
Performance appraisal	No	Yes
Partially or fully meets expectations	82	235
Usually exceeds expectations	357	490
Continually exceeds expectations	63	30

Analyze the data. Do the older employees appear to have lower performance evaluations?

9.24 Shopping at secondhand stores is becoming more popular and has even attracted the attention of business schools. A study of customers' attitudes toward secondhand stores interviewed samples of shoppers at two secondhand stores of the same chain in two cities. The breakdown of the respondents by gender is as follows:[17]

Gender	City 1	City 2
Men	38	68
Women	203	150
Total	241	218

Is there a significant difference between the proportions of women customers in the two cities?

(a) State the null hypothesis, find the sample proportions of women in both cities, do a two-sided z test, and give a P-value using Table A.

(b) Calculate the X^2 statistic and show that it is the square of the z statistic. Show that the P-value from Table F agrees (up to the accuracy of the table) with your result from (a).

(c) Give a 95% confidence interval for the difference between the proportions of women customers in the two cities.

9.25 The study of shoppers in secondhand stores cited in the previous exercise also compared the income distributions of shoppers in the two stores. Here is the two-way table of counts:

Income	City 1	City 2
Under \$10,000	70	62
\$10,000 to \$19,999	52	63
\$20,000 to \$24,999	69	50
\$25,000 to \$34,999	22	19
\$35,000 or more	28	24

Find the X^2 statistic for this table as $X^2 = 3.955$. Give the degrees of freedom and the P-value. Is there good evidence that customers at the two stores have different income distributions?

9.26 You analyzed data from a survey of companies in Exercise 9.11. State appropriate null and alternative hypotheses for these data, give the test statistic with the degrees of freedom and the P-value, and summarize your conclusion.

9.27 Data from a study of three different treatments for cocaine abuse are given in Exercise 9.13. Compare the treatments using a significance test and summarize the results.

9.28 Euthanasia of healthy but unwanted pets by animal shelters is believed to be the leading cause of death for cats and dogs. A study designed to find factors associated with bringing a cat to an animal shelter compared data on cats that were brought to the Humane Society of Saint Joseph County in Mishawaka, Indiana, with controls from the same county that were not brought in.[18] One of the factors examined was the source of the cat; the categories were private owner or breeder, pet store, and other (includes born in home, stray, and obtained from a shelter). This kind of study is called a **case-control study** by epidemiologists. Here are the data:

case-control study

	Source		
Group	Private	Pet store	Other
Cases	124	16	76
Controls	219	24	203

(a) Should you use row or column percents to describe these data? Give reasons for your answer and summarize the pattern that you see in these percents.

(b) Use a significance test to examine the association between the two variables. Summarize the results.

9.29 Refer to the previous exercise. The same researchers did a similar study for dogs.[19] The data are given in the following table:

	Source		
Group	Private	Pet store	Other
Cases	188	7	90
Controls	518	68	142

Analyze the data and write a short report explaining your conclusions.

9.30 In the studies described in the previous two exercises, the *control group data* were obtained by a random-digit-dialing telephone survey and can be considered to be an SRS of households with a cat or a dog in this area. Compare the sources of cats and of dogs in this area. Write a short report on your analysis; include appropriate descriptive statistics, the results of a significance test, and your conclusion.

9.31 Can you increase the response rate for a mail survey by contacting the respondents before they receive the survey? A study designed to address this question compared three groups of subjects.[20] The first group received a preliminary letter about the survey, the second group was phoned, and the third received no preliminary contact. A positive response was defined as returning the survey within two weeks. Here are the counts:

	Intervention		
Response	Letter	Phone call	None
Yes	171	146	118
No	220	68	455
Total	391	214	573

(a) For each intervention find the proportion of positive responses.

(b) Translate the question of interest into appropriate null and alternative hypotheses for this problem.

(c) Give the test statistic, degrees of freedom, and the *P*-value for the significance test. What do you conclude?

9.32 The effect of a prenotification letter on the response rate of a survey was examined in a study of physicians. The response rate of those who received the letters was compared with that of a control group who did not.[21] Here are the data:

Response	Letter	No letter
Yes	2570	2645
No	2448	2384
Total	5018	5029

(a) Give the percents of positive responses for those who received letters and those who did not.

(b) Analyze the two-way table. State the hypotheses, and give the test statistic, degrees of freedom, *P*-value, and your conclusion.

9.33 Psychological and social factors can influence the survival of patients with serious diseases. One study examined the relationship between survival of patients with coronary heart disease (CHD) and pet ownership.[22] Each of 92 patients was classified as having a pet or not and by whether they survived for one year. Here are the data:

	Pet ownership	
Patient status	No	Yes
Alive	28	50
Dead	11	3

(a) Was this study an experiment? Why or why not?

(b) The researchers thought that having a pet might improve survival, so pet ownership is the explanatory variable. Compute appropriate percents to describe the data and state your preliminary findings.

(c) State in words the null hypothesis for this problem. What is the alternative hypothesis?

(d) Find the X^2 statistic, its degrees of freedom, and the *P*-value.

(e) What do you conclude? Do the data give convincing evidence that owning a pet is an effective treatment for increasing the survival of CHD patients?

9.34 The baseball player Reggie Jackson had a reputation for hitting better in the World Series than during the regular season. In his 21-year career, Jackson was at bat 9864 times in regular-season play and had 2584 hits. During World Series games, he was at bat 98 times and had 35 hits. We can view Jackson's regular-season at-bats as a random sample from a population of potential at-bats (he might have batted many more times if the season were longer, for example), and his World Series at-bats as a sample from a second population.

(a) Display the data in a 2×2 table of counts with "regular season" and "World Series" as the column headings, and fill in the marginal sums.

(b) Calculate appropriate percents to compare Jackson's regular-season and World Series performances. Did he hit better in World Series games?

(c) Is there a significant difference between Jackson's regular-season and World Series performances? State hypotheses (in words), and then calculate the X^2 statistic, its degrees of freedom, and its *P*-value. What is your conclusion?

9.35 If the performance of a stock fund is due to the skill of the manager, then we would expect a fund that does well this year to perform well next year also. This is called persistence of fund performance. One study classified funds as losers or winners depending on whether their rate of return was less than or greater than the median of all funds.[23] To examine the question of interest we form a two-way table that classifies each fund as a loser or winner in each of two successive years. Here are the data for one such table:

	Next year	
This year	Winner	Loser
Winner	85	35
Loser	37	83

Is there evidence in favor of persistence of fund performance in this table? Support your conclusion with a complete analysis of the data.

9.36 Refer to the previous exercise. Rerun the analysis using the method for comparing two proportions of Chapter 8. Verify that the X^2 statistic is the square of the z statistic and that the P-values for both analyses are the same.

9.37 In the previous exercise, it is natural to use "this year" to define the two samples. If we drew separate random samples of winners and losers this year and we recorded the outcome next year, we would call this a **prospective study** (forward looking). On the other hand, if we drew separate random samples of winners and losers "next year" and looked back historically to determine if they were winners or losers in the previous year, we would have a **retrospective study** (backward looking). Verify that you get the same value of z (and therefore the same P-value) using these two different approaches.

prospective study

retrospective study

9.38 If we find evidence in favor of an effect in one set of circumstances, it is natural to want to conclude that it holds in many others. Unfortunately, this reasoning can sometimes lead us to incorrect conclusions. For example, here is another table from the study described in Exercise 9.35:

	Next year	
This year	Winner	Loser
Winner	96	148
Loser	145	99

Analyze these data in the same way. What do you conclude? This set of data is for 1987 to 1988; in Exercise 9.35 the years were 1977 to 1978. Many things change with time, as we learned in Chapter 2.

9.39 There is much evidence that high blood pressure is associated with increased risk of death from cardiovascular disease. A major study of this association examined 2676 men with low blood pressure and 3338 men with high blood pressure.[24] During the period of the study, 21 men in the low-blood-pressure and 55 in the high-blood-pressure group died from cardiovascular disease.

(a) What is the explanatory variable? Describe the association in these data numerically and in words.

(b) Do the study data confirm that death rates are higher among men with high blood pressure? State hypotheses, carry out a significance test, and give your conclusions.

(c) Present the data in a two-way table. Is the chi-square test appropriate for the hypotheses you stated in (b)?

(d) Give a 95% confidence interval for the difference between the death rates for the low- and high-blood-pressure groups.

9.40 It is traditional practice in Egypt to withhold food from children with diarrhea. Because it is known that feeding children with this illness reduces mortality, medical authorities undertook a nationwide program designed to promote feeding sick children. To evaluate the impact of the program, surveys were taken before and after the program was implemented.[25] In the first survey, 457 of 1003 surveyed mothers followed the practice of feeding children with diarrhea. For the second survey, 437 of 620 surveyed followed this practice.

(a) Assume that the data come from two independent samples. Test the hypothesis that the program was effective, that is, that the practice of feeding children with diarrhea increased between the time of the first study and the time of the second. State H_0 and H_a, give the test statistic and its P-value, and summarize your conclusion.

(b) Present the data in a two-way table. Can the X^2 statistic test your hypotheses?

(c) Describe the results using a 95% confidence interval for the difference in proportions.

9.41 PTC is a compound that has a strong bitter taste for some people and is tasteless for others. The ability to taste this compound is an inherited trait. Many studies have assessed the proportions of people in different populations who can taste PTC. The following table gives results for samples from several countries:[26]

	Country			
Taster	Ireland	Portugal	Norway	Italy
Yes	558	345	185	402
No	225	109	81	134

Complete the table and describe the data. Do they provide evidence that the proportion of PTC tasters varies among the four countries? Give a complete summary of your analysis.

9.42 There are four major blood types in humans: O, A, B, and AB. In a study conducted using blood specimens from the Blood Bank of Hawaii, individuals were classified according to blood type and ethnic group. The ethnic groups were Hawaiian, Hawaiian-white, Hawaiian-Chinese, and white.[27] Assume that the blood bank specimens are random samples from the Hawaiian populations of these ethnic groups.

	Ethnic group			
Blood type	Hawaiian	Hawaiian-white	Hawaiian-Chinese	White
O	1,903	4,469	2,206	53,759
A	2,490	4,671	2,368	50,008
B	178	606	568	16,252
AB	99	236	243	5,001

Summarize the data numerically and with a graph. Is there evidence to conclude that blood type and ethnic group are related? Explain how you arrived at your conclusion.

9.43 Gastric freezing was once a recommended treatment for ulcers in the upper intestine. A randomized comparative experiment found that 28 of the 82 patients who were subjected to gastric freezing improved, while 30 of the 78 patients in the control group improved. The hypothesis of "no difference" between the two groups can be tested in two ways: using a z statistic or using the X^2 statistic.

(a) State the appropriate hypothesis and a two-sided alternative, and carry out a z test. What is the P-value?

(b) Present the data in a 2×2 table. State the appropriate hypotheses and carry out the chi-square test. What is the P-value? Verify that the X^2 statistic is the square of the z statistic.

(c) What do you conclude about the effectiveness of gastric freezing as a treatment for ulcers? (See Example 3.5 on page 231 for a discussion of gastric freezing.)

9.44 Example 9.9 (page 617) presents artificial data that illustrate Simpson's paradox. The data concern the survival rates of surgery patients at two hospitals.

(a) Apply the chi-square test to the data for all patients combined and summarize the results.

(b) Run separate chi-square analyses for the patients in good condition and for those in poor condition. Summarize these results.

(c) Are the effects that illustrate Simpson's paradox in this example statistically significant?

9.45 The sports goals study described in Example 9.12 (page 620) actually involved *three* categorical variables: gender, social comparison, and mastery. In that example, we combined the social comparison and mastery categories to obtain a two-way table. There are statistical methods for analyzing three-way tables, but these are beyond the scope of this text. We can, however, look at the several two-way tables formed by rearranging the data.

The analyses and discussion of this example presented earlier suggest that the major gender difference is in the social comparison variable. To investigate this idea, rearrange the data into a 2×2 table that classifies the students by gender and social comparison. Analyze this table and summarize the results. Do the same using gender and mastery as the classification variables. Considering these analyses and the analysis presented in Example 9.15 (page 626), what conclusions do you draw?

9.46 A recent study of 865 college students found that 42.5% had student loans.[28] The students were randomly selected from the approximately 30,000 undergraduates enrolled in a large public university. The overall purpose of the study was to examine the effects of student loan burdens on the choice of a career. A student with a large debt may be more likely to choose a field where starting salaries are high so that the loan can more easily be repaid. The following table classifies the students by field of study and whether or not they have a loan:

	Student loan	
Field of study	Yes	No
Agriculture	32	35
Child development and family studies	37	50
Engineering	98	137
Liberal arts and education	89	124
Management	24	51
Science	31	29
Technology	57	71

Carry out a complete analysis of the association between having a loan and field of study, including a description of the association and an assessment of its statistical significance.

9.47 In one part of the study described in the previous exercise, students were asked to respond to some questions regarding their interests and attitudes. Some of these questions form a scale called PEOPLE that measures altruism, or an interest in the welfare of others. Each student was classified as low, medium, or high on this scale. Is there an association between PEOPLE score and field of study? Here are the data:

Field of study	PEOPLE score Low	Medium	High
Agriculture	5	27	35
Child development and family studies	1	32	54
Engineering	12	129	94
Liberal arts and education	7	77	129
Management	3	44	28
Science	7	29	24
Technology	2	62	64

Analyze the data and summarize your results. Are there some fields of study that have very large or very small proportions of students in the high-PEOPLE category?

9.48 In healthy individuals the concentration of various substances in the blood remains within relatively narrow bounds. One such substance is potassium. A person is said to be hypokalemic if the potassium level is too low (less than 3.5 milliequivalents per liter) and hyperkalemic if the level is too high (above 5.5 meq/l). Hypokalemia is associated with a variety of symptoms such as excessive tiredness, while hyperkalemia is generally an indication of a serious problem. Patients being treated with diuretics (pharmaceuticals that help the body to eliminate water) sometimes have abnormal potassium concentrations. In a large study of patients on chronic diuretic therapy, several risk factors were studied to see if they were associated with abnormal potassium levels.[29] Of the 5810 patients studied, 1094 were hypokalemic, 4689 had normal potassium levels, and 27 were hyperkalemic. The following table gives the percents of patients having each of four risk factors in the three potassium groups:

	Potassium group		
	Hypokalemic	Normal	Hyperkalemic
n	1094	4689	27
Hypertension	88.3%	78.1%	40.7%
Heart failure	16.5%	24.7%	55.6%
Diabetes	20.6%	25.5%	29.6%
Gender (% female)	72.5%	68.0%	48.1%

For example, 88.3% of the 1094 hypokalemic patients had hypertension, and 78.1% of the 4689 normal patients had hypertension. For each of the four risk factors, use the percents and n's given to compute the counts for the 2×3 table needed to study the association between the factor and potassium. Then analyze each table using the methods presented in this chapter. Note that there are very few patients in the hyperkalemic group. Therefore, reanalyze the data dropping this category from the tables. Write a short summary explaining what you have found.

9.49 The proportion of women entering many professions has undergone considerable change in recent years. A study of students enrolled in pharmacy programs describes the changes in this field. A random sample of 700 students in their third or higher year of study at colleges of pharmacy was taken in each of nine years.[30] The following table gives the numbers of women in each of these samples:

Year	1970	1972	1974	1976	1978	1980	1982	1984	1986
Women	164	195	226	283	302	342	369	385	412

Use the chi-square test to assess the change in the percent of women pharmacy students over time, and summarize your results. (You will need to calculate the number of male students for each year using the fact that the sample size each year is 700.) Plot the percent of women versus year. Describe the plot. Is it roughly linear? Find the least-squares line that summarizes the relation between time and the percent of women pharmacy students.

9.50 Refer to the previous exercise. Here are the percents of women pharmacy students for the years 1987 to 2000:[31]

Year	1987	1988	1989	1990	1991	1992	1993
Women	60.0%	60.6%	61.6%	62.4%	63.0%	63.4%	63.2%

Year	1994	1995	1996	1997	1998	1999	2000
Women	63.3%	63.4%	63.8%	64.2%	64.4%	64.9%	65.9%

Plot these percents versus year and summarize the pattern. Using your analysis of the data in this and the previous exercise, write a report summarizing the changes that have occurred in the percent of women pharmacy students from 1970 to 2000. Include an estimate of the percent for the year 2010 with an explanation of why you chose this estimate.

9.51 Refer to Exercise 8.81 (page 607) concerning *Castaneda v. Partida*, the case where the Supreme Court review used the phrase "two or three standard deviations" as a criterion for statistical significance. Recall that there were

181,535 persons eligible for jury duty, of whom 143,611 were Mexican Americans. Of the 870 people selected for jury duty, 339 were Mexican Americans. We are interested in finding out if there is an association between being a Mexican American and being selected as a juror. Formulate this problem using a two-way table of counts. Construct the 2×2 table using the variables Mexican American or not and juror or not. Find the X^2 statistic and its P-value. Square the z statistic that you obtained in Exercise 8.81 and verify that the result is equal to the X^2 statistic.

9.52 Refer to Exercise 9.28 (page 646) concerning the case-control study of factors associated with bringing a cat to an animal shelter and the similar study for dogs described in Exercise 9.29. The last category for the source of the pet was given as "Other" and includes born in home, stray, and obtained from a shelter. The following two-way table lists these categories separately for cats:

	Source				
Group	Private	Pet store	Home	Stray	Shelter
Cases	124	16	20	38	18
Controls	219	24	38	116	49

Here is the same breakdown for dogs:

	Source				
Group	Private	Pet store	Home	Stray	Shelter
Cases	188	7	11	23	56
Controls	518	68	20	55	67

Analyze these 2×5 tables and compare the results that you obtained for the 2×3 tables in Exercises 9.28 and 9.29. With a large number of cells, the chi-square test sometimes does not have very much power.

9.53 Refer to Exercises 9.11, 9.21, 9.23, and 9.24. For each, state whether you are comparing two populations (the first model for two-way tables) or testing independence between two categorical variables (the second model).

Prelude: global warming

Is the climate growing warmer? Some very fancy data collection and analysis seek to answer this question. Here's a simpler approach. "The Nenana Ice Classic is a betting contest held in Nenana, Alaska (64°34'N, 149°05'W), in which participants attempt to guess the exact minute in spring that a wooden tripod positioned on the frozen Tanana River will fall through the ice."[1] Because money is at stake, the tripod is watched constantly, and the minute it falls has been recorded by the same standard since 1917.

Sure enough, the ice breakup has moved earlier in the spring over the decades. There is a roughly linear trend but lots of variation from year to year. The slope of the least-squares line of breakup time (in days and fractions of days after the spring equinox) on year is $b = -0.07$ days per year, but r^2 is only 0.07. We want to do more than simply fit a line and note these facts. We want to know if the slope is *statistically significant*. That is, is there good evidence that the negative slope we observed is due to a real change and not merely to random variation in the data? The answer is yes. The P-value is $P = 0.01$, giving reason to conclude that there is really a trend in the moment the tripod falls. In this chapter, devoted to inference about regression, we will learn how to determine this P-value.

CHAPTER 10

Piet Mondrian, *Composition with Red, Yellow, and Blue.*

Inference for Regression

Introduction

Throughout this book we have followed several strategies for working with data. Two of these strategies are (1) descriptive analysis and exploration of data precede formal inference and (2) examination of each variable separately precedes study of relations among quantitative variables. In this chapter we reach the second part of both strategies in the study of formal inference for relations among quantitative variables.

We will describe methods for inference when there is a single response variable and one quantitative explanatory variable. The descriptive tools we learned in Chapter 2—scatterplots, least-squares regression, and correlation—are essential preliminaries to inference and also provide a foundation for confidence intervals and significance tests.

We first met the sample mean $\bar{x}$ in Chapter 1 as a measure of the center of a collection of observations. Later we learned that when the data are a random sample from a population, the sample mean is an estimate of the population mean μ. In Chapters 6 and 7, we used $\bar{x}$ as the basis for confidence intervals and significance tests for inference about μ.

Now we will follow the same program for the problem of fitting straight lines to data. In Chapter 2 we met the least-squares regression line $\hat{y} = a + bx$ as a description of a straight-line relationship between a response variable y and an explanatory variable x. At that point we did not distinguish between sample and population. Now we will think of the least-squares line computed from a sample as an estimate of a *true* regression line for the population, just as $\bar{x}$ is an estimate of the true population mean μ. To emphasize the change in thinking, we introduce new notation. Following the common practice of using Greek letters for population parameters, we will write the population line as $\beta_0 + \beta_1 x$. The least-squares line fitted to sample data is now $b_0 + b_1 x$. This notation reminds us that the intercept b_0 of the fitted line estimates the intercept β_0 of the population line, and the slope b_1 estimates the slope β_1.

We are concerned with inference in the setting of simple linear regression, the familiar case of a single explanatory variable x. By extending the ideas in Chapters 6 and 7, we can give confidence intervals and significance tests for inference about the slope β_1 and the intercept β_0. Because regression lines are often used for prediction, we also consider inference about either the mean response or an individual future observation on y for a given value of the explanatory variable x. Finally, we discuss statistical inference about the correlation between two variables x and y that need not have an explanatory-response relationship.

10.1 Simple Linear Regression

Statistical model for linear regression

Simple linear regression studies the relationship between a response variable y and a single explanatory variable x. We expect that different values of x will produce different mean responses. We encountered a similar but simpler

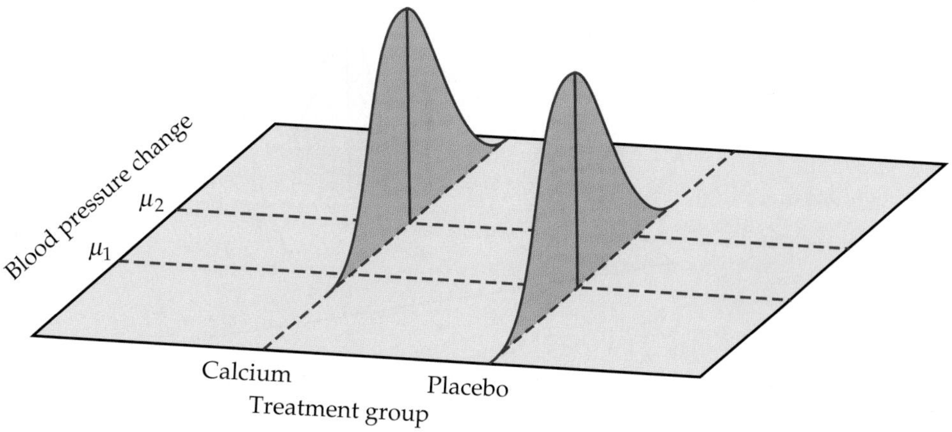

FIGURE 10.1 The statistical model for comparing responses to two treatments; the mean response varies with the treatment.

situation in Chapter 7 when we discussed methods for comparing two population means. Figure 10.1 illustrates the statistical model for a comparison of the blood pressure in two groups of experimental subjects, one group taking a calcium supplement and the other a placebo. In each population the observed change in blood pressure varies according to a normal distribution. The mean change will generally be different for the two populations. In the figure, μ_1 and μ_2 represent the two population means. The two normal curves in Figure 10.1 have the same spread, indicating that the population standard deviations are assumed to be equal. We can think of the treatment (placebo or calcium) as the explanatory variable in this example.

In linear regression the explanatory variable x can have many different values. Imagine, for example, giving different amounts x of calcium to different groups of subjects. We can think of the values of x as defining different **subpopulations,** one for each possible value of x. Each subpopulation consists of all individuals in the population having the same value of x. If you give $x = 1500$ milligrams to everyone, then these are the subjects "who actually receive 1500 milligrams." In this case, the subpopulation would be the same as the sample.

subpopulations

The statistical model for simple linear regression assumes that for each value of x the observed values of the response variable y are normally distributed about a mean that depends on x. We use μ_y to represent these means. Rather than just two means μ_1 and μ_2, we are interested in how the many means μ_y change as x changes. In general the means μ_y can change according to any sort of pattern as x changes. In **simple linear regression** we assume that they all lie on a line when plotted against x. The equation of the line is

simple linear regression

$$\mu_y = \beta_0 + \beta_1 x$$

with intercept β_0 and slope β_1. This is the **population regression line;** it describes how the mean response changes with x. Actually observed y's will vary about these means. The model assumes that this variation, measured by

population regression line

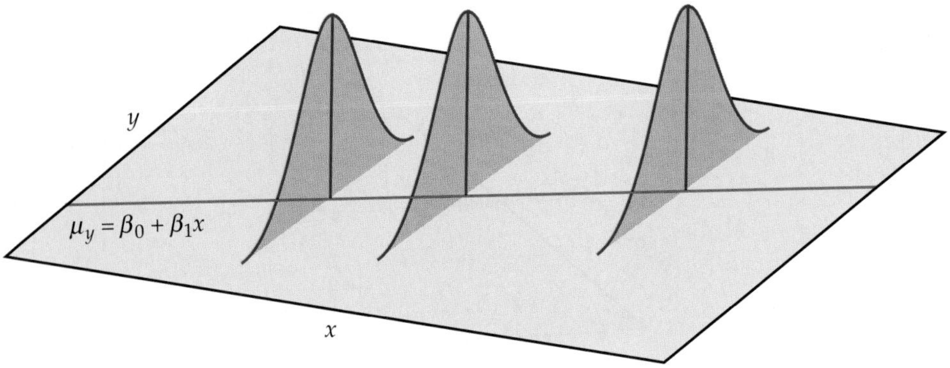

FIGURE 10.2 The statistical model for linear regression; the mean response is a straight-line function of the explanatory variable.

the standard deviation σ, is the same for all values of x. Figure 10.2 displays this statistical model. The line is the population regression line, and the three normal curves show how the response y will vary for three different values of the explanatory variable x.

Data for simple linear regression

The data for a linear regression are observed values of y and x. The model takes each x to be a fixed known quantity. In practice, x may not be exactly known. If the error in measuring x is large, more advanced inference methods are needed. The response y to a given x is a random variable that will take different values if we have several observations with the same x-value. The model describes the mean and standard deviation of the random variable y. These unknown parameters must be estimated from the data.

We will use the following example to explain the fundamentals of simple linear regression. Because regression calculations in practice are always done by statistical software, we will rely on computer output for the arithmetic in this example. In the next section, we give an example that illustrates how to do the work with a calculator if software is unavailable.

EXAMPLE 10.1

"The fat content of the human body has physiological and medical importance. It may influence morbidity and mortality, it may alter the effectiveness of drugs and anesthetics, and it may affect the ability to withstand exposure to cold and starvation."[2]

In practice, fat content is found by measuring body density, the weight per unit volume of the body. High fat content corresponds to low body density. Body density is hard to measure directly—the standard method requires that subjects be weighed underwater. For this reason, scientists have sought variables that are easier to measure and that can be used to predict body density. Research suggests that *skinfold thickness* can accurately predict body density. To measure skinfold thickness, pinch a fold of skin between calipers at four body locations to determine the thickness, and add the four thicknesses. There is a linear relationship between body density and the logarithm of the skinfold thickness measure. The explanatory variable x is the log of the sum of the skinfold measures, and the response variable y is body density.

In the statistical model for predicting body density from skinfold thickness, subpopulations are defined by the x variable. All individuals with the same skinfold thickness x are in the same subpopulation. Their body densities are assumed to be normally distributed about a mean density that has a straight-line relation to skinfold thickness. The mean density for people with log skinfold thickness x is

$$\mu_y = \beta_0 + \beta_1 x$$

This population regression line gives the mean body density for all values of x. We cannot observe this line, because the observed responses y vary about their means. The statistical model for linear regression consists of the population regression line and a description of the variation of y about the line. Figure 10.2 displays the line and the variation in a picture, and the following equation expresses the idea in an equation:

$$\text{DATA} = \text{FIT} + \text{RESIDUAL}$$

The FIT part of the model consists of the subpopulation means, given by the expression $\beta_0 + \beta_1 x$. The RESIDUAL part represents deviations of the data from the line of population means. We assume that these deviations are normally distributed with standard deviation σ. We use ϵ (the Greek letter epsilon) to stand for the RESIDUAL part of the statistical model. A response y is the sum of its mean and a chance deviation ϵ from the mean. The deviations ϵ represent "noise," that is, variation in y due to other causes that prevent the observed (x, y)-values from forming a perfectly straight line on the scatterplot.

Simple Linear Regression Model

Given n observations on the explanatory variable x and the response variable y,

$$(x_1, y_1), (x_2, y_2), \quad \ldots, \quad (x_n, y_n)$$

the **statistical model for simple linear regression** states that the observed response y_i when the explanatory variable takes the value x_i is

$$y_i = \beta_0 + \beta_1 x_i + \epsilon_i$$

Here $\beta_0 + \beta_1 x_i$ is the mean response when $x = x_i$. The deviations ϵ_i are assumed to be independent and normally distributed with mean 0 and standard deviation σ.

The parameters of the model are β_0, β_1, and σ.

Because the means μ_y lie on the line $\mu_y = \beta_0 + \beta_1 x$, they are all determined by β_0 and β_1. Once we have estimates of β_0 and β_1, the linear relationship determines the estimates of μ_y for all values of x. Linear regression allows us to do inference not only for subpopulations for which we have data but also for those corresponding to x's not present in the data. We will learn how to do inference about

- the slope β_1 and the intercept β_0 of the population regression line,
- the mean response μ_y for a given value of x, and
- an individual future response y for a given value of x.

Estimating the regression parameters

The method of least squares presented in Chapter 2 fits a line to summarize a relationship between the observed values of an explanatory variable and a response variable. Now we want to use the least-squares line as a basis for inference about a population from which our observations are a sample. We can do this only when the statistical model just presented holds. In that setting, the slope b_1 and intercept b_0 of the least-squares line

$$\hat{y} = b_0 + b_1 x$$

estimate the slope β_1 and the intercept β_0 of the population regression line.

Using the formulas from Chapter 2 (page 141) and our new notation, the slope of the least-squares line is

$$b_1 = r\frac{s_y}{s_x}$$

and the intercept is

$$b_0 = \bar{y} - b_1\bar{x}$$

Here, r is the correlation between y and x, s_y is the standard deviation of y, and s_x is the standard deviation of x. Some algebra based on the rules for means of random variables (Section 4.4) shows that b_0 and b_1 are unbiased estimators of β_0 and β_1. Furthermore, b_0 and b_1 are normally distributed with means β_0 and β_1 and standard deviations that can be estimated from the data. Normality of these sampling distributions is a consequence of the assumption that the ϵ_i are distributed normally. A general form of the central limit theorem tells us that the distributions of b_0 and b_1 will still be approximately normal even if the ϵ_i are not. On the other hand, outliers and influential observations can invalidate the results of inference for regression.

The predicted value of y for a given value x^* of x is the point on the least-squares line $\hat{y} = b_0 + b_1 x^*$. This is an unbiased estimator of the mean response μ_y when $x = x^*$. The **residuals** are

residuals

$$\begin{aligned} e_i &= \text{observed response} - \text{predicted response} \\ &= y_i - \hat{y}_i \\ &= y_i - b_0 - b_1 x_i \end{aligned}$$

The residuals e_i correspond to the model deviations ϵ_i. The e_i sum to 0, and the ϵ_i come from a population with mean 0.

The remaining parameter to be estimated is σ, which measures the variation of y about the population regression line. Because this parameter is the standard deviation of the model deviations, it should come as no surprise that we use the residuals to estimate it. As usual, we work first with the variance and take the square root to obtain the standard deviation. For simple linear regression the estimate of σ^2 is the average squared residual

$$s^2 = \frac{\sum e_i^2}{n-2} = \frac{\sum (y_i - \hat{y}_i)^2}{n-2}$$

degrees of freedom

We average by dividing the sum by $n - 2$ in order to make s^2 an unbiased estimate of σ^2. The sample variance of n observations uses the divisor $n - 1$ for this same reason. The quantity $n - 2$ is called the **degrees of freedom** for s^2. The estimate of σ is given by

$$s = \sqrt{s^2}$$

In practice, we use software to do the unpleasant arithmetic needed to obtain b_1, b_0, and s. Here are the results for the example of body density and skinfold thickness.

EXAMPLE 10.2

We use the Minitab statistical software package to calculate the regression of body density on log of skinfolds for Example 10.1. In entering the data, we chose the names LSKIN and DEN for the explanatory and response variables. It is good practice to use names, rather than just x and y, to remind yourself which data the output describes. The skinfolds were measured in millimeters (mm) at the biceps, triceps, and the subscapular and suprailiac areas. LSKIN is the log of the sum of these four measures. Density is measured in thousands of kilograms per cubic meter (10^3 kg/m^3). Because the relationship between skinfold and body density varies slightly with age and sex, we look only at data from a sample of 92 males aged 20 to 29. In the discussion that follows, we round off the numbers from the computer output. Software often reports many more digits than are meaningful or useful.

First, we examine the variables and their relationship. Figure 10.3 is a scatterplot of the data. The relationship appears to be linear, and no unusual observations are evident.

We then proceed to fit the least-squares line. Figure 10.4 gives part of the Minitab output for this example. The top part of the output gives the fitted regression equation. We could write this as

$$\hat{y} = 1.16 - 0.0631x$$

Note that this is the line that appears on the scatterplot of Figure 10.3. The middle part of the output gives a table with the regression coefficients and some other quantities that we will discuss later in this chapter. Under the heading Predictor we have two entries: Constant, the term that Minitab uses for the intercept, and LSKIN. The column with the label Coef gives the

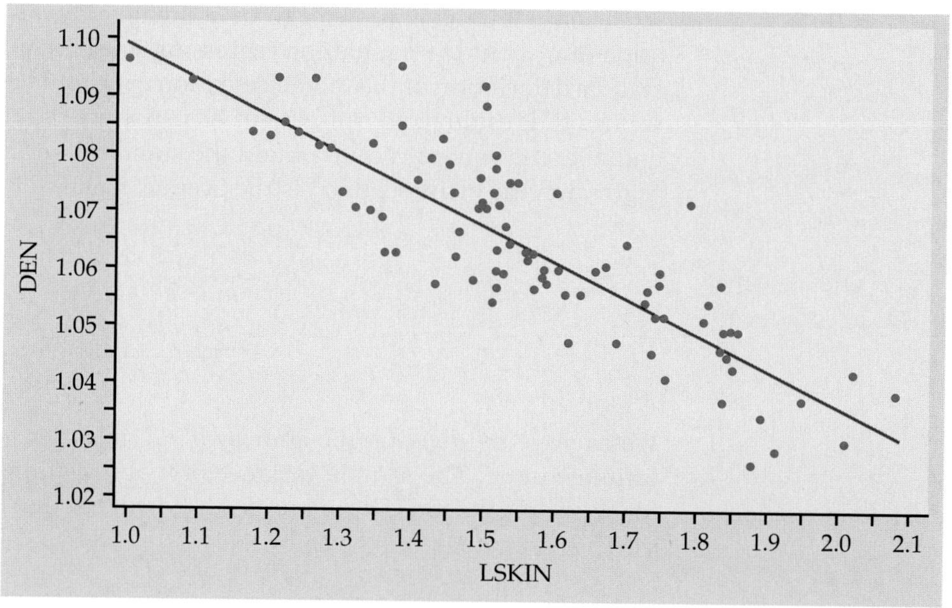

FIGURE 10.3 Scatterplot of DEN versus LSKIN with the least-squares regression line.

```
The regression equation is
DEN = 1.16 - 0.0631 LSKIN

Predictor        Coef        StDev         T        P
Constant      1.16300      0.00656    177.30    0.000
LSKIN       -0.063119     0.004145    -15.23    0.000

S = 0.008539   R-Sq = 72.0%    R-Sq(adj) = 71.7%
```

FIGURE 10.4 Regression output for the body density example.

values of b_0 and b_1. We see that $b_0 = 1.16300$ and $b_1 = -0.063119$. Minitab has rounded these numbers in the top part of the output where the equation is given. The estimated standard deviation s is labeled S and is given as 0.008539. We now have estimates of all of our model parameters β_0, β_1, and σ.

The output contains much other information that we ignore for now. Computer outputs often give more information than we want or need. On the other hand, it is very frustrating to find that a software package does not print out the particular statistics that we want for an analysis. The experienced user of statistical software learns to ignore the parts of the output that are not needed for the problem at hand.

Regression outputs from different software all contain the same basic quantities. Figure 10.5 gives the output for the body density example for Excel, JMP, SPSS, and SAS. You should verify that you can find the regression coefficients and the estimate of σ from these outputs.

Excel

	A	B	C	D	E	F	G
1	SUMMARY OUTPUT						
2							
3	*Regression Statistics*						
4	Multiple R	0.848766779					
5	R Square	0.720405045					
6	Adjusted R Square	0.717298435					
7	Standard Error	0.008539025					
8	Observations	92					
9							
10		*Coefficients*	*Standard Error*	*t Stat*	*P-value*	*Lower 95%*	*Upper 95%*
11	Intercept	1.162999075	0.006559641	177.2962	2.67E-116	1.1499672	1.1760309
12	X Variable 1	-0.063118994	0.00414491	-15.2281	1.22E-26	-0.0713536	-0.0548844
13							

Sheet1 / den /

Ready   Sum=0   NUM

JMP

JMPINTRO - eg10_001- Fit Y by X

File Edit Tables Rows Cols Analyze Graph Tools View Window Help

eg10_001

eg10_001- Fit Y by X

Bivariate Fit of Column3 By Column2

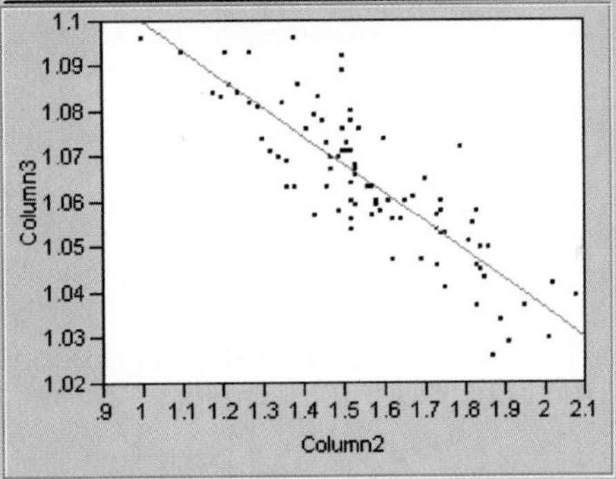

—— Linear Fit

Linear Fit

Column3 = 1.1629991 - 0.063119 Column2

Summary of Fit

RSquare	0.720405
RSquare Adj	0.717298
Root Mean Square Error	0.008539
Mean of Response	1.064033
Observations (or Sum Wgts)	92

Lack Of Fit

Analysis of Variance

Parameter Estimates

Term	Estimate	Std Error	t Ratio	Prob>|t|
Intercept	1.1629991	0.00656	177.30	<.0001
Column2	-0.063119	0.004145	-15.23	<.0001

Ready

FIGURE 10.5 Regression output from Excel, JMP, SPSS, and SAS for the body density example. (*continued*)

SPSS

Model Summary

Model	R	R Square	Adjusted R Square	Std. Error of the Estimate
1	0.849	0.720	0.717	0.00854

a Predictors: (Constant), LSKIN
b Dependent Variable: DEN

Coefficients

	Unstandardized Coefficients B	Std. Error	t	Sig.	95% Confidence Interval for B Lower Bound	Upper Bound
(Constant)	1.163	0.007	177.296	0.000	1.150	1.176
LSKIN	-6.312E-02	0.004	-15.228	0.000	-0.071	-0.055

a Dependent Variable: DEN

SAS

Root MSE	0.00854	R-Square	0.7204
Dependent Mean	1.06403	Adj R-Sq	0.7173
Coeff Var	0.80252		

Parameter Estimates

Variable	DF	Parameter Estimate	Standard Error	t Value	Pr > |t|	95% Confidence Limits	
Intercept	1	1.16300	0.00656	177.30	<0.0001	1.14997	1.17603
LSKIN	1	-0.06312	0.00414	-15.23	<0.0001	-0.07135	-0.05488

FIGURE 10.5 (*continued*) Regression output from Excel, JMP, SPSS, and SAS for the body density example.

Now that we have fitted a line, we should examine the residuals. Plot the residuals both against the case number (especially if this reflects the order in which the observations were taken) and against the explanatory variable. Figure 10.6 is a plot of the residuals versus case number, and Figure 10.7 plots the residuals versus LSKIN. No unusual values or patterns are evident in either plot. Finally, Figure 10.8 (page 668) is a normal quantile plot of the residuals. The plot looks fairly straight, so the assumption of normally distributed deviations appears to be reasonable. That is important for the inference that will follow. We need not make these plots by hand; the software will produce all of them on demand.

For the 92 males aged 20 to 29 in the study, the mean body density was 1.064, and the standard deviation was 0.016. The regression output shows that the standard deviation of body density about the regression line is 0.009, only about half as large. This decrease in the standard deviation is due to the ability of skinfold thickness to explain body density.

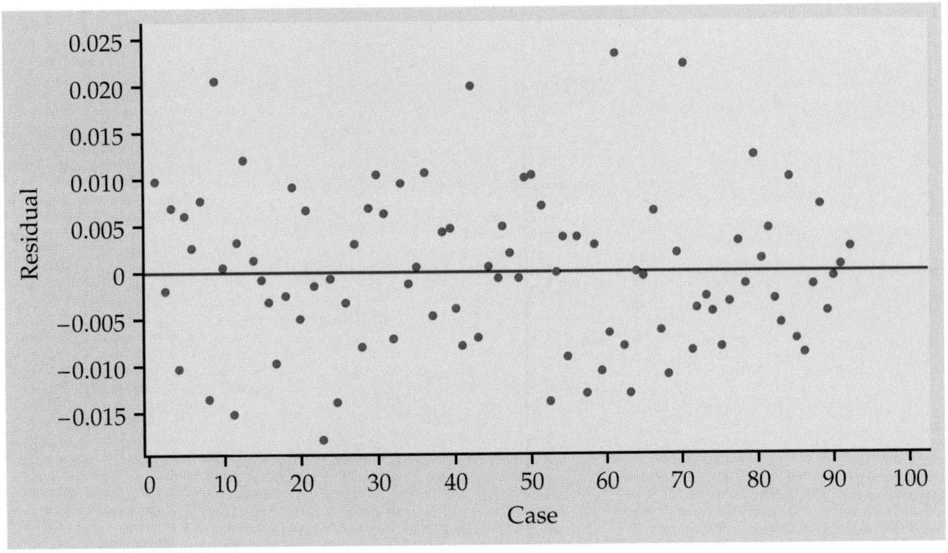

FIGURE 10.6 Plot of residuals versus case number for the body density example.

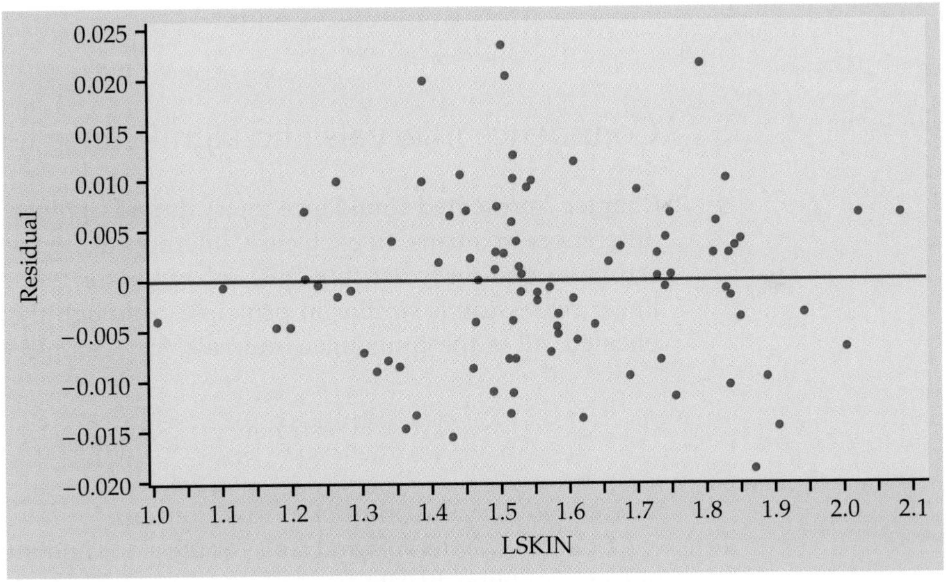

FIGURE 10.7 Plot of residuals versus the explanatory variable for the body density example.

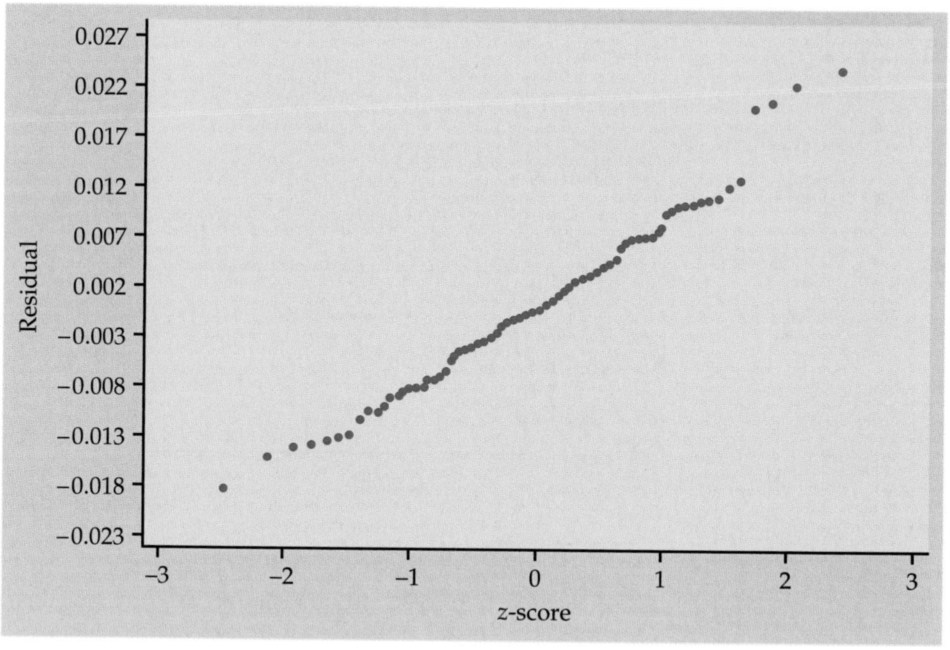

FIGURE 10.8 Normal quantile plot of the residuals for the body density example.

Confidence intervals and significance tests

Chapter 7 presented confidence intervals and significance tests for means and differences in means. In each case, inference rested on the standard errors of estimates and on t distributions. Inference for the intercept and slope in a linear regression is similar in principle, although the recipes are more complicated. All of the confidence intervals, for example, have the form

$$\text{estimate} \pm t^* \text{SE}_{\text{estimate}}$$

where t^* is a critical point of a t distribution.

Confidence intervals and tests for the slope and intercept are based on the normal sampling distributions of the estimates b_1 and b_0. Standardizing these estimates gives standard normal z statistics. The standard deviations of these estimates are multiples of σ, the model parameter that describes the variability about the true regression line. Because we do not know σ, we estimate it by s, the variability of the data about the least-squares line. When we do this, we get t distributions with degrees of freedom $n - 2$, the degrees of freedom of s. We give formulas for the standard errors SE_{b_1} and SE_{b_0} in Section 10.2. For now we will concentrate on the basic ideas and let the computer do the computations.

Confidence Intervals and Significance Tests for Regression Slope and Intercept

A **level C confidence interval for the intercept** β_0 is

$$b_0 \pm t^* \mathrm{SE}_{b_0}$$

A **level C confidence interval for the slope** β_1 is

$$b_1 \pm t^* \mathrm{SE}_{b_1}$$

In these expressions t^* is the value for the $t(n-2)$ density curve with area C between $-t^*$ and t^*.

To test the hypothesis H_0: $\beta_1 = 0$, compute the **test statistic**

$$t = \frac{b_1}{\mathrm{SE}_{b_1}}$$

The **degrees of freedom** are $n-2$. In terms of a random variable T having the $t(n-2)$ distribution, the P-value for a test of H_0 against

H_a: $\beta_1 > 0$ is $P(T \geq t)$

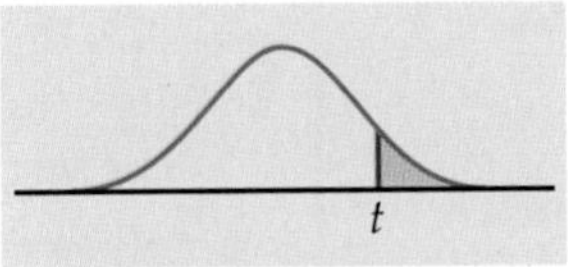

H_a: $\beta_1 < 0$ is $P(T \leq t)$

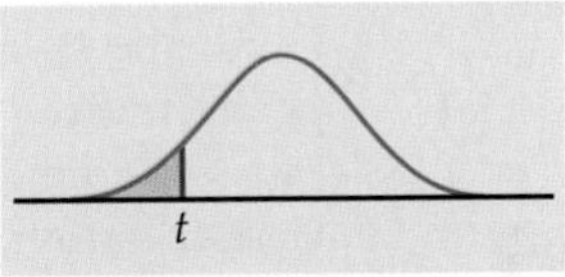

H_a: $\beta_1 \neq 0$ is $2P(T \geq |t|)$

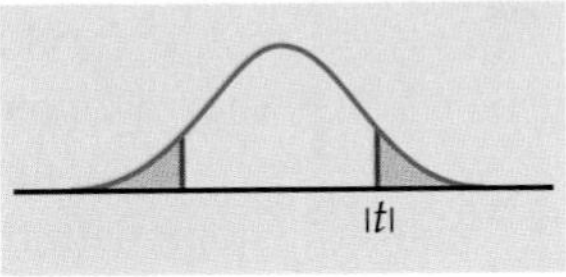

There are similar formulas for significance tests about the intercept β_0, using SE_{b_0} and the $t(n-2)$ distribution. Although computer outputs often include a test of H_0: $\beta_0 = 0$, this information usually has little practical value. From the equation for the population regression line, $\mu_y = \beta_0 + \beta_1 x$, we see that β_0 is the mean response corresponding to $x = 0$. In many practical situations, this subpopulation does not exist or is not interesting.

On the other hand, the test of H_0: $\beta_1 = 0$ is quite useful. When we substitute $\beta_1 = 0$ in the model, the x term drops out and we are left with

$$\mu_y = \beta_0$$

This model says that the mean of y does not vary with x. All of the y's come from a single population with mean β_0, which we would estimate by $\bar{y}$. The hypothesis H_0: $\beta_1 = 0$ therefore says that there is no straight-line relationship between y and x and that linear regression of y on x is of no value for predicting y.

EXAMPLE 10.3

The computer outputs in Figures 10.4 and 10.5 for the body density problem contain the information needed for inference about the regression slope and intercept. Let's look at the Excel output in Figure 10.5. The column labeled Standard Error gives the standard errors of the estimates. The value of SE_{b_1} appears to the right of the estimated slope $b_1 = -0.063118994$; it is 0.00414491. We round these and say that the slope is -0.0631 with a standard error of 0.0041.

The t statistic and P-value for the test of H_0: $\beta_1 = 0$ against the two-sided alternative H_a: $\beta_1 \neq 0$ appear in the columns labeled t Stat and P-value. We can verify the t calculation from the formula for the standardized estimate:

$$t = \frac{b_1}{SE_{b_1}} = \frac{-0.063118994}{0.00414491} = -15.23$$

The P-value is given as 1.22132E$-$26. Because this is a very small number, we have strong evidence against the null hypothesis. We conclude that linear regression on the logarithm of the skinfold measures is useful for predicting body density.

Higher body fat reduces body density and increases skinfold thickness, so we expect a negative association. We might therefore choose the one-sided alternative H_a: $\beta_1 < 0$. This choice does not affect our conclusion for this example, but the P-value for the one-sided alternative is half of the value for the two-sided alternative.

A confidence interval for β_1 requires a critical value t^* from the $t(n-2) = t(90)$ distribution. In Table D there are entries for 80 and 100 degrees of freedom. The values for these rows are very similar. To be conservative, we will use the larger critical value, for 80 degrees of freedom. For a 95% confidence interval we use the value in the 95% column, which is $t^* = 1.99$.

To compute the 95% confidence interval for β_1 we combine the estimate of the slope with the margin of error:

$$\begin{aligned} b_1 \pm t^* SE_{b_1} &= -0.063119 \pm (1.99)(0.00414491) \\ &= -0.0631 \pm 0.0082 \end{aligned}$$

The interval is $(-0.07, -0.05)$. This agrees with the value given by Excel in Figure 10.5. We estimate that an increase of 1 in the logarithm of skinfold thickness is associated with a decrease of between 0.05 and 0.07 times 10^3 kg/m^3 in body density.

Note that the intercept in this example is not of practical interest. It estimates the mean body density when the logarithm of the sum of the four skinfold thicknesses (that's x) is 0, a value that cannot occur. The values of LSKIN in the data range from 1.00 to 2.08. For this reason, we do not compute a confidence interval for β_0.

Confidence intervals for mean response

For any specific value of x, say x^*, the mean of the response y in this subpopulation is given by

$$\mu_y = \beta_0 + \beta_1 x^*$$

To estimate this mean from the sample, we substitute the estimates b_0 and b_1 for β_0 and β_1:

$$\hat{\mu}_y = b_0 + b_1 x^*$$

A confidence interval for μ_y adds to this estimate a margin of error based on the standard error $SE_{\hat{\mu}}$. (The formula for the standard error is given in Section 10.2.)

Confidence Interval for a Mean Response

A **level C confidence interval for the mean response** μ_y when x takes the value x^* is

$$\hat{\mu}_y \pm t^* SE_{\hat{\mu}}$$

where t^* is the value for the $t(n-2)$ density curve with area C between $-t^*$ and t^*.

Many computer programs calculate confidence intervals for the mean response corresponding to each of the x-values in the data. Some can calculate an interval for any value x^* of the explanatory variable.[3]

EXAMPLE 10.4

Figure 10.9 gives more of the output generated by the SAS regression procedure for the data on body density and skinfold measures. We show the output for only the first 10 of the 92 cases.

Obs	LSKIN	Dep Var DEN	Predicted Value	Std Error Mean Predict	95% CL Mean		Residual	Std Error Residual
1	1.27	1.0930	1.0828	0.001522	1.0798	1.0859	0.0102	0.00840
2	1.56	1.0630	1.0645	0.000891	1.0628	1.0663	-0.001533	0.00849
3	1.45	1.0780	1.0715	0.001016	1.0695	1.0735	0.006523	0.00848
4	1.52	1.0560	1.0671	0.000912	1.0652	1.0689	-0.0111	0.00849
5	1.51	1.0730	1.0677	0.000922	1.0659	1.0695	0.005311	0.00849
6	1.51	1.0710	1.0677	0.000922	1.0659	1.0695	0.003311	0.00849
7	1.5	1.0760	1.0683	0.000934	1.0665	1.0702	0.007679	0.00849
8	1.62	1.0470	1.0607	0.000916	1.0589	1.0626	-0.0137	0.00849
9	1.5	1.0890	1.0683	0.000934	1.0665	1.0702	0.0207	0.00849
10	1.75	1.0530	1.0525	0.001167	1.0502	1.0549	0.000459	0.00846

FIGURE 10.9 Computer output giving confidence intervals for subpopulation means for the body density example.

The observed data values y_i appear in the column labeled Dep Var DEN, the predicted values $\hat{\mu}_y$ are given in the column labeled Predicted Value, and the standard errors $SE_{\hat{\mu}}$ of the predicted values appear in the column labeled Std Error Mean Predict. The lower and upper limits for the 95% confidence intervals appear under the heading 95% CL Mean.

Consider the first line of the output. The value of LSKIN for this individual is 1.27. The predicted mean DEN for all individuals with LSKIN = 1.27 is

$$\hat{\mu}_{\mathrm{DEN}(1.27)} = 1.0828 \times 10^3 \text{ kg/m}^3$$

with a standard error of

$$SE_{\hat{\mu}} = 0.002$$

The 95% confidence interval is (1.0798, 1.0859). We conclude with 95% confidence that the mean body density for men aged 20 to 29 whose logarithm of skinfold measures is 1.27 lies between 1.0798 and 1.0859.

We can gain more insight by plotting the upper and lower confidence limits for a range of values of x on a graph with the data and the least-squares line. The 95% confidence limits appear as curved black lines in Figure 10.10. For any x^*, the confidence interval for the mean response extends from the lower black line to the upper black line. The intervals are narrowest for values of x^* near the mean of the observed x's and increase as x^* moves away from $\overline{x}$. To show this effect clearly, we have extended the x axis in the plot far beyond the range of the actual data. In practice, you would avoid extrapolation for such extreme values.

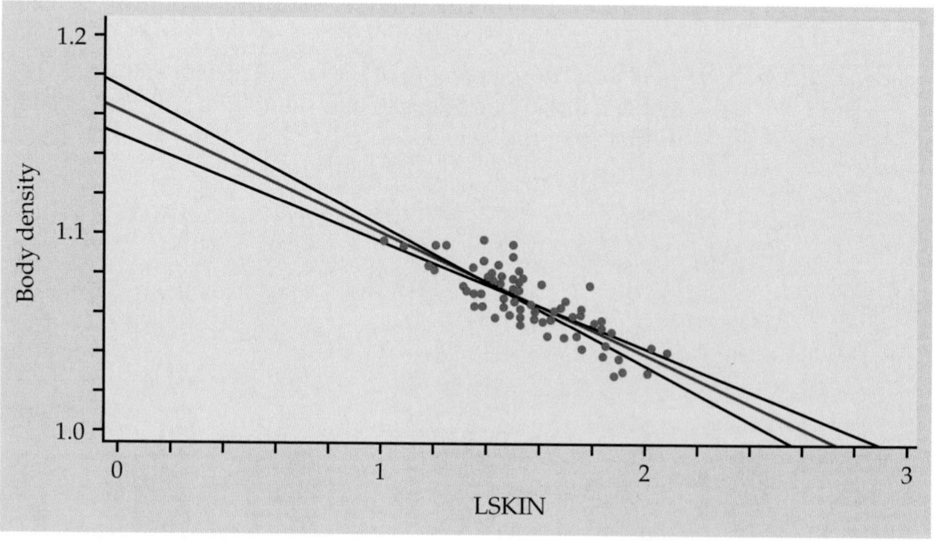

FIGURE 10.10 The 95% confidence limits (black curves) for the mean response for the body density example. The range of LSKIN is extended to show that the intervals grow wider as x^* moves away from the mean of the x observations.

Prediction intervals

In the last example, we predicted the mean body density for all young men with LSKIN equal to 1.27. Suppose that Sean has LSKIN = 1.27. We can also predict Sean's individual body density. The predicted value is the same as if we were predicting the mean of the entire subpopulation, but the margin of error is larger because it is harder to predict one individual value than to predict the mean.

The predicted response y for an individual case with a specific value x^* of the explanatory variable x is

$$\hat{y} = b_0 + b_1 x^*$$

This is the same as the expression for $\hat{\mu}_y$. That is, the fitted line is used both to estimate the mean response when $x = x^*$ and to predict a single future response. We use the two notations $\hat{\mu}_y$ and $\hat{y}$ to remind ourselves of these two distinct uses.

prediction interval

A useful prediction should include a margin of error to indicate its accuracy. The interval used to predict a future observation is called a **prediction interval.** Although the response y that is being predicted is a random variable, the interpretation of a prediction interval is similar to that for a confidence interval. Draw a sample of n observations (x_i, y_i) and then one additional observation (x^*, y). Do this many times, and each time calculate the 95% prediction interval for y based on the sample. Then 95% of the prediction intervals will contain the value of y for the additional observation. In other words, the probability that this method produces an interval that contains the value of a future observation is 0.95.

The form of the prediction interval is very similar to that of the confidence interval for the mean response. The difference is that the standard error $\text{SE}_{\hat{y}}$ used in the prediction interval includes both the variability due to the fact that the least-squares line is not exactly equal to the true regression line *and* the variability of the future response variable y around the subpopulation mean. (The formula for $\text{SE}_{\hat{y}}$ appears in Section 10.2.)

Prediction Interval for a Future Observation

A **level C prediction interval for a future observation** on the response variable y from the subpopulation corresponding to x^* is

$$\hat{y} \pm t^* \text{SE}_{\hat{y}}$$

where t^* is the value for the $t(n-2)$ density curve with area C between $-t^*$ and t^*.

EXAMPLE 10.5

Figure 10.11 gives prediction output from the SAS regression procedure for the first 10 of the 92 cases in the body density versus skinfold data set. The format is similar to Figure 10.9. The 95% prediction intervals are under the heading 95% CL Predict. These limits are wider than the corresponding 95% confidence limits for the mean

Obs	LSKIN	Dep Var DEN	Predicted Value	Std Error Mean Predict	95% CL Predict		Residual	Std Error Residual
1	1.27	1.0930	1.0828	0.001522	1.0656	1.1001	0.0102	0.00840
2	1.56	1.0630	1.0645	0.000891	1.0475	1.0816	-0.001533	0.00849
3	1.45	1.0780	1.0715	0.001016	1.0544	1.0886	0.006523	0.00848
4	1.52	1.0560	1.0671	0.000912	1.0500	1.0841	-0.0111	0.00849
5	1.51	1.0730	1.0677	0.000922	1.0506	1.0848	0.005311	0.00849
6	1.51	1.0710	1.0677	0.000922	1.0506	1.0848	0.003311	0.00849
7	1.5	1.0760	1.0683	0.000934	1.0513	1.0854	0.007679	0.00849
8	1.62	1.0470	1.0607	0.000916	1.0437	1.0778	-0.0137	0.00849
9	1.5	1.0890	1.0683	0.000934	1.0513	1.0854	0.0207	0.00849
10	1.75	1.0530	1.0525	0.001167	1.0354	1.0697	0.000459	0.00846

FIGURE 10.11 Computer output giving prediction intervals for the body density example.

body density that appear in Figure 10.9. (Note that this output does not give $SE_{\hat{y}}$, but only $SE_{\hat{\mu}}$.)

Consider Observation 1, which has LSKIN = 1.27. The 95% prediction interval is 1.0656 to 1.1001. For a male aged 20 to 29 with LSKIN = 1.27, we predict with 95% confidence that his body density will be between 1.0656 and 1.1001 times 10^3 kg/m^3.

Figure 10.12 shows the upper and lower prediction limits for a range of values of x, along with the data and the least-squares line. The 95% prediction limits are indicated by the black curves. Compare this figure with Figure 10.10, which shows the 95% confidence limits drawn to the same scale.

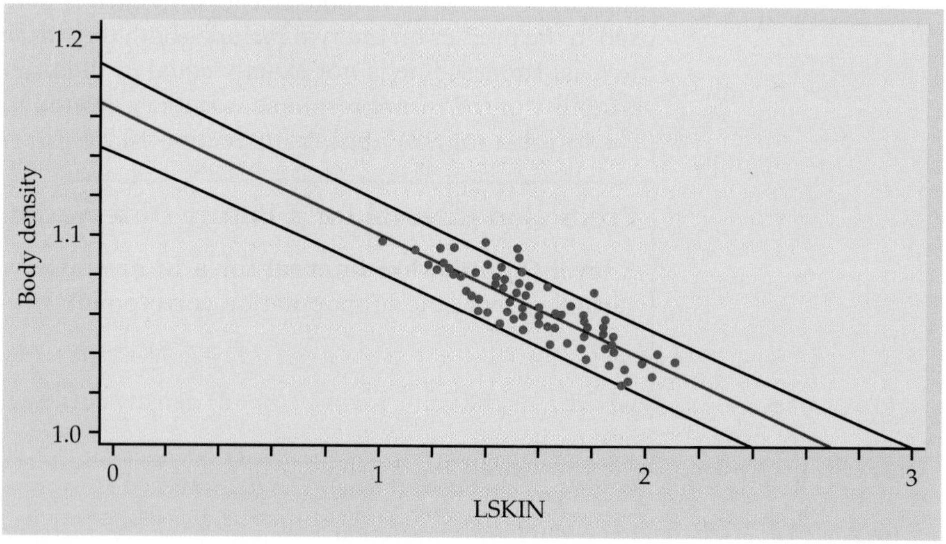

FIGURE 10.12 The 95% prediction limits (black curves) for individual response for the body density example. Compare with Figure 10.10.

The upper and lower limits of the prediction intervals are farther from the least-squares line than are the confidence limits. The figures remind us that the interval for a single future observation must be larger than an interval for the mean of its subpopulation.

Beyond the basics ▶ nonlinear regression

The regression model that we have studied assumes that the relationship between the response variable and the explanatory variable can be summarized with a straight line. When the relationship is not linear, we can sometimes make it linear by a transformation. In other circumstances, we use models that allow for various types of curved relationships. These models are called **nonlinear models.**

nonlinear models

Technical details are much more complicated for nonlinear models. In general we cannot write down simple formulas for the parameter estimates; we use a computer to solve systems of equations to find the estimates. However, the basic principles are those that we have already learned. For example,

$$\text{DATA} = \text{FIT} + \text{RESIDUAL}$$

still applies. The FIT is a nonlinear (curved) function and the residuals are assumed to be an SRS from the $N(0, \sigma)$ distribution. The nonlinear function contains parameters that must be estimated from the data. Approximate standard errors for these estimates are part of the standard output provided by software. Here is an example.

EXAMPLE 10.6

As we age, our bones become weaker and are more likely to break. Osteoporosis (or weak bones) is the major cause of bone fractures in older women. Some researchers have studied this problem by looking at how and when bone mass is accumulated by young women. Understanding the relationship between age and bone mass is an important part of this approach to the problem.

Figure 10.13 displays data for a measure of bone strength, called "total body bone mineral density" (TBBMD), and age for a sample of 256 young women.[4] TBBMD is measured in grams per square centimeter (g/cm^2), and age is recorded in years. The solid curve is the nonlinear fit, and the dashed curves are 95% prediction limits. The fitted nonlinear equation is

$$\hat{y} = 1.162\frac{e^{-1.62+0.28x}}{1 + e^{-1.62+0.28x}}$$

In this equation, $\hat{y}$ is the predicted value of TBBMD, the response variable; and x is age, the explanatory variable. A straight line would not do a very good job of summarizing the relationship between TBBMD and age. At first, TBBMD increases with age, but then it levels off as age increases. The value of the function where it is level is called "peak bone mass"; it is a parameter in the nonlinear model. The estimate is 1.162 and the standard error is 0.008. Software gives the 95% confidence interval as (1.146, 1.178).

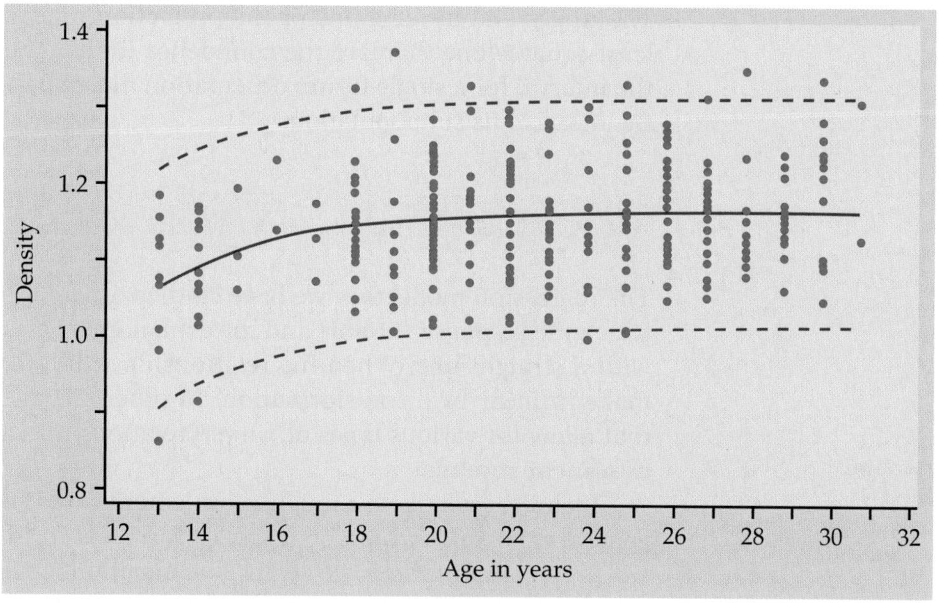

FIGURE 10.13 Plot of total body bone mineral density versus age.

The long-range goals of the researchers who conducted this study include developing intervention programs (exercise and increasing calcium intake have been shown to be effective) for young women that will increase their TBBMD. What age groups should be the target of these interventions? The fitted nonlinear model can be used to obtain estimates of the age (with a standard error) at which any given percent of the peak bone mass is attained. We estimate that the age at which the population reaches 95% of peak bone mass is 16.2 years (SE = 1.1 years). For 99% of peak bone mass, the age is 22.1 years (SE = 2.5 years). Intervention programs should be directed toward high school and college-aged women.

SUMMARY

The statistical model for **simple linear regression** is

$$y_i = \beta_0 + \beta_1 x_i + \epsilon_i$$

where $i = 1, 2, \ldots, n$. The ϵ_i are assumed to be independent and normally distributed with mean 0 and standard deviation σ. The **parameters** of the model are β_0, β_1, and σ.

The intercept and slope β_0 and β_1 are estimated by the intercept and slope of the **least-squares regression line,** b_0 and b_1. The parameter σ is estimated by

$$s = \sqrt{\frac{\sum e_i^2}{n-2}}$$

where the e_i are the **residuals**

$$e_i = y_i - \hat{y}_i$$

A **level C confidence interval for β_1** is

$$b_1 \pm t^* \mathrm{SE}_{b_1}$$

where t^* is the value for the $t(n-2)$ density curve with area C between $-t^*$ and t^*.

The **test of the hypothesis** H_0: $\beta_1 = 0$ is based on the ***t* statistic**

$$t = \frac{b_1}{\mathrm{SE}_{b_1}}$$

and the $t(n-2)$ distribution. There are similar formulas for confidence intervals and tests for β_0, but these are meaningful only in special cases.

The **estimated mean response** for the subpopulation corresponding to the value x^* of the explanatory variable is

$$\hat{\mu}_y = b_0 + b_1 x^*$$

A **level C confidence interval for the mean response** is

$$\hat{\mu}_y \pm t^* \mathrm{SE}_{\hat{\mu}}$$

where t^* is the value for the $t(n-2)$ density curve with area C between $-t^*$ and t^*.

The **estimated value of the response variable** y for a future observation from the subpopulation corresponding to the value x^* of the explanatory variable is

$$\hat{y} = b_0 + b_1 x^*$$

A **level C prediction interval** for the estimated response is

$$\hat{y} \pm t^* \mathrm{SE}_{\hat{y}}$$

where t^* is the value for the $t(n-2)$ density curve with area C between $-t^*$ and t^*.

The following section contains material that you should study if you plan to read Chapter 11 on multiple regression. In addition, the section we just completed assumes that you have access to software or a statistical calculator. If you do not, you now need to study the material on computations in the following optional section. The exercises are given at the end of the chapter.

10.2 More Detail about Simple Linear Regression*

In this section we study three optional topics. The first is analysis of variance for regression. If you plan to study the next chapter on multiple regression, you should study this material. The second topic concerns computations for regression inference. Here we present and illustrate the use of formulas for the inference procedures that we have just studied. Finally, we discuss inference for correlation.

Analysis of variance for regression

analysis of variance

The usual computer output for regression includes additional calculations called **analysis of variance.** Analysis of variance, often abbreviated ANOVA, is essential for multiple regression (Chapter 11) and for comparing several means (Chapters 12 and 13). Analysis of variance summarizes information about the sources of variation in the data. It is based on the

$$\text{DATA} = \text{FIT} + \text{RESIDUAL}$$

framework.

The total variation in the response y is expressed by the deviations $y_i - \bar{y}$. If these deviations were all 0, all observations would be equal and there would be no variation in the response. There are two reasons why the individual observations y_i are not all equal to their mean $\bar{y}$. First, the responses y_i correspond to different values of the explanatory variable x and will differ because of that. The fitted value $\hat{y}_i$ estimates the mean response for the specific x_i. The differences $\hat{y}_i - \bar{y}$ reflect the variation in mean response due to differences in the x_i. This variation is accounted for by the regression line, because the $\hat{y}$'s lie exactly on the line. Second, individual observations will vary about their mean because of variation within the subpopulation of responses to a fixed x_i. This variation is represented by the residuals $y_i - \hat{y}_i$ that record the scatter of the actual observations about the fitted line. The overall deviation of any y observation from the mean of the y's is the sum of these two deviations:

$$(y_i - \bar{y}) = (\hat{y}_i - \bar{y}) + (y_i - \hat{y}_i)$$

For deviations, this equation expresses the idea that DATA = FIT + RESIDUAL.

We have several times measured variation by an average of squared deviations. If we square each of the three deviations above and then sum over all n observations, it is an algebraic fact that the sums of squares add:

$$\sum(y_i - \bar{y})^2 = \sum(\hat{y}_i - \bar{y})^2 + \sum(y_i - \hat{y}_i)^2$$

*This material is optional.

We rewrite this equation as

$$\text{SST} = \text{SSM} + \text{SSE}$$

where

$$\text{SST} = \sum (y_i - \bar{y})^2$$
$$\text{SSM} = \sum (\hat{y}_i - \bar{y})^2$$
$$\text{SSE} = \sum (y_i - \hat{y}_i)^2$$

sum of squares

The SS in each abbreviation stands for **sum of squares,** and the T, M, and E stand for total, model, and error, respectively. ("Error" here stands for deviations from the line, which might better be called "residual" or "unexplained variation.") The total variation, as expressed by SST, is the sum of the variation due to the straight-line model (SSM) and the variation due to deviations from this model (SSE). This partition of the variation in the data between two sources is the heart of analysis of variance.

If H_0: $\beta_1 = 0$ were true, there would be no subpopulations and all of the y's should be viewed as coming from a single population with mean μ_y. The variation of the y's would then be described by the sample variance

$$s_y^2 = \frac{\sum (y_i - \bar{y})^2}{n - 1}$$

degrees of freedom

The numerator in this expression is SST. The denominator is the total **degrees of freedom,** or simply DFT.

Just as the total sum of squares SST is the sum of SSM and SSE, the total degrees of freedom DFT is the sum of DFM and DFE, the degrees of freedom for the model and for the error:

$$\text{DFT} = \text{DFM} + \text{DFE}$$

The model has one explanatory variable x, so the degrees of freedom for this source are DFM $= 1$. Because DFT $= n - 1$, this leaves DFE $= n - 2$ as the degrees of freedom for error. For each source, the ratio of the sum of squares to the degrees of freedom is called the **mean square,** or simply MS. The general formula for a mean square is

mean square

$$\text{MS} = \frac{\text{sum of squares}}{\text{degrees of freedom}}$$

Each mean square is an average squared deviation. MST is just s_y^2, the sample variance that we would calculate if all of the data came from a single population. MSE is familiar to us:

$$\text{MSE} = s^2 = \frac{\sum (y_i - \hat{y}_i)^2}{n - 2}$$

It is our estimate of σ^2, the variance about the population regression line.

Sums of Squares, Degrees of Freedom, and Mean Squares

Sums of squares represent variation present in the responses. They are calculated by summing squared deviations. **Analysis of variance** partitions the total variation between two sources.

The sums of squares are related by the formula

$$\text{SST} = \text{SSM} + \text{SSE}$$

That is, the total variation is partitioned into two parts, one due to the model and one due to deviations from the model.

Degrees of freedom are associated with each sum of squares. They are related in the same way:

$$\text{DFT} = \text{DFM} + \text{DFE}$$

To calculate **mean squares,** use the formula

$$\text{MS} = \frac{\text{sum of squares}}{\text{degrees of freedom}}$$

interpretation of r^2

In Section 2.3 we noted that r^2 is the fraction of variation in the values of y that is explained by the least-squares regression of y on x. The sums of squares make this interpretation precise. Recall that SST = SSM + SSE. It is an algebraic fact that

$$r^2 = \frac{\text{SSM}}{\text{SST}} = \frac{\sum(\hat{y}_i - \bar{y})^2}{\sum(y_i - \bar{y})^2}$$

Because SST is the total variation in y and SSM is the variation due to the regression of y on x, this equation is the precise statement of the fact that r^2 is the fraction of variation in y explained by x.

The ANOVA *F* test

F statistic

The null hypothesis H_0: $\beta_1 = 0$ that y is not linearly related to x can be tested by comparing MSM with MSE. The ANOVA test statistic is an ***F* statistic,**

$$F = \frac{\text{MSM}}{\text{MSE}}$$

When H_0 is true, this statistic has an F distribution with 1 degree of freedom in the numerator and $n-2$ degrees of freedom in the denominator. These degrees of freedom are those of MSM and MSE. The F distributions were introduced in Section 7.3. Figure 7.15 (page 555) shows a typical F density curve. Just as there are many t statistics, there are many F statistics. The ANOVA F statistic is not the same as the F statistic of Section 7.3.

When $\beta_1 \neq 0$, MSM tends to be large relative to MSE. So large values of F are evidence against H_0 in favor of the two-sided alternative.

Analysis of Variance F Test

In the simple linear regression model, the hypotheses

$$H_0: \beta_1 = 0$$
$$H_1: \beta_1 \neq 0$$

are tested by the ***F* statistic**

$$F = \frac{\text{MSM}}{\text{MSE}}$$

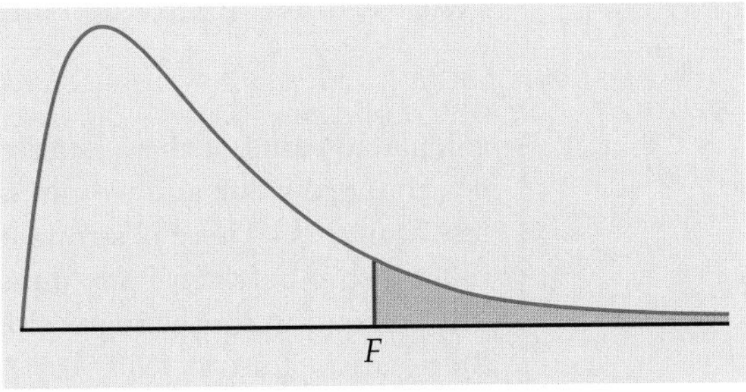

The P-value is the probability that a random variable having the $F(1, n-2)$ distribution is greater than or equal to the calculated value of the F statistic.

The F statistic tests the same null hypothesis as one of the t statistics that we encountered earlier in this chapter, so it is not surprising that the two are related. It is an algebraic fact that $t^2 = F$ in this case. For linear regression with one explanatory variable, we prefer the t form of the test because it more easily allows us to test one-sided alternatives and is closely related to the confidence interval for β_1.

ANOVA table

The ANOVA calculations are displayed in an *analysis of variance table*, often abbreviated **ANOVA table.** Here is the format of the table for simple linear regression:

Source	Degrees of freedom	Sum of squares	Mean square	F
Model	1	$\sum(\hat{y}_i - \bar{y})^2$	SSM/DFM	MSM/MSE
Error	$n - 2$	$\sum(y_i - \hat{y}_i)^2$	SSE/DFE	
Total	$n - 1$	$\sum(y_i - \bar{y})^2$	SST/DFT	

EXAMPLE 10.7

The ANOVA output generated by Minitab for the body density data in Example 10.1 is given in Figure 10.14. Note that Minitab uses the labels Regression, Residual Error, and Total for the three sources of variation. We have called these Model, Error, and Total. Other statistical software packages may use slightly different labels. The

```
S = 0.008539   R-Sq = 72.0%    R-Sq(adj) = 71.7%

Analysis of Variance

Source           DF        SS          MS          F        P
Regression        1   0.016909    0.016909    231.89    0.000
Residual Error   90   0.006562    0.000073
Total            91   0.023471
```

FIGURE 10.14 Regression output with ANOVA table for Example 10.7.

calculated value of the F statistic is 231.89 and the P-value is given as 0.000. This is a rounded value and we can conclude that $P < 0.0005$. (The actual value is much less than this.) There is strong evidence against the null hypothesis that there is no relationship between body density and the skinfold measure. Let's go back to the Minitab output for the regression coefficients given in Figure 10.4. There we see that the t statistic for LSKIN is -15.23. If we square this number we obtain the F statistic. The value of r^2 is also given in the output. Log of the skinfolds explains 71.7% of the variability in body density.

Calculations for regression inference

We recommend using statistical software for regression calculations. With time and care, however, the work is feasible with a calculator. We will use the following example to illustrate how to perform inference for regression analysis using a calculator.

EXAMPLE 10.8

As in many other businesses, technological advances and new methods have produced dramatic results in agriculture. Here are data on the average yield in bushels per acre of U.S. corn for selected years:[5]

Year	1966	1976	1986	1996
Yield	73.1	88.0	119.4	127.1

Here, year is the explanatory variable x and yield is the response variable y. The data and the least-squares line are plotted in Figure 10.15. The strong straight-line pattern suggests that we can use linear regression to model the relationship between yield and time.

We begin our regression calculations by fitting the least-squares line. Fitting the line gives estimates b_1 and b_0 of the model parameters β_1 and β_0. Next we examine the residuals from the fitted line and obtain an estimate s of the remaining parameter σ. These calculations are preliminary to inference. Finally, we use s to obtain the standard errors needed for the various interval estimates and significance tests. Roundoff errors that accumulate during these calculations can ruin the final results. Be sure to carry many significant digits and check your work carefully.

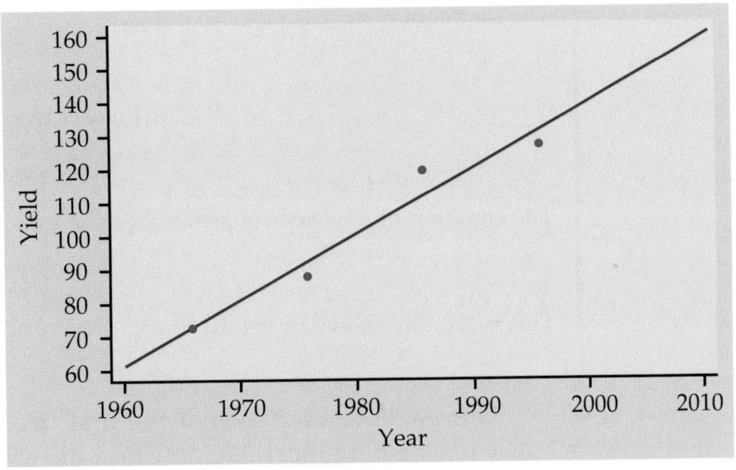

FIGURE 10.15 Data and regression line for Example 10.8.

Preliminary calculations After examining the scatterplot (Figure 10.15) to verify that the data show a straight-line pattern with no unusual points, we begin our calculations.

EXAMPLE 10.9

We start by making a table with the mean and standard deviation for each of the variables, the correlation, and the sample size. These calculations should be familiar from Chapters 1 and 2. Here is the summary:

Variable	Mean	Standard deviation	Correlation	Sample size
Year	$\bar{x} = 1981$	$s_x = 12.91$	$r = 0.97584$	$n = 4$
Yield	$\bar{y} = 101.9$	$s_y = 25.5861$		

These quantities are the building blocks for our calculations.

We will need one additional quantity for the calculations to follow. It is the expression $\sum(x_i - \bar{x})^2$. We obtain this quantity as an intermediate step when we calculate s_x. You could also find it using the fact that $\sum(x_i - \bar{x})^2 = (n-1)s_x^2$. You should verify that the value for our example is

$$\sum(x_i - \bar{x})^2 = 500$$

Our first task is to find the least-squares line. This is easy with the building blocks that we have assembled.

EXAMPLE 10.10

Although a calculator that has a correlation function will also give us b_0 and b_1, we can check the basic facts. The slope of the least-squares line is

$$\begin{aligned} b_1 &= r\frac{s_y}{s_x} \\ &= 0.97584\frac{25.5861}{12.91} \\ &= 1.934 \end{aligned}$$

The intercept is

$$\begin{aligned} b_0 &= \bar{y} - b_1\bar{x} \\ &= 101.9 - (1.934)(1981) \\ &= -3729.354 \end{aligned}$$

The equation of the least-squares regression line is therefore

$$\hat{y} = -3729 + 1.9x$$

This is the line shown in Figure 10.15.

We now have estimates of the first two parameters, β_0 and β_1, of our linear regression model. We now find the estimate of the third parameter σ, the standard deviation s about the fitted line. To do this we need to find the predicted values and then the residuals.

The predicted value for 1966 is

$$\begin{aligned} \hat{y}_1 &= b_0 + b_1x_1 \\ &= -3729.354 + (1.934)(1966) \\ &= 72.89 \end{aligned}$$

and the residual is

$$\begin{aligned} e_1 &= y_1 - \hat{y}_1 \\ &= 73.1 - 72.89 \\ &= 0.21 \end{aligned}$$

The residuals for the other years are calculated in the same way. You should verify that they are -4.23, 7.83, and -3.81. Also check that the four residuals sum to zero.

EXAMPLE 10.11

The estimate of σ^2 is s^2, the sum of the squares of the residuals divided by $n - 2$. The estimated standard deviation about the line is the square root of this quantity.

$$\begin{aligned} s^2 &= \frac{\sum e_i^2}{n-2} \\ &= \frac{(0.21)^2 + (-4.23)^2 + (7.83)^2 + (-3.81)^2}{2} \\ &= 46.8810 \end{aligned}$$

So the estimate of the standard deviation about the line is

$$s = \sqrt{46.8810} = 6.8470$$

Now that we have estimates of the three parameters of our model, we can proceed to the more detailed calculations needed for regression inference.

Inference for slope and intercept Confidence intervals and significance tests for the slope β_1 and intercept β_0 of the population regression line make use of the estimates b_1 and b_0 and their standard errors.

Some algebra using the rules for variances (Section 4.4) establishes that the standard deviation of b_1 is

$$\sigma_{b_1} = \frac{\sigma}{\sqrt{\sum(x_i - \overline{x})^2}}$$

Similarly, the standard deviation of b_0 is

$$\sigma_{b_0} = \sigma\sqrt{\frac{1}{n} + \frac{\overline{x}^2}{\sum(x_i - \overline{x})^2}}$$

To estimate these standard deviations, we need only replace σ by its estimate s.

Standard Errors for Estimated Regression Coefficients

The standard error of the slope b_1 of the least-squares regression line is

$$\text{SE}_{b_1} = \frac{s}{\sqrt{\sum(x_i - \overline{x})^2}}$$

The standard error of the intercept b_0 is

$$\text{SE}_{b_0} = s\sqrt{\frac{1}{n} + \frac{\overline{x}^2}{\sum(x_i - \overline{x})^2}}$$

The plot of the regression line with the data in Figure 10.15 *appears* to show a very strong relationship, but our sample size is very small. We assess the situation with a significance test for the slope.

EXAMPLE 10.12

First we need the standard error of the estimated slope:

$$\begin{aligned}\text{SE}_{b_1} &= \frac{s}{\sqrt{\sum(x_i - \overline{x})^2}}\\ &= \frac{6.8470}{\sqrt{500}}\\ &= 0.3062\end{aligned}$$

To test

$$\begin{aligned}H_0&: \beta_1 = 0\\ H_a&: \beta_1 \neq 0\end{aligned}$$

calculate the t statistic:

$$t = \frac{b_1}{\mathrm{SE}_{b_1}}$$
$$= \frac{1.934}{0.3062} = 6.316$$

Using Table D with $n - 2 = 2$ degrees of freedom, we conclude $P < 0.04$. (The exact value obtained from software is 0.0242.) We conclude that over the period 1966 to 1996 there has been an approximately linear increase in the yield of U.S. corn ($t = 6.32$, df $= 2$, $P < 0.04$).

The significance test tells us that the data provide sufficient information to conclude that there has been an increase. We use the estimate b_1 and its confidence interval to describe what the increase is and how well we have estimated it.

EXAMPLE 10.13

For the corn yield problem, let's find a 95% confidence interval for the slope β_1. The degrees of freedom are $n - 2 = 2$, so t^* from Table D is 4.303. To find the 95% confidence interval for β_1 we compute

$$b_1 \pm t^* \mathrm{SE}_{b_1} = 1.934 \pm (4.303)(0.3062)$$
$$= 1.934 \pm 1.318$$

The interval is (0.62, 3.25). Over the period 1966 to 1996, U.S. corn yields per acre have increased at the rate of about 2 bushels per year. The margin of error for 95% confidence is 1.3 bushels.

Note the effect of the small sample size on the critical value t^*. With one additional observation, it would decrease to 3.182.

In this example, the intercept β_0 does not have a meaningful interpretation. Because β_0 is the mean of y when $x = 0$, β_0 represents corn yield in the year 0. For problems where inference for β_0 is appropriate, the calculations are performed in the same way as those for β_1. Note that there is a different formula for the standard error.

Confidence intervals for the mean response and prediction intervals for a future observation When we substitute a particular value x^* of the explanatory variable into the regression equation and obtain a value of $\hat{y}$, we can view the result in two ways. We have estimated the mean response μ_y, and we have also predicted a future value of the response y. The margins of error for these two uses are quite different. Prediction intervals for an individual response are wider than confidence intervals for estimating a mean response. We now proceed with the details of these calculations. Once again, standard errors are the essential quantities. And once again, these standard errors are multiples of s, our basic measure of the variability of the responses about the fitted line.

Standard Errors for $\hat{\mu}$ and $\hat{y}$

The standard error of $\hat{\mu}$ is

$$SE_{\hat{\mu}} = s\sqrt{\frac{1}{n} + \frac{(x^* - \overline{x})^2}{\sum(x_i - \overline{x})^2}}$$

The standard error for predicting an individual response $\hat{y}$ is[6]

$$SE_{\hat{y}} = s\sqrt{1 + \frac{1}{n} + \frac{(x^* - \overline{x})^2}{\sum(x_i - \overline{x})^2}}$$

Note that the only difference between the recipes for these two standard errors is the extra 1 under the square root sign in the standard error for prediction. This standard error is larger due to the additional variation of individual responses about the mean response. It produces prediction intervals that are wider than the confidence intervals for the mean response.

For problems such as the corn yield example, we often want to estimate values of y in the future. This is a type of forecasting. We use a prediction interval for this purpose.

EXAMPLE 10.14

Assuming that the linear pattern of increasing corn yield continues into the future, let's estimate the yield in the year 2006. We also provide a prediction interval that quantifies the uncertainty in this estimate. For the year 2006, we predict that the corn yield will be

$$\begin{aligned}\hat{y} &= b_0 + b_1x^* \\ &= -3729.354 + (1.934)(2006) = 150.25\end{aligned}$$

The standard error is

$$\begin{aligned}SE_{\hat{y}} &= s\sqrt{1 + \frac{1}{n} + \frac{(x^* - \overline{x})^2}{\sum(x_i - \overline{x})^2}} \\ &= 6.8470\sqrt{1 + \frac{1}{4} + \frac{(2006 - 1981)^2}{500}} \\ &= 6.8470\sqrt{1 + \frac{1}{4} + \frac{625}{500}} = 10.83\end{aligned}$$

To find the 95% prediction interval we compute

$$\begin{aligned}\hat{y} \pm t^*SE_{\hat{y}} &= 150.25 \pm (4.303)(10.83) \\ &= 150.25 \pm 46.60\end{aligned}$$

The interval is (103.65, 196.85). We predict that the U.S. corn yield in the year 2006 will be 150 bushels per acre. The margin of error is 47 bushels per acre for 95% confidence.

Calculations for confidence intervals for the mean response are similar. The only difference is the use of the formula for $SE_{\hat{\mu}}$ in place of $SE_{\hat{y}}$.

Inference for correlation

The correlation coefficient is a measure of the strength and direction of the linear association between two variables. Correlation does not require an explanatory-response relationship between the variables. We can consider the sample correlation r as an estimate of the correlation in the population and base inference about the population correlation on r.

population correlation ρ

The correlation between the variables x and y when they are measured for every member of a population is the **population correlation.** As usual, we use Greek letters to represent population parameters. In this case ρ (the Greek letter rho) is the population correlation. When $\rho = 0$, there is no linear association in the population. In the important case where the two variables x and y are both normally distributed, the condition $\rho = 0$ is equivalent to the statement that x and y are independent. That is, there is no association of any kind between x and y. (Technically, the condition required is that x and y be **jointly normal.** This means that the distribution of x is normal and also that the conditional distribution of y given any fixed value of x is normal.) We therefore may wish to test the null hypothesis that a population correlation is 0.

jointly normal variables

Test for a Zero Population Correlation

To test the hypothesis H_0: $\rho = 0$, compute the t statistic:

$$t = \frac{r\sqrt{n-2}}{\sqrt{1-r^2}}$$

where n is the sample size and r is the sample correlation.

In terms of a random variable T having the $t(n-2)$ distribution, the P-value for a test of H_0 against

H_a: $\rho > 0$ $\quad P(T \geq t)$

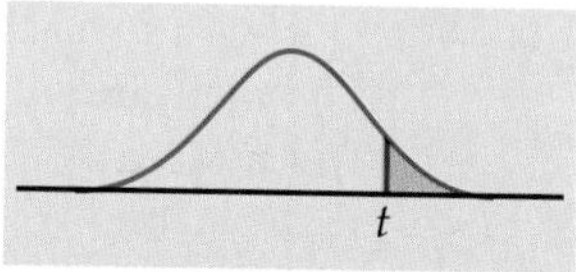

H_a: $\rho < 0$ $\quad P(T \leq t)$

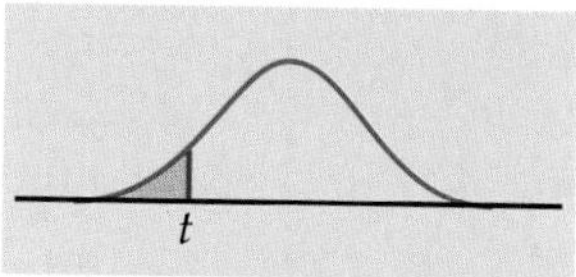

H_a: $\rho \neq 0$ is $\quad 2P(T \geq |t|)$

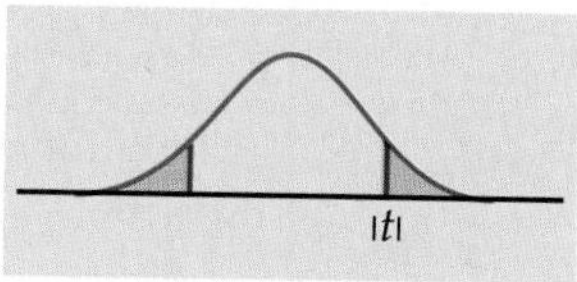

Most computer packages have routines for calculating correlations and some will provide the significance test for the null hypothesis that ρ is zero.

Correlations (Pearson)

```
Correlation of LSKIN and DEN = -0.849, P-Value = 0.000
```

FIGURE 10.16 Correlation output for Example 10.15.

EXAMPLE 10.15

For the body density and skinfold measures problem, the Minitab output appears in Figure 10.16. The sample correlation between body density and log of skinfolds is $r = -0.849$. Minitab calls this a Pearson correlation to distinguish it from other kinds of correlations that it can calculate. The P-value for a two-sided test of H_0: $\rho = 0$ is given as 0.000. This means that the actual P-value is less than 0.0005. To test the one-sided alternative that the population correlation is negative, we divide the P-value in the output by 2, after checking that the sample coefficient is in fact negative. We conclude that there is a negative correlation between body density and log of skinfolds.

If your software does not give the significance test, the computations are not very difficult to do with a calculator.

EXAMPLE 10.16

Figure 10.16 gives the correlation between body density and the log of the skinfold measures. We see that $r = -0.849$. Recall that $n = 92$. The t statistic for testing the null hypothesis that the population correlation is zero is

$$\begin{aligned} t &= \frac{r\sqrt{n-2}}{\sqrt{1-r^2}} \\ &= \frac{-0.849\sqrt{92-2}}{\sqrt{1-(-0.849)^2}} \\ &= -15.5 \end{aligned}$$

From Table D using $n - 2 = 90$ degrees of freedom we have $P < 0.0001$. This agrees with the Minitab output in Figure 10.16, where the P-value is given as 0.000. The data provide clear evidence that body density and the log of skinfolds are negatively related.

There is a close connection between the significance test for a correlation and the test for the slope in a linear regression. Recall that

$$b_1 = r\frac{s_y}{s_x}$$

From this fact we see that if the slope is 0, so is the correlation, and vice versa. It should come as no surprise to learn that the procedures for testing H_0: $\beta_1 = 0$ and H_0: $\rho = 0$ are also closely related. In fact, the t statistics for testing these hypotheses are numerically equal. That is,

$$\frac{b_1}{s_{b_1}} = \frac{r\sqrt{n-2}}{\sqrt{1-r^2}}$$

Check that this holds in both of our examples.

In our examples, the conclusion that there is a significant correlation between the two variables would not come as a surprise to anyone familiar with the meaning of these variables. The significance test simply tells us whether or not there is evidence in the data to conclude that the population correlation is different from 0. The actual size of the correlation is of considerably more interest. We would therefore like to give a confidence interval for the population correlation. Unfortunately, most software packages do not perform this calculation. Because hand calculation of the confidence interval is very tedious, we do not give the method here.[7]

SUMMARY

The **ANOVA table** for a linear regression gives the degrees of freedom, sum of squares, and mean squares for the model, error, and total sources of variation. The **ANOVA *F* statistic** is the ratio MSM/MSE. Under H_0: $\beta_1 = 0$, this statistic has an $F(1,\ n-2)$ distribution and is used to test H_0 versus the two-sided alternative.

The **square of the sample correlation** can be expressed as

$$r^2 = \frac{\text{SSM}}{\text{SST}}$$

and is interpreted as the proportion of the variability in the response variable y that is explained by the explanatory variable x in the linear regression.

The **standard errors** for b_0 and b_1 are

$$\text{SE}_{b_0} = s\sqrt{\frac{1}{n} + \frac{\bar{x}^2}{\sum(x_i - \bar{x})^2}}$$

$$\text{SE}_{b_1} = \frac{s}{\sqrt{\sum(x_i - \bar{x})^2}}$$

The **standard error** that we use for a confidence interval for the estimated mean response for the subpopulation corresponding to the value x^* of the explanatory variable is

$$\text{SE}_{\hat{\mu}} = s\sqrt{\frac{1}{n} + \frac{(x^* - \bar{x})^2}{\sum(x_i - \bar{x})^2}}$$

The **standard error** that we use for a prediction interval for a future observation from the subpopulation corresponding to the value x^* of the explanatory variable is

$$\text{SE}_{\hat{y}} = s\sqrt{1 + \frac{1}{n} + \frac{(x^* - \bar{x})^2}{\sum(x_i - \bar{x})^2}}$$

When the variables y and x are jointly normal, the sample correlation is an estimate of the **population correlation** $\boldsymbol{\rho}$**.** The test of H_0: $\rho = 0$ is based on the ***t*** **statistic**

$$t = \frac{r\sqrt{n-2}}{\sqrt{1-r^2}}$$

which has a $t(n-2)$ distribution under H_0. This test statistic is numerically identical to the t statistic used to test H_0: $\beta_1 = 0$.

CHAPTER 10 EXERCISES

10.1 Returns on common stocks in the United States and overseas appear to be growing more closely correlated as economies become more interdependent. Suppose that this population regression line connects the total annual returns (in percent) of two indexes of stock prices:

$$\text{MEAN OVERSEAS RETURN} = 4.7 + 0.66 \times \text{U.S. RETURN}$$

(a) What is β_0 in this line? What does this number say about overseas returns when the U.S. market is flat (0% return)?

(b) What is β_1 in this line? What does this number say about the relationship between U.S. and overseas returns?

(c) We know that overseas returns will vary during years that have the same return on U.S. common stocks. Write the regression model based on the population regression line given above. What part of this model allows overseas returns to vary when U.S. returns remain the same?

10.2 Suppose that there is a linear relationship between the number of students x in an elementary school and the annual budget y. Write a population regression model to describe this relationship.

(a) Which parameter in your model is the fixed cost in the budget (for example, the salary of the principal and some administrative costs) that does not change as x increases?

(b) Which parameter in your model shows how total cost changes when there are more students in the school? Do you expect this number to be greater than 0 or less than 0?

(c) Actual data from schools will not fit a straight line exactly. What term in your model allows variation among schools of the same size x?

10.3 How is the flow of investors' money into stock mutual funds related to the flow of money into bond mutual funds? Here are data on the net new money flowing into stock and bond mutual funds in the years 1985 to 2000, in billions of dollars.[8] "Net" means that funds flowing out are subtracted from those flowing in. If more money leaves than arrives, the net flow will be negative. To eliminate the effect of inflation, all dollar amounts are in "real dollars" with constant buying power equal to that of a dollar in the year 2000.

Year	1985	1986	1987	1988	1989	1990	1991	1992
Stocks	12.8	34.6	28.8	−23.3	8.3	17.1	50.6	97.0
Bonds	100.8	161.8	10.6	−5.8	−1.4	9.2	74.6	87.1

Year	1993	1994	1995	1996	1997	1998	1999	2000
Stocks	151.3	133.6	140.1	238.2	243.5	165.9	194.3	309.0
Bonds	84.6	−72.0	−6.8	3.3	30.0	79.2	−6.2	−48.0

(a) Make a scatterplot with cash flow into stock funds as the explanatory variable. Find the least-squares line for predicting net bond investments from net stock investments. What do the data suggest?

(b) Is there statistically significant evidence that there is some straight-line relationship between the flows of cash into bond funds and stock funds? (State hypotheses, give a test statistic and its *P*-value, and state your conclusion.)

(c) What fact about the scatterplot explains why the relationship described by the least-squares line is not significant?

10.4 How well does the number of beers a student drinks predict his or her blood alcohol content? Sixteen student volunteers at Ohio State University drank a randomly assigned number of cans of beer. Thirty minutes later, a police officer measured their blood alcohol content (BAC). Here are the data:[9]

Student	1	2	3	4	5	6	7	8
Beers	5	2	9	8	3	7	3	5
BAC	0.10	0.03	0.19	0.12	0.04	0.095	0.07	0.06

Student	9	10	11	12	13	14	15	16
Beers	3	5	4	6	5	7	1	4
BAC	0.02	0.05	0.07	0.10	0.085	0.09	0.01	0.05

The students were equally divided between men and women and differed in weight and usual drinking habits. Because of this variation, many students don't believe that number of drinks predicts blood alcohol well.

(a) Make a scatterplot of the data. Find the equation of the least-squares regression line for predicting blood alcohol from number of beers and add this line to your plot. What is r^2 for these data? Briefly summarize what your data analysis shows.

(b) Is there significant evidence that drinking more beers increases blood alcohol on the average in the population of all students? State hypotheses, give a test statistic and *P*-value, and state your conclusion.

10.5 The capacity (bits) of the largest DRAM (dynamic random access memory) chips commonly available at retail has increased as follows:[10]

Year	1971	1980	1987	1993	1999	2000
Bits	1,024	64,000	1,024,000	16,384,000	256,000,000	512,000,000

(a) Make a scatterplot of the data. Growth is much faster than linear.

(b) Plot the logarithm of DRAM capacity against year. These points are close to a straight line.

(c) Regress the logarithm of DRAM capacity on year. Give a 90% confidence interval for the slope of the population regression line.

10.6 Your scatterplot in Exercise 10.4 shows one unusual point: student number 3, who drank 9 beers.

(a) Does student 3 have the largest residual from the fitted line? (You can use the scatterplot to see this.) Is this observation extreme in the x direction, so that it may be influential?

(b) Do the regression again, omitting student 3. Add the new regression line to your scatterplot. Does removing this observation greatly change predicted BAC? Does r^2 change greatly? Does the *P*-value of your test change greatly? What do you conclude: did your work in the previous problem depend heavily on this one student?

10.7 Metatarsus adductus (call it MA) is a turning in of the front part of the foot that is common in adolescents and usually corrects itself. Hallux abducto valgus (call it HAV) is a deformation of the big toe that is not common in youth and often requires surgery. Perhaps the severity of MA can help predict the severity of HAV. Table 2.3 (page 120) gives data on 38 consecutive patients who came to a medical center for HAV surgery.[11] Using X-rays, doctors measured the angle of deformity for both MA and HAV. They speculated that there is a positive association—more serious MA is associated with more serious HAV.

(a) Make a scatterplot of the data in Table 2.3. (Which is the explanatory variable?)

(b) Describe the form, direction, and strength of the relationship between MA angle and HAV angle. Are there any clear outliers in your graph?

(c) Give a statistical model that provides a framework for asking the question of interest for this problem.

(d) Translate the question of interest into null and alternative hypotheses.

(e) Test these hypotheses and write a short description of the results. Be sure to include the value of the test statistic, the degrees of freedom, the *P*-value, and a clear statement of what you conclude.

10.8 Refer to the previous exercise. Give a 95% confidence interval for the slope. Explain how this interval can tell you what to conclude from a significance test for this parameter.

10.9 We assume that our wages will increase as we gain experience and become more valuable to our employers. Wages also increase because of inflation. By examining a sample of employees at a given point in time, we can look at part of the picture. How does length of service (LOS) relate to wages? Table 10.1 gives data on the LOS in months and wages for 60 women who work in Indiana banks. Wages are yearly total income divided by the number of weeks worked. We have multiplied wages by a constant for reasons of confidentiality.[12]

(a) Plot wages versus LOS. Describe the relationship. There is one woman with relatively high wages for her length of service. Circle this point and do not use it in the rest of this exercise.

(b) Find the least-squares line. Summarize the significance test for the slope. What do you conclude?

(c) State carefully what the slope tells you about the relationship between wages and length of service.

(d) Give a 99% confidence interval for the slope.

TABLE 10.1 Bank wages and length of service

Wages	LOS	Size	Wages	LOS	Size	Wages	LOS	Size
48.3355	94	Large	64.1026	24	Large	41.2088	97	Small
49.0279	48	Small	54.9451	222	Small	67.9096	228	Small
40.8817	102	Small	43.8095	58	Large	43.0942	27	Large
36.5854	20	Small	43.3455	41	Small	40.7000	48	Small
46.7596	60	Large	61.9893	153	Large	40.5748	7	Large
59.5238	78	Small	40.0183	16	Small	39.6825	74	Small
39.1304	45	Large	50.7143	43	Small	50.1742	204	Large
39.2465	39	Large	48.8400	96	Large	54.9451	24	Large
40.2037	20	Large	34.3407	98	Large	32.3822	13	Small
38.1563	65	Small	80.5861	150	Large	51.7130	30	Large
50.0905	76	Large	33.7163	124	Small	55.8379	95	Large
46.9043	48	Small	60.3792	60	Large	54.9451	104	Large
43.1894	61	Small	48.8400	7	Large	70.2786	34	Large
60.5637	30	Large	38.5579	22	Small	57.2344	184	Small
97.6801	70	Large	39.2760	57	Large	54.1126	156	Small
48.5795	108	Large	47.6564	78	Large	39.8687	25	Large
67.1551	61	Large	44.6864	36	Large	27.4725	43	Small
38.7847	10	Small	45.7875	83	Small	67.9584	36	Large
51.8926	68	Large	65.6288	66	Large	44.9317	60	Small
51.8326	54	Large	33.5775	47	Small	51.5612	102	Large

TABLE 10.2 U.S. corn yield

Year	Yield	Year	Yield	Year	Yield	Year	Yield
1957	48.3	1968	79.5	1979	109.5	1990	118.5
1958	52.8	1969	85.9	1980	91.0	1991	108.6
1959	53.1	1970	72.4	1981	108.9	1992	131.5
1960	54.7	1971	88.1	1982	113.2	1993	100.7
1961	62.4	1972	97.0	1983	81.1	1994	138.6
1962	64.7	1973	91.3	1984	106.7	1995	113.5
1963	67.9	1974	71.9	1985	118.0	1996	127.1
1964	62.9	1975	86.4	1986	119.4	1997	126.7
1965	74.1	1976	88.0	1987	119.8	1998	134.4
1966	73.1	1977	90.8	1988	84.6	1999	133.8
1967	80.1	1978	101.0	1989	116.3	2000	136.9

10.10 Refer to the previous exercise. Analyze the data with the outlier included.

(a) How does this change the estimates of the parameters β_0, β_1, and σ?

(b) What effect does the outlier have on the results of the significance test for the slope?

(c) How has the width of the 99% confidence interval changed?

10.11 In Example 10.8 we examined the yield in bushels per acre of corn for the years 1966, 1976, 1986, and 1996. Data for all years between 1957 and 1996 appear in Table 10.2.

(a) Plot the yield versus year. Describe the relationship. Are there any outliers or unusual years?

(b) Perform the regression analysis and summarize the results. How rapidly has yield increased over time?

10.12 Utility companies need to estimate the amount of energy that will be used by their customers. The consumption of natural gas required for heating homes depends on the outdoor temperature. When the weather is cold, more gas will be consumed. A study of one home recorded the average daily gas consumption y (in hundreds of cubic feet) for each month during one heating season. The explanatory variable x is the average number of heating degree-days per day during the month. One heating degree-day is accumulated for each degree a day's average temperature falls below 65° F. An average temperature of 50°, for example, corresponds to 15 degree-days. The data for October through June are given in the following table:[13]

Month	Oct.	Nov.	Dec.	Jan.	Feb.	Mar.	Apr.	May	June
Degree-days	15.6	26.8	37.8	36.4	35.5	18.6	15.3	7.9	0.0
Gas consumption	5.2	6.1	8.7	8.5	8.8	4.9	4.5	2.5	1.1

(a) Find the equation of the least-squares line.

(b) Test the null hypothesis that the slope is zero and describe your conclusion.

(c) Give a 90% confidence interval for the slope.

(d) The parameter β_0 corresponds to natural gas consumption for cooking, hot water, and other uses when there is no demand for heating. Give a 90% confidence interval for this parameter.

10.13 The Leaning Tower of Pisa is an architectural wonder. Engineers concerned about the tower's stability have done extensive studies of its increasing tilt. Measurements of the lean of the tower over time provide much useful information. The following table gives measurements for the years 1975 to 1987. The variable "lean" represents the difference between where a point on the tower would be if the tower were straight and where it actually is. The data are coded as tenths of a millimeter in excess of 2.9 meters, so that the 1975 lean, which was 2.9642 meters, appears in the table as 642. Only the last two digits of the year were entered into the computer.[14]

Year	75	76	77	78	79	80	81	82	83	84	85	86	87
Lean	642	644	656	667	673	688	696	698	713	717	725	742	757

(a) Plot the data. Does the trend in lean over time appear to be linear?

(b) What is the equation of the least-squares line? What percent of the variation in lean is explained by this line?

(c) Give a 99% confidence interval for the average rate of change (tenths of a millimeter per year) of the lean.

10.14 Refer to the previous exercise.

(a) In 1918 the lean was 2.9071 meters. (The coded value is 71.) Using the least-squares equation for the years 1975 to 1987, calculate a predicted value for the lean in 1918. (Note that you must use the coded value 18 for year.)

(b) Although the least-squares line gives an excellent fit to the data for 1975 to 1987, this pattern did not extend back to 1918. Write a short statement explaining why this conclusion follows from the information available. Use numerical and graphical summaries to support your explanation.

10.15 Refer to the previous two exercises.

(a) How would you code the explanatory variable for the year 2005?

(b) The engineers working on the Leaning Tower of Pisa were most interested in how much the tower would lean if no corrective action was taken. Use the least-squares equation to predict the tower's lean in the year 2005.

(c) To give a margin of error for the lean in 2005, would you use a confidence interval for a mean response or a prediction interval? Explain your choice.

10.16 Exercise 10.12 demonstrates that there is a strong linear relationship between household consumption of natural gas and outdoor temperature, measured by heating degree-days. The slope and intercept depend on the particular house and on the habits of the household living there. Data for two heating seasons (18 months) for another household produce the least-squares line $\hat{y} = 2.405 + 0.26896x$ for predicting average daily gas consumption y from average degree-days per day x. The standard error of the slope is $SE_{b_1} = 0.00815$.

(a) Explain briefly what the slope β_1 of the population regression line represents. Then give a 90% confidence interval for β_1.

(b) This interval is based on twice as many observations as the one calculated in Exercise 10.12 for a different household, and the two standard errors are of similar size. How would you expect the margins of error of the two intervals to be related? Check your answer by comparing the two margins of error.

10.17 The standard error of the intercept in the regression of gas consumption on degree-days for the household in the previous exercise is $SE_{b_0} = 0.20351$.

(a) Explain briefly what the intercept represents in this setting. Find a 90% confidence interval for the intercept.

(b) Compare the width of your interval with the one calculated for a different household in Exercise 10.12. Explain why it is narrower.

10.18 Exercise 10.12 gives information about the regression of natural gas consumption on degree-days for a particular household.

(a) What is the t statistic for testing H_0: $\beta_1 = 0$?

(b) For the alternative H_a: $\beta_1 > 0$, what critical value would you use for a test at the $\alpha = 0.05$ significance level? Do you reject H_0 at this level?

(c) How would you report the P-value for this test?

10.19 Can a pretest on mathematics skills predict success in a statistics course? The 102 students in an introductory statistics class took a pretest at the beginning of the semester. The least-squares regression line for predicting the score y on the final exam from the pretest score x was $\hat{y} = 10.5 + 0.73x$. The standard error of b_1 was 0.42.

(a) Test the null hypothesis that there is no linear relationship between the pretest score and the score on the final exam against the two-sided alternative.

(b) Would you reject this null hypothesis versus a two-sided alternative? Explain your answer.

10.20 The human body takes in more oxygen when exercising than when it is at rest. To deliver the oxygen to the muscles, the heart must beat faster. Heart rate is easy to measure, but measuring oxygen uptake requires elaborate equipment. If oxygen uptake (VO2) can be accurately predicted from heart rate (HR), the predicted values can replace actually measured values for various research purposes. Unfortunately, not all human bodies are the same, so no single prediction equation works for all people. Researchers can, however, measure both HR and VO2 for one person under varying sets of exercise conditions and calculate a regression equation for predicting that person's oxygen uptake from heart rate. They can then use predicted oxygen uptakes in place of measured uptakes for this individual in later experiments.[15] Here are data for one individual:

HR	94	96	95	95	94	95	94	104	104	106
VO2	0.473	0.753	0.929	0.939	0.832	0.983	1.049	1.178	1.176	1.292

HR	108	110	113	113	118	115	121	127	131
VO2	1.403	1.499	1.529	1.599	1.749	1.746	1.897	2.040	2.231

(a) Plot the data. Are there any outliers or unusual points?

(b) Compute the least-squares regression line for predicting oxygen uptake from heart rate for this individual.

(c) Test the null hypothesis that the slope of the regression line is 0. Explain in words the meaning of your conclusion from this test.

(d) Calculate a 95% interval for the oxygen uptake of this individual on a future occasion when his heart rate is 96. Repeat the calculation for a heart rate of 115.

(e) From what you have learned in (a), (b), (c), and (d) of this exercise, do you think that the researchers should use predicted VO2 in place of measured VO2 for this individual under similar experimental conditions? Explain your answer.

10.21 Premature infants are often kept in intensive-care nurseries after they are born. It is common practice to measure their blood pressure frequently. The oscillometric method of measuring blood pressure is noninvasive and easy to use. The traditional procedure, called the direct intra-arterial method, is believed to be more accurate but is invasive and more difficult to perform. Several studies have reported high correlations between measurements made by the two methods, ranging from $r = 0.49$ to $r = 0.98$. These correlations are statistically significant. One study that investigated the relation between the two methods reported the regression equation $\hat{y} = 15 + 0.83x$. Here x represents the easy method and y represents the difficult one.[16] The standard error of the slope is 0.065 and the sample size is 81. Calculate the t statistic for testing H_0: $\beta_1 = 0$. Specify an appropriate alternative hypothesis for this problem, and give an approximate P-value for the test. Then explain your conclusion in words a physician can understand. (The authors of the study calculated and plotted prediction intervals. They found the widths to be unacceptably large and concluded that statistical significance does not imply that results are clinically useful.)

10.22 Soil aeration and soil water evaporation involve the exchange of gases between the soil and the atmosphere. Experimenters have investigated the effect of the airflow above the soil on this process. One such experiment varied the speed of the air x and measured the rate of evaporation y. The fitted regression equation based on 18 observations was $\hat{y} = 5.0 + 0.00665x$. The standard error of the slope was reported to be 0.00182.

(a) It is reasonable to suppose that greater airflow will cause more evaporation. State hypotheses to test this belief and calculate the test statistic. Find an approximate *P*-value for the significance test and report your conclusion.

(b) Construct a 90% confidence interval for the additional evaporation experienced when airflow increases by 1 unit.

10.23 In Exercise 10.11 we examined the relationship between time and the yield of corn in the United States. Table 10.3 gives similar data for the yield of soybeans (in bushels per acre). Give a complete analysis of these data. Include a plot of the data, significance test results, examination of the residuals, and your conclusions.

10.24 Refer to Example 10.8 (page 682) and Exercise 10.11 on the changes in the U.S. corn yield over time.

(a) Give a 95% prediction interval for the yield in the year 2006.

(b) Compare this prediction interval with the one given in Example 10.14 (page 687).

(c) Find the margin of error for this prediction interval. Compare it with the margin of error for the prediction interval found in Example 10.14. Why are they different?

TABLE 10.3 U.S. soybean yield

Year	Yield	Year	Yield	Year	Yield	Year	Yield
1957	23.2	1968	26.7	1979	32.1	1990	34.1
1958	24.2	1969	27.4	1980	26.5	1991	34.2
1959	23.5	1970	26.7	1981	30.1	1992	37.6
1960	23.5	1971	27.5	1982	31.5	1993	32.6
1961	25.1	1972	27.8	1983	26.2	1994	41.4
1962	24.2	1973	27.8	1984	28.1	1995	35.3
1963	24.4	1974	23.7	1985	34.1	1996	37.6
1964	22.8	1975	28.9	1986	33.3	1997	38.9
1965	24.5	1976	26.1	1987	33.9	1998	38.9
1966	25.4	1977	30.6	1988	27.0	1999	36.6
1967	24.5	1978	29.4	1989	32.3	2000	38.1

10.25 The corn yields of Table 10.2 (page 695) and the soybean yields of Table 10.3 both vary over time for similar reasons, including improved technology and weather conditions. Let's examine the relationship between the two yields.

(a) Plot the two yields with corn on the x axis and soybeans on the y axis. Describe the relationship.

(b) Find the correlation. How well does it summarize the relation?

(c) Use corn yield to predict soybean yield. Give the equation and the results of the significance test for the slope. This test also tests the null hypothesis that the two yields are uncorrelated.

(d) Obtain the residuals from the model in part (c) and plot them versus time. Describe the pattern.

10.26 The corn yield data of Exercise 10.11 show a larger amount of scatter about the least-squares line for the later years, when the yields are higher. This may be an indication that the standard deviation σ of our model is not a constant but is increasing with time. Take logs of the yields and rerun the analyses. Prepare a short report comparing the two analyses. Include plots, a comparison of the significance test results, and the percent of variation explained by each model.

10.27 Is there a nonzero correlation between net flow of money into stock mutual funds and into bond funds? Use the regression analysis you did in Exercise 10.3 (page 691) to answer this question with no additional calculations.

10.28 Exercise 10.4 (page 692) gives data from measuring the blood alcohol content (BAC) of students 30 minutes after they drank an assigned number of cans of beer. Steve thinks he can drive legally 30 minutes after he drinks 5 beers. The legal limit is BAC = 0.08. Give a 95% confidence interval for Steve's BAC. Can he be confident he won't be arrested if he drives and is stopped?

10.29 How are returns on common stocks in overseas markets related to returns in U.S. markets? Measure U.S. returns by the annual rate of return on the Standard & Poor's 500-stock index and overseas returns by the annual rate of return on the Morgan Stanley EAFE (Europe, Australasia, Far East) index. Both are recorded in percents. Regress the EAFE returns on the S&P 500 returns for the 30 years 1971 to 2000. Here is part of the Minitab output for this regression:

```
The regression equation is
EAFE = 4.76 + 0.663 S&P

Analysis of Variance

Source            DF       SS        MS       F       P
Regression         1   3445.9    3445.9    9.50   0.005
Residual Error
Total             29  13598.3
```

Complete the analysis of variance table by filling in the "Residual Error" row.

10.30 Refer to the previous exercise. What are the values of the regression standard error s and the squared correlation r^2?

10.31 Refer to the previous two exercises. The standard deviation of the S&P 500 returns for these years is 16.45%. From this and your work in the previous exercise, find the standard error for the least-squares slope b_1. Give a 95% confidence interval for the slope β_1 of the population regression line.

10.32 Return to the oxygen uptake and heart rate data given in Exercise 10.20.

(a) Construct the ANOVA table.

(b) What null hypothesis is tested by the ANOVA F statistic? What does this hypothesis say in practical terms?

(c) Give the degrees of freedom for the F statistic and an approximate P-value for the test of H_0.

(d) Verify that the square of the t statistic that you calculated in Exercise 10.20 is equal to the F statistic in your ANOVA table. (Any difference found is due to roundoff error.)

(e) What proportion of the variation in oxygen uptake is explained by heart rate for this set of data?

10.33 A study conducted in the Egyptian village of Kalama examined the relationship between the birth weights of 40 infants and various socioeconomic variables.[17]

(a) The correlation between monthly income and birth weight was $r = 0.39$. Calculate the t statistic for testing the null hypothesis that the correlation is 0 in the entire population of infants.

(b) The researchers expected that higher birth weights would be associated with higher incomes. Express this expectation as an alternative hypothesis for the population correlation.

(c) Determine a P-value for H_0 versus the alternative that you specified in (b). What conclusion does your test suggest?

10.34 Chinese students from public schools in Hong Kong were the subjects of a study designed to investigate the relationship between various measures of parental behavior and other variables. The sample size was 713. The data were obtained from questionnaires filled in by the students. One of the variables examined was parental control, an indication of the amount of control that the parents exercised over the behavior of the students. Another was the self-esteem of the students.[18]

(a) The correlation between parental control and self-esteem was $r = -0.19$. Calculate the t statistic for testing the null hypothesis that the population correlation is 0.

(b) Find an approximate P-value for testing H_0 versus the two-sided alternative and report your conclusion.

10.35 This exercise uses the Gesell Adaptive Score data given in Table 2.9 (page 161). Run a linear regression to predict the Gesell score from age at first word. Use all 21 cases for this analysis and assume that the simple linear regression model holds.

(a) Write the fitted regression equation.

(b) What is the value of the t statistic for testing the null hypothesis that the slope is 0? Give an approximate P-value for the test of H_0 versus the alternative that the slope is negative.

(c) Give a 95% confidence interval for the slope.

(d) What percent of the variation in Gesell scores is explained by the age at first word?

(e) What is s, the estimate of the model standard deviation σ?

(f) In Chapter 2 we noted that Cases 18 and 19 in the Gesell data set were somewhat unusual. The simple linear regression model is therefore suspect when all cases are included. Rerun your analysis with these two cases excluded. Compare your results with those obtained in parts (a) through (e). In particular, to what extent did the conclusions from the first regression run depend on these two cases?

10.36 The data on gas consumption and degree-days from Exercise 10.12 (page 695) are as follows:

Month	Oct.	Nov.	Dec.	Jan.	Feb.	Mar.	Apr.	May	June
Degree-days	15.6	26.8	37.8	36.4	35.5	18.6	15.3	7.9	0.0
Gas consumption	5.2	6.1	8.7	8.5	8.8	4.9	4.5	2.5	1.1

Suppose that the gas consumption for January was incorrectly recorded as 85 instead of 8.5.

(a) Calculate the least-squares regression line for the incorrect set of data.

(b) Find the standard error of b_1.

(c) Compute the test statistic for H_0: $\beta_1 = 0$ and find the P-value. How do the results compare with those for the correct set of data?

10.37 We studied the *Archaeopteryx* data in Example 2.4 (page 110) and Exercise 2.19 (page 131). The data consist of measurements on the lengths in centimeters of the femur (a leg bone) and the humerus (a bone in the upper arm) for the five specimens that preserve both bones:

Femur	38	56	59	64	74
Humerus	41	63	70	72	84

(a) Plot the data and describe the pattern. Is it reasonable to summarize this kind of relationship with a correlation?

(b) Find the correlation and perform the significance test. Summarize the results and report your conclusion.

10.38 Concern about the quality of life for chronically ill patients is becoming as important as treating their physical symptoms. The SF-16, a questionnaire for measuring the health quality of life, was given to 50 patients with chronic obstructive lung disease.[19] A correlation of 0.68 was reported between the component of the questionnaire called general health perceptions (GHP) and a measure of lung function called forced vital capacity (FVC), expressed as a percent of normal. The mean and standard deviation of GHP are 43.5 and 20.3, and for FVC the values are 80.9 and 17.2.

(a) Find the equation of the least-squares line for predicting GHP from FVC.

(b) Give the results of the significance test for the null hypothesis that the slope is 0. (*Hint*: What is the relation between this test and the test for a zero correlation?)

10.39 A study reported a correlation $r = 0.4$ based on a sample size of $n = 20$; another reported the same correlation based on a sample size of $n = 10$. For each, perform the test of the null hypothesis that $\rho = 0$. Describe the results and explain why the conclusions are different.

10.40 Refer to the bank wages data given in Table 10.1 and described in Exercise 10.9 (page 694). The data also include a variable "Size," which classifies the bank as large or small. Obtain the residuals from the regression used to predict wages from LOS, and plot them versus LOS using different symbols for the large and small banks. Include on your plot a horizontal line at 0 (the mean of the residuals). Describe the important features of this plot. Explain what they indicate about wages in this set of data.

10.41 The SAT and the ACT are the two major standardized tests that colleges use to evaluate candidates. Most students take just one of these tests. However, some students take both. Table 10.4 gives the scores of 60 students who did this. How can we relate the two tests?

(a) Plot the data with SAT on the x axis and ACT on the y axis. Describe the overall pattern and any unusual observations.

(b) Find the least-squares regression line and draw it on your plot. Give the results of the significance test for the slope.

(c) What is the correlation between the two tests?

10.42 Refer to the previous exercise. Find the predicted value of ACT for each observation in the data set.

(a) What is the mean of these predicted values? Compare it with the mean of the ACT scores.

(b) Compare the standard deviation of the predicted values with the standard deviation of the actual ACT scores. If least-squares regression

TABLE 10.4 SAT and ACT scores

SAT	ACT	SAT	ACT	SAT	ACT	SAT	ACT
1000	24	870	21	1090	25	800	21
1010	24	880	21	860	19	1040	24
920	17	850	22	740	16	840	17
840	19	780	22	500	10	1060	25
830	19	830	20	780	12	870	21
1440	32	1190	30	1120	27	1120	25
490	7	800	16	590	12	800	18
1050	23	830	16	990	24	960	27
870	18	890	23	700	16	880	21
970	21	880	24	930	22	1020	24
920	22	980	27	860	23	790	14
810	19	1030	23	420	21	620	18
1080	23	1220	30	800	20	1150	28
1000	19	1080	22	1140	24	970	20
1030	25	970	20	920	21	1060	24

is used to predict ACT scores for a large number of students such as these, the average predicted value will be accurate but the variability of the predicted scores will be too small.

(c) Find the SAT score for a student who is one standard deviation above the mean ($z = (x - \overline{x})/s = 1$). Find the predicted ACT score and standardize this score. (Use the means and standard deviations from this set of data for these calculations.)

(d) Repeat part (c) for a student whose SAT score is one standard deviation below the mean ($z = -1$).

(e) What do you conclude from parts (c) and (d)? Perform additional calculations for different z's if needed.

10.43 Refer to the previous two exercises. An alternative to the least-squares method is based on matching standardized scores. Specifically, we set

$$\frac{(y - \overline{y})}{s_y} = \frac{(x - \overline{x})}{s_x}$$

and solve for y. Let's use the notation $y = a_0 + a_1 x$ for this line. The slope is $a_1 = s_x/s_y$ and the intercept is $a_0 = \overline{y} - a_1\overline{x}$. Compare these expressions with the formulas for the least-squares slope and intercept (page 141).

(a) Using the data in Table 10.4, find the values of a_0 and a_1.

(b) Plot the data with the least-squares line and the new prediction line.

(c) Use the new line to find predicted ACT scores. Find the mean and the standard deviation of these scores. How do they compare with the mean and standard deviation of the ACT scores?

10.44 Here are data for 12 perch caught in a lake in Finland:[20]

Weight (grams)	Length (cm)	Width (cm)	Weight (grams)	Length (cm)	Width (cm)
5.9	8.8	1.4	300.0	28.7	5.1
100.0	19.2	3.3	300.0	30.1	4.6
110.0	22.5	3.6	685.0	39.0	6.9
120.0	23.5	3.5	650.0	41.4	6.0
150.0	24.0	3.6	820.0	42.5	6.6
145.0	25.5	3.8	1000.0	46.6	7.6

In this exercise we will examine different models for predicting weight.

(a) Run the regression using length to predict weight. Do the same using width as the explanatory variable. Summarize the results. Be sure to include the value of r^2.

(b) Plot weight versus length and weight versus width. Include the least-squares lines in these plots. Do these relationships appear to be linear? Explain your answer.

10.45 Refer to the previous exercise.

(a) Try to find a better model using a transformation of length. One possibility is to use the square. Make a plot and perform the regression analysis. Summarize the results.

(b) Do the same for width.

10.46 Refer to the previous two exercises.

(a) Create a new variable that is the product of length and width. Do a plot and run the regression using this new variable. Summarize the results.

(b) Write a short report summarizing and comparing the different regression analyses that you performed in this exercise and the previous two exercises.

10.47 How does the fuel consumption of a car change as its speed increases? Here are data for a British Ford Escort. Speed is measured in kilometers per hour, and fuel consumption is measured in liters of gasoline used per 100 kilometers traveled.[21]

Speed (km/h)	Fuel used (liters/100 km)	Speed (km/h)	Fuel used (liters/100 km)
10	21.00	90	7.57
20	13.00	100	8.27
30	10.00	110	9.03
40	8.00	120	9.87
50	7.00	130	10.79
60	5.90	140	11.77
70	6.30	150	12.83
80	6.95		

You were assigned to analyze these data for a team project. (The other two members of your team did not take a statistics course based on this text.) The first team member prepared a draft with the following summary:

> Fuel consumption does not depend on speed for this vehicle ($t = -0.63$, df $= 13$, $P = 0.54$).

The second team member had some trouble with the statistical software. When the data were read, only the first two columns of the data table were used by the software. These are the data for speeds between 10 and 80 km/h. This team member prepared the following draft:

> There is a strong relationship between fuel consumption and speed for this vehicle ($t = -3.63$, df $= 6$, $P = 0.0109$). Speed explains 68.8% of the variation in fuel consumption.

First, verify that your two teammates have computed the quantities that they reported correctly. Then analyze the data and write a summary of your analysis of the relationship between fuel consumption and speed.

10.48 Metabolic rate, the rate at which the body consumes energy, is important in studies of weight gain, dieting, and exercise. The table below gives data on the lean body mass and resting metabolic rate for 12 women and 7 men who are subjects in a study of dieting. Lean body mass, given in kilograms, is a person's weight leaving out all fat. Metabolic rate is measured in calories burned per 24 hours, the same calories used to describe the energy content of foods. The researchers believe that lean body mass is an important influence on metabolic rate.

Subject	Sex	Mass	Rate	Subject	Sex	Mass	Rate
1	M	62.0	1792	11	F	40.3	1189
2	M	62.9	1666	12	F	33.1	913
3	F	36.1	995	13	M	51.9	1460
4	F	54.6	1425	14	F	42.4	1124
5	F	48.5	1396	15	F	34.5	1052
6	F	42.0	1418	16	F	51.1	1347
7	M	47.4	1362	17	F	41.2	1204
8	F	50.6	1502	18	M	51.9	1867
9	F	42.0	1256	19	M	46.9	1439
10	M	48.7	1614				

(a) Make a scatterplot of the data, using different symbols or colors for men and women. Summarize what you see in the plot.

(b) Run the regression to predict metabolic rate from lean body mass for the women in the sample and summarize the results. Do the same for the men.

10.49 Refer to the previous exercise. It is tempting to conclude that there is a strong linear relationship for the women but no relationship for the men. Let's look at this issue a little more carefully.

(a) Find the confidence interval for the slope in the regression equation that you ran for the females. Do the same for the males. What do these suggest about the possibility that these two slopes are the same? (The formal method for making this comparison is a bit complicated and is beyond the scope of this chapter.)

(b) Examine the formula for the standard error of the regression slope given on page 685. The term in the denominator is $\sqrt{\sum(x_i - \overline{x})^2}$. Find this quantity for the females; do the same for the males. How do these calculations help to explain the results of the significance tests?

(c) Suppose you were able to collect additional data for males. How would you use lean body mass in deciding which subjects to choose?

Prelude: Executive Order 11246

Executive Order 11246 signed by President Lyndon B. Johnson in 1965 prohibits discrimination in hiring and employment decisions on the basis of race, color, gender, religion, and national origin. It applies to contractors doing business in excess of $10,000 or more with the U.S. government. The Office of Federal Contract Compliance Programs (OFCCP) is responsible for monitoring compliance with this Executive Order. From the viewpoint of the employee, deciding how much to pay employees is one of the most important employment decisions that a company makes. Not surprisingly, a careful examination of salary is one of the things that OFCCP looks at frequently. Statistical methods are very useful for this purpose. We would construct a statistical model for the process that underlies the determination of salary and ask about discrimination within the context of this model.

Consider gender, for example. We could use the methods of Section 7.2 and compare the average salaries of men and women using a two-sample *t* significance test. This approach would be appropriate in circumstances where all of the employees do approximately the same kind of job and have the same kinds of qualifications and experience. More typically, however, employees do different jobs and they have different qualifications and experiences that can be highly valued by the company. To model this situation, we consider the response variable to be salary. Explanatory variables include the kinds of qualifications and experience that can be shown to be relevant to the performance of the job, as well as variables that distinguish different types of jobs. Finally, we would also consider gender as an explanatory variable. Multiple regression is used to construct a model for explaining salary using the explanatory variables. The gender term in the model addresses the question of whether or not the salaries of men and women are distinguishable when salaries are adjusted for the other explanatory variables. Suppose that the term indicates that men are paid more than women and it is statistically significant. This finding is consistent with a claim of discrimination. The company then needs to find a nondiscriminatory explanation for the disparity or to take some corrective action.

CHAPTER 11

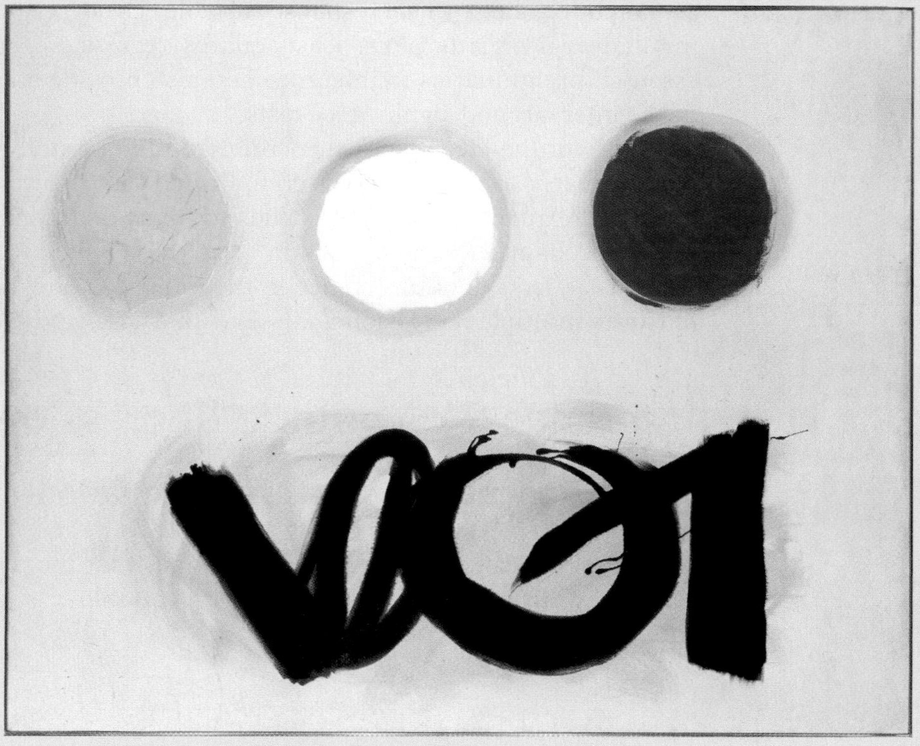

Adolph Gottlieb, *Three Disks*, 1960.

Multiple Regression

Introduction

In Chapter 10 we presented methods for inference in the setting of a linear relationship between a response variable y and a *single* explanatory variable x. With multiple linear regression we use *more than one* explanatory variable to explain or predict a single response variable. The descriptive tools we learned in Chapter 2—scatterplots, least-squares regression, and correlation—are essential preliminaries to inference and also provide a foundation for confidence intervals and significance tests.

Many of the ideas that we encountered in our study of simple linear regression carry over to this more complicated case. However, the introduction of several explanatory variables leads to many additional considerations. In this short chapter we cannot explore all of these issues. Rather, we will outline some basic facts about inference in the multiple regression setting and then illustrate multiple regression analysis with a case study.

11.1 Inference for Multiple Regression

Population multiple regression equation

The simple linear regression model assumes that the mean of the response variable y depends on the explanatory variable x according to a linear equation

$$\mu_y = \beta_0 + \beta_1 x$$

For any fixed value of x, the response y varies normally around this mean and has a standard deviation σ that is the same for all values of x.

In the multiple regression setting, the response variable y depends on not one but p explanatory variables. We will denote these explanatory variables by $x_1, x_2, \ldots, x_p$. The mean response is a linear function of the explanatory variables:

$$\mu_y = \beta_0 + \beta_1 x_1 + \beta_2 x_2 + \cdots + \beta_p x_p$$

population regression equation

This expression is the **population regression equation.** We do not observe the mean response because the observed values of y vary about their means. We can think of subpopulations of responses, each corresponding to a particular set of values for *all* of the explanatory variables $x_1, x_2, \ldots, x_p$. In each subpopulation, y varies normally with a mean given by the population regression equation. The regression model assumes that the standard deviation σ of the responses is the same in all subpopulations.

EXAMPLE 11.1

Our case study uses data collected at a large university on all first-year computer science majors in a particular year.[1] The purpose of the study was to attempt to predict success in the early university years. One measure of success was the cumulative grade point average (GPA) after three semesters. Among the explanatory variables recorded at the time the students enrolled in the university were average high school grades in mathematics (HSM), science (HSS), and English (HSE).

We will use high school grades to predict the response variable GPA. There are $p = 3$ explanatory variables: x_1 = HSM, x_2 = HSS, and x_3 = HSE. The high school grades are coded on a scale from 1 to 10, with 10 corresponding to A, 9 to A−, 8 to B+, and so on. These grades define the subpopulations. For example, the straight-C students are the subpopulation defined by HSM = 4, HSS = 4, and HSE = 4.

One possible multiple regression model for the subpopulation mean GPAs is

$$\mu_{\text{GPA}} = \beta_0 + \beta_1\text{HSM} + \beta_2\text{HSS} + \beta_3\text{HSE}$$

For the straight-C subpopulation of students, the model gives the subpopulation mean as

$$\mu_{\text{GPA}} = \beta_0 + 4\beta_1 + 4\beta_2 + 4\beta_3$$

Data for multiple regression

The data for a simple linear regression problem consist of observations (x_i, y_i) on the two variables. Because there are several explanatory variables in multiple regression, the notation needed to describe the data is more elaborate. In our case study, each observation or case consists of a value for the response variable and for each of the explanatory variables for one student. Call x_{ij} the value of the jth explanatory variable for the ith student. The data are then

Student 1: $(x_{11}, x_{12}, \ldots, x_{1p}, y_1)$

Student 2: $(x_{21}, x_{22}, \ldots, x_{2p}, y_2)$

$\vdots$

Student n: $(x_{n1}, x_{n2}, \ldots, x_{np}, y_n)$

Here, n is the number of cases and p is the number of explanatory variables. Data are often entered into computer regression programs in this format. Each row is a case and each column corresponds to a different variable. The data for Example 11.1, with several additional explanatory variables, appear in this format in the CSDATA data set described in the Data Appendix.

Multiple linear regression model

We combine the population regression equation and assumptions about variation to construct the multiple linear regression model. The subpopulation means describe the FIT part of our statistical model. The RESIDUAL part represents the variation of observations about the means. We will use the same notation for the residual that we used in the simple linear regression model. The symbol ϵ represents the deviation of an individual observation from its subpopulation mean. We assume that these deviations are normally distributed with mean 0 and an unknown standard deviation σ that does not depend on the values of the x variables.

Multiple Linear Regression Model

The **statistical model for multiple linear regression** is

$$y_i = \beta_0 + \beta_1 x_{i1} + \beta_2 x_{i2} + \cdots + \beta_p x_{ip} + \epsilon_i$$

for $i = 1, 2, \ldots, n$.

The **mean response μ_y** is a linear function of the explanatory variables:

$$\mu_y = \beta_0 + \beta_1 x_1 + \beta_2 x_2 + \cdots + \beta_p x_p$$

The **deviations ϵ_i** are independent and normally distributed with mean 0 and standard deviation σ. In other words, they are an SRS from the $N(0, \sigma)$ distribution.

The parameters of the model are $\beta_0, \beta_1, \beta_2, \ldots, \beta_p$, and σ.

The assumption that the subpopulation means are related to the regression coefficients β by the equation

$$\mu_y = \beta_0 + \beta_1 x_1 + \beta_2 x_2 + \cdots + \beta_p x_p$$

implies that we can estimate all subpopulation means from estimates of the β's. To the extent that this equation is accurate, we have a useful tool for describing how the mean of y varies with the x's.

Estimation of the multiple regression parameters

For simple linear regression we used the principle of least squares first described in Section 2.3 to obtain estimators of the intercept and slope of the regression line. For multiple regression the principle is the same but the details are more complicated. Let

$$b_0, \ b_1, \ b_2, \ \ldots, \ b_p$$

denote the estimators of the parameters

$$\beta_0, \ \beta_1, \ \beta_2, \ \ldots, \ \beta_p$$

For the ith observation the predicted response is

$$\hat{y}_i = b_0 + b_1 x_{i1} + b_2 x_{i2} + \cdots + b_p x_{ip}$$

residual

The ith **residual,** the difference between the observed and predicted response, is therefore

$$\begin{aligned} e_i &= \text{observed response} - \text{predicted response} \\ &= y_i - \hat{y}_i \\ &= y_i - b_0 - b_1 x_{i1} - b_2 x_{i2} - \cdots - b_p x_{ip} \end{aligned}$$

method of least squares

The **method of least squares** chooses the values of the b's that make the sum of the squares of the residuals as small as possible. In other words, the parameter estimates $b_0, b_1, b_2, \ldots, b_p$ minimize the quantity

$$\sum (y_i - b_0 - b_1 x_{i1} - b_2 x_{i2} - \cdots - b_p x_{ip})^2$$

The formula for the least-squares estimates is complicated. We will be content to understand the principle on which they are based and to let software do the computations.

The parameter σ^2 measures the variability of the responses about the population regression equation. As in the case of simple linear regression, we estimate σ^2 by an average of the squared residuals. The estimator is

$$\begin{aligned} s^2 &= \frac{\sum e_i^2}{n - p - 1} \\ &= \frac{\sum (y_i - \hat{y}_i)^2}{n - p - 1} \end{aligned}$$

degrees of freedom

The quantity $n - p - 1$ is the **degrees of freedom** associated with s^2. The degrees of freedom equal the sample size n minus $p + 1$, the number of β's we must estimate to fit the model. In the simple linear regression case there is just one explanatory variable, so $p = 1$ and the degrees of freedom are $n - 2$. To estimate σ we use

$$s = \sqrt{s^2}$$

Confidence intervals and significance tests for regression coefficients

We can obtain confidence intervals and significance tests for each of the regression coefficients β_j as we did in simple linear regression. The standard errors of the b's have more complicated formulas, but all are multiples of s. We again rely on statistical software for the calculations.

Confidence Intervals and Significance Tests for β_j

A **level C confidence interval** for β_j is

$$b_j \pm t^* \mathrm{SE}_{b_j}$$

where SE_{b_j} is the standard error of b_j and t^* is the value for the $t(n-p-1)$ density curve with area C between $-t^*$ and t^*.

To test the hypothesis H_0: $\beta_j = 0$, compute the ***t* statistic**

$$t = \frac{b_j}{\mathrm{SE}_{b_j}}$$

In terms of a random variable T having the $t(n-p-1)$ distribution, the P-value for a test of H_0 against

H_a: $\beta_j > 0$ is $P(T \geq t)$

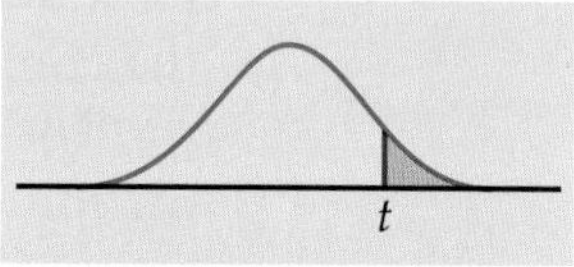

H_a: $\beta_j < 0$ is $P(T \leq t)$

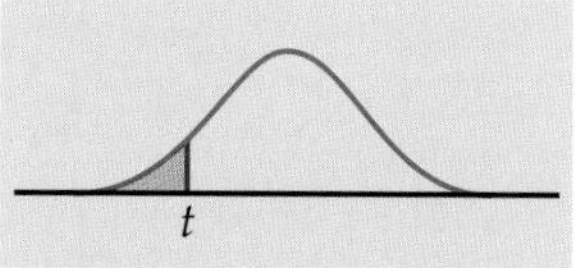

H_a: $\beta_j \neq 0$ is $2P(T \geq |t|)$

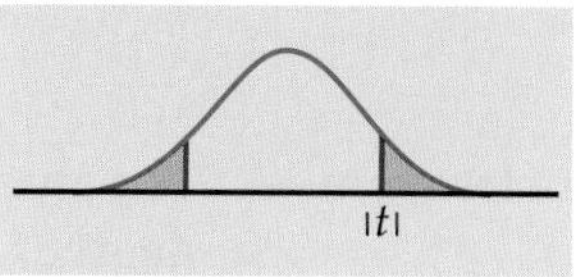

confidence intervals for mean response

prediction intervals

Because regression is often used for prediction, we may wish to construct **confidence intervals for a mean response** and **prediction intervals** for a future observation from multiple regression models. The basic ideas are the same as in the simple linear regression case. In most software systems, the same commands that give confidence and prediction intervals for simple linear regression work for multiple regression. The only difference is that we specify a list of explanatory variables rather than a single variable. Modern software allows us to perform these rather complex calculations without an intimate knowledge of all of the computational details. This frees us to concentrate on the meaning and appropriate use of the results.

ANOVA table for multiple regression

ANOVA F test

In simple linear regression the F test from the ANOVA table is equivalent to the two-sided t test of the hypothesis that the slope of the regression line is 0. For multiple regression there is a corresponding **ANOVA *F* test,** but it tests the hypothesis that *all* of the regression coefficients (with the exception of the intercept) are 0. Here is the general form of the ANOVA table for multiple regression:

Source	Degrees of freedom	Sum of squares	Mean square	F
Model	p	$\sum(\hat{y}_i - \bar{y})^2$	SSM/DFM	MSM/MSE
Error	$n - p - 1$	$\sum(y_i - \hat{y}_i)^2$	SSE/DFE	
Total	$n - 1$	$\sum(y_i - \bar{y})^2$	SST/DFT	

The ANOVA table is similar to that for simple linear regression. The degrees of freedom for the model increase from 1 to p to reflect the fact that we now have p explanatory variables rather than just one. As a consequence, the degrees of freedom for error decrease by the same amount. The sums of squares represent sources of variation. Once again, both the sums of squares and their degrees of freedom add:

$$\text{SST} = \text{SSM} + \text{SSE}$$
$$\text{DFT} = \text{DFM} + \text{DFE}$$

The estimate of the variance σ^2 for our model is again given by the MSE in the ANOVA table. That is, $s^2 = \text{MSE}$.

F statistic

The ratio MSM/MSE is an ***F* statistic** for testing the null hypothesis

$$H_0\colon \beta_1 = \beta_2 = \cdots = \beta_p = 0$$

against the alternative hypothesis

$$H_a\colon \text{at least one of the } \beta_j \text{ is not 0}$$

The null hypothesis says that none of the explanatory variables are predictors of the response variable when used in the form expressed by the multiple regression equation. The alternative states that at least one of them is linearly related to the response. As in simple linear regression, large values of F give evidence against H_0. When H_0 is true, F has the $F(p,\ n - p - 1)$ distribution. The degrees of freedom for the F distribution are those associated with the model and error in the ANOVA table.

Analysis of Variance F Test

In the multiple regression model, the hypothesis

$$H_0: \beta_1 = \beta_2 = \cdots = \beta_p = 0$$

is tested by the analysis of variance F statistic

$$F = \frac{\text{MSM}}{\text{MSE}}$$

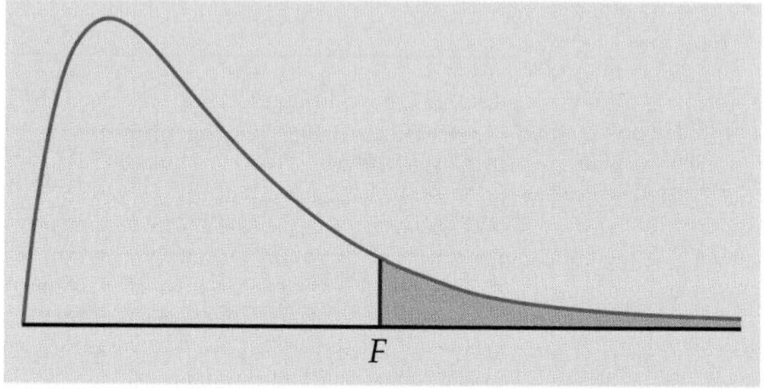

The P-value is the probability that a random variable having the $F(p, n - p - 1)$ distribution is greater than or equal to the calculated value of the F statistic.

Squared multiple correlation R^2

For simple linear regression we noted that the square of the sample correlation could be written as the ratio of SSM to SST and could be interpreted as the proportion of variation in y explained by x. A similar statistic is routinely calculated for multiple regression.

The Squared Multiple Correlation

The statistic

$$R^2 = \frac{\text{SSM}}{\text{SST}} = \frac{\sum(\hat{y}_i - \bar{y})^2}{\sum(y_i - \bar{y})^2}$$

is the proportion of the variation of the response variable y that is explained by the explanatory variables $x_1, x_2, \ldots, x_p$ in a multiple linear regression.

multiple correlation coefficient

Often, R^2 is multiplied by 100 and expressed as a percent. The square root of R^2, called the **multiple correlation coefficient,** is the correlation between the observations y_i and the predicted values $\hat{y}_i$.

11.2 A Case Study

Preliminary analysis

In this section we illustrate multiple regression by analyzing the data from the study described in Example 11.1. The response variable is the cumulative GPA after three semesters for a group of computer science majors at a large university. The explanatory variables previously mentioned are average high school grades, represented by HSM, HSS, and HSE. We also examine the SAT mathematics and SAT verbal scores as explanatory variables. We have data for $n = 224$ students in the study. We use SAS to illustrate the outputs that are given by most software.

The first step in the analysis is to carefully examine each of the variables. Means, standard deviations, and minimum and maximum values appear in Figure 11.1.

The minimum value for the SAT mathematics (SATM) variable appears to be rather extreme; it is $(595 - 300)/86 = 3.43$ standard deviations below the mean. We do not discard this case at this time but will take care in our subsequent analyses to see if it has an excessive influence on our results. The mean for the SATM score is higher than the mean for the verbal score (SATV), as we might expect for a group of computer science majors. The two standard deviations are about the same. The means of the three high school grade variables are similar, with the mathematics grades being a bit higher. The standard deviations for the high school grade variables are very close to each other. The mean GPA is 2.635 on a 4-point scale, with standard deviation 0.779.

Because the variables GPA, SATM, and SATV have many possible values, we could use stemplots or histograms to examine the shapes of their distributions. Normal quantile plots indicate whether or not the distributions look normal. It is important to note that the multiple regression model *does not* require any of these distributions to be normal. Only the deviations of the responses y from their means are assumed to be normal. The purpose of examining these plots is to understand something about each variable alone before attempting to use it in a complicated model. Extreme values of any variable should be noted and checked for accuracy. If found to be correct, the

Variable	N	Mean	Std Dev	Minimum	Maximum
GPA	224	2.6352232	0.7793949	0.1200000	4.0000000
SATM	224	595.2857143	86.4014437	300.0000000	800.0000000
SATV	224	504.5491071	92.6104591	285.0000000	760.0000000
HSM	224	8.3214286	1.6387367	2.0000000	10.0000000
HSS	224	8.0892857	1.6996627	3.0000000	10.0000000
HSE	224	8.0937500	1.5078736	3.0000000	10.0000000

FIGURE 11.1 Descriptive statistics for the computer science student case study.

cases with these values should be carefully examined to see if they are truly exceptional and perhaps do not belong in the same analysis with the other cases. When these data are examined in this way, no obvious problems are evident.

The high school grade variables HSM, HSS, and HSE have relatively few values and are best summarized by giving the relative frequencies for each possible value. The output in Figure 11.2 provides these summaries. The distributions are all skewed, with a large proportion of high grades (10 = A and 9 = A−). Again we emphasize that these distributions need not be normal.

HSM	Frequency	Percent	Cumulative Frequency	Cumulative Percent
2	1	0.4	1	0.4
3	1	0.4	2	0.9
4	4	1.8	6	2.7
5	6	2.7	12	5.4
6	23	10.3	35	15.6
7	28	12.5	63	28.1
8	36	16.1	99	44.2
9	59	26.3	158	70.5
10	66	29.5	224	100.0

HSM	Frequency	Percent	Cumulative Frequency	Cumulative Percent
3	1	0.4	1	0.4
4	7	3.1	8	3.6
5	9	4.0	17	7.6
6	24	10.7	41	18.3
7	42	18.8	83	37.1
8	31	13.8	114	50.9
9	50	22.3	164	73.2
10	60	26.8	224	100.0

HSM	Frequency	Percent	Cumulative Frequency	Cumulative Percent
3	1	0.4	1	0.4
4	4	1.8	5	2.2
5	5	2.2	10	4.5
6	23	10.3	33	14.7
7	43	19.2	76	33.9
8	49	21.9	125	55.8
9	42	23.2	177	79.0
10	47	21.0	224	100.0

FIGURE 11.2 The distributions of the high school grade variables.

Relationships between pairs of variables

The second step in our analysis is to examine the relationships between all pairs of variables. Scatterplots and correlations are our tools for studying two-variable relationships. The correlations appear in Figure 11.3. The output includes the *P*-value for the test of the null hypothesis that the population correlation is 0 versus the two-sided alternative for each pair. Thus, we see that the correlation between GPA and HSM is 0.44, with a *P*-value of 0.0001, whereas the correlation between GPA and SATV is 0.11, with a *P*-value of 0.087. The first is statistically significant by any reasonable standard, and the second is rather marginal.

The high school grades all have higher correlations with GPA than do the SAT scores. As we might expect, math grades have the highest correlation ($r = 0.44$), followed by science grades (0.33) and then English grades (0.29). The two SAT scores have a rather high correlation with each other (0.46), and the high school grades also correlate well with each other (0.45 to 0.58). The SAT mathematics score correlates well with HSM (0.45), less well with HSS (0.24), and rather poorly with HSE (0.11). The correlations of the SAT verbal score with the three high school grades are about equal, ranging from 0.22 to 0.26.

It is important to keep in mind that by examining pairs of variables we are seeking a better understanding of the data. The fact that the correlation of

Pearson Correlation Coefficients / Prob > |R| under Ho: Rho=0 / N = 224

	GPA	SATM	SATV	HSM	HSS	HSE
GPA	1.00000	0.25171	0.11449	0.43650	0.32943	0.28900
	0.0	0.0001	0.0873	0.0001	0.0001	0.0001
SATM	0.25171	1.00000	0.46394	0.45351	0.24048	0.10828
	0.0001	0.0	0.0001	0.0001	0.0003	0.1060
SATV	0.11449	0.46394	1.00000	0.22112	0.26170	0.24371
	0.0873	0.0001	0.0	0.0009	0.0001	0.0002
HSM	0.43650	0.45351	0.22112	1.00000	0.57569	0.44689
	0.0001	0.0001	0.0009	0.0	0.0001	0.0001
HSS	0.32943	0.24048	0.26170	0.57569	1.00000	0.57937
	0.0001	0.0003	0.0001	0.0001	0.0	0.0001
HSE	0.28900	0.10828	0.24371	0.44689	0.57937	1.00000
	0.0001	0.1060	0.0002	0.0001	0.0001	0.0

FIGURE 11.3 Correlations among the case study variables.

a particular explanatory variable with the response variable does not achieve statistical significance does not *necessarily* imply that it will not be a useful (and significant) predictor in a multiple regression.

Numerical summaries such as correlations are useful, but plots are generally more informative when seeking to understand data. Plots tell us whether the numerical summary gives a fair representation of the data. For a multiple regression, each pair of variables should be plotted. For the six variables in our case study, this means that we should examine 15 plots. In general there are $p + 1$ variables in a multiple regression analysis with p explanatory variables, so that $p(p + 1)/2$ plots are required. Multiple regression is a complicated procedure. If we do not do the necessary preliminary work, we are in serious danger of producing useless or misleading results. We leave the task of making these plots as an exercise.

Regression on high school grades

To explore the relationship between the explanatory variables and our response variable GPA, we run several multiple regressions. The explanatory variables fall into two classes. High school grades are represented by HSM, HSS, and HSE, and standardized tests are represented by the two SAT scores. We begin our analysis by using the high school grades to predict GPA. Figure 11.4 gives the multiple regression output.

ANOVA table

The output contains an **ANOVA table,** some additional descriptive statistics, and information about the parameter estimates. When examining any ANOVA table, it is a good idea to first verify the degrees of freedom. This ensures that we have not made some serious error in specifying the model for the program or in entering the data. Because there are $n = 224$ cases, we have $\text{DFT} = n - 1 = 223$. The three explanatory variables give $\text{DFM} = p = 3$ and $\text{DFE} = n - p - 1 = 223 - 3 = 220$.

F statistic

The ANOVA ***F* statistic** is 18.86, with a *P*-value of 0.0001. Under the null hypothesis

$$H_0: \beta_1 = \beta_2 = \beta_3 = 0$$

the *F* statistic has an $F(3, 220)$ distribution. According to this distribution, the chance of obtaining an *F* statistic of 18.86 or larger is 0.0001. We therefore conclude that at least one of the three regression coefficients for the high school grades is different from 0 in the population regression equation.

estimate of σ

R^2

In the descriptive statistics that follow the ANOVA table we find that Root MSE is 0.6998. This value is the square root of the MSE given in the ANOVA table and is s, the **estimate of the parameter σ** of our model. The value of $\boldsymbol{R^2}$ is 0.20. That is, 20% of the observed variation in the GPA scores is explained by linear regression on high school grades. Although the *P*-value is very small, the model does not explain very much of the variation in GPA. Remember, a small *P*-value does not necessarily tell us that we have a large effect, particularly when the sample size is large.

```
Dependent  Variable:  GPA

                        Analysis  of  Variance

                            Sum  of          Mean
Source             DF       Squares          Square       F Value      Prob>F

Model               3       27.71233         9.23744       18.861      0.0001
Error             220      107.75046         0.48977
C Total           223      135.46279

    Root  MSE           0.69984       R-Square        0.2046
    Dep  Mean           2.63522       Adj  R-sq       0.1937
    C.V.               26.55711

                        Parameter  Estimates

                      Parameter         Standard      T for H0:
 Variable    DF        Estimate            Error      Parameter=0      Prob > |T|

 INTERCEP     1        0.589877       0.29424324            2.005          0.0462
 HSM          1        0.168567       0.03549214            4.749          0.0001
 HSS          1        0.034316       0.03755888            0.914          0.3619
 HSE          1        0.045102       0.03869585            1.166          0.2451
```

FIGURE 11.4 Multiple regression output for regression using high school grades to predict GPA.

fitted regression equation

From the Parameter Estimates section we obtain the **fitted regression equation**

$$\widehat{\text{GPA}} = 0.590 + 0.169\text{HSM} + 0.034\text{HSS} + 0.045\text{HSE}$$

t statistics

Recall that the ***t* statistics** for testing the regression coefficients are obtained by dividing the estimates by their standard errors. Thus, for the coefficient of HSM we obtain the *t*-value given in the output by calculating

$$t = \frac{b}{\text{SE}_b} = \frac{0.168567}{0.03549214} = 4.749$$

P-values

The ***P*-values** appear in the last column. Note that these *P*-values are for the two-sided alternatives. HSM has a *P*-value of 0.0001, and we conclude that the regression coefficient for this explanatory variable is significantly different from 0. The *P*-values for the other explanatory variables (0.36 for HSS and 0.25 for HSE) do not achieve statistical significance.

Interpretation of results

The significance tests for the individual regression coefficients seem to contradict the impression obtained by examining the correlations in Figure 11.3. In that display we see that the correlation between GPA and HSS is 0.33 and the correlation between GPA and HSE is 0.29. The *P*-values for both of these correlations are 0.0001. In other words, if we used HSS alone in a regression to predict GPA, or if we used HSE alone, we would obtain statistically significant regression coefficients.

This phenomenon is not unusual in multiple regression analysis. Part of the explanation lies in the correlations between HSM and the other two explanatory variables. These are rather high (at least compared with the other correlations in Figure 11.3). The correlation between HSM and HSS is 0.58, and that between HSM and HSE is 0.45. Thus, when we have a regression model that contains all three high school grades as explanatory variables, there is considerable overlap of the predictive information contained in these variables. *The significance tests for individual regression coefficients assess the significance of each predictor variable assuming that all other predictors are included in the regression equation.* Given that we use a model with HSM and HSS as predictors, the coefficient of HSE is not statistically significant. Similarly, given that we have HSM and HSE in the model, HSS does not have a significant regression coefficient. HSM, however, adds significantly to our ability to predict GPA even after HSS and HSE are already in the model.

Unfortunately, we cannot conclude from this analysis that the *pair* of explanatory variables HSS and HSE contribute nothing significant to our model for predicting GPA once HSM is in the model. The impact of relations among the several explanatory variables on fitting models for the response is the most important new phenomenon encountered in moving from simple linear regression to multiple regression. We can only hint at the many complicated problems that arise.

Residuals

As in simple linear regression, we should always examine the residuals as an aid to determining whether the multiple regression model is appropriate for the data. Because there are several explanatory variables, we must examine several residual plots. It is usual to plot the residuals versus the predicted values $\hat{y}$ and also versus each of the explanatory variables. Look for outliers, influential observations, evidence of a curved (rather than linear) relation, and anything else unusual. Again, we leave the task of making these plots as an exercise. The plots all appear to show more or less random noise around the center value of 0.

If the deviations ϵ in the model are normally distributed, the residuals should be normally distributed. Figure 11.5 presents a normal quantile plot of the residuals. The distribution appears to be approximately normal. There are many other specialized plots that help detect departures from the multiple

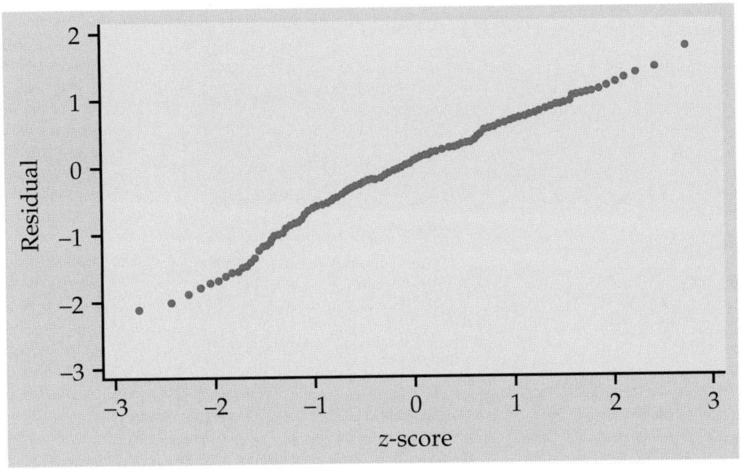

FIGURE 11.5 Normal quantile plot of the residuals from the high school grade model. There are no important deviations from normality.

regression model. Discussion of these, however, is more than we can undertake in this chapter.

Refining the model

Because the variable HSS has the largest P-value of the three explanatory variables (see Figure 11.4) and therefore appears to contribute the least to our explanation of GPA, we rerun the regression using only HSM and HSE as explanatory variables. The SAS output appears in Figure 11.6. The F statistic indicates that we can reject the null hypothesis that the regression coefficients for the two explanatory variables are both 0. The P-value is still 0.0001. The value of R^2 has dropped very slightly compared with our previous run, from 0.2046 to 0.2016. Dropping HSS from the model resulted in the loss of very little explanatory power. The measure s of variation about the fitted equation (Root MSE in the printout) is nearly identical for the two regressions, another indication that dropping HSS loses very little. The t statistics for the individual regression coefficients indicate that HSM is still clearly significant ($P = 0.0001$), while the coefficient of HSE is larger than before (1.747 versus 1.166) and approaches the traditional 0.05 level of significance ($P = 0.082$).

Comparison of the fitted equations for the two multiple regression analyses tells us something more about the intricacies of this procedure. For the first run we have

$$\widehat{\text{GPA}} = 0.590 + 0.169\text{HSM} + 0.034\text{HSS} + 0.045\text{HSE}$$

whereas the second gives us

$$\widehat{\text{GPA}} = 0.624 + 0.183\text{HSM} + 0.061\text{HSE}$$

```
Dependent Variable: GPA

                         Analysis of Variance

                               Sum of          Mean
Source              DF        Squares        Square       F Value       Prob>F

Model                2       27.30349      13.65175        27.894       0.0001
Error              221      108.15930       0.48941
C Total            223      135.46279

     Root MSE            0.69958     R-Square          0.2016
     Dep Mean            2.63522     Adj R-sq          0.1943
     C.V.               26.54718

                         Parameter Estimates

                      Parameter       Standard     T for H0:
 Variable    DF        Estimate          Error   Parameter=0    Prob > |T|

 INTERCEP     1        0.624228     0.29172204         2.140        0.0335
 HSM          1        0.182654     0.03195581         5.716        0.0001
 HSE          1        0.060670     0.03472914         1.747        0.0820
```

FIGURE 11.6 Multiple regression output for regression using HSM and HSE to predict GPA.

Eliminating HSS from the model changes the regression coefficients for all of the remaining variables and the intercept. This phenomenon occurs quite generally in multiple regression. *Individual regression coefficients, their standard errors, and significance tests are meaningful only when interpreted in the context of the other explanatory variables in the model.*

Regression on SAT scores

We now turn to the problem of predicting GPA using the two SAT scores. Figure 11.7 gives the output. The fitted model is

$$\widehat{\text{GPA}} = 1.289 + 0.002283\text{SATM} - 0.000025\text{SATV}$$

The degrees of freedom are as expected: 2, 221, and 223. The F statistic is 7.476, with a P-value of 0.0007. We conclude that the regression coefficients for SATM and SATV are not both 0. Recall that we obtained the P-value 0.0001 when we used high school grades to predict GPA. Both multiple regression equations are highly significant, but this obscures the fact that the two models have quite different explanatory power. For the SAT regression, $R^2 = 0.0634$, whereas for the high school grades model even with only HSM and HSE

Dependent Variable: GPA

Analysis of Variance

Source	DF	Sum of Squares	Mean Square	F Value	Prob>F
Model	2	8.58384	4.29192	7.476	0.0007
Error	221	126.87895	0.57411		
C Total	223	135.46279			

Root MSE	0.75770	R-Square	0.0634
Dep Mean	2.63522	Adj R-sq	0.0549
C.V.	28.75287		

Parameter Estimates

Variable	DF	Parameter Estimate	Standard Error	T for H0: Parameter=0	Prob > |T|
INTERCEP	1	1.288677	0.37603684	3.427	0.0007
SATM	1	0.002283	0.00066291	3.444	0.0007
SATV	1	-0.000024562	0.00061847	-0.040	0.9684

FIGURE 11.7 Multiple regression output for regression using SAT scores to predict GPA.

(Figure 11.6), we have $R^2 = 0.2016$, a value more than three times as large. Stating that we have a statistically significant result is quite different from saying that an effect is large or important.

Further examination of the output in Figure 11.7 reveals that the coefficient of SATM is significant ($t = 3.44$, $P = 0.0007$), and that for SATV is not ($t = -0.04$, $P = 0.9684$). For a complete analysis we should carefully examine the residuals. Also, we might want to run the analysis with SATM as the only explanatory variable.

Regression using all variables

We have seen that either the high school grades or the SAT scores give a highly significant regression equation. The mathematics component of each of these groups of explanatory variables appears to be the key predictor. Comparing the values of R^2 for the two models indicates that high school grades are better predictors than SAT scores. Can we get a better prediction equation using all of the explanatory variables together in one multiple regression?

To address this question we run the regression with all five explanatory variables. The output appears in Figure 11.8. The F statistic is 11.69, with a P-value of 0.0001, so at least one of our explanatory variables has a nonzero

```
Dependent Variable: GPA

                          Analysis of Variance

                              Sum of          Mean
Source              DF       Squares        Square       F Value       Prob>F

Model                5      28.64364       5.72873        11.691       0.0001
Error              218     106.81914       0.49000
C Total            223     135.46279

    Root MSE          0.70000     R-Square        0.2115
    Dep Mean          2.63522     Adj R-sq        0.1934
    C.V.             26.56311

                          Parameter Estimates

                     Parameter         Standard     T for H0:
  Variable   DF       Estimate            Error    Parameter=0     Prob > |T|

  INTERCEP    1       0.326719       0.39999643          0.817         0.4149
  SATM        1       0.000944       0.00068566          1.376         0.1702
  SATV        1      -0.000408       0.00059189         -0.689         0.4915
  HSM         1       0.145961       0.03926097          3.718         0.0003
  HSS         1       0.035905       0.03779841          0.950         0.3432
  HSE         1       0.055293       0.03956869          1.397         0.1637

Test: SAT          Numerator:       0.4657   DF:     2    F value:    0.9503
                   Denominator:   0.489996   DF:   218    Prob>F:     0.3882

Test: HS           Numerator:       6.6866   DF:     3    F value:   13.6462
                   Denominator:   0.489996   DF:   218    Prob>F:     0.0001
```

FIGURE 11.8 Multiple regression output for regression using all variables to predict GPA.

regression coefficient. This result is not surprising given that we have already seen that HSM and SATM are strong predictors of GPA. The value of R^2 is 0.2115, not much higher than the value of 0.2046 that we found for the high school grades regression.

Examination of the t statistics and the associated P-values for the individual regression coefficients reveals that HSM is the only one that is significant ($P = 0.0003$). That is, only HSM makes a significant contribution when it is added to a model that already has the other four explanatory variables. Once again it is important to understand that this result does not necessarily mean that the regression coefficients for the four other explanatory variables are *all* 0.

Figure 11.9 gives the Excel and Minitab multiple regression output for this problem. Although the format and organization of output differ among software packages, the basic results that we need are easy to find.

Excel

SUMMARY OUTPUT

Regression Statistics	
Multiple R	0.459837234
R Square	0.211450282
Adjusted R Square	0.193364279
Standard Error	0.699997195
Observations	224

ANOVA

	df	*SS*	*MS*	*F*	*Signifcance F*
Regression	5	28.64364489	5.728729	11.69138	5.06E-10
Residual	218	106.8191439	0.489996		
Total	223	135.4627888			

	Coefficients	*Standard Error*	*t Stat*	*P-value*	*Lower 95%*	*upper 95%*
intercept	0.326718739	0.399996431	0.816804	0.414932	-0.461636967	1.115074446
HSM	0.14596108	0.039260974	3.717714	0.000256	0.068581358	0.223340801
HSS	0.03590532	0.037798412	0.949916	0.343207	-0.03859183	0.11040247
HSE	0.055292581	0.039568691	1.397382	0.163719	-0.022693622	0.133278785
SATM	0.000843593	0.000685657	1.376187	0.170176	-0.000407774	0.002294959
SATV	0.00040785	0.000591893	-0.68906	0.491518	-0.001574415	0.00075816

Sheet1 Sheet2 e data

Ready Sum=0 CARS

Minitab

```
The regression equation is
GPA = 0.327 + 0.146 HSM + 0.0359 HSS + 0.0553 HSE +0.000944 SATM -
0.000408 SATV

Predictor        Coef       StDev          T        P
Constant       0.3267      0.4000       0.82    0.415
HSM           0.14596     0.03926       3.72    0.000
HSS           0.03591     0.03780       0.95    0.343
HSE           0.05529     0.03957       1.40    0.164
SATM        0.0009436   0.0006857       1.38    0.170
SATV       -0.0004078   0.0005919      -0.69    0.492

S = 0.7000        R-Sq = 21.1%       R-Sq(adj) = 19.3%

Analysis of Variance

Source         DF         SS          MS          F        P
Regression      5    28.6436      5.7287      11.69    0.000
Error         218   106.8191      0.4900
Total         223   135.4628
```

FIGURE 11.9 Excel and Minitab multiple regression output for regression using all variables to predict GPA.

Many statistical software packages provide the capability for testing whether a collection of regression coefficients in a multiple regression model are *all* 0. We use this approach to address two interesting questions about this set of data. We did not discuss such tests in the outline that opened this section, but the basic idea is quite simple.

Test for a collection of regression coefficients

In the context of the multiple regression model with all five predictors, we ask first whether or not the coefficients for the two SAT scores are both 0. In other words, do the SAT scores add any significant predictive information to that already contained in the high school grades? To be fair, we also ask the complementary question—do the high school grades add any significant predictive information to that already contained in the SAT scores?

The answers are given in the last two parts of the output in Figure 11.8. For the first test we see that $F = 0.9503$. Under the null hypothesis that the two SAT coefficients are 0, this statistic has an $F(2, 218)$ distribution and the P-value is 0.39. We conclude that the SAT scores are not significant predictors of GPA in a regression that already contains the high school scores as predictor variables. Recall that the model with just SAT scores has a highly significant F statistic. We now see that whatever predictive information is in the SAT scores can also be found in the high school grades. In this sense, the SAT scores are superfluous.

The test statistic for the three high school grade variables is $F = 13.6462$. Under the null hypothesis that these three regression coefficients are 0, the statistic has an $F(3, 218)$ distribution and the P-value is 0.0001. We conclude that high school grades contain useful information for predicting GPA that is not contained in SAT scores.

Of course, our statistical analysis of these data does not imply that SAT scores are less useful than high school grades for predicting college grades for all groups of students. We have studied a select group of students—computer science majors—from a specific university. Generalizations to other situations are beyond the scope of inference based on these data alone.

Beyond the basics ▸ multiple logistic regression

Many studies have yes/no or success/failure response variables. A surgery patient lives or dies; a consumer does or does not purchase a product after viewing an advertisement. Because the response variable in a multiple regression is assumed to have a normal distribution, this methodology is not suitable for predicting such responses. However, there are models that apply the ideas of regression to response variables with only two possible outcomes.

logistic regression

One type of model that can be used is called **logistic regression.** We think in terms of a binomial model for the two possible values of the response variable and use one or more explanatory variables to explain the probability of success. Details are even more complicated than those for multiple regression but the fundamental ideas are very much the same. Here is an example.

EXAMPLE 11.2

A company uses many different mailed catalogs to sell different types of merchandise. One catalog that features home goods, such as bedspreads and pillows, was mailed to 200,000 people who were not current customers of the catalog.[2] The response variable is whether or not a purchase was made from this catalog. We use logistic regression[3] to model the probability of a purchase, say p, as a function of five explanatory variables. These are x_1, number of purchases within the last 24 months from a home gift catalog; x_2, the proportion of single people in the zip code area based on census data; x_3, number of credit cards; x_4, a variable that distinguishes apartment dwellers from those who live in single-family homes; and x_5, an indicator of whether or not the customer has ever made a purchase from a similar type of catalog.

Similar to the F test in multiple regression, there is a chi-square test for multiple logistic regression that tests the null hypothesis that *all* coefficients of the explanatory variables are zero. The value is $X^2 = 141.34$, and the degrees of freedom are the number of explanatory variables, 5 in this case. (Recall that we studied the chi-square distribution in Chapter 9, where we analyzed two-way tables of counts.) The P-value is reported by software as $P = 0.0001$. (You can verify that it is less than 0.0005 using Table F.) We conclude that not all of the explanatory variables have zero coefficients.

Interpretation of the coefficients is a little more difficult in multiple logistic regression because of the form of the model. For our example, the fitted model is

$$\log\left(\frac{p}{1-p}\right) = -4.80 + 0.11x_1 - 0.11x_2 + 0.14x_3 - 0.43x_4 + 0.22x_5$$

The expression $p/(1-p)$ is the odds of a successful outcome. Logistic regression models the "log odds" as a linear combination of the explanatory variables.

In place of the t tests for individual coefficients in multiple regression, chi-square tests, each with 1 degree of freedom, are used to test whether individual coefficients are zero. The P-values for each of these tests are very small, ranging from 0.0067 to 0.0001. We conclude that all are nonzero. The signs of the coefficients tell us whether the explanatory variable is positively or negatively related to the response in the context of the model. We see that having made purchases in the past and having more credit cards are associated with a higher chance of buying from this catalog, whereas people who live in an apartment and in a neighborhood with a high proportion of single people are less likely to purchase these home goods.

The predicted value for this model allows us to estimate the probability p that a customer will make a purchase from the catalog. In the future, this model will be applied to potential customers who have not yet received the catalog. It will be mailed to those with an estimated probability of making a purchase that is above some cutoff value. From the response rate needed for a mailing to be profitable, an appropriate cutoff can be determined.

SUMMARY

The statistical model for **multiple linear regression** with response variable y and p explanatory variables $x_1, x_2, \ldots, x_p$ is

$$y_i = \beta_0 + \beta_1 x_{i1} + \beta_2 x_{i2} + \cdots + \beta_p x_{ip} + \epsilon_i$$

where $i = 1, 2, \ldots, n$. The ϵ_i are assumed to be independent and normally distributed with mean 0 and standard deviation σ. The **parameters** of the model are $\beta_0, \beta_1, \beta_2, \ldots, \beta_p$, and σ.

The β's are estimated by $b_0, b_1, b_2, \ldots, b_p$, which are obtained by the **method of least squares.** The parameter σ is estimated by

$$s = \sqrt{\text{MSE}} = \sqrt{\frac{\sum e_i^2}{n - p - 1}}$$

where the e_i are the **residuals,**

$$e_i = y_i - \hat{y}_i$$

A **level *C* confidence interval for** β_j is

$$b_j \pm t^* \text{SE}_{b_j}$$

where t^* is the value for the $t(n - p - 1)$ density curve with area C between $-t^*$ and t^*.

The test of the hypothesis H_0: $\beta_j = 0$ is based on the ***t* statistic**

$$t = \frac{b_j}{\text{SE}_{b_j}}$$

and the $t(n - p - 1)$ distribution.

The estimate b_j of β_j and the test and confidence interval for β_j are all based on a specific multiple linear regression model. The results of all of these procedures change if other explanatory variables are added to or deleted from the model.

The **ANOVA table** for a multiple linear regression gives the degrees of freedom, sum of squares, and mean squares for the model, error, and total sources of variation. The **ANOVA *F* statistic** is the ratio MSM/MSE and is used to test the null hypothesis

$$H_0\text{: } \beta_1 = \beta_2 = \cdots = \beta_p = 0$$

If H_0 is true, this statistic has an $F(p, n - p - 1)$ distribution.

The **squared multiple correlation** is given by the expression

$$R^2 = \frac{\text{SSM}}{\text{SST}}$$

and is interpreted as the proportion of the variability in the response variable y that is explained by the explanatory variables $x_1, x_2, \ldots, x_p$ in the multiple linear regression.

CHAPTER 11 EXERCISES

11.1 A multiple regression is used to relate a response variable to a set of 7 explanatory variables. There are 120 observations. Outline the analysis of variance table for this analysis giving the sources of variation and the degrees of freedom.

11.2 A multiple regression analysis of 105 cases was performed with 4 explanatory variables. Suppose that SSM $= 20$ and SSE $= 200$.

(a) Find the value of the F statistic for testing the null hypothesis that the coefficients of all of the explanatory variables are zero.

(b) Use Table E to determine if the result is significant at the 5% level. Is it significant at the 1% level?

11.3 Refer to the previous exercise. What proportion of the variation in the response variable is explained by the explanatory variables?

11.4 An instructor in an introductory statistics class used multiple regression to predict the score on the final exam using the results of 10 quizzes that were given during the semester. The F statistic for this analysis was highly significant ($P < 0.0001$) but none of the t tests for the individual coefficients were significant. Explain this apparent inconsistency.

11.5 Nutrition facts labels provide consumers with information about the nutritional value of food products that they buy. A study of these labels collected data from 152 consumers who were sent information about a frozen chicken dinner. Each subject was asked to give an overall product nutrition score, and also evaluated each of 10 nutrients on a 9-point scale with higher values indicating that the product has a healthy value for the given nutrient. Composite scores for favorable nutrients (such as protein and fiber) and unfavorable nutrients (such as fat and sodium) were used in a multiple regression to predict the overall product nutrition score.[4] The following was reported in a table:

Explanatory variables	b	se	t	Model F	R^2
				33.7**	.31
Unfavorable nutrients	0.82	.12	6.8**		
Favorable nutrients	0.57	.10	5.5**		
Constant	3.33	.13	26.1**		

$^{**}p < .01$

(a) What is the equation of the least-squares line?

(b) Give the null and alternative hypotheses associated with the entry labeled "Model *F*" and interpret this result.

(c) The column labeled "*t*" contains three entries. Explain what each of these means.

(d) What are the degrees of freedom associated with the *t* statistics that you explained in part (c)?

11.6 The product used in the previous exercise was described by the researchers as a poor-nutrition product. The label information for this product had high values for unfavorable nutrients such as fat and low values for favorable nutrients such as fiber. The researchers who conducted this study collected a parallel set of data from subjects who were provided label information for a good-nutrition product. This label had low values for the unfavorable nutrients and high values for the favorable ones. The same type of multiple regression model was run for the 162 consumers who participated in this part of the study. Here are the regression results:

Explanatory variables	b	se	t	Model F	R^2
				44.0**	.36
Unfavorable nutrients	0.86	.12	6.9**		
Favorable nutrients	0.66	.10	6.9**		
Constant	3.96	.12	32.8**		

$^{**}p < .01$

For this analysis, answer the questions in parts (a) to (d) of the previous exercise.

11.7 Refer to the previous two exercises. When the researchers planned these studies, they expected both unfavorable nutrients and favorable nutrients to be positively associated with the overall product nutrition score. They also expected the unfavorable nutrients to have a stronger effect. Examine the regression coefficients and the associated *t* statistics for the two regression models. Then, use this information to discuss how well the researchers' expectations were fulfilled.

The following five exercises use the data given in the next exercise.

11.8 Online stock trading through the Internet has increased dramatically during the past several years. An article discussing this new method of investing provided data on the major Internet stock brokerages who provide this service.[5] Below are some data for the top 10 Internet brokerages. The variables are Mshare, the market share of the firm; Accts, the number of Internet accounts in thousands; and Assets, the total assets in billions of dollars. These firms are not a random sample from any population but we will use multiple regression methods to develop statistical models that relate assets to the other two variables.

ID	Broker	Mshare	Accts	Assets
1	Charles Schwab	27.5	2500	219.0
2	E*Trade	12.9	909	21.1
3	TD Waterhouse	11.6	615	38.8
4	Datek	10.0	205	5.5
5	Fidelity	9.3	2300	160.0
6	Ameritrade	8.4	428	19.5
7	DLJ Direct	3.6	590	11.2
8	Discover	2.8	134	5.9
9	Suretrade	2.2	130	1.3
10	National Discount Brokers	1.3	125	6.8

(a) Plot assets versus accounts and describe the relationship.

(b) Perform a simple linear regression to predict assets from the number of accounts. Give the least-squares line and the results of the hypothesis test for the slope.

(c) Obtain the residuals from part (b) and plot them versus accounts. Describe the plot. What do you conclude?

(d) Construct a new variable that is the square of the number of accounts. Rerun the regression analysis with accounts and the square as explanatory variables. Summarize the results.

11.9 In the multiple regression you performed in the previous exercise, the *P*-value for the number of accounts was 0.8531, while the *P*-value for the square was 0.0070. Unless we have a strong theoretical reason for considering a model with a quadratic term and no linear term, we prefer not to do this. One problem with these two explanatory variables is that they are highly correlated. Here is a way to construct a version of the quadratic term that is uncorrelated with the linear term. We first find the mean for accounts, and then we subtract this value from accounts before squaring. The mean is 793.6, so the new quadratic explanatory variable will be $(\text{Accts} - 793.6)^2$. Run the multiple regression to predict assets using accounts and the new quadratic term. Compare these results with what you found in the previous exercise.

11.10 To one person, the plot of assets versus the number of accounts indicates that the relationship is curved. Another person might see this as a linear relationship with two outliers. Identify the two outliers and rerun the linear regression and the multiple regression with the linear and quadratic terms. Summarize your results.

11.11 Sometimes we attempt to model curved relationships by transforming variables. Take the logarithm of assets and the logarithm of the number of accounts. Does the relationship between the logs appear to be approximately linear? Analyze the data and provide a summary of your results. Be sure to include plots along with the results of your statistical inference.

11.12 Recall that the relationship between an explanatory variable and a response variable can depend on what other explanatory variables are included in the model.

(a) Use a simple linear regression to predict assets using the number of accounts. Give the regression equation and the results of the significance test for the regression coefficient.

(b) Do the same using market share to predict assets.

(c) Run a multiple regression using both the number of accounts and market share to predict assets. Give the multiple regression equation and the results of the significance tests for the two regression coefficients.

(d) Compare the results in parts (a), (b), and (c). If you had to choose one of these three models, which one do you prefer? Give an explanation for your answer.

11.13 Banks charge different interest rates for different loans. A random sample of 2229 loans made for the purchase of new automobiles was studied to identify variables that explain the interest rate charged. All of these loans were made directly by the bank. A multiple regression was run with interest rate as the response variable and 13 explanatory variables.[6]

(a) The F statistic reported is 71.34. State the null and alternative hypotheses for this statistic. Give the degrees of freedom and the P-value for this test. What do you conclude?

(b) The value of R^2 is 0.297. What percent of the variation in interest rates is explained by the 13 explanatory variables?

(c) The researchers report a t statistic for each of the regression coefficients. State the null and alternative hypotheses tested by each of these statistics. What are the degrees of freedom for these t statistics? What values of t will lead to rejection of the null hypothesis at the 5% level?

(d) The following table gives the explanatory variables and the t statistics (these are given without the sign, assuming that all tests are two-sided) for the regression coefficients. Which of the explanatory variables are significantly different from zero in this model?

Variable	b	t
Intercept	15.47	
Loan size (in dollars)	–0.0015	10.30
Length of loan (in months)	–0.906	4.20
Percent down payment	–0.522	8.35
Cosigner (0 = no, 1 = yes)	–0.009	3.02
Unsecured loan (0 = no, 1 = yes)	0.034	2.19
Total payments (borrower's monthly installment debt)	0.100	1.37
Total income (borrower's total monthly income)	–0.170	2.37
Bad credit report (0 = no, 1 = yes)	0.012	1.99
Young borrower (0 = older than 25, 1 = 25 or younger)	0.027	2.85
Male borrower (0 = female, 1 = male)	–0.001	0.89
Married (0 = no, 1 = yes)	–0.023	1.91
Own home (0 = no, 1 = yes)	–0.011	2.73
Years at current address	–0.124	4.21

(e) The signs of many of these coefficients are what we might expect before looking at the data. For example, the negative coefficient of loan size means that larger loans get a smaller interest rate. This is very reasonable. Examine the signs of each of the statistically significant coefficients and give a short explanation of what they mean.

11.14 Refer to the previous exercise. The researchers also looked at loans made indirectly, that is, through an auto dealer. They studied 5664 indirect loans. Multiple regression was used to predict the interest rate using the same set of explanatory variables.

(a) The F statistic reported is 27.97. State the null and alternative hypotheses for this statistic. Give the degrees of freedom and the P-value for this test. What do you conclude?

(b) The value of R^2 is 0.141. What percent of the variation in interest rates is explained by the 13 explanatory variables? Compare this value with the percent explained for direct loans in the previous exercise.

(c) The researchers report a t statistic for each of the regression coefficients. State the null and alternative hypotheses tested by each of these statistics. What are the degrees of freedom for these t statistics? What values of t will lead to rejection of the null hypothesis at the 5% level?

(d) The following table gives the explanatory variables and the t statistics (these are given without the sign, assuming that all tests are two-sided) for the regression coefficients. Which of the explanatory variables are significantly different from zero in this model?

Variable	b	t
Intercept	15.89	
Loan size (in dollars)	−0.0029	17.40
Length of loan (in months)	−1.098	5.63
Percent down payment	−0.308	4.92
Cosigner (0 = no, 1 = yes)	−0.001	1.41
Unsecured loan (0 = no, 1 = yes)	0.028	2.83
Total payments (borrower's monthly installment debt)	−0.513	1.37
Total income (borrower's total monthly income)	0.078	0.75
Bad credit report (0 = no, 1 = yes)	0.039	1.76
Young borrower (0 = older than 25, 1 = 25 or younger)	−0.036	1.33
Male borrower (0 = female, 1 = male)	−0.179	1.03
Married (0 = no, 1 = yes)	−0.043	1.61
Own home (0 = no, 1 = yes)	−0.047	1.59
Years at current address	−0.086	1.73

(e) The signs of many of these coefficients are what we might expect before looking at the data. For example, the negative coefficient of loan size means that larger loans get a smaller interest rate. This is very reasonable. Examine the signs of each of the statistically significant coefficients and give a short explanation of what they mean.

11.15 Refer to the previous two exercises. The authors conclude that banks take higher risks with indirect loans because they do not take into account borrower characteristics when setting the loan rate. Explain how the results of the multiple regressions lead to this conclusion.

11.16 Refer to the educational data for 78 seventh-grade students given in Table 1.6 (page 33). We view GPA as the response variable. IQ, gender, and self-concept are the explanatory variables.

(a) Find the correlation between GPA and each of the explanatory variables. What percent of the total variation in student GPAs can be explained by the straight-line relationship for each of the explanatory variables?

(b) The importance of IQ in explaining GPA is not surprising. The purpose of the study is to assess the influence of self-concept on GPA. So we will include IQ in the regression model and ask, "How much does self-concept contribute to explaining GPA after the effect of IQ on GPA is taken into account?" Give a model that can be used to answer this question.

(c) Run the model and report the fitted regression equation. What percent of the variation in GPA is explained by the explanatory variables in your model?

(d) Translate the question of interest into appropriate null and alternative hypotheses about the model parameters. Give the value of the test statistic and its *P*-value. Write a short summary of your analysis with an emphasis on your conclusion.

11.17 One model for subpopulation means for the computer science study is described in Example 11.1 (page 710) as

$$\mu_{\text{GPA}} = \beta_0 + \beta_1\text{HSM} + \beta_2\text{HSS} + \beta_3\text{HSE}$$

(a) Give the model for the subpopulation mean GPA for students having high school grade scores HSM = 8 (B+), HSS = 9 (A−), and HSE = 8 (B+).

(b) Using the parameter estimates given in Figure 11.4 (page 721), calculate the estimate of this subpopulation mean. Then briefly explain in words what your numerical answer means.

11.18 Use the model given in the previous exercise to do the following:

(a) For students having high school grade scores HSM = 7 (B), HSS = 6 (B−), and HSE = 6 (B), express the subpopulation mean in terms of the parameters β_j.

(b) Calculate the estimate of this subpopulation mean using the b_j given in Figure 11.4 (page 721). Briefly explain the meaning of the number you obtain.

11.19 Consider the regression problem of predicting GPA from the three high school grade variables. The computer output appears in Figure 11.4 (page 721).

(a) Use information given in the output to calculate a 99% confidence interval for the regression coefficient of HSM. Explain in words what this regression coefficient means in the context of this particular model.

(b) Do the same for the explanatory variable HSE.

11.20 Computer output for the regression run for predicting GPA from the high school grade variables HSM and HSE appears in Figure 11.6 (page 724). Answer (a) and (b) of the previous exercise for this regression. Why are the results for this exercise different from those that you calculated for the previous exercise?

11.21 Consider the regression problem of predicting GPA from the three high school grade variables. The computer output appears in Figure 11.4 (page 721).

(a) Write the estimated regression equation.

(b) What is the value of s, the estimate of σ?

(c) State the H_0 and H_a tested by the ANOVA F statistic for this problem. After stating the hypotheses in symbols, explain them in words.

(d) What is the distribution of the F statistic under H_0? What conclusion do you draw from the F test?

(e) What percent of the variation in GPA is explained by these three high school grade variables?

11.22 Figure 11.7 (page 725) gives the output for the regression analysis in which the two SAT scores are used to predict GPA. Answer the questions in the previous exercise for this analysis.

11.23 Multiple regressions are sometimes used in litigation. In the case of *Cargill, Inc. v. Hardin*, the prosecution charged that the cash price of wheat was manipulated in violation of the Commodity Exchange Act. In a statistical study conducted for this case, a multiple regression model was constructed to predict the price of wheat using three supply-and-demand explanatory variables. Data for 14 years were used to construct the regression equation, and a prediction for the suspect period was computed from this equation. The value of R^2 was 0.989. The predicted value was reported as \$2.136 with a standard error of \$0.013. Express the prediction with a 95% interval. (The degrees of freedom were large for this analysis, so use 100 as the df to determine t^*.) The actual price for the period in question was \$2.13. The judge in this case decided that the analysis provided evidence that the price was not artificially depressed, and the opinion was sustained by the court of appeals. Write a short summary of the results of the analysis that relate to the decision and explain why you agree or disagree with it.

The following eight exercises are related to the case study described in this chapter. They require use of the CSDATA data set described in the Data Appendix.

11.24 Use software to make a plot of GPA versus SATM. Do the same for GPA versus SATV. Describe the general patterns. Are there any unusual values?

11.25 Make a plot of GPA versus HSM. Do the same for the other two high school grade variables. Describe the three plots. Are there any outliers or influential points?

11.26 Regress GPA on the three high school grade variables. Calculate and store the residuals from this regression. Plot the residuals versus each of the three predictors and versus the predicted value of GPA. Are there any unusual points or patterns in these four plots?

11.27 Use the two SAT scores in a multiple regression to predict GPA. Calculate and store the residuals. Plot the residuals versus each of the explanatory variables and versus the predicted GPA. Describe the plots.

11.28 It appears that the mathematics explanatory variables are strong predictors of GPA in the computer science study. Run a multiple regression using HSM and SATM to predict GPA.

(a) Give the fitted regression equation.

(b) State the H_0 and H_a tested by the ANOVA F statistic, and explain their meaning in plain language. Report the value of the F statistic, its P-value, and your conclusion.

(c) Give 95% confidence intervals for the regression coefficients of HSM and SATM. Do either of these include the point 0?

(d) Report the t statistics and P-values for the tests of the regression coefficients of HSM and SATM. What conclusions do you draw from these tests?

(e) What is the value of s, the estimate of σ?

(f) What percent of the variation in GPA is explained by HSM and SATM in your model?

11.29 How well do verbal variables predict the performance of computer science students? Perform a multiple regression analysis to predict GPA from HSE and SATV. Summarize the results and compare them with those obtained in the previous exercise. In what ways do the regression results indicate that the mathematics variables are better predictors?

11.30 The variable SEX has the value 1 for males and 2 for females. Create a data set containing the values for males only. Run a multiple regression analysis for predicting GPA from the three high school grade variables for this group. Using the case study in the text as a guide, interpret the results and state what conclusions can be drawn from this analysis. In what way (if any) do the results for males alone differ from those for all students?

11.31 Refer to the previous exercise. Perform the analysis using the data for females only. Are there any important differences between female and male students in predicting GPA?

The following three exercises use the CONCEPT data set described in the Data Appendix.

11.32 Find the correlations between the response variable GPA and each of the explanatory variables IQ, AGE, SEX, SC, and C1 to C6. Of all the explanatory variables, IQ does the best job of explaining GPA in a simple linear regression with only one variable. How do you know this *without* doing all of the regressions? What percent of the variation in GPA can be explained by the straight-line relationship between GPA and IQ?

11.33 Let us look at the role of positive self-concept about one's physical appearance (variable C3). We include IQ in the model because it is a known important predictor of GPA.

(a) Report the fitted regression model

$$\widehat{\text{GPA}} = b_0 + b_1\text{IQ} + b_2\text{C3}$$

with its R^2, and report the t statistic for significance of self-concept about one's physical appearance with its P-value. Does C3 contribute significantly to explaining GPA when added to IQ? How much does adding C3 to the model raise R^2?

(b) Now start with the model that includes overall self-concept SC along with IQ. Does C3 help explain GPA significantly better? Does it raise R^2 enough to be of practical value? In answering these questions, report the fitted regression model

$$\widehat{\text{GPA}} = b_0 + b_1\text{IQ} + b_2\text{C3} + b_3\text{SC}$$

with its R^2 and the t statistic for significance of C3 with its P-value.

(c) Explain carefully in words why the coefficient b_2 for the variable C3 takes quite different values in the regressions of parts (a) and (b). Then explain simply how it can happen that C3 is a useful explanatory variable in part (a) but worthless in part (b).

11.34 A reasonable model explains GPA using IQ, C1 (behavior self-concept), and C5 (popularity self-concept). These three explanatory variables are all significant in the presence of the other two, and no other explanatory variable is significant when added to these three. (You do not have to verify these statements.) Let's use this model.

(a) What is the fitted regression model, its R^2, and the standard deviation s about the fitted model? What GPA does this model predict for a student with IQ 109, C1 score 13, and C5 score 8?

(b) If both C1 and C5 could be held constant, how much would GPA increase for each additional point of IQ score, according to the fitted model? Give a 95% confidence interval for the mean increase in GPA in the entire population if IQ could be increased by 1 point while holding C1 and C5 constant.

(c) Compute the residuals for this model. Plot the residuals against the predicted values and also make a normal quantile plot. Which observation (value of OBS) produces the most extreme residual? Circle this observation in both of your plots of the residuals. For which individual variables does this student have very unusual values? What are these values? (Look at all the variables, not just those in the model.)

(d) Repeat part (a) with this one observation removed. How much did this one student affect the fitted model and the prediction?

The following nine exercises use the CHEESE data set described in the Data Appendix.

11.35 For each of the four variables in the CHEESE data set, find the mean, median, standard deviation, and interquartile range. Display each distribution by means of a stemplot and use a normal quantile plot to assess normality of the data. Summarize your findings. Note that when doing regressions with these data, we do not assume that these distributions are normal. Only the residuals from our model need be (approximately) normal. The careful study of each variable to be analyzed is nonetheless an important first step in any statistical analysis.

11.36 Make a scatterplot for each pair of variables in the CHEESE data set (you will have six plots). Describe the relationships. Calculate the correlation for each pair of variables and report the *P*-value for the test of zero population correlation in each case.

11.37 Perform a simple linear regression analysis using Taste as the response variable and Acetic as the explanatory variable. Be sure to examine the residuals carefully. Summarize your results. Include a plot of the data with the least-squares regression line. Plot the residuals versus each of the other two chemicals. Are any patterns evident? (The concentrations of the other chemicals are lurking variables for the simple linear regression.)

11.38 Repeat the analysis of Exercise 11.37 using Taste as the response variable and H2S as the explanatory variable.

11.39 Repeat the analysis of Exercise 11.37 using Taste as the response variable and Lactic as the explanatory variable.

11.40 Compare the results of the regressions performed in the three previous exercises. Construct a table with values of the F statistic, its *P*-value, R^2, and the estimate s of the standard deviation for each model. Report the three regression equations. Why are the intercepts in these three equations different?

11.41 Carry out a multiple regression using Acetic and H2S to predict Taste. Summarize the results of your analysis. Compare the statistical significance of Acetic in this model with its significance in the model with Acetic alone as a predictor (Exercise 11.37). Which model do you prefer? Give a simple explanation for the fact that Acetic alone appears to be a good predictor of Taste but with H2S in the model, it is not.

11.42 Carry out a multiple regression using H2S and Lactic to predict Taste. Comparing the results of this analysis with the simple linear regressions using each of these explanatory variables alone, it is evident that a better result is obtained by using both predictors in a model. Support this statement with explicit information obtained from your analysis.

11.43 Use the three explanatory variables Acetic, H2S, and Lactic in a multiple regression to predict Taste. Write a short summary of your results, including an examination of the residuals. Based on all of the regression analyses you have carried out on these data, which model do you prefer and why?

The following seven exercises use the corn and soybean data given in Tables 10.2 (page 695) and 10.3 (page 699).

11.44 Run the simple linear regression using year to predict corn yield.

(a) Summarize the results of your analysis, including the significance test results for the slope and R^2 for this model.

(b) Analyze the residuals with a normal quantile plot. Is there any indication in the plot that the residuals are not normal?

(c) Plot the residuals versus soybean yield. Does the plot indicate that soybean yield might be useful in a multiple linear regression with year to predict corn yield?

11.45 Run the simple linear regression using soybean yield to predict corn yield.

(a) Summarize the results of your analysis, including the significance test results for the slope and R^2 for this model.

(b) Analyze the residuals with a normal quantile plot. Is there any indication in the plot that the residuals are not normal?

(c) Plot the residuals versus year. Does the plot indicate that year might be useful in a multiple linear regression with soybean yield to predict corn yield?

11.46 From the previous two exercises, we conclude that year *and* soybean yield may be useful together in a model for predicting corn yield. Run this multiple regression.

(a) Explain the results of the ANOVA F test. Give the null and alternative hypotheses, the test statistic with degrees of freedom, and the P-value. What do you conclude?

(b) What percent of the variation in corn yield is explained by these two variables? Compare it with the percent explained in the simple linear regression models of the previous two exercises.

(c) Give the fitted model. Why do the coefficients of year and soybean yield differ from those in the previous two exercises?

(d) Summarize the significance test results for the regression coefficients for year and soybean yield.

(e) Give a 95% confidence interval for each of these coefficients.

(f) Plot the residuals versus year and versus soybean yield. What do you conclude?

11.47 We need a new variable to model the curved relation that we see between corn yield and year in the residual plot of the last exercise. Let year2 = $(\text{year} - 1978.5)^2$. (When adding a squared term to a multiple regression model, it is a good idea to subtract the mean of the variable being squared before squaring. This avoids all sorts of messy problems that we cannot discuss here.)

(a) Run the multiple linear regression using year, year2, and soybean yield to predict corn yield. Give the fitted regression equation.

(b) Give the null and alternative hypotheses for the ANOVA F test. Report the results of this test, giving the test statistic, degrees of freedom, P-value, and conclusion.

(c) What percent of the variation in corn yield is explained by this multiple regression? Compare this with the model in the previous exercise.

(d) Summarize the results of the significance tests for the individual regression coefficients.

(e) Analyze the residuals and summarize your conclusions.

11.48 Run the model to predict corn yield using year and the squared term year2 defined in the previous exercise.

(a) Summarize the significance test results.

(b) The coefficient of year2 is not statistically significant in this run, but it was highly significant in the model analyzed in the previous exercise. Explain how this can happen.

(c) Obtain the fitted values for each year in the data set and use these to sketch the curve on a plot of the data. Plot the least-squares line on this graph for comparison. Describe the differences between the two regression functions. For what years do they give very similar fitted values; for what years are the differences between the two relatively large?

11.49 Use the simple linear regression model with corn yield as the response variable and year as the explanatory variable to predict the corn yield for the year 2001, and give the 95% prediction interval. Also, use the multiple regression model where year and year2 are both explanatory variables to find another predicted value with the 95% interval. Explain why these two predicted values are so different. The actual yield for 2001 is 138.2 bushels per acre. How well did your models predict this value?

11.50 Repeat the previous exercise doing the prediction for 2002. Compare the results of this exercise with the previous one. Why are they different?

The following three exercises use the bank wages data given in Table 10.1 (page 694). For these exercises we code the size of the bank as 1 if it is large and 0 if it is small. There is one outlier in this data set. Delete it and use the remaining 59 observations in these exercises.

11.51 Use length of service (LOS) to predict wages with a simple linear regression. Write a short summary of your results and conclusions.

11.52 Predict wages using the size of the bank as the explanatory variable. Use the coded values 0 and 1 for this model.

(a) Summarize the results of your analysis. Include a statement of all hypotheses, test statistics with degrees of freedom, P-values, and conclusions.

(b) Calculate the t statistic for comparing the mean wages for the large and the small banks assuming equal standard deviations. Give the degrees of freedom. Verify that this t is the same as the t statistic for the coefficient of size in the regression. Explain why this makes sense.

(c) Plot the residuals versus LOS. What do you conclude?

11.53 Use a multiple linear regression to predict wages from LOS and the size of the bank. Write a report summarizing your work. Include graphs and the results of significance tests.

Prelude: how should we improve the forms?

A university requires its faculty to complete a form three times a year that describes the amount of time spent in various activities. For many years there were complaints that the form was difficult to fill out and that many of the questions were not clearly explained.

A new administration decided to revise the form. Several small groups of faculty members studied the old form and made recommendations for how it should be changed. From these groups, three possible versions of a new form were developed. To determine which of the new versions should be adopted, a randomized experiment was performed. From the list of faculty, 150 names were randomly selected. These were then randomly assigned to receive one of the three proposed forms. Thus, there were 50 faculty who received each form. The faculty were asked to review the proposed form and to rate it on a scale of 1 to 10.

The evaluations for the three proposed forms were compared using the analysis of variance methods described in this chapter. The results gave a clear indication that one of the forms was preferred to the other two. This form is now being used.

CHAPTER 12

Robert Delaunay, *Rhythme I.*

One-Way Analysis of Variance

Introduction

Many of the most effective statistical studies are comparative, whether we compare the salaries of women and men in a sample of firms or the responses to two treatments in a clinical trial. We display comparisons with back-to-back stemplots or side-by-side boxplots, and we measure them with five-number summaries or with means and standard deviations. Now we ask: "Is the difference between groups statistically significant?"

When only two groups are compared, Chapter 7 provides the tools we need. Two-sample t procedures compare the means of two normal populations, and we saw that these procedures, unlike comparisons of spread, are sufficiently robust to be widely useful. Now we will compare any number of means by techniques that generalize the two-sample t and share its robustness and usefulness.

12.1 Inference for One-Way Analysis of Variance

Which of four advertising offers mailed to sample households produces the highest sales in dollars? Which of ten brands of automobile tires wears longest? How long do cancer patients live under each of three therapies for their cancer? In each of these settings we wish to compare several treatments. In each case the data are subject to sampling variability—if we mailed the advertising offers to another set of households, we would get different data. We therefore pose the question for inference in terms of the *mean* response. We compare, for example, the mean tread lifetime of different brands of tires. In Chapter 7 we met procedures for comparing the means of two populations. We are now ready to extend those methods to problems involving more than two populations. The statistical methodology for comparing several means is called **analysis of variance,** or simply **ANOVA.**

ANOVA

We will consider two ANOVA techniques. When there is only one way to classify the populations of interest, we use **one-way ANOVA** to analyze the data. For example, to compare the tread lifetimes of ten specific brands of tires we use one-way ANOVA. This chapter presents the details for one-way ANOVA.

one-way ANOVA

In many practical situations there is more than one way to classify the populations. A mail-order firm might want to compare mailings that offer different discounts and also have different layouts. Will a lower price offered in a plain format draw more sales on the average than a higher price offered in a fancy brochure? Analyzing the effect of price and layout together requires **two-way ANOVA,** and adding yet more factors necessitates higher-way ANOVA techniques. Most of the new ideas in ANOVA with more than one factor already appear in two-way ANOVA, which we discuss in Chapter 13.

two-way ANOVA

Data for a one-way ANOVA

Do two population means differ? If we have random samples from the two populations, we compute a two-sample t statistic and its P-value to assess the statistical significance of the difference in the sample means. We compare several means in much the same way. Instead of a t statistic, ANOVA uses an F statistic and its P-value to evaluate the null hypothesis that all of several population means are equal.

In the sections that follow, we will examine the basic ideas and assumptions that are needed for the analysis of variance. Although the details differ, many of the concepts are similar to those discussed in the two-sample case.

One-way analysis of variance is a statistical method for comparing several population means. We draw a simple random sample (SRS) from each population and use the data to test the null hypothesis that the population means are all equal.

EXAMPLE 12.1

A medical researcher wants to compare the effectiveness of three different treatments to lower the cholesterol of patients with high blood cholesterol levels. He assigns 60 individuals at random to the three treatments (20 to each) and records the reduction in cholesterol for each patient.

EXAMPLE 12.2

An ecologist is interested in comparing the concentration of the pollutant cadmium in five streams. She collects 50 water specimens from each stream and measures the concentration of cadmium in each specimen.

experiment
observational study

These two examples have several similarities. In both there is a single response variable measured on several units; the units are people in the first example and water specimens in the second. We expect the data to be approximately normal and will consider a transformation if they are not. We wish to compare several populations, three treatments in the first example and five streams in the second. But there is also an important difference between the examples. The first is an **experiment** in which patients are randomly assigned to treatments. The second example is an **observational study.** By taking water specimens at randomly chosen locations, the ecologist tries to ensure that her specimens from each stream are a random sample of all water specimens that could be obtained from that stream. In both cases we can use ANOVA to compare the mean responses. The same ANOVA methods apply to data from random samples and randomized experiments. Do keep the data-production method in mind when interpreting the results, however. A strong case for causation is best made by a randomized experiment. We will often use the term *groups* for the populations to be compared in a one-way ANOVA.

Comparing means

To assess whether several populations all have the same mean, we compare the means of samples drawn from each population. Figure 12.1 displays the sample means for Example 12.1. Treatment 2 had the highest average cholesterol reduction. But is the observed difference among the groups just the result of chance variation? We do not expect sample means to be equal even if the population means are all identical. The purpose of ANOVA is to assess whether the observed differences among sample means are *statistically*

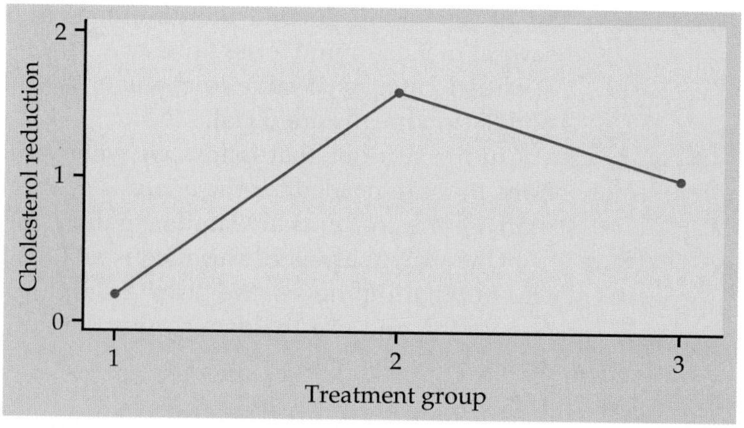

FIGURE 12.1 Mean serum cholesterol reduction in three groups.

significant. In other words, could a variation this large be plausibly due to chance, or is it good evidence for a difference among the population means? This question can't be answered from the sample means alone. Because the standard deviation of a sample mean $\overline{x}$ is the population standard deviation σ divided by $\sqrt{n}$, the answer depends upon both the variation within the groups of observations and the sizes of the samples.

Side-by-side boxplots help us see the within-group variation. Compare Figures 12.2(a) and 12.2(b). The sample medians are the same in both figures, but the large variation within the groups in Figure 12.2(a) suggests that the differences among the sample medians could be due simply to chance variation. The data in Figure 12.2(b) are much more convincing evidence that the populations differ. Even the boxplots omit essential information, however. To assess the observed differences, we must also know how large the samples

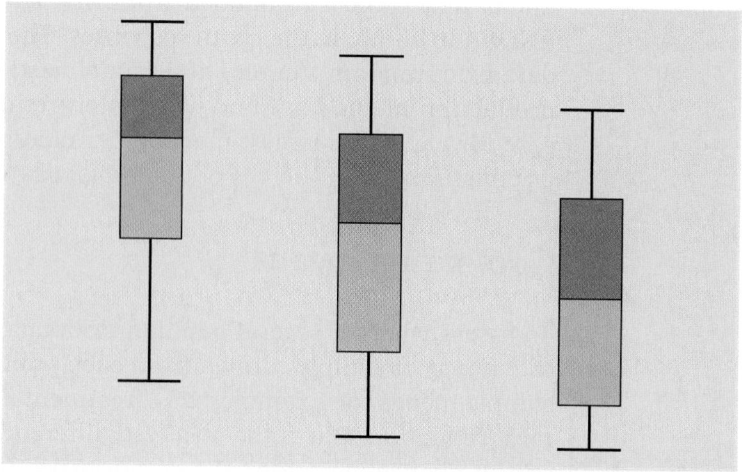

FIGURE 12.2(a) Side-by-side boxplots for three groups with large within-group variation. The differences among centers may be just chance variation.

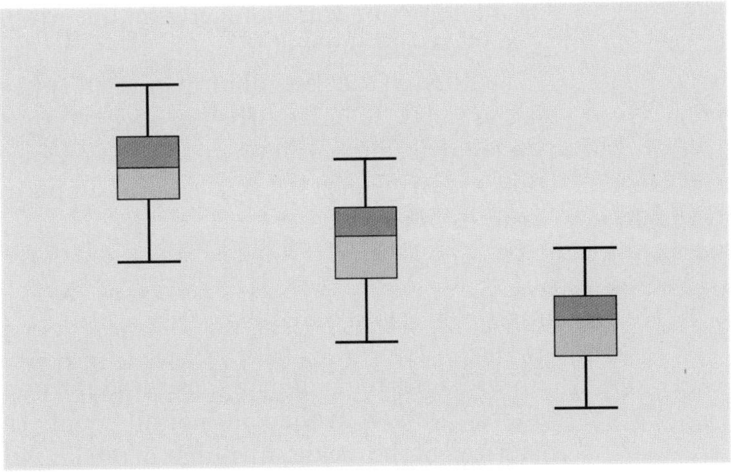

FIGURE 12.2(b) Side-by-side boxplots for three groups with the same centers as in Figure 12.2(a) but with small within-group variation. The differences among centers are more likely to be significant.

are. Nonetheless, boxplots are a good preliminary display of ANOVA data. (ANOVA compares means, and boxplots display medians. If the distributions are nearly symmetric, these two measures of center will be close together.)

The two-sample t statistic

Two-sample t statistics compare the means of two populations. If the two populations are assumed to have equal but unknown standard deviations and the sample sizes are both equal to n, the t statistic is (page 538)

$$t = \frac{\bar{x} - \bar{y}}{s_p\sqrt{\dfrac{1}{n} + \dfrac{1}{n}}} = \frac{\sqrt{\dfrac{n}{2}}(\bar{x} - \bar{y})}{s_p}$$

The square of this t statistic is

$$t^2 = \frac{\dfrac{n}{2}(\bar{x} - \bar{y})^2}{s_p^2}$$

If we use ANOVA to compare two populations, the ANOVA F statistic is exactly equal to this t^2. We can therefore learn something about how ANOVA works by looking carefully at the statistic in this form.

between-group variation

The numerator in the t^2 statistic measures the variation **between** the groups in terms of the difference between their sample means $\bar{x}$ and $\bar{y}$. It includes a factor for the common sample size n. The numerator can be large because of a large difference between the sample means or because the sample sizes are large. The denominator measures the variation **within** groups by s_p^2, the pooled estimator of the common variance. If the within-group variation is

within-group variation

small, the same variation between the groups produces a larger statistic and a more significant result.

Although the general form of the F statistic is more complicated, the idea is the same. To assess whether several populations all have the same mean, we compare the variation *among* the means of several groups with the variation *within* groups. Because we are comparing variation, the method is called *analysis of variance*.

ANOVA hypotheses

ANOVA tests the null hypothesis that the population means are *all* equal. The alternative is that they are not all equal. This alternative could be true because all of the means are different or simply because one of them differs from the rest. This is a more complex situation than comparing just two populations. If we reject the null hypothesis, we need to perform some further analysis to draw conclusions about which population means differ from which others.

The computations needed for an ANOVA are more lengthy than those for the t test. For this reason we generally use computer programs to perform the calculations. Automating the calculations frees us from the burden of arithmetic and allows us to concentrate on interpretation. The following example illustrates the practical use of ANOVA in analyzing data. Later we will explore the technical details.

EXAMPLE 12.3

In the computer science department of a large university, many students change their major after the first year. A detailed study[1] of the 256 students enrolled as first-year computer science majors in one year was undertaken to help understand this phenomenon. Students were classified on the basis of their status at the beginning of their second year, and several variables measured at the time of their entrance to the university were obtained. Here are summary statistics for the SAT mathematics scores:

Second-year major	n	$\bar{x}$	s
Computer science	103	619	86
Engineering and other sciences	31	629	67
Other	122	575	83

Figure 12.3 gives side-by-side boxplots of the SAT data. Compare Figure 12.3 with the plot of the mean scores in Figure 12.4. The means appear to be different, but there is a large amount of overlap in the three distributions. When we perform an ANOVA on these data, we ask a question about the group means. The null hypothesis is that the population mean SAT scores for the three groups are equal, and the alternative is that they are not all equal. The report on the study states that the ANOVA F statistic is 10.35 with $P < 0.001$. There is very strong evidence that the three groups of students do not all have the same mean SAT mathematics scores.

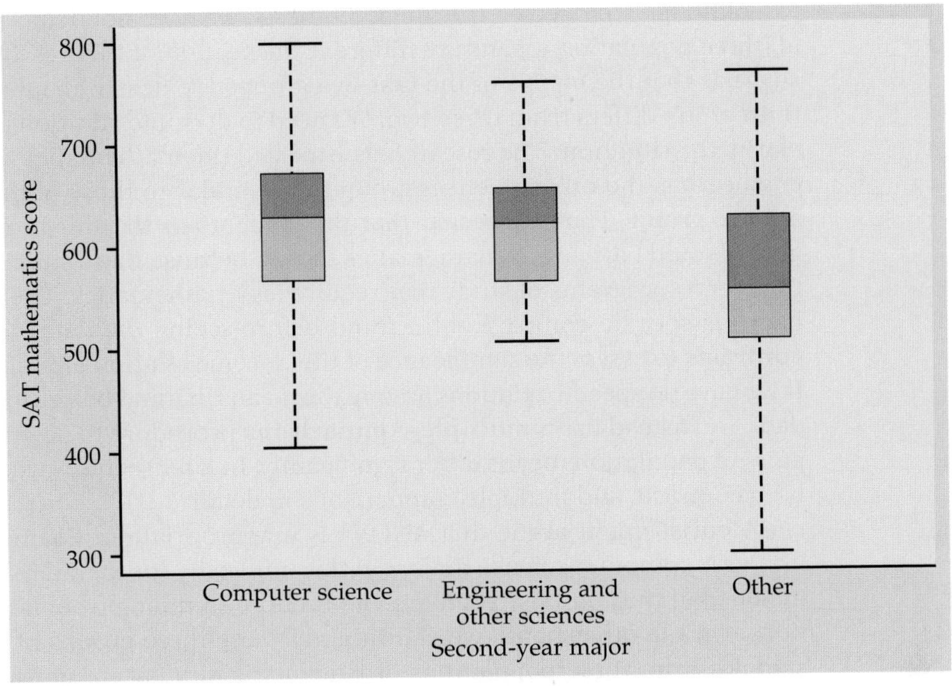

FIGURE 12.3 Side-by-side boxplots of SAT mathematics scores for the change-of-majors study.

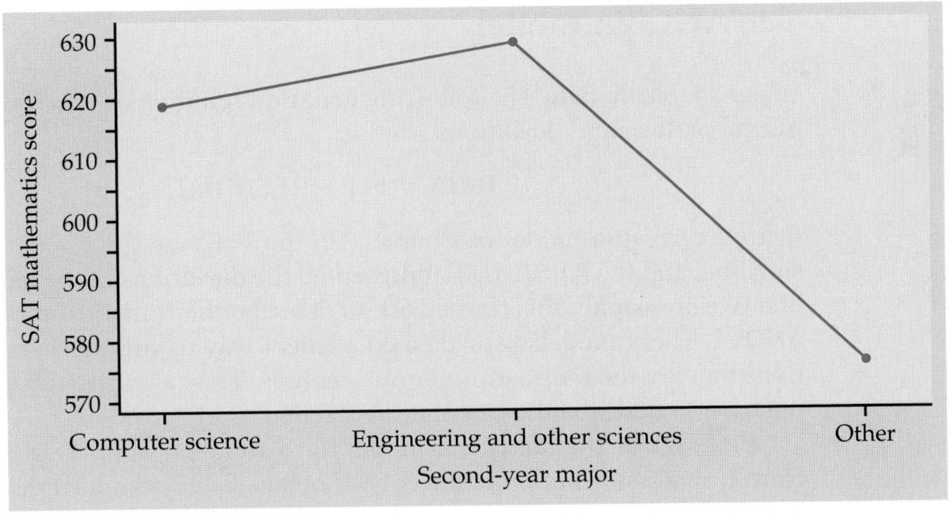

FIGURE 12.4 SAT mathematics score means for the change-of-majors study.

Although we rejected the null hypothesis, we have not demonstrated that all three population means are different. Inspection of Figures 12.3 and 12.4 suggests that the means of the first two groups are nearly equal and that the third group differs from these two. We need to do some additional analysis to clarify the situation. The researchers expected the mathematics scores of the engineering and other sciences group to be similar to those of the computer science group. They suspected that the students in the third (other major) group would have lower mathematics scores, because these students chose to transfer to programs of study that require less mathematics. The researchers had this specific comparison in mind before seeing the data, so they used **contrasts** to assess the significance of this specific relation among the means. If we have no specific relations among the means in mind before looking at the data, we instead use a **multiple-comparisons** procedure to determine which pairs of population means differ significantly. In later sections we will explore both contrasts and multiple comparisons in detail.

contrasts

multiple comparisons

A purist might argue that ANOVA is inappropriate in Example 12.3. *All* first-year computer science majors in this university for the year studied were included in the analysis. There was no random sampling from larger populations. On the other hand, we can think of these three groups of students as samples from three populations of students in similar circumstances. They may be representative of students in the next few years at the same university or other universities with similar programs. Judgments such as this are very important and must be made, not by statisticians, but by people who are knowledgeable about the subject of the study.

The ANOVA model

When analyzing data, the following equation reminds us that we look for an overall pattern and deviations from it:

$$\text{DATA} = \text{FIT} + \text{RESIDUAL}$$

In the regression model of Chapter 10, the FIT was the population regression line, and the RESIDUAL represented the deviations of the data from this line. We now apply this framework to describe the statistical models used in ANOVA. These models provide a convenient way to summarize the assumptions that are the foundation for our analysis. They also give us the necessary notation to describe the calculations needed.

First, recall the statistical model for a random sample of observations from a single normal population with mean μ and standard deviation σ. If the observations are

$$x_1, x_2, \dots, x_n$$

we can describe this model by saying that the x_j are an SRS from the $N(\mu, \sigma)$ distribution. Another way to describe the same model is to think of the x's varying about their population mean. To do this, write each observation x_j as

$$x_j = \mu + \epsilon_j$$

The ϵ_j are then an SRS from the $N(0, \sigma)$ distribution. Because μ is unknown, the ϵ's cannot actually be observed. This form more closely corresponds to our

$$\text{DATA} = \text{FIT} + \text{RESIDUAL}$$

way of thinking. The FIT part of the model is represented by μ. It is the systematic part of the model, like the line in a regression. The RESIDUAL part is represented by ϵ_j. It represents the deviations of the data from the fit and is due to random, or chance, variation.

There are two unknown parameters in this statistical model: μ and σ. We estimate μ by $\overline{x}$, the sample mean, and σ by s, the sample standard deviation. The differences $e_j = x_j - \overline{x}$ are the sample residuals and correspond to the ϵ_j in the statistical model.

The model for one-way ANOVA is very similar. We take random samples from each of I different populations. The sample size is n_i for the ith population. Let x_{ij} represent the jth observation from the ith population. The I population means are the FIT part of the model and are represented by μ_i. The random variation, or RESIDUAL, part of the model is represented by the deviations ϵ_{ij} of the observations from the means.

The One-Way ANOVA Model

The **one-way ANOVA model** is

$$x_{ij} = \mu_i + \epsilon_{ij}$$

for $i = 1, \ldots, I$ and $j = 1, \ldots, n_i$. The ϵ_{ij} are assumed to be from an $N(0, \sigma)$ distribution. The **parameters of the model** are the population means $\mu_1, \mu_2, \ldots, \mu_I$ and the common standard deviation σ.

Note that the sample sizes n_i may differ, but the standard deviation σ is assumed to be the same in all of the populations. Figure 12.5 pictures this model for $I = 3$. The three population means μ_i are different, but the shapes

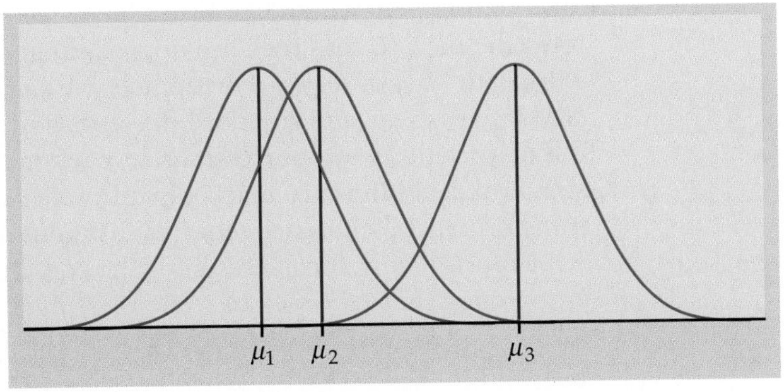

FIGURE 12.5 Model for one-way ANOVA with three groups. The three populations have normal distributions with the same standard deviation.

of the three normal distributions are the same, reflecting the assumption that all three populations have the same standard deviation.

EXAMPLE 12.4 A survey of college students attempted to determine how much money they spend per year on textbooks and to compare students by year of study. From lists of students provided by the registrar, SRSs of size 50 were chosen from each of the four classes (freshmen, sophomores, juniors, and seniors). The students selected were asked how much they had spent on textbooks during the current semester.

There are $I = 4$ populations. The population means μ_1, μ_2, μ_3, and μ_4 are the average amounts spent on textbooks by *all* freshmen, sophomores, juniors, and seniors at this college for this semester. The sample sizes n_i are 50, 50, 50, and 50.

Suppose the first freshman sampled is Eve Brogden. The observation $x_{1,1}$ is the amount spent by Eve. The data for the other freshmen sampled are denoted by $x_{1,2}, x_{1,3}, \ldots, x_{1,50}$. Similarly, the data for the other groups have a first subscript indicating the group and a second subscript indicating the student in the group.

According to our model, Eve's spending is $x_{1,1} = \mu_1 + \epsilon_{1,1}$, where μ_1 is the average for *all* of the students in the freshman class and $\epsilon_{1,1}$ is the chance variation due to Eve's specific needs. We are assuming that the ϵ_{ij} are independent and normally distributed (at least approximately) with mean 0 and standard deviation σ.

Estimates of population parameters

The unknown parameters in the statistical model for ANOVA are the I population means μ_i and the common population standard deviation σ. To estimate μ_i we use the sample mean for the ith group:

$$\overline{x}_i = \frac{1}{n_i}\sum_{j=1}^{n_i} x_{ij}$$

The residuals $e_{ij} = x_{ij} - \overline{x}_i$ reflect the variation about the sample means that we see in the data.

The ANOVA model assumes that the population standard deviations are all equal. If we have unequal standard deviations, we generally try to transform the data so that they are approximately equal. We might, for example, work with $\sqrt{x_{ij}}$ or $\log x_{ij}$. Fortunately, we can often find a transformation that *both* makes the group standard deviations more nearly equal and also makes the distributions of observations in each group more nearly normal. If the standard deviations are markedly different and cannot be made similar by a transformation, inference requires different methods.

Unfortunately, formal tests for the equality of standard deviations in several groups share the lack of robustness against nonnormality that we noted in Chapter 7 for the case of two groups. Because ANOVA procedures are not extremely sensitive to unequal standard deviations, we do *not* recommend a formal test of equality of standard deviations as a preliminary to the ANOVA. Instead, we will use the following rule of thumb.

Rule for Examining Standard Deviations in ANOVA

If the largest standard deviation is less than twice the smallest standard deviation, we can use methods based on the assumption of equal standard deviations and our results will still be approximately correct.[2]

When we assume that the population standard deviations are equal, each sample standard deviation is an estimate of σ. To combine these into a single estimate, we use a generalization of the pooling method introduced in Chapter 7.

Pooled Estimator of σ

Suppose we have sample variances $s_1^2, s_2^2, \ldots, s_I^2$ from I independent SRSs of sizes $n_1, n_2, \ldots, n_I$ from populations with common variance σ^2. The **pooled sample variance**

$$s_p^2 = \frac{(n_1 - 1)s_1^2 + (n_2 - 1)s_2^2 + \cdots + (n_I - 1)s_I^2}{(n_1 - 1) + (n_2 - 1) + \cdots + (n_I - 1)}$$

is an unbiased estimator of σ^2. The **pooled standard deviation**

$$s_p = \sqrt{s_p^2}$$

is the estimate of σ.

Pooling gives more weight to groups with larger sample sizes. If the sample sizes are equal, s_p^2 is just the average of the I sample variances. Note that s_p is *not* the average of the I sample standard deviations.

EXAMPLE 12.5

In the change-of-majors study of Example 12.3 there are $I = 3$ groups and the sample sizes are $n_1 = 103$, $n_2 = 31$, and $n_3 = 122$. The sample means for the SAT mathematics scores are $\bar{x}_1 = 619$, $\bar{x}_2 = 629$, and $\bar{x}_3 = 575$. The sample standard deviations are $s_1 = 86$, $s_2 = 67$, and $s_3 = 83$.

Because the largest standard deviation (86) is less than twice the smallest ($2 \times 67 = 134$), our rule of thumb indicates that we can use the assumption of equal population standard deviations.

The pooled variance estimate is

$$\begin{aligned} s_p^2 &= \frac{(n_1 - 1)s_1^2 + (n_2 - 1)s_2^2 + (n_3 - 1)s_3^2}{(n_1 - 1) + (n_2 - 1) + (n_3 - 1)} \\ &= \frac{(102)(86)^2 + (30)(67)^2 + (121)(83)^2}{102 + 30 + 121} \\ &= \frac{1{,}722{,}631}{253} = 6809 \end{aligned}$$

The pooled standard deviation is

$$s_p = \sqrt{6809} = 82.5$$

This is our estimate of the common standard deviation σ of the SAT mathematics scores in the three populations of students.

Testing hypotheses in one-way ANOVA

Comparison of several means is accomplished by using an F statistic to compare the variation among groups with the variation within groups. We now show how the F statistic expresses this comparison. Calculations are organized in an **ANOVA table,** which contains numerical measures of the variation among groups and within groups.

ANOVA table

First we must specify our hypotheses for one-way ANOVA. As usual, I represents the number of populations to be compared.

Hypotheses for One-Way ANOVA

The **null and alternative hypotheses** for one-way ANOVA are

$$H_0\colon \mu_1 = \mu_2 = \cdots = \mu_I$$
$$H_a\colon \text{not all of the } \mu_i \text{ are equal}$$

The example and discussion that follow illustrate how to do a one-way ANOVA. The calculations are generally performed using statistical software on a computer, so we focus on interpretation of the output.

EXAMPLE 12.6

A study of reading comprehension in children compared three methods of instruction.[3] As is common in such studies, several pretest variables were measured before any instruction was given. One purpose of the pretest was to see if the three groups of children were similar in their comprehension skills. One of the pretest variables was an "intruded sentences" measure, which measures one type of reading comprehension skill. The data for the 22 subjects in each group are given in Table 12.1. The three methods of instruction are called basal, DRTA, and strategies. We use Basal, DRTA, and Strat as values for the categorical variable indicating which method each student received.

Before proceeding with ANOVA, several important preliminaries need attention. Look at the normal quantile plots for the three groups in Figures 12.6(a), 12.6(b), and 12.6(c). The data look reasonably normal; we have no evidence to doubt our assumption that our residuals are normal. Although there are a small number of possible values for the score (the values are all integers between 4 and 17), this should not cause difficulties in using ANOVA. Side-by-side boxplots do not give a particularly good picture of these data. This is a consequence of the relatively small number of observations in each group (22) combined with the fact that the variable has only a few possible values.

TABLE 12.1 Pretest reading scores

Group	Subject	Score	Group	Subject	Score	Group	Subject	Score
Basal	1	4	DRTA	23	7	Strat	45	11
Basal	2	6	DRTA	24	7	Strat	46	7
Basal	3	9	DRTA	25	12	Strat	47	4
Basal	4	12	DRTA	26	10	Strat	48	7
Basal	5	16	DRTA	27	16	Strat	49	7
Basal	6	15	DRTA	28	15	Strat	50	6
Basal	7	14	DRTA	29	9	Strat	51	11
Basal	8	12	DRTA	30	8	Strat	52	14
Basal	9	12	DRTA	31	13	Strat	53	13
Basal	10	8	DRTA	32	12	Strat	54	9
Basal	11	13	DRTA	33	7	Strat	55	12
Basal	12	9	DRTA	34	6	Strat	56	13
Basal	13	12	DRTA	35	8	Strat	57	4
Basal	14	12	DRTA	36	9	Strat	58	13
Basal	15	12	DRTA	37	9	Strat	59	6
Basal	16	10	DRTA	38	8	Strat	60	12
Basal	17	8	DRTA	39	9	Strat	61	6
Basal	18	12	DRTA	40	13	Strat	62	11
Basal	19	11	DRTA	41	10	Strat	63	14
Basal	20	8	DRTA	42	8	Strat	64	8
Basal	21	7	DRTA	43	8	Strat	65	5
Basal	22	9	DRTA	44	10	Strat	66	8

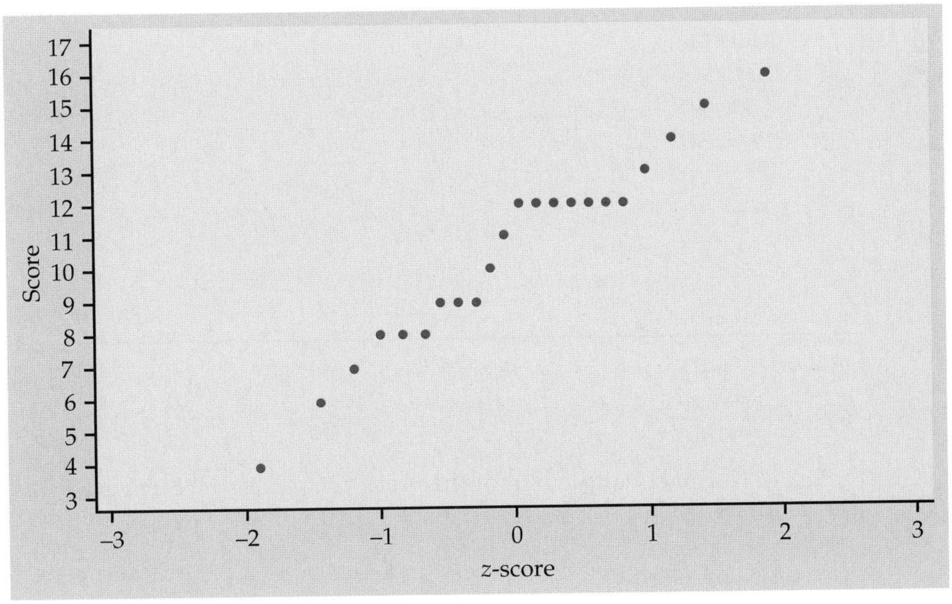

FIGURE 12.6(a) Normal quantile plot of the pretest scores in the Basal group of the reading comprehension study.

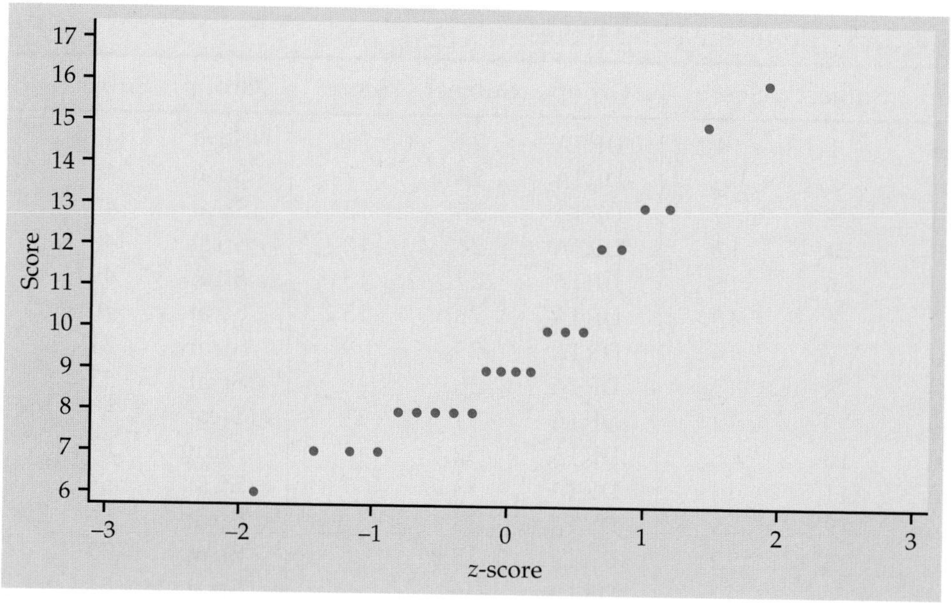

FIGURE 12.6(b) Normal quantile plot of the pretest scores in the DRTA group of the reading comprehension study.

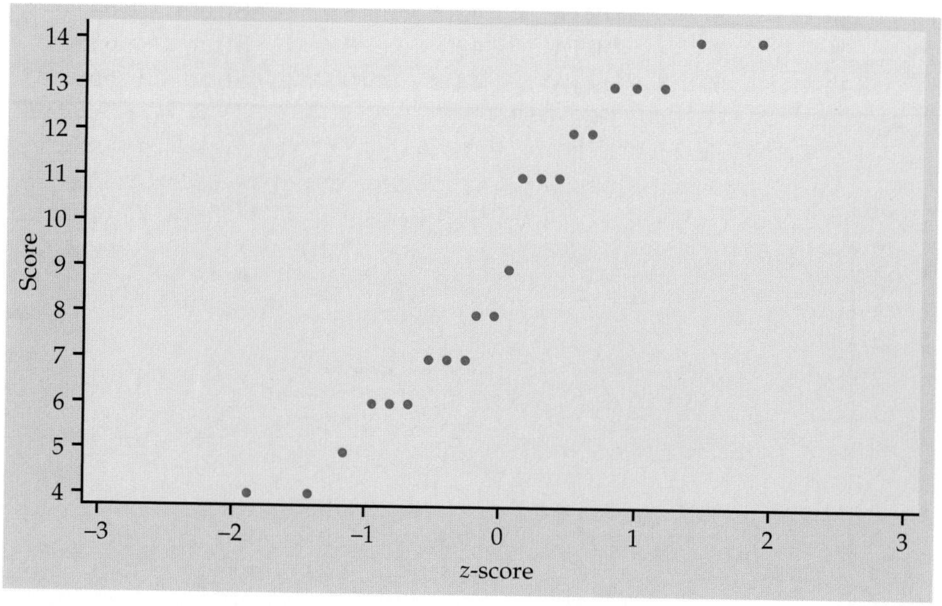

FIGURE 12.6(c) Normal quantile plot of the pretest scores in the Strat group of the reading comprehension study.

```
            Analysis Variable : SCORE

--------------------- GROUP=Basal --------------------

          N           Mean         Std Dev
         22     10.5000000       2.9720924

---------------------- GROUP=DRTA --------------------

          N           Mean         Std Dev
         22      9.7272727       2.6935871

--------------------- GROUP=Strat --------------------

          N           Mean         Std Dev
         22      9.1363636       3.3423039
```

FIGURE 12.7 Summary statistics for the pretest scores in the three groups of the reading comprehension study.

Figure 12.7 gives the summary statistics generated by SAS. Because the largest standard deviation (3.34) is less than twice the smallest ($2 \times 2.69 = 5.38$), our rule of thumb tells us that we need not be concerned about violating the assumption that the three populations have the same standard deviation.

Because the data look reasonably normal and the assumption of equal standard deviations can be used, we proceed with the analysis of variance. The results produced by SAS are shown in Figure 12.8. The pooled standard deviation s_p is given as Root MSE and has the value 3.01. The calculated value of the F statistic appears under the heading F Value, and its P-value is under the heading Pr > F. The value of F is 1.13, with a P-value of 0.3288. That is,

```
                    General Linear Models Procedure

Dependent Variable: SCORE

                                  Sum of             Mean
Source                  DF       Squares           Square   F Value     Pr > F
Model                    2     20.575758        10.287879      1.13     0.3288
Error                   63    572.454545         9.086580
Corrected Total         65    593.030303

                R-Square          C.V.         Root MSE           SCORE Mean
                0.034696      30.79723           3.0144               9.7879
```

FIGURE 12.8 Analysis of variance output for the pretest scores in the reading comprehension study.

an F of 1.13 or larger would occur about 33% of the time by chance when the population means are equal. We have no evidence to reject the null hypothesis that the three populations have equal means.

The ANOVA table

ANOVA table

The information in an analysis of variance is organized in an **ANOVA table.** In the computer output in Figure 12.8, the columns of this table are labeled Source, DF, Sum of Squares, Mean Square, F Value, and Pr > F. The rows are labeled Model, Error, and Corrected Total. These are the three sources of variation in the one-way ANOVA.

variation among groups

The Model row in the table corresponds to the FIT term in our DATA = FIT + RESIDUAL way of thinking. It gives information related to the variation **among** group means. In writing ANOVA tables we will use the generic label "groups" or some other term that describes the factor being studied for this row.

variation within groups

The Error row in the table corresponds to the RESIDUAL term in our DATA = FIT + RESIDUAL model. It gives information related to the variation **within** groups. The term "error" is most appropriate for experiments in the physical sciences, where the observations within a group differ because of measurement error. In business and the biological and social sciences, on the other hand, the within-group variation is often due to the fact that not all firms or plants or people are the same. This sort of variation is not due to errors and is better described as "residual." Finally, the Corrected Total row in the table corresponds to the DATA term in our DATA = FIT + RESIDUAL framework. Often this row is simply labeled Total. So, for analysis of variance,

$$\text{DATA} = \text{FIT} + \text{RESIDUAL}$$

translates into

$$\text{total} = \text{model} + \text{residual}$$

sum of squares

As you might expect, each **sum of squares** is a sum of squared deviations. We use SSG, SSE, and SST for the entries in this column corresponding to groups, error, and total. Each sum of squares measures a different type of variation. SST measures variation of the data around the overall mean, $x_{ij} - \overline{x}$. Variation of the group means around the overall mean $\overline{x}_i - \overline{x}$ is measured by SSG. Finally, SSE measures variation of each observation around its group mean, $x_{ij} - \overline{x}_i$.

The Sum of Squares column in Figure 12.8 gives the values for these three sums of squares. The output gives many more digits than we need. Rounded-off values are SSG = 21, SSE = 572, and SST = 593. In this example it appears that most of the variation is coming from Error, that is, from within groups. You can verify that SST = SSG + SSE for this example. This fact is true in general. The total variation is always equal to the among-group variation plus the within-group variation.

degrees of freedom

Associated with each sum of squares is a quantity called the **degrees of freedom.** Because SST measures the variation of all N observations around the overall mean, its degrees of freedom are DFT $= N - 1$. This is the same as the degrees of freedom for the ordinary sample variance. Similarly, because SSG measures the variation of the I sample means around the overall mean, its degrees of freedom are DFG $= I - 1$. Finally, SSE is the sum of squares of the deviations $x_{ij} - \overline{x}_i$. Here we have N observations being compared with I sample means and DFE $= N - I$.

In our example, we have $I = 3$ and $N = 66$. Therefore,

$$\begin{aligned}\text{DFT} &= N - 1 = 66 - 1 = 65\\ \text{DFG} &= I - 1 = 3 - 1 = 2\\ \text{DFE} &= N - I = 66 - 3 = 63\end{aligned}$$

These are the entries in the DF column of Figure 12.8. Note that the degrees of freedom add in the same way that the sums of squares add. That is, DFT = DFG + DFE.

mean square

For each source of variation, the **mean square** is the sum of squares divided by the degrees of freedom. You can verify this by doing the divisions for the values given on the output in Figure 12.8.

Sums of Squares, Degrees of Freedom, and Mean Squares

Sums of squares represent variation present in the data. They are calculated by summing squared deviations. In one-way ANOVA there are three **sources of variation:** groups, error, and total. The sums of squares are related by the formula

$$\text{SST} = \text{SSG} + \text{SSE}$$

Thus, the total variation is composed of two parts, one due to groups and one due to error.

Degrees of freedom are related to the deviations that are used in the sums of squares. The degrees of freedom are related in the same way as the sums of squares:

$$\text{DFT} = \text{DFG} + \text{DFE}$$

To calculate each **mean square,** divide the corresponding sum of squares by its degrees of freedom.

The pooled variance for Example 12.6 is $572.45/63 = 9.09$, the mean square in the error row in Figure 12.8. It is true in general that

$$s_p^2 = \text{MSE} = \frac{\text{SSE}}{\text{DFE}}$$

That is, the error mean square is an estimate of the within-group variance, σ^2. The output gives s_p under the heading Root MSE.

The F test

If H_0 is true, there are no differences among the group means. The ratio MSG/MSE is a statistic that is approximately 1 if H_0 is true and tends to be larger if H_a is true. This is the ANOVA F statistic. In our example, MSG = 10.29 and MSE = 9.09, so the ANOVA F statistic is

$$F = \frac{\text{MSG}}{\text{MSE}} = \frac{10.29}{9.09} = 1.13$$

When H_0 is true, the F statistic has an F distribution that depends upon two numbers: the *degrees of freedom for the numerator* and the *degrees of freedom for the denominator*. These degrees of freedom are those associated with the mean squares in the numerator and denominator of the F statistic. For one-way ANOVA, the degrees of freedom for the numerator are DFG $= I - 1$ and the degrees of freedom for the denominator are DFE $= N - I$. We use the notation $F(I - 1, N - I)$ for this distribution.

The ANOVA F Test

To test the null hypothesis in a one-way ANOVA, calculate the **F statistic**

$$F = \frac{\text{MSG}}{\text{MSE}}$$

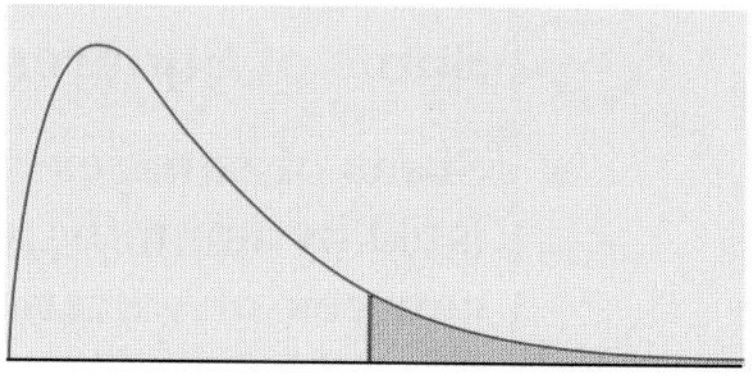

When H_0 is true, the F statistic has the $F(I - 1, N - I)$ distribution. When H_a is true, the F statistic tends to be large. We reject H_0 in favor of H_a if the F statistic is sufficiently large.

The ***P*-value** of the F test is the probability that a random variable having the $F(I - 1, N - I)$ distribution is greater than or equal to the calculated value of the F statistic.

Tables of F critical values are available for use when software does not give the P-value. Table E in the back of the book contains the F critical values for probabilities $p = 0.100$, 0.050, 0.025, 0.010, and 0.001. For one-way ANOVA we use critical values from the table corresponding to $I - 1$ degrees of freedom in the numerator and $N - I$ degrees of freedom in the denominator. We have already seen several examples where the F statistic and its P-value were used to choose between H_0 and H_a.

applet

The *One-Way ANOVA* applet, available on the Web site `www.whfreeman.com/ips` is an excellent way to see how the value of the F statistic and the P-value depend on the variability of the data within the groups and the differences among the means. See Exercises 12.64 and 12.65 for use of this applet.

EXAMPLE 12.7 In the study of computer science students in Example 12.3, $F = 10.35$. There were three populations, so the degrees of freedom in the numerator are DFG $= I - 1 = 2$. For this example the degrees of freedom in the denominator are DFE $= N - I = 256 - 3 = 253$. In Table E we first find the column corresponding to 2 degrees of freedom in the numerator. For the degrees of freedom in the denominator (DFD in the table), we see that there are entries for 200 and 1000. These entries are very close. To be conservative we use critical values corresponding to 200 degrees of freedom in the denominator since these are slightly larger.

p	Critical value
0.100	2.33
0.050	3.04
0.025	3.76
0.010	4.71
0.001	7.15

Because 10.35 is larger than all of the tabulated values, we reject H_0 and conclude that the differences in means are statistically significant with $P < 0.001$.

EXAMPLE 12.8 In the comparison of teaching methods in Example 12.6, $F = 1.13$. Because we are comparing three populations, the degrees of freedom in the numerator are DFG $= I - 1 = 2$, and the degrees of freedom in the denominator are DFE $= N - I = 66 - 3 = 63$. Table E has entries for 2 degrees of freedom in the numerator. We do not find entries for 63 degrees of freedom in the denominator, so we use the slightly larger values corresponding to 60.

p	Critical value
0.100	2.39
0.050	3.15
0.025	3.93
0.010	4.98
0.001	7.77

Because 1.13 is less than any of the tabulated values, we do not have strong evidence in the data against the hypothesis of equal means. We conclude that the differences in sample means are not significant at the 10% level. Statistical software (Figure 12.8) gives the exact value $P = 0.3288$.

Remember that the F test is always one-sided because any differences among the group means tend to make F large. The ANOVA F test shares the robustness of the two-sample t test. It is relatively insensitive to moderate nonnormality and unequal variances, especially when the sample sizes are similar.

The following display shows the general form of a one-way ANOVA table with the F statistic. The formulas in the sum of squares column can be used for calculations in small problems. There are other formulas that are more efficient for hand or calculator use, but ANOVA calculations are usually done by computer software.

Source	Degrees of freedom	Sum of squares	Mean square	F
Groups	$I - 1$	$\sum_{\text{groups}} n_i(\bar{x}_i - \bar{x})^2$	SSG/DFG	MSG/MSE
Error	$N - I$	$\sum_{\text{groups}}(n_i - 1)s_i^2$	SSE/DFE	
Total	$N - 1$	$\sum_{\text{obs}}(x_{ij} - \bar{x})^2$	SST/DFT	

EXAMPLE 12.9

In Example 12.8 we could not reject H_0 that the three groups of students in Example 12.6 had equal mean scores on a pretest of reading comprehension. We now change the data in a way that favors H_a. Add one point to the score of each student in the Basal group, and subtract one point from each of the observations in the Strat group. Leave the observations in the DRTA group as they were. Call the new variable SCOREX. The summary statistics for SCOREX appear in Figure 12.9. Compared with the original means, given in Figure 12.7, the SCOREX means are farther apart. Note that we have not changed any of the standard deviations.

The ANOVA results appear in Figure 12.10. Compare this table with the ANOVA for the original variable SCORE given in Figure 12.8. The degrees of freedom are the same. The values of SSG and SST have increased, but SSE has not changed. In constructing SCOREX, we have added more total variation to the data, and all of this added variation went into the variation between groups. The increase in SSG is reflected in MSG. The new value is about six times the old. Because MSE has not changed, the F statistic is also about six times as large. The P-value has decreased dramatically and we now have evidence to reject H_0 in favor of H_a.

```
Analysis Variable : SCOREX

---------------------- GROUP=Basal ---------------------

   N          Mean           Std Dev
  22    11.5000000         2.9720924

---------------------- GROUP=DRTA ----------------------

   N          Mean           Std Dev
  22     9.7272727         2.6935871

---------------------- GROUP=Strat ---------------------

   N          Mean           Std Dev
  22     8.1363636         3.3423039
```

FIGURE 12.9 Summary statistics for the adjusted scores SCOREX in the reading comprehension study.

General Linear Models Procedure

Dependent Variable: SCOREX

Source	DF	Sum of Squares	Mean Square	F Value	Pr > F
Model	2	124.57576	62.28788	6.85	0.0020
Error	63	572.45455	9.08658		
Corrected Total	65	697.03030			

R-Square	C.V.	Root MSE	SCOREX Mean
0.178724	30.79723	3.0144	9.7879

FIGURE 12.10 Analysis of variance output for SCOREX in the reading comprehension study.

coefficient of determination

One other item in the computer output for ANOVA is worth noting. For an analysis of variance, we define the **coefficient of determination** as

$$R^2 = \frac{\text{SSG}}{\text{SST}}$$

The coefficient of determination plays the same role as the squared multiple correlation R^2 in a multiple regression. In Figure 12.8, R^2 is listed under the heading R-Square. The value is 0.035. This result says that the FIT part of the model (that is, differences among the means of the groups) accounts for 3.5% of the total variation in the data. Compare this value with that for the changed data described in Example 12.9 and given in Figure 12.10. The change has increased the coefficient of determination to 17.9%.

12.2 Comparing the Means

Contrasts

The ANOVA F test gives a general answer to a general question: Are the differences among observed group means significant? Unfortunately, a small P-value simply tells us that the group means are not all the same. It does not tell us specifically which means differ from each other. Plotting and inspecting the means give us some indication of where the differences lie, but we would like to supplement inspection with formal inference.

In the ideal situation, specific questions regarding comparisons among the means are posed before the data are collected. We can answer specific questions of this kind and attach a level of confidence to the answers we give. We now explore these ideas through an example.

EXAMPLE 12.10

Example 12.6 describes a randomized comparative experiment to compare three methods for teaching reading. We analyzed the pretest scores and found no reason to reject H_0 that the three groups had similar population means on this measure.

This was the desired outcome. We now turn to the response variable, a measure of reading comprehension called COMP that was measured by a test taken after the instruction was completed.

Figure 12.11 gives the summary statistics for COMP computed by SAS. Side-by-side boxplots appear in Figure 12.12, and Figure 12.13 plots the group means. The

```
                Analysis Variable : COMP

--------------------- GROUP=Basal ---------------------

                N         Mean          Std Dev
               22    41.0454545       5.6355781

---------------------- GROUP=DRTA ---------------------

                N         Mean          Std Dev
               22    46.7272727       7.3884196

--------------------- GROUP=Strat ---------------------

                N         Mean          Std Dev
               22    44.2727273       5.7667505
```

FIGURE 12.11 Summary statistics for the comprehension scores in the three groups of the reading comprehension study.

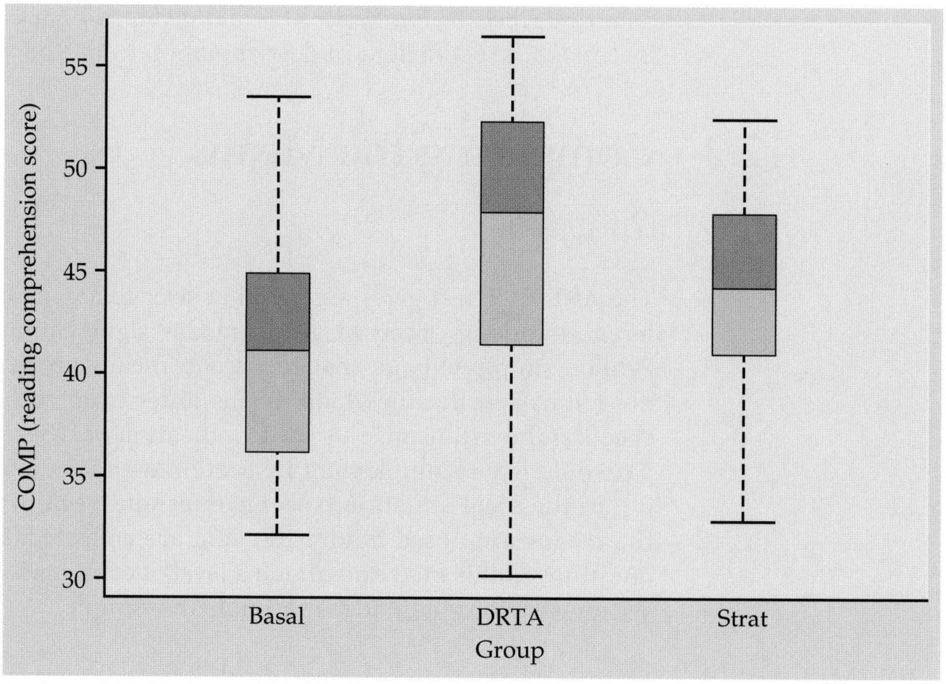

FIGURE 12.12 Side-by-side boxplots of the comprehension scores for the reading comprehension study.

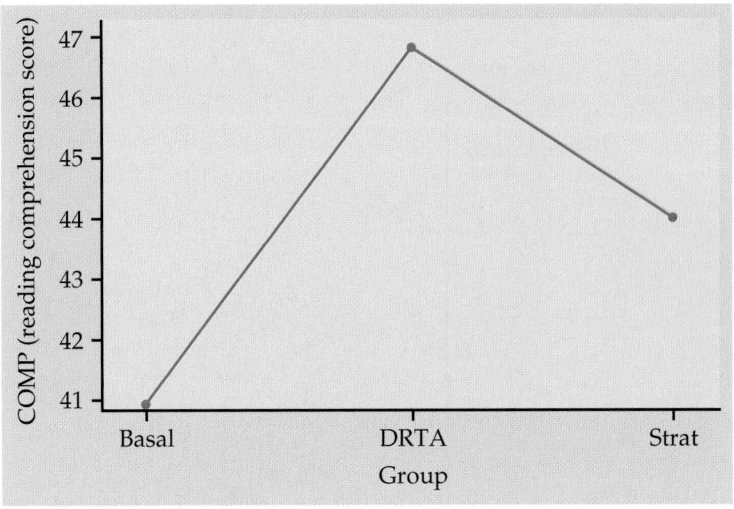

FIGURE 12.13 Comprehension score group means for the reading comprehension study.

ANOVA results generated by SAS are given in Figure 12.14, and a normal quantile plot of the residuals appears in Figure 12.15.

The ANOVA null hypothesis is

$$H_0: \mu_B = \mu_D = \mu_S$$

where the subscripts correspond to the group labels Basal, DRTA, and Strat. Figure 12.14 shows that $F = 4.48$ with degrees of freedom 2 and 63. The P-value is 0.0152. We have good evidence against H_0.

What can the researchers conclude from this analysis? The alternative hypothesis is true if $\mu_B \neq \mu_D$ or if $\mu_B \neq \mu_S$ or if $\mu_D \neq \mu_S$ or if any combination of these statements is true. We would like to be more specific.

```
                    General Linear Models Procedure

Dependent Variable: COMP

                                    Sum of            Mean
Source                 DF          Squares          Square   F Value     Pr > F
Model                   2        357.30303       178.65152      4.48     0.0152
Error                  63       2511.68182        39.86797
Corrected Total        65       2868.98485

                 R-Square             C.V.        Root MSE              COMP Mean
                 0.124540         14.34531          6.3141                 44.015

Contrast               DF      Contrast SS     Mean Square   F Value     Pr > F
B vs D and S            1        291.03030       291.03030      7.30     0.0088
D vs S                  1         66.27273        66.27273      1.65     0.2020
```

FIGURE 12.14 Analysis of variance and contrasts output for the comprehension scores in the reading comprehension study.

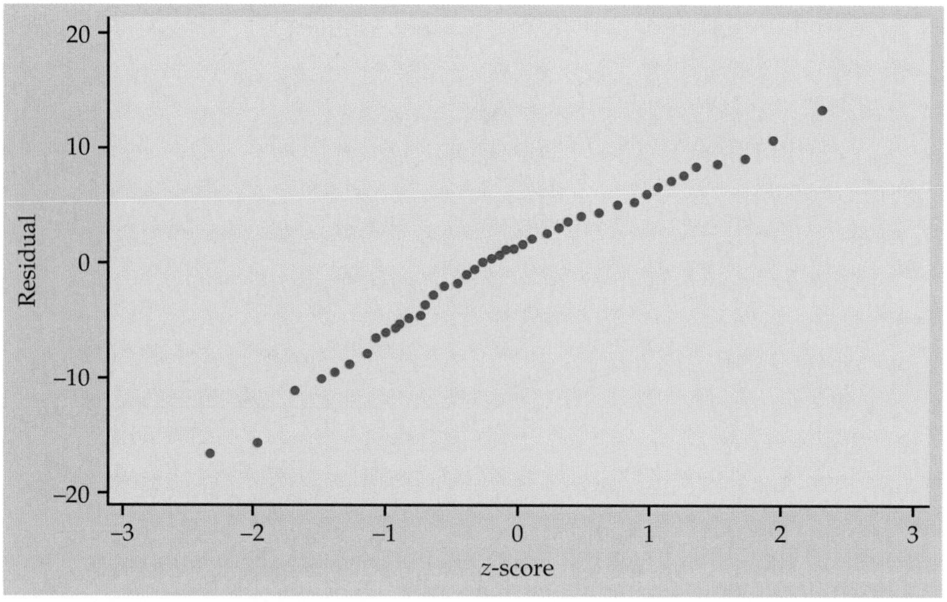

FIGURE 12.15 Normal quantile plot of the residuals for the comprehension scores in the reading comprehension study.

The researchers were investigating a specific theory about reading comprehension. The instruction for the Basal group was the standard method commonly used in schools. The DRTA and Strat groups received innovative methods of teaching that were designed to increase the reading comprehension of the children. The DRTA and Strat methods were not identical, but they both involved teaching the students to use similar comprehension strategies in their reading.

The ANOVA F test demonstrates that there is some difference among the mean responses, but it does not directly answer the researchers' questions.

EXAMPLE 12.11

The primary question can be formulated as a null hypothesis:

$$H_{01}: \frac{1}{2}(\mu_D + \mu_S) = \mu_B$$

with the alternative

$$H_{a1}: \frac{1}{2}(\mu_D + \mu_S) > \mu_B$$

The hypothesis H_{01} compares the average of the two innovative methods (DRTA and Strat) with the standard method (Basal). The alternative is one-sided because the researchers are interested in demonstrating that the new methods are better than the old. We use the subscripts 1 and 2 to distinguish the two sets of hypotheses.

EXAMPLE 12.12

A secondary question involves a comparison of the two new methods. We formulate this as the hypothesis that the methods DRTA and Strat are equally effective:

$$H_{02}: \mu_D = \mu_S$$

versus the alternative

$$H_{a2}: \mu_D \neq \mu_S$$

contrast ψ

Each of H_{01} and H_{02} says that a combination of population means is 0. These combinations of means are called **contrasts.** We use ψ, the Greek letter psi, for contrasts among population means. The two contrasts that arise from our two null hypotheses are

$$\begin{aligned}\psi_1 &= -\mu_\text{B} + \frac{1}{2}(\mu_\text{D} + \mu_\text{S}) \\ &= (-1)\mu_\text{B} + (0.5)\mu_\text{D} + (0.5)\mu_\text{S}\end{aligned}$$

and

$$\psi_2 = \mu_\text{D} - \mu_\text{S}$$

In each case, the value of the contrast is 0 when H_0 is true. Note that we have chosen to define the contrasts so that they will be positive when the alternative of interest (what we expect) is true. Whenever possible, this is a good idea because it makes some computations easier.

sample contrast

A contrast expresses an effect in the population as a combination of population means. To estimate the contrast, form the corresponding **sample contrast** by using sample means in place of population means. Under the ANOVA assumptions, a sample contrast is a linear combination of independent normal variables and therefore has a normal distribution. We can obtain the standard error of a contrast by using the rules for variances given in Section 4.4. Inference is based on t statistics. Here are the details.

Contrasts

A **contrast** is a combination of population means of the form

$$\psi = \sum a_i \mu_i$$

where the coefficients a_i have sum 0. The corresponding **sample contrast** is

$$c = \sum a_i \bar{x}_i$$

The **standard error of c** is

$$\text{SE}_c = s_p \sqrt{\sum \frac{a_i^2}{n_i}}$$

To test the null hypothesis

$$H_0: \psi = 0$$

use the ***t* statistic**

$$t = \frac{c}{\text{SE}_c}$$

with degrees of freedom DFE that are associated with s_p. The alternative hypothesis can be one-sided or two-sided.

A **level C confidence interval for ψ** is

$$c \pm t^* \text{SE}_c$$

where t^* is the value for the t(DFE) density curve with area C between $-t^*$ and t^*.

Because each $\overline{x}_i$ estimates the corresponding μ_i, the addition rule for means tells us that the mean μ_c of the sample contrast c is ψ. In other words, c is an unbiased estimator of ψ. Testing the hypothesis that a contrast is 0 assesses the significance of the effect measured by the contrast. It is often more informative to estimate the size of the effect using a confidence interval for the population contrast.

EXAMPLE 12.13

In our example the coefficients in the contrasts are $a_1 = -1, a_2 = 0.5, a_3 = 0.5$ for ψ_1 and $a_1 = 0, a_2 = 1, a_3 = -1$ for ψ_2, where the subscripts 1, 2, and 3 correspond to B, D, and S. In each case the sum of the a_i is 0. We look at inference for each of these contrasts in turn.

EXAMPLE 12.14

The sample contrast that estimates ψ_1 is

$$\begin{aligned} c_1 &= \overline{x}_B + \frac{1}{2}(\overline{x}_D + \overline{x}_S) \\ &= -41.05 + \frac{1}{2}(46.73 + 44.27) = 4.45 \end{aligned}$$

with standard error

$$\begin{aligned} SE_{c_1} &= 6.314\sqrt{\frac{(-1)^2}{22} + \frac{(0.5)^2}{22} + \frac{(0.5)^2}{22}} \\ &= 1.65 \end{aligned}$$

The t statistic for testing $H_{01}: \psi_1 = 0$ versus $H_{a1}: \psi_1 > 0$ is

$$\begin{aligned} t &= \frac{c_1}{SE_{c_1}} \\ &= \frac{4.45}{1.65} = 2.70 \end{aligned}$$

Because s_p has 63 degrees of freedom, software using the $t(63)$ distribution gives the one-sided P-value as 0.0044. If we used Table D, we would conclude that $P < 0.005$. The P-value is small, so there is strong evidence against H_{01}. The researchers have shown that the new methods produce higher mean scores than the old. The size of the improvement can be described by a confidence interval. To find the 95% confidence interval for ψ_1, we combine the estimate with its margin of error:

$$\begin{aligned} c_1 \pm t^* SE_{c_1} &= 4.45 \pm (2.00)(1.65) \\ &= 4.45 \pm 3.30 \end{aligned}$$

The interval is (1.15, 7.75). We are 95% confident that the mean improvement obtained by using the innovative methods rather than the old method is between 1.15 and 7.75 points.

EXAMPLE 12.15

The second sample contrast, which compares the two new methods, is

$$\begin{aligned} c_2 &= 46.73 - 44.27 \\ &= 2.46 \end{aligned}$$

with standard error

$$\begin{aligned} SE_{c_2} &= 6.314\sqrt{\frac{(1)^2}{22} + \frac{(-1)^2}{22}} \\ &= 1.90 \end{aligned}$$

The t statistic for assessing the significance of this contrast is

$$t = \frac{2.46}{1.90} = 1.29$$

The P-value for the two-sided alternative is 0.2020. We conclude that either the two new methods have the same population means or the sample sizes are not sufficiently large to distinguish them. A confidence interval helps clarify this statement. To find the 95% confidence interval for ψ_2, we combine the estimate with its margin of error:

$$\begin{aligned} c_2 \pm t^* \mathrm{SE}_{c_2} &= 2.46 \pm (2.00)(1.90) \\ &= 2.46 \pm 3.80 \end{aligned}$$

The interval is $(-1.34,\ 6.26)$. With 95% confidence we state that the difference between the population means for the two new methods is between -1.34 and 6.26.

Many statistical software packages report the test statistics associated with contrasts as F statistics rather than t statistics. These F statistics are the squares of the t statistics described above. SAS output for contrasts is given at the bottom of Figure 12.14. The F statistics 7.30 and 1.65 for our two contrasts appear in the F Value column. These are the squares of the t-values 2.70 and 1.29 (up to roundoff error). The degrees of freedom for each F are 1 and 63. The P-values are correct for two-sided alternative hypotheses. They are 0.0088 and 0.2020. To convert the computer-generated results to apply to our one-sided hypothesis concerning ψ_1, simply divide the reported P-value by 2 after checking that the value of c is in the direction of H_a (that is, that c is positive).

Questions about population means are expressed as hypotheses about contrasts. A contrast should express a specific question that we have in mind when designing the study. When contrasts are formulated before seeing the data, *inference about contrasts is valid whether or not the ANOVA H_0 of equality of means is rejected.* Because the F test answers a very general question, it is less powerful than tests for contrasts designed to answer specific questions. Specifying the important questions before the analysis is undertaken enables us to use this powerful statistical technique.

Multiple comparisons

multiple comparisons

In many studies, specific questions cannot be formulated in advance of the analysis. If H_0 is not rejected, we conclude that the population means are indistinguishable on the basis of the data given. On the other hand, if H_0 is rejected, we would like to know which pairs of means differ. **Multiple comparisons** methods address this issue. It is important to keep in mind that multiple comparisons methods are used only *after rejecting* the ANOVA H_0.

EXAMPLE 12.16

Return once more to the reading comprehension data given in Example 12.10. There are three pairs of population means, comparing groups 1 and 2, 1 and 3, and 2 and 3. We can write a t statistic for each of these pairs. For example, the statistic

$$\begin{aligned} t_{12} &= \frac{\overline{x}_1 - \overline{x}_2}{s_p\sqrt{\frac{1}{n_1} + \frac{1}{n_2}}} \\ &= \frac{41.05 - 46.73}{6.31\sqrt{\frac{1}{22} + \frac{1}{22}}} \\ &= -2.99 \end{aligned}$$

compares populations 1 and 2. The subscripts on t specify which groups are being compared.

The t statistics for the other two pairs are

$$\begin{aligned} t_{13} &= \frac{\overline{x}_1 - \overline{x}_3}{s_p\sqrt{\frac{1}{n_1} + \frac{1}{n_3}}} \\ &= \frac{41.05 - 44.27}{6.31\sqrt{\frac{1}{22} + \frac{1}{22}}} \\ &= -1.69 \end{aligned}$$

and

$$\begin{aligned} t_{23} &= \frac{\overline{x}_2 - \overline{x}_3}{s_p\sqrt{\frac{1}{n_2} + \frac{1}{n_3}}} \\ &= \frac{46.73 - 44.27}{6.31\sqrt{\frac{1}{22} + \frac{1}{22}}} \\ &= 1.29 \end{aligned}$$

We performed the last calculation when we analyzed the contrast ψ_2 in the previous section, because that contrast was $\mu_2 - \mu_3$. These t statistics are very similar to the pooled two-sample t statistic for comparing two population means which we described in Chapter 7 (page 538). The difference is that we now have more than two populations, so each statistic uses the pooled estimator s_p from all groups rather than the pooled estimator from just the two groups being compared. This additional information about the common σ increases the power of the tests. The degrees of freedom for all of these statistics are DFE $= 63$, those associated with s_p.

Because we do not have any specific ordering of the means in mind as an alternative to equality, we must use a two-sided approach to the problem of deciding which pairs of means are significantly different.

Multiple Comparisons

To perform a **multiple comparisons procedure,** compute ***t* statistics** for all pairs of means using the formula

$$t_{ij} = \frac{\bar{x}_i - \bar{x}_j}{s_p\sqrt{\frac{1}{n_i} + \frac{1}{n_j}}}$$

If

$$|t_{ij}| \geq t^{**}$$

we declare that the population means μ_i and μ_j are different. Otherwise, we conclude that the data do not distinguish between them. The value of t^{**} depends upon which multiple comparisons procedure we choose.

LSD method

One obvious choice for t^{**} is the upper $\alpha/2$ critical value for the $t(\text{DFE})$ distribution. This choice simply carries out as many separate significance tests of fixed level α as there are pairs of means to be compared. The procedure based on this choice is called the **least-significant differences** method, or simply LSD. LSD has some undesirable properties, particularly if the number of means being compared is large. Suppose, for example, that there are $I = 20$ groups and we use LSD with $\alpha = 0.05$. There are 190 different pairs of means. If we perform 190 t tests, each with an error rate of 5%, our overall error rate will be unacceptably large. We expect about 5% of the 190 to be significant even if the corresponding population means are the same. Since 5% of 190 is 9.5, we expect 9 or 10 false rejections.

The LSD procedure fixes the probability of a false rejection for each single pair of means being compared. It does not control the overall probability of *some* false rejection among all pairs. Other choices of t^{**} control possible errors in other ways. The choice of t^{**} is therefore a complex problem and a detailed discussion of it is beyond the scope of this text. Many choices for t^{**} are used in practice. One major statistical package allows selection from a list of over a dozen choices.

Bonferroni method

We will discuss only one of these, called the **Bonferroni method.** Use of this procedure with $\alpha = 0.05$, for example, guarantees that the probability of *any* false rejection among all comparisons made is no greater than 0.05. This is much stronger protection than controlling the probability of a false rejection at 0.05 for *each separate* comparison.

minimum significant difference, MSD

Many computer programs display the output for multiple comparisons procedures as a table of the group means along with the smallest difference between two means that is statistically significant, called the **minimum significant difference,** or MSD. The MSD depends on the type of multiple comparison we choose. Any two sample means that differ by more than the MSD are significantly different. That is, we conclude that the corresponding population means are different. In our notation, the MSD is

$$\text{MSD} = t^{**}s_p\sqrt{\frac{1}{n_i} + \frac{1}{n_j}}$$

EXAMPLE 12.17

We apply the Bonferroni multiple comparisons procedure with $\alpha = 0.05$ to the data from the reading comprehension study in Example 12.10. The value of t^{**} for this procedure (from software or special tables) is 2.46. Of the statistics $t_{12} = -2.99$, $t_{13} = -1.69$, and $t_{23} = 1.29$ calculated in the beginning of this section, only t_{12} is significant.

The output generated by SAS for Bonferroni comparisons appears in Figure 12.16. The critical value t^{**} is given on the output as Critical Value of T. The MSD is 4.6825. The output shows which group means *are not* significantly different. The letters A under the heading Bon Grouping extend from the DRTA group to the Strat group. This means that these two groups are not distinguishable. Similarly, the letters B extend from the Strat group to the Basal group, indicating that these two groups also cannot be distinguished. There is no collection of letters that connects the DRTA and Basal groups. This means that we can declare these two means to be significantly different.

This conclusion appears to be illogical. If μ_1 is the same as μ_3, and μ_2 is the same as μ_3, doesn't it follow that μ_1 is the same as μ_2? Logically, the answer must be yes.

This apparent contradiction points out dramatically the nature of the conclusions of statistical tests of significance. Some of the difficulty can be resolved by noting the choice of words used above. In describing the inferences, we talk about being able or unable to distinguish between pairs of means. In making the logical statements, we say, "is the same as." There is a big dif-

```
               General Linear Models Procedure

          Bonferroni (Dunn) T tests for variable: COMP

NOTE: This test controls the type I experimentwise error rate,
      but generally has a higher type II error rate than REGWQ.

              Alpha= 0.05  df= 63   MSE= 39.86797
                  Critical Value of T= 2.46
            Minimum Significant Difference= 4.6825

Means with the same letter are not significantly different.

        Bon Grouping             Mean       N   GROUP

                     A         46.727      22   DRTA
                     A
              B      A         44.273      22   Strat
              B
              B                41.045      22   Basal
```

FIGURE 12.16 Bonferroni multiple comparisons output for the comprehension scores in the reading comprehension study.

ference between the two modes of thought. Statistical tests ask, "Do we have adequate evidence to distinguish two means?" It is not illogical to conclude that we have evidence to distinguish μ_1 from μ_3, but not μ_1 from μ_2 or μ_2 from μ_3. It is very unlikely that any two methods of teaching reading comprehension would give *exactly* the same population means, but the data can fail to provide good evidence of a difference.

One way to deal with these difficulties of interpretation is to give confidence intervals for the differences. The intervals remind us that the differences are not known exactly. We want to give **simultaneous confidence intervals,** that is, intervals for *all* differences among the population means at once. Again, we must face the problem that there are many competing procedures—in this case, many methods of obtaining simultaneous intervals.

simultaneous confidence intervals

Simultaneous Confidence Intervals for Differences between Means

Simultaneous confidence intervals for all differences $\mu_i - \mu_j$ between population means have the form

$$(\bar{x}_i - \bar{x}_j) \pm t^{**} s_p \sqrt{\frac{1}{n_i} + \frac{1}{n_j}}$$

The critical values t^{**} are the same as those used for the multiple comparisons procedure chosen.

The confidence intervals generated by a particular choice of t^{**} are closely related to the multiple comparisons results for that same method. If one of the confidence intervals includes the value 0, then that pair of means will not be declared significantly different, and vice versa.

EXAMPLE 12.18

For simultaneous 95% Bonferroni confidence intervals SAS gives the output in Figure 12.17 for the data in Example 12.10. We are 95% confident that *all three* intervals simultaneously contain the true values of the population mean differences.

Software

We have used SAS to illustrate the analysis of the reading comprehension data. Other statistical software gives similar output and you should be able to read it without any difficulty.

EXAMPLE 12.19

Research suggests that customers think that a product is of high quality if it is heavily advertised. An experiment designed to explore this idea collected quality ratings (on a 1 to 7 scale) of a new line of take-home refrigerated entrées based on reading

```
        Bonferroni (Dunn) T tests for variable: COMP

    NOTE: This test controls the type I experimentwise error rate but
          generally has a higher type II error rate than Tukey's for
          all pairwise comparisons.

       Alpha= 0.05  Confidence= 0.95  df= 63  MSE= 39.86797
                    Critical Value of T= 2.45958
               Minimum Significant Difference= 4.6825

Comparisons significant at the 0.05 level are indicated by '***'.

                     Simultaneous              Simultaneous
                         Lower     Difference     Upper
       GROUP          Confidence    Between    Confidence
     Comparison          Limit       Means        Limit

  DRTA  - Strat        -2.228       2.455        7.137
  DRTA  - Basal         0.999       5.682       10.364    ***

  Strat - DRTA         -7.137      -2.455        2.228
  Strat - Basal        -1.455       3.227        7.910

  Basal - DRTA        -10.364      -5.682       -0.999    ***
  Basal - Strat        -7.910      -3.227        1.455
```

FIGURE 12.17 Bonferroni simultaneous confidence intervals output for the comprehension scores in the reading comprehension study.

a magazine ad. Three groups were compared. The first group's ad included information that would undermine (U) the expected positive association between quality and advertising; the second group's ad contained information that would affirm (A) the association; and the third group was a control (C).[4] The data are given in Table 12.2. Outputs from SAS, Excel, Minitab, and the TI-83 are given in Figure 12.18.

TABLE 12.2 Quality ratings in three groups

Group	Quality ratings																			
Undermine	6	5	5	5	4	5	4	6	5	5	5	5	3	3	5	4	5	5	5	4
($n = 55$)	5	4	4	5	4	4	4	5	5	5	4	5	5	5	4	4	5	5	4	5
	5	4	4	5	4	3	4	5	5	5	3	4	4	4	4					
Affirm	4	6	4	6	5	5	5	6	4	5	5	5	4	6	6	5	5	7	4	6
($n = 36$)	6	4	5	4	5	5	6	4	5	5	4	6	4	6	5	5				
Control	5	4	5	6	5	7	5	6	7	5	7	5	4	5	4	4	6	6	5	6
($n = 36$)	5	5	4	5	5	6	6	6	5	6	6	7	6	6	5	5				

SAS

```
General Linear Models Procedure

Dependent Variable: QUALITY
                                   Sum of            Mean
Source                  DF        Squares          Square   F Value     Pr > F
Model                    2      18.828255        9.414127     15,28     0.0001
Error                  124      76.384343        0.616003
Corrected Total        126      95.212598

                  R-Square           C.V.        Root MSE          QUALITY Mean
                  0.197750       15.94832          0.7849                4.9213

Level of         -----------QUALITY-----------
     GROUP            N          Mean                    SD

     affirm          36      5.05555556            0.82615960
     control         36      5.41666667            0.87423436
     under           55      4.50909091            0.69048365
```

Excel

	A	B	C	D	E	F	G
1	Anova Single Factor						
2							
3	SUMMARY						
4	*Groups*	*Count*	*Sum*	*Average*	*Variance*		
5	Column 1	55	248	4.509091	0.476768		
6	Column 2	36	182	5.055556	0.68254		
7	Column 3	36	195	5.416667	0.764286		
8							
9							
10	ANOVA						
11	*Source of Variation*	*SS*	*df*	*MS*	*F*	*P-value*	*F crit*
12	Between Groups	18.82825	2	9.414127	15.28261	1.17E-06	3.069289
13	Within Groups	76.38434	124	0.616003			
14							
15	Total	95.2126	126				

Sheet17 | Sheet18 | Sheet 1 | Sheet 2 | Sheet 3

Ready Sum=0

Minitab

```
One-Way Analysis of Variance

Analysis of Variance
Source     DF        SS        MS        F        P
Factor      2    18.828     9.414    15.28   0.0000
Error     124    76.384     0.616
Total     126    95.123
                                    Individual 95% CIs For Mean
                                    Based on Pooled StDev
Level       N      Mean     StDev   ---+---------+---------+---------+---
Undermin   55    4.5091    0.6905   (-----*----)
Affirm     36    5.0556    0.8262              (-----*------)
Control    36    5.4167    0.8742                      (-----*------)
                                    ---+---------+---------+---------+---
Pooled StDev =   0.7849               4.40      4.80      5.20      5.60
```

FIGURE 12.18 SAS, Excel, and Minitab output for the advertising study. *(continued)*

TI-83

```
One-way ANOVA
 F=15.28260579
 p=1.1673896E-6
 Factor
  df=2
  SS=18.828255
  MS=9.4141275
 Error
  df=124
  SS=76.3843434
  MS=.61600277
 Sxp=.784858439
```

FIGURE 12.18 (*continued*) TI-83 output for the advertising study.

Power*

Recall that the power of a test is the probability of rejecting H_0 when H_a is in fact true. Power measures how likely a test is to detect a specific alternative. When planning a study in which ANOVA will be used for the analysis, it is important to perform power calculations to check that the sample sizes are adequate to detect differences among means that are judged to be important. Power calculations also help evaluate and interpret the results of studies in which H_0 was not rejected. We sometimes find that the power of the test was so low against reasonable alternatives that there was little chance of obtaining a significant F.

In Chapter 7 we found the power for the two-sample t test. One-way ANOVA is a generalization of the two-sample t test, so it is not surprising that the procedure for calculating power is quite similar. Here are the steps that are needed:

1. Specify
   - **(a)** an alternative (H_a) that you consider important; that is, values for the true population means $\mu_1, \mu_2, \ldots, \mu_I$;
   - **(b)** sample sizes $n_1, n_2, \ldots, n_I$; usually these will all be equal to the common value n;
   - **(c)** a level of significance α, usually equal to 0.05; and
   - **(d)** a guess at the standard deviation σ.
2. Find the degrees of freedom DFG $= I - 1$ and DFE $= N - I$ and the critical value that will lead to rejection of H_0. This value, which we denote by F^*, is the upper α critical value for the F(DFG, DFE) distribution.

*This section is optional.

noncentrality parameter

3. Calculate the **noncentrality parameter**[5]

$$\lambda = \frac{\sum n_i(\mu_i - \overline{\mu})^2}{\sigma^2}$$

where $\overline{\mu}$ is a weighted average of the group means,

$$\overline{\mu} = \sum w_i \mu_i$$

and the weights are proportional to the sample sizes,

$$w_i = \frac{n_i}{\sum n_i} = \frac{n_i}{N}$$

noncentral F distribution

4. Find the power, which is the probability of rejecting H_0 when the alternative hypothesis is true, that is, the probability that the observed F is greater than F^*. Under H_a, the F statistic has a distribution known as the **noncentral *F* distribution.** Using the SAS function for this distribution, we find the power as

$$\text{Power} = 1 - \text{PROBF}(F^*, \text{DFG}, \text{DFE}, \lambda)$$

Note that if the n_i are all equal to the common value n, then $\overline{\mu}$ is the ordinary average of the μ_i and

$$\lambda = \frac{n\sum(\mu_i - \overline{\mu})^2}{\sigma^2}$$

If the means are all equal (the ANOVA H_0), then $\lambda = 0$. The noncentrality parameter measures how unequal the given set of means is. Large λ points to an alternative far from H_0, and we expect the ANOVA F test to have high power. Software makes calculation of the power quite easy, but tables and charts are also available.

EXAMPLE 12.20

Suppose that a study on reading comprehension similar to the one described in Example 12.10 has 10 students in each group. How likely is this study to detect differences in the mean responses similar to those observed in the actual study? Based on the results of the actual study, we will use $\mu_1 = 41$, $\mu_2 = 47$, $\mu_3 = 44$, and $\sigma = 7$ in a calculation of power. The n_i are equal, so $\overline{\mu}$ is simply the average of the μ_i:

$$\overline{\mu} = \frac{41 + 47 + 44}{3} = 44$$

The noncentrality parameter is therefore

$$\begin{aligned}\lambda &= \frac{n\sum(\mu_i - \overline{\mu})^2}{\sigma^2} \\ &= \frac{(10)[(41-44)^2 + (47-44)^2 + (44-44)^2]}{49} \\ &= \frac{(10)(18)}{49} = 3.67\end{aligned}$$

Because there are three groups with 10 observations per group, DFG = 2 and DFE = 27. The critical value for $\alpha = 0.05$ is $F^* = 3.35$. The power is therefore

$$1 - \text{PROBF}(3.35, 2, 27, 3.67) = 0.3486$$

The chance that we reject the ANOVA H_0 at the 5% significance level is only about 35%.

If the assumed values of the μ_i in this example describe differences among the groups that the experimenter wants to detect, then we would want to use more than 10 subjects per group. Although H_0 is assumed to be false, the chance of rejecting it is only about 35%. This chance can be increased to acceptable levels by increasing the sample sizes.

EXAMPLE 12.21

To decide on an appropriate sample size for the experiment described in the previous example, we repeat the power calculation for different values of n, the number of subjects in each group. Here are the results:

n	DFG	DFE	F^*	λ	Power
20	2	57	3.16	7.35	0.65
30	2	87	3.10	11.02	0.84
40	2	117	3.07	14.69	0.93
50	2	147	3.06	18.37	0.97
100	2	297	3.03	36.73	≈ 1

With $n = 40$ the experimenters have a 93% chance of rejecting H_0 with $\alpha = 0.05$ and thereby demonstrating that the groups have different means. In the long run, 93 out of every 100 such experiments would reject H_0 at the $\alpha = 0.05$ level of significance. Using 50 subjects per group increases the chance of finding significance to 97%. With 100 subjects per group, the experimenters are virtually certain to reject H_0. The exact power for $n = 100$ is 0.99989. In most real-life situations the additional cost of increasing the sample size from 50 to 100 subjects per group would not be justified by the relatively small increase in the chance of obtaining statistically significant results.

SUMMARY

One-way analysis of variance (ANOVA) is used to compare several population means based on independent SRSs from each population. The populations are assumed to be normal with possibly different means and the same standard deviation.

To do an analysis of variance, first compute sample means and standard deviations for all groups. Side-by-side boxplots give an overview of the data. Examine normal quantile plots (either for each group separately or for the residuals) to detect outliers or extreme deviations from normality. Compute the ratio of the largest to the smallest sample standard deviation. If this ratio is less than 2 and the normal quantile plots are satisfactory, ANOVA can be performed.

The **null hypothesis** is that the population means are *all* equal. The **alternative hypothesis** is true if there are *any* differences among the population means.

ANOVA is based on separating the total variation observed in the data into two parts: variation **among group means** and variation **within groups.** If the variation among groups is large relative to the variation within groups, we have evidence against the null hypothesis.

An **analysis of variance table** organizes the ANOVA calculations. **Degrees of freedom, sums of squares, and mean squares** appear in the table. The ***F* statistic** and its ***P*-value** are used to test the null hypothesis.

Specific questions formulated before examination of the data can be expressed as **contrasts.** Tests and confidence intervals for contrasts provide answers to these questions.

If no specific questions are formulated before examination of the data and the null hypothesis of equality of population means is rejected, **multiple comparisons** methods are used to assess the statistical significance of the differences between pairs of means.

The **power** of the F test depends upon the sample sizes, the variation among population means, and the within-group standard deviation.

CHAPTER 12 EXERCISES

12.1 For each of the following situations, identify the response variable and the populations to be compared, and give I, the n_i, and N.

(a) A company wants to compare three different training programs for its new employees. In a one-month period there are 90 new hires. One-third of these are randomly assigned to each of the three training programs. At the end of the program the employees are asked to rate the effectiveness of the program on a 7-point scale.

(b) A marketing experiment compares six different types of packaging for computer disks. Each package is shown to 50 different potential consumers, who rate the attractiveness of the product on a 1 to 10 scale.

(c) Four different new formulations for a hand lotion have been produced by your research and development group, and you want to decide which of these, if any, to market. Samples of the first new lotion are sent to 100 randomly selected customers who use your regular product. The same procedure is followed for each of the other three new lotions. You ask each customer to compare the new lotion sent to them with the regular product by rating it on a 7-point scale. The middle point of the scale corresponds to no preference, while higher values indicate that the new product is preferred and lower values indicate that the regular product is better.

12.2 For each of the following situations, identify the response variable and the populations to be compared, and give I, the n_i, and N.

(a) A company wants to compare three different water treatment devices that can be attached to a kitchen faucet. From a list of potential customers, they select 225 households who will receive free samples. One-third of the households will receive each of the devices. The household is asked to rate the likelihood that they would buy this kind of device on a 5-point scale.

(b) The strength of concrete depends upon the formula used to prepare it. One study compared five different mixtures. Six batches of each mixture were prepared, and the strength of the concrete made from each batch was measured.

(c) Which of three methods of teaching statistics is most effective? Twenty students are randomly assigned to each of the methods, and their scores on a final exam are recorded.

12.3 Refer to Exercise 12.1. For each situation, give the following:

(a) Degrees of freedom for the model, for error, and for the total.

(b) Null and alternative hypotheses.

(c) Numerator and denominator degrees of freedom for the F statistic.

12.4 Refer to Exercise 12.2. For each situation, give the following:

(a) Degrees of freedom for the model, for error, and for the total.

(b) Null and alternative hypotheses.

(c) Numerator and denominator degrees of freedom for the F statistic.

12.5 An experiment was run to compare three groups. The sample sizes were 10, 12, and 14, and the corresponding estimated standard deviations were 18, 24, and 20.

(a) Is it reasonable to use the assumption of equal standard deviations when we analyze these data?

(b) Give the values of the variances for the three groups.

(c) Find the pooled variance.

(d) What is the value of the pooled standard deviation?

12.6 An experiment was run to compare four groups. The sample sizes were 20, 220, 18, and 15, and the corresponding estimated standard deviations were 62, 40, 52, and 48.

(a) Is it reasonable to use the assumption of equal standard deviations when we analyze these data?

(b) Give the values of the variances for the four groups.

(c) Find the pooled variance.

(d) What is the value of the pooled standard deviation?

(e) Explain why your answer in part (c) is much closer to the standard deviation for the second group than to any of the other standard deviations.

12.7 For each of the following situations find the degrees of freedom for the F statistic and then use Table E to approximate the P-value or use computer software to obtain an exact value.

(a) Three groups are being compared, with 8 observations per group. The value of the F statistic is 5.82.

(b) Six groups are being compared, with 11 observations per group. The value of the F statistic is 2.16.

12.8 For each of the following situations find the degrees of freedom for the F statistic and then use Table E to approximate the P-value or use computer software to obtain an exact value.

(a) Five groups are being compared, with 13 observations per group. The value of the F statistic is 1.61.

(b) Ten groups are being compared, with 4 observations per group. The value of the F statistic is 4.68.

12.9 Return to the change-of-majors study described in Example 12.3 (page 750).

(a) State H_0 and H_a for ANOVA.

(b) Outline the ANOVA table, giving the sources of variation and the degrees of freedom.

(c) What is the distribution of the F statistic under the assumption that H_0 is true?

(d) Using Table E, find the critical value for an $\alpha = 0.05$ test.

12.10 Return to the survey of college students described in Example 12.4 (page 754).

(a) State H_0 and H_a for ANOVA.

(b) Outline the ANOVA table, giving the sources of variation and the degrees of freedom.

(c) What is the distribution of the F statistic under the assumption that H_0 is true?

(d) Using Table E, find the critical value for an $\alpha = 0.05$ test.

12.11 Does bread lose its vitamins when stored? Small loaves of bread were prepared with flour that was fortified with a fixed amount of vitamins. After baking, the vitamin C content of two loaves was measured. Another two loaves were baked at the same time, stored for one day, and then the vitamin C content was measured. In a similar manner, two loaves were stored for three, five, and seven days before measurements were taken. The units are milligrams per hundred grams of flour (mg/100 g).[6] Here are the data:

Condition	Vitamin C (mg/100 g)	
Immediately after baking	47.62	49.79
One day after baking	40.45	43.46
Three days after baking	21.25	22.34
Five days after baking	13.18	11.65
Seven days after baking	8.51	8.13

(a) Give a table with sample size, mean, standard deviation, and standard error for each condition.

(b) Perform a one-way ANOVA for these data. Be sure to state your hypotheses, the test statistic with degrees of freedom, and the *P*-value.

(c) Summarize the data and the means with a plot. Use the plot and the ANOVA results to write a short summary of your conclusions.

12.12 Refer to the previous exercise. Use the Bonferroni or another multiple comparisons procedure to compare the group means. Summarize the results.

12.13 Refer to Exercise 12.11. Measurements of the amounts of vitamin A (β-carotene) and vitamin E in each loaf are given below. Use an analysis of variance to study the data for each of these vitamins.

Condition	Vitamin A (mg/100 g)	Vitamin E (mg/100 g)
Immediately after baking	3.36 3.34	94.6 96.0
One day after baking	3.28 3.20	95.7 93.2
Three days after baking	3.26 3.16	97.4 94.3
Five days after baking	3.25 3.36	95.0 97.7
Seven days after baking	3.01 2.92	92.3 95.1

12.14 Refer to the previous exercise.

(a) Explain why it is inappropriate to perform a multiple comparisons analysis for the vitamin E data.

(b) Perform the Bonferroni or another multiple comparisons procedure for the vitamin A data and summarize the results.

12.15 Refer to Exercises 12.11 to 12.14. Write a report summarizing what happens to vitamins A, C, and E after bread is baked. Include appropriate statistical inference results and graphs.

12.16 If a supermarket product is offered at a reduced price frequently, do customers expect the price of the product to be lower in the future? This question was examined by researchers in a study conducted on students enrolled in an introductory management course at a large midwestern university. For 10 weeks 160 subjects received information about the products. The treatment conditions corresponded to the number of promotions (1, 3, 5, or 7) that were described during this 10-week period. Students were randomly assigned to four groups.[7] Table 12.3 gives the data.

(a) Make a normal quantile plot for the data in each of the four treatment groups. Summarize the information in the plots and draw a conclusion regarding the normality of these data.

(b) Summarize the data with a table containing the sample size, mean, standard deviation, and standard error for each group.

TABLE 12.3 Price promotion data

Number of promotions	Expected price (dollars)									
1	3.78	3.82	4.18	4.46	4.31	4.56	4.36	4.54	3.89	4.13
	3.97	4.38	3.98	3.91	4.34	4.24	4.22	4.32	3.96	4.73
	3.62	4.27	4.79	4.58	4.46	4.18	4.40	4.36	4.37	4.23
	4.06	3.86	4.26	4.33	4.10	3.94	3.97	4.60	4.50	4.00
3	4.12	3.91	3.96	4.22	3.88	4.14	4.17	4.07	4.16	4.12
	3.84	4.01	4.42	4.01	3.84	3.95	4.26	3.95	4.30	4.33
	4.17	3.97	4.32	3.87	3.91	4.21	3.86	4.14	3.93	4.08
	4.07	4.08	3.95	3.92	4.36	4.05	3.96	4.29	3.60	4.11
5	3.32	3.86	4.15	3.65	3.71	3.78	3.93	3.73	3.71	4.10
	3.69	3.83	3.58	4.08	3.99	3.72	4.41	4.12	3.73	3.56
	3.25	3.76	3.56	3.48	3.47	3.58	3.76	3.57	3.87	3.92
	3.39	3.54	3.86	3.77	4.37	3.77	3.81	3.71	3.58	3.69
7	3.45	3.64	3.37	3.27	3.58	4.01	3.67	3.74	3.50	3.60
	3.97	3.57	3.50	3.81	3.55	3.08	3.78	3.86	3.29	3.77
	3.25	3.07	3.21	3.55	3.23	2.97	3.86	3.14	3.43	3.84
	3.65	3.45	3.73	3.12	3.82	3.70	3.46	3.73	3.79	3.94

(c) Is the assumption of equal standard deviations reasonable here? Explain why or why not.

(d) Run the one-way ANOVA. Give the hypotheses tested, the test statistic with degrees of freedom, and the *P*-value. Summarize your conclusion.

12.17 Refer to the previous exercise. Use the Bonferroni or another multiple comparisons procedure to compare the group means. Summarize the results and support your conclusions with a graph of the means.

12.18 Do piano lessons improve the spatial-temporal reasoning of preschool children? In Exercise 7.65 (page 546) we examined this question by comparing the change scores (after treatment minus before treatment) of 34 children who took piano lessons with the scores of 44 children who did not. The latter group actually contained three groups of children: 10 were given singing lessons, 20 had some computer instruction, and 14 received no extra lessons. The data appear in Table 12.4.

(a) Make a table giving the sample size, the mean, the standard deviation, and the standard error for each group.

(b) Analyze the data using one-way analysis of variance. State the null and alternative hypotheses, the test statistic with degrees of freedom, the *P*-value, and your conclusion.

TABLE 12.4 Piano lesson data

Lessons	Scores									
Piano	2	5	7	−2	2	7	4	1	0	7
	3	4	3	4	9	4	5	2	9	6
	0	3	6	−1	3	4	6	7	−2	7
	−3	3	4	4						
Singing	1	−1	0	1	−4	0	0	1	0	−1
Computer	0	1	1	−3	−2	4	−1	2	4	2
	2	2	−3	−3	0	2	0	−1	3	−1
None	5	−1	7	0	4	0	2	1	−6	0
	2	−1	0	−2						

12.19 Refer to the previous exercise. Use the Bonferroni or another multiple comparisons procedure to compare the group means. Summarize the results and support your conclusions with a graph of the means.

12.20 The researchers in Exercise 12.18 based their research on a biological argument for a causal link between music and spatial-temporal reasoning. Therefore, it is natural to test the contrast that compares the mean of the piano lesson group with the average of the three other means. Construct this contrast, perform the significance test, and summarize the results. Note that this is not the test done in Exercise 7.65 (page 546), where we did not differentiate among the three groups who did not receive piano instruction.

12.21 Recommendations regarding how long infants in developing countries should be breast-fed are controversial. If the nutritional quality of the breast milk is inadequate because the mothers are malnourished, then there is risk of inadequate nutrition for the infant. On the other hand, the introduction of other foods carries the risk of infection from contamination. Further complicating the situation is the fact that companies that produce infant formulas and other foods benefit when these foods are consumed by large numbers of customers. One question related to this controversy concerns the amount of energy intake for infants who have other foods introduced into the diet at different ages. Part of one study compared the energy intakes, measured in kilocalories per day (kcal/d), for infants who were breast-fed exclusively for 4, 5, or 6 months.[8] Here are the data:

Breast-fed for:	Energy intake (kcal/d)									
4 months	499	620	469	485	660	588	675	517	649	209
	404	738	628	609	617	704	558	653	548	
5 months	490	395	402	177	475	617	616	587	528	518
	370	431	518	639	368	538	519	506		
6 months	585	647	477	445	485	703	528	465		

(a) Make a table giving the sample size, mean, and standard deviation for each group of infants. Is it reasonable to pool the variances?

(b) Run the analysis of variance. Report the *F* statistic with its degrees of freedom and *P*-value. What do you conclude?

12.22 Refer to the previous exercise.

(a) Examine the residuals. Is the normality assumption reasonable for these data?

(b) Explain why you do not need to use a multiple comparisons procedure for these data.

12.23 Many studies have suggested that there is a link between exercise and healthy bones. Exercise stresses the bones and this causes them to get stronger. One study examined the effect of jumping on the bone density of growing rats.[9] There were three treatments: a control with no jumping, a low-jump condition (the jump height was 30 centimeters), and a high-jump condition (60 centimeters). After 8 weeks of 10 jumps per day, 5 days per week, the bone density of the rats (expressed in mg/cm^3) was measured. Here are the data:

Group	Bone density (mg/cm^3)									
Control	611	621	614	593	593	653	600	554	603	569
Low jump	635	605	638	594	599	632	631	588	607	596
High jump	650	622	626	626	631	622	643	674	643	650

(a) Make a table giving the sample size, mean, and standard deviation for each group of rats. Is it reasonable to pool the variances?

(b) Run the analysis of variance. Report the *F* statistic with its degrees of freedom and *P*-value. What do you conclude?

12.24 Refer to the previous exercise.

(a) Examine the residuals. Is the normality assumption reasonable for these data?

(b) Use the Bonferroni or another multiple comparisons procedure to determine which pairs of means differ significantly. Summarize your results in a short report. Be sure to include a graph.

12.25 Iron-deficiency anemia is the most common form of malnutrition in developing countries, affecting about 50% of children and women and 25% of men. Iron pots for cooking foods had traditionally been used in many of these countries, but they have been largely replaced by aluminum pots, which are cheaper and lighter. Some research has suggested that food cooked in iron pots will contain more iron than food cooked in other types of pots. One study designed to investigate this issue compared the iron content of some Ethiopian foods cooked in aluminum, clay, and iron pots.[10]

One of the foods was *yesiga wet'*, beef cut into small pieces and prepared with several Ethiopian spices. The iron content of four samples of *yesiga wet'* cooked in each of the three types of pots is given below. The units are milligrams of iron per 100 grams of cooked food.

Type of pot	Iron (mg/100 g food)			
Aluminum	1.77	2.36	1.96	2.14
Clay	2.27	1.28	2.48	2.68
Iron	5.27	5.17	4.06	4.22

(a) Make a table giving the sample size, mean, and standard deviation for each type of pot. Is it reasonable to pool the variances?

(b) Run the analysis of variance. Report the F statistic with its degrees of freedom and P-value. What do you conclude?

12.26 Refer to the previous exercise.

(a) Examine the residuals. Is the normality assumption reasonable for these data?

(b) Use the Bonferroni or another multiple comparisons procedure to determine which pairs of means differ significantly. Summarize your results in a short report. Be sure to include a graph.

12.27 One way to repair serious wounds is to insert some material as a scaffold for the body's repair cells to use as a template for new tissue. Scaffolds made from extracellular material (ECM) are particularly promising for this purpose. Because they are made from biological material, they serve as an effective scaffold and are then resorbed. Unlike biological material that includes cells, however, they do not trigger tissue rejection reactions in the body. One study compared 6 types of scaffold material.[11] Three of these were ECMs and the other three were made of inert materials. There were three mice used per scaffold type. The response measure was the percent of glucose phosphated isomerase (Gpi) cells in the region of the wound. A large value is good, indicating that there are many bone marrow cells sent by the body to repair the tissue.

Material	Gpi (%)		
ECM1	55	70	70
ECM2	60	65	65
ECM3	75	70	75
MAT1	20	25	25
MAT2	5	10	5
MAT3	10	15	10

(a) Make a table giving the sample size, mean, and standard deviation for each of the six types of material. Is it reasonable to pool the variances? Note that the sample sizes are small and the data are rounded.

(b) Run the analysis of variance. Report the F statistic with its degrees of freedom and P-value. What do you conclude?

12.28 Refer to the previous exercise.

(a) Examine the residuals. Is the normality assumption reasonable for these data?

(b) Use the Bonferroni or another multiple comparisons procedure to determine which pairs of means differ significantly. Summarize your results in a short report. Be sure to include a graph.

12.29 Refer to the previous two exercises. Use a contrast to compare the three ECM materials with the three other materials. Summarize your conclusions. How do these results compare with those that you obtained from the multiple comparisons procedure in the previous exercise?

12.30 The presence of harmful insects in farm fields is detected by erecting boards covered with a sticky material and then examining the insects trapped on the boards. To investigate which colors are most attractive to cereal leaf beetles, researchers placed six boards of each of four colors in a field of oats in July.[12] The table below gives data on the number of cereal leaf beetles trapped:

Color	Insects trapped					
Lemon yellow	45	59	48	46	38	47
White	21	12	14	17	13	17
Green	37	32	15	25	39	41
Blue	16	11	20	21	14	7

(a) Make a table of means and standard deviations for the four colors, and plot the means.

(b) State H_0 and H_a for an ANOVA on these data, and explain in words what ANOVA tests in this setting.

(c) Using computer software, run the ANOVA. What are the F statistic and its P-value? Give the values of s_p and R^2. What do you conclude?

12.31 Return to the previous exercise. For the Bonferroni procedure with $\alpha = 0.05$, the value of t^{**} is 2.61. Use this multiple comparisons procedure to decide which pairs of colors are significantly different. Summarize your results. Which color would you recommend for attracting cereal leaf beetles?

12.32 A study of the effects of exercise on physiological and psychological variables compared four groups of male subjects. The treatment group (T) consisted of 10 participants in an exercise program. A control group (C) of 5 subjects volunteered for the program but were unable to attend for various reasons. Subjects in the other two groups were selected to be similar to those in the first two groups in age and other characteristics. These were 11 joggers (J) and 10 sedentary people (S) who did not regularly exercise.[13] One of the variables measured at the end of the program was a physical fitness score. Part of the ANOVA table used to analyze these data is given below:

Source	Degrees of freedom	Sum of squares	Mean square	F
Groups	3	104,855.87		
Error	32	70,500.59		
Total				

(a) Fill in the missing entries in the ANOVA table.

(b) State H_0 and H_a for this experiment.

(c) What is the distribution of the F statistic under the assumption that H_0 is true? Using Table E, give an approximate P-value for the ANOVA test. Write a brief conclusion.

(d) What is s_p^2, the estimate of the within-group variance? What is s_p?

12.33 Another variable measured in the experiment described in the previous exercise was a depression score. Higher values of this score indicate more depression. Part of the ANOVA table for these data appears below:

Source	Degrees of freedom	Sum of squares	Mean square	F
Groups	3		158.96	
Error	32		62.81	
Total				

(a) Fill in the missing entries in the ANOVA table.

(b) State H_0 and H_a for this experiment.

(c) What is the distribution of the F statistic under the assumption that H_0 is true? Using Table E, give an approximate P-value for the ANOVA test. What do you conclude?

(d) What is s_p^2, the estimate of the within-group variance? What is s_p?

12.34 The weight gain of women during pregnancy has an important effect on the birth weight of their children. If the weight gain is not adequate, the infant is more likely to be small and will tend to be less healthy. In a study conducted in three countries, weight gains (in kilograms) of women during the third trimester of pregnancy were measured.[14] The results are summarized in the following table:

Country	n	$\bar{x}$	s
Egypt	46	3.7	2.5
Kenya	111	3.1	1.8
Mexico	52	2.9	1.8

(a) Find the pooled estimate of the within-country variance s_p^2. What entry in the ANOVA table gives this quantity?

(b) The sum of squares for countries (groups) is 17.22. Use this information and that given above to complete all the entries in an ANOVA table.

(c) State H_0 and H_a for this study.

(d) What is the distribution of the F statistic under the assumption that H_0 is true? Use Table E to find an approximate P-value for the significance test. Report your conclusion.

(e) Calculate R^2, the coefficient of determination.

12.35 In another part of the study described in the previous exercise, measurements of food intake in kilocalories were taken on many individuals several times during the period of a year. From these data, average daily food intake values were computed for each individual. The results for toddlers aged 18 to 30 months are summarized in the following table:

Country	n	$\bar{x}$	s
Egypt	88	1217	327
Kenya	91	844	184
Mexico	54	1119	285

(a) Find the pooled estimate of the within-country variance s_p^2. What entry in the ANOVA table gives this quantity?

(b) The sum of squares for countries (groups) is 6,572,551. Use this information and that given above to complete all the entries in an ANOVA table.

(c) State H_0 and H_a for this study.

(d) What is the distribution of the F statistic under the assumption that H_0 is true? Use Table E to find an approximate P-value for the significance test. Report your conclusion.

(e) Calculate R^2, the coefficient of determination.

12.36 Refer to the change-of-majors study described in Example 12.3 (page 750). Let μ_1, μ_2, and μ_3 represent the population mean SAT mathematics scores for the three groups.

(a) Because the first two groups (computer science, engineering and other sciences) are majoring in areas that require mathematics skills, it is natural to compare the average of these two with the third group. Write an expression in terms of the μ_i for a contrast ψ_1 that represents this comparison.

(b) Let ψ_2 be a contrast which compares the first two groups. Write an expression in terms of the μ_i for ψ_2 that represents this comparison.

12.37 Return to the survey of college students described in Example 12.4 (page 754). Let μ_1, μ_2, μ_3, and μ_4 represent the population mean expenditures on textbooks for the freshmen, sophomores, juniors, and seniors.

(a) Because juniors and seniors take higher-level courses, which might use more expensive textbooks, we want to compare the average of the freshmen and sophomores with the average of the juniors and seniors. Write a contrast that expresses this comparison.

(b) Write a contrast for comparing the freshmen with the sophomores.

(c) Write a contrast for comparing the juniors with the seniors.

12.38 Return to the SAT mathematics scores for the change-of-majors study in Example 12.3 (page 750). Answer the following questions for the two contrasts that you defined in Exercise 12.36.

(a) For each contrast give H_0 and an appropriate H_a. In choosing the alternatives you should use information given in the description of the problem, but you may not consider any impressions obtained by inspection of the sample means.

(b) Find the values of the corresponding sample contrasts c_1 and c_2.

(c) Using the value $s_p = 82.5$, calculate the standard errors s_{c_1} and s_{c_2} for the sample contrasts.

(d) Give the test statistics and approximate P-values for the two significance tests. What do you conclude?

(e) Compute 95% confidence intervals for the two contrasts.

12.39 The following table presents data on the high school mathematics grades of the students in the change-of-majors study described in Example 12.3 (page 750). The sample sizes in this exercise are different from those in Example 12.3 because complete information was not available on all students. The values have been coded so that 10 = A, 9 = A−, and so on.

Second-year major	n	$\overline{x}$	s
Computer science	90	8.77	1.41
Engineering and other sciences	28	8.75	1.46
Other	106	7.83	1.74

Answer the following questions for the two contrasts that you defined in Exercise 12.36.

(a) For each contrast state H_0 and an appropriate H_a. In choosing the alternatives you should use information given in the description of the problem, but you may not consider any impressions obtained by inspection of the sample means.

(b) Find the values of the corresponding sample contrasts c_1 and c_2.

(c) Using the value $s_p = 1.581$, calculate the standard errors s_{c_1} and s_{c_2} for the sample contrasts.

(d) Give the test statistics and approximate *P*-values for the two significance tests. What do you conclude?

(e) Compute 95% confidence intervals for the two contrasts.

12.40 In the exercise program study described in Exercise 12.32, the summary statistics for physical fitness scores are as follows:

Group	n	$\bar{x}$	s
Treatment (T)	10	291.91	38.17
Control (C)	5	308.97	32.07
Joggers (J)	11	366.87	41.19
Sedentary (S)	10	226.07	63.53

The researchers wanted to address the following questions for the physical fitness scores. In these questions "better" means a higher fitness score. (1) Is T better than C? (2) Is T better than the average of C and S? (3) Is J better than the average of the other three groups?

(a) For each of the three questions, define an appropriate contrast. Translate the questions into null and alternative hypotheses about these contrasts.

(b) Test your hypotheses and give approximate *P*-values. Summarize your conclusions. Do you think that the use of contrasts in this way gives an adequate summary of the results?

(c) You found that the groups differ significantly in the physical fitness scores. Does this study allow conclusions about causation—for example, that a sedentary lifestyle causes people to be less physically fit? Explain your answer.

12.41 Exercise 12.33 gives the ANOVA table for depression scores from the exercise program study described in Exercise 12.32. Here are the summary statistics for the depression scores:

Group	n	$\bar{x}$	s
Treatment (T)	10	51.90	6.42
Control (C)	5	57.40	10.46
Joggers (J)	11	49.73	6.27
Sedentary (S)	10	58.20	9.49

In planning the experiment, the researchers wanted to address the following questions for the depression scores. In these questions "better" means a lower depression score. (1) Is T better than C? (2) Is T better than the average of C and S? (3) Is J better than the average of the other three groups?

(a) For each of the three questions, define an appropriate contrast. Translate the questions into null and alternative hypotheses about these contrasts.

(b) Test your hypotheses and give approximate *P*-values. Summarize your conclusions. Do you think that the use of contrasts in this way gives an adequate summary of the results?

(c) You found that the groups differ significantly in the depression scores. Does this study allow conclusions about causation—for example, that a sedentary lifestyle causes people to be more depressed? Explain your answer.

12.42 Exercise 12.34 gives data on the weight gains of pregnant women in Egypt, Kenya, and Mexico. Computer software gives the critical value for the Bonferroni multiple comparisons procedure with $\alpha = 0.05$ as $t^{**} = 2.41$. Explain in plain language what $\alpha = 0.05$ means in the Bonferroni procedure. Use this procedure to compare the mean weight gains for the three countries. Summarize your conclusions.

12.43 Exercise 12.35 gives summary statistics for the food intake values for toddlers in Egypt, Kenya, and Mexico. Computer software gives the critical value for the Bonferroni multiple comparisons procedure with $\alpha = 0.05$ as $t^{**} = 2.41$. Explain in plain language what $\alpha = 0.05$ means in the Bonferroni procedure. Use this procedure to compare the toddler food intake means for the three countries. What do you conclude?

12.44 Refer to the physical fitness scores for the four groups in the exercise program study discussed in Exercises 12.32 and 12.40. Computer software gives the critical value for the Bonferroni multiple comparisons procedure with $\alpha = 0.05$ as $t^{**} = 2.81$. Use this procedure to compare the mean fitness scores for the four groups. Summarize your conclusions.

12.45 Refer to the depression scores for the four groups in the exercise program study discussed in Exercises 12.33 and 12.41. Computer software gives the critical value for the Bonferroni multiple comparisons procedure with $\alpha = 0.05$ as $t^{**} = 2.81$. Use this procedure to compare the mean depression scores for the four groups. Summarize your conclusions.

12.46 You are planning a study of the weight gains of pregnant women during the third trimester of pregnancy similar to that described in Exercise 12.34. The standard deviations given in that exercise range from 1.8 to 2.5. To perform power calculations, assume that the standard deviation is $\sigma = 2.4$. You have three groups, each with n subjects, and you would like to reject the ANOVA H_0 when the alternative $\mu_1 = 2.6$, $\mu_2 = 3.0$, and $\mu_3 = 3.4$ is true. Use software to make a table of powers against this alternative (similar to the table in Example 12.21, page 780) for the following numbers of women in each group: $n = 50, 100, 150, 175$, and 200. What sample size would you choose for your study?

12.47 Repeat the previous exercise for the alternative $\mu_1 = 2.7$, $\mu_2 = 3.1$, and $\mu_3 = 3.5$. Why are the results the same?

12.48 Analysis of variance methods are often used in clinical trials where the goal is to assess the effectiveness of one or more treatments for a particular medical condition. One such study compared three treatments for dandruff and a placebo. The treatments were 1% pyrithione zinc shampoo (PyrI), the same shampoo but with instructions to shampoo two times (PyrII), 2% ketoconazole shampoo (Keto), and a placebo shampoo (Placebo). After six weeks of treatment, eight sections of the scalp were examined and given a score that measured the amount of scalp flaking on a 0 to 10 scale. The response variable was the sum of these eight scores. An analysis of the baseline flaking measure indicated that randomization of patients to treatments was successful in that no differences were found between the groups. At baseline there were 112 subjects in each of the three treatment groups and 28 subjects in the Placebo group. During the clinical trial, 3 dropped out from the PyrII group and 6 from the Keto group. No patients dropped out of the other two groups. The data are given in the DANDRUFF data set described in the Data Appendix.

(a) Find the mean, standard deviation, and standard error for the subjects in each group. Summarize these, along with the sample sizes, in a table and make a graph of the means.

(b) Run the analysis of variance on these data. Write a short summary of the results and your conclusion. Be sure to include the hypotheses tested, the test statistic with degrees of freedom, and the *P*-value.

12.49 Refer to the previous exercise.

(a) Plot the residuals versus case number (the first variable in the data set). Describe the plot. Is there any pattern that would cause you to question the assumption that the data are independent?

(b) Examine the standard deviations for the four treatment groups. Is there a problem with the assumption of equal standard deviations for ANOVA in this data set? Explain your answer.

(c) Prepare normal quantile plots for each treatment group. What do you conclude from these plots?

(d) Obtain the residuals from the analysis of variance and prepare a normal quantile plot of these. What do you conclude?

12.50 Refer to Exercise 12.48. Use the Bonferroni or another multiple comparisons procedure that your software provides to compare the individual group means in the dandruff study. Write a short summary of your conclusions.

12.51 Refer to Exercise 12.48. There are several natural contrasts in this experiment that describe comparisons of interest to the experimenters. They are

(1) Placebo versus the average of the three treatments; (2) Keto versus the average of the two Pyr treatments; and (3) PyrI versus PyrII.

(a) Express each of these three contrasts in terms of the means (μ's) of the treatments.

(b) Give estimates with standard errors for each of the contrasts.

(c) Perform the significance tests for the contrasts. Summarize the results of your tests and your conclusions.

12.52 Refer to Exercise 12.11, where we studied the effects of storage on the vitamin C content of bread. In this experiment 64 mg/100 g of vitamin C was added to the flour that was used to make each loaf.

(a) Convert the vitamin C amounts (mg/100 g) to percents of the amounts originally in the loaves by dividing the amounts in Exercise 12.11 by 64 and multiplying by 100. Calculate the transformed means, standard deviations, and standard errors and summarize them with the sample sizes in a table.

(b) Explain how you could have calculated the table entries directly from the table you gave in part (a) of Exercise 12.11.

(c) Analyze the percents using analysis of variance. Compare the test statistic, degrees of freedom, P-value, and conclusion you obtain here with the corresponding values that you found in Exercise 12.11.

12.53 Refer to the previous exercise and Exercise 12.13. The flour used to make the loaves contained 5 mg/100 g of vitamin A and 100 mg/100 g of vitamin E. Summarize the effects of transforming the data to percents for the three vitamins.

12.54 Refer to the previous exercise. Can you suggest a general conclusion regarding what happens to the test statistic, degrees of freedom, P-value, and conclusion when you perform analysis of variance on data that have been transformed by multiplying the raw data by a constant and then adding another constant? (That is, if y is the original data, we analyze y^*, where $y^* = a + by$ and a and $b \neq 0$ are constants.)

12.55 Example 12.6 (page 756) describes an experiment to compare three methods of teaching reading. One of the pretest measures is analyzed in that example and one of the posttest measures is analyzed in Example 12.10 (page 765). In the actual study there were two pretest and three posttest measures. The data are given in the READING data set described in the Data Appendix. The pretest variable analyzed in Example 12.6 is Pre1 and the posttest variable analyzed in Example 12.10 is Post3. For this exercise, use the data for the pretest variable Pre2.

(a) Summarize the data with a table of the sample sizes, means, and standard deviations. Plot the means.

(b) Examine the distributions of the scores in the three groups by constructing normal quantile plots. Do the data look normal?

(c) What is the ratio of the largest to the smallest standard deviation? Is it reasonable to proceed with an ANOVA?

(d) Run a one-way ANOVA on these data. State H_0 and H_a for the ANOVA significance test. Report the value of the F statistic and its P-value. Do you have strong evidence against H_0?

(e) Summarize your conclusions from this analysis.

12.56 Refer to the previous exercise. To examine the effect on the ANOVA of changing the means, you will do an analysis similar to that described in Example 12.9. Use the pretest variable Pre2 for this exercise. For each of the observations in the Basal group add two points to the Pre2. For each of the observations in the Strat group, add one point. Leave the observations in the DRTA group unchanged. Call the new variable Pre2X.

(a) Give the sample means and standard deviations for Pre2X in each group. (You can find these from the means and standard deviations for Pre2.)

(b) Run the ANOVA on the variable Pre2X.

(c) Compare the results of this analysis with the ANOVA for Pre2.

(d) Summarize your conclusions.

12.57 Example 12.10 analyzed the scores for the third posttest variable in the reading comprehension study. For this exercise you will analyze the first posttest variable, given as Post1 in the READING data set.

(a) Summarize the data in a table of the sample sizes, means, and standard deviations. Plot the means.

(b) Examine the distributions of the scores in the three groups by constructing normal quantile plots. Do the data look normal?

(c) What is the ratio of the largest to the smallest standard deviation? Is it reasonable to proceed with the ANOVA?

(d) Run the ANOVA and summarize the results.

(e) Following the procedure given in the text in the section on contrasts, analyze the contrast for comparing the Basal group with the average of the other two groups. Test the one-sided H_a and give a 95% confidence interval.

(f) Analyze the contrast that compares the DRTA and the Strat groups. Use a two-sided test and give a 95% confidence interval.

(g) Based on your work in (a), (d), (e), and (f), write a brief nontechnical summary of your conclusions.

12.58 Example 12.10 analyzed the scores for the third posttest variable in the reading comprehension study. For this exercise you will analyze the second posttest variable, given as Post2 in the READING data set.

(a) Summarize the data in a table of the sample sizes, means, and standard deviations. Plot the means.

(b) Examine the distributions of the scores in the three groups by constructing normal quantile plots. Do the data look normal?

(c) What is the ratio of the largest to the smallest standard deviation? Is it reasonable to proceed with the ANOVA?

(d) Run the ANOVA and summarize the results.

(e) Following the procedure given in the text in the section on contrasts, analyze the contrast for comparing the Basal group with the average of the other two groups. Test the one-sided H_a and give a 95% confidence interval.

(f) Analyze the contrast that compares the DRTA and the Strat groups. Use a two-sided test and give a 95% confidence interval.

(g) Based on your work in (a), (d), (e), and (f), write a brief nontechnical summary of your conclusions.

12.59 Refer to the color attractiveness experiment described in Exercise 12.30 (page 789). Suppose that when entering the data into the computer, you accidentally entered the first observation as 450 rather than 45.

(a) Run the ANOVA with the incorrect observation. Summarize the results.

(b) Compare this run with the results obtained with the correct data set. What does this illustrate about the effect of outliers in an ANOVA?

(c) Compute a table of means and standard deviations for each of the four treatments using the incorrect data. How would this table have helped you to detect the incorrect observation?

12.60 Refer to the color attractiveness experiment described in Exercise 12.30 (page 789). The square root transformation is often used for variables that are counts, such as the number of insects trapped in this example. In many cases data transformed in this way will conform more closely to the assumptions of normality and equal standard deviations. Rerun the ANOVA using the square roots of the original counts of insects. Answer the questions given in Exercise 12.30. Compare these results with those obtained by analyzing the raw data.

12.61 You are planning a study of the SAT mathematics scores of four groups of students. From Example 12.3, we know that the standard deviations of the three groups considered in that study were 86, 67, and 83. In Example 12.5, we found the pooled standard deviation to be 82.5. Since the power of the F test decreases as the standard deviation increases, use $\sigma = 86$ for the calculations in this exercise. This choice will lead to sample sizes that are perhaps a little larger than we need but will prevent us from choosing sample sizes that are too small to detect the effects of interest. You would like to conclude that the population means are different when $\mu_1 = 620$, $\mu_2 = 600$, $\mu_3 = 580$, and $\mu_4 = 560$.

(a) Pick several values for n (the number of students that you will select from each group) and calculate the power of the ANOVA F test for each of your choices.

(b) Plot the power versus the sample size. Describe the general shape of the plot.

(c) What choice of n would you choose for your study? Give reasons for your answer.

12.62 Refer to the previous exercise. Repeat all parts for the alternative $\mu_1 = 610$, $\mu_2 = 600$, $\mu_3 = 590$, and $\mu_4 = 580$.

12.63 Refer to the price promotion study that we examined in Exercise 12.16. The explanatory variable in this study is the number of price promotions in a ten-week period, with possible values of 1, 3, 5, and 7. When using analysis of variance we treat the explanatory variable as categorical. An alternative analysis is to use simple linear regression. Perform this analysis and summarize the results. Plot the residuals from the regression model versus the number of promotions. What do you conclude?

applet

12.64 The *One-Way ANOVA* applet lets you see how the *F* statistic and the *P*-value depend on the variability of the data within groups and the differences among the means.

(a) The black dots are at the means of the three groups. Move these up and down until you get a configuration that gives a *P*-value of about 0.01. What is the value of the *F* statistic?

(b) Now increase the variation within the groups by dragging the mark on the pooled standard error scale to the right. Describe what happens to the *F* statistic and the *P*-value. Explain why this happens.

applet

12.65 Set the pooled standard error for the *One-Way ANOVA* applet at a middle value. Drag the black dots so that they are approximately equal.

(a) What is the *F* statistic? Give its *P*-value.

(b) Drag the mean of the second group up and the mean of the third group down. Describe the effect on the *F* statistic and its *P*-value. Explain why they change in this way.

Prelude: should we use a one-way ANOVA or a two-way ANOVA?

A nutrition researcher was interested in studying the effect of calcium intake on the blood pressure of rats. In consultation with a statistician, she decided to use three groups of rats with diets that were identical except for the amount of calcium. The levels of calcium were selected to include a low, a medium, and a high level. Using reports of previous studies on rats, a rough guess at the standard deviation was obtained. From this information, she determined that there would be sufficient power to detect differences that would be important if each diet was fed to 15 rats. So the experiment was planned. A total of 45 rats would be ordered and they would be randomly assigned to the three diets. The data would be analyzed using a one-way ANOVA.

At the end of the consultation, the researcher mentioned that she also intended to do a similar study on the effect of magnesium in the diet. She was pleased with the design of the calcium study and said that she would use the same design for the magnesium study. A one-way ANOVA with three levels would then be used to analyze the data. At this point, the statistician suggested that the discussion of these designs should continue for a bit. Unlike the calcium study, it was expected that the magnesium should have relatively little effect on the rats' blood pressure.

After some discussion, the researcher decided to use a single study to investigate the effects of both calcium and magnesium on the blood pressure of rats. Three levels of each of these factors would be used. Each level of calcium would be paired with each level of magnesium, giving a total of 3×3, or 9, diets. For each of these diets, 9 rats would be used. A total of 81 rats would be randomly assigned to the 9 treatment combinations. The experiment was performed and the data were analyzed using two-way ANOVA methods. As expected, there was an effect from calcium but no effect from magnesium. By combining the experiments, the total number of rats needed was reduced from 90 (45 for each) to 81. In addition, for the calcium question, there were now 9×3, or 27, rats for each level rather than only 15 in the proposed one-way design. This gave considerably more power at essentially no additional cost.

CHAPTER 13

Josef Albers, *Study for Oscillating*, 1940–1945.

Two-Way Analysis of Variance

Introduction

The t procedures of Chapter 7 compare the means of two populations. We generalized these in Chapter 12 so that we could compare the means of several populations.

Two-way ANOVA compares the means of populations that are classified in two ways or the mean responses in two-factor experiments. Many of the key concepts are similar to those of one-way ANOVA, but the presence of more than one factor also introduces some new ideas. We once more assume that the data are approximately normal and that groups may have different means but do have the same standard deviation; we again pool to estimate the variance; and we again use F statistics for significance tests. The major difference between one-way and two-way ANOVA is in the FIT part of the model. We will carefully study this term, and we will find much that is both new and useful.

13.1 The Two-Way ANOVA Model

We begin with a discussion of the advantages of the two-way ANOVA design and illustrate these with some examples. Then we discuss the model and the assumptions.

Advantages of two-way ANOVA

In one-way ANOVA, we classify populations according to one categorical variable, or factor. In the two-way ANOVA model, there are two factors, each with its own number of levels. When we are interested in the effects of two factors, a two-way design offers great advantages over several single-factor studies. Several examples will illustrate these advantages.

EXAMPLE 13.1

In an experiment on the influence of dietary minerals on blood pressure, rats receive diets prepared with varying amounts of calcium and varying amounts of magnesium, but with all other ingredients of the diets the same. Calcium and magnesium are the two factors in this experiment.

As is common in such experiments, high, normal, and low values for each of the two minerals were selected for study. So there are three levels for each of the factors and a total of nine diets, or treatments. The following table summarizes the factors and levels for this experiment:

	Calcium		
Magnesium	L	M	H
L	1	2	3
M	4	5	6
H	7	8	9

For example, Diet 2 contains magnesium at its low level combined with the normal (medium) level of calcium. Each diet is fed to 9 rats, giving a total of 81 rats used in this experiment. The response variable is the blood pressure of a rat after some time on the diet.[1]

The researcher in this example was primarily interested in the effect of calcium on the blood pressure. She could have conducted a single-factor experiment with magnesium set at its normal value and three levels of calcium, using one-way ANOVA to analyze the responses. Had she then decided to study the effect of magnesium, she could have performed another experiment with calcium set at its normal value and three levels of magnesium. Suppose that the budget allowed for only 80 to 90 rats in all. For the first experiment, the researcher might assign 15 rats to each of the three calcium diets. Similarly, she could assign 15 rats to each of the three magnesium diets in the second experiment. This would require a total of 90 rats, 9 more than the two-factor design in Example 13.1.

Let us compare the two-way experiment with the two one-way experiments. In the two-way experiment 27 rats are assigned to each of the three calcium diets and 27 rats to each of the three magnesium diets. In the one-way experiments there are only 15 rats per diet. Therefore, the two-way experiment provides more information to estimate the effects of both calcium and magnesium. This fact can be expressed in terms of standard deviations. If σ is the standard deviation of blood pressure for rats fed the same diet, then the standard deviation of a sample mean for n rats is $\sigma/\sqrt{n}$. For the two-way experiment this is $\sigma/\sqrt{27} = 0.19\sigma$, whereas for the two one-way experiments we have $\sigma/\sqrt{15} = 0.26\sigma$. The sample mean responses for any one level of either calcium or magnesium are less variable in the two-way design.

When the actual experiment was analyzed using two-way ANOVA, no effect of magnesium on blood pressure was found. On the other hand, the data clearly demonstrated that the high-calcium diet had the beneficial effect of lowering the average blood pressure. Means for the three calcium levels were reported with standard errors based on 27 rats per group.

The conclusions about the effect of calcium on blood pressure are also more general in the two-way experiment. In the one-way experiments, calcium is varied only for the normal level of magnesium. All levels of calcium are combined with all levels of magnesium in the two-way design, so that conclusions about calcium are not confined to one level of magnesium. The same increased generality also applies to conclusions about magnesium. So the conclusions are both more precise and more general, yet the two-way design requires fewer rats. Experiments with several factors are an efficient use of resources.

EXAMPLE 13.2

A group of students and administrators concerned about abuse of alcohol is planning an education and awareness program for the student body at a large university. To aid in the design of their program they decide to gather information on the attitudes of undergraduate students toward alcohol.

Initially, they consider taking an SRS of all the students on the campus. They soon realize that the attitudes of freshmen, sophomores, juniors, and seniors may be quite different. A stratified sample is therefore more appropriate. The study planners decide to use year in school as a factor with four levels and to take separate random samples from each class of students.

Further discussion suggests that men and women may have different attitudes toward alcohol. The team therefore decides to add gender as a second factor. The final sampling design has two factors: year with four levels, and gender with two levels. An SRS of size 50 is drawn from each of the eight populations. The eight groups for this study are shown in the following table:

	Year			
Gender	Fr.	So.	Jr.	Sr.
Male	1	2	3	4
Female	5	6	7	8

Example 13.2 illustrates another advantage of two-way designs analyzed by two-way ANOVA. In the one-way design with year as the only factor, all four random samples would include both men and women. If a gender difference exists, the one-way ANOVA would assign this variation to the RESIDUAL (within groups) part of the model. In the two-way ANOVA, gender is included as a factor, and therefore this variation is included in the FIT part of the model. Whenever we can move variation from RESIDUAL to FIT, we reduce the σ of our model and increase the power of our tests.

EXAMPLE 13.3

A textile researcher is interested in how four different colors of dye affect the durability of fabrics. Because the effects of the dyes may be different for different types of cloth, he applies each dye to five different kinds of cloth. The two factors in this experiment are dyes with four levels and cloth types with five levels. Six fabric specimens are dyed for each of the 20 dye-cloth combinations, and all 120 specimens are tested for durability.

The researcher in Example 13.3 could have used only one type of cloth and compared the four dyes using a one-way analysis. However, based on knowledge of the chemistry of textiles and dyes, he suspected that some of the dyes might have a chemical reaction with some types of cloth. Such a reaction could have a strong negative effect on the durability of the cloth. He wanted his results to be useful to manufacturers who would use the dyes on a variety of cloths. He chose several cloth types for the study to represent the different kinds that would be used with these dyes.

If one or more specific dye-cloth combinations produced exceptionally bad or exceptionally good durability measures, the experiment should discover this combined effect. Effects of the dyes that may differ for different types of cloth are represented in the FIT part of a two-way model as **interactions.** In contrast, the average values for dyes and cloths are represented as **main effects.** The two-way model represents FIT as the sum of a main effect for each of the two factors *and* an interaction. One-way designs that vary a single factor and hold other factors fixed cannot discover interactions. We will discuss interactions more fully in a later section.

interactions

main effects

These examples illustrate several reasons why two-way designs are preferable to one-way designs.

Advantages of Two-Way ANOVA

1. It is more efficient to study two factors simultaneously rather than separately.
2. We can reduce the residual variation in a model by including a second factor thought to influence the response.
3. We can investigate interactions between factors.

These considerations also apply to study designs with more than two factors. We will be content to explore only the two-way case. The choice of sampling or experimental design is fundamental to any statistical study. Factors and levels must be carefully selected by an individual or team who understands both the statistical models and the issues that the study will address.

The two-way ANOVA model

When discussing two-way models in general, we will use the labels A and B for the two factors. For particular examples and when using statistical software, it is better to use names for these categorical variables that suggest their meaning. Thus, in Example 13.1 we would say that the factors are calcium and magnesium. The numbers of levels of the factors are often used to describe the model. Again using Example 13.1 as an illustration, we would call this a 3×3 ANOVA. Similarly, Example 13.2 illustrates a 2×4 ANOVA. In general, factor A will have I levels and factor B will have J levels. Therefore, we call the general two-way problem an $I \times J$ ANOVA.

In a two-way design every level of A appears in combination with every level of B, so that $I \times J$ groups are compared. The sample size for level i of factor A and level j of factor B is n_{ij}.[2]

The total number of observations is

$$N = \sum n_{ij}$$

Assumptions for Two-Way ANOVA

We have independent SRSs of size n_{ij} from each of $I \times J$ normal populations. The population means μ_{ij} may differ, but all populations have the same standard deviation σ. The μ_{ij} and σ are unknown parameters.

Let x_{ijk} represent the kth observation from the population having factor A at level i and factor B at level j. The statistical model is

$$x_{ijk} = \mu_{ij} + \epsilon_{ijk}$$

for $i = 1, \ldots, I$ and $j = 1, \ldots, J$ and $k = 1, \ldots, n_{ij}$, where the deviations ϵ_{ijk} are from an $N(0, \sigma)$ distribution.

The FIT part of the model is the means μ_{ij}, and the RESIDUAL part is the deviations ϵ_{ijk} of the individual observations from their group means. To estimate a population mean μ_{ij} we use the sample mean of the observations in the samples from this group:

$$\overline{x}_{ij} = \frac{1}{n_{ij}} \sum_k x_{ijk}$$

The k below the $\sum$ means that we sum the n_{ij} observations that belong to the (i, j)th sample.

The RESIDUAL part of the model contains the unknown σ. We calculate the sample variances for each SRS and pool these to estimate σ^2:

$$s_p^2 = \frac{\sum(n_{ij} - 1)s_{ij}^2}{\sum(n_{ij} - 1)}$$

Just as in one-way ANOVA, the numerator in this fraction is SSE and the denominator is DFE. Also as in the one-way analysis, DFE is the total number of observations minus the number of groups. That is, DFE $= N - IJ$. The estimator of σ is s_p.

Main effects and interactions

In this section we will explore in detail the FIT part of the two-way ANOVA, which is represented in the model by the population means μ_{ij}. The two-way design gives some structure to the set of means μ_{ij}.

Because we have independent samples from each of $I \times J$ groups, we can think of the problem initially as a one-way ANOVA with IJ groups. Each population mean μ_{ij} is estimated by the corresponding sample mean $\overline{x}_{ij}$. We can calculate sums of squares and degrees of freedom as in one-way ANOVA. In accordance with the conventions used by many computer software packages, we use the term *model* when discussing the sums of squares and degrees of freedom calculated as in one-way ANOVA with IJ groups. Thus, SSM is a model sum of squares constructed from deviations of the form $\overline{x}_{ij} - \overline{x}$, where $\overline{x}$ is the average of all of the observations and $\overline{x}_{ij}$ is the mean of the (i, j)th group. Similarly, DFM is simply $IJ - 1$.

In two-way ANOVA, the terms SSM and DFM are broken down into terms corresponding to a main effect for A, a main effect for B, and an AB interaction. Each of SSM and DFM is then a sum of terms:

$$\text{SSM} = \text{SSA} + \text{SSB} + \text{SSAB}$$

and

$$\text{DFM} = \text{DFA} + \text{DFB} + \text{DFAB}$$

The term SSA represents variation among the means for the different levels of the factor A. Because there are I such means, DFA $= I - 1$ degrees of freedom.

Similarly, SSB represents variation among the means for the different levels of the factor B, with DFB $= J - 1$.

Interactions are a bit more involved. We can see that SSAB, which is SSM − SSA − SSB, represents the variation in the model that is not accounted for by the main effects. By subtraction we see that its degrees of freedom are

$$\text{DFAB} = (IJ - 1) - (I - 1) - (J - 1) = (I - 1)(J - 1)$$

There are many kinds of interactions. The easiest way to study them is through examples.

EXAMPLE 13.4

The Bureau of Labor Statistics collects data on earnings of workers in the United States classified according to various characteristics. Here are the median weekly earnings in dollars of men and women in two age groups who were working full-time in the first quarter of 1997:[3]

Age	Women	Men	Mean
16–19	239	264	251.5
20–24	302	333	317.5
Mean	270.5	298.5	284.5

The table includes averages of the means in the rows and columns. For example, the first entry in the far right margin is the average of the salaries for women and men aged 16 to 19:

$$\frac{239 + 264}{2} = 251.5$$

Similarly, the average of women's salaries in the two age groups is

$$\frac{239 + 302}{2} = 270.5$$

marginal means

These averages are called **marginal means** (because of their location at the *margins* of such tabulations).

Figure 13.1 is a plot of the group means. From the plot it is clear that men earn more than women. In statistical language, there is a main effect for gender. We also see that workers aged 20 to 24 earn more than those aged 16 to 19. This is the main effect of age. These main effects can be described by differences between the marginal means. For example, the mean for men is \$298.50, which is \$28 higher than \$270.50, the mean for women.

What about the interaction between gender and age? An interaction is present if the main effects provide an incomplete description of the data. The marginal means for gender show that the salary difference for men versus women is \$28. This is the main effect of gender. The body of the table shows that the gender salary differences for the two age groups are \$25 and \$31.

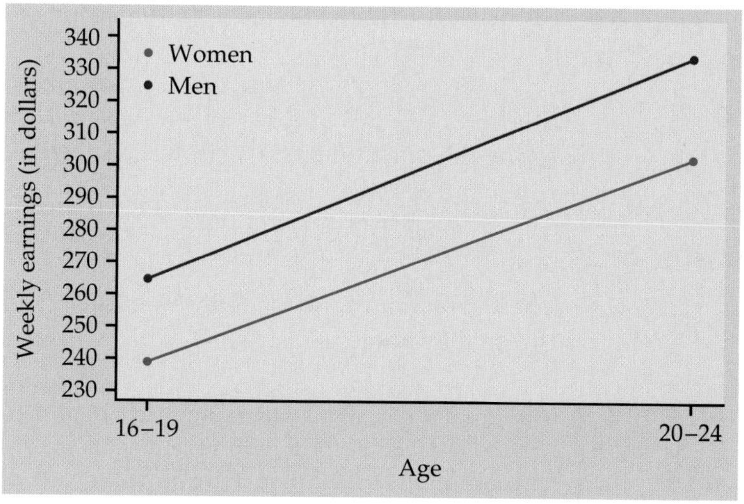

FIGURE 13.1 Plot of median weekly earnings of men and women for two age groups, for Example 13.4.

Although the difference is a bit higher for the older group, most of the information in the table concerning the gender difference is summarized by the main effect, the marginal difference of \$28. Similarly, the main effect of age, \$251.50 versus \$317.50, summarizes the fact that the older group has higher earnings. The difference is similar for men and women. In this example, the interaction is small.

The lack of interaction can be seen in Figure 13.1. The line connecting the earnings for women and the line connecting the earnings for men are roughly parallel. This happens because the difference between the lines is the difference between the men's and women's salaries for each age group. As we have seen, this difference is approximately the same for the two groups.

In the next example, we add four more age groups to our data. We will see that in this case a very strong interaction is present.

EXAMPLE 13.5

The survey described in the previous example also gave weekly earnings for groups of older workers. Adding these gives the following table:

Age	Women	Men	Mean
16–19	239	264	251.5
20–24	302	333	317.5
25–34	422	514	468.0
35–44	475	639	557.0
45–54	496	692	594.0
55–64	426	660	543.0
Mean	393	517	455.0

Including the additional age groups changes the marginal means for gender and the overall mean of all groups. Figure 13.2 is a plot of the group means.

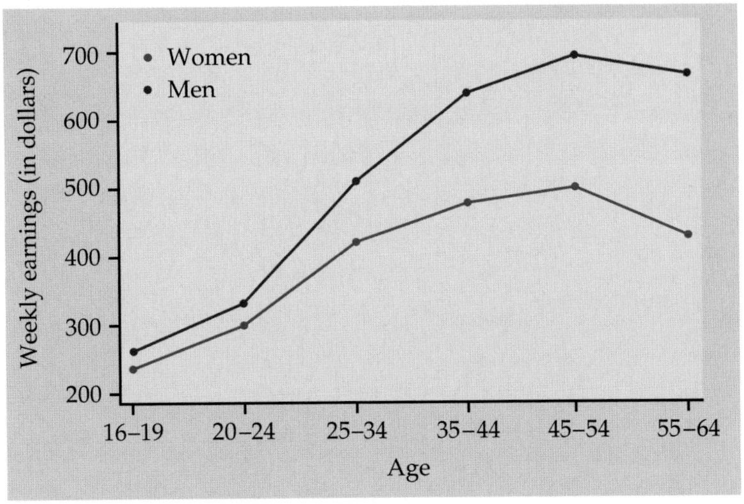

FIGURE 13.2 Plot of median weekly earnings of men and women for six age groups, for Example 13.5.

The two lines in Figure 13.2 are clearly not parallel. The gender difference in earnings depends on which age group we examine. The difference increases with age. This is an interaction between gender and age. The marginal means alone do not provide an adequate summary of the information in this table.

The presence of an interaction does not necessarily mean that the main effects are uninformative. It is still clear that men earn more than women, and this is evident in the marginal means for the two genders. Similarly, the marginal means for age show that earnings increase with age until the last group, where there is a decrease. We see this pattern in the plot for both men and women.

Interactions come in many shapes and forms. When we find them, a careful examination of the means is needed to properly interpret the data. Simply stating that interactions are significant tells us little. Plots of the group means are very helpful.

EXAMPLE 13.6

A study of the energy expenditure of farmers in Burkina Faso (formerly known as Upper Volta) collected data during the wet and dry seasons.[4] The farmers grow millet during the wet season. In the dry season, there is relatively little activity because the ground is too hard to grow crops. The mean energy expended (in calories) by men and women in Burkina Faso during the wet and dry seasons is given in the following table:

Season	Men	Women	Mean
Dry	2310	2320	2315
Wet	3460	2890	3175
Mean	2885	2605	2745

Figure 13.3 is the plot of the group means.

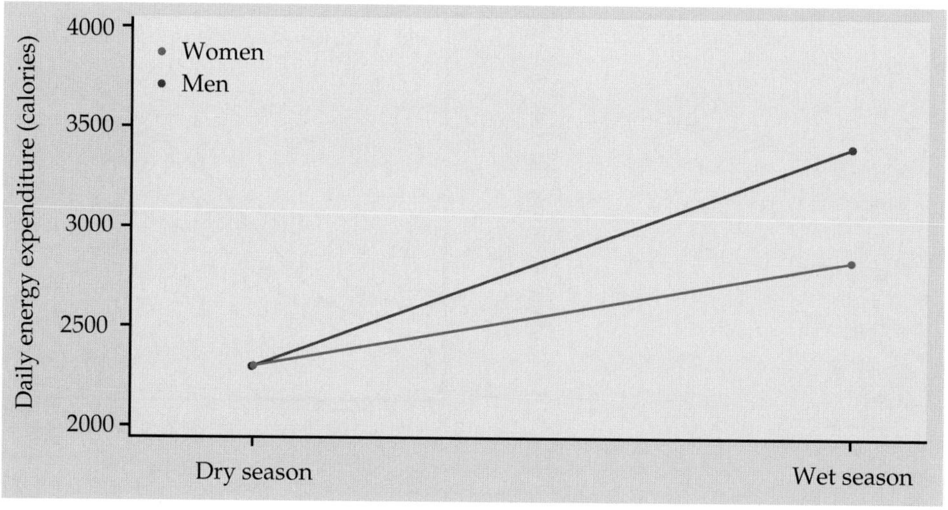

FIGURE 13.3 Plot of mean energy expenditures of farmers in Burkina Faso by sex and season, for Example 13.6.

During the dry season both men and women use about the same number of calories. When the wet season arrives, both genders use more energy. The men, who by social convention do most of the field work, expend much more energy than the women. The amount of energy used by the men in the wet season is very high by any reasonable standard. Such values are sometimes found in developed countries for people engaged in coal mining and lumberjacking.

In a statistical analysis, the pattern of means shown in Figure 13.3 produced significant main effects for season and gender in addition to a season-by-gender interaction. The main effects record that men use more energy than women and that the wet season is associated with greater activity than the dry season. This clearly does not tell the whole story. We need to discuss the men and women in each of the two seasons to fully understand how energy is used in Burkina Faso.

A different kind of interaction is present in the next example. Here, we must be very cautious in our interpretation of the main effects since one of them can lead to a distorted conclusion.

EXAMPLE 13.7

An experiment to study how noise affects the performance of children tested second-grade hyperactive children and a control group of second graders who were not hyperactive.[5] One of the tasks involved solving math problems. The children solved problems under both high-noise and low-noise conditions. Here are the mean scores:

Group	High noise	Low noise	Mean
Control	214	170	192
Hyperactive	120	140	130
Mean	167	155	161

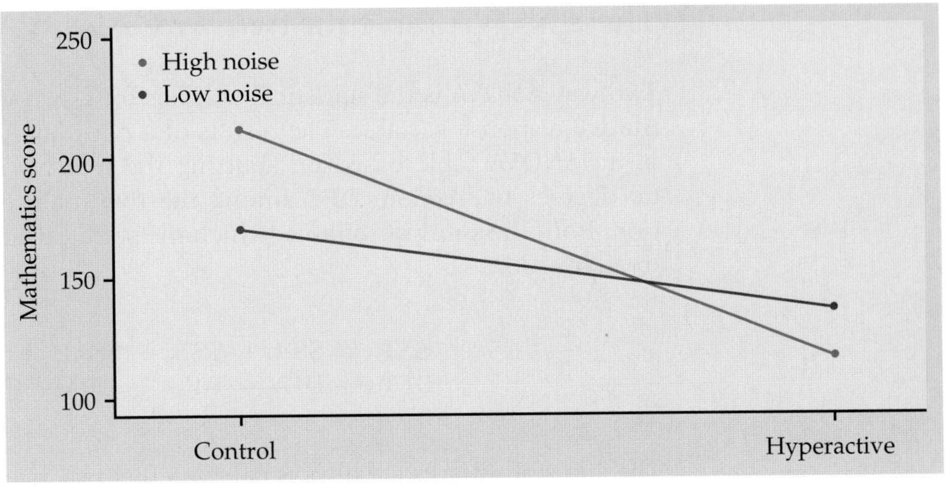

FIGURE 13.4 Plot of mean math scores of control and hyperactive children under low- and high-noise conditions, for Example 13.7.

The means are plotted in Figure 13.4. In the analysis of this experiment, both of the main effects and the interaction were statistically significant. How are we to interpret these results?

What catches our eye in the plot is that the lines cross. The hyperactive children did better in low noise than in high, but the control children performed better in high noise than in low. This interaction between noise and type of children is an important result. Which noise level is preferable depends on which type of children are in the class. The main effect for type of children is straightforward: the normal children did better than the hyperactive children at both levels of noise. However, the main effect for noise could easily be misinterpreted if we do not look at the means carefully. Under high-noise conditions the marginal mean was 167, compared with a low-noise mean of 155. This suggests that high noise is, on the average, a more favorable condition than low noise. In fact, high noise is more favorable only for the control group. The hyperactive children performed better under the low-noise condition. Because of the interaction, we must know which type of children we have in order to say which noise level is better for working math problems.

13.2 Inference for Two-Way ANOVA

Inference for two-way ANOVA involves F statistics for each of the two main effects and an additional F statistic for the interaction. As with one-way ANOVA, the calculations are organized in an ANOVA table.

The ANOVA table for two-way ANOVA

Two-way ANOVA is the statistical analysis for a two-way design with a quantitative response variable. The results of a two-way ANOVA are summarized in an ANOVA table based on splitting the total variation SST and the total degrees of freedom DFT among the two main effects and the interaction. Both the sums of squares (which measure variation) and the degrees of freedom add:

$$\begin{aligned} \text{SST} &= \text{SSA} + \text{SSB} + \text{SSAB} + \text{SSE} \\ \text{DFT} &= \text{DFA} + \text{DFB} + \text{DFAB} + \text{DFE} \end{aligned}$$

The sums of squares are always calculated in practice by statistical software. When the n_{ij} are not all equal, some methods of analysis can give sums of squares that do not add. From each sum of squares and its degrees of freedom we find the mean square in the usual way:

$$\text{mean square} = \frac{\text{sum of squares}}{\text{degrees of freedom}}$$

The significance of each of the main effects and the interaction is assessed by an F statistic that compares the variation due to the effect of interest with the within-group variation. Each F statistic is the mean square for the source of interest divided by MSE. Here is the general form of the two-way ANOVA table:

Source	Degrees of freedom	Sum of squares	Mean square	F
A	$I - 1$	SSA	SSA/DFA	MSA/MSE
B	$J - 1$	SSB	SSB/DFB	MSB/MSE
AB	$(I - 1)(J - 1)$	SSAB	SSAB/DFAB	MSAB/MSE
Error	$N - IJ$	SSE	SSE/DFE	
Total	$N - 1$	SST	SST/DFT	

There are three null hypotheses in two-way ANOVA, with an F test for each. We can test for significance of the main effect of A, the main effect of B, and the AB interaction. It is generally good practice to examine the test for interaction first, since the presence of a strong interaction may influence the interpretation of the main effects. Be sure to plot the means as an aid to interpreting the results of the significance tests.

Significance Tests in Two-Way ANOVA

To test the main effect of A, use the F statistic

$$F_A = \frac{\text{MSA}}{\text{MSE}}$$

To test the main effect of B, use the F statistic

$$F_B = \frac{\text{MSB}}{\text{MSE}}$$

To test the interaction of A and B, use the F statistic

$$F_{AB} = \frac{\text{MSAB}}{\text{MSE}}$$

If the effect being tested is zero, the calculated F statistic has an F distribution with numerator degrees of freedom corresponding to the effect and denominator degrees of freedom equal to DFE. Large values of the F statistic lead to rejection of the null hypothesis. The P-value is the probability that a random variable having the corresponding F distribution is greater than or equal to the calculated value.

The following example illustrates how to do a two-way ANOVA. As with the one-way ANOVA, we focus our attention on interpretation of the computer output.

EXAMPLE 13.8

A study of cardiovascular risk factors compared runners who averaged at least 15 miles per week with a control group described as "generally sedentary." Both men and women were included in the study.[6] The design is a 2×2 ANOVA with the factors group and gender. There were 200 subjects in each of the four combinations. One of the variables measured was the heart rate after 6 minutes of exercise on a treadmill. SAS computer analysis produced the outputs in Figure 13.5 and Figure 13.6.

We begin with the usual preliminary examination. From Figure 13.5 we see that the ratio of the largest to the smallest standard deviation is less than 2. Therefore, we are not concerned about violating the assumption of equal population standard deviations. Normal quantile plots (not shown) do not reveal any outliers, and the data appear to be reasonably normal.

The ANOVA table at the top of the output in Figure 13.6 is in effect a one-way ANOVA with four groups: female control, female runner, male control, and male runner. In this analysis Model has 3 degrees of freedom, and Error has 796 degrees of freedom. The F test and its associated P-value for this analysis refer to the hypothesis that all four groups have the same population mean. We are interested in the main effects and interaction, so we ignore this test.

The sums of squares for the group and gender main effects and the group-by-gender interaction appear at the bottom of Figure 13.6 under the heading

```
Analysis Variable : HR

 GROUP=Control GENDER=Female
  N           Mean        Std Dev      Std Error
-------------------------------------------------
200         148.00          16.27           1.15
-------------------------------------------------

 GROUP=Control GENDER=Male
  N           Mean        Std Dev      Std Error
-------------------------------------------------
200         130.00          17.10           1.21
-------------------------------------------------

 GROUP=Runners GENDER=Female
  N           Mean        Std Dev      Std Error
-------------------------------------------------
200         115.99          15.97           1.13
-------------------------------------------------

 GROUP=Runners GENDER=Male
  N           Mean        Std Dev      Std Error
-------------------------------------------------
200         103.98          12.50           0.88
-------------------------------------------------
```

FIGURE 13.5 Summary statistics for heart rates in the four groups of a 2×2 ANOVA, for Example 13.8.

```
General Linear Models Procedure
Dependent Variable: HR
                                 Sum of          Mean
Source                  DF      Squares        Square   F Value      Pr > F
Model                    3    215256.09      71752.03    296.35      0.0001
Error                  796    192729.83        242.12
Corrected Total        799    407985.92

              R-square             C.V.      Root MSE               HR Mean
              0.527607         12.49924        15.560                124.49

Source                  DF    Type I SS   Mean Square   F Value      Pr > F
GROUP                    1    168432.08     168432.08    695.65      0.0001
GENDER                   1     45030.00      45030.00    185.98      0.0001
GROUP*GENDER             1      1794.01       1794.01      7.41      0.0066
```

FIGURE 13.6 Two-way analysis of variance output for heart rates, for Example 13.8.

Type I SS. These sum to the sum of squares for Model. Similarly, the degrees of freedom for these sums of squares sum to the degrees of freedom for Model. Two-way ANOVA splits the variation among the means (expressed by the Model sum of squares) into three parts that reflect the two-way layout.

Because the degrees of freedom are all 1 for the main effects and the interaction, the mean squares are the same as the sums of squares. The F statistics for the three effects appear in the column labeled F Value, and the P-values are under the heading Pr > F. For the group main effect, we verify the calculation of F as follows:

$$F = \frac{\text{MSG}}{\text{MSE}} = \frac{168{,}432}{242.12} = 695.65$$

All three effects are statistically significant. The group effect has the largest F, followed by the gender effect and then the group-by-gender interaction. To interpret these results, we examine the plot of means with bars indicating one standard error in Figure 13.7. Note that the standard errors are quite small due to the large sample sizes. The significance of the main effect for group is due to the fact that the controls have higher average heart rates than the runners for both genders. This is the largest effect evident in the plot.

The significance of the main effect for gender is due to the fact that the females have higher heart rates than the men in both groups. The differences are not as large as those for the group effect, and this is reflected in the smaller value of the F statistic.

The analysis indicates that a complete description of the average heart rates requires consideration of the interaction in addition to the main effects. The two lines in the plot are not parallel. This interaction can be described in two ways. The female-male difference in average heart rates is greater for the controls than for the runners. Alternatively, the difference in average heart

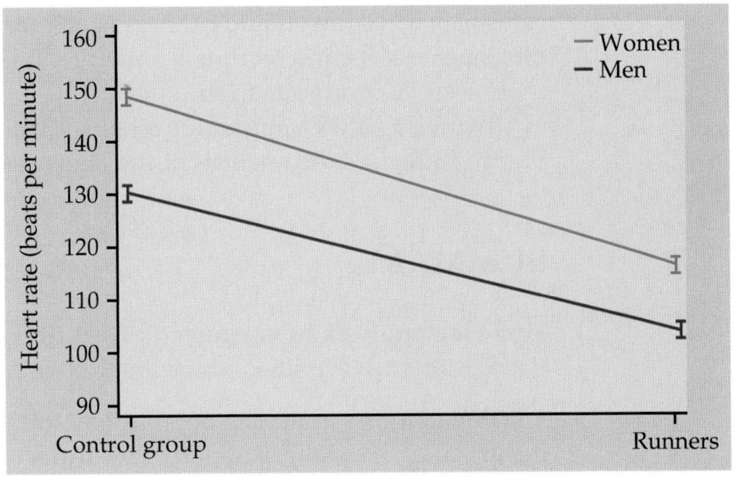

FIGURE 13.7 Plot of the group means with standard errors for heart rates in the 2 × 2 ANOVA, for Example 13.8.

Excel

	F	G	H	I	J	K	L
1	Anova: Two Factor With Replication						
2							
3	SUMMARY	Control	Runners	Total			
4	*Female*						
5	Count	200	200	400			
6	Sum	29600	23197	52797			
7	Average	148	115.985	131.9925			
8	Variance	264.7437	255.0902	516.1478			
9							
10	*Male*						
11	Count	200	200	400			
12	Sum	26000	20795	46795			
13	Average	130	103.975	116.9875			
14	Variance	292.4221	156.2356.2	393.5161			
15							
16	*Total*						
17	Count	400	400				
18	Sum	55600	43992				
19	Average	139	109.98				
20	Variance	359.0877	241.2978				
21							
22							
23	ANOVA						
24	*Source of Variation*	*SS*	*df*	*MS*	*F*	*P-value*	*F crit*
25	Sample	45030.01	1	45030.01	185.9799	3.29E-38	3.85316
26	Columns	168432.1	1	168432.1	695.647	1.10E-110	3.85316
27	Interaction	1794.005	1	1794.005	7.409481	0.00663	3.85316
28	Within	192729.8	796	242.1229			
29							
30	Total	407985.9	799				
31							

hrate1

Ready Sum=0 NUM

Minitab

Analysis of Variance for HR

Source	DF	SS	MS
Group	1	168432	168432
Gender	1	45030	45030
Interaction	1	1794	1794
Error	796	192730	242
Total	799	407986	

FIGURE 13.8 Excel and Minitab two-way analysis of variance output for the heart rate study, for Example 13.8.

rates between controls and runners is greater for women than for men. As the plot suggests, the interaction is not large. It is statistically significant because there were 800 subjects in the study.

Two-way ANOVA output for other software is similar to that given by SAS. Figure 13.8 gives the analysis of the heart rate data using Excel and Minitab.

SUMMARY

Two-way analysis of variance is used to compare population means when populations are classified according to two factors.

ANOVA assumes that the populations are normal with possibly different means and the same standard deviation and that independent SRSs are drawn from each population.

As with one-way ANOVA, preliminary analysis includes examination of means, standard deviations, and normal quantile plots. **Marginal means** are calculated by taking averages of the cell means across rows and columns. Pooling is used to estimate the within-group variance.

ANOVA separates the total variation into parts for the **model** and **error.** The model variation is separated into parts for each of the **main effects** and the **interaction.**

The calculations are organized into an **ANOVA table.** F statistics and P-values are used to test hypotheses about the main effects and the interaction.

Careful inspection of the means is necessary to interpret significant main effects and interactions. Plots are a useful aid.

CHAPTER 13 EXERCISES

13.1 Each of the following situations is a two-way study design. For each case, identify the response variable and both factors, and state the number of levels for each factor (I and J) and the total number of observations (N).

(a) A company wants to compare three different training programs for its new employees. Each of these programs takes 8 hours to complete. The training can be given for 8 hours on one day or for 4 hours on two consecutive days. The next 120 employees that the company hires will be the subjects for this study. After the training is completed, the employees are asked to evaluate the effectiveness of the program on a 7-point scale.

(b) A marketing experiment compares four different types of packaging for computer disks. Each type of packaging can be presented in three different colors. Each combination of package type with a particular color is shown to 40 different potential customers, who rate the attractiveness of the product on a 1 to 10 scale.

(c) Five different formulations for your hand lotion product have been produced by your research and marketing group, and you want to decide which of these, if any, to market. The lotions can be made with three different fragrances. Samples of each formulation-by-fragrance lotion are sent to 120 randomly selected customers who use your regular product. You ask each customer to compare the new lotion with the regular product by rating it on a 7-point scale. The middle point of the scale corresponds to no preference, while higher values indicate that the new product is preferred and lower values indicate that the regular product is better.

13.2 Each of the following situations is a two-way study design. For each case, identify the response variable and both factors, and state the number of levels for each factor (I and J) and the total number of observations (N).

(a) A study of smoking classifies subjects as nonsmokers, moderate smokers, or heavy smokers. Samples of 80 men and 80 women are drawn from each group. Each person reports the number of hours of sleep he or she gets on a typical night.

(b) The strength of concrete depends upon the formula used to prepare it. An experiment compares six different mixtures. Nine specimens of concrete are poured from each mixture. Three of these specimens are subjected to 0 cycles of freezing and thawing, three are subjected to 100 cycles, and three specimens are subjected to 500 cycles. The strength of each specimen is then measured.

(c) Four methods for teaching sign language are to be compared. Sixteen students in special education and sixteen students majoring in other areas are the subjects for the study. Within each group they are randomly assigned to the methods. Scores on a final exam are compared.

13.3 For each part of Exercise 13.1, outline the ANOVA table, giving the sources of variation and the degrees of freedom. (Do not compute the numerical values for the sums of squares and mean squares.)

13.4 For each part of Exercise 13.2, outline the ANOVA table, giving the sources of variation and the degrees of freedom. (Do not compute the numerical values for the sums of squares and mean squares.)

13.5 A two-way ANOVA model was used to analyze an experiment with three levels of one factor, four levels of a second factor, and 6 observations per treatment combination.

(a) For each of the main effects and the interaction, give the degrees of freedom for the corresponding F statistic.

(b) Using Table E or statistical software, find the value that each of these F statistics must exceed for the result to be significant at the 5% level.

(c) Answer part (b) for the 1% level.

13.6 A two-way ANOVA model was used to analyze an experiment with two levels of one factor, three levels of a second factor, and 6 observations per treatment combination.

(a) For each of the main effects and the interaction, give the degrees of freedom for the corresponding F statistic.

(b) Using Table E or statistical software, find the value that each of these F statistics must exceed for the result to be significant at the 5% level.

(c) Answer part (b) for the 1% level.

13.7 In the course of a clinical trial of measures to prevent coronary heart disease, blood pressure measurements were taken on 12,866 men. Individuals were classified by age group and race.[7] The means for systolic blood pressure are given in the following table:

	35–39	40–44	45–49	50–54	55–59
White	131.0	132.3	135.2	139.4	142.0
Nonwhite	132.3	134.2	137.2	141.3	144.1

(a) Plot the group means, with age on the x axis and blood pressure on the y axis. For each racial group connect the points for the different ages.

(b) Describe the patterns you see. Does there appear to be a difference between the two racial groups? Does systolic blood pressure appear to vary with age? If so, how does it vary? Is there an interaction?

(c) Compute the marginal means. Then find the differences between the white and nonwhite mean blood pressures for each age group. Use this information to summarize numerically the patterns in the plot.

13.8 The means for diastolic blood pressure recorded in the clinical trial described in the previous exercise are:

	35–39	40–44	45–49	50–54	55–59
White	89.4	90.2	90.9	91.6	91.4
Nonwhite	91.2	93.1	93.3	94.5	93.5

(a) Plot the group means with age on the x axis and blood pressure on the y axis. For each racial group connect the points for the different ages.

(b) Describe the patterns you see. Does there appear to be a difference between the two racial groups? Does diastolic blood pressure appear to vary with age? If so, how does it vary? Is there an interaction between race and age?

(c) Compute the marginal means. Find the differences between the white and nonwhite mean blood pressures for each age group. Use this information to summarize numerically the patterns in the plot.

13.9 The amount of chromium in the diet has an effect on the way the body processes insulin. In an experiment designed to study this phenomenon, four diets were fed to male rats. There were two factors. Chromium had two levels: low (L) and normal (N). The rats were allowed to eat as much as they wanted (M) or the total amount that they could eat was restricted (R). We call the second factor Eat. One of the variables measured was the amount of an enzyme called GITH.[8] The means for this response variable appear in the following table:

	Eat	
Chromium	M	R
L	4.545	5.175
N	4.425	5.317

(a) Make a plot of the mean GITH for these diets, with the factor Chromium on the x axis and GITH on the y axis. For each Eat group connect the points for the two Chromium means.

(b) Describe the patterns you see. Does the amount of chromium in the diet appear to affect the GITH mean? Does restricting the diet rather than letting the rats eat as much as they want appear to have an effect? Is there an interaction?

(c) Compute the marginal means. Compute the differences between the M and R diets for each level of Chromium. Use this information to summarize numerically the patterns in the plot.

13.10 The Chapin Social Insight Test measures how well people can appraise others and predict what they may say or do. A study administered this test to different groups of people and compared the mean scores.[9] Some of the results are given in the table below. Means for males and females who were psychology graduate students (PG) and liberal arts undergraduates (LU) are presented. The two factors are labeled Gender and Group.

	Group	
Gender	PG	LU
Males	27.56	25.34
Females	29.25	24.94

Plot the means and describe the essential features of the data in terms of main effects and interactions.

13.11 The change-of-majors study described in Example 12.3 (page 750) classified students into one of three groups depending upon their major in the sophomore year. In this exercise we also consider the gender of the students. There are now two factors: major with three levels, and gender with two levels. The mean SAT mathematics scores for the six groups appear in the following table. For convenience we use the labels CS for computer science majors, EO for engineering and other science majors, and O for other majors.

	Major		
Gender	CS	EO	O
Males	628	618	589
Females	582	631	543

Describe the main effects and interaction using appropriate graphs and calculations.

13.12 The mean high school mathematics grades for the students in the previous exercise are summarized in the following table. The grades have been coded so that 10 = A, 9 = A−, etc.

Gender	Major		
	CS	EO	O
Males	8.68	8.35	7.65
Females	9.11	9.36	8.04

Summarize the results of this study using appropriate plots and calculations to describe the main effects and interaction.

13.13 One way to repair serious wounds is to insert some material as a scaffold for the body's repair cells to use as a template for new tissue. Scaffolds made from extracellular material (ECM) are particularly promising for this purpose. Because they are made from biological material, they serve as an effective scaffold and are then resorbed. Unlike biological material that includes cells, however, they do not trigger tissue rejection reactions in the body. One study compared 6 types of scaffold material.[10] Three of these were ECMs and the other three were made of inert materials. There were three mice used per scaffold type. The response measure was the percent of glucose phosphated isomerase (Gpi) cells in the region of the wound. A large value is good, indicating that there are many bone marrow cells sent by the body to repair the tissue. In Exercise 12.27 (page 788) we analyzed the data for rats whose tissues were measured 4 weeks after the repair. The experiment included additional groups of rats who received the same types of scaffold but were measured at different times. Here are the data for 4 weeks and 8 weeks after the repair:

Material	Gpi (%)					
	4 weeks			8 weeks		
ECM1	55	70	70	60	65	65
ECM2	60	65	65	60	70	60
ECM3	75	70	75	70	80	70
MAT1	20	25	25	15	25	25
MAT2	5	10	5	10	5	5
MAT3	10	15	10	5	15	10

(a) Make a table giving the sample size, mean, and standard deviation for each of the material-by-time combinations. Is it reasonable to pool the variances? Because the sample sizes in this experiment are very small, we expect a large amount of variability in the sample standard deviations. Although they vary more than we would prefer, we will proceed with the ANOVA.

(b) Make a plot of the means. Describe the main features of the plot.

(c) Run the analysis of variance. Report the F statistics with degrees of freedom and P-values for the main effects and the interaction. What do you conclude? Write a short paragraph summarizing the results of your analysis.

13.14 Refer to the previous exercise. Here are the data that were collected at 2 weeks, 4 weeks, and 8 weeks:

	Gpi (%)								
Material	2 weeks			4 weeks			8 weeks		
ECM1	70	75	65	55	70	70	60	65	65
ECM2	60	65	70	60	65	65	60	70	60
ECM3	80	60	75	75	70	75	70	80	70
MAT1	50	45	50	20	25	25	15	25	25
MAT2	5	10	15	5	10	5	10	5	5
MAT3	30	25	25	10	15	10	5	15	10

Rerun the analyses that you performed in the previous exercise. How does the addition of the data for 2 weeks change the conclusions? Write a summary comparing these results with those in the previous exercise.

13.15 Refer to the previous exercise. Analyze the data for each time period separately using a one-way ANOVA. Use a multiple comparison procedure where needed. Summarize the results.

13.16 Iron-deficiency anemia is the most common form of malnutrition in developing countries, affecting about 50% of children and women and 25% of men. Iron pots for cooking foods had traditionally been used in many of these countries, but they have been largely replaced by aluminum pots, which are cheaper and lighter. Some research has suggested that food cooked in iron pots will contain more iron than food cooked in other types of pots. One study designed to investigate this issue compared the iron content of some Ethiopian foods cooked in aluminum, clay, and iron pots.[11] In Exercise 12.25, we analyzed the iron content of *yesiga wet'*, beef cut into small pieces and prepared with several Ethiopian spices. The researchers who conducted this study also examined the iron content of *shiro wet'*, a legume-based mixture of chickpea flour and Ethiopian spiced pepper, and *ye-atkilt allych'a*, a lightly spiced vegetable casserole. In the table below, these three foods are labeled meat, legumes, and vegetables. Four samples of each food were cooked in each type of pot. The iron in the food is measured in milligrams of iron per 100 grams of cooked food. Here are the data:

	Iron content											
Type of pot	Meat				Legumes				Vegetables			
Aluminum	1.77	2.36	1.96	2.14	2.40	2.17	2.41	2.34	1.03	1.53	1.07	1.30
Clay	2.27	1.28	2.48	2.68	2.41	2.43	2.57	2.48	1.55	0.79	1.68	1.82
Iron	5.27	5.17	4.06	4.22	3.69	3.43	3.84	3.72	2.45	2.99	2.80	2.92

(a) Make a table giving the sample size, mean, and standard deviation for each type of pot. Is it reasonable to pool the variances? Although the standard deviations vary more than we would like, this is partially due to the small sample sizes and we will proceed with the analysis of variance.

(b) Plot the means. Give a short summary of how the iron content of foods depends upon the cooking pot.

(c) Run the analysis of variance. Give the ANOVA table, the *F* statistics with degrees of freedom and *P*-values, and your conclusions regarding the hypotheses about main effects and interactions.

13.17 Refer to the previous exercise. Although there is a statistically significant interaction, do you think that these data support the conclusion that foods cooked in iron pots contain more iron than foods cooked in aluminum or clay pots? Discuss.

13.18 Refer to Exercise 13.16. Rerun the analysis as a one-way ANOVA with 9 groups and 4 observations per group. Report the results of the *F* test. Examine differences in means using a multiple comparison procedure. Summarize your results and compare them with those you obtained in Exercise 13.16.

13.19 One step in the manufacture of large engines requires that holes of very precise dimensions be drilled. The tools that do the drilling are regularly examined and are adjusted to ensure that the holes meet the required specifications. Part of the examination involves measurement of the diameter of the drilling tool. A team studying the variation in the sizes of the drilled holes selected this measurement procedure as a possible cause of variation in the drilled holes. They decided to use a designed experiment as one part of this examination. Some of the data are given in Table 13.1. The diameters in millimeters (mm) of five tools were measured by the same operator at three times (8:00 A.M., 11:00 A.M., and 3:00 P.M.). Three measurements were taken on each tool at each time. The person taking the measurements could not tell which tool was being measured, and the measurements were taken in random order.[12]

(a) Make a table of means and standard deviations for each of the 5×3 combinations of the two factors.

(b) Plot the means and describe how the means vary with tool and time. Note that we expect the tools to have slightly different diameters. These will be adjusted as needed. It is the process of measuring the diameters that is important.

(c) Use a two-way ANOVA to analyze these data. Report the test statistics, degrees of freedom, and *P*-values for the significance tests.

(d) Write a short report summarizing your results.

13.20 Refer to the previous exercise. Multiply each measurement by 0.04 to convert from millimeters to inches. Redo the plots and rerun the ANOVA using the transformed measurements. Summarize what parts of the analysis have changed and what parts have remained the same.

TABLE 13.1 Tool diameter data

Tool	Time	Diameter		
1	1	25.030	25.030	25.032
1	2	25.028	25.028	25.028
1	3	25.026	25.026	25.026
2	1	25.016	25.018	25.016
2	2	25.022	25.020	25.018
2	3	25.016	25.016	25.016
3	1	25.005	25.008	25.006
3	2	25.012	25.012	25.014
3	3	25.010	25.010	25.008
4	1	25.012	25.012	25.012
4	2	25.018	25.020	25.020
4	3	25.010	25.014	25.018
5	1	24.996	24.998	24.998
5	2	25.006	25.006	25.006
5	3	25.000	25.002	24.999

13.21 How does the frequency that a supermarket product is promoted at a discount affect the price that customers expect to pay for the product? Does the percent reduction also affect this expectation? These questions were examined by researchers in a study conducted on students enrolled in an introductory management course at a large midwestern university. For 10 weeks 160 subjects received information about the products. The treatment conditions corresponded to the number of promotions (1, 3, 5, or 7) that were described during this 10-week period and the percent that the product was discounted (10%, 20%, 30%, and 40%). Ten students were randomly assigned to each of the $4 \times 4 = 16$ treatments.[13] Table 13.2 gives the data.

(a) Summarize the means and standard deviations in a table and plot the means. Summarize the main features of the plot.

(b) Analyze the data with a two-way ANOVA. Report the results of this analysis.

(c) Using your plot and the ANOVA results, prepare a short report explaining how the expected price depends on the number of promotions and the percent of the discount.

13.22 Refer to the previous exercise. Rerun the analysis as a one-way ANOVA with $4 \times 4 = 16$ treatments. Summarize the results of this analysis. Use the Bonferroni multiple comparisons procedure to describe combinations of number of promotions and percent discounts that are similar or different.

TABLE 13.2 Expected price data

Number of promotions	Percent discount	Expected price									
1	40	4.10	4.50	4.47	4.42	4.56	4.69	4.42	4.17	4.31	4.59
1	30	3.57	3.77	3.90	4.49	4.00	4.66	4.48	4.64	4.31	4.43
1	20	4.94	4.59	4.58	4.48	4.55	4.53	4.59	4.66	4.73	5.24
1	10	5.19	4.88	4.78	4.89	4.69	4.96	5.00	4.93	5.10	4.78
3	40	4.07	4.13	4.25	4.23	4.57	4.33	4.17	4.47	4.60	4.02
3	30	4.20	3.94	4.20	3.88	4.35	3.99	4.01	4.22	3.70	4.48
3	20	4.88	4.80	4.46	4.73	3.96	4.42	4.30	4.68	4.45	4.56
3	10	4.90	5.15	4.68	4.98	4.66	4.46	4.70	4.37	4.69	4.97
5	40	3.89	4.18	3.82	4.09	3.94	4.41	4.14	4.15	4.06	3.90
5	30	3.90	3.77	3.86	4.10	4.10	3.81	3.97	3.67	4.05	3.67
5	20	4.11	4.35	4.17	4.11	4.02	4.41	4.48	3.76	4.66	4.44
5	10	4.31	4.36	4.75	4.62	3.74	4.34	4.52	4.37	4.40	4.52
7	40	3.56	3.91	4.05	3.91	4.11	3.61	3.72	3.69	3.79	3.45
7	30	3.45	4.06	3.35	3.67	3.74	3.80	3.90	4.08	3.52	4.03
7	20	3.89	4.45	3.80	4.15	4.41	3.75	3.98	4.07	4.21	4.23
7	10	4.04	4.22	4.39	3.89	4.26	4.41	4.39	4.52	3.87	4.70

13.23 Return to the chromium-insulin experiment described in Exercise 13.9. Here is part of the ANOVA table for these data:

Source	Degrees of freedom	Sum of squares	Mean square	F
A (Chromium)		0.00121		
B (Eat)		5.79121		
AB		0.17161		
Error		1.08084		
Total				

(a) In all, 40 rats were used in this experiment. Fill in the missing values in the ANOVA table.

(b) What is the value of the F statistic to test the null hypothesis that there is no interaction? What is its distribution when the null hypothesis is true? Using Table E, find an approximate P-value for this test.

(c) Answer the questions in part (b) for the main effect of Chromium and the main effect of Eat.

(d) What is s_p^2, the within-group variance? What is s_p?

(e) Using what you have learned in this exercise and your answer to Exercise 13.9, summarize the results of this experiment.

13.24 Return to the Chapin Social Insight Test study described in Exercise 13.10. Part of the ANOVA table for these data is given below:

Source	Degrees of freedom	Sum of squares	Mean square	F
A (Gender)		62.40		
B (Group)		1,599.03		
AB				
Error		13,633.29		
Total		15,458.52		

(a) There were 150 individuals tested in each of the groups. Fill in the missing values in the ANOVA table.

(b) What is the value of the F statistic to test the null hypothesis that there is no interaction? What is its distribution when the null hypothesis is true? Using Table E, find an approximate P-value for this test.

(c) Answer the questions in part (b) for the main effect of Gender and the main effect of Group.

(d) What is s_p^2, the within-group variance? What is s_p?

(e) Using what you have learned in this exercise and your answer to Exercise 13.10, summarize the results of this study.

13.25 Do left-handed people live shorter lives than right-handed people? A study of this question examined a sample of 949 death records and contacted next of kin to determine handedness.[14] Note that there are many possible definitions of "left-handed." The researchers examined the effects of different definitions on the results of their analysis and found that their conclusions were not sensitive to the exact definition used. For the results presented here, people were defined to be right-handed if they wrote, drew, and threw a ball with the right hand. All others were defined to be left-handed. People were classified by gender (female or male) and handedness (left or right), and a 2 × 2 ANOVA was run with the age at death as the response variable. The F statistics were 22.36 (handedness), 37.44 (gender), and 2.10 (interaction). The following marginal mean ages at death (in years) were reported: 77.39 (females), 71.32 (males), 75.00 (right-handed), and 66.03 (left-handed).

(a) For each of the F statistics given above find the degrees of freedom and an approximate P-value. Summarize the results of these tests.

(b) Using the information given, write a short summary of the results of the study.

13.26 Scientists believe that exposure to the radioactive gas radon is associated with some types of cancers in the respiratory system. Radon from natural sources is present in many homes in the United States. A group of researchers decided to study the problem in dogs because dogs get similar types of cancers and are exposed to environments similar to those of their owners. Radon detectors are available for home monitoring but the researchers wanted to obtain actual measures of the exposure of a sample of dogs. To do this they placed the detectors in holders and attached them to the collars of the dogs. One problem was that the holders might in some way affect the radon readings. The researchers therefore devised a laboratory experiment to study the effects of the holders. Detectors from four series of production were available, so they used a two-way ANOVA design (series with 4 levels and holder with 2, representing the presence or absence of a holder). For each combination of these factors, there were 8 detectors. All detectors were exposed to the same radon source and the radon measure in picocuries per liter was recorded.[15] The F statistic for the effect of series is 7.02, for holder it is 1.96, and for the interaction it is 1.24.

(a) Using Table E or statistical software find approximate P-values for the three test statistics. Summarize the results of these three significance tests.

(b) The mean radon readings for the four series were 330, 303, 302, and 295. The results of the significance test for series were of great concern to the researchers. Explain why.

13.27 The PLANTS1 data set in the Data Appendix gives the percent of nitrogen in four different species of plants grown in a laboratory. The species are *Leucaena leucocephala*, *Acacia saligna*, *Prosopis juliflora*, and *Eucalyptus citriodora*. The researchers who collected these data were interested in commercially growing these plants in parts of the country of Jordan where there is very little rainfall. To examine the effect of water, they varied the amount per day from 50 millimeters (mm) to 650 mm in 100 mm increments. There were nine plants per species-by-water combination. Because the plants are to be used primarily for animal food, with some parts that can be consumed by people, a high nitrogen content is very desirable.

(a) Find the means for each species-by-water combination. Plot these means versus water for the four species, connecting the means for each species by lines. Describe the overall pattern.

(b) Find the standard deviations for each species-by-water combination. Is it reasonable to pool the standard deviations for this problem? Note that with sample sizes of size 9, we expect these standard deviations to be quite variable.

(c) Run the two-way analysis of variance. Give the results of the hypothesis tests for the main effects and the interaction.

13.28 Refer to the previous exercise. Examine the residuals. Are there any unusual patterns or outliers? If you think that there are one or more points that are somewhat extreme, rerun the two-way analysis without these observations. Does this change the results in any substantial way?

13.29 Refer to Exercise 13.27. Run a separate one-way analysis of variance for each water level. If there is evidence that the species are not all the same, use a multiple comparisons procedure to determine which pairs of species are significantly different. In what way, if any, do the differences appear to vary by water level? Write a short summary of your conclusions.

13.30 Refer to Exercise 13.27. Run a separate one-way analysis of variance for each species and summarize the results. Since the amount of water is a quantitative factor, we can also analyze these data using regression. Run simple linear regressions separately for each species to predict nitrogen percent from water. Use plots to determine whether or not a line is a good way to approximate this relationship. Summarize the regression results and compare them with the one-way ANOVA results.

13.31 Refer to Exercise 13.27. Additional data collected by the same researchers according to a similar design are given in the PLANTS2 data set in the Data Appendix. Here, there are two response variables. They are fresh biomass and dry biomass. High values for both of these variables are desirable. The same four species and seven levels of water are used for this experiment. Here, however, there are four plants per species-by-water combination. Analyze each of the response variables in the PLANTS2 data set using the outline from Exercise 13.27.

13.32 Perform the tasks described in Exercise 13.28 for the two response variables in the PLANTS2 data set.

13.33 Perform the tasks described in Exercise 13.29 for the two response variables in the PLANTS2 data set.

13.34 Perform the tasks described in Exercise 13.30 for the two response variables in the PLANTS2 data set.

13.35 Refer to the data given for the change-of-majors study in the data set MAJORS described in the Data Appendix. Analyze the data for SATV, the SAT verbal score. Your analysis should include a table of sample sizes, means, and standard deviations; normal quantile plots; a plot of the means; and a two-way ANOVA using sex and major as the factors. Write a short summary of your conclusions.

13.36 Refer to the data given for the change-of-majors study in the data set MAJORS described in the Data Appendix. Analyze the data for HSS, the high school science grades. Your analysis should include a table of sample sizes, means, and standard deviations; normal quantile plots; a plot of the means; and a two-way ANOVA using sex and major as the factors. Write a short summary of your conclusions.

13.37 Refer to the data given for the change-of-majors study in the data set MAJORS described in the Data Appendix. Analyze the data for HSE, the high school English grades. Your analysis should include a table of sample sizes, means, and standard deviations; normal quantile plots; a plot of the means; and a two-way ANOVA using sex and major as the factors. Write a short summary of your conclusions.

13.38 Refer to the data given for the change-of-majors study in the data set MAJORS described in the Data Appendix. Analyze the data for GPA, the college grade point average. Your analysis should include a table of sample sizes, means, and standard deviations; normal quantile plots; a plot of the means; and a two-way ANOVA using sex and major as the factors. Write a short summary of your conclusions.

DATA APPENDIX

Some of the computer exercises in the text refer to nine relatively large data sets that are on the CD that accompanies this text. The CD also contains data from many other exercises and examples.

Background information for each of the nine data sets is presented below. For each, the first five cases are given here.

1 CHEESE

Text Reference: Exercises 1.139 and 11.35–11.43

As cheddar cheese matures, a variety of chemical processes take place. The taste of matured cheese is related to the concentration of several chemicals in the final product. In a study of cheddar cheese from the LaTrobe Valley of Victoria, Australia, samples of cheese were analyzed for their chemical composition and were subjected to taste tests.

Data for one type of cheese-manufacturing process appear below. The variable "Case" is used to number the observations from 1 to 30. "Taste" is the response variable of interest. The taste scores were obtained by combining the scores from several tasters.

Three of the chemicals whose concentrations were measured were acetic acid, hydrogen sulfide, and lactic acid. For acetic acid and hydrogen sulfide (natural) log transformations were taken. Thus the explanatory variables are the transformed concentrations of acetic acid ("Acetic") and hydrogen sulfide ("H2S") and the untransformed concentration of lactic acid ("Lactic"). These data are based on experiments performed by G. T. Lloyd and E. H. Ramshaw of the CSIRO Division of Food Research, Victoria, Australia. Some results of the statistical analyses of these data are given in G. P. McCabe, L. McCabe, and A. Miller, "Analysis of taste and chemical composition of cheddar cheese, 1982–83 experiments," CSIRO Division of Mathematics and Statistics Consulting Report VT85/6; and in I. Barlow et al., "Correlations and changes in flavour and chemical parameters of cheddar cheeses during maturation," *Australian Journal of Dairy Technology*, 44 (1989), pp. 7–18. The table below gives the data for the first five cases.

Case	Taste	Acetic	H2S	Lactic
01	12.3	4.543	3.135	0.86
02	20.9	5.159	5.043	1.53
03	39.0	5.366	5.438	1.57
04	47.9	5.759	7.496	1.81
05	5.6	4.663	3.807	0.99

2 CONCEPT

Text Reference: Exercises 11.32–11.34

Darlene Gordon of the Purdue University School of Education provided the data in the CONCEPT data set. The data were collected on 78 seventh-grade students in a rural midwestern school. The research concerned the relationship between the students' "self-concept" and their academic performance. The variables are OBS, a subject identification number; GPA, grade point average; IQ, score on an IQ test; AGE, age in years; SEX, female (1) or male (2); SC, overall score on the Piers-Harris Children's Self-Concept Scale; and C1 to C6, "cluster scores" for specific aspects of self-concept: C1 = behavior, C2 = school status, C3 = physical appearance, C4 = anxiety, C5 = popularity, and C6 = happiness. The following table gives the first five cases:

OBS	GPA	IQ	AGE	SEX	SC	C1	C2	C3	C4	C5	C6
001	7.940	111	13	2	67	15	17	13	13	11	9
002	8.292	107	12	2	43	12	12	7	7	6	6
003	4.643	100	13	2	52	11	10	5	8	9	7
004	7.470	107	12	2	66	14	15	11	11	9	9
005	8.882	114	12	1	58	14	15	10	12	11	6

3 CSDATA

Text Reference: Exercises 1.140, 2.132, 3.72, 3.73, 7.123, Example 11.1 and Exercises 11.24–11.31

The computer science department of a large university was interested in understanding why a large proportion of their first-year students failed to graduate as computer science majors. An examination of records from the registrar indicated that most of the attrition occurred during the first three semesters. Therefore, they decided to study all first-year students entering their program in a particular year and to follow their progress for the first three semesters.

The variables studied included the grade point average after three semesters and a collection of variables that would be available as students entered their program. These included scores on standardized tests such as the SATs and high school grades in various subjects. The individuals who conducted the study were also interested in examining differences between men and women in this program. Therefore, sex was included as a variable.

Data on 224 students who began study as computer science majors in a particular year were analyzed. A few exceptional cases were excluded, such as students who did not have complete data available on the variables of interest (a few students were admitted who did not take the SATs). Data for the first five students appear below. There are eight variables for each student. OBS is a variable used to identify the student. The data files kept by the registrar identified students by social security number, but for this study they were simply given a number from 1 to 224. The grade point average after three semesters is the variable GPA. This university uses a four-point scale, with A corresponding to 4, B to 3, C to 2, etc. A straight-A student has a 4.00 GPA.

The high school grades included in the data set are the variables HSM, HSS, and HSE. These correspond to average high school grades in math, science, and English. High schools use different grading systems (some high schools have a grade higher than A for honors courses), so the university's task in constructing these variables is not easy. The researchers were willing to accept the university's judgment and used its values. High school grades were recorded on a scale from 1 to 10, with 10 corresponding to A, 9 to A−, 8 to B+, etc.

The SAT scores are SATM and SATV, corresponding to the mathematics and verbal parts of the SAT. Gender was recorded as 1 for men and 2 for women. This is an arbitrary code. For software packages that can use alphanumeric variables (values do not have to be numbers), it is more convenient to use M and F or Men and Women as values for the sex variable. With this kind of user-friendly capability, you do not have to remember who are the 1's and who are the 2's.

Results of the study are reported in P. F. Campbell and G. P. McCabe, "Predicting the success of freshmen in a computer science major," *Communications of the ACM*, 27 (1984), pp. 1108–1113. The table below gives data for the first five students.

OBS	GPA	HSM	HSS	HSE	SATM	SATV	SEX
001	3.32	10	10	10	670	600	1
002	2.26	6	8	5	700	640	1
003	2.35	8	6	8	640	530	1
004	2.08	9	10	7	670	600	1
005	3.38	8	9	8	540	580	1

4 DANDRUFF

Text Reference: Exercises 12.48–12.51

The DANDRUFF data set is based on W. L. Billhimer et al., "Results of a clinical trial comparing 1% pyrithione zinc and 2% ketoconazole shampoos," *Cosmetic Dermatology*, 9 (1996), pp. 34–39. The study reported in this paper is a clinical trial that compared three treatments for dandruff and a placebo. The treatments were 1% pyrithione zinc shampoo (PyrI), the same shampoo but with instructions to shampoo two times (PyrII), 2% ketoconazole shampoo (Keto), and a placebo shampoo (Placebo). After six weeks of treatment, eight sections of the scalp were examined and given a score that measured the amount of scalp flaking on a 0 to 10 scale. The response variable was the sum of these eight scores. An analysis of the baseline flaking measure indicated that randomization of patients to treatments was successful in that no differences were found between the groups. At baseline there were 112 subjects in each of the three treatment groups and 28 subjects in the Placebo group. During the clinical trial 3 dropped out from the PyrII group and 6 from the Keto group. No patients dropped out of the other two groups. Summary statistics given in the paper were used to generate random data that give the same conclusions. Here are the first five cases:

OBS	Treatment	Flaking
001	PyrI	17
002	PyrI	16
003	PyrI	18
004	PyrI	17
005	PyrI	18

5 INDIVIDUALS

Text Reference: Exercises 1.42–1.44, 1.137, 1.138, and 2.44

Each March, the Bureau of Labor Statistics carries out an Annual Demographic Supplement to its monthly Current Population Survey. The BLS collects detailed data about the 50,000 randomly selected households in its sample. The data file *individuals.dat* contains data about 55,899 people from the March 2000 survey. We included all people between the ages of 25 and 65 who have worked but whose main work experience is not in agriculture. Moreover, we combined the 16 levels of education in the BLS survey to form 6 levels.

There are five variables in the data set. Age is age in years. Education is the highest level of education a person has reached, with the following values: 1 = did not reach high school; 2 = some high school, but no high school diploma; 3 = high school diploma; 4 = some college but no bachelor's degree (this includes people with an associate degree); 5 = bachelor's degree; 6 = postgraduate degree (master's, professional, or doctorate). Sex is gender coded as 1 = male and 2 = female. Total income is income from all sources. Note that income can be less than zero in some cases. Job class is a categorization of the person's main work experience, with 5 = private sector (outside households); 6 = government; 7 = self-employed. Here are the first five cases:

Age	Education	Sex	Total income	Job class
25	2	2	7234	5
25	5	1	37413	5
25	4	2	29500	5
25	3	2	13500	5
25	4	1	17660	6

The first individual in the file is a 25-year-old female who did not graduate from high school, works in the private sector, and had $7234 of income.

6 MAJORS

Text Reference:
Exercises 13.35–13.38

See the description of the CSDATA data set for background information on the study for this data set. In this data file, the variables described for CSDATA are given with an additional variable "Maj" that specifies the student's major field of study at the end of three semesters. The codes 1, 2, and 3 correspond to Computer Science, Engineering and Other Sciences, and Other. All available data were used in the analyses performed, which resulted in sample sizes that were unequal in the six sex-by-major groups.

For a one-way ANOVA this causes no particular problems. However, for a two-way ANOVA several complications arise when the sample sizes are unequal. A detailed discussion of these complications is beyond the scope of this text. To avoid these difficulties and still use these interesting data, simulated data based on the results of this study are given on the data disk. ANOVA based on these simulated data gives the same qualitative conclusions as those obtained with the original data. Here are the first five cases:

OBS	SEX	Maj	SATM	SATV	HSM	HSS	HSE	GPA
001	1	1	640	530	8	6	8	2.35
002	1	1	670	600	9	10	7	2.08
003	1	1	600	400	8	8	7	3.21
004	1	1	570	480	7	7	6	2.34
005	1	1	510	530	6	8	8	1.40

7 READING

Text Reference:
Exercises 7.124, 12.55–12.58, Examples 12.6 and 12.8–12.18

Jim Baumann and Leah Jones of the Purdue University School of Education conducted a study to compare three methods of teaching reading comprehension. The 66 students who participated in the study were randomly assigned to the methods (22 to each). The standard practice of comparing new methods with a traditional one was used in this study. The traditional method is called Basal and the two innovative methods are called DRTA and Strat.

In the data set the variable Subject is used to identify the individual students. The values are 1 to 66. The method of instruction is indicated by the variable Group, with values B, D, and S, corresponding to Basal, DRTA, and Strat. Two pretests and three posttests were given to all students. These are the variables Pre1, Pre2, Post1, Post2, and Post3. Data for the first five subjects are given below.

Subject	Group	Pre1	Pre2	Post1	Post2	Post3
01	B	4	3	5	4	41
02	B	6	5	9	5	41
03	B	9	4	5	3	43
04	B	12	6	8	5	46
05	B	16	5	10	9	46

8 PLANTS1

Text Reference:
Exercises 13.27–13.31

These data were collected by Maher Tadros from the Purdue University Department of Forestry and Natural Resources under the direction of his major professor Andrew Gillespie. Maher is from Jordan, a Middle Eastern country where there is very little rainfall in many areas. His research concerns four species of plants that may be suitable for commercial development in his country. Products produced by these species can be used as feed for animals and in some cases for humans. The four species of plants are *Leucaena leucocephala*, *Acacia saligna*, *Prosopis juliflora*, and *Eucalyptus citriodora*. A major research question concerns how well these species can tolerate drought stress.

PLANTS1 gives data for a laboratory experiment performed by Maher at Purdue University. Seven different amounts of water were given daily to plants of each species. For each of the species-by-water combinations, there were nine plants. The response variable is the percent of the plant that consists of nitrogen. A high nitrogen content is desirable for plant products that are used for food.

The actual experiment performed to collect these data had an additional factor that is not given in the data set. The 4 × 7, species-by-water combinations were actually run with three plants per combination. This design was then repeated three times. The repeat factor is often called a replicate, or rep, and is a standard part of most well-designed experiments of this type. For our purposes we ignore this additional factor and analyze the design as a 4 × 7 two-way ANOVA with 9 observations per treatment combination. The first five cases are listed below. The four species, *Leucaena leucocephala*, *Acacia saligna*, *Prosopis juliflora*, and *Eucalyptus citriodora*, are coded 1 to 4. The water levels, 50 mm, 150 mm, 250 mm, 350 mm, 450 mm, 550 mm, and 650 mm, are coded 1 to 7.

OBS	Species	Water	pctnit
001	1	1	3.644
002	1	1	3.500
003	1	1	3.509
004	1	1	3.137
005	1	1	3.100

9 PLANTS2

Text Reference:
Exercises 13.32–13.34

PLANTS2 gives data for a second experiment conducted by Maher Tadros in a lab at Purdue University. As in PLANTS1, there are the same four species of plants and the same seven levels of water. Here, however, there are four plants per species-by-water combination. The two response variables in the data set are fresh biomass and dry biomass. High values for these response variables indicate that the plants of the given species are resistant to drought at the given water level. Here are the first five cases:

OBS	Species	Water	fbiomass	dbiomass
001	1	1	105.13	37.65
002	1	1	138.95	48.85
003	1	1	90.05	38.85
004	1	1	102.25	36.91
005	1	1	207.90	74.35

TABLES

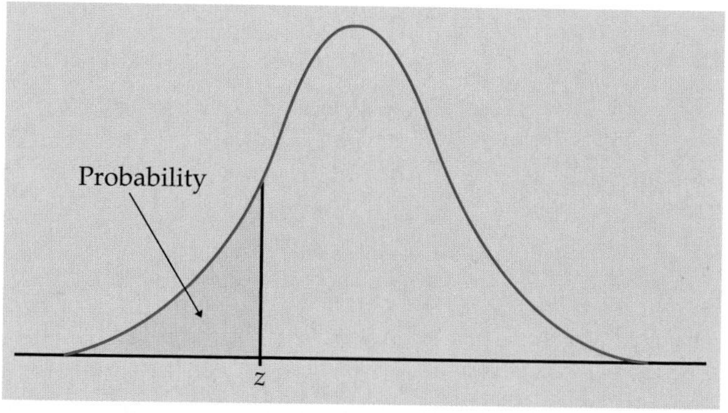

Table entry for z is the area under the standard normal curve to the left of z.

TABLE A Standard normal probabilities

z	.00	.01	.02	.03	.04	.05	.06	.07	.08	.09
−3.4	.0003	.0003	.0003	.0003	.0003	.0003	.0003	.0003	.0003	.0002
−3.3	.0005	.0005	.0005	.0004	.0004	.0004	.0004	.0004	.0004	.0003
−3.2	.0007	.0007	.0006	.0006	.0006	.0006	.0006	.0005	.0005	.0005
−3.1	.0010	.0009	.0009	.0009	.0008	.0008	.0008	.0008	.0007	.0007
−3.0	.0013	.0013	.0013	.0012	.0012	.0011	.0011	.0011	.0010	.0010
−2.9	.0019	.0018	.0018	.0017	.0016	.0016	.0015	.0015	.0014	.0014
−2.8	.0026	.0025	.0024	.0023	.0023	.0022	.0021	.0021	.0020	.0019
−2.7	.0035	.0034	.0033	.0032	.0031	.0030	.0029	.0028	.0027	.0026
−2.6	.0047	.0045	.0044	.0043	.0041	.0040	.0039	.0038	.0037	.0036
−2.5	.0062	.0060	.0059	.0057	.0055	.0054	.0052	.0051	.0049	.0048
−2.4	.0082	.0080	.0078	.0075	.0073	.0071	.0069	.0068	.0066	.0064
−2.3	.0107	.0104	.0102	.0099	.0096	.0094	.0091	.0089	.0087	.0084
−2.2	.0139	.0136	.0132	.0129	.0125	.0122	.0119	.0116	.0113	.0110
−2.1	.0179	.0174	.0170	.0166	.0162	.0158	.0154	.0150	.0146	.0143
−2.0	.0228	.0222	.0217	.0212	.0207	.0202	.0197	.0192	.0188	.0183
−1.9	.0287	.0281	.0274	.0268	.0262	.0256	.0250	.0244	.0239	.0233
−1.8	.0359	.0351	.0344	.0336	.0329	.0322	.0314	.0307	.0301	.0294
−1.7	.0446	.0436	.0427	.0418	.0409	.0401	.0392	.0384	.0375	.0367
−1.6	.0548	.0537	.0526	.0516	.0505	.0495	.0485	.0475	.0465	.0455
−1.5	.0668	.0655	.0643	.0630	.0618	.0606	.0594	.0582	.0571	.0559
−1.4	.0808	.0793	.0778	.0764	.0749	.0735	.0721	.0708	.0694	.0681
−1.3	.0968	.0951	.0934	.0918	.0901	.0885	.0869	.0853	.0838	.0823
−1.2	.1151	.1131	.1112	.1093	.1075	.1056	.1038	.1020	.1003	.0985
−1.1	.1357	.1335	.1314	.1292	.1271	.1251	.1230	.1210	.1190	.1170
−1.0	.1587	.1562	.1539	.1515	.1492	.1469	.1446	.1423	.1401	.1379
−0.9	.1841	.1814	.1788	.1762	.1736	.1711	.1685	.1660	.1635	.1611
−0.8	.2119	.2090	.2061	.2033	.2005	.1977	.1949	.1922	.1894	.1867
−0.7	.2420	.2389	.2358	.2327	.2296	.2266	.2236	.2206	.2177	.2148
−0.6	.2743	.2709	.2676	.2643	.2611	.2578	.2546	.2514	.2483	.2451
−0.5	.3085	.3050	.3015	.2981	.2946	.2912	.2877	.2843	.2810	.2776
−0.4	.3446	.3409	.3372	.3336	.3300	.3264	.3228	.3192	.3156	.3121
−0.3	.3821	.3783	.3745	.3707	.3669	.3632	.3594	.3557	.3520	.3483
−0.2	.4207	.4168	.4129	.4090	.4052	.4013	.3974	.3936	.3897	.3859
−0.1	.4602	.4562	.4522	.4483	.4443	.4404	.4364	.4325	.4286	.4247
−0.0	.5000	.4960	.4920	.4880	.4840	.4801	.4761	.4721	.4681	.4641

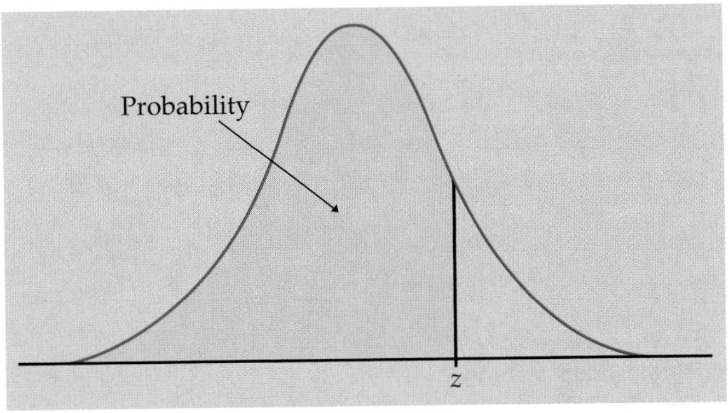

Table entry for *z* is the area under the standard normal curve to the left of *z*.

TABLE A Standard normal probabilities (*continued*)

z	.00	.01	.02	.03	.04	.05	.06	.07	.08	.09
0.0	.5000	.5040	.5080	.5120	.5160	.5199	.5239	.5279	.5319	.5359
0.1	.5398	.5438	.5478	.5517	.5557	.5596	.5636	.5675	.5714	.5753
0.2	.5793	.5832	.5871	.5910	.5948	.5987	.6026	.6064	.6103	.6141
0.3	.6179	.6217	.6255	.6293	.6331	.6368	.6406	.6443	.6480	.6517
0.4	.6554	.6591	.6628	.6664	.6700	.6736	.6772	.6808	.6844	.6879
0.5	.6915	.6950	.6985	.7019	.7054	.7088	.7123	.7157	.7190	.7224
0.6	.7257	.7291	.7324	.7357	.7389	.7422	.7454	.7486	.7517	.7549
0.7	.7580	.7611	.7642	.7673	.7704	.7734	.7764	.7794	.7823	.7852
0.8	.7881	.7910	.7939	.7967	.7995	.8023	.8051	.8078	.8106	.8133
0.9	.8159	.8186	.8212	.8238	.8264	.8289	.8315	.8340	.8365	.8389
1.0	.8413	.8438	.8461	.8485	.8508	.8531	.8554	.8577	.8599	.8621
1.1	.8643	.8665	.8686	.8708	.8729	.8749	.8770	.8790	.8810	.8830
1.2	.8849	.8869	.8888	.8907	.8925	.8944	.8962	.8980	.8997	.9015
1.3	.9032	.9049	.9066	.9082	.9099	.9115	.9131	.9147	.9162	.9177
1.4	.9192	.9207	.9222	.9236	.9251	.9265	.9279	.9292	.9306	.9319
1.5	.9332	.9345	.9357	.9370	.9382	.9394	.9406	.9418	.9429	.9441
1.6	.9452	.9463	.9474	.9484	.9495	.9505	.9515	.9525	.9535	.9545
1.7	.9554	.9564	.9573	.9582	.9591	.9599	.9608	.9616	.9625	.9633
1.8	.9641	.9649	.9656	.9664	.9671	.9678	.9686	.9693	.9699	.9706
1.9	.9713	.9719	.9726	.9732	.9738	.9744	.9750	.9756	.9761	.9767
2.0	.9772	.9778	.9783	.9788	.9793	.9798	.9803	.9808	.9812	.9817
2.1	.9821	.9826	.9830	.9834	.9838	.9842	.9846	.9850	.9854	.9857
2.2	.9861	.9864	.9868	.9871	.9875	.9878	.9881	.9884	.9887	.9890
2.3	.9893	.9896	.9898	.9901	.9904	.9906	.9909	.9911	.9913	.9916
2.4	.9918	.9920	.9922	.9925	.9927	.9929	.9931	.9932	.9934	.9936
2.5	.9938	.9940	.9941	.9943	.9945	.9946	.9948	.9949	.9951	.9952
2.6	.9953	.9955	.9956	.9957	.9959	.9960	.9961	.9962	.9963	.9964
2.7	.9965	.9966	.9967	.9968	.9969	.9970	.9971	.9972	.9973	.9974
2.8	.9974	.9975	.9976	.9977	.9977	.9978	.9979	.9979	.9980	.9981
2.9	.9981	.9982	.9982	.9983	.9984	.9984	.9985	.9985	.9986	.9986
3.0	.9987	.9987	.9987	.9988	.9988	.9989	.9989	.9989	.9990	.9990
3.1	.9990	.9991	.9991	.9991	.9992	.9992	.9992	.9992	.9993	.9993
3.2	.9993	.9993	.9994	.9994	.9994	.9994	.9994	.9995	.9995	.9995
3.3	.9995	.9995	.9995	.9996	.9996	.9996	.9996	.9996	.9996	.9997
3.4	.9997	.9997	.9997	.9997	.9997	.9997	.9997	.9997	.9997	.9998

TABLE B Random digits

Line								
101	19223	95034	05756	28713	96409	12531	42544	82853
102	73676	47150	99400	01927	27754	42648	82425	36290
103	45467	71709	77558	00095	32863	29485	82226	90056
104	52711	38889	93074	60227	40011	85848	48767	52573
105	95592	94007	69971	91481	60779	53791	17297	59335
106	68417	35013	15529	72765	85089	57067	50211	47487
107	82739	57890	20807	47511	81676	55300	94383	14893
108	60940	72024	17868	24943	61790	90656	87964	18883
109	36009	19365	15412	39638	85453	46816	83485	41979
110	38448	48789	18338	24697	39364	42006	76688	08708
111	81486	69487	60513	09297	00412	71238	27649	39950
112	59636	88804	04634	71197	19352	73089	84898	45785
113	62568	70206	40325	03699	71080	22553	11486	11776
114	45149	32992	75730	66280	03819	56202	02938	70915
115	61041	77684	94322	24709	73698	14526	31893	32592
116	14459	26056	31424	80371	65103	62253	50490	61181
117	38167	98532	62183	70632	23417	26185	41448	75532
118	73190	32533	04470	29669	84407	90785	65956	86382
119	95857	07118	87664	92099	58806	66979	98624	84826
120	35476	55972	39421	65850	04266	35435	43742	11937
121	71487	09984	29077	14863	61683	47052	62224	51025
122	13873	81598	95052	90908	73592	75186	87136	95761
123	54580	81507	27102	56027	55892	33063	41842	81868
124	71035	09001	43367	49497	72719	96758	27611	91596
125	96746	12149	37823	71868	18442	35119	62103	39244
126	96927	19931	36089	74192	77567	88741	48409	41903
127	43909	99477	25330	64359	40085	16925	85117	36071
128	15689	14227	06565	14374	13352	49367	81982	87209
129	36759	58984	68288	22913	18638	54303	00795	08727
130	69051	64817	87174	09517	84534	06489	87201	97245
131	05007	16632	81194	14873	04197	85576	45195	96565
132	68732	55259	84292	08796	43165	93739	31685	97150
133	45740	41807	65561	33302	07051	93623	18132	09547
134	27816	78416	18329	21337	35213	37741	04312	68508
135	66925	55658	39100	78458	11206	19876	87151	31260
136	08421	44753	77377	28744	75592	08563	79140	92454
137	53645	66812	61421	47836	12609	15373	98481	14592
138	66831	68908	40772	21558	47781	33586	79177	06928
139	55588	99404	70708	41098	43563	56934	48394	51719
140	12975	13258	13048	45144	72321	81940	00360	02428
141	96767	35964	23822	96012	94591	65194	50842	53372
142	72829	50232	97892	63408	77919	44575	24870	04178
143	88565	42628	17797	49376	61762	16953	88604	12724
144	62964	88145	83083	69453	46109	59505	69680	00900
145	19687	12633	57857	95806	09931	02150	43163	58636
146	37609	59057	66967	83401	60705	02384	90597	93600
147	54973	86278	88737	74351	47500	84552	19909	67181
148	00694	05977	19664	65441	20903	62371	22725	53340
149	71546	05233	53946	68743	72460	27601	45403	88692
150	07511	88915	41267	16853	84569	79367	32337	03316

TABLE B Random digits (*continued*)

Line								
151	03802	29341	29264	80198	12371	13121	54969	43912
152	77320	35030	77519	41109	98296	18984	60869	12349
153	07886	56866	39648	69290	03600	05376	58958	22720
154	87065	74133	21117	70595	22791	67306	28420	52067
155	42090	09628	54035	93879	98441	04606	27381	82637
156	55494	67690	88131	81800	11188	28552	25752	21953
157	16698	30406	96587	65985	07165	50148	16201	86792
158	16297	07626	68683	45335	34377	72941	41764	77038
159	22897	17467	17638	70043	36243	13008	83993	22869
160	98163	45944	34210	64158	76971	27689	82926	75957
161	43400	25831	06283	22138	16043	15706	73345	26238
162	97341	46254	88153	62336	21112	35574	99271	45297
163	64578	67197	28310	90341	37531	63890	52630	76315
164	11022	79124	49525	63078	17229	32165	01343	21394
165	81232	43939	23840	05995	84589	06788	76358	26622
166	36843	84798	51167	44728	20554	55538	27647	32708
167	84329	80081	69516	78934	14293	92478	16479	26974
168	27788	85789	41592	74472	96773	27090	24954	41474
169	99224	00850	43737	75202	44753	63236	14260	73686
170	38075	73239	52555	46342	13365	02182	30443	53229
171	87368	49451	55771	48343	51236	18522	73670	23212
172	40512	00681	44282	47178	08139	78693	34715	75606
173	81636	57578	54286	27216	58758	80358	84115	84568
174	26411	94292	06340	97762	37033	85968	94165	46514
175	80011	09937	57195	33906	94831	10056	42211	65491
176	92813	87503	63494	71379	76550	45984	05481	50830
177	70348	72871	63419	57363	29685	43090	18763	31714
178	24005	52114	26224	39078	80798	15220	43186	00976
179	85063	55810	10470	08029	30025	29734	61181	72090
180	11532	73186	92541	06915	72954	10167	12142	26492
181	59618	03914	05208	84088	20426	39004	84582	87317
182	92965	50837	39921	84661	82514	81899	24565	60874
183	85116	27684	14597	85747	01596	25889	41998	15635
184	15106	10411	90221	49377	44369	28185	80959	76355
185	03638	31589	07871	25792	85823	55400	56026	12193
186	97971	48932	45792	63993	95635	28753	46069	84635
187	49345	18305	76213	82390	77412	97401	50650	71755
188	87370	88099	89695	87633	76987	85503	26257	51736
189	88296	95670	74932	65317	93848	43988	47597	83044
190	79485	92200	99401	54473	34336	82786	05457	60343
191	40830	24979	23333	37619	56227	95941	59494	86539
192	32006	76302	81221	00693	95197	75044	46596	11628
193	37569	85187	44692	50706	53161	69027	88389	60313
194	56680	79003	23361	67094	15019	63261	24543	52884
195	05172	08100	22316	54495	60005	29532	18433	18057
196	74782	27005	03894	98038	20627	40307	47317	92759
197	85288	93264	61409	03404	09649	55937	60843	66167
198	68309	12060	14762	58002	03716	81968	57934	32624
199	26461	88346	52430	60906	74216	96263	69296	90107
200	42672	67680	42376	95023	82744	03971	96560	55148

TABLE C Binomial probabilities

Entry is $P(X = k) = \binom{n}{k} p^k (1-p)^{n-k}$

n	k	p = .01	.02	.03	.04	.05	.06	.07	.08	.09
2	0	.9801	.9604	.9409	.9216	.9025	.8836	.8649	.8464	.8281
	1	.0198	.0392	.0582	.0768	.0950	.1128	.1302	.1472	.1638
	2	.0001	.0004	.0009	.0016	.0025	.0036	.0049	.0064	.0081
3	0	.9703	.9412	.9127	.8847	.8574	.8306	.8044	.7787	.7536
	1	.0294	.0576	.0847	.1106	.1354	.1590	.1816	.2031	.2236
	2	.0003	.0012	.0026	.0046	.0071	.0102	.0137	.0177	.0221
	3				.0001	.0001	.0002	.0003	.0005	.0007
4	0	.9606	.9224	.8853	.8493	.8145	.7807	.7481	.7164	.6857
	1	.0388	.0753	.1095	.1416	.1715	.1993	.2252	.2492	.2713
	2	.0006	.0023	.0051	.0088	.0135	.0191	.0254	.0325	.0402
	3			.0001	.0002	.0005	.0008	.0013	.0019	.0027
	4									.0001
5	0	.9510	.9039	.8587	.8154	.7738	.7339	.6957	.6591	.6240
	1	.0480	.0922	.1328	.1699	.2036	.2342	.2618	.2866	.3086
	2	.0010	.0038	.0082	.0142	.0214	.0299	.0394	.0498	.0610
	3		.0001	.0003	.0006	.0011	.0019	.0030	.0043	.0060
	4						.0001	.0001	.0002	.0003
	5									
6	0	.9415	.8858	.8330	.7828	.7351	.6899	.6470	.6064	.5679
	1	.0571	.1085	.1546	.1957	.2321	.2642	.2922	.3164	.3370
	2	.0014	.0055	.0120	.0204	.0305	.0422	.0550	.0688	.0833
	3		.0002	.0005	.0011	.0021	.0036	.0055	.0080	.0110
	4					.0001	.0002	.0003	.0005	.0008
	5									
	6									
7	0	.9321	.8681	.8080	.7514	.6983	.6485	.6017	.5578	.5168
	1	.0659	.1240	.1749	.2192	.2573	.2897	.3170	.3396	.3578
	2	.0020	.0076	.0162	.0274	.0406	.0555	.0716	.0886	.1061
	3		.0003	.0008	.0019	.0036	.0059	.0090	.0128	.0175
	4				.0001	.0002	.0004	.0007	.0011	.0017
	5								.0001	.0001
	6									
	7									
8	0	.9227	.8508	.7837	.7214	.6634	.6096	.5596	.5132	.4703
	1	.0746	.1389	.1939	.2405	.2793	.3113	.3370	.3570	.3721
	2	.0026	.0099	.0210	.0351	.0515	.0695	.0888	.1087	.1288
	3	.0001	.0004	.0013	.0029	.0054	.0089	.0134	.0189	.0255
	4			.0001	.0002	.0004	.0007	.0013	.0021	.0031
	5							.0001	.0001	.0002
	6									
	7									
	8									

TABLE C Binomial probabilities (*continued*)

Entry is $P(X = k) = \binom{n}{k} p^k (1-p)^{n-k}$

		p								
n	*k*	.10	.15	.20	.25	.30	.35	.40	.45	.50
2	0	.8100	.7225	.6400	.5625	.4900	.4225	.3600	.3025	.2500
	1	.1800	.2550	.3200	.3750	.4200	.4550	.4800	.4950	.5000
	2	.0100	.0225	.0400	.0625	.0900	.1225	.1600	.2025	.2500
3	0	.7290	.6141	.5120	.4219	.3430	.2746	.2160	.1664	.1250
	1	.2430	.3251	.3840	.4219	.4410	.4436	.4320	.4084	.3750
	2	.0270	.0574	.0960	.1406	.1890	.2389	.2880	.3341	.3750
	3	.0010	.0034	.0080	.0156	.0270	.0429	.0640	.0911	.1250
4	0	.6561	.5220	.4096	.3164	.2401	.1785	.1296	.0915	.0625
	1	.2916	.3685	.4096	.4219	.4116	.3845	.3456	.2995	.2500
	2	.0486	.0975	.1536	.2109	.2646	.3105	.3456	.3675	.3750
	3	.0036	.0115	.0256	.0469	.0756	.1115	.1536	.2005	.2500
	4	.0001	.0005	.0016	.0039	.0081	.0150	.0256	.0410	.0625
5	0	.5905	.4437	.3277	.2373	.1681	.1160	.0778	.0503	.0313
	1	.3280	.3915	.4096	.3955	.3602	.3124	.2592	.2059	.1563
	2	.0729	.1382	.2048	.2637	.3087	.3364	.3456	.3369	.3125
	3	.0081	.0244	.0512	.0879	.1323	.1811	.2304	.2757	.3125
	4	.0004	.0022	.0064	.0146	.0284	.0488	.0768	.1128	.1562
	5		.0001	.0003	.0010	.0024	.0053	.0102	.0185	.0312
6	0	.5314	.3771	.2621	.1780	.1176	.0754	.0467	.0277	.0156
	1	.3543	.3993	.3932	.3560	.3025	.2437	.1866	.1359	.0938
	2	.0984	.1762	.2458	.2966	.3241	.3280	.3110	.2780	.2344
	3	.0146	.0415	.0819	.1318	.1852	.2355	.2765	.3032	.3125
	4	.0012	.0055	.0154	.0330	.0595	.0951	.1382	.1861	.2344
	5	.0001	.0004	.0015	.0044	.0102	.0205	.0369	.0609	.0937
	6			.0001	.0002	.0007	.0018	.0041	.0083	.0156
7	0	.4783	.3206	.2097	.1335	.0824	.0490	.0280	.0152	.0078
	1	.3720	.3960	.3670	.3115	.2471	.1848	.1306	.0872	.0547
	2	.1240	.2097	.2753	.3115	.3177	.2985	.2613	.2140	.1641
	3	.0230	.0617	.1147	.1730	.2269	.2679	.2903	.2918	.2734
	4	.0026	.0109	.0287	.0577	.0972	.1442	.1935	.2388	.2734
	5	.0002	.0012	.0043	.0115	.0250	.0466	.0774	.1172	.1641
	6		.0001	.0004	.0013	.0036	.0084	.0172	.0320	.0547
	7				.0001	.0002	.0006	.0016	.0037	.0078
8	0	.4305	.2725	.1678	.1001	.0576	.0319	.0168	.0084	.0039
	1	.3826	.3847	.3355	.2670	.1977	.1373	.0896	.0548	.0313
	2	.1488	.2376	.2936	.3115	.2965	.2587	.2090	.1569	.1094
	3	.0331	.0839	.1468	.2076	.2541	.2786	.2787	.2568	.2188
	4	.0046	.0185	.0459	.0865	.1361	.1875	.2322	.2627	.2734
	5	.0004	.0026	.0092	.0231	.0467	.0808	.1239	.1719	.2188
	6		.0002	.0011	.0038	.0100	.0217	.0413	.0703	.1094
	7			.0001	.0004	.0012	.0033	.0079	.0164	.0312
	8					.0001	.0002	.0007	.0017	.0039

TABLE C Binomial probabilities (*continued*)

Entry is $P(X = k) = \binom{n}{k} p^k (1-p)^{n-k}$

		p								
n	*k*	.01	.02	.03	.04	.05	.06	.07	.08	.09
9	0	.9135	.8337	.7602	.6925	.6302	.5730	.5204	.4722	.4279
	1	.0830	.1531	.2116	.2597	.2985	.3292	.3525	.3695	.3809
	2	.0034	.0125	.0262	.0433	.0629	.0840	.1061	.1285	.1507
	3	.0001	.0006	.0019	.0042	.0077	.0125	.0186	.0261	.0348
	4			.0001	.0003	.0006	.0012	.0021	.0034	.0052
	5						.0001	.0002	.0003	.0005
	6									
	7									
	8									
	9									
10	0	.9044	.8171	.7374	.6648	.5987	.5386	.4840	.4344	.3894
	1	.0914	.1667	.2281	.2770	.3151	.3438	.3643	.3777	.3851
	2	.0042	.0153	.0317	.0519	.0746	.0988	.1234	.1478	.1714
	3	.0001	.0008	.0026	.0058	.0105	.0168	.0248	.0343	.0452
	4			.0001	.0004	.0010	.0019	.0033	.0052	.0078
	5					.0001	.0001	.0003	.0005	.0009
	6									.0001
	7									
	8									
	9									
	10									
12	0	.8864	.7847	.6938	.6127	.5404	.4759	.4186	.3677	.3225
	1	.1074	.1922	.2575	.3064	.3413	.3645	.3781	.3837	.3827
	2	.0060	.0216	.0438	.0702	.0988	.1280	.1565	.1835	.2082
	3	.0002	.0015	.0045	.0098	.0173	.0272	.0393	.0532	.0686
	4		.0001	.0003	.0009	.0021	.0039	.0067	.0104	.0153
	5				.0001	.0002	.0004	.0008	.0014	.0024
	6							.0001	.0001	.0003
	7									
	8									
	9									
	10									
	11									
	12									
15	0	.8601	.7386	.6333	.5421	.4633	.3953	.3367	.2863	.2430
	1	.1303	.2261	.2938	.3388	.3658	.3785	.3801	.3734	.3605
	2	.0092	.0323	.0636	.0988	.1348	.1691	.2003	.2273	.2496
	3	.0004	.0029	.0085	.0178	.0307	.0468	.0653	.0857	.1070
	4		.0002	.0008	.0022	.0049	.0090	.0148	.0223	.0317
	5			.0001	.0002	.0006	.0013	.0024	.0043	.0069
	6						.0001	.0003	.0006	.0011
	7								.0001	.0001
	8									
	9									
	10									
	11									
	12									
	13									
	14									
	15									

TABLE C Binomial probabilities (*continued*)

		p								
n	*k*	.10	.15	.20	.25	.30	.35	.40	.45	.50
9	0	.3874	.2316	.1342	.0751	.0404	.0207	.0101	.0046	.0020
	1	.3874	.3679	.3020	.2253	.1556	.1004	.0605	.0339	.0176
	2	.1722	.2597	.3020	.3003	.2668	.2162	.1612	.1110	.0703
	3	.0446	.1069	.1762	.2336	.2668	.2716	.2508	.2119	.1641
	4	.0074	.0283	.0661	.1168	.1715	.2194	.2508	.2600	.2461
	5	.0008	.0050	.0165	.0389	.0735	.1181	.1672	.2128	.2461
	6	.0001	.0006	.0028	.0087	.0210	.0424	.0743	.1160	.1641
	7			.0003	.0012	.0039	.0098	.0212	.0407	.0703
	8				.0001	.0004	.0013	.0035	.0083	.0176
	9						.0001	.0003	.0008	.0020
10	0	.3487	.1969	.1074	.0563	.0282	.0135	.0060	.0025	.0010
	1	.3874	.3474	.2684	.1877	.1211	.0725	.0403	.0207	.0098
	2	.1937	.2759	.3020	.2816	.2335	.1757	.1209	.0763	.0439
	3	.0574	.1298	.2013	.2503	.2668	.2522	.2150	.1665	.1172
	4	.0112	.0401	.0881	.1460	.2001	.2377	.2508	.2384	.2051
	5	.0015	.0085	.0264	.0584	.1029	.1536	.2007	.2340	.2461
	6	.0001	.0012	.0055	.0162	.0368	.0689	.1115	.1596	.2051
	7		.0001	.0008	.0031	.0090	.0212	.0425	.0746	.1172
	8			.0001	.0004	.0014	.0043	.0106	.0229	.0439
	9					.0001	.0005	.0016	.0042	.0098
	10							.0001	.0003	.0010
12	0	.2824	.1422	.0687	.0317	.0138	.0057	.0022	.0008	.0002
	1	.3766	.3012	.2062	.1267	.0712	.0368	.0174	.0075	.0029
	2	.2301	.2924	.2835	.2323	.1678	.1088	.0639	.0339	.0161
	3	.0852	.1720	.2362	.2581	.2397	.1954	.1419	.0923	.0537
	4	.0213	.0683	.1329	.1936	.2311	.2367	.2128	.1700	.1208
	5	.0038	.0193	.0532	.1032	.1585	.2039	.2270	.2225	.1934
	6	.0005	.0040	.0155	.0401	.0792	.1281	.1766	.2124	.2256
	7		.0006	.0033	.0115	.0291	.0591	.1009	.1489	.1934
	8		.0001	.0005	.0024	.0078	.0199	.0420	.0762	.1208
	9			.0001	.0004	.0015	.0048	.0125	.0277	.0537
	10					.0002	.0008	.0025	.0068	.0161
	11						.0001	.0003	.0010	.0029
	12								.0001	.0002
15	0	.2059	.0874	.0352	.0134	.0047	.0016	.0005	.0001	.0000
	1	.3432	.2312	.1319	.0668	.0305	.0126	.0047	.0016	.0005
	2	.2669	.2856	.2309	.1559	.0916	.0476	.0219	.0090	.0032
	3	.1285	.2184	.2501	.2252	.1700	.1110	.0634	.0318	.0139
	4	.0428	.1156	.1876	.2252	.2186	.1792	.1268	.0780	.0417
	5	.0105	.0449	.1032	.1651	.2061	.2123	.1859	.1404	.0916
	6	.0019	.0132	.0430	.0917	.1472	.1906	.2066	.1914	.1527
	7	.0003	.0030	.0138	.0393	.0811	.1319	.1771	.2013	.1964
	8		.0005	.0035	.0131	.0348	.0710	.1181	.1647	.1964
	9		.0001	.0007	.0034	.0116	.0298	.0612	.1048	.1527
	10			.0001	.0007	.0030	.0096	.0245	.0515	.0916
	11				.0001	.0006	.0024	.0074	.0191	.0417
	12					.0001	.0004	.0016	.0052	.0139
	13						.0001	.0003	.0010	.0032
	14								.0001	.0005
	15									

TABLE C Binomial probabilities (*continued*)

n	k	.01	.02	.03	.04	.05	.06	.07	.08	.09
20	0	.8179	.6676	.5438	.4420	.3585	.2901	.2342	.1887	.1516
	1	.1652	.2725	.3364	.3683	.3774	.3703	.3526	.3282	.3000
	2	.0159	.0528	.0988	.1458	.1887	.2246	.2521	.2711	.2818
	3	.0010	.0065	.0183	.0364	.0596	.0860	.1139	.1414	.1672
	4		.0006	.0024	.0065	.0133	.0233	.0364	.0523	.0703
	5			.0002	.0009	.0022	.0048	.0088	.0145	.0222
	6				.0001	.0003	.0008	.0017	.0032	.0055
	7						.0001	.0002	.0005	.0011
	8								.0001	.0002
	9									
	10									
	11									
	12									
	13									
	14									
	15									
	16									
	17									
	18									
	19									
	20									

n	k	.10	.15	.20	.25	.30	.35	.40	.45	.50
20	0	.1216	.0388	.0115	.0032	.0008	.0002	.0000	.0000	.0000
	1	.2702	.1368	.0576	.0211	.0068	.0020	.0005	.0001	.0000
	2	.2852	.2293	.1369	.0669	.0278	.0100	.0031	.0008	.0002
	3	.1901	.2428	.2054	.1339	.0716	.0323	.0123	.0040	.0011
	4	.0898	.1821	.2182	.1897	.1304	.0738	.0350	.0139	.0046
	5	.0319	.1028	.1746	.2023	.1789	.1272	.0746	.0365	.0148
	6	.0089	.0454	.1091	.1686	.1916	.1712	.1244	.0746	.0370
	7	.0020	.0160	.0545	.1124	.1643	.1844	.1659	.1221	.0739
	8	.0004	.0046	.0222	.0609	.1144	.1614	.1797	.1623	.1201
	9	.0001	.0011	.0074	.0271	.0654	.1158	.1597	.1771	.1602
	10		.0002	.0020	.0099	.0308	.0686	.1171	.1593	.1762
	11			.0005	.0030	.0120	.0336	.0710	.1185	.1602
	12			.0001	.0008	.0039	.0136	.0355	.0727	.1201
	13				.0002	.0010	.0045	.0146	.0366	.0739
	14					.0002	.0012	.0049	.0150	.0370
	15						.0003	.0013	.0049	.0148
	16							.0003	.0013	.0046
	17								.0002	.0011
	18									.0002
	19									
	20									

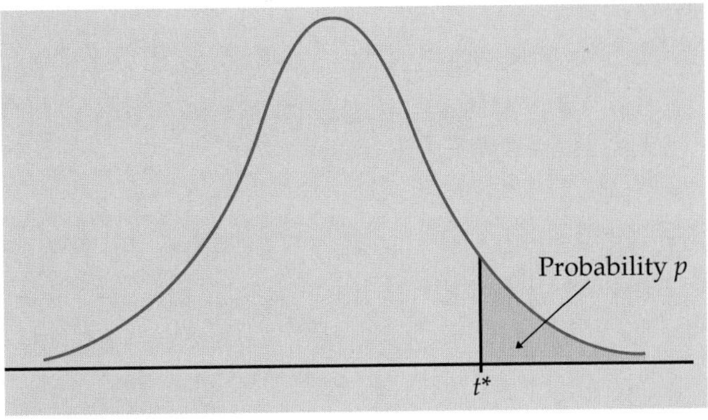

Table entry for p and C is the critical value t^* with probability p lying to its right and probability C lying between $-t^*$ and t^*.

TABLE D *t* distribution critical values

	Upper tail probability p											
df	.25	.20	.15	.10	.05	.025	.02	.01	.005	.0025	.001	.0005
1	1.000	1.376	1.963	3.078	6.314	12.71	15.89	31.82	63.66	127.3	318.3	636.6
2	0.816	1.061	1.386	1.886	2.920	4.303	4.849	6.965	9.925	14.09	22.33	31.60
3	0.765	0.978	1.250	1.638	2.353	3.182	3.482	4.541	5.841	7.453	10.21	12.92
4	0.741	0.941	1.190	1.533	2.132	2.776	2.999	3.747	4.604	5.598	7.173	8.610
5	0.727	0.920	1.156	1.476	2.015	2.571	2.757	3.365	4.032	4.773	5.893	6.869
6	0.718	0.906	1.134	1.440	1.943	2.447	2.612	3.143	3.707	4.317	5.208	5.959
7	0.711	0.896	1.119	1.415	1.895	2.365	2.517	2.998	3.499	4.029	4.785	5.408
8	0.706	0.889	1.108	1.397	1.860	2.306	2.449	2.896	3.355	3.833	4.501	5.041
9	0.703	0.883	1.100	1.383	1.833	2.262	2.398	2.821	3.250	3.690	4.297	4.781
10	0.700	0.879	1.093	1.372	1.812	2.228	2.359	2.764	3.169	3.581	4.144	4.587
11	0.697	0.876	1.088	1.363	1.796	2.201	2.328	2.718	3.106	3.497	4.025	4.437
12	0.695	0.873	1.083	1.356	1.782	2.179	2.303	2.681	3.055	3.428	3.930	4.318
13	0.694	0.870	1.079	1.350	1.771	2.160	2.282	2.650	3.012	3.372	3.852	4.221
14	0.692	0.868	1.076	1.345	1.761	2.145	2.264	2.624	2.977	3.326	3.787	4.140
15	0.691	0.866	1.074	1.341	1.753	2.131	2.249	2.602	2.947	3.286	3.733	4.073
16	0.690	0.865	1.071	1.337	1.746	2.120	2.235	2.583	2.921	3.252	3.686	4.015
17	0.689	0.863	1.069	1.333	1.740	2.110	2.224	2.567	2.898	3.222	3.646	3.965
18	0.688	0.862	1.067	1.330	1.734	2.101	2.214	2.552	2.878	3.197	3.611	3.922
19	0.688	0.861	1.066	1.328	1.729	2.093	2.205	2.539	2.861	3.174	3.579	3.883
20	0.687	0.860	1.064	1.325	1.725	2.086	2.197	2.528	2.845	3.153	3.552	3.850
21	0.686	0.859	1.063	1.323	1.721	2.080	2.189	2.518	2.831	3.135	3.527	3.819
22	0.686	0.858	1.061	1.321	1.717	2.074	2.183	2.508	2.819	3.119	3.505	3.792
23	0.685	0.858	1.060	1.319	1.714	2.069	2.177	2.500	2.807	3.104	3.485	3.768
24	0.685	0.857	1.059	1.318	1.711	2.064	2.172	2.492	2.797	3.091	3.467	3.745
25	0.684	0.856	1.058	1.316	1.708	2.060	2.167	2.485	2.787	3.078	3.450	3.725
26	0.684	0.856	1.058	1.315	1.706	2.056	2.162	2.479	2.779	3.067	3.435	3.707
27	0.684	0.855	1.057	1.314	1.703	2.052	2.158	2.473	2.771	3.057	3.421	3.690
28	0.683	0.855	1.056	1.313	1.701	2.048	2.154	2.467	2.763	3.047	3.408	3.674
29	0.683	0.854	1.055	1.311	1.699	2.045	2.150	2.462	2.756	3.038	3.396	3.659
30	0.683	0.854	1.055	1.310	1.697	2.042	2.147	2.457	2.750	3.030	3.385	3.646
40	0.681	0.851	1.050	1.303	1.684	2.021	2.123	2.423	2.704	2.971	3.307	3.551
50	0.679	0.849	1.047	1.299	1.676	2.009	2.109	2.403	2.678	2.937	3.261	3.496
60	0.679	0.848	1.045	1.296	1.671	2.000	2.099	2.390	2.660	2.915	3.232	3.460
80	0.678	0.846	1.043	1.292	1.664	1.990	2.088	2.374	2.639	2.887	3.195	3.416
100	0.677	0.845	1.042	1.290	1.660	1.984	2.081	2.364	2.626	2.871	3.174	3.390
1000	0.675	0.842	1.037	1.282	1.646	1.962	2.056	2.330	2.581	2.813	3.098	3.300
z^*	0.674	0.841	1.036	1.282	1.645	1.960	2.054	2.326	2.576	2.807	3.091	3.291
	50%	60%	70%	80%	90%	95%	96%	98%	99%	99.5%	99.8%	99.9%
	Confidence level C											

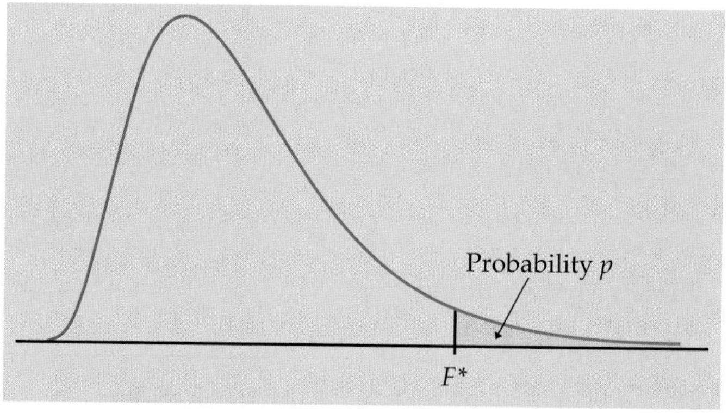

Table entry for p is the critical value F^* with probability p lying to its right.

TABLE E *F* critical values

Degrees of freedom in the denominator	p	Degrees of freedom in the numerator: 1	2	3	4	5	6	7	8	9
1	.100	39.86	49.50	53.59	55.83	57.24	58.20	58.91	59.44	59.86
	.050	161.45	199.50	215.71	224.58	230.16	233.99	236.77	238.88	240.54
	.025	647.79	799.50	864.16	899.58	921.85	937.11	948.22	956.66	963.28
	.010	4052.2	4999.5	5403.4	5624.6	5763.6	5859.0	5928.4	5981.1	6022.5
	.001	405284	500000	540379	562500	576405	585937	592873	598144	602284
2	.100	8.53	9.00	9.16	9.24	9.29	9.33	9.35	9.37	9.38
	.050	18.51	19.00	19.16	19.25	19.30	19.33	19.35	19.37	19.38
	.025	38.51	39.00	39.17	39.25	39.30	39.33	39.36	39.37	39.39
	.010	98.50	99.00	99.17	99.25	99.30	99.33	99.36	99.37	99.39
	.001	998.50	999.00	999.17	999.25	999.30	999.33	999.36	999.37	999.39
3	.100	5.54	5.46	5.39	5.34	5.31	5.28	5.27	5.25	5.24
	.050	10.13	9.55	9.28	9.12	9.01	8.94	8.89	8.85	8.81
	.025	17.44	16.04	15.44	15.10	14.88	14.73	14.62	14.54	14.47
	.010	34.12	30.82	29.46	28.71	28.24	27.91	27.67	27.49	27.35
	.001	167.03	148.50	141.11	137.10	134.58	132.85	131.58	130.62	129.86
4	.100	4.54	4.32	4.19	4.11	4.05	4.01	3.98	3.95	3.94
	.050	7.71	6.94	6.59	6.39	6.26	6.16	6.09	6.04	6.00
	.025	12.22	10.65	9.98	9.60	9.36	9.20	9.07	8.98	8.90
	.010	21.20	18.00	16.69	15.98	15.52	15.21	14.98	14.80	14.66
	.001	74.14	61.25	56.18	53.44	51.71	50.53	49.66	49.00	48.47
5	.100	4.06	3.78	3.62	3.52	3.45	3.40	3.37	3.34	3.32
	.050	6.61	5.79	5.41	5.19	5.05	4.95	4.88	4.82	4.77
	.025	10.01	8.43	7.76	7.39	7.15	6.98	6.85	6.76	6.68
	.010	16.26	13.27	12.06	11.39	10.97	10.67	10.46	10.29	10.16
	.001	47.18	37.12	33.20	31.09	29.75	28.83	28.16	27.65	27.24
6	.100	3.78	3.46	3.29	3.18	3.11	3.05	3.01	2.98	2.96
	.050	5.99	5.14	4.76	4.53	4.39	4.28	4.21	4.15	4.10
	.025	8.81	7.26	6.60	6.23	5.99	5.82	5.70	5.60	5.52
	.010	13.75	10.92	9.78	9.15	8.75	8.47	8.26	8.10	7.98
	.001	35.51	27.00	23.70	21.92	20.80	20.03	19.46	19.03	18.69
7	.100	3.59	3.26	3.07	2.96	2.88	2.83	2.78	2.75	2.72
	.050	5.59	4.74	4.35	4.12	3.97	3.87	3.79	3.73	3.68
	.025	8.07	6.54	5.89	5.52	5.29	5.12	4.99	4.90	4.82
	.010	12.25	9.55	8.45	7.85	7.46	7.19	6.99	6.84	6.72
	.001	29.25	21.69	18.77	17.20	16.21	15.52	15.02	14.63	14.33

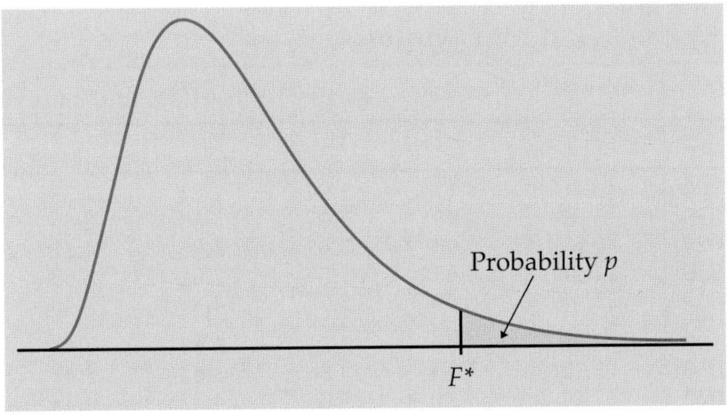

Table entry for p is the critical value F^* with probability p lying to its right.

TABLE E *F* critical values (*continued*)

Degrees of freedom in the numerator

10	12	15	20	25	30	40	50	60	120	1000
60.19	60.71	61.22	61.74	62.05	62.26	62.53	62.69	62.79	63.06	63.30
241.88	243.91	245.95	248.01	249.26	250.10	251.14	251.77	252.20	253.25	254.19
968.63	976.71	984.87	993.10	998.08	1001.4	1005.6	1008.1	1009.8	1014.0	1017.7
6055.8	6106.3	6157.3	6208.7	6239.8	6260.6	6286.8	6302.5	6313.0	6339.4	6362.7
605621	610668	615764	620908	624017	626099	628712	630285	631337	633972	636301
9.39	9.41	9.42	9.44	9.45	9.46	9.47	9.47	9.47	9.48	9.49
19.40	19.41	19.43	19.45	19.46	19.46	19.47	19.48	19.48	19.49	19.49
39.40	39.41	39.43	39.45	39.46	39.46	39.47	39.48	39.48	39.49	39.50
99.40	99.42	99.43	99.45	99.46	99.47	99.47	99.48	99.48	99.49	99.50
999.40	999.42	999.43	999.45	999.46	999.47	999.47	999.48	999.48	999.49	999.50
5.23	5.22	5.20	5.18	5.17	5.17	5.16	5.15	5.15	5.14	5.13
8.79	8.74	8.70	8.66	8.63	8.62	8.59	8.58	8.57	8.55	8.53
14.42	14.34	14.25	14.17	14.12	14.08	14.04	14.01	13.99	13.95	13.91
27.23	27.05	26.87	26.69	26.58	26.50	26.41	26.35	26.32	26.22	26.14
129.25	128.32	127.37	126.42	125.84	125.45	124.96	124.66	124.47	123.97	123.53
3.92	3.90	3.87	3.84	3.83	3.82	3.80	3.80	3.79	3.78	3.76
5.96	5.91	5.86	5.80	5.77	5.75	5.72	5.70	5.69	5.66	5.63
8.84	8.75	8.66	8.56	8.50	8.46	8.41	8.38	8.36	8.31	8.26
14.55	14.37	14.20	14.02	13.91	13.84	13.75	13.69	13.65	13.56	13.47
48.05	47.41	46.76	46.10	45.70	45.43	45.09	44.88	44.75	44.40	44.09
3.30	3.27	3.24	3.21	3.19	3.17	3.16	3.15	3.14	3.12	3.11
4.74	4.68	4.62	4.56	4.52	4.50	4.46	4.44	4.43	4.40	4.37
6.62	6.52	6.43	6.33	6.27	6.23	6.18	6.14	6.12	6.07	6.02
10.05	9.89	9.72	9.55	9.45	9.38	9.29	9.24	9.20	9.11	9.03
26.92	26.42	25.91	25.39	25.08	24.87	24.60	24.44	24.33	24.06	23.82
2.94	2.90	2.87	2.84	2.81	2.80	2.78	2.77	2.76	2.74	2.72
4.06	4.00	3.94	3.87	3.83	3.81	3.77	3.75	3.74	3.70	3.67
5.46	5.37	5.27	5.17	5.11	5.07	5.01	4.98	4.96	4.90	4.86
7.87	7.72	7.56	7.40	7.30	7.23	7.14	7.09	7.06	6.97	6.89
18.41	17.99	17.56	17.12	16.85	16.67	16.44	16.31	16.21	15.98	15.77
2.70	2.67	2.63	2.59	2.57	2.56	2.54	2.52	2.51	2.49	2.47
3.64	3.57	3.51	3.44	3.40	3.38	3.34	3.32	3.30	3.27	3.23
4.76	4.67	4.57	4.47	4.40	4.36	4.31	4.28	4.25	4.20	4.15
6.62	6.47	6.31	6.16	6.06	5.99	5.91	5.86	5.82	5.74	5.66
14.08	13.71	13.32	12.93	12.69	12.53	12.33	12.20	12.12	11.91	11.72

TABLE E *F* critical values (*continued*)

Degrees of freedom in the denominator	p	Degrees of freedom in the numerator 1	2	3	4	5	6	7	8	9
8	.100	3.46	3.11	2.92	2.81	2.73	2.67	2.62	2.59	2.56
	.050	5.32	4.46	4.07	3.84	3.69	3.58	3.50	3.44	3.39
	.025	7.57	6.06	5.42	5.05	4.82	4.65	4.53	4.43	4.36
	.010	11.26	8.65	7.59	7.01	6.63	6.37	6.18	6.03	5.91
	.001	25.41	18.49	15.83	14.39	13.48	12.86	12.40	12.05	11.77
9	.100	3.36	3.01	2.81	2.69	2.61	2.55	2.51	2.47	2.44
	.050	5.12	4.26	3.86	3.63	3.48	3.37	3.29	3.23	3.18
	.025	7.21	5.71	5.08	4.72	4.48	4.32	4.20	4.10	4.03
	.010	10.56	8.02	6.99	6.42	6.06	5.80	5.61	5.47	5.35
	.001	22.86	16.39	13.90	12.56	11.71	11.13	10.70	10.37	10.11
10	.100	3.29	2.92	2.73	2.61	2.52	2.46	2.41	2.38	2.35
	.050	4.96	4.10	3.71	3.48	3.33	3.22	3.14	3.07	3.02
	.025	6.94	5.46	4.83	4.47	4.24	4.07	3.95	3.85	3.78
	.010	10.04	7.56	6.55	5.99	5.64	5.39	5.20	5.06	4.94
	.001	21.04	14.91	12.55	11.28	10.48	9.93	9.52	9.20	8.96
11	.100	3.23	2.86	2.66	2.54	2.45	2.39	2.34	2.30	2.27
	.050	4.84	3.98	3.59	3.36	3.20	3.09	3.01	2.95	2.90
	.025	6.72	5.26	4.63	4.28	4.04	3.88	3.76	3.66	3.59
	.010	9.65	7.21	6.22	5.67	5.32	5.07	4.89	4.74	4.63
	.001	19.69	13.81	11.56	10.35	9.58	9.05	8.66	8.35	8.12
12	.100	3.18	2.81	2.61	2.48	2.39	2.33	2.28	2.24	2.21
	.050	4.75	3.89	3.49	3.26	3.11	3.00	2.91	2.85	2.80
	.025	6.55	5.10	4.47	4.12	3.89	3.73	3.61	3.51	3.44
	.010	9.33	6.93	5.95	5.41	5.06	4.82	4.64	4.50	4.39
	.001	18.64	12.97	10.80	9.63	8.89	8.38	8.00	7.71	7.48
13	.100	3.14	2.76	2.56	2.43	2.35	2.28	2.23	2.20	2.16
	.050	4.67	3.81	3.41	3.18	3.03	2.92	2.83	2.77	2.71
	.025	6.41	4.97	4.35	4.00	3.77	3.60	3.48	3.39	3.31
	.010	9.07	6.70	5.74	5.21	4.86	4.62	4.44	4.30	4.19
	.001	17.82	12.31	10.21	9.07	8.35	7.86	7.49	7.21	6.98
14	.100	3.10	2.73	2.52	2.39	2.31	2.24	2.19	2.15	2.12
	.050	4.60	3.74	3.34	3.11	2.96	2.85	2.76	2.70	2.65
	.025	6.30	4.86	4.24	3.89	3.66	3.50	3.38	3.29	3.21
	.010	8.86	6.51	5.56	5.04	4.69	4.46	4.28	4.14	4.03
	.001	17.14	11.78	9.73	8.62	7.92	7.44	7.08	6.80	6.58
15	.100	3.07	2.70	2.49	2.36	2.27	2.21	2.16	2.12	2.09
	.050	4.54	3.68	3.29	3.06	2.90	2.79	2.71	2.64	2.59
	.025	6.20	4.77	4.15	3.80	3.58	3.41	3.29	3.20	3.12
	.010	8.68	6.36	5.42	4.89	4.56	4.32	4.14	4.00	3.89
	.001	16.59	11.34	9.34	8.25	7.57	7.09	6.74	6.47	6.26
16	.100	3.05	2.67	2.46	2.33	2.24	2.18	2.13	2.09	2.06
	.050	4.49	3.63	3.24	3.01	2.85	2.74	2.66	2.59	2.54
	.025	6.12	4.69	4.08	3.73	3.50	3.34	3.22	3.12	3.05
	.010	8.53	6.23	5.29	4.77	4.44	4.20	4.03	3.89	3.78
	.001	16.12	10.97	9.01	7.94	7.27	6.80	6.46	6.19	5.98
17	.100	3.03	2.64	2.44	2.31	2.22	2.15	2.10	2.06	2.03
	.050	4.45	3.59	3.20	2.96	2.81	2.70	2.61	2.55	2.49
	.025	6.04	4.62	4.01	3.66	3.44	3.28	3.16	3.06	2.98
	.010	8.40	6.11	5.19	4.67	4.34	4.10	3.93	3.79	3.68
	.001	15.72	10.66	8.73	7.68	7.02	6.56	6.22	5.96	5.75

TABLE E *F* critical values (*continued*)

Degrees of freedom in the numerator

10	12	15	20	25	30	40	50	60	120	1000
2.54	2.50	2.46	2.42	2.40	2.38	2.36	2.35	2.34	2.32	2.30
3.35	3.28	3.22	3.15	3.11	3.08	3.04	3.02	3.01	2.97	2.93
4.30	4.20	4.10	4.00	3.94	3.89	3.84	3.81	3.78	3.73	3.68
5.81	5.67	5.52	5.36	5.26	5.20	5.12	5.07	5.03	4.95	4.87
11.54	11.19	10.84	10.48	10.26	10.11	9.92	9.80	9.73	9.53	9.36
2.42	2.38	2.34	2.30	2.27	2.25	2.23	2.22	2.21	2.18	2.16
3.14	3.07	3.01	2.94	2.89	2.86	2.83	2.80	2.79	2.75	2.71
3.96	3.87	3.77	3.67	3.60	3.56	3.51	3.47	3.45	3.39	3.34
5.26	5.11	4.96	4.81	4.71	4.65	4.57	4.52	4.48	4.40	4.32
9.89	9.57	9.24	8.90	8.69	8.55	8.37	8.26	8.19	8.00	7.84
2.32	2.28	2.24	2.20	2.17	2.16	2.13	2.12	2.11	2.08	2.06
2.98	2.91	2.85	2.77	2.73	2.70	2.66	2.64	2.62	2.58	2.54
3.72	3.62	3.52	3.42	3.35	3.31	3.26	3.22	3.20	3.14	3.09
4.85	4.71	4.56	4.41	4.31	4.25	4.17	4.12	4.08	4.00	3.92
8.75	8.45	8.13	7.80	7.60	7.47	7.30	7.19	7.12	6.94	6.78
2.25	2.21	2.17	2.12	2.10	2.08	2.05	2.04	2.03	2.00	1.98
2.85	2.79	2.72	2.65	2.60	2.57	2.53	2.51	2.49	2.45	2.41
3.53	3.43	3.33	3.23	3.16	3.12	3.06	3.03	3.00	2.94	2.89
4.54	4.40	4.25	4.10	4.01	3.94	3.86	3.81	3.78	3.69	3.61
7.92	7.63	7.32	7.01	6.81	6.68	6.52	6.42	6.35	6.18	6.02
2.19	2.15	2.10	2.06	2.03	2.01	1.99	1.97	1.96	1.93	1.91
2.75	2.69	2.62	2.54	2.50	2.47	2.43	2.40	2.38	2.34	2.30
3.37	3.28	3.18	3.07	3.01	2.96	2.91	2.87	2.85	2.79	2.73
4.30	4.16	4.01	3.86	3.76	3.70	3.62	3.57	3.54	3.45	3.37
7.29	7.00	6.71	6.40	6.22	6.09	5.93	5.83	5.76	5.59	5.44
2.14	2.10	2.05	2.01	1.98	1.96	1.93	1.92	1.90	1.88	1.85
2.67	2.60	2.53	2.46	2.41	2.38	2.34	2.31	2.30	2.25	2.21
3.25	3.15	3.05	2.95	2.88	2.84	2.78	2.74	2.72	2.66	2.60
4.10	3.96	3.82	3.66	3.57	3.51	3.43	3.38	3.34	3.25	3.18
6.80	6.52	6.23	5.93	5.75	5.63	5.47	5.37	5.30	5.14	4.99
2.10	2.05	2.01	1.96	1.93	1.91	1.89	1.87	1.86	1.83	1.80
2.60	2.53	2.46	2.39	2.34	2.31	2.27	2.24	2.22	2.18	2.14
3.15	3.05	2.95	2.84	2.78	2.73	2.67	2.64	2.61	2.55	2.50
3.94	3.80	3.66	3.51	3.41	3.35	3.27	3.22	3.18	3.09	3.02
6.40	6.13	5.85	5.56	5.38	5.25	5.10	5.00	4.94	4.77	4.62
2.06	2.02	1.97	1.92	1.89	1.87	1.85	1.83	1.82	1.79	1.76
2.54	2.48	2.40	2.33	2.28	2.25	2.20	2.18	2.16	2.11	2.07
3.06	2.96	2.86	2.76	2.69	2.64	2.59	2.55	2.52	2.46	2.40
3.80	3.67	3.52	3.37	3.28	3.21	3.13	3.08	3.05	2.96	2.88
6.08	5.81	5.54	5.25	5.07	4.95	4.80	4.70	4.64	4.47	4.33
2.03	1.99	1.94	1.89	1.86	1.84	1.81	1.79	1.78	1.75	1.72
2.49	2.42	2.35	2.28	2.23	2.19	2.15	2.12	2.11	2.06	2.02
2.99	2.89	2.79	2.68	2.61	2.57	2.51	2.47	2.45	2.38	2.32
3.69	3.55	3.41	3.26	3.16	3.10	3.02	2.97	2.93	2.84	2.76
5.81	5.55	5.27	4.99	4.82	4.70	4.54	4.45	4.39	4.23	4.08
2.00	1.96	1.91	1.86	1.83	1.81	1.78	1.76	1.75	1.72	1.69
2.45	2.38	2.31	2.23	2.18	2.15	2.10	2.08	2.06	2.01	1.97
2.92	2.82	2.72	2.62	2.55	2.50	2.44	2.41	2.38	2.32	2.26
3.59	3.46	3.31	3.16	3.07	3.00	2.92	2.87	2.83	2.75	2.66
5.58	5.32	5.05	4.78	4.60	4.48	4.33	4.24	4.18	4.02	3.87

TABLE E *F* critical values (*continued*)

Degrees of freedom in the denominator	*p*	1	2	3	4	5	6	7	8	9
		Degrees of freedom in the numerator								
18	.100	3.01	2.62	2.42	2.29	2.20	2.13	2.08	2.04	2.00
	.050	4.41	3.55	3.16	2.93	2.77	2.66	2.58	2.51	2.46
	.025	5.98	4.56	3.95	3.61	3.38	3.22	3.10	3.01	2.93
	.010	8.29	6.01	5.09	4.58	4.25	4.01	3.84	3.71	3.60
	.001	15.38	10.39	8.49	7.46	6.81	6.35	6.02	5.76	5.56
19	.100	2.99	2.61	2.40	2.27	2.18	2.11	2.06	2.02	1.98
	.050	4.38	3.52	3.13	2.90	2.74	2.63	2.54	2.48	2.42
	.025	5.92	4.51	3.90	3.56	3.33	3.17	3.05	2.96	2.88
	.010	8.18	5.93	5.01	4.50	4.17	3.94	3.77	3.63	3.52
	.001	15.08	10.16	8.28	7.27	6.62	6.18	5.85	5.59	5.39
20	.100	2.97	2.59	2.38	2.25	2.16	2.09	2.04	2.00	1.96
	.050	4.35	3.49	3.10	2.87	2.71	2.60	2.51	2.45	2.39
	.025	5.87	4.46	3.86	3.51	3.29	3.13	3.01	2.91	2.84
	.010	8.10	5.85	4.94	4.43	4.10	3.87	3.70	3.56	3.46
	.001	14.82	9.95	8.10	7.10	6.46	6.02	5.69	5.44	5.24
21	.100	2.96	2.57	2.36	2.23	2.14	2.08	2.02	1.98	1.95
	.050	4.32	3.47	3.07	2.84	2.68	2.57	2.49	2.42	2.37
	.025	5.83	4.42	3.82	3.48	3.25	3.09	2.97	2.87	2.80
	.010	8.02	5.78	4.87	4.37	4.04	3.81	3.64	3.51	3.40
	.001	14.59	9.77	7.94	6.95	6.32	5.88	5.56	5.31	5.11
22	.100	2.95	2.56	2.35	2.22	2.13	2.06	2.01	1.97	1.93
	.050	4.30	3.44	3.05	2.82	2.66	2.55	2.46	2.40	2.34
	.025	5.79	4.38	3.78	3.44	3.22	3.05	2.93	2.84	2.76
	.010	7.95	5.72	4.82	4.31	3.99	3.76	3.59	3.45	3.35
	.001	14.38	9.61	7.80	6.81	6.19	5.76	5.44	5.19	4.99
23	.100	2.94	2.55	2.34	2.21	2.11	2.05	1.99	1.95	1.92
	.050	4.28	3.42	3.03	2.80	2.64	2.53	2.44	2.37	2.32
	.025	5.75	4.35	3.75	3.41	3.18	3.02	2.90	2.81	2.73
	.010	7.88	5.66	4.76	4.26	3.94	3.71	3.54	3.41	3.30
	.001	14.20	9.47	7.67	6.70	6.08	5.65	5.33	5.09	4.89
24	.100	2.93	2.54	2.33	2.19	2.10	2.04	1.98	1.94	1.91
	.050	4.26	3.40	3.01	2.78	2.62	2.51	2.42	2.36	2.30
	.025	5.72	4.32	3.72	3.38	3.15	2.99	2.87	2.78	2.70
	.010	7.82	5.61	4.72	4.22	3.90	3.67	3.50	3.36	3.26
	.001	14.03	9.34	7.55	6.59	5.98	5.55	5.23	4.99	4.80
25	.100	2.92	2.53	2.32	2.18	2.09	2.02	1.97	1.93	1.89
	.050	4.24	3.39	2.99	2.76	2.60	2.49	2.40	2.34	2.28
	.025	5.69	4.29	3.69	3.35	3.13	2.97	2.85	2.75	2.68
	.010	7.77	5.57	4.68	4.18	3.85	3.63	3.46	3.32	3.22
	.001	13.88	9.22	7.45	6.49	5.89	5.46	5.15	4.91	4.71
26	.100	2.91	2.52	2.31	2.17	2.08	2.01	1.96	1.92	1.88
	.050	4.23	3.37	2.98	2.74	2.59	2.47	2.39	2.32	2.27
	.025	5.66	4.27	3.67	3.33	3.10	2.94	2.82	2.73	2.65
	.010	7.72	5.53	4.64	4.14	3.82	3.59	3.42	3.29	3.18
	.001	13.74	9.12	7.36	6.41	5.80	5.38	5.07	4.83	4.64
27	.100	2.90	2.51	2.30	2.17	2.07	2.00	1.95	1.91	1.87
	.050	4.21	3.35	2.96	2.73	2.57	2.46	2.37	2.31	2.25
	.025	5.63	4.24	3.65	3.31	3.08	2.92	2.80	2.71	2.63
	.010	7.68	5.49	4.60	4.11	3.78	3.56	3.39	3.26	3.15
	.001	13.61	9.02	7.27	6.33	5.73	5.31	5.00	4.76	4.57

TABLE E *F* critical values (*continued*)

Degrees of freedom in the numerator

10	12	15	20	25	30	40	50	60	120	1000
1.98	1.93	1.89	1.84	1.80	1.78	1.75	1.74	1.72	1.69	1.66
2.41	2.34	2.27	2.19	2.14	2.11	2.06	2.04	2.02	1.97	1.92
2.87	2.77	2.67	2.56	2.49	2.44	2.38	2.35	2.32	2.26	2.20
3.51	3.37	3.23	3.08	2.98	2.92	2.84	2.78	2.75	2.66	2.58
5.39	5.13	4.87	4.59	4.42	4.30	4.15	4.06	4.00	3.84	3.69
1.96	1.91	1.86	1.81	1.78	1.76	1.73	1.71	1.70	1.67	1.64
2.38	2.31	2.23	2.16	2.11	2.07	2.03	2.00	1.98	1.93	1.88
2.82	2.72	2.62	2.51	2.44	2.39	2.33	2.30	2.27	2.20	2.14
3.43	3.30	3.15	3.00	2.91	2.84	2.76	2.71	2.67	2.58	2.50
5.22	4.97	4.70	4.43	4.26	4.14	3.99	3.90	3.84	3.68	3.53
1.94	1.89	1.84	1.79	1.76	1.74	1.71	1.69	1.68	1.64	1.61
2.35	2.28	2.20	2.12	2.07	2.04	1.99	1.97	1.95	1.90	1.85
2.77	2.68	2.57	2.46	2.40	2.35	2.29	2.25	2.22	2.16	2.09
3.37	3.23	3.09	2.94	2.84	2.78	2.69	2.64	2.61	2.52	2.43
5.08	4.82	4.56	4.29	4.12	4.00	3.86	3.77	3.70	3.54	3.40
1.92	1.87	1.83	1.78	1.74	1.72	1.69	1.67	1.66	1.62	1.59
2.32	2.25	2.18	2.10	2.05	2.01	1.96	1.94	1.92	1.87	1.82
2.73	2.64	2.53	2.42	2.36	2.31	2.25	2.21	2.18	2.11	2.05
3.31	3.17	3.03	2.88	2.79	2.72	2.64	2.58	2.55	2.46	2.37
4.95	4.70	4.44	4.17	4.00	3.88	3.74	3.64	3.58	3.42	3.28
1.90	1.86	1.81	1.76	1.73	1.70	1.67	1.65	1.64	1.60	1.57
2.30	2.23	2.15	2.07	2.02	1.98	1.94	1.91	1.89	1.84	1.79
2.70	2.60	2.50	2.39	2.32	2.27	2.21	2.17	2.14	2.08	2.01
3.26	3.12	2.98	2.83	2.73	2.67	2.58	2.53	2.50	2.40	2.32
4.83	4.58	4.33	4.06	3.89	3.78	3.63	3.54	3.48	3.32	3.17
1.89	1.84	1.80	1.74	1.71	1.69	1.66	1.64	1.62	1.59	1.55
2.27	2.20	2.13	2.05	2.00	1.96	1.91	1.88	1.86	1.81	1.76
2.67	2.57	2.47	2.36	2.29	2.24	2.18	2.14	2.11	2.04	1.98
3.21	3.07	2.93	2.78	2.69	2.62	2.54	2.48	2.45	2.35	2.27
4.73	4.48	4.23	3.96	3.79	3.68	3.53	3.44	3.38	3.22	3.08
1.88	1.83	1.78	1.73	1.70	1.67	1.64	1.62	1.61	1.57	1.54
2.25	2.18	2.11	2.03	1.97	1.94	1.89	1.86	1.84	1.79	1.74
2.64	2.54	2.44	2.33	2.26	2.21	2.15	2.11	2.08	2.01	1.94
3.17	3.03	2.89	2.74	2.64	2.58	2.49	2.44	2.40	2.31	2.22
4.64	4.39	4.14	3.87	3.71	3.59	3.45	3.36	3.29	3.14	2.99
1.87	1.82	1.77	1.72	1.68	1.66	1.63	1.61	1.59	1.56	1.52
2.24	2.16	2.09	2.01	1.96	1.92	1.87	1.84	1.82	1.77	1.72
2.61	2.51	2.41	2.30	2.23	2.18	2.12	2.08	2.05	1.98	1.91
3.13	2.99	2.85	2.70	2.60	2.54	2.45	2.40	2.36	2.27	2.18
4.56	4.31	4.06	3.79	3.63	3.52	3.37	3.28	3.22	3.06	2.91
1.86	1.81	1.76	1.71	1.67	1.65	1.61	1.59	1.58	1.54	1.51
2.22	2.15	2.07	1.99	1.94	1.90	1.85	1.82	1.80	1.75	1.70
2.59	2.49	2.39	2.28	2.21	2.16	2.09	2.05	2.03	1.95	1.89
3.09	2.96	2.81	2.66	2.57	2.50	2.42	2.36	2.33	2.23	2.14
4.48	4.24	3.99	3.72	3.56	3.44	3.30	3.21	3.15	2.99	2.84
1.85	1.80	1.75	1.70	1.66	1.64	1.60	1.58	1.57	1.53	1.50
2.20	2.13	2.06	1.97	1.92	1.88	1.84	1.81	1.79	1.73	1.68
2.57	2.47	2.36	2.25	2.18	2.13	2.07	2.03	2.00	1.93	1.86
3.06	2.93	2.78	2.63	2.54	2.47	2.38	2.33	2.29	2.20	2.11
4.41	4.17	3.92	3.66	3.49	3.38	3.23	3.14	3.08	2.92	2.78

TABLE E *F* critical values (*continued*)

Degrees of freedom in the denominator		Degrees of freedom in the numerator								
	p	1	2	3	4	5	6	7	8	9
28	.100	2.89	2.50	2.29	2.16	2.06	2.00	1.94	1.90	1.87
	.050	4.20	3.34	2.95	2.71	2.56	2.45	2.36	2.29	2.24
	.025	5.61	4.22	3.63	3.29	3.06	2.90	2.78	2.69	2.61
	.010	7.64	5.45	4.57	4.07	3.75	3.53	3.36	3.23	3.12
	.001	13.50	8.93	7.19	6.25	5.66	5.24	4.93	4.69	4.50
29	.100	2.89	2.50	2.28	2.15	2.06	1.99	1.93	1.89	1.86
	.050	4.18	3.33	2.93	2.70	2.55	2.43	2.35	2.28	2.22
	.025	5.59	4.20	3.61	3.27	3.04	2.88	2.76	2.67	2.59
	.010	7.60	5.42	4.54	4.04	3.73	3.50	3.33	3.20	3.09
	.001	13.39	8.85	7.12	6.19	5.59	5.18	4.87	4.64	4.45
30	.100	2.88	2.49	2.28	2.14	2.05	1.98	1.93	1.88	1.85
	.050	4.17	3.32	2.92	2.69	2.53	2.42	2.33	2.27	2.21
	.025	5.57	4.18	3.59	3.25	3.03	2.87	2.75	2.65	2.57
	.010	7.56	5.39	4.51	4.02	3.70	3.47	3.30	3.17	3.07
	.001	13.29	8.77	7.05	6.12	5.53	5.12	4.82	4.58	4.39
40	.100	2.84	2.44	2.23	2.09	2.00	1.93	1.87	1.83	1.79
	.050	4.08	3.23	2.84	2.61	2.45	2.34	2.25	2.18	2.12
	.025	5.42	4.05	3.46	3.13	2.90	2.74	2.62	2.53	2.45
	.010	7.31	5.18	4.31	3.83	3.51	3.29	3.12	2.99	2.89
	.001	12.61	8.25	6.59	5.70	5.13	4.73	4.44	4.21	4.02
50	.100	2.81	2.41	2.20	2.06	1.97	1.90	1.84	1.80	1.76
	.050	4.03	3.18	2.79	2.56	2.40	2.29	2.20	2.13	2.07
	.025	5.34	3.97	3.39	3.05	2.83	2.67	2.55	2.46	2.38
	.010	7.17	5.06	4.20	3.72	3.41	3.19	3.02	2.89	2.78
	.001	12.22	7.96	6.34	5.46	4.90	4.51	4.22	4.00	3.82
60	.100	2.79	2.39	2.18	2.04	1.95	1.87	1.82	1.77	1.74
	.050	4.00	3.15	2.76	2.53	2.37	2.25	2.17	2.10	2.04
	.025	5.29	3.93	3.34	3.01	2.79	2.63	2.51	2.41	2.33
	.010	7.08	4.98	4.13	3.65	3.34	3.12	2.95	2.82	2.72
	.001	11.97	7.77	6.17	5.31	4.76	4.37	4.09	3.86	3.69
100	.100	2.76	2.36	2.14	2.00	1.91	1.83	1.78	1.73	1.69
	.050	3.94	3.09	2.70	2.46	2.31	2.19	2.10	2.03	1.97
	.025	5.18	3.83	3.25	2.92	2.70	2.54	2.42	2.32	2.24
	.010	6.90	4.82	3.98	3.51	3.21	2.99	2.82	2.69	2.59
	.001	11.50	7.41	5.86	5.02	4.48	4.11	3.83	3.61	3.44
200	.100	2.73	2.33	2.11	1.97	1.88	1.80	1.75	1.70	1.66
	.050	3.89	3.04	2.65	2.42	2.26	2.14	2.06	1.98	1.93
	.025	5.10	3.76	3.18	2.85	2.63	2.47	2.35	2.26	2.18
	.010	6.76	4.71	3.88	3.41	3.11	2.89	2.73	2.60	2.50
	.001	11.15	7.15	5.63	4.81	4.29	3.92	3.65	3.43	3.26
1000	.100	2.71	2.31	2.09	1.95	1.85	1.78	1.72	1.68	1.64
	.050	3.85	3.00	2.61	2.38	2.22	2.11	2.02	1.95	1.89
	.025	5.04	3.70	3.13	2.80	2.58	2.42	2.30	2.20	2.13
	.010	6.66	4.63	3.80	3.34	3.04	2.82	2.66	2.53	2.43
	.001	10.89	6.96	5.46	4.65	4.14	3.78	3.51	3.30	3.13

TABLE E *F* critical values (*continued*)

Degrees of freedom in the numerator										
10	12	15	20	25	30	40	50	60	120	1000
1.84	1.79	1.74	1.69	1.65	1.63	1.59	1.57	1.56	1.52	1.48
2.19	2.12	2.04	1.96	1.91	1.87	1.82	1.79	1.77	1.71	1.66
2.55	2.45	2.34	2.23	2.16	2.11	2.05	2.01	1.98	1.91	1.84
3.03	2.90	2.75	2.60	2.51	2.44	2.35	2.30	2.26	2.17	2.08
4.35	4.11	3.86	3.60	3.43	3.32	3.18	3.09	3.02	2.86	2.72
1.83	1.78	1.73	1.68	1.64	1.62	1.58	1.56	1.55	1.51	1.47
2.18	2.10	2.03	1.94	1.89	1.85	1.81	1.77	1.75	1.70	1.65
2.53	2.43	2.32	2.21	2.14	2.09	2.03	1.99	1.96	1.89	1.82
3.00	2.87	2.73	2.57	2.48	2.41	2.33	2.27	2.23	2.14	2.05
4.29	4.05	3.80	3.54	3.38	3.27	3.12	3.03	2.97	2.81	2.66
1.82	1.77	1.72	1.67	1.63	1.61	1.57	1.55	1.54	1.50	1.46
2.16	2.09	2.01	1.93	1.88	1.84	1.79	1.76	1.74	1.68	1.63
2.51	2.41	2.31	2.20	2.12	2.07	2.01	1.97	1.94	1.87	1.80
2.98	2.84	2.70	2.55	2.45	2.39	2.30	2.25	2.21	2.11	2.02
4.24	4.00	3.75	3.49	3.33	3.22	3.07	2.98	2.92	2.76	2.61
1.76	1.71	1.66	1.61	1.57	1.54	1.51	1.48	1.47	1.42	1.38
2.08	2.00	1.92	1.84	1.78	1.74	1.69	1.66	1.64	1.58	1.52
2.39	2.29	2.18	2.07	1.99	1.94	1.88	1.83	1.80	1.72	1.65
2.80	2.66	2.52	2.37	2.27	2.20	2.11	2.06	2.02	1.92	1.82
3.87	3.64	3.40	3.14	2.98	2.87	2.73	2.64	2.57	2.41	2.25
1.73	1.68	1.63	1.57	1.53	1.50	1.46	1.44	1.42	1.38	1.33
2.03	1.95	1.87	1.78	1.73	1.69	1.63	1.60	1.58	1.51	1.45
2.32	2.22	2.11	1.99	1.92	1.87	1.80	1.75	1.72	1.64	1.56
2.70	2.56	2.42	2.27	2.17	2.10	2.01	1.95	1.91	1.80	1.70
3.67	3.44	3.20	2.95	2.79	2.68	2.53	2.44	2.38	2.21	2.05
1.71	1.66	1.60	1.54	1.50	1.48	1.44	1.41	1.40	1.35	1.30
1.99	1.92	1.84	1.75	1.69	1.65	1.59	1.56	1.53	1.47	1.40
2.27	2.17	2.06	1.94	1.87	1.82	1.74	1.70	1.67	1.58	1.49
2.63	2.50	2.35	2.20	2.10	2.03	1.94	1.88	1.84	1.73	1.62
3.54	3.32	3.08	2.83	2.67	2.55	2.41	2.32	2.25	2.08	1.92
1.66	1.61	1.56	1.49	1.45	1.42	1.38	1.35	1.34	1.28	1.22
1.93	1.85	1.77	1.68	1.62	1.57	1.52	1.48	1.45	1.38	1.30
2.18	2.08	1.97	1.85	1.77	1.71	1.64	1.59	1.56	1.46	1.36
2.50	2.37	2.22	2.07	1.97	1.89	1.80	1.74	1.69	1.57	1.45
3.30	3.07	2.84	2.59	2.43	2.32	2.17	2.08	2.01	1.83	1.64
1.63	1.58	1.52	1.46	1.41	1.38	1.34	1.31	1.29	1.23	1.16
1.88	1.80	1.72	1.62	1.56	1.52	1.46	1.41	1.39	1.30	1.21
2.11	2.01	1.90	1.78	1.70	1.64	1.56	1.51	1.47	1.37	1.25
2.41	2.27	2.13	1.97	1.87	1.79	1.69	1.63	1.58	1.45	1.30
3.12	2.90	2.67	2.42	2.26	2.15	2.00	1.90	1.83	1.64	1.43
1.61	1.55	1.49	1.43	1.38	1.35	1.30	1.27	1.25	1.18	1.08
1.84	1.76	1.68	1.58	1.52	1.47	1.41	1.36	1.33	1.24	1.11
2.06	1.96	1.85	1.72	1.64	1.58	1.50	1.45	1.41	1.29	1.13
2.34	2.20	2.06	1.90	1.79	1.72	1.61	1.54	1.50	1.35	1.16
2.99	2.77	2.54	2.30	2.14	2.02	1.87	1.77	1.69	1.49	1.22

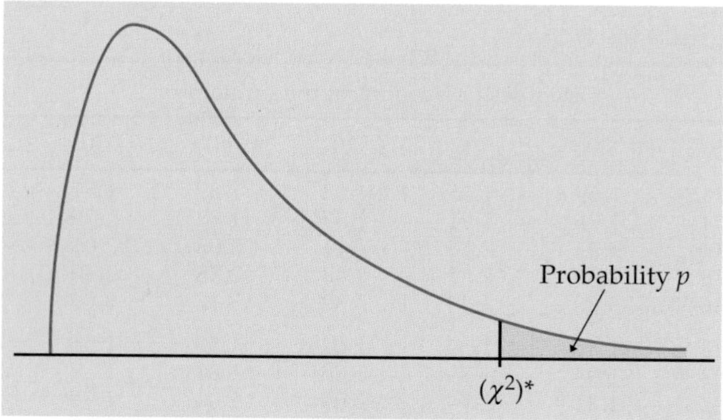

Table entry for p is the critical value $(\chi^2)^*$ with probability p lying to its right.

TABLE F χ^2 distribution critical values

	Tail probability p											
df	.25	.20	.15	.10	.05	.025	.02	.01	.005	.0025	.001	.0005
1	1.32	1.64	2.07	2.71	3.84	5.02	5.41	6.63	7.88	9.14	10.83	12.12
2	2.77	3.22	3.79	4.61	5.99	7.38	7.82	9.21	10.60	11.98	13.82	15.20
3	4.11	4.64	5.32	6.25	7.81	9.35	9.84	11.34	12.84	14.32	16.27	17.73
4	5.39	5.99	6.74	7.78	9.49	11.14	11.67	13.28	14.86	16.42	18.47	20.00
5	6.63	7.29	8.12	9.24	11.07	12.83	13.39	15.09	16.75	18.39	20.51	22.11
6	7.84	8.56	9.45	10.64	12.59	14.45	15.03	16.81	18.55	20.25	22.46	24.10
7	9.04	9.80	10.75	12.02	14.07	16.01	16.62	18.48	20.28	22.04	24.32	26.02
8	10.22	11.03	12.03	13.36	15.51	17.53	18.17	20.09	21.95	23.77	26.12	27.87
9	11.39	12.24	13.29	14.68	16.92	19.02	19.68	21.67	23.59	25.46	27.88	29.67
10	12.55	13.44	14.53	15.99	18.31	20.48	21.16	23.21	25.19	27.11	29.59	31.42
11	13.70	14.63	15.77	17.28	19.68	21.92	22.62	24.72	26.76	28.73	31.26	33.14
12	14.85	15.81	16.99	18.55	21.03	23.34	24.05	26.22	28.30	30.32	32.91	34.82
13	15.98	16.98	18.20	19.81	22.36	24.74	25.47	27.69	29.82	31.88	34.53	36.48
14	17.12	18.15	19.41	21.06	23.68	26.12	26.87	29.14	31.32	33.43	36.12	38.11
15	18.25	19.31	20.60	22.31	25.00	27.49	28.26	30.58	32.80	34.95	37.70	39.72
16	19.37	20.47	21.79	23.54	26.30	28.85	29.63	32.00	34.27	36.46	39.25	41.31
17	20.49	21.61	22.98	24.77	27.59	30.19	31.00	33.41	35.72	37.95	40.79	42.88
18	21.60	22.76	24.16	25.99	28.87	31.53	32.35	34.81	37.16	39.42	42.31	44.43
19	22.72	23.90	25.33	27.20	30.14	32.85	33.69	36.19	38.58	40.88	43.82	45.97
20	23.83	25.04	26.50	28.41	31.41	34.17	35.02	37.57	40.00	42.34	45.31	47.50
21	24.93	26.17	27.66	29.62	32.67	35.48	36.34	38.93	41.40	43.78	46.80	49.01
22	26.04	27.30	28.82	30.81	33.92	36.78	37.66	40.29	42.80	45.20	48.27	50.51
23	27.14	28.43	29.98	32.01	35.17	38.08	38.97	41.64	44.18	46.62	49.73	52.00
24	28.24	29.55	31.13	33.20	36.42	39.36	40.27	42.98	45.56	48.03	51.18	53.48
25	29.34	30.68	32.28	34.38	37.65	40.65	41.57	44.31	46.93	49.44	52.62	54.95
26	30.43	31.79	33.43	35.56	38.89	41.92	42.86	45.64	48.29	50.83	54.05	56.41
27	31.53	32.91	34.57	36.74	40.11	43.19	44.14	46.96	49.64	52.22	55.48	57.86
28	32.62	34.03	35.71	37.92	41.34	44.46	45.42	48.28	50.99	53.59	56.89	59.30
29	33.71	35.14	36.85	39.09	42.56	45.72	46.69	49.59	52.34	54.97	58.30	60.73
30	34.80	36.25	37.99	40.26	43.77	46.98	47.96	50.89	53.67	56.33	59.70	62.16
40	45.62	47.27	49.24	51.81	55.76	59.34	60.44	63.69	66.77	69.70	73.40	76.09
50	56.33	58.16	60.35	63.17	67.50	71.42	72.61	76.15	79.49	82.66	86.66	89.56
60	66.98	68.97	71.34	74.40	79.08	83.30	84.58	88.38	91.95	95.34	99.61	102.7
80	88.13	90.41	93.11	96.58	101.9	106.6	108.1	112.3	116.3	120.1	124.8	128.3
100	109.1	111.7	114.7	118.5	124.3	129.6	131.1	135.8	140.2	144.3	149.4	153.2

SOLUTIONS TO ODD-NUMBERED EXERCISES

Chapter 1

1.1 **(a)** Gender is categorical. **(b)** Age is quantitative. **(c)** Race is categorical. **(d)** Smoker is categorical. **(e)** Systolic blood pressure is quantitative. **(f)** Calcium level is quantitative.

1.3 The individuals are the different mutual funds. The variables are category, net assets, year-to-date return, and largest holding. Category and largest holding are categorical, and net assets and year-to-date return are quantitative.

1.5 You might measure such quantities as "number of sit-ups in 2 minutes" or "maximum weight that can be lifted" or "time in the 50-yard dash." These would require minimal equipment, such as a stopwatch. You might also want to measure "body mass index" or "oxygen consumption during a standard exercise." These would require more sophisticated machines.

1.7 The number of deaths from cancer can increase just because the population is growing. Because cancer is mainly a disease of old age, the percent of deaths from cancer can rise even if cancer treatment improves, simply because more people live to advanced ages. Progress against infections that once killed many people, for example, leaves more people to die of cancer later in life. Survival time will increase if the disease is detected earlier even if treatment is no more effective.

1.9 We need to compare the rates of deaths per number of riders. If many more children ride in autos than in buses, then it would be reasonable that more would die in cars.

1.11 **(a)** Yes. The total of all the categories is 12,299,000, which is 1000 more than the figure given for the total.

1.15 **(a)** 10.9% and 11.0%. **(b)** The shape is rather symmetric. The center of the distribution is about 13.9%. The spread is from about 11% to 16%.

1.17 The distribution is quite symmetric. The center is between 6 and 7. Most of the values are within ± 3 or 4 of the center.

1.19 The distribution seems to be in two parts. There are schools costing less than $6000 and schools costing more than $12,000. It is possible that this is the difference between private and public schools.

1.21 Skewed to the right. Main peak is around 40, and the maximum is around 200.

1.23 **(a)** One woman had a score of 200, which might be an outlier. There is no obvious outlier for the men. **(b)** The midpoint for the women is around 140. For the men, the midpoint is around 110. The range of the women (omitting the possible outlier) is 70 points. The range for the men is 110 points.

1.25 **(b)** Midpoint for men = 120. Midpoint for women = 180.

1.27 From the histogram, estimate 5.5 or 5.6. From the stemplot, estimate 5.44.

1.29 **(b)** -26.6. Roughly symmetric with center near 2 and spread from about -20 to 20.

1.31 **(a)** There are 4 variables (not counting OBS). Gender is the only categorical variable. The others are quantitative. **(c)** The shape is skewed left. The center is around 7.5. Most of the observations lie between 3.5 and 10. There are no suspected outliers because the 3 smallest values seem to trail down smoothly from the previous values. **(d)** Both distributions seem to peak around 7.5. The scores for males (on the right) may be more spread out, but this may also be because there are more males in the data set.

1.33 The center is around 60. The shape is skewed to the left. Most of the values lie between 35 and 73.

1.35 A class $20,000 wide would be 1/4 as tall as a class $5000 wide.

1.37 **(b)** 1974, 1981, 1984, 1989. Not clear how to treat 1995 or 2000. **(c)** 1981. After that, generally downward.

1.41 There are several gaps in the data, as can be seen from the histogram.

Min.	Q_1	M	Q_3	Max.
2508	13,385	17,022.5	23,552	26,241

1.43 5th percentile occurs at about the 550th person and the value is about \$10K. 95th percentile occurs at about the 10,441st person and the value is about \$110K.

1.45 **(a)** \$20 is the most common value. **(b)** $IQR = 4$ and $Q_3 = 22$, so anything over 28 is a possible outlier. There are several values over \$30.

1.47 Estimating the heights of the bars, it appears that 50% is just about the total for one-, two-, and three-letter words. The median is 3 or 4, depending on whether these add to just over 50% or just under 50%. Min. = 1, $Q_1 = 2$, $Q_3 = 5$, max. (given) = 12.

1.49 **(b)** For women, the mean is 141.06 and the median is 138.50. For men, the mean is 121.25 and the median is 114.50. This indicates that the distributions are skewed to the right. **(c)** Five-number summary is given below. The $1.5 \times IQR$ rule says that for women, anything over 180.5 is a possible outlier. The maximum for women is 200, which is the only observation over 180.5.

	Min.	Q_1	M	Q_3	Max.
Women	101	126	138.5	154	200
Men	70	98	114.5	143	187

1.53 **(a)**

Mean	St. dev.	Min.	Q_1	M	Q_3	Max.
234.3	57.62	154	200	224.5	246	412

(b) Mean and St. dev. do not reveal skewness.

1.55

Type	Min.	Q_1	M	Q_3	Max.	Low	High
Beef	111	140	152.5	178.5	190	82.25	236.25
Meat	107	138.5	153	180.5	195	75.5	243.5
Poultry	86	100.5	129	143.5	170	36	208

1.57

Type	Min.	Q_1	M	Q_3	Max.	Mean	Low	High
Beef	253	320.5	380.5	478	645	401.2	84.25	714.25
Meat	144	379	405	501	545	418.5	196	684
Poultry	357	379	430	535	588	459	145	769

1.59 $\bar{x}$ = \$62,500. Median = \$25,000. All but the owner earned less than the mean.

1.61 The mean rises to \$87,500, while the median is unchanged.

1.65 **(a)** Choose all four to be the same, such as {1, 1, 1, 1}. **(b)** We want the greatest spread; {0, 0, 10, 10} is the only answer.

1.69

Min.	Q_1	M	Q_3	Max.
43	82.5	102.5	151.5	598

1.71 100, 107, 112, and 119.
1.73 **(a)** $a = -98.6, b = 1$. **(b)** $a = 0, b = 100/120$.
1.75 Multiply by 6.4516.
1.79 **(a)** 1/2. **(b)** 1/2. **(c)** 0.4.
1.81 **(a)** Mean is C; median is B. **(b)** Mean and median are A. **(c)** Mean is A; median is B.
1.83 **(a)** 327 to 345 days. **(b)** 0.16.
1.85 Within one standard deviation of the mean is 95.75 to 122.09. Two standard deviations is 82.58 to 135.26. Three standard deviations is 69.41 to 148.43. 70.5% are within one standard deviation of the mean. 93.5% are within two standard deviations of the mean. All are within three standard deviations of the mean.
1.87 Cobb: $z = 4.15$. Williams: $z = 4.26$. Brett: $z = 4.07$.
1.89 **(a)** 0.0122. **(b)** 0.9878. **(c)** 0.0384. **(d)** $0.9878 - 0.0384 = 0.9494$.
1.91 **(a)** 0.84. **(b)** 0.38 or 0.39.
1.93 **(a)** 0.0793. **(b)** 0.2727.
1.95 0.3121. 0.4428.
1.97 **(a)** About 3%. **(b)** 1%.
1.99 **(a)** 0.0521. **(b)** 0.5466. **(c)** 279.4 days.
1.101 31st.
1.103 129.75. 98.25.
1.105 124.675. 134.95.
1.107 **(a)** -1.28 and $+1.28$. **(b)** 245.52 and 286.48.
1.109 The sharp rise at the right end.
1.111 There is no important deviation from normality.
1.113 No outliers are noted. Mean $= 56.96$, Min. $= 20$, $Q_1 = 51$, $M = 59.5$, $Q_3 = 66$, Max. $= 80$.
1.115 The normal plot is close to a line, except at the high end.
1.117 Categorical: Gender, automobile preference. Quantitative: age, household income.
1.119 **(a)** The other 95% of incomes will contain many more observations, which would make comparison difficult. **(b)** High-income families have a smaller proportion of 1s than do the others. High-income families also have a higher proportion of 4s than do the others.
1.121 Not necessarily.
1.123 **(a)** The records for 1942 to 1945 are somewhat lower than those for surrounding years. **(b)** Ruth's records were much higher than those in earlier years. McGwire's records (except for 70) were not remarkably higher.
1.125 Yes, the percents given add up to 84%.
1.127 **(a)** $x_{new} = -50 + 2x$. **(b)** $x_{new} = -49.09 + (20/11)x$. **(c)** David is higher within his grade. **(d)** 0.6179. 0.3594.
1.129 Because the data are multiples of 10, the plots show granularity. Otherwise, they appear normal.
1.131 **(b)** OL are the heaviest. DB and R are the lightest, but K are low with a high spread. **(c)** No outliers are obvious.
1.133 The distribution is highly right-skewed. The mean is much larger than the median. The median and quartiles are a better description.

Mean	Median	St. dev.	Q_1	Q_3
3,541,674	1,600,000	4,031,430	270,000	6,500,000

1.135 **(a)** Francs $= 1.72 \times$ dollars. **(b)** Multiply by 1.72. **(c)** $IQR = 10{,}715{,}600$.

Mean	Median	St. dev.	Q_1	Q_3
6,091,679	2,752,000	6,934,060	464,400	11,180,000

1.137 Summary statistics (up to \$100K) are given below. The distribution is somewhat right-skewed, as evidenced by the average being somewhat higher than the median.

Min.	Q_1	M	Q_3	Max.	Mean	St. dev.
–24,998	17,000	29,717	46,503	425,510	37,864	36,158

1.139 H2S is slightly right-skewed. LACTIC is more symmetric and more spread out. The five-number summary is more appropriate for H2S, while mean and standard deviation are more appropriate for LACTIC.

	Mean	St. dev.	Min.	Q_1	M	Q_3	Max.
H2S	5.942	2.127	2.996	3.912	5.329	7.601	10.199
LACTIC	1.442	0.303	0.860	1.250	1.450	1.680	2.010

Chapter 2

2.1 **(a)** Grade would be response, and time of study would be explanatory. **(b)** Relationship. **(c)** Rain would be explanatory, and yield would be response. **(d)** Relationship. **(e)** Income would be explanatory, and education would be response.

2.3 **(a)** When individual income is higher, household income will also be higher. Household income will be higher when there are multiple wage earners in the family. Also, income per person is diluted by non-wage earners. **(b)** Since we are dealing with the *median* household income but the *mean* income per person, some high outliers could cause the mean of one measure to be greater than the median of the other. **(c)** Except for the District of Columbia, the relationship is fairly linear with a positive slope. **(d)** Large households raise household income relative to personal income. Many very high earners (Wall Street) raise mean personal income.

2.5 **(a)** The number returning could increase as pairs increase even if the percent surviving goes down. **(b)** Yes, there is a general downward trend.

2.7 **(b)** The pattern is fairly linear, with a negative slope. The relationship is fairly strong. **(c)** Negative. This means that higher levels of alcohol consumption are associated with lower death rates. This does not mean there is a causal effect. There could be lurking variables that explain the effect. Higher alcohol consumption could be linked to higher incomes, which would allow better medical care.

2.9 **(b)** The relationship is fairly linear, with a positive slope. The strength is high.

2.11 **(b)** Overall, the relationship is fairly strong and linear, with a positive slope. The relationship for the men does not appear as strong, but there are fewer data points. As a group, the men are higher in both variables.

2.13 The improvement seems to be rather linear except for the last data point, which shows a very minor improvement for the time period involved.

2.15 **(a)** Both sexes are steadily improving. On this scale, the women's improvement appears to be rather linear. The women's times are higher than the men's but drop more rapidly. **(b)** The faster improvement is evident. It also appears that the men's times are faster.

2.17 **(b)** The mean returns were high in sectors 2 and 4, Financial services and Utilities. **(c)** No, sector is not a quantitative variable.

2.19

	Femur	Humerus	z (femur)	z (humerus)	Product
	38	41	−1.530	−1.573	2.408
	56	63	−0.167	−0.189	0.031
	59	70	0.061	0.252	0.015
	64	72	0.439	0.378	0.166
	74	84	1.197	1.133	1.356
Mean	58.2	66		Sum =	3.977
St. dev.	13.198	15.890		Sum/4 =	0.994

2.21 **(a)** Correlation is probably positive but not large. **(b)** $r = +0.6$. **(c)** Unchanged. Correlation does not help answer the question of whether the men are taller. **(d)** Unchanged. **(e)** $+1$.

2.23 Correlation is $+0.2$. The relationship is strong but not linear. Correlation only measures linear relationships.

2.25 **(a)** The Insight appears to fall on the line. **(b)** With the Insight, $r = 0.988$. Without the Insight, $r = 0.958$. Since the Insight appears to fit the pattern but is far higher, including it increases the correlation.

2.27 Yes, there is a straight-line pattern. Correlation is $+0.9$. There are no apparent outliers.

2.29 **(a)** The graph will look essentially the same. The correlation (-0.2) is unchanged. **(b)** The correlation is now -0.1.

2.31 1. Perfect straight-line relationship.

2.33 The paper is interpreting zero correlation as if it were negative correlation. The proper explanation of the report is that there is no (linear) association between teaching and research. Faculty who are good teachers may be good researchers or may be poor researchers.

2.35 **(b)** \$2500. **(c)** $y = 500 + 200x$.

2.37 **(a)** $100 + 40x$. **(c)** When $x = 104$ weeks, $y = 4260$ grams, or about 9.4 lbs.

2.39 **(b)** At weeks = 1, pH = 5.4247. At weeks = 150, pH = 4.6350. **(c)** The slope is -0.0053. This says that for every week that passes, the pH will decrease by 0.0053.

2.41 For run 8903, Length = $-2.39 + 0.158$ Time. For run 8905, Length = $-1.45 + 0.0911$ Time.

2.43 **(a)** Return = $16.2352 - 0.5727$ T-bill. Assuming that a 5% return corresponds to T-bill = 5 produces Return = 13.4. **(b)** The slope means that for each increase in T-bill rates, Return will decrease by 0.5727. This would indicate that high interest rates are bad for stock returns. **(c)** The prediction would not be very accurate. For a given value of T-bill, there is a wide range of returns.

2.45 **(a)** In the following, Year is taken as given, that is, between 1900 and 2000. For women, Record = $41{,}373 - 19.9$ Year. For men, Record = $8166 - 3.29$ Year. **(b)** The record times are decreasing more rapidly for women (19.9 per year) than for men (3.29 per year). **(c)** The two times should be equal in about 1999.

2.47 For length, mean = 12.66, st. dev. = 8.50. For time, mean = 95.0, st. dev. = 53.385. Correlation = 0.996. Slope is $b = 0.1585$. Intercept is -2.39.

2.49 Slope is $b = 0.5 \times 2.7/2.5 = 0.54$. Intercept is $a = 68.5 - 0.54 \times 64.5 = 33.67$. For wife = 67, husband's predicted height is 69.85.

2.51 **(a)** The units of mean and st. dev. for time are minutes. The units for length are centimeters. Correlation is 0.996.

	Mean	St. dev.
Time	95	53.4
Length	12.66	8.5

(b) Mean and StDev will increase by 2.54. Correlation is not changed. **(c)** The slope in terms of inches will be 1/2.54 of the slope in terms of centimeters. The slope in centimeters is 0.158483. The slope in inches is 0.062395.

2.53 +0.4.

2.55 **(a)** Relationship appears to be linear, positive, and strong. **(b)** Equation is women = 2.93 + 1.41 men. **(c)** The three countries with the highest residuals are Yemen, Guinea-Bissau, and Togo. Lesotho has the most negative residual.

2.59 **(b)** A line is only a poor approximation of the relationship. **(c)** There is a very distinct pattern in the residual plot.

2.61 **(a)** The influential observation is at Round1 = 105, Round2 = 89. **(b)** The steeper line omits the influential observation because including the influential observation would tend to flatten the line, given its position.

2.63 **(a)** The general relationship is quite weak and negative. The correlation is −0.369. **(b)** The regression line is NAEP = 233.6 − 0.4658 Poverty. The underachievers are MS, HI, and AR. The overachievers are NC, MA, and TX.

2.65 **(a)** $r^2 = 99.8\%$. **(b)** Stride = 1.766 + 0.0803 Speed. **(d)** The shape suggests that the relationship is not perfectly linear.

2.69 **(a)** The line shows a general decrease but not a strong relationship. **(b)** The general pattern is concave upward (bowl shaped). This might strengthen the conclusion to avoid hospitals that treat few heart attacks.

2.71 The only test scores you have are for people you hired. Presumably, these are all competent people, and you would expect their performance to be at least acceptable. Probably the boss wants to distinguish between who to hire and who not to hire. But you don't have performance ratings for those you didn't hire.

2.73 **(a)** Yes, the relationship appears to be linear and strong: City = −2.58 + 1.0935 Rural, and $r^2 = 95.1\%$. **(b)** With respect to x, there are two suspicious points. One of the points may be influential, and the other has a large positive residual. **(c)** Observation at time = 7 is most influential. It does not have the largest residual. **(d)** 98.8. **(e)** The residuals do not appear to be normal.

2.75 This could occur if salaries go up in each group but go down between groups.

2.77 **(a)** Low. **(b)** The residual plot is curved. The salaries of players in the major leagues for 8 years are generally underestimated because the residuals are positive.

2.79 Students with experience in music may come from wealthier families or families that are more involved with their children. Such experience is certainly not causative.

2.81 This is an observational study, not an experiment. Minority children who take algebra and geometry in high school may already be among the better-performing students.

2.85 Confounders: parental involvement, family values (value attached to education), attention span, socioeconomic level (to the extent that children in poor families may have the TV as a babysitter).

2.87 The explanatory variable is studying a foreign language. The response variable is score on the English achievement test. The lurking variable could be that better students take a foreign language. These students are expected to score better on the English exam.

2.89 We would need information on the type of surgery and on the age, sex, and condition of the patient.

2.91 Sine increases monotonically for $0 < t < 90$ and decreases monotonically for $90 < t < 180$.

2.93 **(a)** Exponential. **(b)** Activity = exp(−0.0116 Days). **(c)** The model fits too well for experimental data.

2.95 **(a)** Larger values of length are associated with larger weights. Except for the two shortest fish, the relationship appears to be rather linear. **(b)** Assuming that the fish are the same proportions, the weight should be a function of length^3. This plot is quite linear, even for the shorter fish.

2.97 One hour: 16 bacteria. Five hours = 1,048,576 bacteria.

2.99 **(a)**

Year	0	1	2	3	4	5
Value	500.00	537.50	577.81	621.15	667.73	717.81

Year	6	7	8	9	10
Value	771.65	829.52	891.74	958.62	1030.52

2.101 **(b)** The logarithm is more nearly linear than Population.
2.103 **(a)** $y = \exp[-2.00 + 2.42\ln(x)] = e^{-2} \times x^{2.42} = 0.1353x^{2.42}$. **(b)** $y = 508.08$.
2.105 The filled slope is 0.22. For $x = 65$, predict 14.4 years.
2.107 Being a million times as large makes you only 10,000 times as strong.
2.109 **(a)** The graph is slightly concave down, so $p < 1$. **(b)** Square root is best.
2.111 The graph of log(Pop) versus Year is quite linear, which would suggest exponential growth over this time period.
2.113 $r^2 = 0.34$ is only moderately strong.
2.115 A correlation of 0.217 is not particularly strong.
2.117 **(a)** Both relationships appear to be linear. There are no influential observations and no apparent outliers. **(b)** Before: +0.189. After: +0.157. **(c)** Before: 7.704. After: 6.348. **(d)** $40.35.
2.119 **(a)** Yield is the response variable and Weeds are the explanatory variable. **(b)** As Weeds increase, Yield seems to decrease. **(c)** Since the slope is negative, we see that Yield decreases as Weeds increase. **(d)** 159.89.
2.121 **(d)** The two lines fit the data quite well. I would ask the experts what differences there are between samples less than 9 inches and greater than 9 inches in length.
2.125 64%.
2.127 **(a)** There is a general decline. **(b)** In 1970, between the elections of 1968 and 1972, the number of eligible voters jumped, due to inclusion of those 18 to 21 years old. If the number of voters remains roughly constant, having a larger eligible population will reduce the percent who vote.
2.129 For a linear model, $r^2 = 82.3\%$. For exponential, $r^2 = 98.8\%$. An exponential model fits much better.
2.131 log(aysym) is a linear function of log(breed). $r^2 = 89.8\%$.

Chapter 3

3.1 This is anecdotal evidence. It was not the result of a designed study. We do not know what other factors might be present. Her acquaintances may have several factors in common that would raise the chance that two of them would develop tumors. Also, she may count someone as an "acquaintance" if she heard that he or she had a brain tumor, while she might take no notice of them otherwise.
3.3 This is an observational study. No action was taken by the experimenter that would affect the outcome. The explanatory variable is cell phone use. The response variable is the occurrence of brain cancer.
3.5 Experiment. Viewing the animation or using the text are the treatments. The explanatory variable is which treatment was used. The response variable is increase in test scores.
3.7 This is an experiment. The treatment is the exercise. The response variable is the change in metabolic rate before the exercise and 12 hours later. The explanatory variable is the exercise.
3.9 Units are the individual trees. Factor is the amount of light. Treatments are full light and reduced light. Response variable is the weight of the trees.
3.11 The units are the individuals who were called. One factor is what information is offered. Treatments are (1) giving name, (2) identifying university, (3) both of these. The second factor is offering to send a copy of the results. The treatments are either offering or not offering. The response is whether the interview was completed.
3.17 In design 1, one group consists of women who have already chosen to use bee pollen. These could have some notion that they have potential health problems. Or

they could be more health conscious and therefore also have other behaviors that would distinguish them from the control group. In design 2, randomly assigning treatments to subjects ensures that there will be no systematic difference between the groups.

3.19 "Significantly higher" means that these returns are higher than would be expected due to chance. "Not significantly different from zero" means that, although the sample may not be exactly zero, the difference from zero is within the range that can occur simply due to chance.

3.21 **(a)** "Randomized" means that patients were randomly assigned to either Saint-John's-wort or the placebo. "Placebo controlled" means that we will compare the results for the group using Saint-John's-wort with those for the group that received the placebo. Double blind: see page 237.

3.29 Select 15 right-handed people. Toss a coin. If it is heads, have them use their right hand first to squeeze the scale, and then have them squeeze the scale with their left hand. If the coin comes up tails, they use their left hand first and right hand second.

3.31 In each field have two boards for each color. One method of randomization would be to assign the labels 1 to 8 to each pole in field 1, then select from line 105: 5 and 2 for blue, 4 and 7 for green, 6 and 1 for white, and 3 and 8 for yellow. Proceeding on from there, in the second field we assign 1 and 4 for blue, 8 and 6 for green, 7 and 5 for white, and 2 and 3 for yellow.

3.33 **(b)** One ethical issue is that some subjects will receive less medical care (presumably those who have to pay a larger share themselves). A practical concern would be the bias that would result if a subject had an extreme situation and was allowed to leave the study because of it.

3.35 **(a)** False. **(b)** True. **(c)** False.

3.37 The population is all the words in Tom Clancy's novels. The sample is the 250 words selected. The variable is the length of the words.

3.39 The variable is whether or not adults approve of the president's job performance. The population that was desired was all adults in the country. The sample was the 1210 adults who were reached by phone. The sample could be biased because only people with telephones could be contacted. There could also be bias due to the types of people who would respond to a telephone poll.

3.41 Numbering 1 to 28 as given and using pairs from Table B will give 04-Bowman, 10-Fonseca, 17-Lujan, 19-Nunez, 12-Goel, 13-Gomez.

3.45 If there is any "unusual" pattern in the line you consistently use, then this will show up in all the samples you take.

3.47 Using row 147 for students: 08-Golomb, 19-Moran, 20-Moskowitz, 23-Pearl. Using row 111 for faculty: 8-Rabinowitz and 1-Besicovitch.

3.49 Midsize: 417 494 322 247 097. Small: 3698 1452 2605 2480 3716.

3.51 **(a)** People without telephones or with unlisted phone numbers would be omitted. Many of those without phones cannot afford one. People with unlisted numbers might include prominent citizens who do not want their numbers published. **(b)** Those with unlisted numbers will be dialed anyway by this scheme.

3.53 **(a)** 29777. **(b)** There's nothing to prevent a person from answering several times. Also, the respondents were only those who went to that Web site and took the time to respond. We cannot define "nonresponse" in this situation. **(c)** The results are slanted toward the opinions of men, who might be less likely to feel that female athletes should earn as much as men.

3.55 **(a)** The wording is clear. The question is somewhat slanted in favor of warning labels. **(b)** The question is clear, but it is clearly slanted in favor of national health insurance by asserting it would reduce administrative costs. **(c)** The question could be clearer by using simpler language. It is slanted in favor of incentives by starting out discussing environmental degradation.

3.57 The two questions appear to be rather opposite. The first question mentions "higher performance," which is an admirable goal and probably led to its high support. The second question mentions another admirable goal, which is to avoid major differences in economic distribution. This also drew support. Neither question mentions the other side to its proposal, which would have diluted support for either issue.

3.59 2.503 cm is a parameter; 2.515 cm is a statistic.

3.61 Both 35% and 14% are statistics.

3.63 A higher sample size will have less variability.

3.65 **(a)** The variability of the estimate depends on the size of the sample, not the population. **(b)** Yes, if we take a fixed proportion of the state's population, we will have a much larger sample size for California than for Wyoming. The estimate for California will be much less variable than the estimate for Wyoming.

3.69 **(a)** Let 0 and 1 represent egg masses being present and the other digits represent no egg masses. Three of the 10 yards had egg masses, so we would estimate 30%. **(b)** The mean of the sampling distribution should be close to 0.2. The distribution will be somewhat skewed to the right.

3.75 It is an experiment because the researcher attempts to affect the outcome by showing different price histories. The explanatory variable is whether the student sees a steady price or temporary price cuts. The response variable is the price the student would expect to pay.

3.77 **(a)** "How much alcohol did you consume over the past weekend?" **(b)** "Do students who bring their texts to class get better grades?" **(c)** "Does playing music in class improve performance on a simple arithmetic quiz?"

3.79 Assume we have several courses, one level and one or more that are steeper. The rider would ride each course 3 times, using a different gear each time, and the times to complete the course would be recorded. Randomize the order of the gears and, if feasible, randomize the order of courses. It is preferable to intersperse runs on different courses.

3.81 Divide the players into two groups: one to receive oxygen and the other without oxygen during the rest period. Match the players so each player receiving oxygen has a corresponding player of similar speed who does not receive oxygen.

3.83 **(a)** Simply being included in the study would raise their awareness of safe-sex issues.

3.85 **(a)** One possible population is all full-time undergraduates in the fall term. **(b)** A stratified sample might include 125 students from each year. **(c)** Mailed questionnaires might have a high nonresponse rate. Telephone interviews would not reach those students without phones, and there might be problems in reaching students at home.

3.87 **(a)** The factors are storage and cooking. Storage is either fresh, a month at room temperature, or a month refrigerated. Cooking is either cooked immediately or cooked after an hour at room temperature. The response variables are the ratings of the color and flavor. **(c)** The different combinations of treatments will be presented in a random order.

3.89 **(a)** Choose 10 subjects at random for 70° temperature and the remaining for 90°. **(b)** Have each subject perform in each temperature. Use the temperatures in a random order.

3.91 Women chose for themselves whether or not to respond. If many of the letters described unfeeling treatment by men, this may suggest that women with this experience are more likely to write in. The 72% figure is likely higher than in the overall population.

3.93 All three graphs are centered at $p = 0.6$, but the larger sample sizes have smaller variability.

Chapter 4

4.5 This means that in a large number of hands, about 1/50 of the hands will contain three of a kind. It is not necessarily the case that in every 50 hands exactly 1 will be three of a kind. It is possible to deal several hands of three of a kind in a row, but it is extremely unlikely.

4.11 **(a)** The sample space would be a time in the range 0 to 24 hours. **(b)** The sample space would be a whole number of cents from 0 on up. **(c)** At most schools, the sample space would be {A, B, C, D, F}. **(d)** The question could be posed so the answer is Yes or No. Or the question could be posed so that the sample space is 0, 1, 2 for how many years the student took math.

4.13 **(a)** 4%. **(b)** 56%.
4.15 **(a)** Legitimate. **(b)** Not legitimate. **(c)** Not legitimate.
4.17 **(a)** The probabilities all lie between 0 and 1 and they sum to 1. **(b)** 0.41. **(c)** 0.38.
4.19 **(a)** 0.295. **(b)** 0.348. **(c)** 0.121104.
4.21 A and B are not independent. $P(\text{A and B}) \neq P(\text{A})\,P(\text{B})$.
4.23 **(a)** 1/38. **(b)** 0.474. **(c)** 0.316.
4.25 The probability for each face should be 1/4, if all the faces are the same size and shape. Alternatively, if the die is as symmetrical as possible, then there should be no reason to favor any one side. Hence, all four probabilities should be the same and sum to 1.
4.27 **(a)** 10,000. **(b)** 3439/10,000.
4.29 **(a)** Abby and Deborah, Abby and Mei-Ling, Abby and Sam, Abby and Roberto, Deborah and Mei-Ling, Deborah and Sam, Deborah and Roberto, Mei-Ling and Sam, Mei-Ling and Roberto, Sam and Roberto.
(b) 1/10. **(c)** 4/10. **(d)** 6/10.
4.31 One big battle is preferred.
4.33 No. It is unlikely that "college" and "laborers" are independent.
4.35 **(a)** 0.2746. **(b)** 0.35. **(c)** 0.545.
4.37 **(a)** B, O. **(b)** $P(\text{O}) = 1/4$. $P(\text{B}) = 3/4$.
4.39 **(a)** $P(\text{O}) = 1/4$. **(b)** $P(\text{3 children with type O}) = 1/64$. $P(\text{first is type O and next two are not}) = 9/64$.
4.41 $P(< 3) = P(1 \text{ or } 2) = 2/6 = 1/3$.
4.43 Owned units tend to have 5 to 7 rooms and rented units tend to have 3 or 4 rooms.
4.45 **(a)** $\{X \geq 5\}$ or $\{X > 4\}$. $P = 0.868$. **(b)** $\{X > 5\}$ means the unit has more than 5 rooms (but not 5). The probability is 0.658.
4.47 **(b)** $\{X > 1\}$ means "people who own at least one car." It is everyone except those who don't own a vehicle. $P = 0.91$. **(c)** This is $P(X > 2) = 0.20$.
4.49 Y takes on the values 1 to 12, each with probability 1/12.
4.51 **(a)** 0.5. **(b)** 0.5.
4.53 **(a)** 1/2. **(b)** 1/2. **(c)** 0.4. **(d)** 0.6.
4.55 **(a)** 0. **(b)** 0.0150. **(c)** 0.9700.
4.57 Owner-occupied: 6.284. Renter-occupied: 4.187.
4.59 Owner-occupied: 1.63992. Renter-occupied: 1.30768.
4.61 **(a)**

Amount	P
\$1690	1/4
\$975	1/2
\$562.5	1/4

(b) \$1050.63.
4.63 **(a)** First die, 4.5. Second die, 2.5. **(b)**

2	3	4	5	6	7	8	9	10	11	12
1/36	2/36	3/36	4/36	5/36	6/36	5/36	4/36	3/36	2/36	1/36

Mean = 7.

4.65 **(a)** Not independent: Knowing the total of the first two cards tells us something about the total for the three cards. **(b)** Independent: Separate rolls of the dice should be independent.
4.67 **(a)** 31 seconds. **(b)** No. **(c)** No.
4.69 Independent, 4.4721. Correlated, 4.9800.
4.73 **(a)** $\sigma_Y = 138.65$ units. **(b)** $\sigma_{X+Y} = 2796.29$ units. **(c)** $\sigma_Z = \$5{,}606{,}738$.
4.75 **(a)** Mean = 0.001 g. Standard deviation = 0.002236 g. **(b)** Mean = 2.0005 g. $\sigma_Z = 0.001118$ g.
4.77 With only 12 policies, the total income from policies would be about \$3000. If any one of the 12 houses should burn down, then the cost would be much more than

\$3000. However, if you sell thousands of policies, the number of payouts is more predictable and the company can manage the risk.

4.79 \$303.353 is the mean.

4.81 9707.6 is the standard deviation.

4.83 His new mean is 1.23. His new standard deviation is 4.584.

4.85 The mean is 1.034. The standard deviation is 3.819.

4.87 0.31496.

4.89 0.8.

4.91 **(a)** 0.3. **(b)** 0.3. **(c)** 0.2. **(d)** 0.2.

4.93 **(a)** 70.13086. **(b)** 0.348 is the proportion of women who are *married* among those that are *age 18 to 29*.

4.95 **(a)** 0.44256. **(b)** 0.68989. **(c)** $P(\text{A and B}) = 0.30532$.

4.97 Did not finish high school = 7.73%. High school but no college = 4.6%. Less than bachelor's degree = 4.07%. College graduate = 2.74%.

4.99 $P(\text{graduate} \mid \text{employed}) = 0.3167$. $P(\text{employed} \mid \text{graduate}) = 36{,}259/47{,}371$.

4.101 0.08.

4.103 0.75.

4.105 **(a)** Each parent has to have *a* in order for her brother to be albino. However, her parents are not albino, so neither of them can be *aa*. Therefore, each parent has to be *aA*. **(b)** Types *aa* and *AA* each have probability 1/4, while *aA* has probability 1/2. **(c)** $P(AA \mid \text{not albino}) = 1/3$. $P(aA \mid \text{not albino}) = 2/3$.

4.107 1.

4.109 0.93333.

4.113 $P(\text{better}) = 0.03$. $P(\text{not worthless}) = 0.50$.

4.115 **(a)** $P(\text{nothing}) = 0.9389$. **(b)** Mean = \$1.30. **(c)** Standard deviation = \$12.9734.

4.117 The mean of the total is 10.5. The standard deviation of the total is 2.958.

4.119 **(a)** and **(b)**

Ann	Bob	X
A	A	0
A	B	2
A	C	−3
A	D	0
B	A	−2
B	B	0
B	C	0
B	D	3
C	A	3
C	B	0
C	C	0
C	D	−4
D	A	0
D	B	−3
D	C	4
D	D	0

(c)

X	P
−4	1/16
−3	2/16
−2	1/16
0	8/16
2	1/16
3	2/16
4	1/16

(d) The mean is 0, so the game is fair. Standard deviation is 2.1794.

4.121 1/4.
4.123 0.84.
4.125 **(b)** 0.015925. **(c)** 0.627.

Chapter 5

5.1 **(a)** Binomial if we assume no identical twins or multiple births (this would violate independence) and that the probability of a female is the same for all births. **(b)** No, not a fixed number of trials. **(c)** No, opinions of husband and wife are not necessarily independent.
5.3 **(a)** Yes, it is reasonable to assume that the probability of passing is the same for all students and they are independent. **(b)** No, the additional instruction is given if you miss a problem. This violates the assumption that getting one question right is independent of getting an earlier question right (or wrong). **(c)** No, probability of dissolving is not the same for each trial.
5.5 **(a)** 0.17. **(b)** 0.2968.
5.7 **(a)** 4.5. **(b)** 1.775. **(c)** With $p = 0.1$, $\sigma = 1.1619$. With $p = 0.01$, $\sigma = 0.3854$. This shows that the standard deviation gets smaller as p gets smaller. There is less variation when the probability of success is small.
5.9 **(a)** 0.4095. **(b)** Mean is 4.
5.11 **(a)** 59.044. **(b)** $Z = 1.516$. $P(Z > 1.516)$ is between 0.0655 and 0.0643. **(c)** Essentially zero.
5.13 **(a)** 62%. **(b)** 0.1438.
5.15 **(a)** 0.9429. **(b)** As p moves away from 0.5 (in either direction), the probability of being within ±3% increases, since the variability decreases.
5.17 **(a)** $n = 4, p = 1/4$. **(b)**

X	0	1	2	3	4
P	0.316	0.422	0.211	0.047	0.004

(c) Mean $= 4 \times 1/4 = 1$.
5.19 **(a)** $p = 1/4$. **(b)** $p = 0.0139$. **(c)** Mean = 5. Standard deviation = 1.9365. **(d)** Not binomial because the trials are not independent since the total of each shape is known to be 5.
5.21 **(a)** Mean = 200. Standard deviation = 12.65. **(b)** Mean = 0.2. Standard deviation = 0.01265. **(c)** 0.00078. **(d)** 0.2295.
5.23 **(a)** 0.11367. **(b)** Mean $= 1500 \times 0.11367 = 170.5$. **(c)** Nearly 1/2.
5.25 **(a)** 0.124. **(b)** 0.034. **(c)** 4 times as many. **(d)** Yes, the values involved are different, but the factor of 4 still applies.
5.27 **(a)** 5/36. **(b)** 25/216. **(c)** $5^{k-1}/6^k$.
5.29 **(a)** "Although the average may be higher or lower than the mean, μ, the amounts higher will tend to balance the amounts lower." **(b)** Since the variability of the average is less for larger sample sizes, the average will tend to be closer to the true mean when the sample size is larger.
5.31 **(a)** Higher than 300: 1/2. Higher than 335: 0.1587. **(b)** Higher than 300 is still 1/2. Higher than 335: 0.0228.
5.33 **(a)** $\mu = 2.45$. $\sigma = 1.099$. **(b)** $\mu = 2.45$. $\sigma = 0.1554$ **(c)** For a single student, 0.3084. For the average of 50 students, 0.0002.
5.35 **(a)** 0.0668. **(b)** 0.00135.
5.37 133.225.
5.39 There is virtually no probability of exceeding this.
5.41 **(a)** With only 12 policies, the standard deviation of the average claim is \$288. This is greater than the mean and suggests that the central limit theorem hasn't kicked in yet. With more policies, the standard deviation decreases, so that with 1000 policies, the standard deviation of the average is \$31.62. This means that virtually all claims will lie within ±3 standard deviations, or from about \$150 to \$350. **(b)** 0.0062.
5.43 12.513.

5.45 **(a)** Mean = 25 g. Standard deviation = 16.62. **(b)** The distribution of $\overline{x}$ is normal with mean 360 and standard deviation $55/\sqrt{20}$ = 12.298. The distribution of $\overline{y}$ is normal with mean 385 and standard deviation $50/\sqrt{20}$ = 11.18. The distribution of $\overline{y} - \overline{x}$ is also normal with mean and standard deviation from part (a). **(c)** 1/2.

5.47 **(a)** $\overline{y}$ has a normal distribution with mean μ_Y and standard deviation $\sigma_Y/\sqrt{m}$. $\overline{x}$ has a normal distribution with mean μ_X and standard deviation $\sigma_X/\sqrt{n}$. **(b)** The distribution of $\overline{y} - \overline{x}$ is normal with mean $\mu_Y - \mu_X$ and standard deviation $\sqrt{\sigma_Y^2/m + \sigma_X^2/n}$.

5.49 0.07706.

5.51 **(a)** Yes, this result does not rely on independence. **(b)** No, this formula is correct only when X and Y are independent. It is likely that husbands' and wives' incomes are positively correlated.

5.53 0.0989.

5.55 **(a)** Mean = 3025 and standard deviation = 51.565. **(b)** Zero.

5.57 No. The number of trials is not fixed. In fact, the random variable is what is being measured.

5.59 0.9537.

5.61 **(a)** No, not a fixed number of trials. **(b)** No, X has to be an integer and cannot be negative. **(c)** Approximate distribution is normal with mean 1.5 and standard deviation 0.02835. **(d)** 0.1039.

5.63 0.0123.

5.65 −2.6256 to 3.0324.

5.67 **(a)** The distribution of $\hat{p}_1$ is approximately normal with mean 0.82 and standard deviation 0.02817. The distribution of $\hat{p}_2$ is approximately normal with mean 0.95 and standard deviation 0.01422. **(b)** The distribution of $\hat{p}_2 - \hat{p}_1$ is approximately normal with mean 0.95 − 0.82 = 0.07 and standard deviation 0.03156. **(c)** 0.1709.

Chapter 6

6.1 **(a)** 11.3%. **(b)** No, with such a small response rate, there are many other things that can be going on in the population.

6.3 **(a)** 94.32 to 125.68. **(b)** No, this is a confidence interval for the mean.

6.5 **(a)** 20. **(b)** 40. **(c)** 40.

6.7 **(a)** 119.6 to 128.0. **(b)** 118.7 to 128.9. **(c)** 117.1 to 130.5. **(d)** As confidence increases, the interval gets wider.

6.9 474.83 to 605.17. Wider.

6.11 **(a)** 3.071 to 3.729. **(b)** 3.210 to 3.590.

6.13 31.84, 38.34.

6.17 Sample size will be smaller because you need fewer data to produce a wider interval. n = 245.86, rounded up to 246.

6.19 **(a)** 2.352. **(b)** 7.438. **(c)** Take n = 61.46 or n = 62.

6.21 7.517, so use n = 8.

6.23 **(a)** No, only 95% confident. **(b)** This interval is based on methods that produce correct results (contain the true percentage) 95% of the time. **(c)** 0.0153. **(d)** No, it only accounts for random fluctuation.

6.25 Probably greater because the estimate for four-person families in Michigan is based on a much smaller sample.

6.27 **(a)** 51% to 57%. **(b)** Yes, even the lower end of the interval is greater than 1/2.

6.29 **(a)** The actual percent may be 3% higher or lower (with 95% confidence). **(b)** The method used produces an interval such that 95% of the time the population percent will lie within the interval. **(c)** There were fewer men.

6.31 **(a)** H_0: $\mu = 1.4$ versus H_a: $\mu \neq 1.4$. **(b)** H_0: $\mu = 0$ versus H_a: $\mu < 0$. **(c)** H_0: $\mu = 60$ versus H_a: $\mu \neq 60$.

6.33 **(a)** H_0: $p_m = p_f$ versus H_a: $p_m = p_f$, where p_m is the percent of males who like math and p_f is the percent of females who like math. **(b)** H_0: $\mu_A = \mu_B$ versus H_a: $\mu_A > \mu_B$, where μ_A is the mean for group A and μ_B is the mean for group B. **(c)** H_0: $\rho = 0$ versus H_a: $\rho > 0$, where ρ is the population correlation between income and percent of disposable income saved.

6.35 **(a)** 0.0359. **(b)** 0.9641. **(c)** 0.0718.

6.37 **(a)** Assuming a two-sided alternative, we cannot reject $\mu = 34$ at the 5% level, because 34 is inside the 95% confidence interval. **(b)** Since 36 is outside the interval, we can reject $\mu = 36$ at the 5% level.

6.39 r^2 measures the percent variation that is explained by the regression model. When it is large, this means that the model "explains" most of the variation in the data. $P < 0.01$ means that the value of r^2 in this case would occur by chance alone with probability less than 1%.

6.41 While some difference between church attenders and nonattenders is not unexpected, the size of the difference that was observed has a less than 5% chance of occurring by chance alone.

6.43 The difference between men and women was so large that there is only a small chance (3.8%) that it is due to random fluctuation. The difference between the races was of a size (47.6%) that could well have occurred simply due to chance.

6.45 **(a)** $z = 3.01$. The P-value is 0.0013; therefore, we reject H_0 and conclude that older students do have a higher (mean) score. **(b)** We assume that the 20 students are an SRS and that the distribution of scores is nearly normal. The SRS assumption is more important.

6.47 **(a)** $z = 2.548$. The P-value is 0.0054, which is strong evidence that the mean is greater than 20. **(b)** Randomly assign some students to take the course and compare their mean score to that of those who did not take the course.

6.49 **(a)** H_0: $\mu = 32$, H_a: $\mu > 32$. **(b)** $z = 1.86$. The P-value is 0.0314. This is fairly good evidence that these scores are higher than the national average. There is only a 3% chance of an average this high simply due to chance.

6.53 **(a)** Since $z > 1.645$, the result is significant at the 5% level. **(b)** Since $z > 2.326$, it is also significant at 1%.

6.55 To be significant at the 1% level, the results must be such that there is less than 1% chance of the results occurring due to random fluctuations. If the chance is less than 1%, then it is also <5%, so the result is also significant at the 5% level.

6.57 If $2.58 < |z| < 2.81$, then the (two-sided) result will be significant at the 1% level but not at the 0.5% level.

6.59 More than 25%.

6.61 **(a)** Between 0.15 and 0.10. Table A: 0.1314.

6.63 **(a)** 99.04 to 109.23. **(b)** H_0: $\mu = 105$, H_a: $\mu \neq 105$. Since 105 is in the confidence interval, we do not have enough evidence to reject H_0.

6.65 **(a)** H_0: $\mu = 7$, H_a: $\mu \neq 7$. Since 7 is not in the given interval, we can reject H_0. **(b)** No, since 5 is in the interval.

6.67 **(a)** H_0: $p = 0.5$, H_a: $p > 0.5$. **(b)** Binomial, $p = 0.5$, $n = 5$. **(c)** $P(X > 4) \geq 0.1875$.

6.69 **(a)** $z = 1.64$. This is just less than 1.645, so it is not significant at 5%. **(b)** $z = 1.65$. This is over 1.645, so it *is* significant at 5%.

6.71 If you perform a large number of tests, you would expect about 5% of these tests to be significant even when there is no real effect. It would be better to use these variables in a follow-up study that tests only this small set of variables.

6.73 There is evidence that vitamin C is effective, but we cannot say if the effect is strong. The small P-value could be the result of a large sample size.

6.75 **(a)** 452.24 to 503.76. **(b)** 469.85 to 486.15. **(c)** 475.42 to 480.58.

6.77 Since the data are based on voluntary response rather than an SRS, the methods of this section cannot be used. The interval does not apply to the whole population.

6.79 The fourth and sixth tests would be significant at 5%.

6.81 **(a)** Binomial with $n = 77$ and $P = 0.05$. **(b)** $P(X > 2) = 0.9027$.

6.83 64%.

6.85 **(a)** 0.977. **(b)** 1.

6.87 **(a)** 1/2. **(b)** 0.212. **(c)** 0.008.

6.89 P(Type I) $= 0.01$. P(Type II) $= 0.36$.

6.91 **(a)** H_0: patient is ill (should see a doctor) and H_a: patient is well (does not need to see a doctor). Type I error is a false-negative, concluding that a patient is well when he should be seen by a doctor. Type II error is a false-positive, having the patient see the doctor unnecessarily. **(b)** If you lower the probability of a false-negative, you ensure that almost all sick patients are seen by a doctor. If you lower the probability of a false-positive, you reduce costs because the doctor sees fewer well patients.

6.93 4.76 to 5.97.

6.95 **(b)** 26.06 to 34.74. **(c)** $z = 2.44$ and $P = 0.007$.

6.97 \$782.82 ± 6.022.

6.99 12.3 to 13.5.

6.101 **(b)** The *P*-value is slightly over 10%. **(c)** Not significant at the 5% level.

6.103 H_0: $\mu = 18/38$, H_a: $\mu \neq 18/38$.

6.105 Yes. Statistical significance allows us to distinguish results that could have arisen by chance from those that are too extreme to be reasonably attributed simply to chance variation.

6.107 The results are essentially the same as in Exercise 6.104.

6.109 **(b)** 19.6. **(d)** All of the possibilities from 46 through 49 have probability at least 10%.

Chapter 7

7.1 **(a)** 25.298. **(b)** 9.

7.3 Larger. The interval is \$445.92 to \$616.08.

7.5 **(a)** 2.55 is between the columns for 0.01 and 0.005, *P*-value is between 0.02 and 0.01. **(b)** *P*-value is between 0.10 and 0.05.

7.7 **(a)** $|t| > 2.576$. **(b)** The z^* row. t approaches $N(0,1)$ as df increases.

7.9 **(b)** The 95% confidence interval is from 20.00 to 27.12.

7.13 **(a)** 49. **(b)** With df = 50, $1.299 < 1.65 < 1.676$. **(c)** The *P*-value is between 20% and 10%. **(d)** It is significant at neither 10% nor 5%. **(e)** 0.1053.

7.15 $\bar{x} = 544.75$, $s = 79.701$, SE = 39.85. This is not a random sample.

7.17 **(b)** $\bar{x} = 856.2$, $s = 306.7$, SE = 51.1. **(c)** Confidence interval is 752.4 to 960.0.

7.19 **(b)** $t = -5.40$. **(d)** $t = -8.68$ with df = 35. Same conclusion.

7.21 $t = 0.83$ with df = 49. Two-sided *P*-value = 0.41.

7.23 \$224.76 to \$276.84 per year.

7.25 **(a)** \$254,742.50. **(b)** \$216,250 to \$293,235.

7.27 87.60 to 104.4 days.

7.29 **(b)** $\bar{x} = 3.618$, $s = 3.055$, SE = 0.524. **(c)** Interval is 2.552 to 4.684.

7.31 $t = 31.24$ with df = 4

7.33 **(a)** $t = 47.3$. **(b)** \$541.54 to \$588.46. **(c)** The sample size is very large. **(d)** Make the offer to 250 customers and keep 250 other customers as a control group.

7.35 $t = 2.086$ with df = 5. The *P*-value is 0.046.

7.37 $t = -0.32$ with df = 11. The *P*-value is 0.76.

7.39 **(b)** $t = 4.96$ with df = 26. The *P*-value is < 0.0005. **(c)** The confidence intervals are given below. Note that the interval for Change cannot be determined from the other two.

Factory	40.955	44.749
Haiti	36.553	38.484
Change	3.12	7.54

7.41 7.019 to 10.441.

7.43 $t = 1.581$ and the *P*-value is 0.074.

7.45 There is no sampling involved.

7.47 **(a)** With df = 100, $t^* = 2.364$ (one-sided). **(b)** Reject H_0: $\bar{x} > 63.1188$. **(c)** The power is between 0.999 and 0.9995, from Table D.

7.49 **(b)** Either $P(\geq 28$ improvements) or $P(\leq 4$ negatives). $z = 3.944$.

7.51 With base 10 logarithms, 2.4929 to 2.6175. With natural logarithms, 5.7402 to 6.0270.

7.53 **(a)** Two-sided. **(b)** If we use the smaller of $n_1 - 1$ and $n_2 - 1$, then df = 29. Software may determine another value. **(c)** Between 0.005 and 0.002.

7.55 We now have a paired design.

7.57 It is possible that new employees will come in batches, so that we might end up with more employees in one department (with different needs) who get the flat screens.

7.59 **(a)** H_0: $\mu_1 = \mu_2$, H_a: $\mu_1 < \mu_2$. **(b)** $t = 2.025$ with df $= 9$. The P-value is 0.037. **(c)** No, only that the *mean* for one-bedrooms is less than the *mean* for two-bedrooms.

7.61 **(a)** We now have a paired design. **(b)** $t = 49.83$ and the P-value is 0.006. The interval is 23.5 to 30.32.

7.63 With smaller samples, the SE is larger and the critical values for t are a bit larger, so it is more "difficult" to reject H_0. In Exercise 7.60, there is a large difference in means and a small P-value. In Exercise 7.62, the difference in means is small (and in the wrong direction) and we do not reject H_0.

7.65 **(b)**

	n	Mean	St. dev.	SE
Piano	34	3.618	3.06	0.5240
Control	44	0.386	2.42	0.3653

(c) $t = 5.06$ with df $= 33$. The P-value is < 0.0005.

7.69 **(a)** $t = 8.23$ with df $= 26$. The P-value is < 0.0001. This is significant at almost any level. **(b)** The study was based on college faculty, who are known not to be representative of the general population.

7.71 **(a)** $s_1 = 0.22$ and $s_2 = 0.19$. **(b)** $t = 17.8$ and df $= 44$.

7.73 **(a)** -1.274 to 7.274.

7.75 **(a)** $t = 2.454$ with df $= 29$. The P-value is 0.01022. This is significant at 5%. The t statistic is slightly less than the 1% table value, so it is just barely not significant at 1%. **(b)** 0.632 to 10.832.

7.77 **(a)** $t = 1.604$ with df $= 9$. The P-value is 0.0716. **(b)** The 95% interval is -0.76 to 11.30.

7.79 **(a)** 11.35. 15.61. **(b)** $t = 1.249$ with df $= 9$. The P-value is 0.2431. **(c)** The 95% interval is -15.8 to 54.8.

7.81 **(a)** Using $t^* = 1.660$ (df $= 100$), the interval is \$412.68 to \$635.58. Using $t^* = 1.6473$ (df $= 620$), the interval is \$413.54 to \$634.72. Using $t^* = 1.6461$ (df $= 1249.2$), the interval is \$413.62 to \$634.64.

7.83 **(a)** Unpooled, $t_{\text{crit}} = 6.314$. **(b)** Pooled $t_{\text{crit}} = 2.920$. **(c)** The increased df means that values of the statistic between 2.920 and 6.314 are significant at the 5% level for the pooled test.

7.87 **(a)** $t = 1.663$ and the P-value is 0.052. **(b)** -0.1938 to 1.994.

7.89 2.16.

7.91 **(a)** Not significant at the 5% level. **(b)** Between 10% and 5%.

7.93 $F(9, 9) = 6.612$. $P < 0.01$.

7.95 $F = 1.258$. $P = 0.4033$.

7.97 **(b)** $F = 1.544$. **(c)** $P > 0.10$.

7.99 0.8267.

7.101

Size	20	40	60	80	100	120
Margin	285	201	164	142	127	116

7.103 64.55 to 92.09.

7.105 For GPA, $t = 0.91$. For IQ, $t = 1.64$.

7.107 $t = 3.650$. $P = 0.0002$. A 99% confidence interval for the difference is 0.477 to 2.903.

7.109 $t = -0.7586$.

7.111 **(b)** 55.71 to 63.47.

7.113 $t = -0.3533$ and the P-value is 0.3621.

7.115 No, not a random sample.

7.117 $t = 1.173$. $P = 0.1265$. **(b)** -14.57 to 52.57. **(c)** 165.53 to 220.47. **(d)** It is assumed that we have SRSs from each group and that the distributions are at least approximately normal, with no outliers. It is unlikely that the pet dogs are an SRS.

7.119 Omit the two zeros and $z = -2.45$, $P = 0.0113$.

7.121 The 95% confidence interval for mean difference in abdomen skinfold thickness is 11.62 to 15.38 (df $= 103.6$). For mean thigh difference, the interval is 9.738 to 12.86.

7.123 **(a)** $t = 4.0$, $P = 0.0001$. Interval = 23.7 to 69.8. There appears to be a difference. **(b)** $t = 0.94$, $P = 0.35$. Interval = −13.4 to 38. No difference.

7.125 $n = 820$.

7.127 As df grows, the critical value gets closer to the normal value of 1.96.

7.129

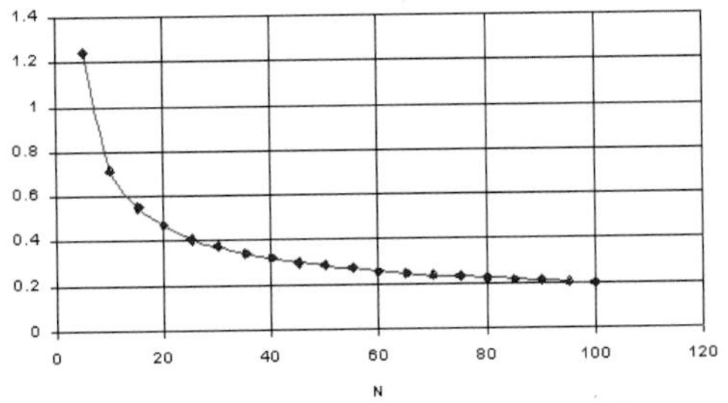

7.131 **(b)** $t = -3.33$, $P = 0.002$. Three-bedroom houses cost less (on average) than four-bedroom houses. **(c)** Yes, we would expect that smaller houses would cost less. **(d)** \$105,258 to \$25,538.

Chapter 8

8.1 For $\hat{p}$: 0.1366 to 0.4134. For Wilson estimate: 0.1606 to 0.4303.

8.3 **(a)** $\tilde{p} = 7/9 = 0.7778$. **(b)** Margin of error = 0.3644. The 95% interval is 0.5062 to 1.0494, with the upper limit replaced by 1.

8.5 No. Some of those who lied about their degree might also have lied about their major. Someone may have claimed to have a degree in business but was an English major who did not graduate.

8.7 0.1996 to 0.2516.

8.9 For $\hat{p}$: 0.8885 to 0.9615. For Wilson estimate: 0.8787 to 0.9546.

8.11 For $\hat{p}$: 0.6124 to 0.7609. For Wilson estimate: 0.6083 to 0.7554.

8.13 **(a)** No. **(b)** Yes. **(c)** No. **(d)** Yes.

8.15 **(b)** $z = 0.9317$, which is still not significant.

8.17 For $\hat{p}$: 0.1014 to 0.2452. For Wilson estimate: 0.1173 to 0.2625.

8.19 **(a)** $z = 1.34$ and the two-sided *P*-value is 0.18. **(b)** The 95% interval is 0.4969 to 0.5165.

8.21 **(b)** $z = 2.0997$. **(c)** Reject at 5% because $z > 1.645$. The actual *P*-value is 0.018. **(d)** The 90% interval for her percent is 0.39705 to 0.65057.

8.23 384.

8.25 288.12. Round up to 289.

8.27 144.06. 150.06.

8.29

P	0.10	0.20	0.30	0.40	0.50	0.60	0.70	0.80	0.90
Margin	0.1859	0.2479	0.2840	0.3036	0.3099	0.3036	0.2840	0.2479	0.1859

8.31 For high-tech–non-high-tech: 0.0081 to 0.230.

8.33 0.0299 to 0.0799.

8.35 **(a)** The Wilson estimate for females is 0.7903 with SE = 0.051. The estimate for males is 0.3955 with SE = 0.0419. **(b)** The 90% interval for the difference is 0.0668 to 0.285.

8.37 −0.1296 to 0.0943.

8.39 $z = 5.5287$.

8.41 **(b)** $z = 1.222$ and the *P*-value is 0.22. This is not significant. **(c)** −0.036 to 0.154.

8.43 **(a)** 0.5404. **(b)** 0.0786. **(d)** $z = 0.2436$. The one-sided *P*-value is 0.404.

8.45 **(a)** Tippecanoe County: 0.5106. Benton County: 0.40845. **(b)** 0.02935. **(c)** 0.0447 to 0.1597.

8.47 $z = 1.823$ and the one-sided *P*-value is 0.034.

8.49 **(a)** $z = 4.921$. This is highly significant. **(b)** 0.053 to 0.1247.

8.51 **(a)** 72.2% for plasterboard and 50% for glass. The Wilson estimates are 0.7 and 0.5. **(b)** −0.09724 to 0.49724. **(c)** $z = 1.3675$ and the one-sided P-value is 0.086.

8.53 **(a)** $z = 1.934$ and the one-sided P-value is 0.0266. **(b)** The same proportions based on a larger sample size result in smaller P-values (that is, more significant results).

8.55 Sample proportion: $z = -1.50$ and $P = 0.133$. Wilson estimate: $z = -1.640$ and $P = 0.101$.

8.57 $z = 3.309$ and the two-sided P-value is 0.0009.

8.59 **(a)** $z = 6.985$. **(b)** 0.1143 to 0.2019.

8.61 $z = 1.34$ and the two-sided P-value is 0.18. The 95% interval is −0.116 to 0.0615.

8.63 $z = 8.945$, which is extremely significant. The interval is 0.3666 to 0.5553.

8.65 0.02615.

8.67 Reject H_0 for all but Text 3. The last three are not different.

Text	1	2	3	4	5	6	7	8	9	10
$\hat{p}$	0.872	0.900	0.537	0.674	0.935	0.688	0.643	0.647	0.710	0.876
z	4.64	6.69	0.82	5.31	5.90	5.20	3.02	2.10	6.60	9.05

8.69 No, it is a voluntary response sample.

8.71 0.0279 to 0.0788.

8.73 $z = -2.15$ and the P-value is 0.0316.

8.75 British: The 95% interval is −0.0154 to 0.0088; $z = -0.5$; $P = 0.308$. American: $z = 5.00$.

8.77

n	10	20	40	60	80	100	500
Z	0.9129	1.2910	1.8257	2.2361	2.5820	2.8868	6.4550
P	0.1967	0.0679	0.0253	0.0098	0.0039	0.0000	

8.79

n	16	26	56	86	106	506
Margin	0.3465	0.2718	0.1852	0.1494	0.1346	0.0616

8.81 **(a)** $p_0 = 0.7911$. **(b)** $z = -29.1$. **(c)** $z = -29.2$.

Chapter 9

9.1 **(a)** The tables below are in percents.

	25–34	35–54	55+
Did not complete high school	3.2	5.5	9.6
Completed high school	8.4	14.5	11.0
College, 1–3 years	7.0	12.0	5.8
College, 4 years	6.2	11.9	4.8

(b)

	25–34	35–54	55+
Total	24.9	43.9	31.3

(c)

Did not complete high school	18.3
Completed high school	33.9
College, 1–3 years	24.8
College, 4 years	23.0

9.3

	25–34	35–54	55+
Did not complete high school	17.5	30.0	52.6
Completed high school	24.9	42.6	32.5
College, 1–3 years	28.3	48.3	23.4
College, 4 years	27.1	52.0	20.9

9.5 The marginal distribution for type is the bottom row divided by the overall total (14,050).

9.7

	2-yr, FT	2-yr, PT	4-yr, FT	4-yr, PT	Total
<18	1.8	2.8	1.2	1.4	7.3
18–21	55.8	20.5	54.5	8.7	139.5
22–34	31.4	45.3	37.8	54.0	168.5
35 and over	11.0	31.4	6.5	35.8	84.7

9.11 **(a)** 51.1% nonresponse. **(b)** Small = 41.7%, medium = 51.7%, large = 60.0%, total = 51.1%. **(d)** Small = 39.8%, medium = 33.0%, large = 27.3%. **(e)** No. There are nearly 50% more responses from Small businesses than from Large businesses.

9.13 Percent of nonrelapses: desipramine = 58.3, lithium = 25.0, placebo = 16.7.

9.15 **(a)** The difference is surely due to roundoff. **(b)** Never married = 19.4, married = 59.2, widowed = 11.1, divorced = 10.3.

(c)

	Never married	Married	Widowed	Divorced
18–24	73.6	24.1	0.2	2.1
40–64	6.3	72.7	6.0	15.0

9.17 **(a)**

	Yes	No	Total
Male	490	310	800
Female	400	300	700
Total	890	610	1500

(b) For males, the admission rate is 61.3%. For females, it is 57.1%.

(c)

	Business	Law
Male	66.7	45
Female	66.7	50

(d) Males apply in higher numbers (600 to 200) to the business school, which has a higher admission rate. Thus, although they have a lower rate in the law school, this does not hurt their overall numbers at the university.

9.19 **(a)** For Alaska Airlines, delayed flights were 15.3%. For America West, delays were 10.9%. **(b)** The percents for each city are given on the following page. **(c)** Most of America West's flights are from Phoenix, where the weather is good. Most of Alaska Airlines flights are from Seattle, where the weather is worse.

	Alaska	America West
Los Angeles	11.1	14.4
Phoenix	5.2	7.9
San Diego	8.6	14.5
San Francisco	16.9	28.7
Seattle	14.2	23.3

9.21 **(a)**

	Women	Men	Total
Label	63	27	90
Nonlabel	233	224	457
Total	296	251	547

(b) $X^2 = 10.9493$ with df = 1. This is greater than the 0.001 entry in Table F, so this is very strong evidence of a difference in label preference among men and women. **(c)** $z = 3.30897$.

9.23 $X^2 = 57.86$ with df = 2. The P-value is essentially 0. The two age groups differ in their performance.

9.25 $X^2 = 3.95527$ and df = 4. P-value > 0.25.

9.27 $X^2 = 10.5$ with df = 2. The P-value is slightly greater than 0.005.

9.29 $X^2 = 26.939$ with df = 2. This is highly significant.

9.31 **(a)**

	Letter	Phone	None
Yes	43.7	68.2	20.6
No	56.3	31.8	79.4
Total	100.0	100.0	100.0

(c) $X^2 = 163.4$ with df = 2. This is highly significant.

9.33 **(a)** This study was not an experiment. There was no treatment applied. **(b)**

	No	Yes
Alive	71.8	94.3
Dead	28.2	5.7
Total	100.0	100.0

(d) $X^2 = 8.85$ with df = 1. The P-value is between 0.005 and 0.0025. **(e)** This is good evidence that the two rates are different.

9.35 $X^2 = 38.41$ with df = 1. This is extremely significant. This is extremely strong (overwhelming) evidence that the two rates are different.

9.39 **(a)** Blood pressure. **(b)** $z = -2.9773$ and the (one-sided) P-value is 0.0014. **(c)**

	Low BP	High BP	Total
Died	21	55	76
Not dead	2655	3283	5938
Total	2676	3338	6014

X^2 not appropriate for a one-sided alternative. **(d)** The 95% interval (for High − Low) is 0.00302 to 0.01409.

9.41 $X^2 = 5.957$ with df = 3. The P-value is between 0.15 and 0.10.

9.43 **(a)** $z = -0.56766$ (Freezing − Control) and the two-sided P-value is 57%. The two proportions do not appear to be different. **(b)**

	Freezing	Control	Total
Relief	28	30	58
No relief	54	48	102
Total	82	78	160

$X^2 = 0.32211$ with df $= 1$. **(c)** We conclude that there is no difference in relief rate between the Freezing group and the Control group. This is not an effective treatment.

9.45 SC: $X^2 = 23.45$ with df $= 1$. This is extremely significant. This is extremely strong evidence of a difference between female and male. Mastery: $X^2 = 0.03$ with df $= 1$. This is extremely insignificant. There is no evidence of a difference between female and male.

9.47 $X^2 = 43.4875$ with df $= 12$. This is extremely significant. There is a relationship between PEOPLE score and field of study.

9.49 $X^2 = 359.677$ with df $= 8$. This is very significant. The regression line is $-4448 + 2.27(\text{time})$.

9.51 $X^2 = 852$ with df $= 1$.

9.53 9.11: Comparing populations. 9.21: Independence. 9.22: Independence. 9.24: Independence.

Chapter 10

10.1 **(a)** 4.7. **(b)** 0.66. **(c)** Overseas $= 4.7 + 0.66 \times$ U.S. return $+ \epsilon$.

10.3 **(b)** The t statistic is -1.27 with df $= 14$. The P-value is 0.226.

10.5 **(c)** 90% interval for slope $= 0.446390 + 2.132(0.006856)$, or 0.431773 to 0.461007. $\ln(\text{bits}) = -873 + 0.446 \times$ year.

10.7 **(c)** HAV $= \beta_0 + \beta_1$MA $+ \epsilon$. **(e)** The t statistic is 1.90 with df $= 36$. The P-value is 0.065.

10.9 **(a)** Employee 15 is the outlier. **(b)** $y = 43.4 + 0.0733x$. The t statistic is 2.85 with df $= 57$. The P-value is 0.006. **(d)** 0.004748 to 0.14175.

10.11 **(b)** The equation is $y = -3545 + 1.84x$. The t statistic is 13.06 with df $= 38$.

10.13 **(b)** $y = -61.1 + 9.32x$. $r^2 = 98.8\%$. **(c)** 8.356 to 10.281.

10.15 **(a)** 105. **(b)** The prediction is $-61.1 + 9.32(105) = 917.5$, or a lean of 2917.5. **(c)** We should use a prediction interval.

10.17 **(a)** This is the gas used in nonheating applications, such as cooking. The 90% interval is 2.0497 to 2.7603. **(b)** The width of the interval in this problem is 0.71066. In Exercise 10.12, the width is 1.0838.

10.19 **(a)** $t = 1.738$ with df $= 100$. The two-sided P-value is between 0.05 and 0.10.

10.21 $t = 12.77$ with df $= 79$.

10.23 $y = -659 + 0.348x$. The t statistic is 11.03 with df $= 38$.

10.25 **(b)** The correlation is 0.9334. **(c)** $y = 12.2 + 0.183x$. $t = 16.04$.

10.27 $t = -1.27$, the same as for slope.

10.29 DF $= 28$, SS $= 10152.4$, MS $= 362.5857$.

10.31 -12.103 to 13.429.

10.33 **(a)** $t = 2.611$. **(c)** 0.0064.

10.35 **(a)** Score $= 110 - 1.13$ Age. **(b)** $t = -3.63$. The one-sided P-value is close to 0.001. **(c)** -1.776 to -0.0478. **(d)** 41.0%. **(e)** 11.02. **(f)** The equation is only slightly changed: $y = 108 - 1.05x$. The confidence interval is -1.933 to -0.167. $r^2 = 27\%$; $s = 8.831$.

10.37 **(b)** Correlation $= 0.994$. $t = 15.94$ with df $= 3$.

10.39 When $n = 20$, $t = 1.85164$. When $n = 10$, $t = 1.23443$.

10.41 **(b)** ACT $= 1.626 + 0.021374$ SAT. $t = 10.78$ with df $= 58$. **(c)** 81.7%.

10.43 **(a)** $a_1 = 0.02617$ and $a_0 = -2.7522$. **(c)** Same as for original.

10.45 **(a)** $y = -118 + 0.497x$. **(b)** $y = -99.0 + 18.7x$.

10.49 **(a)** For women, the 95% interval is 14.726 to 33.326. For men, the interval is -9.46 to 42.96. **(b)** For women, the term in the denominator is 389,955. For men, it is 214,870.

Chapter 11

11.1

Source	df	SS	MS	F
Model	7	SSM	SSM/7	MSM/MSE
Error	112	SSE	SSE/112	
Total	119	SST		

11.3 0.0909.

11.5 **(a)** Overall = 3.33 + 0.57 × Favorable + 0.82 × Unfavorable. **(b)** $H_0 : \beta_1 = \beta_2 = 0$ versus H_a : not all slopes = 0. df = (2, 149). The value of 33.7 is significant well beyond the 0.001 level. **(c)** These are the t statistics for testing each coefficient separately. The t for Constant is for testing if the intercept is zero. All of the values are large, indicating that both slopes contribute to the model. **(d)** Each t statistic will have df given by the denominator df for F, namely, $152 - 1 - 2 = 149$.

11.7 In both cases, the two slopes are positive, so the expectation of positive association is met. Assuming that the scales of the unfavorable and favorable nutrient scores are comparable, the fact that the unfavorable nutrient coefficient is large in each case indicates a stronger effect.

11.9 R^2 is the same, but now t for Accts is significant. Assets = −13.6 + 0.0488 Accts + 0.000034(Accts − 793.6)^2.

11.11 $y = -5.06 + 1.29x$. $t = 6.96$.

11.13 **(a)** The P-value is very close to 0. **(b)** 29.7%. **(c)** $|t| > 1.96$, two-sided.

11.15 The intercept for 11.14 is larger.

11.17 2.608.

11.19 **(a)** 0.07714 to 0.259995. **(b)** −0.0546 to 0.14478.

11.21 **(a)** GPA = 0.589877 + 0.168567 HSM + 0.034316 HSS + 0.045102 HSE. **(b)** 0.6998. **(d)** The P-value for the F test is quite small, indicating that GPA depends on at least some of the given variables. **(e)** 20.46%.

11.25 Not much pattern in any graph. HSM might suggest a positive association.

11.27 GPA = 1.29 + 0.00228 SATM − 0.000025 SATV.

11.29 $F = 10.34$. The t statistic for SATV has a P-value of 0.481.

11.31 $F = 8.39$. $R^2 = 25.1\%$. GPA = 0.648 + 0.205 HSM + 0.0018 HSS + 0.0324 HSE. Best model uses only HSM.

11.33 **(a)** GPA = −2.83 + 0.0822 IQ + 0.163 C3. $R^2 = 45.9\%$. The t statistic for C3 is 2.83. **(b)** $R^2 = 47.5$. GPA = −3.49 + 0.0761 IQ + 0.0670 C3 + 0.0369 SC. **(c)** Coefficients are sensitive to changes in the model.

11.35

Variable	Mean	Median	St. dev.	IQR
Taste	24.53	20.95	16.26	24.58
Acetic	5.498	5.425	0.571	0.713
H2S	5.942	5.329	2.127	3.766
Lactic	1.442	1.45	0.3035	0.4625

11.37 $R^2 = 30.2\%$. $y = -61.5 + 15.6x$.

11.39 $R^2 = 49.6\%$. $y = -29.9 + 37.7x$.

11.41 Taste = −26.9 + 3.80 Acetic + 5.15 H2S. $R^2 = 58.2\%$. Acetic does not add to the model.

11.43 Taste = −28.9 + 0.33 Acetic + 3.91 H2S + 19.7 Lactic. $R^2 = 65.2\%$. The preferred model would include H2S and Lactic, but not Acetic.

11.45 The t statistic for slope is 16.04, which is highly significant. $R^2 = 86.8\%$. The regression equation is corn = −46.2 + 4.76 soybean.

11.47 **(a)** corn = 1002 − 0.512 year − 0.0451 year2 + 3.90 soybean. **(b)** $F = 233.05$ with df = (3, 36) and $P < 0.001$. **(c)** $R^2 = 95.1\%$. **(d)** The t statistic for year is not significant. The others are very significant.

11.49 Using linear model: Fit = 136.41. 95.0% PI = 114.20 to 158.61. Using quadratic model: Fit = 127.14. 95.0% PI = 101.82 to 152.45.

11.51 The t statistic is 2.85 and the P-value is 0.006. $R^2 = 12.5\%$. Wages = 43.4 + 0.0733 LOS.

11.53 The F statistic is 11.50 and the P-value < 0.001. Both t statistics are significant. $R^2 = 29.1\%$. Wages = 37.6 + 0.0829 LOS + 8.92 Size.

Chapter 12

12.1 **(a)** The response variable is the effectiveness rating. There are $I = 3$ groups. $N = 90$ and $n_i = 30$ for each group. **(b)** Response is the attractiveness rating. $I = 6$, $N = 50 \times 6 = 300$, and $n_i = 50$ for each group. **(c)** Response is the preference rating. $I = 4$, $N = 100 \times 4 = 400$, and $n_i = 100$ for each group.

12.3 **(a)** Training effectiveness: Model df = 3 − 1 = 2. Error df = 90 − 3 = 87. Total df = $N - 1$ = 89. H_0: all methods equally effective, H_a: not all methods equally effective. Numerator df = Model df = 2. Denominator df = Error df = 87.
(b) Computer disks: Model df = 6 − 1 = 5. Error df = 300 − 6 = 294. Total df = $N - 1$ = 299. H_0: all packages equally attractive, H_a: not all packages equally attractive. Numerator df = Model df = 5. Denominator df = Error df = 294.
(c) Lotion preference: Model df = 4 − 1 = 3. Error df = 400 − 4 = 396. Total df = $N - 1$ = 399. H_0: all formulations equally preferred, H_a: not all formulations equally preferred. Numerator df = Model df = 3. Denominator df = Error df = 396.

12.5 **(a)** The ratio of largest to smallest standard deviation is 24/18 < 2. **(b)** 324, 576, and 400. **(c)** 437.9394. **(d)** 20.927.

12.7 **(a)** Numerator df = $I - 1$ = 3 − 1 = 2. Denominator df = $N - I$ = 24 − 3 = 21. P-value = 0.00975 (software). **(b)** Numerator df = $I - 1$ = 6 − 1 = 5. Denominator df = $N - I$ = 66 − 6 = 60. P-value = 0.07047 (software).

12.9 **(a)** H_0: $\mu_1 = \mu_2 = \ldots = \mu_I$ vs. H_a: not all μ_I equal.
(b)

Source	df	SS	MS	F
Groups	2	SSG	SSG/2	
Error	253	SSE	SSE/253	
Total	255			

(c) F distribution with df = (2, 253). **(d)** Table entry for df = (2, 200) = 3.04.

12.11 **(b)** Reject H_0.

Source	DF	SS	MS	F	P
Cond.	4	2565.72	641.43	367.74	0.000
Error	5	8.72	1.74		
Total	9	2574.44			

12.13 $F = 12.09$. $P = 0.009$.
12.17 All four groups are significantly different.
12.19 Piano is different, but no other comparison is.
12.21 $F = 2.72$, $P = 0.078$.
12.23 $F = 7.98$, $P = 0.002$.
12.25 $F = 31.16$, $P < 0.001$.
12.27 $F = 137.94$, $P < 0.001$.
12.29 $t = 7.407$ with df = 12.
12.31 Blue and white are not significantly different. Lemon is better than all other colors.
12.33 **(a)**

Source	df	SS	MS	F
Groups	3	476.88	158.96	2.53081
Error	32	2009.92	62.81	
Total	35	2486.8		

(c) $0.05 < P$-value < 0.10. **(d)** Variance $= 62.81$. Standard deviation $= 7.953$.
12.35 **(a)** 72412. **(d)** <0.001. **(e)** 28.3%.

Source	df	SS	MS	F
Groups	2	6,572,551	3,286,276	45.383
Error	230	16,654,788	72,412	
Total	232	23,227,339		

12.37 **(a)** $\psi_1 = (\mu_1 + \mu_2)/2 - (\mu_3 + \mu_4)/2$. **(b)** $\psi_2 = \mu_1 - \mu_2$. **(c)** $\psi_2 = \mu_3 - \mu_4$.

12.39 **(b)** $c_1 = 0.93$ and $c_2 = 0.02$. **(c)** $S_{c_1} = 0.2299$ and $S_{c_2} = 0.3421$. **(d)** $t_1 = 4.045$ (df $= 221$). $t_2 = 0.0585$. **(e)** For ψ_1, 0.474 to 1.386. For ψ_2, -0.659 to 0.699.

12.41 **(a)** $\psi_1 = \mu_T - \mu_C$, $H_0\colon \psi_1 = 0$ versus $H_a\colon \psi_1 < 0$. $\psi_2 = \mu_T - (\mu_C + \mu_S)/2$, $H_0\colon \psi_2 = 0$ versus $H_a\colon \psi_2 < 0$. $\psi_3 = \mu_J - (\mu_T + \mu_C + \mu_S)/3$, $H_0\colon \psi_3 = 0$ versus $H_a\colon \psi_3 < 0$. **(b)** $t_1 = -1.27$, $t_2 = -1.78$, $t_3 = -2.09$.

12.43 To compare 1 to 2, $t = 9.27$. To compare 1 to 3, $t = 2.1067$. To compare 2 to 3, $t = 5.949$.

12.45 To compare T to C, $t = -1.27$. To compare T to J, $t = -0.63$. To compare T to S, $t = 1.78$. To compare C to J, $t = -1.79$. To compare C to S, $t = 0.18$. To compare J to S, $t = 2.44$.

12.47 Table is the same.

12.51 **(a)** $\psi_1 = (\mu_1 + \mu_2 + \mu_3)/3 - \mu_4$, $\psi_2 = (\mu_1 + \mu_2)/2 - \mu_3$, $\psi_3 = \mu_1 - \mu_2$.
(b)

	1	2	3
C	−12.51	1.269	0.191
SE	0.2355	0.1413	0.1609
t	−53.17	8.98	1.19
P-value	<0.0005	<0.0005	0.2359

12.53 For vitamin A, change all the columns except N by the same factor, 100/5. This will also change the SS and MS columns by $(100/5)^2$ but will not change F, df, or the P-value. For vitamin E, this will have no effect at all either on the means and standard deviations or on the ANOVA table. In all three cases, the F statistic is unchanged, as are the conclusions.

12.55 $F = 0.11$, $P = 0.895$.

12.57 **(d)** $F = 5.32$. $P = 0.007$. **(e)** $t = -2.51$ with df $= 63$. **(f)** $t = 2.0504$.

12.59 **(a)** $F = 2.00$.

12.61 **(a)**

Size	25	50	75	100
Power	0.554	0.8776	0.9752	0.9959

12.63 Response $= 4.36 - 0.116$ Treatment.

Chapter 13

13.1 **(a)** Response variable: Program Effectiveness. Factors: Program ($I = 3$) and Days ($J = 2$). $N = 120$. **(b)** Response variable: Attractiveness. Factors: Type of Package ($I = 4$) and Color ($J = 3$). $N = 40 \times 4 \times 3 = 480$. **(c)** Response variable: Lotion Comparison. Factors: Formulation ($I = 5$) and Fragrance ($J = 3$). $N = 120$.

13.3 **(a)** Program Effectiveness (df $= 2$), Days (df $= 1$), Interaction (df $= 2$), Error (df $= 119 - 2 - 1 - 2 = 114$), Total (df $= 119$). **(b)** Type of Package (df $= 3$), Color (df $= 2$), Interaction (df $= 6$), Error (df $= 479 - 3 - 2 - 6 = 468$), Total (df $= 479$). **(c)** Formulation (df $= 4$), Fragrance (df $= 2$), Interaction (df $= 8$), Error (df $= 119 - 4 - 2 - 8 = 105$), Total (df $= 119$).

13.5 **(a)** Total $N = 72$ and total df $= 71$. Factor 1, df $= 2$. Factor 2, df $= 3$. Interaction, df $= 6$. Error df $= 71 - 6 - 3 - 2 = 60$. So df for F for Factor 1 $= (2, 60)$. For

Factor 2, df = (3, 60). For Interaction, df = (6, 60). **(b)** Critical value for Factor 1 is 3.15. For Factor 2, critical value is 2.76. For Interaction, critical value is 2.25. **(c)** For 1% level, critical value for Factor 1 is 4.98. For Factor 2, critical value is 4.13. For Interaction, critical value is 3.12.

13.7 **(b)** For both races, blood pressure increases with age. Nonwhite pressure is higher than white pressure at each age. The difference appears to be constant with age, so there does not appear to be an interaction. **(c)** The difference in the averages is 1.84. Except for age 35, this value appears to be a typical difference.

13.9 **(b)** There seems to be a large difference between the means based on how much the rats were allowed to eat, but not very much difference based on the chromium level. There may be an interaction. **(c)** L mean: 4.86, N mean: 4.871. M mean: 4.485, R mean: 5.246. LR − LM = 0.63. NR − NM = 0.892.

13.11 The Other category had the lowest mean SATM score for both genders. Males had higher mean scores in CS and O, while females are slightly higher in EO. This seems to be an interaction. Overall, the marginal means are 611.7 (males) and 585.3 (females).

13.13 **(a)** ECM1/4 weeks has a rather larger standard deviation. The rest are within a factor of 2.

Material	Time	Mean	St. dev.
ECM1	4	65.00	8.660
ECM1	8	63.33	2.887
ECM2	4	63.33	2.887
ECM2	8	63.33	5.774
ECM3	4	73.33	2.887
ECM3	8	73.33	5.774
MAT1	4	23.33	2.887
MAT1	8	21.67	5.774
MAT2	4	6.67	2.887
MAT2	8	6.67	2.887
MAT3	4	11.67	2.887
MAT3	8	10.00	5.000

(c) F for Material = 251.26, df = (5, 24). Time = 0.29, df = (1, 24). Interaction = 0.06, df = (5,24).

13.15 Two weeks, F = 50.71. Four weeks, F = 137.94. Eight weeks, F = 115.54. All significant.

13.17 Yes, the F for Pot is so much larger. It would appear that Iron is still much higher than the others.

13.19 **(a)**

Tool	Time	Mean	St. dev.
1	1	25.031	0.00115
1	2	25.028	0.00000
1	3	25.026	0.00000
2	1	25.017	0.00115
2	2	25.020	0.00200
2	3	25.016	0.00000
3	1	25.006	0.00153
3	2	25.013	0.00115
3	3	25.009	0.00115
4	1	25.012	0.00000
4	2	25.019	0.00115
4	3	25.014	0.00400
5	1	24.997	0.00115
5	2	25.006	0.00000
5	3	25.000	0.00153

(c) Tool: $F = 412.94$, df = (4, 30). Time: $F = 43.60$, df = (2, 30). Interaction: $F = 7.65$, df = (8, 30). All significant.

13.21 **(a)** The mean price decreases as discount increases and also as the number of promotions increases.

Promo	Discount	Mean	St. dev.
1	10	4.423	0.18476
1	20	4.225	0.38561
1	30	4.689	0.23307
1	40	4.920	0.15202
3	10	4.284	0.20403
3	20	4.097	0.23462
3	30	4.524	0.27073
3	40	4.756	0.24291
5	10	4.058	0.17599
5	20	3.890	0.16289
5	30	4.251	0.26485
5	40	4.393	0.26854
7	10	3.780	0.21437
7	20	3.760	0.26179
7	30	4.094	0.24075
7	40	4.269	0.26992

(b) Promo: $F = 47.73$, df = (3, 144). Discount: $F = 47.42$, df = (3, 144). Interaction: $F = 0.44$, df = (9, 144). Promo and Discount are significant.

13.23 **(a)**

Source	df	SS	MS	F
Chromium	1	0.00121	0.00121	0.0403
Eat	1	5.79121	5.79121	192.89
Interaction	1	0.17161	0.17161	5.72
Error	36	1.08084	0.03002	
Total	39	7.04487		

(b) $F = 5.72$. $P = 0.02216$. **(c)** $F = 0.0403$. **(d)** Variance = 0.03002. Standard deviation = 0.1732.

13.25 **(a)** All three F-values have df = (1, 945). The P-values are <0.001, <0.001, and 0.1477.

13.27 **(a) (b)**

Level	N	Mean	St. dev.
11	9	3.2543	0.2287
12	9	2.7636	0.0666
13	9	2.8429	0.2333
14	9	2.9362	0.0709
15	9	3.0519	0.0909
16	9	3.0963	0.0815
17	9	3.3334	0.2482
21	9	2.4216	0.1654
22	9	2.0502	0.1454
23	9	2.0524	0.1481
24	9	1.9673	0.2203
25	9	1.9560	0.1571
26	9	1.9839	0.2895
27	9	2.2184	0.1238
31	9	3.0589	0.1525
32	9	3.1541	0.3324
33	9	3.2003	0.2341
34	9	3.1419	0.2965
35	9	3.3956	0.2533
36	9	3.4961	0.3437
37	9	3.5437	0.3116
41	9	1.4230	0.1738
42	9	1.3037	0.2661
43	9	1.1253	0.1230
44	9	1.0087	0.1310
45	9	1.2584	0.2489
46	9	1.2712	0.0795
47	9	0.9788	0.2090

(c) Plant: $F = 1301.32$, df $= (3, 224)$. Rain: $F = 9.76$, df $= (6, 224)$. Interaction: $F = 5.97$, df $= (18, 224)$.

13.29

	F
R1	184.05
R2	115.93
R3	208.87
R4	218.37
R5	220.01
R6	174.14
R7	230.17

13.31 Fresh: Plant $F = 81.45$, Rain $F = 43.71$, Interaction $F = 1.79$.
Dry: Plant $F = 79.93$. Rain $F = 44.79$, Interaction: $F = 2.22$.

13.33

	Fresh	Dry
R1	15.88	8.14
R2	11.81	26.26
R3	62.08	22.58
R4	10.83	11.86
R5	22.62	21.38
R6	8.20	14.77
R7	10.81	8.66

13.35 Sex $F = 0.47$, Major $F = 9.32$, Interaction $F = 1.81$. Only Major is significant.
13.37 Sex $F = 50.32$, Major $F = 1.40$, Interaction $F = 1.33$. Only Sex is significant.

NOTES AND DATA SOURCES

PREFACE NOTES

1. The committee's report is George Cobb, "Teaching statistics," in L. A. Steen (ed.), *Heeding the Call for Change: Suggestions for Curricular Action*, Mathematical Association of America, 1990, pp. 3–43. A summary has been endorsed by the ASA Board of Directors.

2. See, for example, the evaluation of the current state of the first course in statistics, with discussion by leading statisticians from industry as well as academia, in David S. Moore and discussants, "New pedagogy and new content: the case of statistics," *International Statistical Review*, 65 (1997), pp. 123–165.

3. A detailed discussion appears in George Cobb and David S. Moore, "Mathematics, statistics, and teaching," *American Mathematical Monthly*, 104 (1997), pp. 801–823. Readers interested in relations between mathematics and statistics as fields might also look at David S. Moore and George Cobb, "Statistics and mathematics: tension and cooperation," *American Mathematical Monthly*, 107 (2000), pp. 615–630.

4. A. Agresti and B. A. Coull, "Approximate is better than 'exact' for interval estimation of binomial proportions," *The American Statistician*, 52 (1998), pp. 119–126. Alan Agresti and Brian Caffo, "Simple and effective confidence intervals for proportions and differences of proportions result from adding two successes and two failures," *The American Statistician*, 45 (2000), pp. 280–288.

5. Lawrence D. Brown, Tony Cai, and Anirban DasGupta, "Confidence intervals for a binomial proportion and asymptotic expansions," *Annals of Statistics*, 30 (2002), pp. 160–201.

CHAPTER 1 NOTES

1. Based on news articles in the *New York Times* of April 18, 1992, and September 10, 1992.

2. Data for 1999, collected by the Current Population Survey and reported in the 2000 *Statistical Abstract of the United States*.

3. Newcomb's data and the background information about his work appear in S. M. Stigler, "Do robust estimators work with real data?" *Annals of Statistics*, 5 (1977), pp. 1055–1078.

4. From National Highway Traffic Safety Administration, *Traffic Safety Facts 1999*, December 2000.

5. Ruth, McGwire, and Sammy Sosa of the Chicago Cubs are the only players to hit 50 or more home runs in four different seasons. Barry Bonds's high before his record 73 in 2001 was 49.

6. From the Census Bureau Web site, `www.census.gov`.

7. From the Consumer Price Index home page at the Bureau of Labor Statistics Web site, `stats.bls.gov/cpihome`.

8. Data in (a) and (b) from K. D. Kochanek, B. L. Smith, and R. N. Anderson, *National Vital Statistics Reports*, 49, No. 3, "Deaths: preliminary data for 1999," Centers for Disease

Control and Prevention, 2001. Survival rates in (c) from the updated tables for *Health, United States, 2000*, National Center for Health Statistics, 2000.

9. *Digest of Education Statistics, 2000*, National Center for Education Statistics, 2001.

10. See the first source in Note 8.

11. U.S. Census Bureau, *Profiles of General Demographic Characteristics: 2000*, 2001.

12. *The Condition of Education, 2001*, National Center for Education Statistics, 2001.

13. Data from Gary Community School Corporation, courtesy of Celeste Foster, Department of Education, Purdue University.

14. J. Marcus Jobe and Hutch Jobe, "A statistical approach for additional infill development," *Energy Exploration and Exploitation*, 18 (2000), pp. 89–103.

15. Data provided by Maribeth Cassidy Schmitt, from her Ph.D dissertation, "The effects of an elaborated directed reading activity on the metacomprehension skills of third graders," Purdue University, 1987.

16. See the Note 3 reference for Cavendish's data and background information about his work.

17. *Statistical Abstract of the United States, 2000*.

18. T. Bjerkedal, "Acquisition of resistance in guinea pigs infected with different doses of virulent tubercle bacilli," *American Journal of Hygiene*, 72 (1960), pp. 130–148.

19. Data provided by Darlene Gordon, Purdue University.

20. Stanley B. Goldenberg et al., "The recent increase in Atlantic hurricane activity: causes and implications," *Science*, 293 (2001), pp. 474–479.

21. Returns from the Web site of Professor Kenneth French of Dartmouth, `mba.tuck.dartmouth.edu/pages/faculty/ken.french/data_library.html`.

22. Environmental Protection Agency, *2001 Fuel Economy Guide*, `www.fueleconomy.gov`.

23. Data from the August 2000 supplement to the Current Population Survey, from the Census Bureau Web site, `www.census.gov`.

24. *Consumer Reports*, June 1986, pp. 366–367.

25. The density curve in Figure 1.35 was produced by a kernel density estimator. See B. W. Silverman, *Density Estimation for Statistics and Data Analysis*, Chapman and Hall, 1986.

26. Stephen Jay Gould, "Entropic homogeneity isn't why no one hits .400 anymore," *Discover*, August 1986, pp. 60–66.

27. Results for 1988 to 1991 from a large sample survey, reported in National Center for Health Statistics, *Health, United States, 1995*, 1996.

28. Reed Abelson, "Companies turn to grades, and employees go to court," *New York Times*, March 19, 2000.

29. Graham Hurlbut, Eldon Glade, and John McLaughlin, "Teaching attitudes and study attitudes of Indian education students," *Journal of American Indian Education*, 29, No. 3 (1990), pp. 12–18.

30. Data provided by Charles Hicks, Purdue University.

31. Data from the March 2000 Current Population Survey, available on the Census Bureau Web site, `www.census.gov`.

32. James W. Grier, "Ban of DDT and subsequent recovery of reproduction of bald eagles," *Science*, 218 (1982), pp. 1232–1235.

33. From the Florida State University athletics Web site, `seminoles.fansonly.com`.

34. These are a random sample from the more than 66,000 randomly chosen housing units in the 1999 American Housing Survey, from data available at the Census Bureau Web site, `www.census.gov`.

CHAPTER 2 NOTES

1. Figure 2.1 is based on data from the College Board Web site, `www.collegeboard.org`.

2. A sophisticated treatment of improvements and additions to scatterplots is W. S. Cleveland and R. McGill, "The many faces of a scatterplot," *Journal of the American Statistical Association*, 79 (1984), pp. 807–822.

3. Marilyn A. Houck et al., "Allometric scaling in the earliest fossil bird, *Archaeopteryx lithographica*," *Science*, 247 (1990), pp. 195–198. The authors conclude from a variety of evidence that all specimens represent the same species.

4. W. L. Colville and D. P. McGill, "Effect of rate and method of planting on several plant characters and yield of irrigated corn," *Agronomy Journal*, 54 (1962), pp. 235–238.

5. The motorcycle crash test data are often used to test scatterplot smoothers. They appear in, for example, W. Hardle, *Applied Nonparametric Regression*, Cambridge University Press, 1990. The smoother used to produce Figure 2.5 is called "lowess." A full account appears in section 4.6 of J. M. Chambers, W. S. Cleveland, B. Kleiner, and P. A. Tukey, *Graphical Methods for Data Analysis*, Wadsworth, 1983.

6. The data, from the March 2000 supplement to the Current Population Survey, appear in the file *individuals.dat*, described in the Data Appendix. We have combined the 16 levels of education in the original data to form 6 levels.

7. Figure 2.8 is based on data from the Web site of Professor Kenneth French. See Note 21 for Chapter 1.

8. Christer G. Wiklund, "Food as a mechanism of density-dependent regulation of breeding numbers in the merlin *Falco columbarius*," *Ecology*, 82 (2001), pp. 860–867.

9. M. H. Criqui, University of California, San Diego, reported in the *New York Times*, December 28, 1994.

10. Alan S. Banks et al., "Juvenile hallux abducto valgus association with metatarsus adductus," *Journal of the American Podiatric Medical Association* 84 (1994), pp. 219–224.

11. National Center for Education Statistics, `nces.ed.gov/nationsreportcard/`.

12. Based on T. N. Lam, "Estimating fuel consumption from engine size," *Journal of Transportation Engineering*, 111 (1985), pp. 339–357. The data for 10 to 50 km/h are measured; those for 60 and higher are calculated from a model given in the paper and are therefore smoothed.

13. N. Maeno et al., "Growth rates of icicles," *Journal of Glaciology*, 40 (1994), pp. 319–326.

14. From the *Runner's World* Web site, `www.runnersworld.com`.

15. Data provided by Matthew Moore.

16. Compiled from Fidelity data by the *Fidelity Insight* newsletter, `fidelity.kobren.com`.

17. Modified from M. C. Wilson and R. E. Shade, "Relative attractiveness of various luminescent colors to the cereal leaf beetle and the meadow spittlebug," *Journal of Economic Entomology*, 60 (1967), pp. 578–580.

18. A careful study of this phenomenon is W. S. Cleveland, P. Diaconis, and R. McGill, "Variables on scatterplots look more highly correlated when the scales are increased," *Science*, 216 (1982), pp. 1138–1141.

19. From the performance data for the fund presented at the Vanguard Group Web site, `personal.vanguard.com`. The EAFE returns differ from those given in Table 2.8. For reasons unknown to us, we often find minor variations in EAFE returns reported by different sources. We have elected to use Vanguard's reported values in this exercise.

20. Data provided by Linda McCabe and the Agency for International Development.

21. E. P. Hubble, "A relation between distance and radial velocity among extra-galactic nebulae," *Proceedings of the National Academy of Sciences*, 15 (1929), pp. 168–173.

22. William M. Lewis and Michael C. Grant, "Acid precipitation in the western United States," *Science*, 207 (1980), pp. 176–177.

23. From a presentation by Charles Knauf, Monroe County (New York) Environmental Health Laboratory.

24. The U.S. returns are for the Standard & Poor's 500-stock index. The overseas returns are for the Morgan Stanley Europe, Australasia, Far East (EAFE) index.

25. Illiteracy data from the Social Indicators home page of the United Nations Statistical Division, `www.un.org/Depts/unsd/`.

26. American Heart Association statement "Cardiovascular disease in women," found at `www.americanheart.org`.

27. Data provided by Peter Cook, Department of Mathematics, Purdue University.

28. These data were originally collected by L. M. Linde of UCLA but were first published by M. R. Mickey, O. J. Dunn, and V. Clark, "Note on the use of stepwise regression in detecting outliers," *Computers and Biomedical Research*, 1 (1967), pp. 105–111. The data have been used by several authors. We found them in N. R. Draper and J. A. John, "Influential observations and outliers in regression," *Technometrics*, 23 (1981), pp. 21–26.

29. This example is drawn from M. Goldstein, "Preliminary inspection of multivariate data," *The American Statistician*, 36 (1982), pp. 358–362.

30. The facts in Example 2.22 come from Nancy W. Burton and Leonard Ramist, *Predicting Success in College: Classes Graduating Since 1980*, Research Report No. 2001-2, The College Board, 2001.

31. Zeinab E. M. Afifi, "Principal components analysis of growth of Nahya infants: size, velocity and two physique factors," *Human Biology*, 57 (1985), pp. 659–669.

32. D. A. Kurtz (ed.), *Trace Residue Analysis*, American Chemical Society Symposium Series No. 284, 1985, Appendix.

33. State poverty rates from *Poverty in the United States, 1999*, U.S. Census Bureau Current Population Report P60-210, 2000. To reduce sampling errors, the rates are averages for the years 1997, 1998, and 1999.

34. R. C. Nelson, C. M. Brooks, and N. L. Pike, "Biomechanical comparison of male and female distance runners," in P. Milvy (ed.), *The Marathon: Physiological, Medical, Epidemiological, and Psychological Studies*, New York Academy of Sciences, 1977, pp. 793–807.

35. Frank J. Anscombe, "Graphs in statistical analysis," *The American Statistician*, 27 (1973), pp. 17–21.

36. Data provided by Matthew Moore.

37. P. Velleman, *ActivStats 2.0*, Addison Wesley Interactive, 1997.

38. This analysis was done by Crystal Richard.

39. Laura L. Calderon et al., "Risk factors for obesity in Mexican-American girls: dietary factors, anthropometric factors, physical activity, and hours of television viewing," *Journal of the American Dietetic Association*, 96 (1996), pp. 1177–1179.

40. Saccharin appears on the National Institutes of Health's list of suspected carcinogens but remains in wide use. In October 1997, an expert panel recommended by a 4 to 3 vote to keep it on the list, despite recommendations from other scientific groups that it be removed.

41. A detailed study of this correlation appears in E. M. Remolona, P. Kleinman, and D. Gruenstein, "Market returns and mutual fund flows," *Federal Reserve Bank of New York Economic Policy Review*, 3, No. 2 (1997), pp. 33–52.

42. M. S. Linet et al., "Residential exposure to magnetic fields and acute lymphoblastic leukemia in children," *New England Journal of Medicine*, 337 (1997), pp. 1–7.

43. *The Health Consequences of Smoking: 1983*, U.S. Public Health Service, 1983.

44. David E. Bloom and David Canning, "The health and wealth of nations," *Science*, 287 (2000), pp. 1207–1208.

45. Contributed by Marigene Arnold of Kalamazoo College.

46. D. E. Powers and D. A. Rock, *Effects of Coaching on SAT I: Reasoning Test Scores*, Educational Testing Service Research Report 98-6, College Entrance Examination Board, 1998.

47. L. E. Moses and F. Mosteller, "Safety of anesthetics," in J. M. Tanur et al. (eds.), *Statistics: A Guide to the Unknown*, 3rd ed., Wadsworth, 1989, pp. 15–24.

48. Data from G. A. Sacher and E. F. Staffelt, "Relation of gestation time to brain weight for placental mammals: implications for the theory of vertebrate growth," *American Naturalist*, 108 (1974), pp. 593–613. We found these data in F. L. Ramsey and D. W. Schafer, *The Statistical Sleuth: A Course in Methods of Data Analysis*, Duxbury, 1997.

49. There are several mathematical ways to show that $\log t$ fits into the power family at $p = 0$. Here's one. For powers $p \neq 0$, the indefinite integral $\int t^{p-1} dt$ is a multiple of t^p. When $p = 0$, $\int t^{-1} dt$ is $\log t$.

50. Data from the World Bank's *1999 World Development Indicators*. Life expectancy is estimated for 1997 and GDP per capita (purchasing power parity basis) for 1998.

51. Data from the Energy Information Administration, recorded in Robert H. Romer, *Energy: An Introduction to Physics*, W. H. Freeman, 1976, for 1880 to 1972, and from the EIA's *International Energy Annual* for later years.

52. From the Intel Web site, `www.intel.com/research/silicon/mooreslaw.htm`.

53. Data on a sample of 12 of 56 perch in a data set contributed to the *Journal of Statistics Education* data archive (`www.amstat.org/publications/jse/`) by Juha Puranen of the University of Helsinki.

54. From Joel Best, *Damned Lies and Statistics: Untangling Numbers from the Media, Politicians, and Activists*, University of California Press, 2001.

55. Similar experiments are described by A. P. J. Trinci, "A kinetic study of the growth of *Aspergillus nidulans* and other fungi," *Journal of General Microbiology*, 57 (1969), pp. 11–24. These data were provided by Thomas Kuczek, Purdue University.

56. Jérôme Chave, Bernard Riéra, and Marc-A. Dubois, "Estimation of biomass in a neotropical forest of French Guiana: spatial and temporal variability," *Journal of Tropical Ecology*, 17 (2001), pp. 79–96.

57. The power law connecting heart rate with body weight was found online at "The Worldwide Anaesthetist," `www.anaesthetist.com`. Anesthesiologists are interested in power laws because they must judge how drug doses should increase in bigger patients.

58. Data from Stillman Drake, *Galileo at Work*, University of Chicago Press, 1978. We found these data in D. A. Dickey and J. T. Arnold, "Teaching statistics with data of historic significance," *Journal of Statistics Education*, 3 (1995), `www.amstat.org/publications/jse/`.

59. Data from many studies compiled in D. F. Greene and E. A. Johnson, "Estimating the mean annual seed production of trees," *Ecology*, 75 (1994), pp. 642–647.

60. Todd W. Anderson, "Predator responses, prey refuges, and density-dependent mortality of a marine fish," *Ecology*, 81 (2001), pp. 245–257.

61. Oskar Kindvall, "Habitat heterogeneity and survival in a bush cricket metapopulation," *Ecology*, 77 (1996), pp. 207–214.

62. J. T. Dwyer et al., "Memory of food intake in the distant past," *American Journal of Epidemiology*, 130 (1989), pp. 1033–1046.

63. Data provided by Robert Dale, Purdue University.

64. Data provided by Samuel Phillips, Purdue University.

65. See Note 34.

66. Data provided by Jim Bateman and Michael Hunt, Purdue University.

67. "Dancing in step," *Economist*, March 22, 2001.

68. Gary Smith, "Do statistics test scores regress toward the mean?" *Chance*, 10, No. 4 (1997), pp. 42–45.

69. S. Chatterjee and B. Price, *Regression Analysis by Example*, Wiley, 1977.

70. From several editions of the *Statistical Abstract of the United States*.

71. Ji X. He and Donald J. Stewart, "Age and size at first reproduction of fishes: predictive models based only on growth trajectories," *Ecology*, 82 (2001), pp. 784–792. The data are available from the electronic archive maintained by this journal. The data include 235 populations, often separately for male and female fish. We have used only the first values reported for each species and only data for females if the sexes are listed separately.

CHAPTER 3 NOTES

1. "Acadian ambulance officials, workers flood call-in poll," *Baton Rouge Advocate*, January 22, 1999.

2. M. R. Burt et al., *Homelessness: Programs and the People They Serve*, The Urban Institute, 1999. Found at `www.huduser.org/publications/homeless/homelessness/`.

3. S. G. Stolberg, "Link found between behavioral problems and time in child care," *New York Times*, April 19, 2001. The study is the National Institute of Child Health and Human Development Study of Early Child Care. See also S. G. Stolberg, "Science, studies and motherhood," *New York Times*, April 22, 2001. Details about this large government-sponsored study are available at `www.nichd.nih.gov/about/crmc/secc/`.

4. J. E. Muscat et al., "Handheld cellular telephone use and risk of brain cancer," *Journal of the American Medical Association*, 284 (2000), pp. 3001–3007.

5. K. J. Mukamal et al., "Prior alcohol consumption and mortality following acute myocardial infarction," *Journal of the American Medical Association*, 285 (2001), pp. 1965–1970.

6. L. E. Moses and F. Mosteller, "Safety of anesthetics," in J. M. Tanur et al. (eds.), *Statistics: A Guide to the Unknown*, 3rd ed., Wadsworth, 1989, pp. 15–24.

7. Steering Committee of the Physicians' Health Study Research Group, "Final report on the aspirin component of the ongoing Physicians' Health Study," *New England Journal of Medicine*, 321 (1989), pp. 129–135.

8. L. L. Miao, "Gastric freezing: an example of the evaluation of medical therapy by randomized clinical trials," in J. P. Bunker, B. A. Barnes, and F. Mosteller (eds.), *Costs, Risks, and Benefits of Surgery*, Oxford University Press, 1977, pp. 198–211.

9. The study is described in G. Kolata, "New study finds vitamins are not cancer preventers," *New York Times*, July 21, 1994. Look in the *Journal of the American Medical Association* of the same date for the details.

10. Based on the Electronic Encyclopedia of Statistical Examples and Exercises (EESEE) story "Surgery in a Blanket," found on the IPS Web site `www.whfreeman.com/ips`.

11. K. Wang, Y. Li, and J. Erickson, "A new look at the Monday effect," *Journal of Finance*, 52 (1997), pp. 2171–2186.

12. Carol A. Warfield, "Controlled-release morphine tablets in patients with chronic cancer pain," *Cancer*, 82 (1998), pp. 2299–2306.

13. R. C. Shelton et al., "Effectiveness of St. John's wort in major depression," *Journal of the American Medical Association*, 285 (2001), pp. 1978–1986.

14. Based on Evan H. DeLucia et al., "Net primary production of a forest ecosystem with experimental CO_2 enhancement," *Science*, 284 (1999), pp. 1177–1179. The investigators used the block design.

15. E. M. Peters et al., "Vitamin C supplementation reduces the incidence of postrace symptoms of upper-respiratory tract infection in ultramarathon runners," *American Journal of Clinical Nutrition*, 57 (1993), pp. 170–174.

16. Reported by D. Horvitz in his contribution to "Pseudo-opinion polls: SLOP or useful data?" *Chance*, 8, No. 2 (1995), pp. 16–25.

17. The information in this example is taken from *The ASCAP Survey and Your Royalties*, ASCAP, undated.

18. The most recent account of the design of the CPS appears in the "Technical Notes" to the Bureau of Labor Statistics' report *Employment and Earnings*, available online at `stats.bls.gov/cpstn3.htm`. The account here omits many complications, such as the need to separately sample "group quarters" like college dormitories.

19. For more detail on the material of this section and complete references, see P. E. Converse and M. W. Traugott, "Assessing the accuracy of polls and surveys," *Science*, 234 (1986), pp. 1094–1098.

20. The nonresponse rate for the CPS comes from "Technical notes to household survey data published in *Employment and Earnings*," found on the Bureau of Labor Statistics Web site at `stats.bls.gov/cpshome.htm`. The General Social Survey reports its response rate on its Web site, `www.norc.org/projects/gensoc.asp`. The Pew study is described in Gregory Flemming and Kimberly Parker, "Race and reluctant respondents: possible consequences of non-response for pre-election surveys," Pew Research Center for the People and the Press, 1997, found at `www.people-press.org`.

21. For more detail on the limits of memory in surveys, see N. M. Bradburn, L. J. Rips, and S. K. Shevell, "Answering autobiographical questions: the impact of memory and inference on surveys," *Science*, 236 (1987), pp. 157–161.

22. The responses on welfare are from a *New York Times*/CBS News Poll reported in the *New York Times*, July 5, 1992. Those for Scotland are from "All set for independence?" *Economist*, September 12, 1998. Professors and colleges and many other examples appear in T. W. Smith, "That which we call welfare by any other name would smell sweeter," *Public Opinion Quarterly*, 51 (1987), pp. 75–83.

23. Giuliana Coccia, "An overview of non-response in Italian telephone surveys," *Proceedings of the 99th Session of the International Statistical Institute, 1993*, Book 3, pp. 271–272.

24. From *CIS Boletin 9, Spaniards' Economic Awareness*, found online at `www.cis.es/ingles/opinion/economia.htm`.

25. The survey question and result are reported in Trish Hall, "Shop? Many say 'Only if I must,' " *New York Times*, November 28, 1990.

26. Warren McIsaac and Vivek Goel, "Is access to physician services in Ontario equitable?" Institute for Clinical Evaluative Sciences in Ontario, October 18, 1993.

27. Based on Christopher Anderson, "Measuring what works in health care," *Science*, 263 (1994), pp. 1080–1082.

28. NIMH Multisite HIV Prevention Trial Group, "The NIMH multisite HIV prevention trial: reducing HIV sexual risk behavior," *Science*, 280 (1998), pp. 1889–1894.

29. This exercise is based on the EESEE story "What makes a pre-teen popular?" found on the IPS Web site `www.whfreeman.com/ips`.

CHAPTER 4 NOTES

1. An informative and entertaining account of the origins of probability theory is Florence N. David, *Games, Gods and Gambling*, Charles Griffin, London, 1962.

2. You can find a mathematical explanation of Benford's Law in Ted Hill, "The first-digit phenomenon," *American Scientist*, 86 (1996), pp. 358–363; and Ted Hill, "The difficulty of faking data," *Chance*, 12, No. 3 (1999), pp. 27–31. Applications in fraud detection are discussed in the second paper by Hill and in Mark A. Nigrini, "I've got your number," *Journal of Accountancy*, May 1999, available online at `www.aicpa.org/pubs/jofa/joaiss.htm`.

3. From the Dupont Automotive North America Color Popularity Survey, reported at `www.dupont.com/automotive/`.

4. Corey Kilgannon, "When New York is on the end of the line," *New York Times*, November 7, 1999.

5. We use $\bar{x}$ both for the random variable, which takes different values in repeated sampling, and for the numerical value of the random variable in a particular sample. Similarly, s and $\hat{p}$ stand both for random variables and for specific values. This notation is mathematically imprecise but statistically convenient.

6. We will consider only the case in which X takes a finite number of possible values. The same ideas, implemented with more advanced mathematics, apply to random variables with an infinite but still countable collection of values.

7. From `www.ncsu.edu/class/grades/`.

8. From the Census Bureau's 1998 American Housing Survey.

9. The mean of a continuous random variable X with density function $f(x)$ can be found by integration:

$$\mu_X = \int x f(x)\,dx$$

This integral is a kind of weighted average, analogous to the discrete-case mean

$$\mu_X = \sum x P(X\bar{x})$$

The variance of a continuous random variable X is the average squared deviation of the values of X from their mean, found by the integral

$$\sigma_X^2 = \int (x - \mu)^2 f(x)\,dx$$

10. See A. Tversky and D. Kahneman, "Belief in the law of small numbers," *Psychological Bulletin*, 76 (1971), pp. 105–110, and other writings of these authors for a full account of our misperception of randomness.

11. Probabilities involving runs can be quite difficult to compute. That the probability of a run of three or more heads in 10 independent tosses of a fair coin is $(1/2) + (1/128) = 0.508$ can be found by clever counting, as can the other results given in the text. A general treatment using advanced methods appears in Section XIII.7 of William Feller, *An Introduction to Probability Theory and Its Applications*, vol. 1, 3rd ed., Wiley, 1968.

12. R. Vallone and A. Tversky, "The hot hand in basketball: on the misperception of random sequences," *Cognitive Psychology*, 17 (1985), pp. 295–314. A later series of articles that debate the independence question is A. Tversky and T. Gilovich, "The cold facts about the 'hot hand' in basketball," *Chance*, 2, No. 1 (1989), pp. 16–21; P. D. Larkey, R. A. Smith, and J. B. Kadane, "It's OK to believe in the 'hot hand,'" *Chance*, 2, No. 4 (1989), pp. 22–30; and A. Tversky and T. Gilovich, "The 'hot hand': statistical reality or cognitive illusion?" *Chance*, 2, No. 4 (1989), pp. 31–34.

13. See Note 21 for Chapter 1.

14. Returns data from several sources, especially the *Fidelity Insight* newsletter, `fidelity.kobren.com`.

15. These probabilities come from studies by the sociologist Harry Edwards, reported in the *New York Times*, February 25, 1986.

16. This example is modeled on Benjamin A. Barnes, "An overview of the treatment of end-stage renal disease and a consideration of some of the consequences," in J. P. Bunker, B. A. Barnes, and F. W. Mosteller (eds.), *Costs, Risks and Benefits of Surgery*, Oxford University Press, 1977, pp. 325–341. The probabilities are recent estimates based on data from the United Network for Organ Sharing (www.unos.org) and Rebecca D. Williams, "Living day-to-day with kidney dialysis," Food and Drug Administration, www.fda.gov.

17. Projections by the National Center for Education Statistics, from the *Digest of Education Statistics, 2000*, National Center for Education Statistics, 2001, available at nces.ed.gov.

18. Data for fall 1997, from the source cited in Note 17.

19. Based loosely on M. C. Weinstein, J. S. Pliskin, and W. B. Stason, "Coronary artery bypass surgery: decision and policy analysis," in J. P. Bunker, B. A. Barnes, and F. W. Mosteller (eds.), *Costs, Risks and Benefits of Surgery*, Oxford University Press, 1977, pp. 342–371.

20. Probabilities from trials with 2897 people known to be free of HIV antibodies and 673 people known to be infected, reported in J. Richard George, "Alternative specimen sources: methods for confirming positives," 1998 Conference on the Laboratory Science of HIV, found online at the Centers for Disease Control and Prevention, www.cdc.gov.

CHAPTER 5 NOTES

1. Henry Wechsler et al., *Binge Drinking on America's Colleges Campuses*, Harvard School of Public Health, 2001.

2. David W. Moore and Joseph Carroll, "Most Americans call their physical and mental health 'good' or 'excellent,' " www.gallup.com/poll/releases/, November 28, 2001.

3. Lydia Saad, "Hold the gravy! Six in ten Americans want to lose weight," www.gallup.com/poll/releases/, November 21, 2001.

4. Office of Technology Assessment, *Scientific Validity of Polygraph Testing: A Research Review and Evaluation*, Government Printing Office, 1983.

5. Statistical methods for dealing with time-to-failure data, including the Weibull model, are presented in Wayne Nelson, *Applied Life Data Analysis*, Wiley, 1982.

6. From www.ncsu.edu/class/grades/.

7. Based on data from over 7000 students, reported at www.act.org/explore/newscale/summary.html.

8. From the Web site of the Finnish National Public Health Institute, www.ktl.fi/monica/. This institute is the data center for MONICA.

CHAPTER 6 NOTES

1. The standard reference here is Bradley Efron and Robert J. Tibshirani, *An Introduction to the Bootstrap*, Chapman Hall, 1993. A less technical overview is in Bradley Efron and Robert J. Tibshirani, "Statistical data analysis in the computer age," *Science*, 253 (1991), pp. 390–395.

2. P. H. Lewis, "Technology" column, *New York Times*, May 29, 1995.

3. Jon E. Keeley, C. J. Fotheringham, and Marco Morais, "Reexamining fire suppression impacts on brushland fire regimes," *Science*, 284 (1999), pp. 1829–1831.

4. From R. M. Lyle et al., "Blood pressure and metabolic effects of calcium supplementation in normotensive white and black men," *Journal of the American Medical Association*, 257 (1987), pp. 1772–1776.

5. From a study by M. R. Schlatter et al., Division of Financial Aid, Purdue University.

6. Data provided by Diana Schellenberg, Purdue University School of Health Sciences.

7. From D. L. Shankland et al., "The effect of 5-thio-D-glucose on insect development and its absorption by insects," *Journal of Insect Physiology*, 14 (1968), pp. 63–72.

8. From William T. Robinson, "Sources of market pioneer advantages: the case of industrial goods industries," *Journal of Marketing Research*, 25 (1988), pp. 87–94.

9. R. A. Fisher, "The arrangement of field experiments," *Journal of the Ministry of Agriculture of Great Britain*, 33 (1926), p. 504, quoted in Leonard J. Savage, "On rereading R. A. Fisher," *The Annals of Statistics*, 4 (1976), p. 471.

10. Fisher's work is described in a biography by his daughter: Joan Fisher Box, *R. A. Fisher: The Life of a Scientist*, Wiley, 1978.

11. T. D. Sterling, "Publication decisions and their possible effects on inferences drawn from tests of significance—or vice versa," *Journal of the American Statistical Association*, 54 (1959), pp. 30–34. Related comments appear in J. K. Skipper, A. L. Guenther, and G. Nass, "The sacredness of 0.05: a note concerning the uses of statistical levels of significance in social science," *American Sociologist*, 1 (1967), pp. 16–18.

12. This example is cited by William Feller, "Are life scientists overawed by statistics?" *Scientific Research*, February 3, 1969, p. 26.

13. Robert J. Schiller, "The volatility of stock market prices," *Science*, 235 (1987), pp. 33–36.

14. Data provided by Joan M. Susic.

15. From a study by Mugdha Gore and Joseph Thomas, Purdue University School of Pharmacy.

16. Barbara A. Almanza, Richard Ghiselli, and William Jaffe, "Foodservice design and aging baby boomers: importance and perception of physical amenities in restaurants," *Foodservice Research International*, 12 (2000), pp. 25–40.

CHAPTER 7 NOTES

1. From C. Don Wiggins, "The legal perils of 'underdiversification'—a case study," *Personal Financial Planning*, 1, No. 6 (1999), pp. 16–18.

2. These data are from "Results report on the vitamin C pilot program," prepared by SUSTAIN (Sharing United States Technology to Aid in the Improvement of Nutrition) for the U.S. Agency for International Development. The report was used by the Committee on International Nutrition of the National Academy of Sciences/Institute of Medicine (NAS/IOM) to make recommendations on whether or not the vitamin C content of food commodities used in U.S. food aid programs should be increased. The program was directed by Peter Ranum and Françoise Chomé.

3. Data provided by Joseph A. Wipf, Department of Foreign Languages and Literatures, Purdue University.

4. These recommendations are based on extensive computer work. See, for example, Harry O. Posten, "The robustness of the one-sample t-test over the Pearson system," *Journal of Statistical Computation and Simulation*, 9 (1979), pp. 133–149; and E. S. Pearson and N. W. Please, "Relation between the shape of population distribution and the robustness of four simple test statistics," *Biometrika*, 62 (1975), pp. 223–241.

5. The data and the use of the log transformation come from Thomas J. Lorenzen, "Determining statistical characteristics of a vehicle emissions audit procedure," *Technometrics*, 22 (1980), pp. 483–493.

6. Data from the August 2000 supplement to the Current Population Survey, from the Census Bureau Web site, `www.census.gov`.

7. You can find a practical discussion of distribution-free inference in Myles Hollander and Douglas A. Wolfe, *Nonparametric Statistical Methods*, 2nd ed., Wiley, 1999.

8. Based on R. G. Hood, "Results of the prices received by farmers for grain—quality assurance project," USDA Report SRB-95-07, 1995.

9. Based on A. N. Dey, "Characteristics of elderly home health care users," National Center for Health Statistics, 1996.

10. Data from F. H. Rauscher et al., "Music training causes long-term enhancement of preschool children's spatial-temporal reasoning," *Neurological Research*, 19 (1997), pp. 2–8.

11. Data are from Joan M. Susic, "Dietary phosphorus intakes, urinary and peritoneal phosphate excretion and clearance in continuous ambulatory peritoneal dialysis patients," M.S. thesis, Purdue University, 1985.

12. Data provided by Diana Schellenberg, Purdue University School of Health Sciences.

13. From S. C. Hand and E. Gnaiger, "Anaerobic dormancy quantified in *Artemia* embryos," *Science*, 239 (1988), pp. 1425–1427.

14. Data provided by Timothy Sturm, Purdue University.

15. Data provided by Joseph A. Wipf, Department of Foreign Languages and Literatures, Purdue University.

16. Data from Wayne Nelson, *Applied Life Data Analysis*, Wiley, 1982, p. 471.

17. Detailed information about the conservative t procedures can be found in Paul Leaverton and John J. Birch, "Small sample power curves for the two sample location problem," *Technometrics*, 11 (1969), pp. 299–307; in Henry Scheffé, "Practical solutions of the Behrens-Fisher problem," *Journal of the American Statistical Association*, 65 (1970), pp. 1501–1508; and in D. J. Best and J. C. W. Rayner, "Welch's approximate solution for the Behrens-Fisher problem," *Technometrics*, 29 (1987), pp. 205–210.

18. This example is adapted from Maribeth C. Schmitt, "The effects of an elaborated directed reading activity on the metacomprehension skills of third graders," Ph.D. dissertation, Purdue University, 1987.

19. See the extensive simulation studies in Harry O. Posten, "The robustness of the two-sample t test over the Pearson system," *Journal of Statistical Computation and Simulation*, 6 (1978), pp. 295–311.

20. From H. G. Gough, *The Chapin Social Insight Test*, Consulting Psychologists Press, 1968.

21. This example is loosely based on D. L. Shankland, "Involvement of spinal cord and peripheral nerves in DDT-poisoning syndrome in albino rats," *Toxicology and Applied Pharmacology*, 6 (1964), pp. 197–213.

22. This study is reported in Roseann M. Lyle et al., "Blood pressure and metabolic effects of calcium supplementation in normotensive white and black men," *Journal of the American Medical Association*, 257 (1987), pp. 1772–1776. The individual measurements in Table 7.4 were provided by Dr. Lyle.

23. The data were provided by Helen Park. See Helen Park et al., "Fortifying bread with each of three antioxidants," *Cereal Chemistry*, 74 (1997), pp. 202–206.

24. The data are from F. H. Rauscher et al., "Music training causes long-term enhancement of preschool children's spatial-temporal reasoning," *Neurological Research*, 19 (1997), pp. 2–8.

25. Based on C. Papoulias and P. Theodossiou, "Analysis and modeling of recent business failures in Greece," *Managerial and Decision Economics*, 13 (1992), pp. 163–169.

26. Based on A. H. Ismail and R. J. Young, "The effect of chronic exercise on the personality of middle-aged men," *Journal of Human Ergology*, 2 (1973), pp. 47–57.

27. Data from a study conducted at the Medical University of South Carolina in 1989.

28. See Note 8.

29. From M. F. Picciano and R. H. Deering, "The influence of feeding regimens on iron status during infancy," *American Journal of Clinical Nutrition*, 33 (1980), pp. 746–753.

30. From M. A. Zlatin and R. A. Koenigsknecht, "Development of the voicing contrast: a comparison of voice onset time in stop perception and production," *Journal of Speech and Hearing Research*, 19 (1976), pp. 93–111.

31. Based on studies conducted by Marvin Schlatter, Division of Financial Aid, Purdue University.

32. The problem of comparing spreads is difficult even with advanced methods. Common distribution-free procedures do not offer a satisfactory alternative to the *F* test, because they are sensitive to unequal shapes when comparing two distributions. A good introduction to the available methods is W. J. Conover, M. E. Johnson, and M. M. Johnson, "A comparative study of tests for homogeneity of variances, with applications to outer continental shelf bidding data," *Technometrics*, 23 (1981), pp. 351–361. Modern resampling procedures often work well. See Dennis D. Boos and Colin Brownie, "Bootstrap methods for testing homogeneity of variances," *Technometrics*, 31 (1989), pp. 69–82.

33. G. E. P. Box, "Non-normality and tests on variances," *Biometrika*, 40 (1953), pp. 318–335. The quote appears on page 333.

34. This exercise is based on events that are real. The data and details have been altered to protect the privacy of the individuals involved.

35. See Note 16.

36. These data were provided by Professor Sebastian Heath, School of Veterinary Medicine, Purdue University.

37. Based loosely on D. R. Black et al., "Minimal interventions for weight control: a cost-effective alternative," *Addictive Behaviors*, 9 (1984), pp. 279–285.

38. Based on G. Salvendy, "Selection of industrial operators: the one-hole test," *International Journal of Production Research*, 13 (1973), pp. 303–321.

39. Data from R. A. Berner and G. P. Landis, "Gas bubbles in fossil amber as possible indicators of the major gas composition of ancient air," *Science*, 239 (1988), pp. 1406–1409.

40. From J. W. Marr and J. A. Heady, "Within- and between-person variation in dietary surveys: number of days needed to classify individuals," *Human Nutrition: Applied Nutrition*, 40A (1986), pp. 347–364.

41. From V. D. Bass, W. E. Hoffmann, and J. L. Dorner, "Normal canine lipid profiles and effects of experimentally induced pancreatitis and hepatic necrosis on lipids," *American Journal of Veterinary Research*, 37 (1976), pp. 1355–1357.

42. From M. L. Pollock et al., "Body composition of elite class distance runners," in P. Milvey (ed.), *The Marathon: Physiological, Medical, Epidemiological, and Psychological Studies*, New York Academy of Sciences, 1977, p. 366.

43. Data taken from the Web site `www.assist-2-sell.com` on July 9, 2001.

CHAPTER 8 NOTES

1. This case is discussed in D. H. Kaye and M. Aickin (eds.), *Statistical Methods in Discrimination Litigation*, Marcel Dekker, 1986; and D. C. Baldus and J. W. L. Cole, *Statistical Proof of Discrimination*, McGraw-Hill, 1980.

2. Details of exact binomial procedures can be found in Myles Hollander and Douglas Wolfe, *Nonparametric Statistical Methods*, 2nd ed., Wiley, 1999.

3. See A. Agresti and B. A. Coull, "Approximate is better than 'exact' for interval estimation of binomial proportions," *The American Statistician*, 52 (1998), pp. 119–126. A detailed theoretical study is Lawrence D. Brown, Tony Cai, and Anirban DasGupta, "Confidence intervals for a binomial proportion and asymptotic expansions," *Annals of Statistics*, 30 (2002), pp. 160–201.

4. Results of this survey are reported in Henry Wechsler et al., "Health and behavioral consequences of binge drinking in college," *Journal of the American Medical Association*, 272 (1994), pp. 1672–1677.

5. This example is adapted from a survey directed by Professor Joseph N. Uhl of the Department of Agricultural Economics, Purdue University. The survey was sponsored by the Indiana Christmas Tree Growers Association and was conducted in April 1987.

6. The poll is part of the "American Express Retail Index Project" and is reported in *Stores*, December 2000, pp. 38–40.

7. Data provided by Jude M. Werra & Associates, Brookfield, Wisconsin.

8. Data from Guohua Li and Susan P. Baker, "Alcohol in fatally injured bicyclists," *Accident Analysis and Prevention*, 26 (1994), pp. 543–548.

9. See Alan Agresti and Brian Caffo, "Simple and effective confidence intervals for proportions and differences of proportions result from adding two successes and two failures," *The American Statistician*, 45 (2000), pp. 280–288. The Wilson interval is a bit conservative (true coverage probability is higher than the confidence level) when p_1 and p_2 are equal and close to 0 or 1, but the traditional interval is much less accurate and has the fatal flaw that the true coverage probability is *less* than the confidence level.

10. The study and data are described in J. Stamler, "The mass treatment of hypertensive disease: defining the problem," in *Mild Hypertension: To Treat or Not to Treat*, New York Academy of Sciences, 1978, pp. 333–358.

11. Based on Greg Clinch, "Employee compensation and firms' research and development activity," *Journal of Accounting Research*, 29 (1991), pp. 59–78.

12. Sara J. Solnick and David Hemenway, "Complaints and disenrollment at a health maintenance organization," *The Journal of Consumer Affairs*, 26 (1992), pp. 90–103.

13. From Monica Macaulay and Colleen Brice, "Don't touch my projectile: gender bias and stereotyping in syntactic examples," *Language*, 73, No. 4 (1997), pp. 798–825. The first part of the title is a direct quote from one of the texts.

14. See S. W. Lagakos, B. J. Wessen, and M. Zelen, "An analysis of contaminated well water and health effects in Woburn, Massachusetts," *Journal of the American Statistical Association*, 81 (1986), pp. 583–596, and the following discussion. This case is the basis for the movie *A Civil Action*.

15. Data from H. A. Witkin et al., "Criminality in XYY and XXY men," *Science*, 193 (1976), pp. 547–555.

16. From William S. Fields et al., "Controlled trial of aspirin in cerebral ischemia," *Stroke*, 8 (1977), pp. 301–315.

17. Based on Elray M. Roper and Charles G. Wright, "German cockroach (Orthoptera: Blatellidae) mortality on various surfaces following application of diazinon," *Journal of Economic Entomology*, 78 (1985), pp. 733–737.

18. National Institute for Occupational Safety and Health, *Stress at Work*, 2000, `www.cdc.gov/niosh/stresswk.html`.

19. Results of this survey were reported in *Restaurant Business*, September 15, 1999, pp. 45–49.

20. Marsha A. Dickson, "Utility of no sweat labels for apparel customers: profiling label users and predicting their purchases," *The Journal of Consumer Affairs*, 35 (2001), pp. 96–119.

21. Based on Robert T. Driescher, "A quality swing with Ping," *Quality Progress*, August 2001, pp. 37–41.

22. From Karin Weber and Weley S. Roehl, "Profiling people searching for and purchasing travel products on the world wide web," *Journal of Travel Research*, 37 (1999), pp. 291–298.

23. From Dennis N. Bristow and Richard J. Sebastian, "Holy cow! Wait till next year! A closer look at the brand loyalty of Chicago Cubs baseball fans," *Journal of Consumer Marketing*, 18 (2001), pp. 256–275.

24. Based on a Gallup poll conducted April 6–8, 2001.

25. Based on Deborah Roedder John and Ramnath Lakshmi-Ratan, "Age differences in children's choice behavior: the impact of available alternatives," *Journal of Marketing Research*, 29 (1992), pp. 216–226.

26. Data provided by Chris Melby and David Goldflies, Department of Physical Education, Health, and Recreation Studies, Purdue University.

27. The first study is reported in an article in the *New York Times* of January 30, 1988; the second was described in the *New York Times* on January 27, 1988.

28. See the references given in Note 1.

CHAPTER 9 NOTES

1. C. M. Ryan, C. A. Northrup-Clewes, B. Knox, and D. I. Thurnham, "The effect of in-store music on consumer choice of wine," *Proceedings of the Nutrition Society*, 57 (1998), p. 1069A.

2. You can find a clear and comprehensive discussion of numerical measures of association for categorical data in Chapter 3 of A. M. Liebetrau, *Measures of Association*, Sage Publications, 1983.

3. National Science Board, *Science and Engineering Indicators, 1991*, U.S. Government Printing Office, 1991. The detailed data appear in Appendix Table 3-5, p. 274.

4. This study is reported in Joan L. Duda, "The relationship between goal perspectives, persistence and behavioral intensity among male and female recreational sport participants," *Leisure Sciences*, 10 (1988), pp. 95–106.

5. When the expected cell counts are small, it is best to use a test based on the exact distribution rather than the chi-square approximation, particularly for 2×2 tables. Many statistical software systems offer an "exact" test as well as the chi-square test for 2×2 tables.

6. The full report of the study appeared in George H. Beaton et al., "Effectiveness of vitamin A supplementation in the control of young child morbidity and mortality in developing countries," United Nations ACC/SCN State-of-the-Art Series, Nutrition Policy Discussion Paper No. 13, 1993.

7. These data were taken from Ray H. Rosenman et al., "A 4-year prospective study of the relationship of different habitual vocational physical activity to risk and incidence of ischemic heart disease in volunteer male federal employees," in P. Milvey (ed.), *The Marathon: Physiological, Medical, Epidemiological, and Psychological Studies*, New York Academy of Sciences, 301 (1977), pp. 627–641.

8. An alternative formula that can be used for hand or calculator computations is

$$X^2 = \sum \frac{(\text{observed})^2}{\text{expected}} - n$$

9. See, for example, Alan Agresti, *Categorical Data Analysis*, Wiley, 1990.

10. From F. D. Blau and M. A. Ferber, "Career plans and expectations of young women and men," *Journal of Human Resources*, 26 (1991), pp. 581–607.

11. Data from D. M. Barnes, "Breaking the cycle of addiction," *Science*, 241 (1988), pp. 1029–1030.

12. Data from National Science Foundation Science Resources Studies Division, *Data Brief* No. 12 (1996), p. 1.

13. Although these data are fictitious, similar though less simple situations occur. See P. J. Bickel and J. W. O'Connell, "Is there a sex bias in graduate admissions?" *Science*, 187 (1975), pp. 398–404.

14. These data, from reports submitted by airlines to the Department of Transportation, appear in A. Barnett, "How numbers can trick you," *Technology Review*, October 1994, pp. 38–45.

15. See Note 1.

16. Marsha A. Dickson, "Utility of no sweat labels for apparel customers: profiling label users and predicting their purchases," *The Journal of Consumer Affairs*, 35 (2001), pp. 96–119.

17. William D. Darley, "Store-choice behavior for pre-owned merchandise," *Journal of Business Research*, 27 (1993), pp. 17–31.

18. From Gary J. Patronek et al., "Risk factors for relinquishment of cats to an animal shelter," *Journal of the American Veterinary Medical Association*, 209 (1996), pp. 582–588.

19. From Gary J. Patronek et al., "Risk factors for relinquishment of dogs to an animal shelter," *Journal of the American Veterinary Medical Association*, 209 (1996), pp. 572–581.

20. Data from James E. Stafford, "Influence of preliminary contact on mail returns," *Journal of Marketing Research*, 3 (1966), pp. 410–411.

21. These data are from Patricia H. Shiono and Mark A. Klebanoff, "The effect of two mailing strategies on the response to a survey of physicians," *American Journal of Epidemiology*, 134 (1991), pp. 539–542.

22. From Erika Friedmann et al., "Animal companions and one-year survival of patients after discharge from a coronary care unit," *Public Health Reports*, 96 (1980), pp. 307–312.

23. These data are from Burton G. Malkeil, "Returns from investing in equity mutual funds, 1971 to 1991," *Journal of Finance*, 50 (1995), pp. 549–572.

24. Data taken from J. Stamler, "The mass treatment of hypertensive disease: defining the problem," in *Mild Hypertension: To Treat or Not to Treat*, New York Academy of Sciences, 1978, pp. 333–358.

25. Data taken from O. M. Galal et al., "Feeding the child with diarrhea: a strategy for testing a health education message within the primary health care system in Egypt," *Socio-economic Planning Sciences*, 21 (1987), pp. 139–147.

26. Data from A. E. Mourant et al., *The Distribution of Human Blood Groups and Other Polymorphisms*, Oxford University Press, 1976.

27. See Note 26.

28. Data provided by Susan Prohofsky, from her Ph.D. dissertation, "Selection of undergraduate major: the influence of expected costs and expected benefits," Purdue University, 1991.

29. Data taken from William M. Tierney, Clement J. McDonald, and George P. McCabe, "Serum potassium testing in diuretic-treated outpatients," *Medical Decision Making*, 5 (1985), pp. 91–104.

30. The data are based on *Seventh Report to the President and Congress on the Status of Health Personnel in the United States*, Public Health Service, 1990.

31. Data provided by Dr. Susan Meyer, Senior Vice President of the American Association of Colleges of Pharmacy.

CHAPTER 10 NOTES

1. From Raphael Sagarin and Fiorenza Micheli, "Climate change in nontraditional data sets," *Science*, 294 (2001), p. 811.

2. The quote and the example are taken from J. V. G. A. Durnin and J. Womersley, "Body fat assessed from total body density and its estimation from skinfold thicknesses:

measurements on 481 men and women aged 16 to 72 years," *British Journal of Nutrition*, 32 (1974), pp. 77–97. The data analyzed in the examples that follow have been generated by computer simulation in such a way that they give the same statistical results reported in the article.

3. Minitab does this directly in response to the PREDICT subcommand of the REGRESS command. Some software packages, such as SAS, use an indirect method: Define an additional observation with LSKIN equal to the value of x^* and DEN missing. The software ignores the additional case in computing the least-squares line but computes and prints the predicted value and associated statistics.

4. The data were provided by Linda McCabe and were collected as part of a large study of women's bone health and another study of calcium kinetics, both directed by Professor Connie Weaver of the Foods and Nutrition Department, Purdue University. Results related to this example are reported in Dorothy Teegarden et al., "Peak bone mass in young women," *Journal of Bone and Mineral Research*, 10 (1995), pp. 711–715.

5. These data were provided by Robert Dale of Purdue University.

6. This quantity is the estimated standard deviation of $\hat{y} - y$, not the estimated standard deviation of $\hat{y}$ alone.

7. The method is described in Chapter 10 of G. W. Snedecor and W. G. Cochran, *Statistical Methods*, 8th ed., Iowa State University Press, 1989.

8. Net cash flow data from Sean Collins, *Mutual Fund Assets and Flows in 2000*, Investment Company Institute, 2001. Found online at `www.ici.org`. The raw data were converted to real dollars using annual average values of the CPI.

9. These are part of the data from the EESEE story "Blood Alcohol Content," found on the IPS Web site `www.whfreeman.com/ips`.

10. Data for the first four years from Manuel Castells, *The Rise of the Network Society*, 2nd ed., Blackwell, 2000, p. 41.

11. Alan S. Banks et al., "Juvenile hallux abducto valgus association with metatarsus adductus," *Journal of the American Podiatric Medical Association* 84 (1994), pp. 219–224.

12. The data were provided by Professor Shelly MacDermid, Department of Child Development and Family Studies, Purdue University, from a study reported in S. M. MacDermid et al., "Is small beautiful? Work-family tension, work conditions, and organizational size," *Family Relations*, 44 (1994), pp. 159–167.

13. Data were provided by Professor Robert Dale of the Purdue University Agronomy Department.

14. These data are from G. Geri and B. Palla, "Considerazioni sulle più recenti osservazioni ottiche alla Torre Pendente di Pisa," *Estratto dal Bollettino della Società Italiana di Topografia e Fotogrammetria*, 2 (1988), pp. 121–135. Professor Julia Mortera of the University of Rome provided valuable assistance with the translation.

15. These data are from experiments conducted in Don Corrigan's laboratory at Purdue University and were provided by Paul Waldsmith.

16. From John A. Wareham et al., "Prediction of arterial blood pressure on the premature neonate using the oscillometric method," *American Journal of Diseases of Children*, 141 (1987), pp. 1108–1110.

17. This study is reported in M. El-Kholy, F. Shaheen, and W. Mahmoud, "Relationship between socioeconomic status and birth weight, a field study in a rural community in Egypt," *Journal of the Egyptian Public Health Association*, 61 (1986), pp. 349–358.

18. This study is reported in S. Lau and P. C. Cheung, "Relations between Chinese adolescents' perception of parental control and organization and their perception of parental warmth," *Developmental Psychology*, 23 (1987), pp. 726–729.

19. From Donald A. Mahler and John I. Mackowiak, "Evaluation of the short-form 36-item questionnaire to measure health-related quality of life in patients with COLD," *Chest*, 107 (1995), pp. 1585–1589.

20. Data on a sample of 12 of 56 perch in a data set contributed to the *Journal of Statistics Education* data archive (`www.amstat.org/publications/jse/`) by Juha Puranen of the University of Helsinki.

21. Based on T. N. Lam, "Estimating fuel consumption from engine size," *Journal of Transportation Engineering*, 111 (1985), pp. 339–357. The data for 10 to 50 km/h are measured; those for 60 and higher are calculated from a model given in the paper and are therefore smoothed.

CHAPTER 11 NOTES

1. Results of the study are reported in P. F. Campbell and G. P. McCabe, "Predicting the success of freshmen in a computer science major," *Communications of the ACM*, 27 (1984), pp. 1108–1113.

2. This example describes an analysis of real data performed by Christine Smiley of Kesenbaum & Company, Chicago, Illinois.

3. For more information on logistic regression see Chapter 14 of J. Neter et al., *Applied Linear Statistical Models*, 4th ed., Irwin, 1996.

4. This study is reported in Scot Burton et al., "Implications of accurate usage of nutrition facts panel information for food product evaluations and purchase intentions," *Journal of the Academy of Marketing Science*, 27 (1999), pp. 470–480.

5. The data are given in Alan Levinsohn, "Online brokerage, the new core account?" *ABA Banking Journal*, September 1999, pp. 34–42.

6. From Michael E. Staten et al., "Information costs and the organization of credit markets: a theory of indirect lending," *Economic Inquiry*, 28 (1990), pp. 508–529.

7. A description of this case as well as other examples of the use of statistics in legal settings is given in Michael O. Finkelstein, *Quantitative Methods in Law*, Free Press, 1978.

CHAPTER 12 NOTES

1. Results of the study are reported in P. F. Campbell and G. P. McCabe, "Predicting the success of freshmen in a computer science major," *Communications of the ACM*, 27 (1984), pp. 1108–1113.

2. This rule is intended to provide a general guideline for deciding when serious errors may result by applying ANOVA procedures. When the sample sizes in each group are very small, the sample variances will tend to vary much more than when the sample sizes are large. In this case, the rule may be a little too conservative. For unequal sample sizes, particular difficulties can arise when a relatively small sample size is associated with a population having a relatively large standard deviation. Careful judgment is needed in all cases. By considering P-values rather than fixed level α testing, judgments in ambiguous cases can more easily be made; for example, if the P-value is very small, say 0.001, then it is probably safe to reject H_0 even if there is a fair amount of variation in the sample standard deviations.

3. This example is based on data provided from a study conducted by Jim Baumann and Leah Jones of the Purdue University School of Education.

4. This example is based on Amna Kirmani and Peter Wright, "Money talks: perceived advertising expense and expected product quality," *Journal of Consumer Research*, 16 (1989), pp. 344–353.

5. Several different definitions for the noncentrality parameter of the noncentral F distribution are in use. When $I = 2$, the λ defined here is equal to the square of the noncentrality parameter δ that we used for the two-sample t test in Chapter 7. Many authors prefer $\phi = \sqrt{\lambda/I}$. We have chosen to use λ because it is the form needed for the SAS function PROBF.

6. These data were provided by Helen Park; see H. Park et al., "Fortifying bread with each of three antioxidants," *Cereal Chemistry*, 74 (1997), pp. 202–206.

7. Based on M. U. Kalwani and C. K. Yim, "Consumer price and promotion expectations: an experimental study," *Journal of Marketing Research*, 29 (1992), pp. 90–100.

8. Based on a study by J. E. Stuff and B. L. Nichols reported in Chelsea Lutter, "Recommended length of exclusive breast-feeding, age of introduction of complementary foods and the weaning dilemma," World Health Organization, 1992.

9. Data provided by Jo Welch of the Purdue University Department of Foods and Nutrition.

10. Based on A. A. Adish et al., "Effect of consumption of food cooked in iron pots on iron status and growth of young children: a randomised trial," *The Lancet*, 353 (1999), pp. 712–716.

11. See S. Badylak et al., "Marrow-derived cells populate scaffolds composed of xenogeneic extracellular matrix," *Experimental Hematology*, 29 (2001), pp. 1310–1318.

12. Modified from M. C. Wilson and R. E. Shade, "Relative attractiveness of various luminescent colors to the cereal leaf beetle and the meadow spittlebug," *Journal of Economic Entomology*, 60 (1967), pp. 578–580.

13. Data provided by Dennis Lobstein, from his Ph.D. dissertation, "A multivariate study of exercise training effects on beta-endorphin and emotionality in psychologically normal, medically healthy men," Purdue University, 1983.

14. These data were taken from Collaborative Research Support Program in Food Intake and Human Function, *Management Entity Final Report*, University of California, Berkeley, 1988.

CHAPTER 13 NOTES

1. This experiment was conducted by Connie Weaver, Department of Foods and Nutrition, Purdue University.

2. We present the two-way ANOVA model and analysis for the general case in which the sample sizes may be unequal. If the sample sizes vary a great deal, serious complications can arise. There is no longer a single standard ANOVA analysis. Most computer packages offer several options for the computation of the ANOVA table when cell counts are unequal. When the counts are approximately equal, all methods give essentially the same results.

3. The data in Examples 13.4 and 13.5 are from `stats.bls.gov`, the U.S. Bureau of Labor Statistics Web site.

4. This example is based on a study described in P. Payne, "Nutrition adaptation in man: social adjustments and their nutritional implications," in K. Blaxter and J. C. Waterlow (eds.), *Nutrition Adaptation in Man*, Libbey, 1985.

5. Example 13.7 is based on a study described in S. S. Zentall and J. H. Shaw, "Effects of classroom noise on performance and activity of second-grade hyperactive and control children," *Journal of Educational Psychology*, 72 (1980), pp. 830–840.

6. Example 13.8 is based on a study described in P. D. Wood et al., "Plasma lipoprotein distributions in male and female runners," in P. Milvey (ed.), *The Marathon: Physiological, Medical, Epidemiological, and Psychological Studies*, New York Academy of Sciences, 1977.

7. Data from W. M. Smith et al., "The multiple risk factor intervention trial," in H. M. Perry, Jr., and W. M. Smith (eds.), *Mild Hypertension: To Treat or Not to Treat*, New York Academy of Sciences, 1978, pp. 293–308.

8. Data provided by Julie Hendricks and V. J. K. Liu of the Department of Foods and Nutrition, Purdue University.

9. This exercise is based on results reported in H. G. Gough, *The Chapin Social Insight Test*, Consulting Psychologists Press, 1968.

10. See S. Badylak et al., "Marrow-derived cells populate scaffolds composed of xenogeneic extracellular matrix," *Experimental Hematology*, 29 (2001), pp. 1310–1318.

11. Based on A. A. Adish et al., "Effect of consumption of food cooked in iron pots on iron status and growth of young children: a randomised trial," *The Lancet*, 353 (1999), pp. 712–716.

12. Based on a problem from Renée A. Jones and Regina P. Becker, Department of Statistics, Purdue University.

13. Based on M. U. Kalwani and C. K. Yim, "Consumer price and promotion expectations: an experimental study," *Journal of Marketing Research*, 29 (1992), pp. 90–100.

14. For a summary of this study and other research in this area, see Stanley Coren and Diane F. Halpern, "Left-handedness: a marker for decreased survival fitness," *Psychological Bulletin*, 109 (1991), pp. 90–106.

15. Data provided by Neil Zimmerman of the Purdue University School of Health Sciences.

DATA TABLE INDEX

BEYOND THE BASICS INDEX

INDEX

CREDITS

p. 3	Max Ernst, *The Hat Makes the Man.* The Museum of Modern Art, NY.
p. 103	Laure Cuvillier, monotype, *Human Body in Motion*. Laure Cuvillier.
p. 221	Francis Picabia, *Machines Turn Quickly*. ©2002 Artists Rights Society (ARS), New York/ADAGP, Paris.
p. 281	Wassily Kandinsky, *Yellow, Red, and Blue*. Musée National d'Art Moderne, Centre Georges Pompidou, Paris/Superstock.
p. 365	Robert Indiana, *The Small Diamond Deluth Five*. ©2002 Morgan Art Foundation Ltd./Artists Rights Society (ARS), New York.
p. 415	Hans Hoffman, *Golden Splendor*. Howard Gallery/Art Resource, NY.
p. 491	Frank Stella, *Flin Flon III*. ©2002 Artists Rights Society (ARS), New York.
p. 571	Johannes Itten, *Kind Light*. ©2002 Artists Rights Society (ARS), New York/ProLitteris, Zurich.
p. 611	Gene Davis, *Prince Albert*. Smithsonian American Art Museum, Washington D.C./Art Resource, NY.
p. 657	Piet Mondrian, *Composition with Red, Yellow, and Blue*. The Tate Gallery, London/Art Resource, NY.
p. 709	Adolph Gottlieb, *Three Disks*. Adolph and Esther Gottlieb Foundation/Licensed by VAGA, New York, NY.
p. 745	Robert Delaunay, *Rhythme I*. Christie's Images/Superstock.
p. 801	Josef Albers, *Study for Oscillating*. ©2002 the Josef and Anni Albers Foundation/Artists Rights Society (ARS), New York.

TABLE B Random digits

Line								
101	19223	95034	05756	28713	96409	12531	42544	82853
102	73676	47150	99400	01927	27754	42648	82425	36290
103	45467	71709	77558	00095	32863	29485	82226	90056
104	52711	38889	93074	60227	40011	85848	48767	52573
105	95592	94007	69971	91481	60779	53791	17297	59335
106	68417	35013	15529	72765	85089	57067	50211	47487
107	82739	57890	20807	47511	81676	55300	94383	14893
108	60940	72024	17868	24943	61790	90656	87964	18883
109	36009	19365	15412	39638	85453	46816	83485	41979
110	38448	48789	18338	24697	39364	42006	76688	08708
111	81486	69487	60513	09297	00412	71238	27649	39950
112	59636	88804	04634	71197	19352	73089	84898	45785
113	62568	70206	40325	03699	71080	22553	11486	11776
114	45149	32992	75730	66280	03819	56202	02938	70915
115	61041	77684	94322	24709	73698	14526	31893	32592
116	14459	26056	31424	80371	65103	62253	50490	61181
117	38167	98532	62183	70632	23417	26185	41448	75532
118	73190	32533	04470	29669	84407	90785	65956	86382
119	95857	07118	87664	92099	58806	66979	98624	84826
120	35476	55972	39421	65850	04266	35435	43742	11937
121	71487	09984	29077	14863	61683	47052	62224	51025
122	13873	81598	95052	90908	73592	75186	87136	95761
123	54580	81507	27102	56027	55892	33063	41842	81868
124	71035	09001	43367	49497	72719	96758	27611	91596
125	96746	12149	37823	71868	18442	35119	62103	39244
126	96927	19931	36089	74192	77567	88741	48409	41903
127	43909	99477	25330	64359	40085	16925	85117	36071
128	15689	14227	06565	14374	13352	49367	81982	87209
129	36759	58984	68288	22913	18638	54303	00795	08727
130	69051	64817	87174	09517	84534	06489	87201	97245
131	05007	16632	81194	14873	04197	85576	45195	96565
132	68732	55259	84292	08796	43165	93739	31685	97150
133	45740	41807	65561	33302	07051	93623	18132	09547
134	27816	78416	18329	21337	35213	37741	04312	68508
135	66925	55658	39100	78458	11206	19876	87151	31260
136	08421	44753	77377	28744	75592	08563	79140	92454
137	53645	66812	61421	47836	12609	15373	98481	14592
138	66831	68908	40772	21558	47781	33586	79177	06928
139	55588	99404	70708	41098	43563	56934	48394	51719
140	12975	13258	13048	45144	72321	81940	00360	02428
141	96767	35964	23822	96012	94591	65194	50842	53372
142	72829	50232	97892	63408	77919	44575	24870	04178
143	88565	42628	17797	49376	61762	16953	88604	12724
144	62964	88145	83083	69453	46109	59505	69680	00900
145	19687	12633	57857	95806	09931	02150	43163	58636
146	37609	59057	66967	83401	60705	02384	90597	93600
147	54973	86278	88737	74351	47500	84552	19909	67181
148	00694	05977	19664	65441	20903	62371	22725	53340
149	71546	05233	53946	68743	72460	27601	45403	88692
150	07511	88915	41267	16853	84569	79367	32337	03316

TABLE B Random digits (*continued*)

Line								
151	03802	29341	29264	80198	12371	13121	54969	43912
152	77320	35030	77519	41109	98296	18984	60869	12349
153	07886	56866	39648	69290	03600	05376	58958	22720
154	87065	74133	21117	70595	22791	67306	28420	52067
155	42090	09628	54035	93879	98441	04606	27381	82637
156	55494	67690	88131	81800	11188	28552	25752	21953
157	16698	30406	96587	65985	07165	50148	16201	86792
158	16297	07626	68683	45335	34377	72941	41764	77038
159	22897	17467	17638	70043	36243	13008	83993	22869
160	98163	45944	34210	64158	76971	27689	82926	75957
161	43400	25831	06283	22138	16043	15706	73345	26238
162	97341	46254	88153	62336	21112	35574	99271	45297
163	64578	67197	28310	90341	37531	63890	52630	76315
164	11022	79124	49525	63078	17229	32165	01343	21394
165	81232	43939	23840	05995	84589	06788	76358	26622
166	36843	84798	51167	44728	20554	55538	27647	32708
167	84329	80081	69516	78934	14293	92478	16479	26974
168	27788	85789	41592	74472	96773	27090	24954	41474
169	99224	00850	43737	75202	44753	63236	14260	73686
170	38075	73239	52555	46342	13365	02182	30443	53229
171	87368	49451	55771	48343	51236	18522	73670	23212
172	40512	00681	44282	47178	08139	78693	34715	75606
173	81636	57578	54286	27216	58758	80358	84115	84568
174	26411	94292	06340	97762	37033	85968	94165	46514
175	80011	09937	57195	33906	94831	10056	42211	65491
176	92813	87503	63494	71379	76550	45984	05481	50830
177	70348	72871	63419	57363	29685	43090	18763	31714
178	24005	52114	26224	39078	80798	15220	43186	00976
179	85063	55810	10470	08029	30025	29734	61181	72090
180	11532	73186	92541	06915	72954	10167	12142	26492
181	59618	03914	05208	84088	20426	39004	84582	87317
182	92965	50837	39921	84661	82514	81899	24565	60874
183	85116	27684	14597	85747	01596	25889	41998	15635
184	15106	10411	90221	49377	44369	28185	80959	76355
185	03638	31589	07871	25792	85823	55400	56026	12193
186	97971	48932	45792	63993	95635	28753	46069	84635
187	49345	18305	76213	82390	77412	97401	50650	71755
188	87370	88099	89695	87633	76987	85503	26257	51736
189	88296	95670	74932	65317	93848	43988	47597	83044
190	79485	92200	99401	54473	34336	82786	05457	60343
191	40830	24979	23333	37619	56227	95941	59494	86539
192	32006	76302	81221	00693	95197	75044	46596	11628
193	37569	85187	44692	50706	53161	69027	88389	60313
194	56680	79003	23361	67094	15019	63261	24543	52884
195	05172	08100	22316	54495	60005	29532	18433	18057
196	74782	27005	03894	98038	20627	40307	47317	92759
197	85288	93264	61409	03404	09649	55937	60843	66167
198	68309	12060	14762	58002	03716	81968	57934	32624
199	26461	88346	52430	60906	74216	96263	69296	90107
200	42672	67680	42376	95023	82744	03971	96560	55148

Table entry for p and C is the critical value t^* with probability p lying to its right and probability C lying between $-t^*$ and t^*.

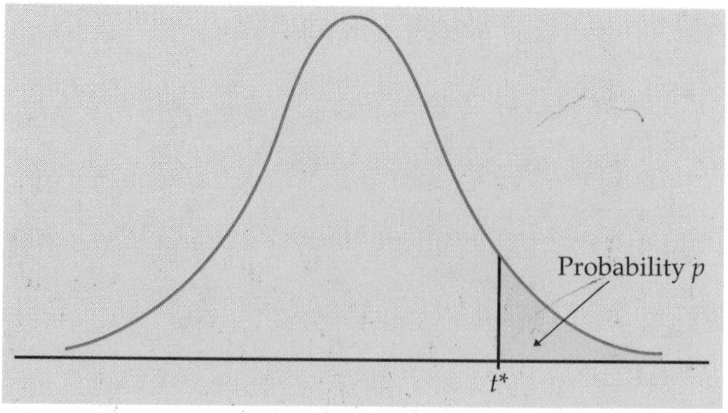

TABLE D t distribution critical values

	Upper tail probability p											
df	.25	.20	.15	.10	.05	.025	.02	.01	.005	.0025	.001	.0005
1	1.000	1.376	1.963	3.078	6.314	12.71	15.89	31.82	63.66	127.3	318.3	636.6
2	0.816	1.061	1.386	1.886	2.920	4.303	4.849	6.965	9.925	14.09	22.33	31.60
3	0.765	0.978	1.250	1.638	2.353	3.182	3.482	4.541	5.841	7.453	10.21	12.92
4	0.741	0.941	1.190	1.533	2.132	2.776	2.999	3.747	4.604	5.598	7.173	8.610
5	0.727	0.920	1.156	1.476	2.015	2.571	2.757	3.365	4.032	4.773	5.893	6.869
6	0.718	0.906	1.134	1.440	1.943	2.447	2.612	3.143	3.707	4.317	5.208	5.959
7	0.711	0.896	1.119	1.415	1.895	2.365	2.517	2.998	3.499	4.029	4.785	5.408
8	0.706	0.889	1.108	1.397	1.860	2.306	2.449	2.896	3.355	3.833	4.501	5.041
9	0.703	0.883	1.100	1.383	1.833	2.262	2.398	2.821	3.250	3.690	4.297	4.781
10	0.700	0.879	1.093	1.372	1.812	2.228	2.359	2.764	3.169	3.581	4.144	4.587
11	0.697	0.876	1.088	1.363	1.796	2.201	2.328	2.718	3.106	3.497	4.025	4.437
12	0.695	0.873	1.083	1.356	1.782	2.179	2.303	2.681	3.055	3.428	3.930	4.318
13	0.694	0.870	1.079	1.350	1.771	2.160	2.282	2.650	3.012	3.372	3.852	4.221
14	0.692	0.868	1.076	1.345	1.761	2.145	2.264	2.624	2.977	3.326	3.787	4.140
15	0.691	0.866	1.074	1.341	1.753	2.131	2.249	2.602	2.947	3.286	3.733	4.073
16	0.690	0.865	1.071	1.337	1.746	2.120	2.235	2.583	2.921	3.252	3.686	4.015
17	0.689	0.863	1.069	1.333	1.740	2.110	2.224	2.567	2.898	3.222	3.646	3.965
18	0.688	0.862	1.067	1.330	1.734	2.101	2.214	2.552	2.878	3.197	3.611	3.922
19	0.688	0.861	1.066	1.328	1.729	2.093	2.205	2.539	2.861	3.174	3.579	3.883
20	0.687	0.860	1.064	1.325	1.725	2.086	2.197	2.528	2.845	3.153	3.552	3.850
21	0.686	0.859	1.063	1.323	1.721	2.080	2.189	2.518	2.831	3.135	3.527	3.819
22	0.686	0.858	1.061	1.321	1.717	2.074	2.183	2.508	2.819	3.119	3.505	3.792
23	0.685	0.858	1.060	1.319	1.714	2.069	2.177	2.500	2.807	3.104	3.485	3.768
24	0.685	0.857	1.059	1.318	1.711	2.064	2.172	2.492	2.797	3.091	3.467	3.745
25	0.684	0.856	1.058	1.316	1.708	2.060	2.167	2.485	2.787	3.078	3.450	3.725
26	0.684	0.856	1.058	1.315	1.706	2.056	2.162	2.479	2.779	3.067	3.435	3.707
27	0.684	0.855	1.057	1.314	1.703	2.052	2.158	2.473	2.771	3.057	3.421	3.690
28	0.683	0.855	1.056	1.313	1.701	2.048	2.154	2.467	2.763	3.047	3.408	3.674
29	0.683	0.854	1.055	1.311	1.699	2.045	2.150	2.462	2.756	3.038	3.396	3.659
30	0.683	0.854	1.055	1.310	1.697	2.042	2.147	2.457	2.750	3.030	3.385	3.646
40	0.681	0.851	1.050	1.303	1.684	2.021	2.123	2.423	2.704	2.971	3.307	3.551
50	0.679	0.849	1.047	1.299	1.676	2.009	2.109	2.403	2.678	2.937	3.261	3.496
60	0.679	0.848	1.045	1.296	1.671	2.000	2.099	2.390	2.660	2.915	3.232	3.460
80	0.678	0.846	1.043	1.292	1.664	1.990	2.088	2.374	2.639	2.887	3.195	3.416
100	0.677	0.845	1.042	1.290	1.660	1.984	2.081	2.364	2.626	2.871	3.174	3.390
1000	0.675	0.842	1.037	1.282	1.646	1.962	2.056	2.330	2.581	2.813	3.098	3.300
z^*	0.674	0.841	1.036	1.282	1.645	1.960	2.054	2.326	2.576	2.807	3.091	3.291
	50%	60%	70%	80%	90%	95%	96%	98%	99%	99.5%	99.8%	99.9%
	Confidence level C											